Oncothermia: Principles and Practices

Andras Szasz · Nora Szasz · Oliver Szasz

Oncothermia: Principles and Practices

Springer

Andras Szasz
Department of Biotechnics
Faculty of Engineering
St. Istvan University
Pater K. u. 1
2103 Godollo
Hungary
Szasz.Andras@gek.szie.hu

Nora Szasz
McKinsey & Co.
Park Plaza 75
02116 Boston
MA, USA
Nora_Szasz@mckinsey.com

Oliver Szasz
Oncotherm Inc.
Ibolya u. 2
2071 Paty
Hungary
SzaszOliver@oncotherm.org

ISBN 978-90-481-9497-1 e-ISBN 978-90-481-9498-8
DOI 10.1007/978-90-481-9498-8
Springer Dordrecht Heidelberg London New York

Printed on acid-free paper

Springer is part of Springer Science+Business Media (www.springer.com)

Contents

Objective of the Book

Oncothermia is a treatment modality for malignant tumor diseases.

Cancer has been a permanent fear of human kind since ancient times. It was incurable, fatal and from ancient times we have been in a continuous war against these malignant diseases. Despite all our sustained efforts to cure cancer this has not become a reality yet.

Hyperthermia is an ancient treatment. The fire (Sun) had symbolic significance in ancient human cultures, so heat delivery was naturally at the top of curative possibilities. Ancient heat-delivery methods were of course ineffective, but modern electromagnetic heating techniques were able to renew this methodology. Heating up the whole body or a part or local volume began to rapidly develop in modern oncotherapeutic practices. Selective energy absorption has several favorable physiological and cellular effects promoting direct and indirect tumor destruction without notable toxicity. Its main success lies in its complementary applications. Oncological hyperthermia is an ideal combination therapy; it provides synergies with most of the conventional treatment modalities, boosts their efficacy, and helps in desensitizing previously non-effective treatments. Hyperthermia in oncology has been debated in an increasing number of books and high-ranking clinical publications.

Contrary to its long history, the state of oncological hyperthermia today is similar to that of therapies in their infancy. Like many early-stage therapies, it lacks adequate treatment experience and long-range, comprehensive statistics that can help us optimize its use for all indications. Nevertheless, we will present a wealth of information about the mechanisms and effects of hyperthermia in oncology, showing the clinical efficacy in a wide range of malignancies.

This relatively simple, physical-physiological method has a phoenix-like history with some bright successes and many deep disappointments. What we have in hand? Is it a brilliant, miraculous, non-toxic treatment or a quackery of some charlatans? In our present book we try to answer these questions too.

Oncothermia employs many modern scientific achievements in order to assure and enhance the clinical success of hyperthermia in oncology. Oncothermia is a certain further development of hyperthermia. Of course it also cannot offer a miracle; it is only a new weapon to be applied. Oncothermia uses many natural processes to reach the curative effect, initializing and supporting the complex human system to make a successful defense against cancer and promote the probable cure or at least the palliative treatment.

To explain oncothermia, we present updated scientific results, complex approaches using the state-of-art in biophysics and medicine. We use a wide range of inter-disciplinary facilities: from physics we employ thermodynamics, statistical mechanics, quantum mechanics, electrodynamics together with biophysics achievements in fractal physiology, stress adaptation, cellular signaling, and add a huge technical arsenal including the radiofrequency technique, impedance control, and antenna fitting.

There are a great number of books, published to date, devoted to the efficacy and power of hyperthermia in oncology [1–19]. The method is a part of the universal and well-accepted oncology knowledge [20] discussed in detail in large text-books of radiology/radiotherapy [21] and general oncology [22] also. We would like to demonstrate the force and perspectives of oncothermia, as a highly specialized hyperthermia in clinical oncology. Our aim is to prove the ability of oncothermia to be a candidate to become a widely accepted modality of standard cancer care. We would like to show the proofs and challenges of oncothermia applications, to provide the presently available data and summarize the knowledge on the topic. We concentrate mainly on local/regional hyperthermia with its non-invasive electromagnetic applications, so whole-body hyperthermia and the RF-ablation techniques will not receive much attention and neither will ultrasound or other heating techniques.

We would like to show a new paradigm for oncological thermal treatment and give its perspectives for the future. The rich reference list and discussions of "hot topics" will help specialists both in labs and clinical practices to use the book as a handy reference.

For the numerous questions that could be addressed you may find answers in this book to the following:

- Hyperthermia has a long history in oncology – however, it has no acceptance. Why?
- Oncologic hyperthermia has a large number of publications – but the results are often contradictory. Why?
- Huge investigation efforts are made in hyperthermia in oncology worldwide – yet there is no complete understanding of the underlying mechanism. Why?
- Hyperthermia has many convicted enthusiastic oncology practitioners – using contradictory explanations. Why?
- There are several possible microscopic effects that could be used to control the hyperthermia process – but only the temperature is used generally. Why?
- Hyperthermia is basically nontoxic and widely compatible with other oncotherapies – but it is less recognized than much more toxic treatments. Why?
- What is against the active use of hyperthermia in oncology?
- Has hyperthermia a future in oncology at all?

After these questions obviously one more arises: what do we have in hand? We would like to offer the physical effects of oncothermia, and present the results, which could make harmonizing the disciplines easier, exploring the heart of oncological hyperthermia as an inter-disciplinary topic.

The aim of this book is to point out the challenges, set the right questions (starting with the apparent contradiction between the oldest method in cancer treatment being at the same time the youngest method in modern oncology), and try to generate the right answers. The open questions need further investigation and we strive to motivate the kind reader.

In summary: we propose a change of paradigm for hyperthermia in oncology. The old paradigm, the conventional, classical hyperthermia, (we use further the term "hyperthermia"), while the proposed new paradigm is referred to as "oncothermia."

This book is an invitation to share the challenge of the new method, share the excitement of applying a new effective treatment, and share enjoyment in the results.

Chapter 1
Oncology – Treatments and Their Limits

The fight against cancer is as old as medicine itself. Despite the huge achievements of human kind, and its triumph over many fatal diseases; the war against cancer has not yet been won.

1.1 Cancer – Short History and Efforts to Cure

1.1.1 Historical Notes

Naturally, the development of oncological hyperthermia cannot be separated from the general progress of oncotherapies. One of the very first written vestiges referring to cancer cases and their possible handling described was hyperthermia, the treatment of tumors with heat. The development of hyperthermia ran parallel to the development of tumor treatment in human history; in fact, it is the oldest method used for curative purposes in oncology.

With the advancement of our understanding of malignant diseases the generic paradigm of oncology also advanced from its inception through of "untouchable" tumor to the modern use of genetic technologies.

Some detailed works have been devoted to the history of cancer [23–27] and there have been numerous others reports on actual results.

The widely accepted paradigm of cancer development states that the malignant tumor is derived from a "renegade" cell [28]. This cell and its daughter cells grow without control and cancer starts to develop. Cancer cells have damaged DNA which cannot be repaired by the usual mechanisms of the system.

The loss of the healthy control of cell cycles is not a simple process. Numerous (at least five, [29]) deviations have to be active to develop such a cancerous situation. The normal, healthy cells are under the control of others (social control, a collective state), they grow, divide, and die in a definite way, which depends on the person and the body part (age, sex, etc.) as well. In a healthy process cells divide only to replace the age-worn or dying cells and to repair injuries. The entire process, as well as the normal functioning of the cells is strictly controlled by the system; the cells exhibit energy and material consumption in a definite regular form depending on

A. Szasz et al., *Oncothermia: Principles and Practices*,
DOI 10.1007/978-90-481-9498-8_1,

their role assigned by labor division within the body. Cancer cells are different: they grow without control, their energy and material exchange is limited only by availability and is not affected by any regular control. Cancer has its own growth factor, and is not sensitive to growth inhibitors, it can avoid apoptosis and has unlimited replicating potential in addition to enhanced angiogenetic and dissemination (invasive) potential. They are autonomic, instead of collective; they have a competitive driving force to survive among the shrinking sources and diminishing availability of survival. This competitive position means cancer cells evolve to be different in their interactions, and makes it possible for cancer cells to develop special defense mechanisms against by alongside stimulation the ebbing resources (missing adequate nutrition) neoangiogenesis, intensive anaerobic metabolism, false information exchange, etc. Cancer cells often disseminate through systemic pathways (lymph and/or blood stream) forming metastases, replacing or limiting the normal tissue in the distant area. However, not all tumors are cancers, some of them are benign, and do not produce metastases.

Cancer forms solid tumors in most cases; however, some cancers, like leukemia, do not form tumors, circulating with the blood flow all over the body.

A powerful new theory [30] alongside connected observations [31] has been established, showing other mechanisms may be at work, that is, a cancerous cell developed from stem cells. Although up to now, the widely accepted cancer model was based on equal tumorigenesis of all cancer cells, in the stem-cell model only a few cells are able to form new cancer cells. The main change in the conventional explanations is the denegation of malignant proliferation, adducing a stem-cell disorder for the dissemination of the disease. Moreover, the new theory challenges the genetic origin of unregulated growth – instead, the disruption of stem-cell renewal mechanisms are considered to be responsible for the autonomy of cancer cells.

Also, a new picture is drawn for carcinogenesis from the study of chronic inflammation, hypothesizing a direct connection to cancer formation and development [32, 33].

Cancer kills by its invasion of surrounding tissues and it is metastasized in the body by spread of disseminated cells. The main causes of death by cancer can be divided into four groups by frequency [34]. The most frequent is organ failure due to malignant invasion (CNS, Lungs, Liver, Renal, etc.); the next is the infarction of lungs or heart, the third is bleeding, and the fourth is carcinomatosis (metabolic/electrolyte disorder).

1.1.2 The "War" Against Cancer

Richard Nixon, the president of the United States declared a "war" in 1971 against one of the greatest challenges of medical science for centuries, against cancer.

Enormous economic and human resources are involved in this field, but according to epidemic data the solution is still eagerly awaited. Analyzing the evidence-based clinical data of 5-year survivals, the conclusion is [35], that 5-year survivals changed only a little from 1950 to 1995, and these changes depended more on better diagnosis than on therapy. The contribution of curative and adjuvant cytotoxic chemotherapy to 5-year survival in adults (counting 22 different localizations) was estimated to be 2.3% in Australia and 2.1% in USA [36] in 2004 over the previous 20 years. This is a minor contribution to the observed 5-year survival rate, which is over 50% for the same time period.

The progress is of course debated: "We are losing the war against cancer" [37], which was immediately corrected to a broader view [38], taking into account the successes in pediatric cancer and in the quality-of-life (QoL) of patients during the curative and palliative treatments. This picture was a little diluted: "Perhaps not lost, but certainly not won." [39]. This was also supported 10-years later [40]. This emotional aggravation has induced very hurtful opinions like the double Nobel-laureate L. Pauling who formulated "Everyone should know that the 'war on cancer' is largely a fraud" [41]. A more immoderate opinion was formulated on the "failed war" by editorial of People Against Cancer [42].

Filtering out extreme opinions, new statistical data [43] supports the shadowed picture: the mortality data from 1975 to 2000 are fairly constant, while the incidence (morbidity) slightly grows over the same time interval. (Interestingly, the incidence has a definite peak in the first half of the 1990s in the group of males, but the mortality does not follow suit.)

The report of Princeton University (USA) [44], formulates directly: "In 1971 President Nixon declared a war against cancer. Thirty years later, many have declared the war a failure: the age-adjusted mortality rate from cancer in 2000 was essentially the same as in the early 1970s. Meanwhile the age-adjusted mortality rate from cardiovascular disease fell dramatically."

However the war is expensive, its costs have rapidly grown. The person-years of life lost due to cancer were 7,725,600 counting the 22 most-frequent cancer types for all races and both sexes in 2003 [45]. Almost a third of this loss (2,403,100 years) was from lung&bronchus cancer. The monetary expense was also enormous. In the USA, the National Cancer Institute (NCI) alone supports each year some 5,000 principal investigators, as well as funding cancer centers, research teams, education, etc. [46]. The budget increase of NCI is growing by a square of the elapsed years ($R^2 = 0.956$), (see Fig. 1.1), which causes a linear decrease of mortality among the patients (relative to incidence, $R^2 = 0.9247$, see Fig. 1.1).

Note, the ration of the mortality to the incidence is not correct, because the morbidity comes of course years earlier than the mortality (most patients fortunately do not die in the same year as the cancer develops, was diagnosed). However, this shift of the data in proportion does not make for huge errors, because the decrease is slow.

The general trend shows about 3.8% improvement of survival among cancer patients for every 10 years in the USA [47], and a further decrease in mortality looks more and more difficult. The growing number of complications and challenges load the budget increasingly with every subsequent improvement. The picture is even more disturbing, if we take into account the efforts of the connected industries (mainly the pharmaceutical industry), which is of course not calculated as part of

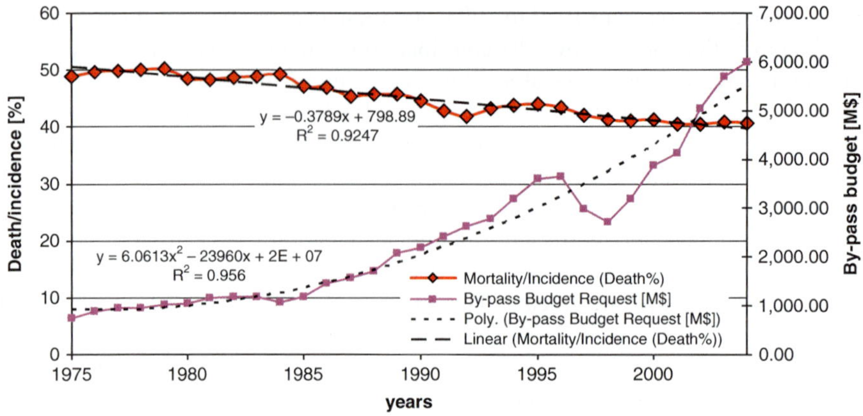

Fig. 1.1 The decrease of the mortality ratio to the cancer incidence in USA, (according to SEER), and the increase of the NCI budget (USA) by years

the NCI budget, but is probably more than an order of magnitudes higher in the US, than the NCI budget capacity (the overall estimated cancer-related costs in the US was $210 billion in 2005 [48], which is more than 30-times higher than the $6.2 billion NCI budget in the same year. Again, this sum does not include the investments within industrial/market spheres). Not only at the sacrifice of increasingly growing monetary funds, but the remarkable (approximately linear, $R^2 = 0.9319$, see Fig. 1.2) growth of supported projects is a good indicator of the difficulty and complexity of the task at hand.

The industrial revenue of the pharmaceutical "cancer market" grows in a very dynamic fashion [49], while the other cancer-connected (medical devices, medical accessories, etc.) industrial revenues are not calculated.

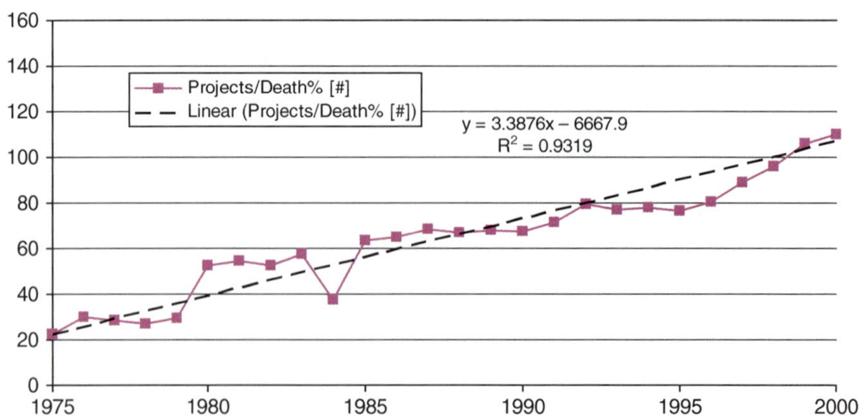

Fig. 1.2 NCI project number relative to the death ratio from SEER and NCI data

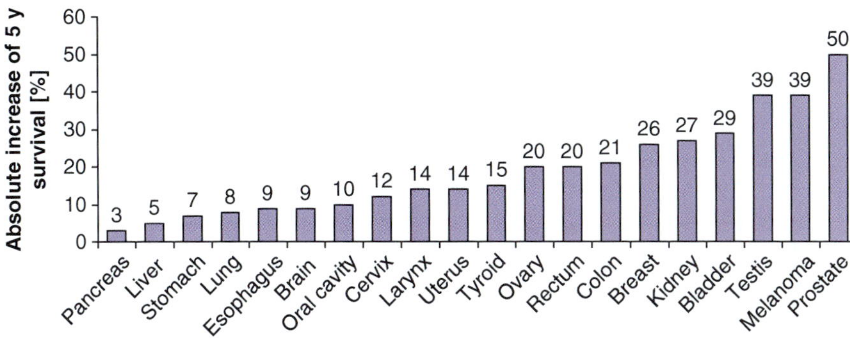

Fig. 1.3 The 5-year survival gain in % from the interval 1950–1954 to 1989–1995

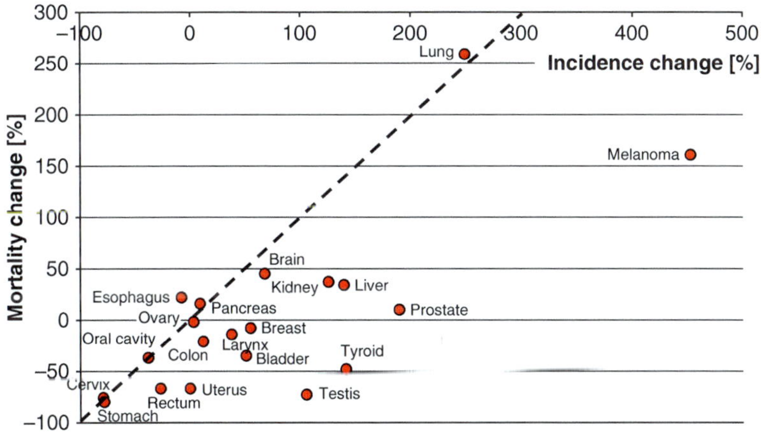

Fig. 1.4 Mortality rates vs. incidence of some common primary cancers (the *dashed line* shows the equal growth correspondence)

However, the massive revenue finances the rapidly growing demand in the R&D investments for drug approvals [50].

The absolute increase of the 5-year survival for the 1950–1954 to 1989–1995 interval is positive for many common primary cancers (see Fig. 1.3 [35]), and also the growth of incidence of these localizations mostly exceed the corresponding mortalities (see Fig. 1.4).

Unfortunately, neither the incidence rate (see Fig. 1.5) nor the mortality rate (see Fig. 1.6) correlate with the 5-year survival [35] for the same localizations. This shows the present imperfectness that cancers with high incidence- and high mortality-rate growths, like lung, liver, brain, and pancreas, have only a low gain in their 5-year survival. This is the essence of the negative answer to the question [35]: "Are increasing 5-year survival rates evidence of success against cancer?"

The real problem with the mortality is of course not the primary tumor, but its metastases, which can block some essential life functions in the attacked organ.

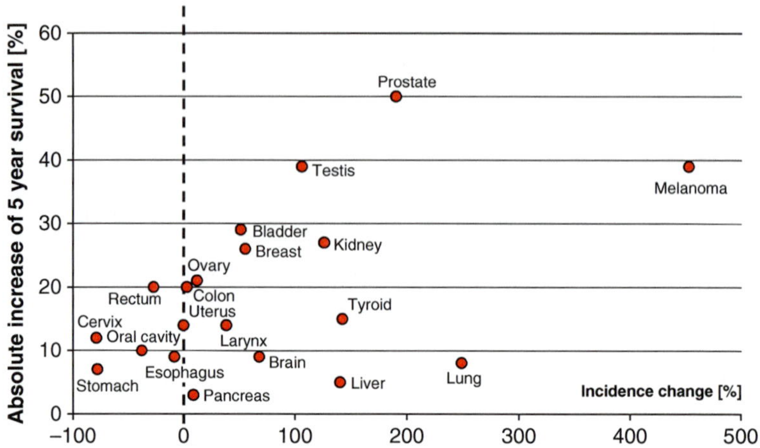

Fig. 1.5 Absolute increase of the 5-year survival vs. incidence change for some primary tumors (the *dashed line* marks the unchanged incidence)

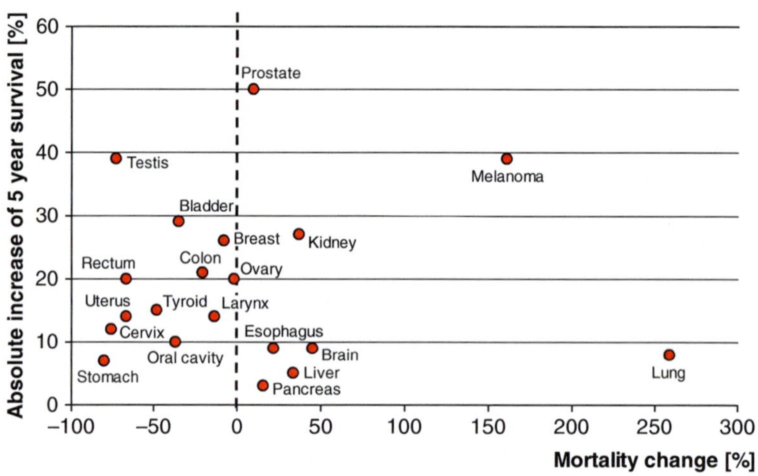

Fig. 1.6 Absolute increase of the 5-year survival vs. mortality change for some primary tumors (the *dashed line* marks the unchanged incidence)

Cancer is the second leading cause of death in the United States – just 16% less than a "competitor": the heart diseases [51] (the third "place," taken by celebrovascular diseases [strokes], is 75% behind the "winner.") Statistically, one-half and one-third of all men and women in the US will develop cancer during their lifetimes, respectively. According to statistical data [52], cancer became the number one killer in England & Wales; with an increase over heart disease of more than 22%. Cancer mortality grew from 15 to 27% and from 16 to 23% for men and women, respectively between 1950 and 1999. More than 30% growth of mortality was observed in Italy from 1970 till 2000 while the new and prevalent cases were increasing at about the same ratio [48]. The 5-year survival in Europe (22 countries) is behind the same in the USA [53], counting 42 cancer localizations.

Presently millions of people all over the world are suffering from cancer, and we have to take great pride in those millions who were successfully cured. Today, millions of people are living with cancer or have had cancer. In spite of the sad statistical data, the rapidly developing prevention networks and changes in human lifestyles (abandonee smoking, caring about environmental pollution, healthy diets, etc.) moderate the rapid development of the mortality, and even stop it in some tumor entities. General preventive actions are mainly based on reduced carcinogenic agents in the environment as well as eliminating cancer-causing bad life-habits (e.g. smoking, drinking, etc.). Also healthy eating is one of the popular forms of prevention. Besides the effective preventive strategies, the prophylactics, the appropriate and effective regular check-ups are essential in the "cancer war."

It is shocking data that about a quarter of all children presently born will die from cancer and more of them confront this disease personally in their lifetime. It is not a surprise, that fear of cancer is very common amongst people [54, 55]. This fear originates not only from the very frequent morbidity and believed "incurability," but also from the long and severe suffering during treatment and dying (the same origins for fear as in AIDS, [56]). Death is a normal process for humans, but it could be earlier than normal, and the fear is mainly based on the path to death, which due to the long and severe suffering is seen as against of human ethical values.

Nowadays, oncology has become one of the most inter-disciplinary research fields: including biology, biophysics, biochemistry, genetics, environmental sciences, epidemiology, immunology, microbiology, pathology, physiology, pharmacology, psychology, virology, etc. Moreover, a wide range of diagnostic and treatment methods are available to identify and destroy the malignant tissue. However, without doubt, a method with low toxicity and rare complications has been a dream for a long time and at the same time represents one of the greatest challenges of oncology. This is the point, when the physical methods have good perspectives for the curative processes of the malignant cancer diseases.

1.2 Paradigm and Challenges of Oncotherapies

Certainly, cancer is not the first and probably not the last among diseases for which – despite exceptional human efforts – a cure has not been found for a long time. One of the most deadly European epidemics, which killed more than 50% of inhabitants in most of the crowded European cities, was the "Black Death," namely the Pest. It was one of the most obvious examples of inadequate medical knowledge and its consequences at the time: "... the disease (Pest) deeply shaming the physicians because they were not able to give any help..." (Guy de Chauliac, 1349, [57]). The development of medical knowledge in most cases follows on from critical situations and crises with the aim of preparing medical science to avoid a future crisis of the same nature.

Most current cancer treatments are inherently the opposite of ancient practice. Instead of the ancient "untouchable" position the general cancer curative strategy is based on distortion, the aim of which is to eliminate the cancer cells by their entire liquidation, and clean the system from the malignancy by a drastic destruction [58]. No paradigm (no tool) has been developed with the strategic aim of re-establishing the original, tumor-free state, of reversing tumor development without killing the cells. The distortion strategy offers a proper solution if the method is effective and selective enough (destroys all and only the cancer cells), and the destruction does not expand into a dangerously large part of the body or paralyzes essential body functions.

The treatment strategy, of course, has to be based on our actual knowledge of the disease and it is heavily modified by certain conditions of the actual cases (e.g. the accuracy of the diagnosis, accuracy of the localization, complications caused by accompanying illnesses or unusual sensitivities of the individual, etc.). Presently there are no reliable strategies available for many cases. The physician makes decisions based in these situations more on general experience then through confident actions where the improvement would be certain. The ancients recognized that there was no curative treatment once a cancer had spread and that intervention might be more harmful than no treatment at all. We are, in a much better situation today, although success and long survival are not guaranteed in all cases yet.

There are generally three major forms ("gold standards") which compose present cancer therapies: surgery, radiotherapy, and chemotherapy. These were all developed firstly with palliation as the primary goal, and only later were curative applications introduced. Also, their introduction showed the most common features of emerging therapies: initial skepticism and often rejection from the medical community, and later an overheated optimism, with expectations of the final solution to the "cancer enigma." Both approaches of course were not realistic, and it was a long and hard learning procedure to realize the actual limits and benefits. Naturally, development adds more and more practical and principal details which place these methods in a definite place within the structure of oncotherapy, as well as further developing the methods on their own and towards collective combinative solutions. Human beings as well as their illnesses are very complex issues, and especially cancer is one of the most complex problems in medicine today. This fact supports the general opinion of the specialist that one kind of treatment modality is not satisfactory for achieving a cure, and multi-modal curative processes are spreading in the medical community. However, what is the final optimal strategy, and which new modalities have to be introduced to "win the war" are open questions at this time, and they look like remaining so for a long time yet.

1.3 Limitations of Oncotherapies – The Quest for a Step Forward

There are numerous challenges to the old paradigm. We list a few of them:

- The first of these is the renaissance of Warburg's theory [59]: that malignancy is a metabolic (mitochondrial) dysfunction (See Section 3.2.5).

- Oncogenes and proto-oncogenes are present not only in malignant cells [60], but have a role in various reparative and growth processes, like in pregnancy [61], in embryonic development [62, 63], in wound healing [64], and in the synthesis of growth factors [65].
- Much cancer development is connected to chronic inflammation [66–69];
- Oncogenes show a wide range of apoptotic functions in cells that have a role in wound healing [70, 69].

Presently two new paradigms are rapidly developing:

1. A definite metabolic difference between the metabolic processes of malignant and healthy cells was observed [71] and honored by the Nobel prize to Otto Warburg. His idea has been revised [72, 73], and is presently undergoing a renaissance [74] and "returns in a New Theory of Cancer" [75], and new hypotheses are formed on this basis [76, 77].
2. The stem cell hypothesis is a completely different approach [78–80] to the conventional well-established "renegade cell" [81] concept. There are numerous clinical concepts [82–85], but presently there are many more questions than answers [86, 87, 83].

Medical practice in prehistoric and ancient times used three different approaches in combination to provide the most complex solution for the actual disease: the psychology-based therapeutic (sacral, ritual, religion derived) methods, chemistry-based therapeutic (herbal, dietetic, etc.), and the physics-based therapeutic (massage, surgery, etc.) methods. In ancient Egypt lettuce and onion, surgery and positive (sacral) thinking were the bases of the cure [57]. They simultaneously applied herbal medicaments, physical treatments, and ritual incantations. This complex approach has been permanently present for millennia in the development of medicine with some fluctuations on the emphases. The complicated, complex human system was treated by an inter-disciplinary approach, which was sometimes muddled, but the goal was never forgotten: to cure the individual as a complex unit. They knew that the health problem manifested somewhere locally, but it needs systemic care, taking into account the individual completely, together with the psycho-factors. This harmony was partly broken by rapid technical-chemical-biological development and connected economic improvements in the past century.

Modern medical solutions break the complex, inter-disciplinary balance; overemphasizing pharmaceutical (chemical) solutions, even in cases where the replacing or conjunctly used applications could be equally or more effective. Today, psycho-factors have become an important part of cancer care (including prophylactics and preventive care). The physical approaches are presented in the widely applied surgery and in the (photon and particle) ionizing radiation technologies. Hyperthermia is a completion of the physical methods, using non-ionizing radiation or convective/conductive methods to energize the treated area.

One of the main disadvantages of the onco-treatments is their disturbance of the normal functions of the organs, which causes various levels of side effects. The wish to cure without unwanted side effects is presently just a dream. Any of our curative interventions basically might hurt the normal functions of the human body. The actual shortfall in our medical knowledge and uncertainties in the applications support unscientific charlatanism and the aim to find "alternative" solutions.

Modern oncology applies highly effective methods and treatments, but their side effects and, in consequence, the impairment of QoL are also remarkable. Patients are treated with chemo- and radiotherapy to their toxicity limits in order to achieve maximal tumor destruction. However, these treatments are often not enough. In general, the tolerable toxic level limits the applications. The actually expected tumor destruction would require higher doses than tolerable in addition to the point that the debris of the tumor liberates toxins so the actual treatment has to be a compromise in terms of the accepted level of side effects. This is a definite therapeutic gap.

The gap between the toxic tolerance and the desirable destruction has to be bridged by a method, such as hyperthermia: based on physical and physiological effects, its stress has no chemical origin or serious toxicity. In this way hyperthermia is an ideal combination therapy. It has low toxicity, mild side effects, and has been shown to provide synergies with many of the traditional treatment modalities.

In addition to toxicity, developed resistance against the actually applied treatments could also limit the efficacy of the applied methods. While the first treatment is able to suppress the tumor to below the point where it is detectable (clinical outcome is a complete remission); some malignant cells remain behind undetected causing possible relapse and/or dissemination. At the same time, a more serious problem arises: some of the still-present malignant cells might become resistant to the actual treatment, so its next application might not be as successful as the one before, and in successive steps we lose this treatment modality. In this way the observed and hopefully complete remission in most cases is only a temporary success [88]. Hyperthermia can be helpful in these cases as well because it may re-sensitize the malignant cells and enable them to be destroyed.

1.3.1 Medical Challenge of Oncotherapies

The special challenge in hyperthermia is tightly connected to a general one. The subject of human medicine is to study and control the biosystem, but it is too complex to understand the processes in detail with our present knowledge. Ignoring the challenge of the complexity of human medicine, leads us in a false direction, "where the medicine went wrong" [89].

The social point of view also led to the formulation of some critical problems, "the death of human medicine" [90], which had a warm evaluation in its foreword by the editor of *The Lancet* [91]. The huge challenge of medicine is well indicated by the large number of patients turning to the "alternatives" and many of these are closeted in spiritual, sacral images instead of enlisting the help of (using their terminology: "scholastic") medicine. These problems inspired many authors to formulate their rather pessimistic view on science (like "End of Science" [92], "Forbidden Science" [93]). Bitter skeptic formulations criticize medicine for its business orientation, for neglecting many ethical points [94–97].

1.3.2 Ethical Challenge of Oncotherapies

Safety and medical ethics represent an inherent problem. The very acceptable ethical attitude formed by Hippocrates: "Primum nil nocere" ["*First, do not harm*"] could be formally fulfilled if cancer therapy does nothing at all, like Gallen proposed. But the medical ethics of cancer therapy dictates completely differently: "Primum succurrere!" ["*First, hasten to help!*"]. Every cancer treatment potentially harms, and maximizing the benefit/risk (benefit/harm) could be the aim instead of the "nil nocere" principle. This apparent contradiction is solved by the general ethical principle "apply the necessary help as much as possible." This principle makes the evaluation of the actual therapy more complex than usual. The clinical response together with the QoL has to be evaluated, bringing into focus the survival time in a disease having presently a high mortality rate. Nowadays many discussions are devoted to this topic [98] also discussing the role of the US Food & Drug Administration (FDA) [99], and working out the rules for clinical trials according to evidence-based principles [100].

The risk/benefit ratio could be evaluated only in the light of the attained results together with the path taken ("the end does not justify the means"). The goals could be measured by clinical responses (how the actual diagnosis shows the changes). This later could refer to direct responses (complete or partial remission, stable disease [or no change], progressive disease; CR, PR, SD, PD, respectively). Also the disease-free survival, relapse time, local control and some other parameters could identify success. These all, however, have to be evaluated in the light of the survival rate (measured over a definite interval, conventionally 5 years is regarded as the disease-free time), which together with the QoL is more realistic from the patient's point of view. There is another factor which is not a medical point at all: the cost/benefit ratio. Limited financial resources could unfortunately modify the above actions distributing the possible help to all of the suffering patients.

1.3.3 The Challenge of Evaluating the Results

To evaluate the success parameters above, challenges also exist on the conditions and treatment guidelines:

- The proper definition of the dose, and its reproducible effect.
- The proper selection of the treatment area (if possible microscopic, cellular).
- Warrant of high-level safety.
- Proper control of the actions and reproducibility among actual treatment conditions.
- Collection of evidence-based proofs.

Hyperthermia is a complementary therapy. The modern complementary therapies are based on supplementing conventional therapies and not contradicting them. The aim is completing, supporting and helping, and not at all isolation from well-established conventional treatments. We do not think that alternatives exist in medicine: it is a unique possibility to cure the patients, and the so-called alternatives have a role only when conventional treatments have failed to offer a solution. Such treatments are not alternatives; hyperthermia is a complementary method. Hyperthermia is a part of human medical knowledge; it is "scholastic" in the meaning of a part of university education. Of course it is in its infancy (no satisfactory amount of proofs yet), but necessarily all and any medical methods started from this position, irrespective of its very "scholastic" position today. In fact, in Hippocratic terms, no alternatives to human medicine exist.

Medical therapies need verification, obtaining a solid proof of their overall benefit is mandatory. The established method to find proofs is the so-called "evidence-based medicine," (EBM) [101]. The definition of EBM is [102]: "the conscientious explicit and judicious use of the current best evidence in making decisions about the care of individual patients." The EBM proofs could be obtained from various sources, differing in their level of evidence [103]. The evaluation of clinical evidences is based on biostatistics [104], which uses hypothesis checks of the measured data. The positive evaluation of an applied method enables us to work out the generally accepted conventions in the actual method. The therapies of this category are the conventional methods. Two other categories are applied in medical practice: complementary and alternative medicine. Complementary medicine supports or expands the effects of the conventional methods, improves the results and/or personalizes them for the actual patient and/or suppresses the side effects, improving the QoL of the patient [105].

Alternative medicine offers an alternative solution of the treatment, and replaces the conventional and complementary methods by another option [106]. These methods in most cases are not categorized as EBM, they are still being proved in many cases, and they have no demand and/or facility for proof using rigorous biostatistical methods. These methods do have a place in the medical spectrum [107]; but their application needs special care. The primary objective of this care is obviously to cause no direct or indirect harm to the patient. The direct harm has to be evaluated on the well-accepted risk/benefit balance, while the indirect harm is more complex. The real danger of indirect harm is the isolation of the patient from the well-proven EBM methods, blocking their conventional treatment and increasing the risk of unknown and uncontrolled factors. This approach is too risky for any serious applications in such a fatal disease as cancer; and is only acceptable if conventional treatments are not available, not applicable, or not effective. In this sense, alternative methods in oncology are mainly used for palliation.

There are some ethical debates about EBM [108–110], and the problems of EBM and the observational studies (OS) have not yet been finally solved [111, 112]. Public trust has become unbalanced in relation to clinical trials [113], pointing to the statement of Frederici Di Trocchio, researcher of life of Casanova: "Swindle was used to an art. Nowadays it became a science too..." [114]. This "swindle-science" appeared to be a real danger, questioning real science and its solid proofs, it unsettles the general public and acceptance of the apparent, unproved "evidences" with oversimplified popular explanations became rife. These inaccurate approaches misinterpret and falsely explain even the solid facts and lead to fallacy and misconclusion in personal decisions.

The problems of the possible accuracy of data collection, the creditability of data evaluation, the fidelity and truthfulness of the interpretation intensifies the general medical challenges. The mandatory request of EBM, which is devoted to the handling of data collection and evaluation, also faces an immanent contradiction. The common tool of statistical evaluation is the hypothesis check with well-proven methods [115]. The significance level of the data is measured with the comparison of the distributions, which have to be based on normal- (*Gaussian-*) distribution. This postulation could be satisfied by identical participants, which is as a matter of course, not grantable. It is very unlikely that two genetically identical humans exist (except twins from a single ovum), not to mention the need for identical raising, identical conditions, and identical disease history to complete this perfect identification. This problem could be reduced by the central-limit theorem of statistics [116], which guarantees that the sum of a large number of stochastic variables obeys a Gaussian distribution. Regarding the human's complex system, cohorts could be constructed in this way. To make the cohort unified during treatment, chemo-studies fix the toxicity limit, and target the tumor of the individuals by this unified dose, irrespectively of the size of the tumor and ignoring other personal differences as well. This makes the study person-independent: all the doses and handling depend on the general human parameters (e.g. heights, weight, disease, surface, etc.), and are not modified by the actual tumor size or any other personal differences in the cohort group. This is the point, where a good protocol (strict inclusion and exclusion criteria) is obligatory. The strong, subjective and in most cases positive placebo effects [117] modify further the individual results, which (despite its positive influence on the patients recovery) then need to be eliminated to measure objectively the efficacy of the applied drug or process. These possible negative biases could be reduced drastically by proper randomization and by possible double-blind and placebo-compared grouping [118, 119]. Because of these complexities, clinical trials can not be multicentered without strict control of identical protocols and treatment conditions. This strict control introduces a new challenge: a trial effect exists [120]. The transfer of obtained data from one place to another, needs exceptional caution. To make EBM for essentially personalized and condition-dependent treatments could be extremely difficult and in many cases an unacceptably long process. After a successful trial the transfer of the obtained efficacy data from the highly equipped and experienced, well-controlled location of the clinical study to a general hospital with different conditions and different cultures worldwide is again a great challenge, emphasizing the complex and strict control of the medical establishments.

The usually popular endpoint of clinical studies, measuring the response rate (RR) could be misleading as well, because the median survival time (MST) could be unchanged or even worsened with definite growth of RR. For example, advanced colorectal cancer, stage IV, the trial for oxaliplatine/5FU/LV vs. 5FU/LV shows 53 and 16% remission rates, together with 19.4 and 19.9 months MSTs, respectively [121]. Another negative example of the RR is in a study of advanced breast cancer, stage IV, a trial shows 14.3% RR with 15.2 months MST [122]; while another trial [123] shows 25% RR with 11.5 months MST.

Negative bias could be introduced by various society conditions. Good examples for this bias are that every fourth clinical study is not published [124]; the proven advantages of a pharmaceutical are four times more frequent if it is sponsored by the producer rather than by others [125]; the significant results against placebo are three times more frequent than the non-significant results [126]. The evaluation of the data could be misleading due to personal conflicts of the evaluator: in a strictly competitive market the opinions are not independent and objective [127] and of course the conflict of interest could cause considerable bias [128, 129].

To formulate the challenge with the words of Berthold Brecht: "the aim is not to open a door to the infinite wisdom, but to circumscribe the infinite fallacy. ... The main reason of the poverty in science is the conceited property." [130].

It is not a question: EBM is an important factor to stop fraud and direct or casual fallacy by providing an objective and scientific approach, probing the actions with probability theory. However, EBM is not an overall omnipotent solution for the actual problems of cancer treatments. The experience and wisdom of the physicians in the actual case, counting the personal, individual character of the studied disease, considering the locally studied, treated organ not as a separated study object but as belonging to a human individual, these are the real modifications to the overall applicability of EBM in everyday medical practice. The narrow-minded limiting of medicine to only EBM leads to an automatism of medical practice ignoring the patients' personal demands. If this automatism were a winning concept, then one could not differentiate between an experienced physician with some 10 years practice and one who had just started his career and who learned how to find the "prescribed recipes" of the EBM. We have to stop this tendency to narrow down medicine to EBM making human connections secondary, degrading medicine to the use of fact-books with the applicable results of EBM. On the other hand, in parallel we have to stop the growing charlatanism and cheating of patients offering them non-proven, non-safe and non-legal treatments.

All of these above complications and challenges of EBM emphasize that the method is not a new science of philosophy to make human medicine objective, but "it is a continuously evolving tool to optimize the clinical practice" [131]. Some new statistical methods, breaking away from the hypothesis-check paradigm, (e.g. Bayesian survival statistics [132–134]), as well as single arm [135] and small trials [136]) are intensively developing and are at the center of various recent debates.

The evaluation of hyperthermia results bears all the hallmarks of the challenges of EBM and has some peculiarities also. The "heating procedure" is not identical with the "cooking process in the kitchen," making a definite "cook book" impossible. This method is very much personalized, and can not handle the "identical" patients with strictly unified treatment. No automatism exists, the actual conditions are important (e.g. patient extremely sensitive to heat, can not tolerate the chosen treatment, patient has severe side effects from previous treatments blocking the usually applied hyperthermia protocol, etc.). No real automatism exists in the parameters and treatment conditions; the maximal tolerable dose (personal decision) has to be given. The physician will not increase the power over the tolerance, even in the case when the treatment parameters do not reach the prescribed (defined) values, and moreover, if the patient tolerates higher power and more heat in the target area, it is desirable to achieve this if we are sure of the focusing, i.e. not producing a hot spot somewhere in the healthy volumes. The main challenge, however, for hyperthermia is in its late application, introducing it only in advanced, and often hopeless cases. Hyperthermia is applied on third or higher lines, for what to make any EBM has extreme complications. Even for simple drug studies treatments are not common, due to the inclusion and evaluation difficulties. This is again a factor, which makes the hyperthermia evaluation so complex and difficult.

Chapter 2
Hyperthermia Results and Challenges

Hyperthermia is not a widely acknowledged treatment, and there is no consensus even among its users. Its effects are mostly acknowledged, but the clinical studies have many challenging problems. Numerous supporters believe hyperthermia is the future miracle of oncology, and more believe the complete opposite, regarding hyperthermia as ineffective and a dead-end among the methods of oncology. Both approaches are basically wrong. Hyperthermia is one of the tools of oncology, having many problems and requesting detailed research in labs and in clinics. Both believers, positive and negative are annoying: believers must not be the basis of any serious medical approach. The facts are necessary! In this book we try to collect these.

2.1 Hyperthermia Approach

2.1.1 Definition of Hyperthermia in Oncology

Looking up various dictionaries, hyperthermia is mainly defined as a special disease or disease-inducing action. From Medicine.net (a medical dictionary) [137]: Hyperthermia: overheating of the body. This may be due to extreme weather conditions. Unrelieved hyperthermia can lead to collapse and death, particularly in the elderly. It is also known as heatstroke or heat prostration. Prevention via air conditioning, ventilation, and drinking extra water is the key for vulnerable persons. In emergency cases, i.v. infusion solution and rapid cooling of the body may be needed.

The free dictionary formulates the treatment and also [138] defined hyperthermia: An abnormally high body temperature, usually resulting from infection, certain drugs and medications, or head injury. Hyperthermia is sometimes created intentionally to treat diseases, especially some cancers.

The National Cancer Institute (USA) [139] fixes both variants: Abnormally high body temperature. This may be caused as part of treatment, by an infection, or by exposure to heat.

A. Szasz et al., *Oncothermia: Principles and Practices*,
DOI 10.1007/978-90-481-9498-8_2, © Springer Science+Business Media B.V. 2011

Oncology Encyclopedia concentrates on the treatment of course [140]: Hyperthermia is the use of therapeutic heat to treat various cancers on and inside the body.

Hyperthermia might appear to be a very new, modern therapy in oncology. However, completely the opposite is true: hyperthermia is the oldest identified weapon against cancer.

Despite the trivial and mandatory demand of the strict definition of the topic with which we are working, this is one of the most sensitive points among hyperthermia users. The above definitions use the "heat" and the "temperature" like synonyms. This is the clue for the disorientation. The emphasis in the definition on the heat or on the temperature could change the entire paradigm of the application and fundamentally could modify the technical tools of the area.

Interpretation of the word "hyperthermia" starts from the composition of words "hyper" (Greek: over, above, extensive . . .) and "therme" (Greek: heat). According to the free medical dictionary [141], hyperthermia is "An abnormal elevation of body temperature, usually as a result of a pathologic process." This definition is different to that of "fever" (Latin origin; "febris"). The latter definitely describes a pathophysiological process, having various origins of unusual interventions (viruses, bacterial or other toxins, incompatible proteins, extended necrosis, inflammation, etc.), or might be induced by failures in the temperature-regulating system.

Hyperthermia is a "modern" wording of heat therapy, however, it is misleading. The word originated from malignant hyperthermia, which is not at all connected to oncology. Malignant hyperthermia is a failure of thermal regulation. The only point, in which it may be equivalent, is whole-body hyperthermia; where the overall temperature is the factor. Local and regional heating could not be connected to the malignant hyperthermia phenomena.

In spite of the definitive difference, active overheating (hyperthermia) and higher body temperature (fever) are often mismatched and used like synonyms. This "glossary problem" is a source of many misunderstandings and false explanations. The heat (which is the origin of the hyperthermia) and the temperature (which is definitively connected to fever) are basically different categories (see Section 3.2.1). Embroilment of heat and temperature is the basis of many "belief" discussions among medical experts. One of the main reasons for calling the special hyperthermia, which is developed for oncology, "oncothermia" is to distinguish heat transfer from simple temperature elevation. However, naturally the increasing temperature of tissue targeted by heat is one of the numerous consequences of constrained heat transfer.

Mismatch of heat and temperature is supported by some of the early steps of curative hyperthermia, where the artificially elevated body temperature was the source of heating (see later), which is used even today for some applications (passive hyperthermia [142]). Also the definite temperature symptom of malignant hyperthermia supports the jumble of definitions. The term *malignant hyperthermia* is by definition [143] an *inherited disease, that causes a rapid rise of body temperature (fever) and severe muscle contractions when the affected person undergoes general anesthesia.*

This systemic irregular homeostasis of the body is categorically different from our present topic: oncological hyperthermia.

To bring clarity to this basic definition problem we have to start from the goal of hyperthermic oncology. The common aim of all oncological treatments is undoubtedly to destroy completely and selectively malignant cells. The only differences between the methods are the applied tools for this task. Oncological hyperthermia uses heat energy to fulfill its function, and applies this special energetic tool to solve the challenge. The definition has to be centered on the heat, energy and work, not on the temperature. The energy source can be from outside (active hyperthermia, like local or regional heating) or make use of the resources of the living object (passive hyperthermia, fever therapy) induced by toxins or other fever-inducing agents. The temperature changes in consequence of the energy intake. Its intensive nature of course tries to eliminate the differences, the possible gradients which provide real work, gradually disappear. The temperature character is the homogeneity, and the dose measures its homogeneity as a definite request in the targeted area. The main postulate to make a proper selection remains a dream, the selection made blindly, supposing the major boundaries of the tumor.

The problem of the traditional hyperthermia definition on average (by the temperature) is historical, originally applied in malignant hyperthermia. However, fever is systemic, and the system is in equilibrium (homeostasis), which is easy to characterize by the average (temperature). Nonetheless, if the system is only locally different from the whole, the temperature has no such characteristic meaning. The temperature is not the parameter, which creates and keeps the actual local conditions. The feature is the absorbed energy, which is converted to heat, and through this can be characterized the temperature. Nevertheless, if the transmitted energy does nothing else, but increases the average energy of all the parts of the system, what are we doing the treatment for? The goal of the treatment (like the general paradigm in oncology) is to destroy the malignant cells. The distortion changes the structure and the chemical composition of the tissue. The average energy (temperature) does not describe a certain distortional ability, but defines the average energizing only. In this sense, the temperature measures the energy which is distributed all over the target, but does not have a definite task to act. In this meaning and as the most trivial example if we provide only heat and in consequence higher temperature for a living system, it never can use this energy for its metabolism, which always requires certain chemical energy (nutrition) which addresses the energy to the chosen chemical reactions.

Discussion of the mechanisms of oncological hyperthermia is a permanent task of the medical community [144] leading to an increasing number of international hyperthermia conferences, books [145, 146, 17], and journals [147]. Publications and an increasing number of clinical trials also started to appear in the top, highly prestigious medical and scientific journals [148–150].

The growth of peer-reviewed (PubMed registered) publications has continued for three decades. The cumulative number of publications in oncological hyperthermia reached 9,000, and about 8% of these are clinical trials (containing many randomized, controlled, prospective ones) [151] (see Fig. 2.1).

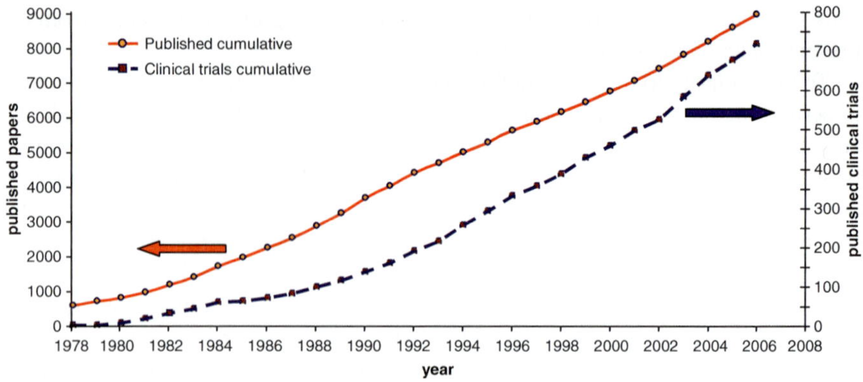

Fig. 2.1 The cumulative number of oncological hyperthermia publications. Search profile: (cancer OR tumor OR oncology OR neoplasm OR malignant) AND (hyperthermia OR heat-therapy OR thermotherapy) NOT (malignant-hyperthermia OR fever), the clinical trials were searched by the limits: (clinical trial OR randomized controlled trial)

At present, the acceptance of hyperthermia is not complete, but large handbooks comprehensively summarizing oncological radiology [21], and oncological therapies in general [152, 22] devote reasonable space to discussing this method.

2.1.2 Basic Concepts of Oncological Hyperthermia

To categorize the various heating systems first we group the treatments by their basic heating mechanisms in the tumor (see Table 2.1). The differentiation of heating mechanisms is based on the role of the blood, which could work like a heater (e.g. in whole-body treatment) or like a cooler (e.g. in local deep heating).

The systemic (whole body) application heats the body (and the tumor) by directly heated blood. Blood perfusion transfers the heat over the entire body. The heat source in the deep-seated tumor is the blood flow.

Local/regional hyperthermia works by energy/heat absorption in the targeted volume, and the rest of the body is not treated. Consequently, the role of the blood in this method category is just the opposite: the blood remains at the body temperature, and the blood flow is a cooling medium in the heated volume.

Table 2.1 Basic categories of hyperthermia in oncology

Active physical effect	Example
Heat delivery	Conduction, convection, radiation, bioactive
Energy source	Chemical, biological, mechanical, electromagnetic
Invasivity	Noninvasive, semi-invasive, invasive

Table 2.2 Typical physical parameters of the basic hyperthermia treatment categories

Basic hyperthermia categories	Typical absorbed energy density (SAR) (W/kg)	Typical operating temperature (°C)	Typical treated mass (kg)
Systemic (blood heating)	5–10	38–42	40–100
Local/regional (tissue heating)	10–50	40–45	1–25
Ablation (tissue burning)	5,000–25,000	60–250	0.001–0.02

There are some hyperthermia methods where the applied energy density is so high that the process is very quick and has the surgical effect of drastic ablation. In these processes the physiology of blood-flow has a marginal or no role.

The typical specific energy absorption rate (SAR), the typical operating temperature and the treated mass are summarized in Table 2.2. The requested heat-flow intensity through the skin in the case of external (noninvasive) applications are typically in the range of 0.03–0.1 and 0.1–1 W/cm^2 in systemic and in local/regional treatments, respectively. The ablation methods work on a typical necrosis basis, and the energy is typically provided by impedance heating with minimally invasive electrode insertion; no heat flow exists through the skin. The modern laser ablative techniques work with ultrafast pulses with ultra large energy-density flow. Depending on the pulse duration it can be a few 100 W/cm^2 and in *ns* intervals can go to 10^7–10^8 W/cm^2 [153], but the provided energy in total reaches a maximum of a few tens of watts.

The same forwarded energy exposition with identical energy flow (W/m^2) can cause different energy absorptions depending on the given conditions [154, 155], the actual organ [156], and the actual frequency [157].

There are clear differences between systemic whole-body hyperthermia (WBH) and local/regional hyperthermia (LRH). Both treatments have a lot of proven advantages and in parallel, both of these have effects that are not entirely clear, and are therefore surrounded by much doubt and skepticism. Both treatment modalities are more and more accepted in the oncological-radiological communities, but they are also both battling to move from the biomedical experiment status to the clinically proven one [158, 159]. Considerations of the risk/benefit or cost/benefit ratios on one side, and the evaluation of the limited amount of controlled randomized clinical studies on the other place different weights into the decisional balance. The complementary curative and palliative applications of hyperthermia are an excellent mode to enhance the effect of the conventional therapies and reduce their possible side effects.

Systemic treatment uses blood circulation to heat up the whole body. The oldest such method is contact heating, immersing the patient into a hot bath, but due to its numerous disadvantages this method is rarely in use today. Mainly two direct methods are available in modern medicine to carry out systemic hyperthermia: the less frequently used extra-corporal blood heating, that is, transport of the blood in a continuous flow through a definite arterial outlet, and pumping the externally heated blood back into the patient. The other method, in situ heating, relies on heating up the blood in the capillary bed of the corium and sub cutis, which heats up the entire blood stream and in this way the whole body. This method also has some special solutions: direct conductive heating on the epidermis by water steam (a sauna-like solution) as well as deep subcutaneous heating by radiation. The method has many early descriptions [160–163]; but the dominant systemic hyperthermia method is based on infrared radiation by multi-reflecting filtering [164, 165] or by water filtering [166–168].

Local/regional hyperthermia also has large technical variability. The old direct heating methods (hot solids or liquids in the area or in the nearest body cavity) were not effective enough to reach the required deep heating of the local area without damaging the surface layers. A real technical solution was only available after the discovery of electromagnetic heating. The appropriately chosen electromagnetic waves could penetrate into the body deeply, and it is possible to keep these absorptions in a desired locality. These methods are limited of course by the cooling effect of the actual blood flow from the unheated area of the body. The main local/regional systems work by radiative [149, 169, 170], or by capacitive [171–173] technical solutions.

Thermodynamically, systemic and local/regional treatment differ in the amount of energy intake, but the natural (biological) cooling volume is identical. Naturally, in order to heat up and reach equal temperature in the whole body, more energy is necessary, than to heat up only a local volume, which will never be in thermal equilibrium with its neighboring areas.

The two basic kinds of heating methods also differ by their physiological limitations: the systemic treatment of course modifies the entire physiology of the organism, which can limit the applied energy absorption and body temperature (see Table 2.3).

Systemic treatment uses the thermal equilibrium phase (homeostasis, plateau) for the treatment, where the temperature is an exact and comparable parameter to control the treatment. Such treatment is not selective, heats the whole body. Its main goal is not direct cell destruction, it makes completion for other oncotherapies.

Table 2.3 Differences and limits of the technical conditions of the heating methods

	Hyperthermia	
Technical parameters	Local/regional	Systemic
Time to reach the stationer state	5–50 min	30–180 min
Complete treatment time	30–90 min	120–720 min
Maximal safe temperature (in the tumor)	Not limited (incl. ablation)	42°C
Systemic safety control (temperature, pulse rate, pO2, ECG, blood pressure, electrolytes, etc.)	Not relevant	Necessary
Local safety control	Necessary	Not relevant

2.1.3 Technical Variations of Hyperthermia in Oncology

To heat up the malignant tumors is apparently a simple task, however doing it at the desired depth, making accurate selection (focused on the malignancy only), and reaching the energy delivery demand into the target is not a simple technical task. This complication was the basis of the phrase: "the physics are against us [158].

Not only the complexity of living objects and their various complicated mechanisms make it difficult to find a solution, but the variety of different methods make a balk and form the situation tanglesome. We can be sure that the highly necrotizing treatments like ablations (e.g. RF or laser), or concentrated mechanical (ultrasound) energy (e.g. High-Intensity Focused Ultrasound, HIFU) have different reaction mechanisms. It is even more likely that the minimally invasive interstitial, or the semi-invasive intra-luminar/intra-cavital methods differ in their action, and also, they definitely work on another principle to the non-invasive, non-ionizing radiations or radio-frequency conductions. Not only the methods are expected to darken our understanding, but the local, regional, part-body and whole-body (systemic) applications probably have different effects due to the activation of a different fraction of the whole system.

A tremendous number of different methods exist to apply active hyperthermia [174]. Simple natural heating methods like the sauna and hot bath are accompanied by sophisticated mechanical (like ultrasound) or electrical (like non-ionizing radiation) techniques, as well as their various combinations. Because of a missing definitive dose concept, the methods using different set doses for particular treatments are not comparable. Moreover, because of the lack of a clear explanation of the mechanisms, many methods are ineffective or even contra-effective, but are still in practice. For example, some treatments are carried out in a sauna or infra-cabin, imitating whole-body hyperthermia. However, due the missing increase of body temperature these methods probably have no effect on cancer. Just the very opposite might even happen, as the living system, in order to maintain homeostasis, increases the pulse rate and acts in various other ways to regulate body temperature. Some of these might actually promote metastases.

The active physical parameters could be differentiated by the heat-delivery method, by the character of the energy source, and by the invasivity of the method (see Table 2.4).

One of the decisional parameters involved in choosing the appropriate method from the various technical solutions is the location of the target tissue (see Table 2.5), and the optimal energy production (see Table 2.6) has to be chosen for proper treatment.

Table 2.4 Main technical parameters of the oncological hyperthermia methods

Active physical effect	Example
Heat delivery	Conduction, convection, radiation, bioactive
Energy source	Chemical, biological, mechanical, electromagnetic
Invasivity	Noninvasive, semi-invasive, invasive

Table 2.5 Technical solution of oncological hyperthermia depending on the location of the target tissue

Location of the target	Technical solution
Superficial	Infrared, microwave, surface currents, conduction
Interstitial/intraluminar/intracavital	Electromagnetic energy delivery, laser-infrared, conductive/convective heating
Deep-seated target	Electromagnetic or mechanical energy delivery, invasive methods
Whole body	Conductive/convective/radiative heating

Table 2.6 The energy-production variations for oncological hyperthermia

Energy production	Example
Contact methods	Convective/conductive heat transfer
Chemical methods	Nutrition for higher/driven metabolism
Biological methods	Pyretotherapy or pyrotherapy (could be chemical, bacterial, virus, etc.)
Mechanical heat production	Ultrasound (HIFU/sono-hyperthermia)
Electromagnetic heat production	Radiative/electric/magnetic type of generation

The contact methods use convective or conductive heat transfers.

Each method has various subdivisions which are realized in different technical solutions. The contact methods (see Table 2.7) work with local conductive mechanisms (e.g. heated intra-luminar or intra-cavital applications), in which the heating mechanism makes use of natural tissue heat conduction and the heated blood in the area. The heat-conduction processes are widely used in our everyday life and in hyperthermia as well. It is technically simple, can be implemented by using different methods including simple hot water or other environmental heat sources. The

Table 2.7 The methods of contact heating

Contact method	Example of technical solution
Conduction	
Hot catheters	Electromagnetically or mechanically or circulatorily heated
Hot bath	Hot water, hot wax, hot mud, hot sand, etc.
Hot cabins	Heat- and steam-boxes/cabins, non-wet infracabins, blankets
Convection	
Hot fluids	Hydro-colon therapies, drinking therapies, other gastrointestinal
Extra-corporeal fluid heating	Heat-exchange of body fluids extracorporally
Intra-peritoneal heating	Open- or minimally invasive intra-peritoneal hot bath
Mechanical excitation	
Ultrasound heating	Focused heating by ultrasound, including HIFU, intracavitary, etc.

heat energy is transferred into the body through heat diffusion using a direct contact, conductive heating method. Fundamentally, these conductive practices work by the heat of the blood: the outer skin or the internal epithelium is heated up, and the heat is then distributed by heat flow (convection) further. The physiological affluence of blood creates a perfect heat exchanger, which results in a suitably effective heat transfer. A physiological positive feedback promotes the heat transfer, the blood perfusion is increased via the increasing temperature. This blood-flow-supported conduction because of its relatively high absorption of energy has a high heating rate. Direct heating is not optimal for local treatments, it is mainly applicable to the systematic warming, to the whole-body hyperthermia. In this case of course a large surface area needs to be heated to avoid the burn caused by the large heat flow through the given surface area. Local warming up by this heat transfer can not be controlled. Therefore, its possible application is the treatment of non-deep-seated areas in a very restricted way. It should be noted that the possible heat insulation features of adipose tissues may screen the required effect. Nevertheless, this kind of hyperthermia is still being used for the treatment of rheumatics and muscle spasms. However, its oncological use is ineffective, and due to the uncontrollable processes in fact this method is no longer used.

In conductive/convective hyperthermia applications, hot-water cavity heating (hot water catheters, intra-peritoneal hyperthermia by using either minimal invasive or complete invasive intervention, extra-corporal blood heating etc.) is the most common.

Very common conductive methods include use of electric currents through the tissues or body part. Practically, this electro-conductive method applies the principle of convective heat transfer by ions of body electrolytes, as well as heat conduction by heat diffusion from the place where the heat is originally generated. It could be galvanic (direct current, DC) or alternating current (AC) conductivity of tissues at relatively low frequencies avoiding physiological stimulation of the nerves. Long- and medium-wave (<500 kHz) radio frequency (RF) applications are common, and this is called *impedance heating*. Technically, the current is introduced into the body by using invasive or semi-invasive (intraluminar or intracavital) methods. Significant advantages of RF impedance heating applications are the proper ability to focus by the position of the inter-stitial or intra-luminar applications and the relatively precise adjustment of energy input. This method in many cases requires more attention and sometimes also an interventional radiologist or a minimally invasive surgical intervention. The significant advantage of impedance methods is the intensive effect on natural biological heat production in the treated tissues. This natural support makes real hyperthermia with low power (eg. electro cancer therapies (ECT)).

Ultrasound heating is a method, heating with a mechanical wave source. It has an ablative solution (HIFU) also.

The conductive and convective heating processes are in fact not distinguishable in living objects, where a combined heat-transfer mechanism is active. For example: during conductive heating of the surface the blood stream continues the transfer by convection, and on the other hand convective heating of the peritoneum (e.g. intra-peritoneal hyperthermia) heats the tissues by diffusion, so using conduction.

Table 2.8 The methods of radiative heating

Radiative method	Example of technical realization
Electromagnetic near-field	*(Wavelength)>> (target distance from the source)*
Electric field	Capacitive coupling
Magnetic field	Inductive coupling
Pointing vector (electromagnetic energy flux)	Radiative-antenna coupling (below microwave frequencies)
Electromagnetic far-field	*(Wavelength)<< (target distance from the source)*
Pointing vector (electromagnetic energy flux)	Absorbing directed radiation (mostly microwave)
Some resonance methods	Biophotonic and bioresonance actions
Mixed situations	
Phase-array radiation	(Wavelength) ≈ (target distance from the source)
Body resonance methods	(Wavelength) ≈ (target size)

In the early days, simple heat diffusion was utilized using – as we can see above – hot water or wax baths and heated objects [175]. Today, focused and unfocused energy delivery using electromagnetic fields are used (see Table 2.8).

These radiative methods have many additive utilized effects, most commonly some resonances: like microwave (Holt's method [176, 177]), ionic cyclotron resonance [178], fluctuations/noises [179], etc.

An older and nowadays not widely used method category is the bioactivation of hyperthermia (see Table 2.9), which could be applied systemically as well as locally. The use of electric stimulation however is expanding among the experts, using other basic effects of the electric fields [180].

Selectivity (see Table 2.10) is obviously not applied for systemic treatment alone, but in complementary applications, together with other therapies (local ionizing

Table 2.9 The methods of biologically stimulated heating

Bioactive method	Example of realization
Injection of pyrogenic material	Organic or inorganic compounds
Infections	Bacteria, viruses (toxin production)
Stimulations	Electric, mechanical, or chemical stimuli

Table 2.10 The selectivity of the various hyperthermia techniques

Technical solution	Selectivity (local/regional)
Bioactive	By its placing, injection, etc.
Conductive	By the heat-source placement, quick spreading
Convective	By the heat-source placement, quick spreading
Radiative	Mostly selective by artificial or by self-focusing

radiation, local or selective chemotherapies, selective hormone therapies, etc.) it is selective. There the opposite direction of enhancement of selectivity exists: locally focused hyperthermia can activate only locally heat/temperature-sensitive drugs or other therapy modalities. The most exciting and selective tumor targeting of this type is temperature-sensitive liposome-jacketed drug delivery [181], as well as the drug delivery with temperature-activated nanoparticle complexes [182].

The invasivity of the methods (see Table 2.11) is also very different. Obviously, the most popular among the patients are the non-invasive methods; however the physicians choose the modality in which they are most experienced (and one that is actually available) to attain the optimal result.

There are definite technical expectations from the various heating techniques (see Table 2.12). The requirements are mostly ethically and medically motivated, make the (anyway very simple) hyperthermia techniques delicate, taking it actually acceptable in the human medicine.

Table 2.11 The invasivity of various hyperthermia techniques

Selected methods	Invasivity
Antenna-phase array heating	Noninvasive (except – temperature measurement)
Antenna microwave applications	Noninvasive (except – temperature measurement)
Capacitive impedance heating	Noninvasive (energy control)
Capacitive radiative heating	Noninvasive (except – temperature measurement)
Extra-corporeal body-fluid heating	Minimally invasive
Extra-corporeal non-body-fluid heating	Semi-invasive (fluids in body lumens/cavities)
Inter-stitial heating (incl. needle-galvanic therapy)	Minimally invasive
Inductive heating (incl. suspensions, seeds, rods, etc.)	Minimally invasive (insertion and remove the materials)
Inductive (Eddy current) heating	Noninvasive (except – temperature measurement)
Intra-peritoneal hyperthermia	Invasive (open surgery) or minimally invasive
Infra- and other local heating	Noninvasive (except – temperature measurement)
Intra-luminar catheter-galvanic therapy	Semi-invasive
Intra-cavitary/intra-luminar impedance heating	Semi-invasive
Intra-cavitary/intra-luminar heat-conductive catheters	Semi-invasive
Microwave catheter applications	Semi-invasive
Thermoablation (incl. microwave, laser, etc.)	Minimally invasive (impedance control)
Ultrasound (incl. focused array, HIFU, etc.)	Noninvasive (except – temperature measurement)
Ultrasound catheter applications	Semi-invasive (except – temperature measurement)

Table 2.12 The most important technical demands for treatment techniques

Treatment demand from the technique	Explanation
General demands	
Reproducibility/stability	Stable, reliable, trustable (and understandable) technique
Control/dose	Clear treatment parameters, dosing, comparability, unitary method
Personalization/flexibility	Modify/optimize for the particular patient and the actual stage of the tumor
Quality guidelines/protocol	Fixed, accepted, (possibly consensus-based) protocols, treatment guidelines
Safety/low risk	Fits to medical standards, approved, low side effects, low (risk/benefit) ratio
Demands for systemic treatments	
Quick heat-up process	Reasonable heating time to keep the complications low
Stable plateau	A constant and suitably long plateau phase at the desired temperature
Demands for local/regional treatments	
Deepness/penetration	Reach the deep-seated target, the tumor at depth
Selection/focus	Select the tumor only, do not affect the normal surroundings

Technically different solutions are applied for all the treatment types: the categories of systemic-, local/regional-, ablative- and some unclassifiable treatments have various subcategories and those are subdivided in more detail as well.

For whole-body hyperthermia the extra-corporeal, intra-corporeal and intra-cavitary treatments are the most common ones (see Table 2.13). The intra-cavitary treatment mode [183], is not permeated among practitioners, because of possible complications due to inhalation injury.

The deep-heating local methods have different technical characters (see Table 2.14). All the categories have many subgroups, involving various technical solutions.

The ablative solutions are the transition methods between surgery and hyperthermia. The main idea is the entire coagulation, ablation of the tissue. Its technical variants are summarized in Table 2.15.

Table 2.13 Typical technical solutions of systemic hyperthermia in oncology

Systemic (whole-body) treatment	Technical character
Extracorporeal	External heat exchanger for the blood flow
Intracorporeal	Infrared or radiofrequency heating of the capillary bed
Intracavitational	Breathing of hot-gas mixture

Table 2.14 The technical characters of the various local deep-hyperthermia methods

Local deep-heat treatment	Technical character
Noninvasive	Electromagnetic or mechanical (ultrasound) energy transfer
Catheter operated	Electromagnetic, mechanical or hot-material contact
Interstitial, minimally invasive	Electromagnetic or hot-material contact

Table 2.15 Technical variations of the ablative hyperthermia solutions

Ablative treatment	Technical character
Minimally invasive	Locally large electromagnetic energy density
Noninvasive	Locally large mechanical (ultrasound) energy density

Table 2.16 Summary of the mixed superficial hyperthermia techniques

Unclassifiable, miscellaneous heating	Technical character
External surface heating	Local infrared or conductive/convective heating
Internal surface heating	Intra-luminar/cavital conductive/convective heating

There is a large category of hyperthermia techniques (see Table 2.16) which are partly based on some solutions of previous categories, and are partly used as independent hyperthermia applications. They can be applied in whole-body treatment (heating the blood in the capillary bed) and in many local solutions (external or internal [catheter] arrangements) too. These are dominantly superficial (external or internal surfaces of the body) effects. External superficial treatment is active on the skin, characteristically the deepest action is in the dermis and in the subcutaneous capillary bed, the hypodermis (superficial- and deep fascia) has no active role, but by heat conduction they are also part of the treatment.

Despite the large number of heating methods, presently the electromagnetic methods dominate in the oncological hyperthermia field. Through concentration of electromagnetic solutions the wide-ranging heating modes abate. The division via their important parameters is shown in Table 2.17.

Table 2.17 Division of electromagnetic heating methods

Division by electromagnetic heat-delivery methods	Technical character
By frequency	Infra, microwave, radiofrequency, low frequency
By radiation/conduction	Radiative, conductive, mixture
By target distance	Far-field, near-field
By intensity (SAR)	Ablative, active heating, stimulating
By wave phase	Phase-dependent, independent
By electrolyte conduction	Electrolyte-selective, not selective

Table 2.18 The spectrum of frequencies and their connection to hyperthermia

Division by electromagnetic frequency	Methods	Frequencies (f) (Hz)	Wavelengths (in vacuum) (λ) (m)
X-rays, γ-rays	Not in use for hyperthermia	$f>3\cdot10^{18}$ (>7,500 PHz)	$\lambda<1\cdot10^{-10}$ ($\lambda<0.1$ nm)
Ultraviolet	Not in use for hyperthermia	$3\cdot10^{18}>f>7.5\cdot10^{14}$ (7,500 PHz>f>750 THz)	$1\cdot10^{-10}<\lambda<4\cdot10^{-7}$ (0.1 nm<λ<400 nm)
Visible	Not in use for hyperthermia	$7.5\cdot10^{14}>f>4\cdot10^{14}$ (750 THz>f>400 THz)	$4\cdot10^{-7}<\lambda<7.5\cdot10^{-7}$ (400 nm<λ<750 nm)
Near-infrared	IR-A, IR-B, IR-C	$4\cdot10^{14}>f>2.1\cdot10^{14}$ (400 THz>f>210 THz)	$7.5\cdot10^{-7}<\lambda<1.4\cdot10^{-6}$ (750 μm<λ<1.4 μm)
Far-infrared	Far-infrared, terahertz	$2.1\cdot10^{14}>f>1\cdot10^{12}$ (210 THz>f>1 THz)	$1.4\cdot10^{-6}<\lambda<3\cdot10^{-4}$ (1.4 μm<λ<300 μm)
Microwave (μw)	Far-μw, near-μw	$1\cdot10^{12}>f>5\cdot10^{8}$ (1 THz>f>0.5 GHz)	$3\cdot10^{-4}<\lambda<0.6$ (300 μm<λ<60 cm)
Radiofrequency (RF)	High-RF, middle-RF, low-RF	$5\cdot10^{8}>f>2\cdot10^{4}$ (0.5 GHz>f>20 kHz)	$0.6<\lambda<1.5\cdot10^{4}$ (60 cm<λ<15 km)
"Audio" range frequency	Not in use for heat production (TENS)	$2\cdot10^{4}>f>1\cdot10^{2}$ (20 kHz>f>100 Hz)	$1.5\cdot10^{4}<\lambda<3\cdot10^{6}$ (15 km<λ<3,000 km)
Very low frequency (mains)	AC	$1\cdot10^{2}>f>10$ (100 Hz>f>10 Hz)	$3\cdot10^{6}<\lambda<3\cdot10^{7}$ (3,000 km<λ< 30,000 km)
Extremely low frequency	Not in use for hyperthermia	$10>f>0$ (10 Hz>f>0)	$3\cdot10^{7}<\lambda$ (30,000 km<λ)
Direct current (no frequency, battery)	Galvanic technique (stimulated heat)	Not applied (zero)	Not applied (infinite)

The heat delivery is strongly frequency-dependent (see Table 2.18). The applied frequencies are: low frequency (impedance heating), radiowave, microwave, and infrared radiation.

The highest frequencies in use for hyperthermia are the infrared regime, between 1 and 400 THz. The natural infraregion of the sunlight has mostly the so-called IR-A, IR-B, and IR-C radiations [184] (see Table 2.19).

Table 2.19 The infrared radiation with medical relevance

Methods (IR) $7.5\cdot10^{-7}<\lambda<6\cdot10^{-6}$ (750 nm< λ<6 μm)	Frequencies (f) (Hz)	Wavelengths (λ) (m)
IR-A	400 THz >f> 210 THz	$7.5\cdot10^{-7}<\lambda<1.4\cdot10^{-6}$ (750 nm < λ < 1.4 μm)
IR-B	210 THz >f> 100 THz	$1.4\cdot10^{-7}<\lambda<3\cdot10^{-6}$ (1.4 μm < λ < 3 μm)
IR-C	100 THz >f> 50 THz	$3\cdot10^{-6}<\lambda<6\cdot10^{-6}$ (3 μm < λ < 6 μm)

Table 2.20 The percentages of energy absorption in the layers of human skin

Methods (IR)	Absorption into the surface region (%) Epidermis (0–0.5 mm)	Dermis (0.5–3 mm)	Subcutis (3–25 mm)	Wavelength (μm)
IR-A	35	48	17	0.8–1.4
IR-B	72	20	8	1.4–3
IR-C	100	0	0	3–6

In this region of the electromagnetic waves the penetration depth is very small, except the so called "water window" (IR-A region), where a considerable part of the energy can reach the deep subcutaneous layers to heat up the blood in the capillaries – see Table 2.20.

This IR-A water-window effect is intensively used for whole-body treatments by multi-reflection filtering [185], by water-filter solution [186], and by multi-layer reflection filtering [187]. The facility is used in local hyperthermia treatments [188] as well. The present book concentrates on oncothermia, which is a local/regional method; consequently the whole-body hyperthermia treatment is not emphasized in the following.

The extensively used oncological hyperthermia is presently connected to radiofrequency applications. Three characteristically different electromagnetic heating techniques exist in this regime: electric-field-coupled energy transfer (capacitive coupling), magnetic-field-coupled energy transfer (inductive coupling), and radiative energy transfer (antenna coupling). We place less emphasis on the emerging ablation technologies, which are more and more becoming a tool of surgery and interventional radiology, isolated a little from general hyperthermia practices. Their intensive necrosis and burning-type radical solution differ from the gentler hyperthermia effects. In general hyperthermia, burning is not usual because of the less effective focusing, therefore, these methods generally avoid the radical burn-necrosis regarding it as a kind of toxicity (hot spots and adipose burn [fat necrosis]).

The most important physical parameters used for these techniques are connected to the basic modern treatment facilities of local/regional hyperthermia (see Table 2.21).

Numerous heating techniques exist for local hyperthermia. We will focus our discussions on electromagnetic heat deliveries for the targeted treatment of any malignancy.

Electromagnetic, "deep thermal" (not convective and not conductive, not invasive) heating can be considered significantly better than convective and conductive methods. The achieved effect depends on the intensity, frequency; and phase of the applied electromagnetic effect. The technical solution of the various coupling categories are unequivocally different in terms of their position within the electronic resonant circuit (see Fig. 2.2) where the technical realizations also definitely differ.

Table 2.21 The active parameters of electromagnetic heat delivery

Active parameters at electromagnetic heat delivery	Denotation (unit)
Electric field	E (V/m)
Magnetic field	B ($T = \text{Vs/m}^2$) (in vacuum: H (A/m))
Electromagnetic radiation (energy flux)	Poynting vector, S (W/m^2)
Electromagnetic photon energy	Photon-quant $E = \text{hf}$, $h = 6.63{*}10^{-34}$ (Js)
Electromagnetic field energy	Energy density (W/m^3)
Dielectric constant (permittivity)	Ratio to vacuum ($\varepsilon_0 = 8.86{*}10^{-12}$ (F/m))
Magnetic permeability	Ratio to vacuum ($\mu_0 = 4\pi{*}10^{-7}$ (N/A^2))
Electrical conductivity	σ (S/m)
Electrical impedance	Z (Ω) (complex value)

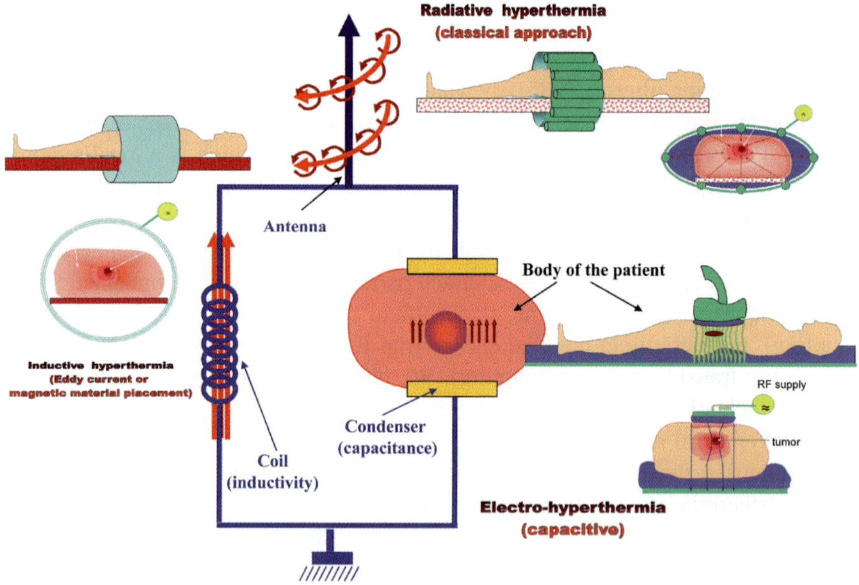

Fig. 2.2 The main versions of electromagnetic heat delivery methods

A complete comparison of the most important electromagnetic hyperthermia methods is given in Table 2.22. The conductive type represented by capacitive coupling (capacitor) and magnetic/inductive coupling (coil) work in the so-called "near-field" regime, where the source–target distance is considerably less than the wavelength of the actual field. For radiatively coupled electromagnetic fields (antenna couplings, radiative capacitor coupling) the far-field effects are important to calculate the dosing accurately. Most microwave applications represent the far-field approximation, since the applied wavelength is smaller than the target distance. The division of the RF-ranges (see Table 2.23) points out the application area as well. For low-frequency microwave or high-frequency radiofrequency

Table 2.22 A comparison of the main parameters, focusing ability, frequency, and transmitter, for electromagnetic hyperthermia

Parameters, conditions for treatments of humans	Type of method							
	Antenna		Capacitor		Coil		Resistance	
	Low frequency	High frequency	Low frequency	High frequency	Low frequency	High frequency	Low frequency	High frequency
Dominant effect	Near-field	Far-field (waves)	Change influence	Capacitive admittance	Magnetic effect	Inductive effect	Resistance	Impedance
Dominant parameter/field	**E** [V/m], **B** [T], En [J/m^3]	**S** [W/m^2]	**E** [V/m]	**E** [V/m], S [W/m^2]	**B** [T]	**B** [T], En [J/m^3]	Q [C] (change)	**I** [A] (current)
Area of dominant heat-loss	Entire tissue	Near-surface area	Extracellular matrix	Near-surface area	Entire tissue	Entire tissue	Extracellular matrix	Extracellular matrix
Usual conditional transmitter	Air	Air	Water-bolus	Various dielectricums	Air	Air	Conductive material	Conductive material
Usual internal transmitter	None	None	None	None	Magnetic material	Non	Non	Non
Usual forwarded power [W]	150–2000	500–3000	1–200	20–2000	50–100	50–500	0.01–100	5–200
Usual selection	Parametrical focus	Parametrical focus	Not selective	Self-focusing	Parametrical focus	Not selective	Positional focus	Positional focus
Usual invasivity	Non-invasive	Non-invasive	Non-invasive	Non-invasive	Invasive (transmitter)	Non-invasive	(Minimally-) invasive	(Minimally-) invasive
Usual safety control	Temperature	Temperature	Energy flux (skin)	Energy flux (skin)	Material-character	Skin-temperature	Resistance/charge	Impedance/temperature
Usual dose control	CEM43°CT_{90}	CEM43°CT_{91}	Energy dose [J/kg]	CEM43°CT_{91}	CEM43°CT_{91}	CEM43°CT_{91}	Charge [C]	Energy dose [J/kg]

Table 2.23 The division of the RF-applications by frequency

Radiofrequency $5 \cdot 10^9 > f > 2 \cdot 10^4$ (0.5 GHz$>f>$20 kHz)	Methods	Frequencies (f) (Hz)	Wavelengths (λ) (m)
High-RF	Radiative heating	$5 \cdot 10^3 > f > 5 \cdot 10^r$ (0.5 GHz $> f >$ 50 MHz)	$0,6 < \lambda < 6$ (60 cm $< \lambda < 6$ m)
Middle-RF	Capacitive and impedance heating	$5 \cdot 10^r > f > 1 \cdot 10^5$ (50 MHz $> f >$ 1 MHz)	$6 < \lambda < 300$ (6 m $< \lambda < 300$ m)
Low-RF	Impedance heating and ablations	$5,1 10^5 > f > 2 \cdot 10^4$ (1 MHz $> f >$ 20 kHz)	$300 < \lambda < 1.5 \cdot 10^4$ (300 m $< \lambda < 15$ km)

(100–200 MHz) the wavelength and the source–target distance are comparable, therefore focusing requires numerical approximations with limited accuracy.

The wavelength of the applied RF is the longest in a vacuum, and shortens by the dielectric permittivity and conduction of the media through which it travels.

Conductive (resistance) heating was the very first "modern" method of oncological hyperthermia, which started in the late nineteenth century, and was called "galvanocautery" [189]. The method was further developed by D'Arsonval introducing the impedance (alternating current [AC], later higher frequencies, even spark-generated currents) calling it "Arsonvalization" [190], and later a more modernized form was "fulguration" [191].

This method was developed in three branches: interstitial hyperthermia, including galvanic heat stimulation (*electrochemical cancer treatment*), the ablation techniques, and capacitive coupling. The first capacitive-coupled device on a conductive basis was the "Universal Thermoflux." It was launched on to the market by Siemens, a giant of the electric industry at that time, and was later further developed, and the new device called "Radiotherm" was launched on to the market in the early 1930s. The beginning of the new capacitive-coupling technologies was in 1976 by LeVeen [192] and they have been widely applied since [193, 194, 173]. Most hyperthermia devices use capacitive coupling since it requires no extra shielding and the energy deposition is easy to control. Capacitive coupling has less contraindication, and can be used for such sensitive common tumors like the lung and brain. Its efficacy was discussed and proven in the relevant literature in its time [195–199, 57]. However, it became more and more clear, that knowledge of the radiofrequency technique alone was not enough for successful development, and refining of the system for physiological effects is the clue to success [171]. The frequency typically chosen for capacitive coupling is in the 5–30 MHz radiofrequency range. In this range the so-called "free frequencies" are preferred. These are RF-waves that are kept for industrial/medical use; their exploitation does not need extra permission from the authorities controlling communication frequencies.

Treatments with coils (magnetic and inductive) are relatively rarely used due to the negligible magnetic permeability of living systems [200]. In order to improve the magnetic energy absorption within the target tissue, magnetic materials, such as microparticles [201] and ferrite rods [202], are usually injected into the targeted area [203]. Ferrite rods (seeds) have also been used for non-oncological ablative therapies [204]. Using the same idea, a new "intra-cellular hyperthermia" method was developed [205], however, the efficacy of this treatment is still debated [206]. There is however an emerging field of application of magnetic treatments using nanoparticle magnetic suspensions [207] and other magnetic liquids.

Another type of inductive heating is typically achieved without inclusion of extra magnetic material into the tumor, and uses only induction of Eddy currents [208–210]. This method has low efficacy, but its penetration crosses all over the body. Because of the problems of tumor selection this method is not popular.

A widely used method for electromagnetic energy delivery is antenna-array coupling [211]. Its subsequent developments the annular phase array [212], the matched phase array [213], the Sigma60 [214], and Sigma-Eye [215], applicators use high-frequency RF (60–150 MHz). The body is ringed by the antenna array and delivers a chosen field intensity with controlled phase and frequency. The higher frequency used in this method is necessary for accurate focusing, however, these frequencies lie outside the electromagnetic compatibility standards for free frequencies, therefore requiring shielding (Faraday cage). Nevertheless, multiple controlled clinical trials have shown the efficacy of this method [216–218, 149].

2.2 Effects of Hyperthermia

The general physiology is naturally temperature-dependent. It is determined by the thermal homeostasis. The physiologic regulations depend to a significant extent on the temperature, all the systemic networks (blood flow, lymph network, and nerve system) are a reaction to the temperature. Some of the well established and widely accepted mechanisms described below characterize the successes of heat treatments in oncology. We list below the most important mechanism.

2.2.1 Higher Baseline Temperature

The rapid growth and higher metabolism of tumors typically yields higher tumor temperatures than the surrounding healthy baseline temperature [219]. Hyperthermia increases biochemical reaction rates [220] and therefore the metabolic rate as well. The metabolic rate (q_m) grows by the temperature gain ΔT: $q_m \sim 1.1^{\Delta T}$ [221]; therefore, in the case of a 6°C increase the amount of growth will be 1.8-times higher than the lower counterpart. The metabolic heat production of a tumor depends on the doubling time of its volume (see Fig. 2.3 [222]), so it is determined by the speed of development.

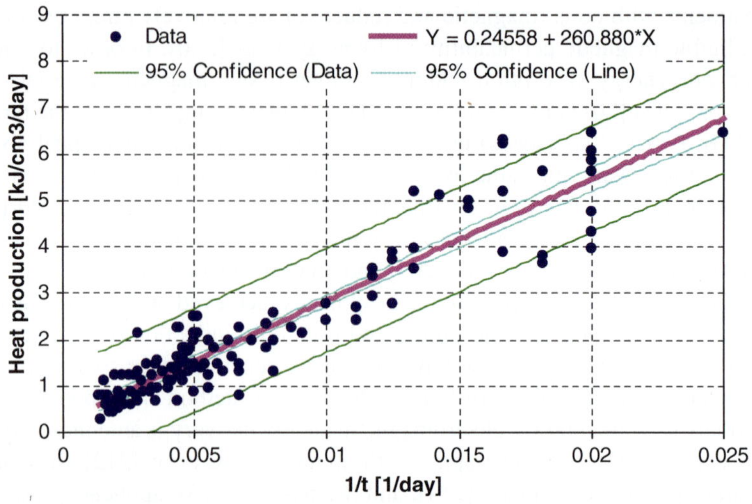

Fig. 2.3 Metabolic heat production of breast cancer lesions (tumor size: 0.6–4.0 cm), shown as a function of the reciprocal value of volume doubling time. The rapid procedures (large $1/t$ values) are more intensive heat producers, (1 kJ/liter/day≈11.6 mW/liter)

2.2.2 Vascular Changes

Sure the blood flow, which is one of the heat exchangers in the body, is the most sensitive for any local or systemic temperature changes. The blood stream has a central role to maintain homeostasis, and regulates well the heat exchange to ensure the proper functional conditions of the targeted area. The blood stream tries to compensate for overheating by intensive perfusion and regulation of the flow capacity of the vessels.

It has been shown that an increase in temperature can cause vasoconstriction in certain tumors leading to decreased blood perfusion and heat conduction [223–225], while causing vasodilatation in the healthy tissues leading to increased relative blood perfusion and heat conduction in this region [226, 227]. Yet, blood perfusion of the tumor relative to the surrounding healthy tissue is always lower [227] providing an effective heat trap [228].

An effective vascular response to heating could be observed [224, 225], which over a tumor-specific threshold (from about 38°C) differentiates between the malignant tumor and normal/benign tissues, suppressed and increased blood perfusion, respectively. This effect functions as an effective heat trap [229] for the tumor. However, this effect is strong only in the neo-vascularized area (the epithelium of new vessels differs from the normal [230].)

The emphasized large difference between the absolute blood flows of the tumor and of healthy tissue is recognized [231–235] (e.g. Fig. 2.4, [227]); also the relative change of the blood flow by temperature is well described in other works [236, 237]. From these studies we learned that the relative blood flow

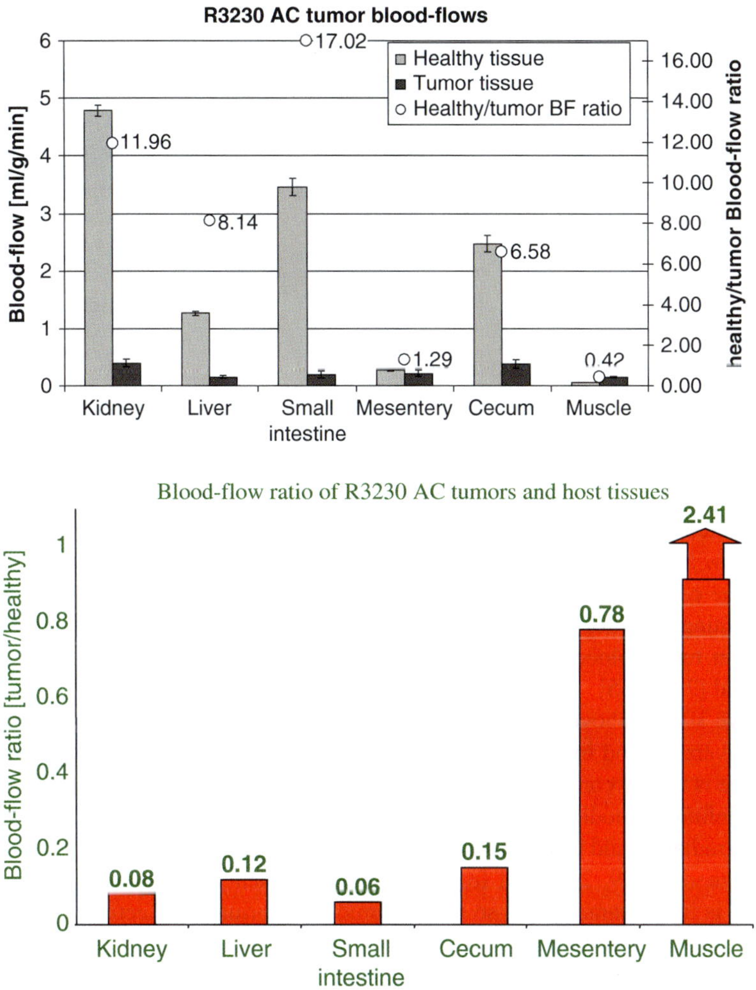

Fig. 2.4 Absolute and relative blood flows in the tumor compared to its healthy counterpart in a R3230 AC tumor [227]

changes in healthy and in tumor tissues are parallel below the specific threshold, starting at 38°C). The maximal threshold value in the literature is 42.5°C. The limit anyway corresponds well with the believed cellular phase transition observed around 42.5°C [238]. This however surprisingly accurately fits to experimental results of in vitro studies by Arrhenius plot [239, 240]. However, above this limit the blood flow suddenly splits, the tumor and the healthy tissue are characteristically down- and up-regulated, respectively. The blood flow further increases in healthy tissue, while to the contrary, in tumor tissue it becomes down-regulated.

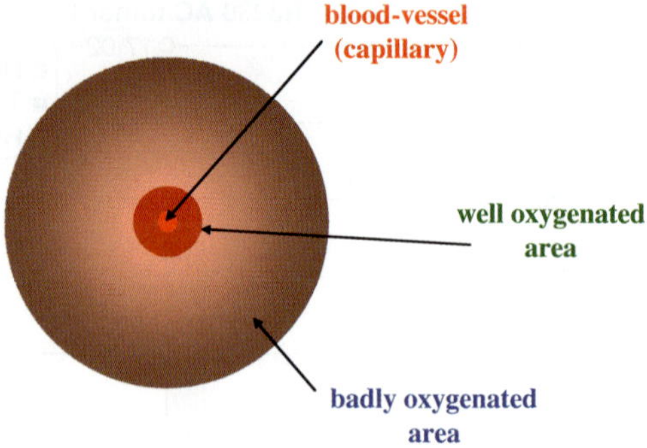

Fig. 2.5 The microscopic difference in the vicinity of a capillary vessel [257]

The tumor blood flow depends on the tumor weight through a negative logarithmic function [241]; which is a further modifying factor in the development of the tumor mass.

The microcirculation of the tumor and its changes through hyperthermia have been studied in detail [242–248, 224, 228]. The physiologic effects connected to the blood flow are considered important and also have been studied in detail [249–252, 227]. The central fact is: change in the blood flow is temperature-dependent and has a turning point for tumor lesions. Detailed reviews and discussions on the latest results on tumor blood flow affecting the applied temperature have recently been published [253–255].

As we shown above, the capillary vessels have a special role in heat delivery. The delivered blood by capillaries is the source of oxygen, so the oxygenation of the tissue is tightly connected to capillary blood perfusion (See Fig. 2.5 [256, 257]); so it is definitely modified by hyperthermia.

The gradient could be guessed in equilibrium to 50 μm [258]. According to direct measurements [259, 260] and from density experiments [261], the inter-capillary distance is in the same range, so the dynamic equilibrium is effective in the whole mass of the living tumor (the necrotic part is excluded).

The characteristic oxygen distribution (see Fig. 2.5) is effective around a capillary at any time, its existence is irrespective of the hyperthermia applications, only its actual value could differ in therapeutically effective periods. However, it is not only the character of the oxygenation, but the behavior of all the diffusion-derived phenomena (pH distribution, drug distribution, and in general all the chemical species delivered in- or out by blood) has the same character. The gradient direction will not change through the applied heat, only the gradient value will be modified. (See Fig. 2.6).

2.2.3 Cellular Membrane Changes

It has long been known that hyperthermia can cause a softening or melting of the lipid bilayer [262, 263, 250], it can change lipid–protein interactions [264], and it can denature proteins [265]. All of these events can significantly disrupt a tumor cell's capacity to divide.

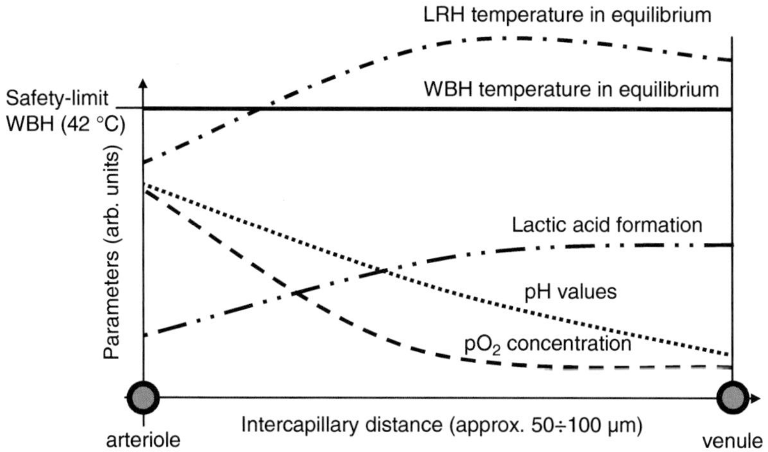

Fig. 2.6 Schematics of the main physiologic parameters in the thermodynamic equilibrium of hyperthermia in the microscopic range. The capillary destruction above 42°C is not considered in this scheme (see later). The values are in arbitrary units, only the trends [259] are shown; relative comparison of the values is impossible

Heat treatment causes structural alteration in trans-membrane proteins causing a change in active membrane transport and membrane capacity [266] leading to substantial changes in potassium, calcium, and sodium ion gradients [267], membrane potential [268, 269], cellular function [270, 271], and causing thermal block of electrically excitable cells [272, 220]. The thermo tolerable cells have significantly higher (~15%) membrane potential than the naïve cells [269], and the difference rapidly grows by the elapsed time at 45°.

2.2.4 Lactic Acid Formation

Hyperthermia increases biochemical reaction rates [220] and therefore the metabolic rate as well. However, there is hypoxia [220] and anaerobe metabolism producing lactate [237] (see Fig. 2.7), as well as pH drops by hyperthermia, suppressing the relative survival of the cells in most tumors.

2.2.5 ATP Depletion

Increased metabolism significantly decreases cellular ATP stores leading to increased cell destruction [237] (see Fig. 2.8). The cellular ATP level at the beginning of hyperthermia (43°C [237]) increases from 3.0 femtomole/cell to 4.2 femtomole/cell (gain from start: 40%, slope: 0.12 femtomole/min) in the first 10 min due to heat-promoted energization. Afterwards it decreases well linearly by time to

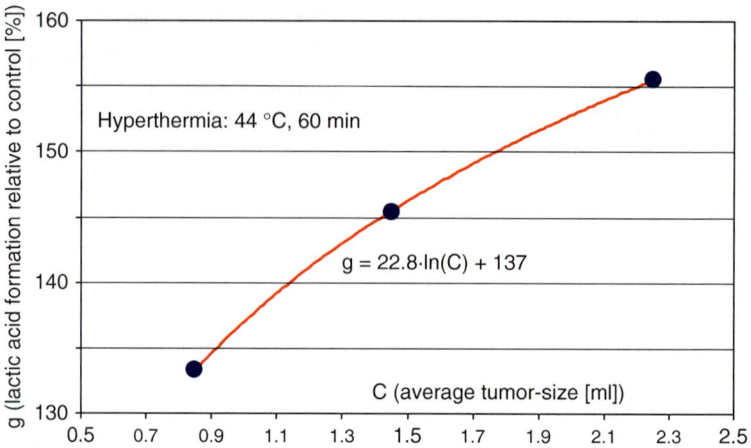

Fig. 2.7 Relative amount (%) of global lactate concentrations compared to the control. Hyperthermia at 44°C/60 min for DS sarcomas in three different volume categories. The best fit is logarithmic which is shown

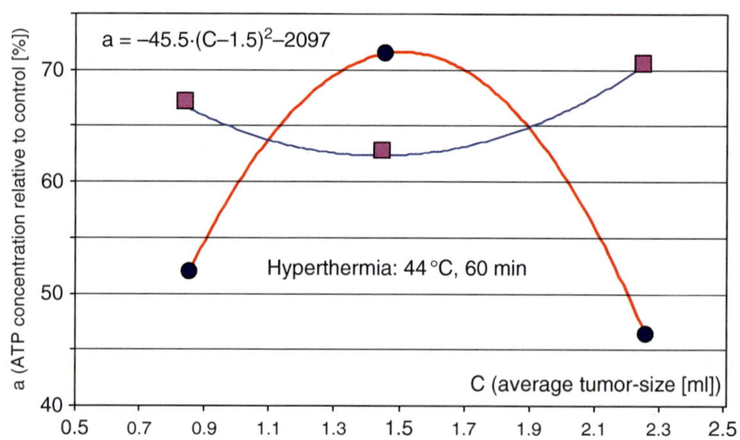

Fig. 2.8 Development of the global ATP concentrations ($a =$ ATP} compared to the control (C), (44°C/60 min) (•) for DS-sarcomas of three different volume classes [the peak at 15 ml is probably misleading. It is the consequence of the reference ATP concentrations (■)]

2.7 femtomole/cell (decrease from top: −35%; slope: −0.03 femtomole/min) during the full time (end at 60 min) of the treatment. The ATP depletion produces a heavy ionic imbalance in cells [273]. Moreover, the ATP deprivation causes protein aggregation in cytosol, the cytoskeleton collapses, the plasma membrane becomes destabilized, and the cell becomes necrotic [273].

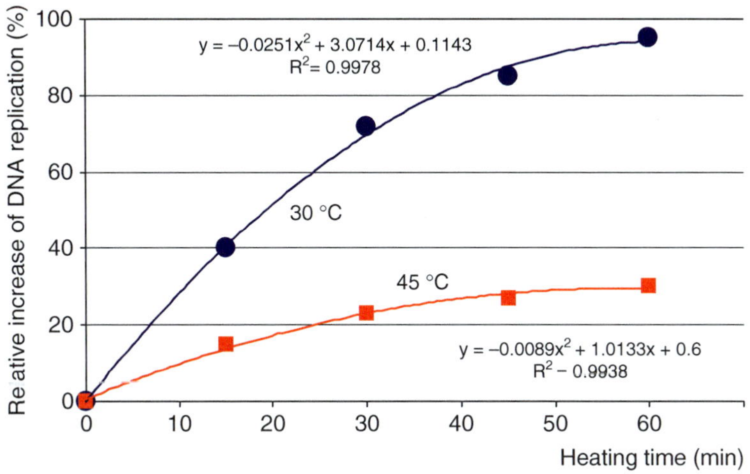

Fig. 2.9 DNA reproduction is suppressed by the temperature (the best fits are shown for eye orientation)

2.2.6 Altered DNA Replication

Increased temperatures can slow down or even block DNA replication (see Fig. 2.9) [274, 275], acting mainly in the S-phase of mitosis [276]. This has been hypothesized to have a sensitizing effect on radiotherapy [277].

2.2.7 Enhanced Immune Reaction

Hyperthermia has been shown to stimulate the immune system [274] with observed increases in natural killer cell activity [278]. Moreover, the elevated temperature distributes tumor-specific antigens on the surface of various tumor cells [279] and assists in their secretion into the extra-cellular fluid [280], triggering immune reactions against the malignant cells (see Fig. 2.10 [274]).

2.2.8 Pain Reduction

Certain electric fields (TENS) are used regularly to reduce pain [281]. Hyperthermia, and especially electric-field-induced hyperthermia, have also shown significant pain reduction during treatments (see Fig. 2.11 [282]). Various authors observed a drastic pain relief from hyperthermia [282].

2.2.9 Selective Gain of the Heat Resistance

In chemo-thermo therapies the role of chaperone proteins is important. Chaperones [stress- or heat-shock proteins (HSP)] are highly conserved proteins, which are

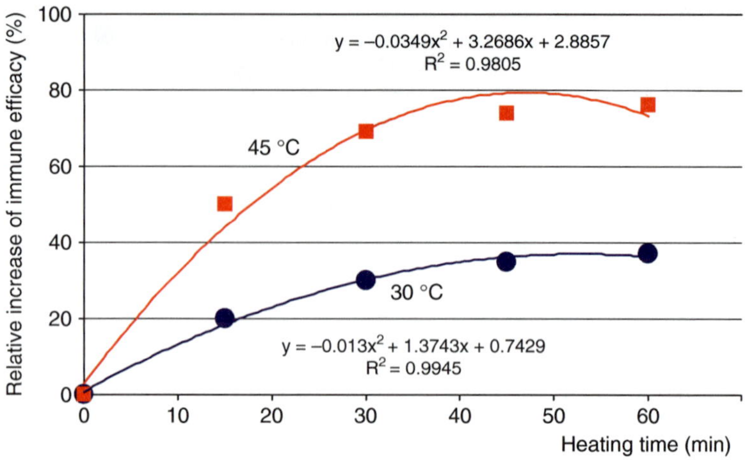

Fig. 2.10 Immune efficacy grows with the temperature and treatment time (the best fits are shown for eye orientation)

Fig. 2.11 Remarkable increase of the quality-of-life could be observed by the hyperthermia applications even as long duration as 3 months from hyperthermia treatments [282]

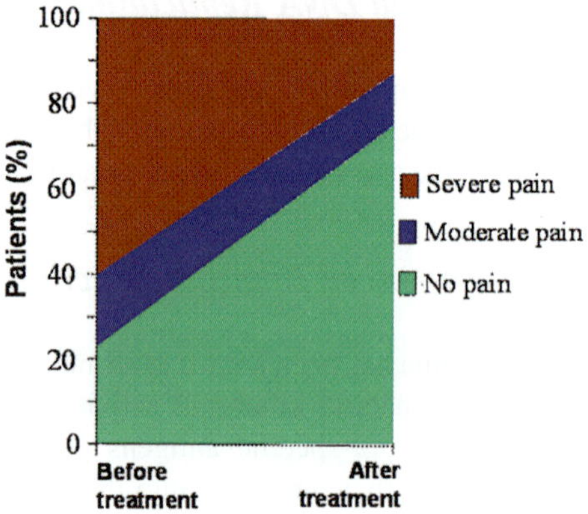

vital in almost every living cell and on their surfaces during their whole lifetime, regardless of their stage in evolution [283]. Any kind of change in the dynamic equilibrium of cell life (environmental stresses, various pathogen processes, diseases, etc.) activates their synthesis [284]. Excretion of the chaperones is the "stress response" of the cells to accommodate themselves to the new challenges. As a consequence of the stressful "life" of malignant cells, molecular chaperones are present in all cancerous cells [285–287] for adaptation to the actual stress conditions to help tumor cell survival. Moreover, shock proteins are induced by every oncological treatment method that is devoted to eliminating the malignancy: after conventional hyperthermia [288], after chemotherapy [289], after radiotherapy [290], or even after phototherapy [291] intense HSP synthesis was shown. Through this stress adaptation the induction or over-expression of stress proteins generally provides effective protection for the cell against apoptosis [292], but their extra-cellular expression acts in an opposite fashion: providing a signal to the immune system on the defect of the actual cell [293]. Furthermore, induction of various HSPs (HSP27, HSP70, and HSP90) was observed in numerous metastases and the HSP90 homologue, GRP94 may act as a mediator of metastasis generation. HSPs generally degrade the effect of the hyperthermia therapy because it may increase tumor cell survival, and its massive induction may generate tumor thermotolerance and in parallel drug- and radio-tolerance. Heat treatment can also lead to multi-drug resistance [294].

Non-temperature-dependent effects (mainly field stresses) could also produce chaperone synthesis [295]. The HSP manifestation in the biopsies could give a good clinical indication for the treatment response [296].

On the other hand, the chaperone HSP70 assists in freezing the actual dynamic equilibrium (the "status quo") and so tries to re-establish the cellular communication in the extra-cellular electrolyte [293]. It is shown that their expression on the cell membrane increases apoptotic signals and enhances immune reactions [293]. HSP participates in the activation of the p53 tumor suppresser [297] and has been associated with the tumor-suppresser retinoblastoma protein [298].

Recently, numerous scientific theories have also concentrated on the significance of thermally induced non-thermal effects, such as HSP production [299, 300]. Development of thermotolerance is one of the suppressors of hyperthermia efficacy [288]. From the point of view of thermotolerance one of the most prominent chaperone proteins is HSP72. The concentration of this HSP is 5–10-times lower in healthy cells than in malignant ones, and increases in both through heat treatment (see Fig. 2.12 [301]). However, the response to the heat treatment varied as the

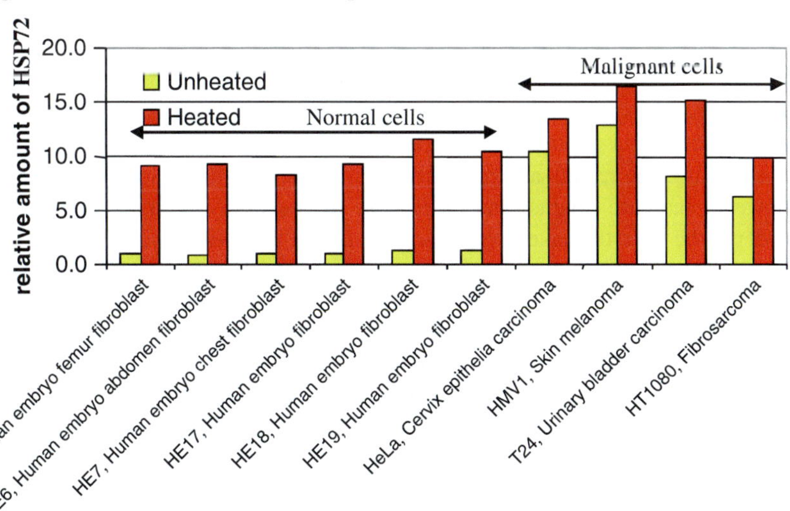

Fig. 2.12 Development of thermotolerance-inducing HSP72 chaperone protein in healthy and malignant cell cultures [288]

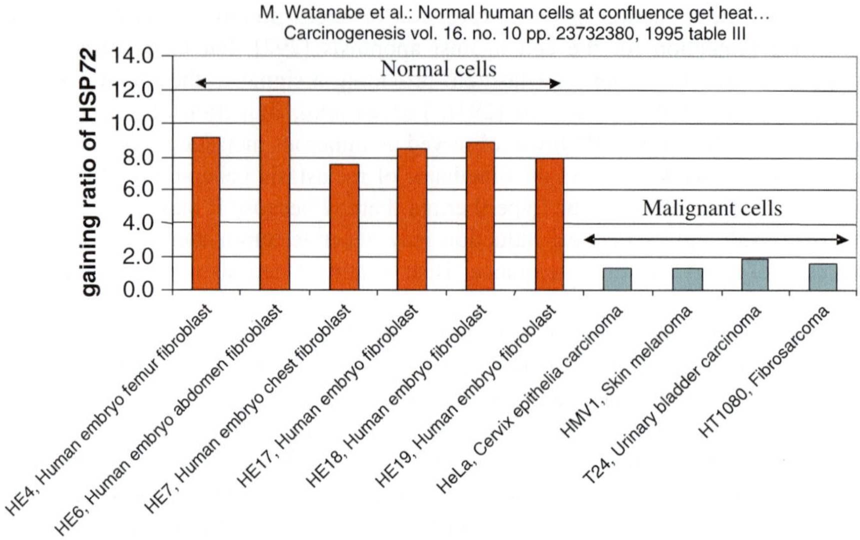

Fig. 2.13 The relative change of HSP72 in healthy and malignant tissues. Despite that the absolute values are higher in malignant tissue, the relative change is much smaller, and therefore the healthy tissue is relatively more protected. This is a kind of selective sensibility to the heating

concentration in healthy cells multiplied 8–10 times, while during the same heat treatment the HSP72 multiplication was only 1.2–1.5-times higher in malignant cells, see Fig. 2.13 [288].

2.3 Clinical Oncological Hyperthermia

Both clinical experience and studies have revealed a positive outcome in most of the hyperthermia applications. Depending on their endpoints and the investigated malignancy, the results are generally significantly positive. Some opposing results however increase doubts, and emphasize the necessity of specifying the dose in order to be able to make comparisons between the treatments. The clinical experience, together with the importance of temperature and its active time, show the homogeneity of the treated volume/area. The latest temperature-based dose considers the temperature, time, and part of the treated area which has such temperature conditions. In a hypothetic homogeneous tumor the volume idea is trivial: the effect is linearly proportional with the perfectly treated volume/area. However, in reality tumors are far from homogeneous, and their boundaries are not characteristic either (in the case of malignant tumors only). So the portion of the volume can be misleading, the conditions are more regulated by tissue nonhomogeneities than the treatment itself.

There are also concerns about the possible contradiction between the direct clinical results (response rates) and the survival rates. The direct local response does not necessarily affect the survival, because the metastasis potential could be increased. The discussions on this topic are very wide ranging among experts.

The diversity of the clinical trials can be observed from the diversity of the applied methods. A direct comparison of divergent technical solutions (for example whole-body treatment with RF-ablation in the liver) is not possible.

Another important factor in the clinical trials is the quality-of-life (QoL). One of the huge advantages of hyperthermia is the relatively low toxicity (side effects) and the good support of QoL, including pain reduction.

2.3.1 Local and Whole-Body Heating

In spite of the possible equal temperature in the tumor, local and whole-body treatment has significant differences; these methods act differently even at the same temperature.

Local/regional treatment has no safety temperature limit in the tumor (we only have to avoid hot spots in the healthy neighborhood). The non-uniform heating of LRH characterizes the dynamic equilibrium [302]. In the inter-capillary region under the 42°C threshold the radio-efficacy has the same trend as oxygenation, but the chemo-efficacy changes. It is enhanced well by the increasing temperature (and blood perfusion; see Fig. 2.14).

However, when LRH exceeds the threshold, the broken perfusion suppresses the radio-efficacy and increases greatly the chemo-effect in the region, where the drug is trapped (see Fig. 2.15).

Macroscopically the local and whole-body heating techniques also have differences. In the therapeutically important region the systemic treatment has

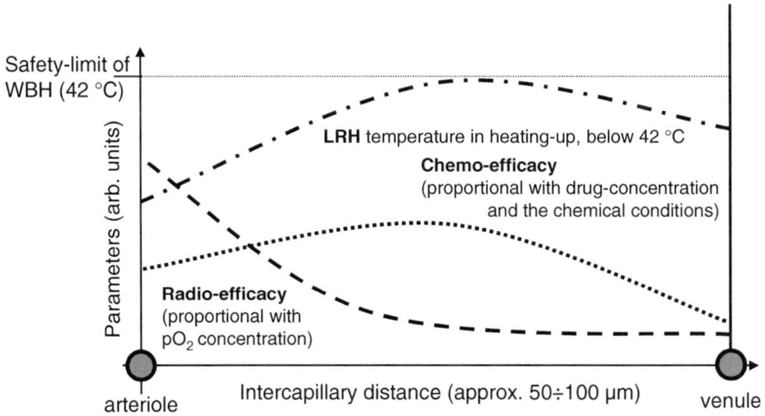

Fig. 2.14 Schematics of the main complementary actions in the heating-up period (below 42°C) of LRH in the microscopic range. The values are in arbitrary units, only the trends are shown; relative comparison of the values is impossible

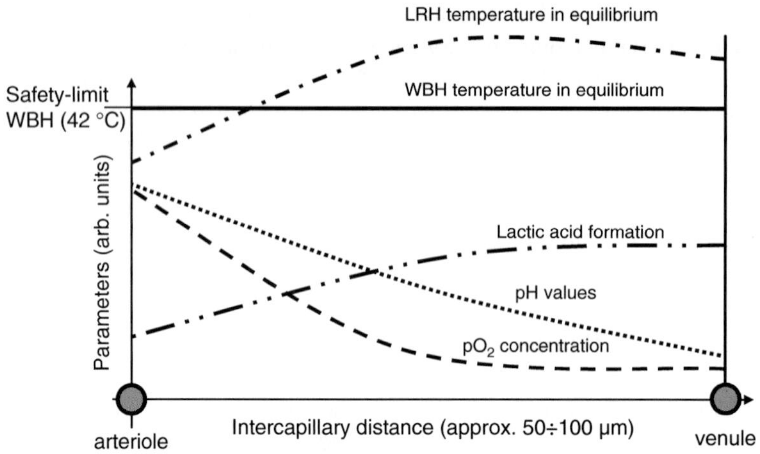

Fig. 2.15 Schematics of the main complementary actions of LRH above 42°C in the microscopic range. The values are in arbitrary units, only the trends are shown; relative comparison of the values is impossible

homogeneous heating all over the body, with no selection of the tumor. However, the local/regional treatment concentrates the heat on the tumor and its connected neighborhood.

The whole body systemically can be heated up by WBH to the physiological limit at 42°C and the thermal distribution in the tumor is homogeneous, well controllable. No hot spots exist, no questions arise about isotherms; the physiologically extreme temperature could be fixed all over the body. Contrary to this clear advantage, the physiological risk is higher than in the LRH treatment, where the systemic physiological factors are unchanged.

Suppress the risk; a decreased treatment temperature (moderate WBH/whole-body fever-range thermal therapy) is also applied. The application of lower temperatures for a longer time period (fever therapy, or mild hyperthermia) also showed surprisingly good efficacy for whole-body hyperthermia treatments [303–306]. Whole-body hyperthermia and even its fever-range version offer an effective immune support [307–310], which could be a very important factor for patients with weak immune system.

These physiological changes and trends modify the efficacy of the conventional radio- and chemotherapies as well. Both heating approaches (systemic and local) provide an excellent possibility of synergy with conventional therapies for different reasons. In the temperature region below 42°C, the local/regional treatment targets mainly the hypoxic area, while the systemic treatment promotes the conventional therapies in the well blood-perfunded microregions.

The chemodrugs are delivered to the tumor by the blood stream. Because of this fact, the drug delivery to the tumor has a gradient from the source of the capillaries to the tissue. In systemic treatment the temperature gradient and the concentration

Table 2.24 Blood-perfusion trends and its effect on complementary treatments

	Effect/efficacy	
Treatment modality/effect	Near to capillary	Far from capillary
Local/regional hyperthermia (LRH)	⇩	⇩
Systemic hyperthermia (WBH)	⇧	⇧
Chemotherapy	⇧	⇩
Radiotherapy	⇧	⇩

gradient are parallel, while in local treatment due to the relatively cold capillaries in the heating period, the temperature gradient is the opposite. The hot drug is more reactive [311], so the reaction of the given pharmaceutics for systemic treatment is near to the capillaries, while the local treatment mainly supports a reaction deeper in the tissue.

The microscopic action of radiation therapy is also different for the two heating procedures. Radiotherapy has high efficacy in the well-oxygenated areas, and rapidly loses its effect in hypoxia [312]. This effect microscopically prefers the near-arterial capillary regions. Because below 42°C the hottest areas are the most oxygenated, these are the most radio-sensitive as well. However, for local treatment the most oxygenated area is the relatively coldest under 42°C. The effective blood perfusion and their consequences on the complementary treatments are shown in Table 2.24.

The application of hyperthermia with radiotherapy has a definite sequence [313]: heat before radiation gave a significantly higher thermal enhancement ratio than applied hyperthermia after radiation. This is typical in cases where the hyperthermia enhances the blood supply (under 42°C), and with this supports the subsequent radiation, while afterwards this effect is not active.

Macroscopically it also has differences: the well blood-perfunded peripheries of the tumor are hotter than their central counterpart in the systemic- and it is directly the opposite in the local/regional treatments (see Fig. 2.16).

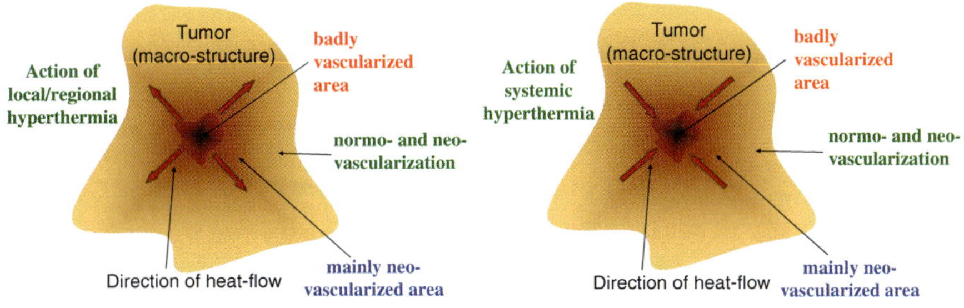

Fig. 2.16 The macroscopic heating difference of systemic and local/regional hyperthermia. Note the direction of the heat flow, which is directly opposite in the two methods

This is the consequence of the heat delivery: in systemic treatment the blood delivers the heat to the tumor, in the local/regional case the blood cools the tumor, because the blood remains at the temperature of the rest of the body. (In WBH at thermal equilibrium [plateau phase] a temperature difference no longer exists between the blood and the tissues, while this difference remains constant in the case of local/regional heating.) The different heat flow forms the main difference between the methods.

In the heating-up period, in both the systemic and local/regional cases, the processes are in a non-equilibrium state till the dynamic equilibrium is reached ("plateau" situation). But the process to reach the equilibrium state differs entirely in the two techniques. The whole-body treatment loses energy only on the body surface, the equilibrium could be regulated by the surface energy exchange. The local/regional treatment has a bulky energy loss, all the capillary veins sink the energy and the "fresh" artery blood has to be heated, which also represents a permanent energy loss. In this case a microscopic temperature gradient permanently exists in the local/regional case also in the dynamic equilibrium phase, while the tumor temperature is homogeneous in the equilibrium phase for systemic heating, (see Fig. 2.17).

2.3.2 Hyperthermia as a Complementary Method

Hyperthermia in most cases is a complementary method. Its synergy with other therapies is dominantly temperature-dependent in these cases. The primary goal of complementing other therapies is to enhance the temperature for promoting the treatment applied, while abandoning (or at least reducing) the thermal-distortion idea; the increased temperature is the task. Direct cytotoxic effects are not expected in this application, hyperthermia has a conditional task by accelerating the effects of the other treatments through increased temperature. The aim is to complete the other methods, which are probably more radical at higher temperature conditions. Here the hyperthermia ensures the optimal conditions for the other treatments, where the temperature could be an ideal parameter (while the role of the temperature differs in the distortion mechanisms).

Hyperthermia provides an excellent possibility for synergy with different actions of conventional therapies (see Fig. 2.18). The local/regional treatment targets mainly

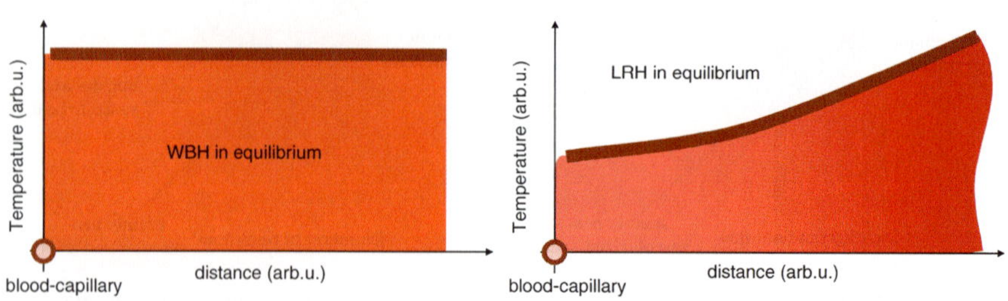

Fig. 2.17 Dynamic equilibrium heating difference for systemic and local/regional hyperthermia

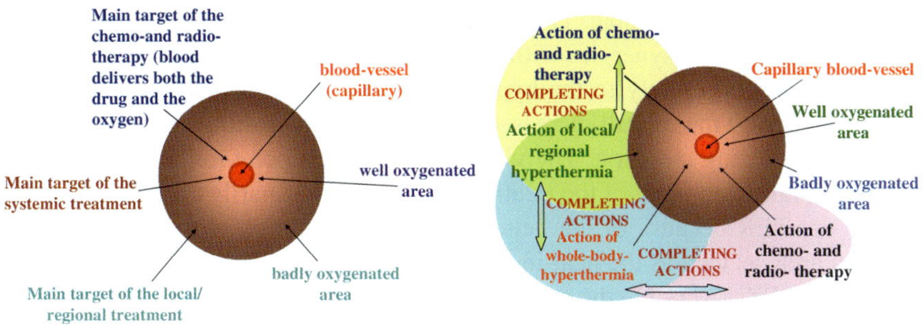

Fig. 2.18 The completion of conventional methods by systemic and local/regional hyperthermia in microregions

the hypoxic area, while the conventional radio- and chemo-therapies are active on the well blood-perfunded microregions: the efficacy of the radiotherapy is oxygen-dependent, while the chemotherapy depends on the blood-delivered drug.

The most active regions of a tumor and regions far from blood supply are usually severely hypoxic, and therefore radiation has reduced efficacy in these regions. The possible vasodilatation caused by hyperthermia aids the synergy via the overall increased blood perfusion (oxygenation) [314], creating considerable sensitization to ionizing radiation. This approach was one of the first lesions of the modern hyperthermic effect. Its characterization was introduced by the Thermal Enhancement Ratio (TER) [315, 257]. TER measures the gain of the efficacy. Hyperthermia, however, speeds up cellular metabolism with possible accentuation of the hypoxia, which works in the opposite fashion: reducing the efficacy of radiotherapy. Here both approaches could be used: a reduced radiotherapy dose which is promoted by hyperthermia using the complex TER factor [316–318].

Chemotherapy drugs are delivered into the tumor through the blood circulation; therefore, it is most effective in the regions near arterioles. In this respect, chemotherapy is similar to radiation therapy in that it primarily targets regions of high blood perfusion due to the oxygen-rich conditions. Nevertheless, the region which is more distant from the fresh blood perfusion is less cooled, so treated effectively by hyperthermia, completing the treatment of the chemotherapy-treated volume. Moreover, an increase in temperature accelerates the pharmaco-kinetics (improves the reaction rates). This effect could be accompanied by the temperature sensitivity of the actual drug also [319]. The thermo-chemotherapy results in a better therapeutic effect and increases the target specificity as well as reducing systemic side effects [320, 217]. In some cases low-dose chemotherapy could be used [321, 322] with hyperthermia promotion, also it is applied in low-dose metronomic chemoregulation [323].

Many current conventional treatments for cancer are difficult to tolerate due to their high toxicity levels. In general, patients are treated with chemo- and/or radiotherapy to their toxicity limits in order to achieve maximal tumor destruction. However, these treatments are often not enough, or the patient develops resistance, or even worse, kidney or liver failure develops. Hyperthermia is an ideal treatment to combine with other therapies [149]. It has low toxicity, mild side effects, and has been shown to provide synergies with many of the traditional treatment modalities.

Furthermore, oppositely to the hypoxia, the vascular changes described above aid the synergy by the overall increased blood perfusion creating considerable sensitization to ionizing radiation. The visualization effects point to the importance of individual evaluation in all of the treatment cases. The primarily applied heat could enhance the TER because of the better blood flow and higher oxygen concentration in an area, as well as the heat-induced decrease of the DNA-dependent protein kinase (DNA-PK) [277]. On the other hand it could have the opposite effect also: the heat accelerates the metabolism, the tissue becomes hypoxic and the ionizing radiation afterwards remains ineffective. This Janus-face behavior has to be evaluated in all cases. The action of oncothermia at the beginning is dominantly not heat based, but governed by the field and the thermodynamic flows. In this case its application before radiology is desired, however, after a long period of oncothermia, when the equilibrium has been constructed, and hypoxia becomes dominant, the radiation treatment is preferred first. (This last is legally also clearer: the patient receives the classical evidence-based treatment first, and the helpful promotion afterwards.)

Sensitization of classical ionizing radiation by hyperthermia has been well known [7, 21] for a long time, and different review articles have summarized this knowledge [324–327, 315]. The advantage of combining heat treatment with classical ionizing radiation is unambiguous [1, 13, 16, 17], the synergy between the methods is well known [328, 329] and successfully applied [148–150]. The primary basis for the synergy is the complementary targets of the two treatments (see Table 2.25). Briefly, ionizing radiation is most effective in the M and G_1 phase, in relatively alkaline, well-oxygenated regions. Hyperthermia on the other hand exerts the greatest effects in the S phase [330], in relatively acidic, hypoxic regions.

The synergy between heat treatment and many of the chemotherapies is well established [331, 16, 17, 217]. The effect of chemotherapy is more reactive beside the arteries delivering the treating drug. Consequently the chemotherapies (systemic, regional, or local) can be complemented by hyperthermia in the same way as with radiotherapy. Chemotherapy drugs are specific to cells in the M and G_2 phase and show little or no efficacy against cells in the G_0 phase. Hyperthermia reduces the average time spent in the G0 phase making them susceptible to chemotherapy (see Table 2.26) [14].

Moreover, a robust synergy prefers the combination of chemotherapy with heat: the thermally increased metabolism, (enhanced chemometabolism), increased absorption of cytotoxines [332, 333]. The cellular chemo-penetration is promoted

Table 2.25 Complementary effects of radiotherapy and hyperthermia

Effect/method	Ionizing radiation	Hyperthermia
Cell cycle specificity	Acts in M+G1 phase	Acts in S phase
pH-dependence	Acts in relatively alkaline regions	Acts in relatively acidic regions
Oxygen specificity	Acts in well-oxygenated tissues	Acts in hypoxic tissue

Table 2.26 Complementary effects of chemotherapy and hyperthermia

Effects/method	Chemotherapy	Hyperthermia
Place of primary activity	Near to arteries	Far from arteries
Reaction rate	Normal	Enhanced
Chemo penetration	Low, due to high pressure	Enhanced transport by electro-osmosis
Chemometabolism	Normal	Enhanced
Chemo selection	By chemical reactions	Definite local enhancement
Cell division	Acts in M+G_2 phase	Acts in S phase
Activity	No activity in G_0 phase	Decreased time in G_0 phase
Treatment failure	Blood/organ failure, developed tolerance	Resensitizes, decreases the load on organs and blood stream

strongly by the non-equilibrium heat flows (electro-osmosis) [334]. Also, for pure chemical reasons, the reaction rate of drugs increases with the gain in temperature (Boltzmann law) [335]. In addition to the reaction acceleration expected also is a decrease in the activation energy (Arrhenius fit) [238], which fits surprisingly accurately to experimental results [239, 240] and changes via the chemotherapies [336].

The promoted, optimized chemo-intake helps to overcome the possible failing of chemotherapies due to patient intolerance (when large doses of drugs are not possible, for example because of renal or liver insufficiency, insufficient blood composition, etc.). In these cases the same results may be achieved by the combination of decreased chemo-dose and heat therapy [337].

Anyway the applied local heat selects the heat-targeted tissue (which is hopefully the tumor) and it results in a better therapeutic index increasing target specificity and reducing systemic side effects [338].

The differences between the normal and neoplasm tissues in their blood-flow reaction cause selective chemo wash-out from the areas, so the locally applied chemo is quickly passed from into the healthy tissue while being trapped in the tumor. This effect has a positive feedback through the increasing energy intake.

Applications of chemo-thermotherapy involve many factors not only in the physiological context but the pharmaco-kinetic behavior of the drug also has an important modification role in the treatment strategy. Because of the very different kinetics of drug efficacy with time (and temperature if any) for various drugs, in proper treatment we have to fit the drug kinetics to the kinetics of the oncothermia treatment. This requires harmonization of the time of the maximum of the chemometabolism, with the maximum of the hyperthermia energy.

Complex applications (radio-chemo-thermotherapy) also became popular [339]. Complementing hyperthermia methods with classical oncological modalities is very promising. Magnifying a blood vessel in the tumor, its vicinity is relatively well oxygenated compared to tissues further away from the vessel wall. Both the radio- and chemotherapies mainly act in the vessel neighborhood, because of their higher activity in oxygen-rich tissues and via blood-delivered drug diffusion, respectively.

Hyperthermia has also been found to have pronounced advantages for surgical interventions. Through hyperthermia-induced inhibition of angiogenesis and heat entrapment, the outline of the tumor often becomes pronounced and the size of the tumor often shrinks making previously dangerous operations possible [340]. The feasibility of pre-operative application for locally advanced rectal cancer is shown well by a Phase II clinical trial [341]. Post-operative application of hyperthermia has also been thought to prevent relapses and metastatic processes [342]. Intra-operative radiofrequency ablation [343] and local hyperthermia [344] has also been used to improve surgical outcomes.

The combination of hyperthermia with gene therapy looks very promising also, as shown by the successful combination of hyperthermia and HSP-promoter-mediated gene therapy in advanced breast cancer patients [345]. Hyperthermia improved the results of the HSP-promoter gene therapy by inducing local HSP production and by enhancing the local rate of release from liposome [346]; also helpful in the double suicide gene transfer into prostate carcinoma cells [347]. It was shown that this combination therapy was highly selective for mammary carcinoma cells. Also heat-induced gene expression could be an excellent tool in targeted cancer gene therapy [348].

Combination with hormone therapies is also a vivid method, applied for prostate [349] and in melanoma treatment [350]. Enzyme therapy [351], photodynamic therapy [352], gene therapy [353], immune- [354] and other supportive therapies [355] are also used in combination with hyperthermia.

2.4 Hyperthermia Successes

There are numerous books published about hyperthermia in oncology, [1–19]. We would like to give here a brief and far not complete outlook about the results, which we selected as most important. The local hyperthermia method (capacitive coupled) was widely applied in Japan and has seen great successes in the last 30 years. The response rates [195] (see Figs. 2.19 and 2.20), and the 2-year survivals are considerably improved [195] (see Fig. 2.21) and proven by other studies also in combination with chemotherapies for deep-seated tumors [331].

Similar success was presented in combination with radiotherapy, see Fig. 2.22 [195], with the advantages also shown by other trials in combination with radiotherapy for deep-seated tumors [356, 357, 198]. The multi-national Radiation Therapy Oncology Group (RTOG) also evaluated the method as feasible [358].

Other summaries were also published, showing the significant advantages both for shallow-seated (see Fig. 2.23) [359] and deep-seated [356] (see Fig. 2.24) tumors.

The QoL is significantly improved together with the above clinical responses (see Fig. 2.25) [195].

Some results of local/regional hyperthermia, which are achieved where conventional therapies are less successful (brain, pancreas, lung, liver) are shown below.

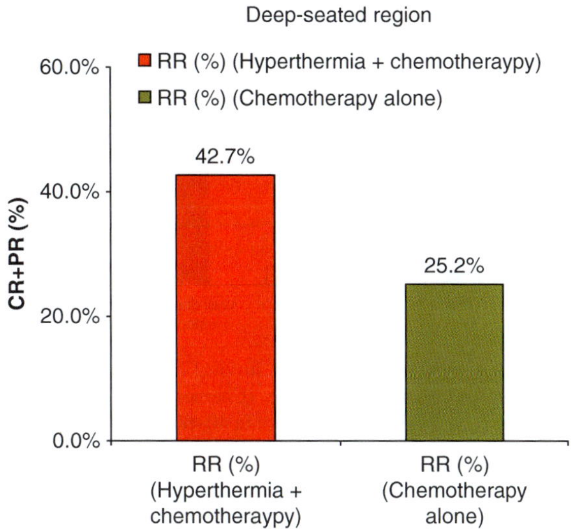

Fig. 2.19 General outcome of capacitive hyperthermia applied in Japan for various tumors and stages. The response rate (CR+PR,%) is shown for hyperthermia plus chemotherapy ($n = 592$) and chemotherapy alone ($n = 270$) as a comparison

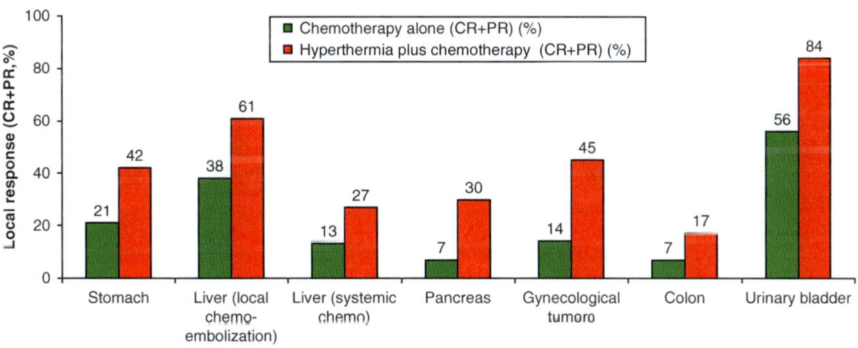

Fig. 2.20 Local response of hyperthermia plus chemotherapy compared to chemotherapy alone. (Chemotherapy is mostly Adriamycin, Bleomycin, Cisplatin, Mitomycin, and 5FU), (Hyperthermia 40–60 min, capacitive, 8 MHz, 4–16 lesions)

Unfortunately all the relevant literature can not be referred to here. There are tremendous references available from MedLine (PubMed) databases as well as in the numerous monographs listed above. Below we show some special localizations that have been treated by hyperthermia successfully: brain glioms, pancreas, lung, liver, and other localizations.

2.4.1 Brain Tumor Treatment by Hyperthermia

The increased intra-cranial temperature, in general, brings a disadvantage to hyperthermia treatment of brain tumors; namely it could increase edema and brain

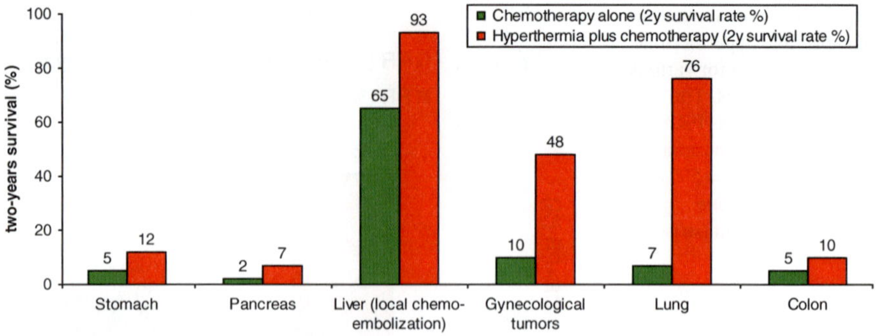

Fig. 2.21 Comparison of 2-year survival after treatment with hyperthermia plus chemotherapy compared to chemotherapy alone. (Chemotherapy is mostly Adriamycin, Bleomycin, Cisplatin, Mitomycin, and 5FU), (Hyperthermia 40–60 min, 8 MHz, 4–16 lesions)

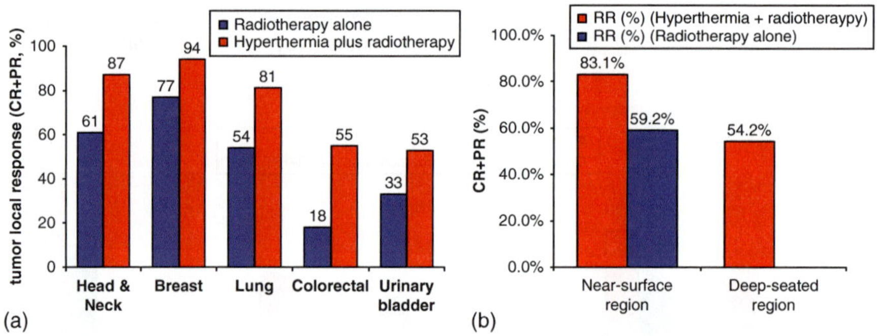

Fig. 2.22 Summary of the results obtained in Japan by capacitive hyperthermia combined with radiotherapy

pressure, which could be fatal. Because of the lack of effective traditional therapies, classical hyperthermia could be an important target to improve treatment facilities. There is a relatively high interest in studying the effects of heat on the brain [360–362]. Despite the risk of edema creation it was challenged in many studies. Avoid the severe edema the proper localization of the incident energy is essential. The requirement is such hyperthermia method, which does not induce or even reduces edema could be one modality for brain tumor intra-cranial treatment. This safe method must not increase the temperature all over the brain, but should act only locally. Numerous, very local, invasive (ablative, interstitial) hyperthermia treatments combined with local irradiation has been applied for glioms [363–372], combined also with laser techniques [373, 374], in addition implant applications [375, 376] and nanoparticle magnetic heating [377] are also in use. A post-operative application was also published [378]. The combination of inter-stitial hyperthermia with external radiation has also been tried [379]. One of these methods, inter-stitial (minimally invasive) hyperthermia was applied in a randomized, controlled double-armed (with and without hyperthermia) clinical study [380]. It showed surprisingly

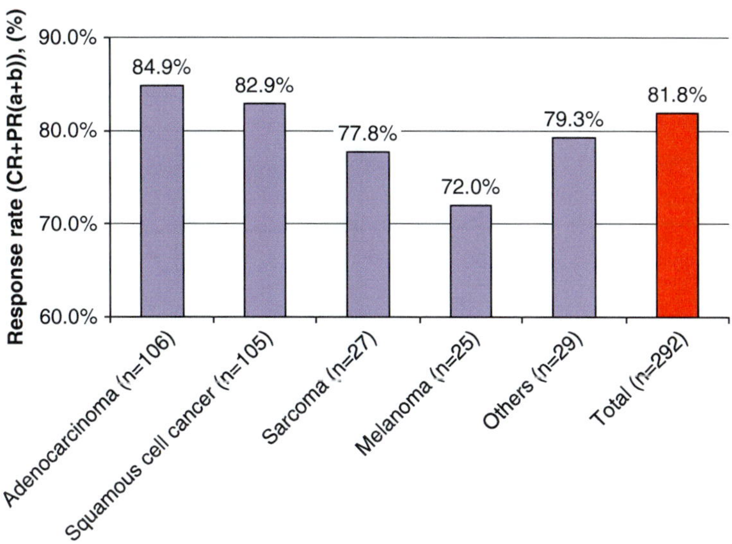

Fig. 2.23 Summary of the RTT of shallow-seated tumors in Japan

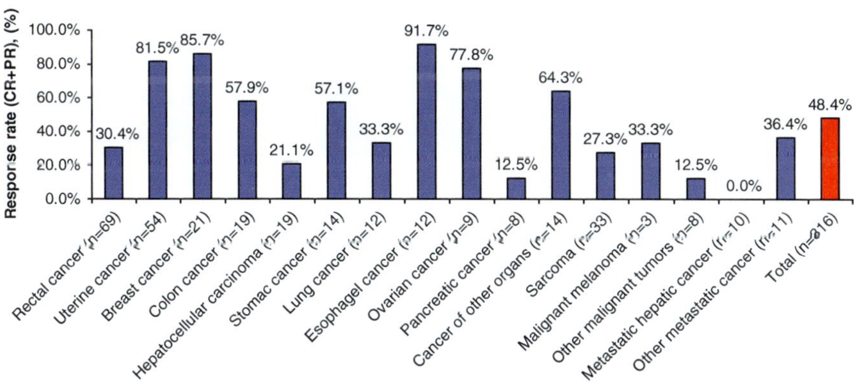

Fig. 2.24 Summary of the RTT of deep-seated tumors in Japan

good efficacy for brain glioms: the median survival increased from 76 to 85 weeks, and the 2-year survival increased to 31% from 15% (see Fig. 2.26). In consequence the FDA certified brain inter-stitial hyperthermia.

Radiofrequency hyperthermia was also applied intra- and extracranially [381–383], and ultrasound hyperthermia was also tried [384]. It was also shown that electric capacitive coupling (known as Electric Capacitive Transference) could be effective transcranially [385]. The distribution and number of clinical trails for hyperthermia in the brain is shown in Fig. 2.27.

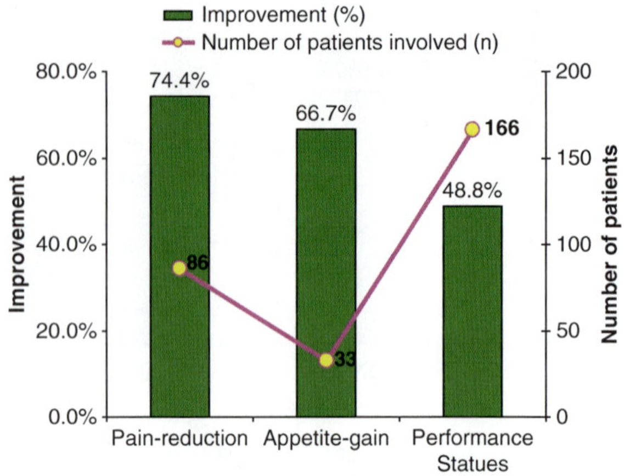

Fig. 2.25 Analysis of the QoL improved by capacitive hyperthermia treatment [195]

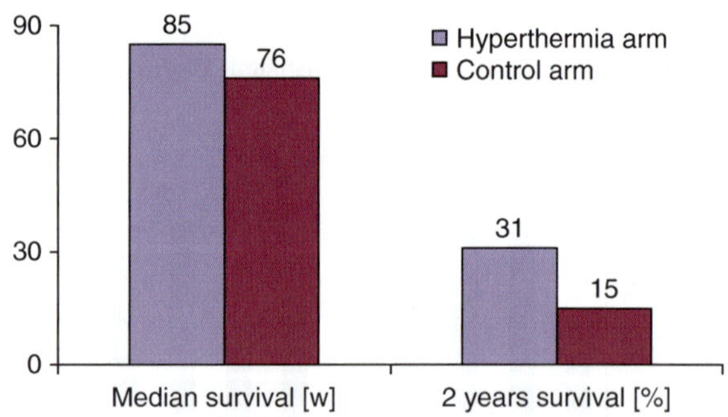

Fig. 2.26 Results of prospective, controlled, randomized clinical study for glioblastoma multiform treated by inter-stitial hyperthermia

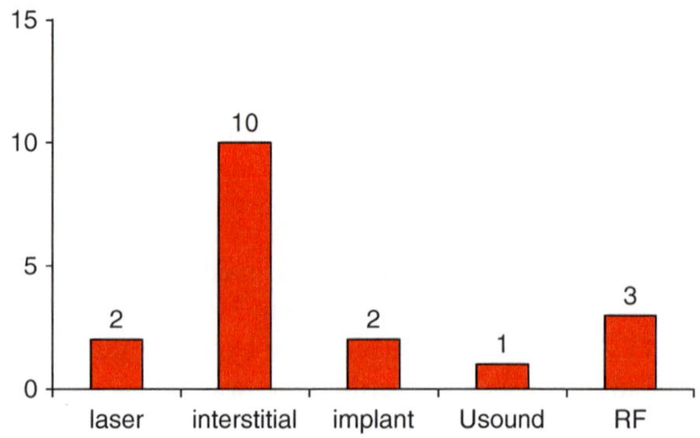

Fig. 2.27 Number of conventional hyperthermia clinical trials for brain glioms by treatment kind. None were negative or unsuccessful

2.4.2 Pancreas Tumor Treatment by Hyperthermia

A few clinical studies were carried out for pancreatic cancer with hyperthermia. One of the earliest extended therapy studies for pancreas [386], studied $n = 77$ patients with resectable and non-resectable adenocarcinoma. The method used was capacitive coupling at 13.56 MHz, and mainly the immune support of the patients was targeted. Patients that received hyperthermia received a lower chemo-dose. Two groups were compared: with and without addition of selective immune stimulation. (The first-year survival percentages and the median survival time are collected in Table 2.27). The leading chemotherapies [387–389], give 25, 22, 28% 1-year survival ratios and 6, 7, and 6.5-months median survival time, respectively. The hyperthermia results indicate the feasibility of the use of hyperthermia in conjunction with immune support in pancreatic cancer.

The hyperthermia could in any case be well combined with the gold standard for pancreatic treatment, i.e. surgery. The unresectable pancreas could be treated intraoperatively; results are shown in Table 2.28 [390]. Other intra-operative applications show the success of hyperthermia in combination with radiation [391] and chemotherapy [392, 344].

The minimally invasive ablative hyperthermia technique is also successfully applied for unresectable pancreatic carcinomas [393].

Pancreatic hyperthermia treatment together with a complex supportive therapy (high-dose enzyme therapy, endocrine combination therapy and vit. A) is applied successfully ($n = 46$) (see Table 2.29 [338]) with remarkable gain in quality-of-life (QoL, see Table 2.30).

Table 2.27 First-year survival percentages and MST of pancreas hyperthermia treatments (capacitive, 13.56 MHz)

Additive therapy	All (%)	Far-advanced diseased (%)	No response on conventional (%)	Operated (bypass or resected) (%)	Median survival time (MST) (m)
Immune stimulation	35	13.3	34.1	48.7	8
No immune stimulation	6	0	0	11.1	6

Table 2.28 Intra-operative radiation combined hyperthermia (capacitive, 13.56 MHz) results. The median survival time was not significantly different, but was higher in the hyperthermia-treated group

Survival percentages	Hyperthermia ($n = 14$) (%)	Control ($n = 55$) (%)	Gain (%)
First year	21.4	12.7	68.5
Second year	7.1	1.8	294.4

Table 2.29 Local hyperthermia (capacitive, 13.56 MHz) results on inoperable pancreatic carcinoma

Survival from ...	1-year (%)	2-year (%)	3-year (%)	4-year (%)	5-year (%)
... First diagnosis	41	20	13	9	9
... First hyperthermia	22	9	7	4	4

Table 2.30 The improvement in QoL by local hyperthermia for inoperable pancreatic cancer patients

QoL	Free of pain (%)	Marked pain relief (%)	Normal appetite (%)	Improved appetite (%)
Improvement	68	24	24	40

Systemic hyperthermia treatment in pancreatic cancer has also been applied, and has an ongoing clinical trial, presently recruiting patients [394], but other publications could not find a benefit for treatment with whole-body hyperthermia on local pancreatic cancer [395].

2.4.3 Lung and Bronchus

Some successful clinical trials in combination with radiotherapy have shown the feasibility of the hyperthermia method for non-small-cell lung cancer (NSCLC). Most of these are combined with radiotherapy, with a 14–70 Gy dose in a given cycle. The measured response rate (RR) was surprisingly high RR=75%, ($n=12$, [396], and *RR*=100% ($n=13$, [397]). In this last study the MST was 15 months (mean value 17 months) for the tumors with an average size of 22 cm^3. The total dose average was 60 Gy, heating time average 52.3 min, and total sessions average in the cycle was 27. Others had a comparison to a control arm (not randomized), studying the response rate (RR); (see Table 2.31).

Table 2.31 Response and survival rates in controlled double-arm studies for NSCLC by capacitive hyperthermia (8 MHz). (Karasawa [398] and Sukurai [399]) (RR – remission rate)

	Response rate by arms, RR (%)		First-year survival ratio (%)		Second-year survival ratio (%)	
Study	Hyperthermia	Control	Hyperthermia	Control	Hyperthermia	Control
Karasawa [398]	94.7 ($n=19$)	70 ($n=30$)	55 ($n=19$)	30 ($n=30$)	35 ($n=19$)	15 ($n=30$)
Sukurai [399]	76.9 ($n=13$)	53.8 ($n=13$)			44.4 ($n=13$)	15.4 ($n=13$)

Table 2.32 The results of radio-thermotherapy for advanced NSCLC patients. (CR – complete remission/response, PR – partial remission/response, NR – no response; IIB and IV are the staging categories according to WHO)

1-year survival rate (%)		MST (m)		Objective response			Pain relief		
IIIB	IV	IIIB	IV	CR	PR	NR	complete	partial	no relief
63.0	40.0	13.5	10.0	15.4	69.2	15.4	15.4	76.9	7.7

A study involving advanced NSCLC patients ($n = 13$, capacitive coupling, $f = 8$ MHz) to control local chest invasion [400], showed similarly good results, and the pain relief was also surprisingly good (see Table 2.32).

Also locally advanced NSCLC was studied ($n = 32$) with fractional radiation (180–300 cGy/fraction, 5 fraction/week, median dose 5.58 Gy) [401] (Table 2.33). Results indicate differences, but were not significant.

The 13.5-month median survival of the historical control was increased by post-operative (lobectomy or pneumonectomy) application of intra-thoracic chemo-thermotherapy to 17.5 m, by capacitive-coupled hyperthermia [248, 342]. The post-operative application was successful in another study also [402, 342].

The survivals for the treatment group (resection + post-operative intra-thoracic chemo-thermotherapy; [PICT], $n = 32$) and the historical control groups: resection only ($n = 20$) or exploratory thoracotomy ($n = 11$) had median survival rates of 25, 15, and 10 months, respectively [248]. The survival curve of the treatment group was significantly better than those of the control groups. Also the local relapse free survival for the treatment group (resection + PICT, $n = 32$) and the historical control group (resection only $n = 20$) was drastically and significantly increased, having more than double relapsed-free survival in the hyperthermia-treated group than in its historical counterpart after 48 months.

Table 2.33 The measured data in locally advanced non-small-cell lung cancer patients. (CR – complete remission/response, PR – partial remission/response, NR – no response; TRT# - number of TRT treatments)

		Clinical response (%)			2-year survival (%)			Survival mean (months)		
Arm	Number of patients	CR	PR	NR	all	TRT# <10	TRT# >10	all	TRT# <10	TRT# >10
Radiotherapy (RT)	18	11.1	55.6	33.3	0			8.1		
Thermo-radiotherapy (TRT)	14	0	85.7	14.3	7.1	7.4	40	10.5	18.2	40

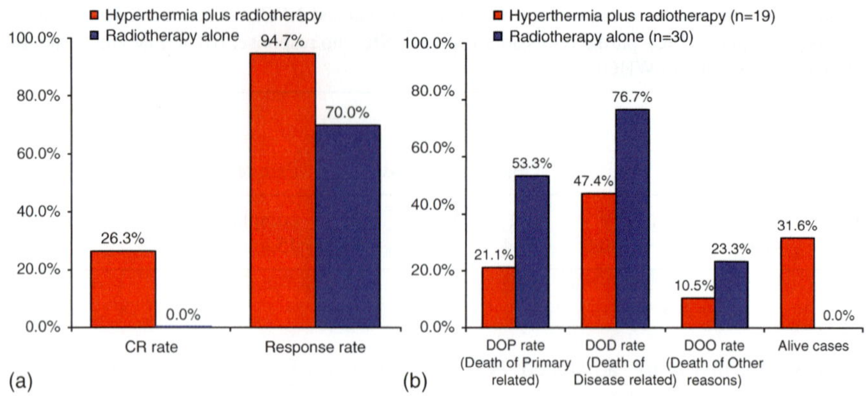

Fig. 2.28 Local clinical responses [complete remission (CR) and total response (CR+PR)] (**a**), and death analysis (**b**)

Capacitive hyperthermia in combination with radiotherapy was also successful for locally advanced non-small lung cancer [398], see Fig. 2.28. The complete remission rate was 26% and 0% with and without hyperthermia, respectively, while the full response rate was 95 and 70%, respectively.

Others also had good results for complementary radiotherapy with capacitive hyperthermia [397].

One of the latest results ($n = 80$) on NSCLC shows no significant differences between the active (hyperthermia plus radiotherapy) and control (radiotherapy alone) arms [403], in the local response rate and in the overall survival. The local progression-free survival was however significantly better ($p = 0.036$) in the hyperthermia arm.

The chemo-thermotherapy combination was also investigated for NSCLC with success. In pre-clinical trials Cisplatin was shown to be effective [404], so the clinical studies concentrated on this drug and its combinations. The synergy between Gemcitabine and hyperthermia in NSCLC was shown in vitro, and in vivo on a nude-mice xenograft model [405]. The decrease in tumor size and a significant inhibitory effect of growth are shown, and hyperthermia supports Gemcitabine-induced apoptosis. A special case report has shown the feasibility [402] of hyperthermia. The median survival was measured in PICT, and measured a definite gain (from 15 ($n = 20$) to 25 ($n = 32$) months), (capacitive coupling, 8 MHz, and 13.56 MHz). The MST was 10, 15, and 25 months for the treatment group (resection + PICT) and historical control groups (resection only and exploratory thoracotomy only), respectively. The survival curve of the treatment group was significantly better than those of the control groups. Measuring the local relapse-free survival for the treatment group (resection + PICT) and the historical control group (resection only) showed also a significant benefit for hyperthermia treatment. Another chemo-thermotherapy study [406] shows pretty good results: MST is 19.2 months, RR = 73%, and 1-year survival is 75%. The 5-year median survival after operation was measured in another study [342], showing a rather high number (24.5%) for patients with N0+N1 status ($n = 14$).

Whole-body hyperthermia has also been applied for advanced lung cancer [407]. This study ($n = 49$) also showed the effective benefit of hyperthermia, which was more effective in elderly (>60 year) patients. The remission rate was 50%, the MST is 7 months (mean is 12.7 months) in primary and 5.5 months for metastatic diseases.

Percutaneous ablation by radiofrequency [408, 409] and by laser-induced interstitial thermotherapy [410] are also in use for pulmonary tumors. Intra-pleural hyperthermia by perfusion is also in use in clinical practice [411]. The breathable perfluorochemical liquid used in convective hyperthermia looks also to be feasible for lung-cancer treatment [412].

2.4.4 Hepatocellular Carcinoma and Metastatic Tumors of the Liver

The liver is one of the most problematic organs for cancer, because of its own tumors, (hepatocellular carcinoma, HCC) and the very frequent metastases (non-HCC) from various other localizations. The liver can be successfully treated by local chemotherapy (chemoembolization), which is one of the most popular and successful chemo treatments. Hyperthermia is excellent in synergism with chemoembolization, increasing the remission rate by more than 12% [413], see Fig. 2.29. The result is remarkable for larger sized tumors. Others have supported these results ([414, 415], see Fig. 2.30).

Hyperthermia works in synergy with numerous different therapies, and all achieved good results in HCC and also in non-HCC studies as well [416] (see

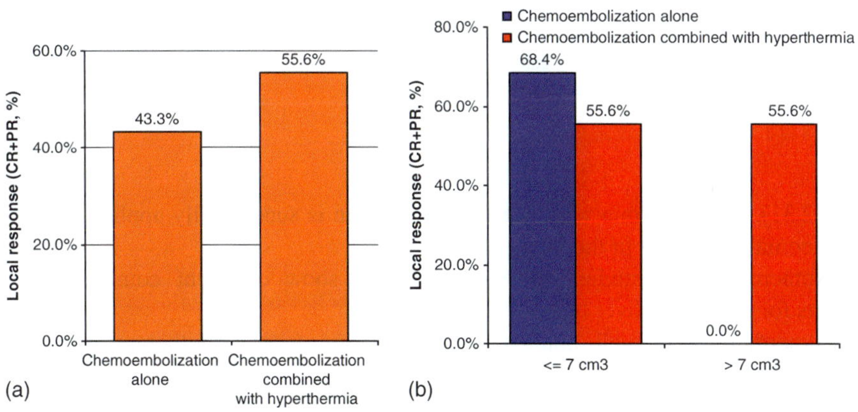

Fig. 2.29 Chemoembolization for non-resectable hepatocellular carcinoma using degradable starch microspheres showing local response rate (CR+PR) (**a**) and response by tumor size (**b**)

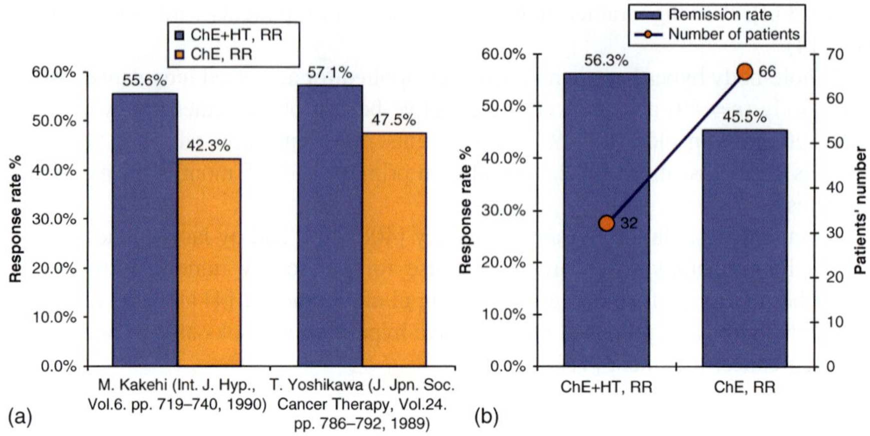

Fig. 2.30 Local remission rates in two different studies for HCC with and without chemoembolization (**a**) and their common meta-analysis (**b**)

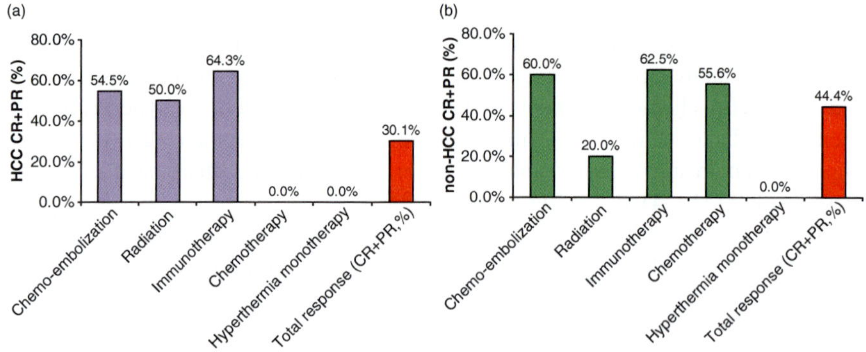

Fig. 2.31 Results on liver tumors, categorized by primary [hepatocellular carcinoma (HCC) (**a**)], and metastatic (non-HCC) (**b**) tumors

Fig. 2.31). Hyperthermia with chemoembolization is successfully applied also for melanoma metastases in the liver [417].

Numerous other studies ([418–420, 1371]) show excellent complementary results for metastatic liver tumors (Fig. 2.32).

Hyperthermia is well applied in combination with radiotherapy for non-resectable cases [421], and also there are many aggressive ablation hyperthermia methods applied to eliminate the liver tumor. This can be done by laser [422], laser in combination with chemoembolization [423], or with the RF ablation technique [424].

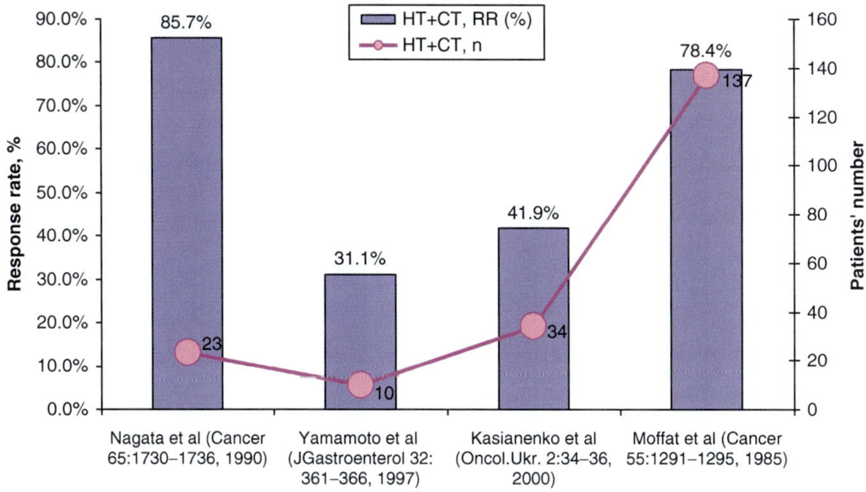

Fig. 2.32 Comparison of publications on capacitive-coupled hyperthermia studies for metastatic liver tumors [IAC = Intra-hepatoarterial catheter (infusion)]

2.4.5 Colo-Rectal Tumors

Radiotherapy combined with capacitive hyperthermia for recurrent or non-resectable colo-rectal tumors is spectacular, having only two cases of progressive disease from $n = 44$ patients. [425], Fig. 2.33. Similar results were obtained with other studies [426–428] as well. Comparison [427] of the active group ($n = 35$) to a control one ($n = 36$) showed the clear advantage of hyperthermia, Fig. 2.34. Success could also be obtained with hyperthermia applied together with chemotherapy in the case of pre-radiated treatments [429].

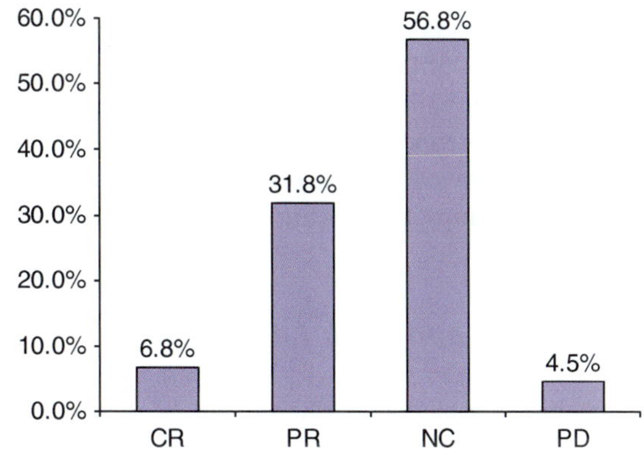

Fig. 2.33 Colo-rectal and rectal cancer ($n = 44$) treatment by capacitive hyperthermia (8 MHz)

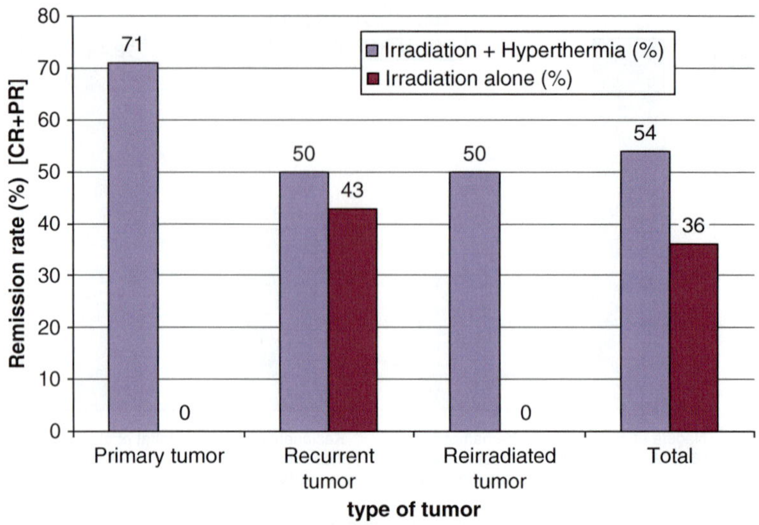

Fig. 2.34 There is a clear difference between the radiation alone and combined with hyperthermia

Pre-operative hyperthermia applications were also successful in trimodal (chemotherapy, radiotherapy, and hyperthermia combination) approaches, [430–432] even intraoperatively [433].

2.4.6 Esophagus

Good results have been obtained for esophagus carcinoma treatment by capacitive-coupled (intraluminar, 13.56 MHz) hyperthermia [434, 435]. A histopathology examination revealed the treatment effect of each type of pre-operative adjuvant therapy. The effective rate was 68.8% in the hyperthermochemoradiotherapy (HCR) group and 44.1% in the chemo-radiotherapy (CR) group ($P < 0.05$). The survival rates were 50.4% in the hyperthermochemoradiotherapy (HCR) group and 24.2% in the chemo-radiotherapy (CR) group. Results are shown in comparison with other studies ([436–438]) (see Fig. 2.35). The treatment efficacy shows a difference also in comparison with and without hyperthermia (see Fig. 2.36), and it is feasible to apply it also preoperatively [439].

Results showed advantages of hyperthermia in recurrent esophagus carcinoma with radiotherapy [440], (Fig. 2.37) and also combined with chemo-radiotherapy [441–445].

2.4.7 Head and Neck Localizations

Capacitive-coupled hyperthermia in head and neck carcinoma also has definite advantages [446], see Fig. 2.38. Curative resection after locally applied radiotherapy with hyperthermia is also feasible [447].

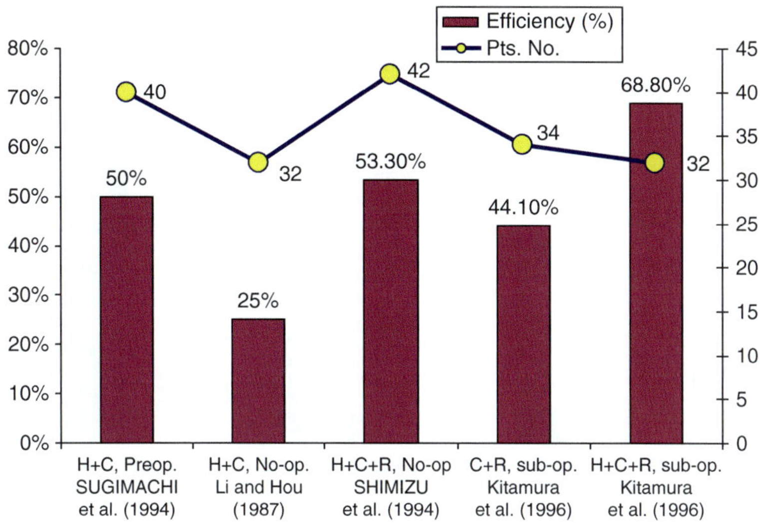

Fig. 2.35 Combination of various oncotherapies (C = chemotherapy; op = surgery; R = radiation; preop = adjuvant or neoadjuvant treatment) with thermotherapy

An important result for radiotherapy combined with hyperthermia is the observation that hyperthermia synergy is much higher in the advanced stages than in lower staged cases [448], A randomized study [449] and a summary of some clinical studies [450] directly show the definite feasibility of using hyperthermia combined with radiotherapy (even for trimodal applications with Cisplatin [451]), and this was confirmed by the long-term (5 year) follow-up also [452].

2.4.8 *Gastric Tumors*

The efficacy of intra-peritoneal chemohyperthermia for gastric cancer patients with peritoneal carcinomatosis was better for the hyperthermia group, but the results were not significant [453]. The radiotherapy combined treatments were effective in most of the trials [454], see Fig. 2.39. Both the pre-operative [455] and post-operative [456] treatments were successfully applied, and radiotherapy [454], chemotherapy [457, 453] or trimodal therapy [458] were also feasible.

2.4.9 *Breast Tumors*

The breast tumor is also frequently and successfully treated by hyperthermia in combination with radiotherapy showing significant advantages compared to radiotherapy alone [459–461]. The results of five controlled clinical trials were collected showing the feasibility of hyperthermia in breast cancer [462]. Capacitive-coupled

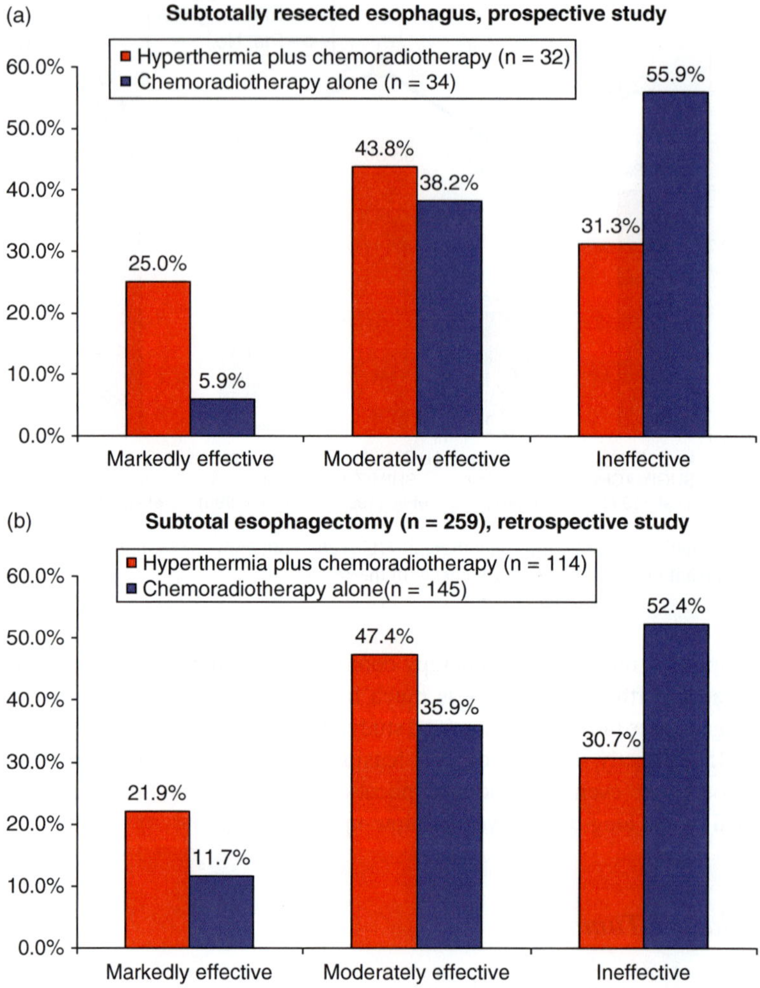

Fig. 2.36 Esophagus cancer in two-arm studies with trimodal application (chemo-radiotherapy compared to hyperthermia combined chemo-radiotherapy). Prospective study (**a**) and retrospective one (**b**) is shown for comparison on a large number of patients by capacitive-coupled (intraluminar 13.56 MHz) hyperthermia

hyperthermia in combination with radiotherapy compared to radiotherapy alone [463] was studied in recurrent and advanced cases also [464, 463]. It is clearly shown [465], when the tumor is larger, that the local response is better (the gain in efficacy to radiotherapy alone is 13.7% when the tumor is smaller than 100 cm^3 and 22.6% when it is larger). The 4-year overall survival is increased to almost 4-times higher [465]. The advantages can also be seen well in the local control and local response rate (see Fig. 2.40).

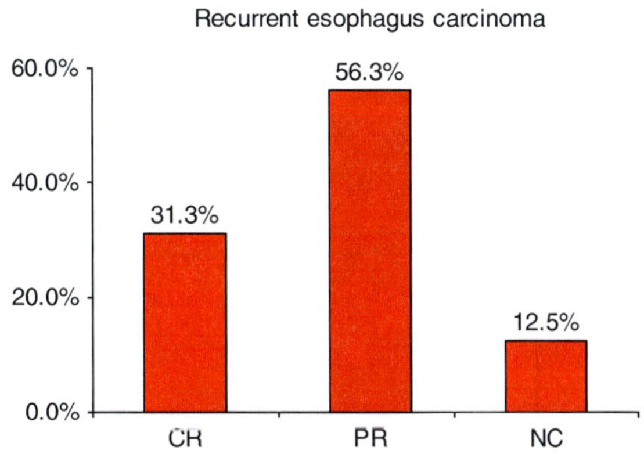

Fig. 2.37 Esophagus results

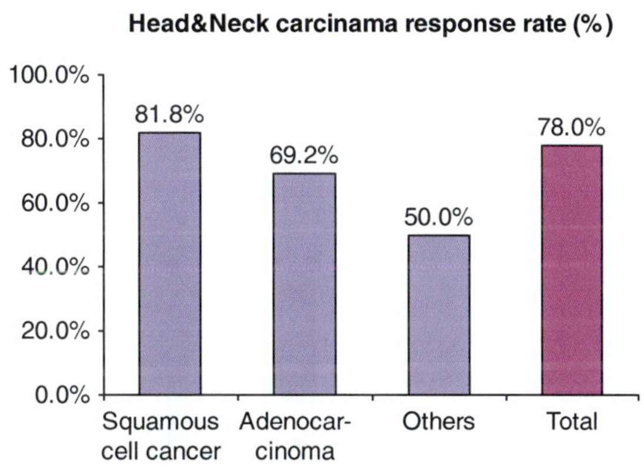

Fig. 2.38 Remission rate of HT+RT for various kinds of head and neck tumors

With chemotherapy (liposomal doxorubicin) [466], and with trimodal therapy [467], a good effect was also stated, except for inflammatory cases [339]. The ablation technique is also applied with success [468]. The question was also formulated "Is metastatic breast cancer refractory to usual therapy curable?" [469] The answer is of course not a full cure, but remarkably good results for metastatic breast cancer have been seen, see Fig. 2.41. The complete remission in these advanced cases was 40% (n = 59); [469].

2.4.10 Other Localizations Treated by Hyperthermia

Peritoneal carcinomatosis, pelvic and abdominal tumors were also successfully treated by hyperthermia in combination with radiotherapy [470, 471] as well as in combination with platinum derivatives (Oxalyplatine [472–474]; Oxalyplatine + Irinotecan [475].

Superficial tumors are also widely applied with great success mainly in combination with radiotherapy [465]. The advantages can be seen well in the local control and local response rate also (see Fig. 2.42) [476–484, 359].

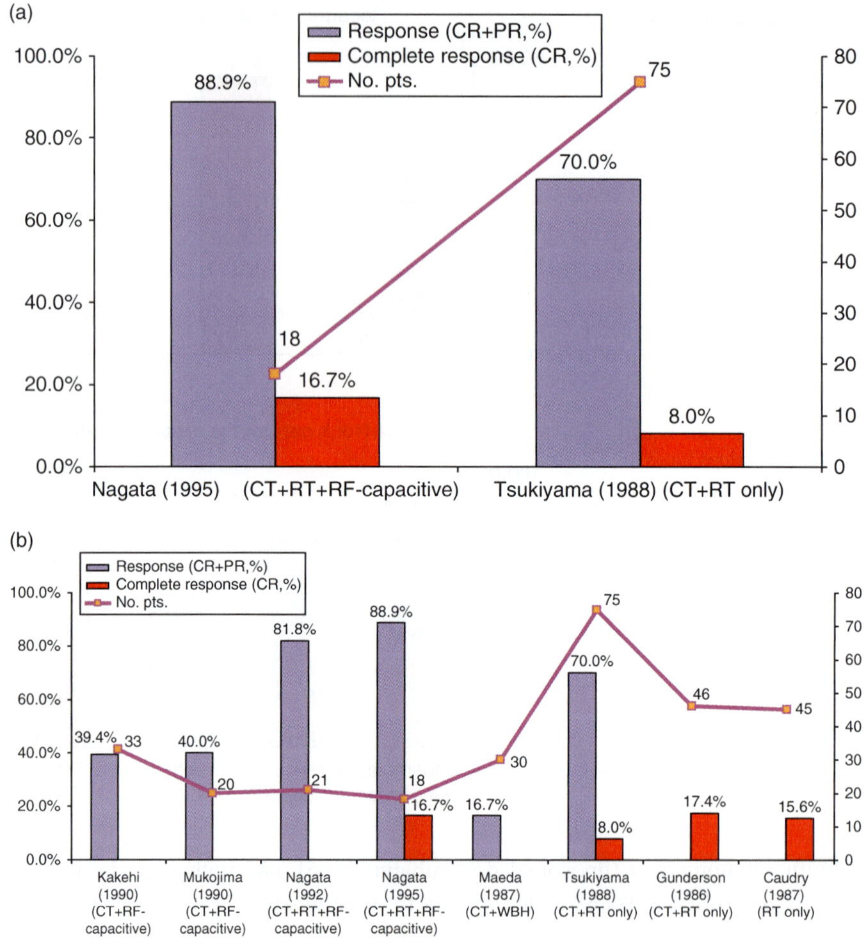

Fig. 2.39 Results for gastric tumors. Its local response comparison with the only radiotherapy publication (**a**) and comparison with other available literature (**b**)

Hyperthermia treatments are popular in gynecological applications [486, 487]. These center on radiotherapy combinations [57], Fig. 2.43; showing highly significant benefit of hyperthermia in overall survival, disease-free survival, and local-relapse-free survival in a randomized trial [57]. A successful large randomized controlled clinical trial for radiohyperthermia has been published in the Lancet [148]. This was regarded as a breakthrough publication.

The chemotherapy combination (Cisplatin + hyperthermia for previously radiated cases) also shows feasibility [488] as well as trimodal applications for cervix [489–491]. There are large debates in regard to this topic [492], with counterpoints [493], and contras [494].

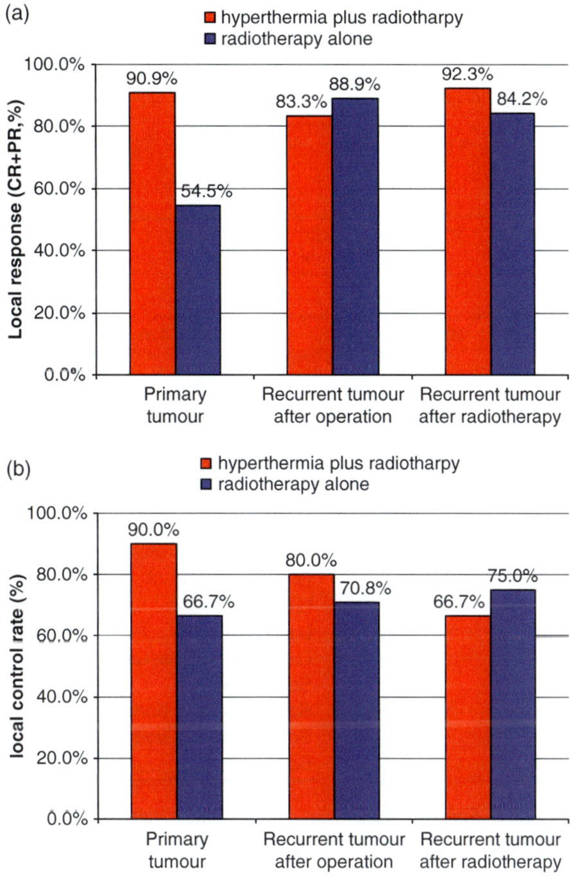

Fig. 2.40 Local response rate (CR+PR) (**a**) and volume dependence (**b**) for hyperthermia plus radiotherapy and for radiotherapy alone in advanced breast cancer

Treatment of prostate tumors by hyperthermia is one of the most rapid developments in heat therapies (see Fig. 2.44). Numerous technical solutions have been developed for prostate heat therapy:

- High-intensity focused ultrasound heating (HIFU);
- Inter-stitial heating (RF-ablation techniques);
- Inter-stitial cooling (cryoablation techniques);
- Laser ablation (vaporization) techniques;
- Hot water trans-urethral heating;
- Metal rods, seeds, or ferromagnetic suspension heating;
- External local/regional heating techniques;
- External systemic (whole body) hyperthermia;
- Trans-urethral microwave heating;
- Trans-urethral oncothermia.

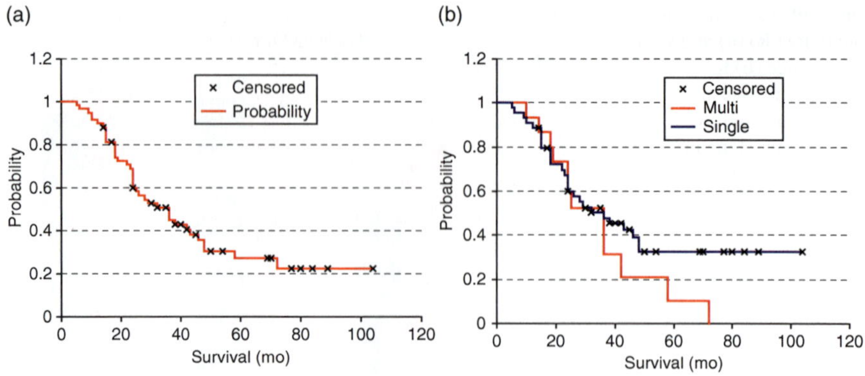

(c)

Metastasis	Number of patients
Liver	6
Liver and soft tissue	2
Brain	2
Brain and lung	2
Brain and bone	1
Lung	9
Lung and node	1
Lung and soft tissue	1
Soft tissue	13
Soft tissue and bone	1
Soft tissue and node	7
Bone	14

Fig. 2.41 Survival of metastatic breast cancer. (**a**) overall survival (**b**) division for single and multi metastases ($p = 0.29$) (**c**) distribution of metastases in the studied group of patients [469]

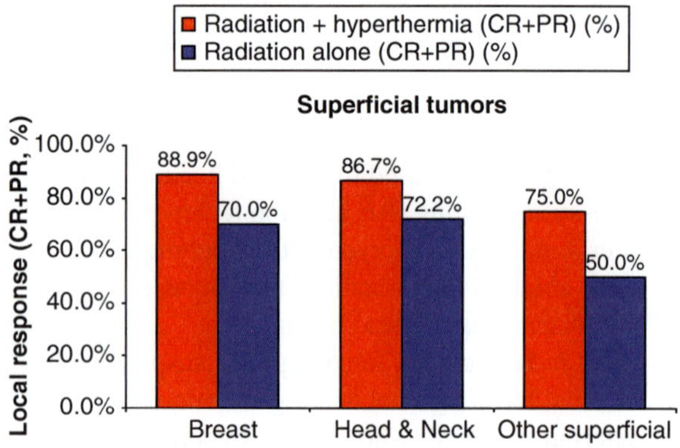

Fig. 2.42 Local clinical response results for some superficial tumors [485]

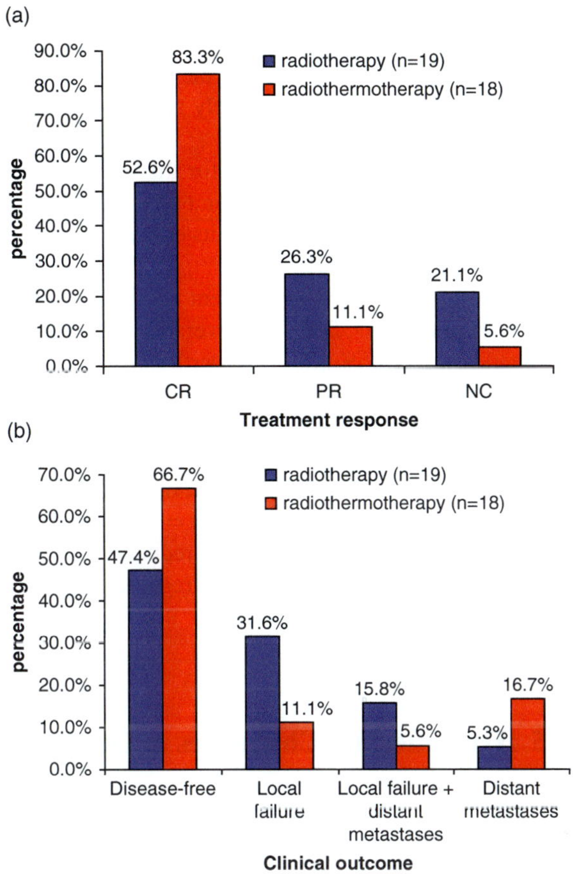

Fig. 2.43 Cervical (uteri) carcinoma results [57] by local response (**a**) and clinical outcome (**b**)

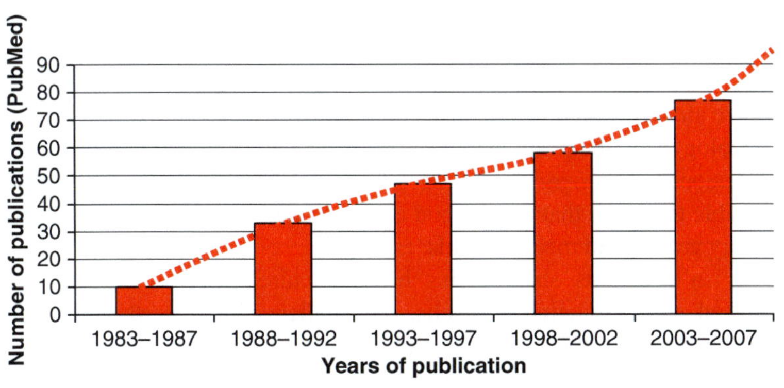

Fig. 2.44 Development of hyperthermia treatment of prostate cancer from published data (PubMed search profile: "prostate AND hyperthermia AND (cancer OR malignant OR tumor)"

Hyperthermia in combination with radiotherapy [495–497] was successful, and a randomized controlled clinical study proved its efficacy for benign prostatic hypertrophy (BPH) [498], measured over long-term follow-up also [499]. Its cost-effectiveness was positively analyzed compared to trans-urethral resection [500].

Hyperthermia can be applied for the urinary bladder [501]. Survival rates are much longer (over 60%) in its combination with radiotherapy compared to radiotherapy alone [502]. It has shown its efficacy in high-risk cases also in combination with chemotherapy [503].

Hyperthermia has given excellent results in soft-tissue malignancies [504] especially sarcomas [505, 506]. The overall survival rate for long-term follow-up (over 10 years) is over 85% [507]. Hyperthermia could be applied preoperatively [508, 509], and intraoperatively [510], and its whole-body application together with combined chemotherapies (Ifosfamide, Carboplatin, and Etoposide) has also been published [511–514]. The topic was the focus of a very large extended research and clinical investigation carried out by Prof. Issels and his group [515–522, 218] which ended in a successful large ($n = 341$) randomized controlled Phase III clinical study [523], showing 29.2% risk reduction (after 5.7-year median follow-up) for local progression or death by local/regional hyperthermia combined with surgery and radiotherapy.

2.5 Hyperthermia Challenges in Oncology

The challenge is trivial: hyperthermia was the first ever cancer treatment; hyperthermia is more than two thousand years old; and hyperthermia is less accepted nowadays than other "young" (less than a hundred years old) therapies. Why is this so?

What are the reasons for these challenges, apparently contradicting the extremely long-term history and the strong evidences of positive useful effects?

The problems originate from the same root: inadequate techniques to allow treatments to be reproduced in the clinics in such a successful way as seen in the lab.

We have already seen how effective and promising the hyperthermia applications are in oncology; and the positive impact that can be attained in combination with all the conventional and emerging new oncotherapies. The picture is very positive and it is plausible to expect that the method is in the focus of interest among specialists in oncology and related medical fields.

But it isn't! Doubts shadow this bright picture. Despite the large number of published excellent clinical results, the challenge of hyperthermia in oncology is plausible from the perspective of its few-thousand-year history. Medicine faces unsolved problems in hyperthermia, mainly in relation to the controversial results obtained from the very beginning. In addition to the unexplained mechanisms of hyperthermia its control for efficacy and for safety remains unsolved as well.

It is probable that hyperthermia is one of the subjects which has the greatest number of questions in the published literature. Numerous definite questions have been formulated, such as:

- Is the community radiation oncologist ready for clinical hyperthermia? [524];
- Is there a future for hyperthermia in cancer treatment? [158];
- Is heating the patient a promising approach? [159];
- Hyperthermia: has its time come? [525];
- What is against the acceptance of hyperthermia? [526];
- Progress in hyperthermia? [527];
- Prostate cancer: hot, but hot enough? [528];
- What happened to hyperthermia and what is its current status in cancer treatment? [529];
- Where there's smoke, is there fire? [530],
- Should inter-stitial thermometry be used for deep hyperthermia? [531];
- If we can't define the quality, can we assure it? [532].

Questions pile up but satisfactory answers are as yet missing. The results are sometimes very promising and significantly show the healing power of hyperthermia, but there are disappointing clinical trials as well. The real challenge is of course the controversial results, addressing many further questions and raising doubts. This is exactly what happened in a cervical cancer study where the results were very promising [148], and a control study [533] was disappointing. The explanation may be simple: a reference point was missing [534].

Despite the many promises and proven effects, hyperthermia faces many serious challenges in oncology. We do not understand clearly the underlying mechanisms, the possible risks and safety issues, and the limits of its applications. The most problematic issues are connected to

- the controversial clinical results;
- the unstable reproducibility;
- the missing consensus-based relevant dose definition (quantifiable process).

In consequence of the missing consensus and widely accepted treatment guidelines the requests are:

- mandatory prospective, randomized, controlled clinical trials requested by evidence-based-medicine;
- wide social support and funding of the treatments;
- forceful educational activity, and public relations.

The complications in detail show many practically applied technical solutions, and no consensus of their comparison. The actually obtained temperature is the only criteria, which has serious principal and practical problems (see later).

The opposition opinions can accept the "sometimes possible" positive effect of hyperthermia, but the central question is of course: could hyperthermia harm the patient? In connection to this is a basic doubt has been formulated: hyperthermia may increase the risk of distant metastases. Indeed, hyperthermia was applied as a monotherapy in laboratory experiments and an increased rate of metastases was found for mammary carcinoma of C3H mice [535] and for rat's sarcoma [536] as well. However, other animal experiments show the opposite: a prevention of distant metastases of mouse sarcomas was observed in two independent studies [537, 538]. Nevertheless in clinical studies on dogs the induction of spontaneous distant metastases was observed in local [539], and in systemic hyperthermia [540]. This area of possible metastatic tumor genesis by hyperthermia is challenging. The increased carcinogenesis was shown in an animal model in combination with X-rays [541]. Hyperthermia induced transformation of the Bleomycin and Cisplatin to carcinogenic was observed in vitro [542]; while others provided contrary proofs to these negative experiments [543, 544].

Some further challenges follow from the effects of the combined therapies, because hyperthermia enhancement of the effects of radio- and especially chemotherapies could possibly cause long-term side effects. Clinical observations support the extra toxicity problems of thermo-radiotherapy [545, 546], but these were not confirmed by others [547].

Considerations on thermo-resistance are also controversial. The heat-shock-induced chaperone proteins could develop a multi-stress-resistant state, decreasing the chances of further treatments. Also the blood-flow reduction in the tumor through hyperthermia could decrease or even block the effect of chemotherapy (i.e. insufficient drug delivery into the tumor; or severe hypoxia could block X-ray efficacy). We could discuss these last objections. The physiological problems could be solved by a perfect treatment protocol, taking into account the actually interacting physiological changes, and choosing the right sequence of complementary treatments.

All of these challenging contradictory observations could be cleared-up with a definite change in the overall survival, which would demonstrate the long-term effect, and may possibly negate the negative observations. However, these data are also not comfortably conclusive. The obviously noted survival increase in radiotherapy with additional hyperthermia in cervical cancer is not accompanied by the same increase in the rectum and bladder in the same pelvic study [148].

All of the above challenges of hyperthermia practice, however, are of a more technical kind than really medical. The problems of dose application, selection/focusing solutions, and safety predominantly request technical solutions. Well-controlled mechanisms could provide a comfortable answer to all the doubts. The missing acceptance of hyperthermia by the medical community is basically the consequence of inadequate deep heating (focusing, selective heating), the indefinite dose, and missing reproducible treatment protocols.

Many researchers in the field share the opinion of the editorial comment of the European Journal of Cancer in 2001: the biological effects are impressive, but physically the heat delivery is problematic. They formulated this opinion as: *The biology is with us, the physics are against us* [158]. In the latest oncological hyperthermia consensus meeting, the physics was less problematic. However, in accordance with the many complex physiological effects, a modification was proposed: *The biology and the physics are with us, but the physiology is against us* [548]. The present situation apparently supports this opinion. However we would like to suggest such physical effects, which could make easier the harmony of the two disciplines and both could be "with us" to win this battle of the war against cancerous diseases. We are convinced that this clear definition will help to solve many problems in oncological hyperthermia, helping the development of the process, giving a definite positive answer to the question: *Heating the patient: a promising approach?* [159].

Only a few highly ranked oncologists and radiotherapists regard hyperthermia as a stable complementary solution or a useful sensitizing process, and even less is the number of those, who actually use the method, or at least recommend it for patients in their actual need.

The challenge of hyperthermia lies also in its relatively complicated practice, as despite its simple principle the required permanent and intensive care from the medical personnel on duty takes a very long time in comparison to other (radio- or chemo-) therapies; in addition the preparatory work is also not less than that for conventional therapies.

Hyperthermia today, like many early-stage therapies, lacks adequate treatment experience and long-range, comprehensive statistics that could help us optimize its use for all indications. Nevertheless, we will present a wealth of information about the mechanisms and effects of hyperthermia from the scientific literature and our own experience with the hope of proving hyperthermia's worth for further research.

The present state of oncological hyperthermia is similar to that of radiology at its infancy. When ionizing radiation was first discovered, many hypothesized its usefulness in oncology, yet its exact dose, contraindications, limits, and the conditions of optimal treatment were not determined until several decades later.

One of the modern technical outgrowths of the development of oncological hyperthermia is oncothermia, devoted to improving the technical conditions of hyperthermia, and providing a better solution to the ancient problems of selectivity, reproducibility, and safety.

There is a definite group of physicians who believe in the curative force of oncological hyperthermia, and there also exists a group, which may be larger than the first, that believe the opposite. Of course the positive and negative believers are not helpful to the aim of bringing clarity to the situation. We need data, scientific analyses, and hypotheses to proceed with our topic. And we need healthy and well-established doubts with good and relevant questions. We think the questions could never be unscientific, only the answers could sometimes be such. We are free to address questions and very careful and cautious in answering them. Naturally we will make great efforts to provide answers to the best of presently available knowledge, as well as providing our final "answer" which would be the development of a technical facility (oncothermia) which is devoted to finding solutions to the controversial problems, and providing a stable and predictable treatment with a new hyperthermia paradigm in oncology.

2.5.1 Challenge of Selection and Focus

The distortion paradigm of oncotherapies has a huge cost in practice: considerable side effects could be generated by improper selection of the distortion and/or by the liberated toxic species of the destroyed cells. The natural tolerance, adaptation, and protective mechanisms of the cells against lethal effects also limit the efficacy of the distortion mechanisms. To provide the best possible treatment an adequate selection has to be made, to destroy the malignant cells only.

The main difficulty of proper selection is the inhomogeneity of the cancer and the lack of definite boundaries. The solitary malignance is only apparently local, it is systemic in the meaning of dissemination of cells, and in regard to those multiple complex changes which makes the malignant tumor systemic (e.g. immune changes, observable changes in laboratory results, fatigue syndrome, weight loss, etc.). (Anyway these systemic effects distinguish the malignancy from benign tumors, which are really local.)

Such an apparently exact selection like resection in open surgery does not offer a satisfactory solution in most cases, due to the lack of a boundary for the malignancy. Complete remission, as a direct clinical response, could be achieved, but later relapses are very likely to occur. The remaining cells are likely to disseminate and form metastases, rapidly suppressing the survival time. Accurate selection has to be cellular including the cells in a large area/volume in and around the malignant tumor.

Consequently artificial (made by macro images) focusing has some characteristic problems:

- To focus at depth in the living system is difficult due to the aqueous electrolytes (the healthy tissues) through which the energy must be transported. These healthy living tissues (covering the deep-seated tumors) absorb the energy which heats them up. Their cooling means a very considerable energy loss.
- The unwanted absorption does not occur when magnetic seeds or other magnetic energy-absorbing materials (rods, suspensions, nanoparticle dose, etc.) orient the field to the tumor. The moving magnetic field interacts almost exclusively with the high magnetic permeability materials, which have to be placed invasively where the energy absorption is expected. Also the invasive ablation techniques do not face unwanted absorption of the middle layer.
- The focusing has to properly follow the geometrical symmetry of the tumor, which means it has to have spherical symmetry. It is in most cases not possible, because the focusing mechanism produces a translational (layer-like) or cylindrical (belt-like) symmetry, which can not cover exclusively a ball-like structure.
- The focusing needs measured feedback to adjust and concentrate on the selected area. This needs a method to measure the energy parameter (in most cases the temperature or SAR). This request produces many technical complications, which are mainly connected to the previously described problems of temperature

measurement. The main challenge is the not strong correlation between the SAR and the temperature developed by it.
- If focusing would be perfect and very accurate that is also not enough for treatment. The tumor is not a solid ball with a special structure and (most importantly) it has no definite boundary. (This indefinite boundary makes it malignant; with a definite size it would be benign.) So a proper focus means the maximum just like in surgery, where the problem of the "boundary-less" tumor is also the main problem of success.
- Some invasive (eg. ablative, interstitial, intraluminar, etc) solutions have such small volume to destroy, which could be well controlled by the surface of the applicator in the tumor. The focus here is approximated centrally around the applicator.

Naturally the main challenge of hyperthermia selection is technical: how to select the tumor cells properly in the treated volume without affecting the healthy parts of the volume and without the measurement causing a safety risk. This challenge is the real task of oncothermia (see Section 2.5.6).

2.5.2 The Challenge of Temperature

Despite the very practical challenges in hyperthermia, in connection to temperature the problems are more conceptual than technical. We know that temperature and heat are different in their conceptions, their idea and definitions cover different thermodynamic values. Unequivocally, temperature and heat are categorically distinct qualities; their physical and biophysical differences basically govern the principle of oncothermia. (This will be discussed in Section 3.2). In practical applications using the terminology "heat dose," and "heat exposure," instead of *temperature* or vice-versa is incorrect, and could be misleading to the treatment aims and the evaluation of the results.

The mismatch between temperature and heat is the central problem of hyperthermia dosing and treatment standardization also, which is still a significant problem in hyperthermia principles and practices. There are considerable discussions in regard to the relevant treatment parameters, treatment optimization, and guidelines. Discussions concentrate on the temperature: its role, its measurement and its medical effects and technical realizations. To control heat transfer into and within the body, and provide the same reproducible heat dose within the target tissue is technically very difficult, and it is even insoluble and inextricable if we do not clearly determine the goals and do not understand the underlying mechanisms.

To decide which parameter hyperthermia uses, we have to define the task of hyperthermia and the aim that we would like to achieve; and naturally understanding the processes is obligatory.

It is without doubt that the most common and requested parameter of hyperthermia treatments is the temperature. However, measurement of temperature in the malignant area is not a simple task. Numerous conditions make it difficult to determine the temperature in a living object. The temperature is very much nonhomogenous site by site. The various size scales have different modifications on the temperature distribution which makes it inherently inhomogeneous:

- at the cellular level: the metabolic rate;
- at the cluster level: necrotization;
- at the tissue level: blood perfusion;
- at the level of organs: the function and interconnections.

In addition to these biological modifications many physical, material factors are added to form the final temperature distribution, including conduction, convection, and radiative processes as well as the intensive and permanent connection with the environment. This causes a smearing of the actual temperature and could seriously modify the heating pattern.

This is the first point casting serious doubt on temperature omnipotence. The measurement of temperature in a definite local point in the tissue can not give general information; it depends greatly on the actual sensor size and position as well as on the actual tumor/tissue status and connections. Consequently, the usual contact temperature point-measurement could give realistic control, except if an appropriate average from many of the measuring points is observed [549].

The disadvantage with multiple invasive temperature sensors is derived from problems of serious discomfort, pain, possible infections, ulcers, and even some metastasizing through release of tumor cells into the blood flow. Technically the point measurement is also problematic. The metallic sensors, like a receiver antenna, may collect a lot of noise, which complicates its filtering. Moreover, the extra energy absorption heats up the sensor and its wires differently than the measurable media. Application of optical wire sensing [550], is an accurate point measurement, but the inadequate volumetric sensing remain unsolved. The most frequently used types of temperature measurements are summarized in Table 2.34.

Non-invasive methods are very much desired, but their application is limited by serious technical complications as well as their complex phenomena and expensive techniques. The complications of these non-invasive methods have various sources. MRI or other imaging systems measure the chemical or structural fingerprints of the temperature. Its accuracy depends on the phantom to calibrate the actual measurement. If the phantom has no adequate material data, physiology-equivalence, and also energy-consuming distortion, than the calibration will be insufficient. The only water calibration (which is the practice in most cases [551]) validates only the temperature change, but the energy, which makes the distortion job is out of the calibration. When we heat up the tumor cell like we do for a water phantom, the treatment efficacy is doubtful. The paradigm is the cell-destruction, and not the simple increase of the temperature.

The most popular methods and their complications are listed in Table 2.35.

In many practical cases the temperature is measured with intra-luminar or intracavitary catheters. These semi-invasive applications use technically one of the invasive methods to measure the temperature on the surface of the catheter. The

Table 2.34 Some typical temperature measurement techniques in hyperthermia

#	Thermometry	Typical solution (example)	Invasivity	RF-incompatibility	Measurement complication	Calibration, validation
1	Thermocouple	Pt-Pt (Rh)	Invasive	Yes	Very local (point)	Simple technical
2	Electrical resistance	Thermistor, Pt-wire	Invasive	Yes	Very local (point)	Simple technical
3	Fiberoptic solution	Fluorescent signal	Invasive	No	Very local (point)	Simple technical
4	Electric-impedance (local)	Point measurement	Invasive	Yes	Very local (point)	Simple technical
5	Electric-impedance tomography	Multi-pair measurement	Noninvasive	No	Difficult interpretation	Phantom
6	Thermo-radiometry	1–20 GHz radiation	Noninvasive	No	Low resolution in depth	Phantom
7	X-ray thermoden-sitometry	X-ray analysis	Noninvasive	No	Low resolution in volume	Phantom
8	MRI-temperature tomography	Time shift (T_1 & T_2) measurements	Noninvasive	Yes/no	Multi-dependent parameters	Phantom
	Ultrasound	Ultrasound echo analysis	Noninvasive	Yes	Limits by bone and air	Phantom
9	Doppler thermometry	Sensitive Doppler apparatus	Noninvasive	No	Blood-flow-dependent	Technical
10	Infrared thermometry	Thermocamera (5–12 μm range)	Noninvasive	No	Surface measurement	Technical

Table 2.35 Complications of popular non-invasive temperature measurements

Non-invasive method	Main complications	Technical condition
Infrared thermometry	Measures only on the surface. The internal temperature could be visible only in open surgery. Some special locations (e.g. breast) could be measured deep, due to heat-condition/convection characteristics	Thermocamera or equivalent visualization is necessary optimally in the range of 5–12 micron
MRI-temperature tomography	Measures the changes in the chemical bonds, which are temperature-dependent. However, no method to distinguish between the origin of the change, only the phantom calibration could decide. The phantom must follow the same complexity as the biomaterial, including the physiological and biochemical conditions	MRI-device with specialized software is necessary, together with a MRI compatible hyperthermia device
Thermo-radiometry	Depth information could be followed but with low resolution. Its calibration is also a problem, because the resolution construction depends on the actual densities and electromagnetic parameters of the tissues	Multi-frequency passive-radar technique has to be applied. Horizontal resolution is complicated
Electric-impedance tomography	Difficult to evaluate the very general mixture of the impedance conditions from the obtained pattern. It has low resolution and a possible phantom effect could be generated	Multi-frequency mode and software with accurate material data are mandatory

measurement, however, in a lumen (esophagus, rectum, vagina, etc.) is not accurate enough to be sure of the focusing and safety (avoiding hot spots), and definitely not accurate enough to conduct a treatment far from the lumen.

The minimal dose is the "success parameter," and personalized hyperthermia treatment uses the highest tolerable dose for the treatment guidelines. During temperature control we try to exceed the minimal requested temperature and reach the maximal tolerable temperature above this point. Temperature measurement is necessary for hyperthermia treatment not only for dose-fixing, but for safety also. Temperature control could avoid hot spots and help in artificial focusing. This role of temperature is sometimes more important than the dose itself, because the tolerability (toxicity = burning) limits the action anyway. The role of temperature

Table 2.36 The main parameters connected to temperature applications

Parameters	Safety	Quality guideline	Control, reproducibility
Challenge	Hot spot	Give proper dose	Patient's physiology & movements
Dose	Temperature	"Thermal dose" ($CAM43°CT_{30}$)	Focus by temperature
Technique	Temperature measurement	Space distribution	Cylindrical action on spherical tumor

and its technical requirements are presented in Table 2.36. At the point of treatment planning many physical, physiological, and biochemical parameters have to be taken into consideration to improve the efficiency of focusing and calculate the heat conduction/convection within the body.

Some of the applications (ablative hyperthermia) perform a local burn to destroy the unwanted tissue. In these cases, the temperature has only a role in coagulation as a threshold (T>60°C). To sense the threshold the phase transition [the coagulation (ablation)] is measured not necessarily by the temperature. In most cases this is done by impedance measurement, and also down-regulation in the case of overheating (carbonization) is controlled by a change in the impedance [18].

One of the most polarizing challenges among specialists is the question: has hyperthermia any non-temperature-derived (this is often incorrectly termed "non-thermal") effects, or is it purely temperature-dependent (again with incorrect terminology i.e. "thermal")? We will explain this in detail in Section 3.2.6.

The thermal equilibrium (which could be characterized by temperature alone) can not be the solution of such a dynamism, which is required for the desired distortional changes. The "non-thermal" "chemical machinery" must be effective in causing something more than only an increase in temperature in the actual volume. We should never forget what the aim of the treatment is: distortion of the malignant cells! The temperature is a possible tool for this among other cooperative ones. The distortion-free increase of temperature also approaches the task of course. It could provide optimal conditions and efficacy enhancements for other complementary therapies. In this case, the task of distortion is carried out by the other complementary method (like chemo- or radiotherapy) and hyperthermia tailors optimal conditions and boosts their effect.

According to everyday use, the observed thermal effects are temperature-dependent. Naturally, the non-temperature-dependent but heat-connected effects (like melting) are also thermal, (coming from the original meaning of the word *therm*) but due to the acceptance of the vulgar terminology in the science as well, we will discuss the purely heat-dependent effects as nonthermal. (It is very similar to considerations relating to the steam engine: the water is heated until the boiling temperature of the water is reached and an appropriate amount of steam is developed. After this point no temperature change occurs in the system, all the heat energy from the fuel is used for the desired mechanical movements and to replace the employed energy in the system. This effect, from the point when the dynamical equilibrium (stationary process) was reached, is absolutely not temperature-dependent (nonthermal), only heat-dependent (we have to pump heat energy in to keep the process going, without any observable change in temperature). We will keep this unfortunately very sloppy formulation of thermal and non-thermal conditions to be in harmony with the widely accepted terminology of hyperthermia.

To clearly formulate hyperthermia categories using this language: if the goal is direct cell destruction, then hyperthermia has to use non-thermal effects. In this context the question has to concentrate on the aim: is the energy absorption the main effect, and the temperature is it a consequence, or do we need the temperature as a primary condition for the other effects. The first process is nonthermal, the second however is a thermal category. When hyperthermia performs the destruction, it isn't thermal, but if it enhances other methods to provide them with optimal conditions, then that hyperthermia is definitely thermal. Of course, none of the categories are purely distinguishable from the other, hyperthermia is a complex effect, consequently only the dominant action is denoted with the thermal categories.

Using this terminology the whole-body (systemic) treatment is thermal and the local/regional is treatment is nonthermal in their dominant effects.

The application of hyperthermia faces two different tasks:

1. Set hyperthermia conditions that directly destroy the malignant cells, according to the oncologic paradigm. The definitive energy absorption has to cover the biochemical reactions, and only the part, which is unused, is used to increase the temperature of the whole treated volume. On pumping energy into the target to perform certain work there we have to form inequalities (gradients) which would be the driving force of the work. For this we need working energy in the target, not an equal temperature which prefers equilibrium and excludes gradients. (The picture is naturally different when we form microscopic temperature gradients. This case will not necessarily increase the overall temperature or at least by much less than in the equilibrium case.).
2. The point is that the temperature is an important initializing parameter to push the system from the actual dynamic equilibrium to a different one. If the task, which we expect to perform, is not a direct distortion of the malignant cells, but the creation of better (optimal) conditions for more drastic other methods, than the temperature would be the active parameter. The higher temperature could help to increase the reactivity of drug reactions in chemotherapy or provide better conditions for radiotherapy due to the promoted blood perfusion (oxygenation). However, the selection on the cellular level would be important again to promote selectively the actions of the complementary treatment.

In both cases modified complex physiology interactions have to be considered: the temperature is modified by conductive and convective heat transfers as well as active biological heat production.

Nowadays, clinical practice uses the patient categories of "heatable" or "unheatable" to select them (inclusion criteria) into hyperthermia clinical trials [552]. The selection is based on the possibility of temperature increase in the given patient in the actual localization by local/regional treatment. Should hyperthermia be used as a distortion tool, this method of selection places patients into incorrect categories when measured by the temperature. The distortion is an energy-dependent mechanism, and a missing measurable temperature increase does not necessarily mean that the absorbed energy is not used for the distortion job. If the "heatability" categorization is made to select patients for the optimal completion of another therapy, it could be a correct selection. For this distinguishing a clear terminology is mandatory.

The challenge, which urges us to choose between "heat or temperature" is not accurate. It is better to say "heat and temperature." However, it is mandatory to realize that the measured temperature or heat alone cannot characterize the actual processes. (Simple example: it is impossible to measure the invested energy of freely falling rocks by the temperature of the ground where they fell and lost their energy.)

2.5.3 Medical Challenges of Hyperthermia in Oncology

The problems in hyperthermia adequately focus our attention on the technical demands, which are the key points to provide for proper treatment. Reproducibility and dosing are also among the technical challenges to be solved. For these basic knowledge about the mechanisms of the hyperthermia action is mandatory; however, the debate about the exact mechanism has not yet finished in the scientific community:

- Widely prevalent opinion conceives the single important quantity is the temperature. The observed phase-transition-type behavior (at 42.5°C) [238] and the experimentally perfectly fitting Arrhenius line [239] serve as proof for these opinions.
- There are also opinions suggesting the heat quantity as the main driving parameter [551]. This idea is confirmed by the definite time-duration dependence of the treatment effect, which is a clear dosage-type effect. Also the lower efficacy of homogeneous high-temperature heating (whole-body hyperthermia) compared to the inhomogeneous local/regional one challenges the temperature concept, and supports the pro-heat opinions.
- There are various considerations to formulate special non-thermal effects, mainly with direct electromagnetic interactions [553, 176, 299]. Supporting these arguments are the observed data of not thermally generated HSP synthesis [300]. The applied electromagnetic effect may cause in itself an HSP increase and undesired tolerance formation without any heat input and temperature increase [554], and may have some effect on the reactions of the immune system as well [555]. The activation of programmed necrocytosis [296] and the blockage of angiogenesis [556] also could be generated non-thermally.

- The synergy of electromagnetic heat effects with radiation- and chemotherapy as well as with surgical interventions (success of pre- and post-operative treatments), indicates a long-term (memory-like) effect of hyperthermia [310], which is effective after the heating has ceased, so is not directly connected to the temperature itself.
- A treatment of complex metamorphosis requires a complex parameter ensemble. It is not expected that such a complex system as the human being could be successfully controlled by one or two parameters. There are some parameters to control the technical application, but it must not be understand as a control of the processes.

The main technical challenge of hyperthermia is of course the focused deep heat input. Misfocused heat input involves the risk of necrocytosis of healthy tissues, as well as the increased blood perfusion which may increase metastasis. The appropriate selection of heating techniques and the construction of actual applicators play a crucial role in the application.

To deliver the heat three fundamental processes exist: conductive and convective heating and the various kinds of radiation.

2.5.4 Challenge of Quality Control and Dosimetry of Hyperthermia

Quality control is one of the central problems of modern medicine in general.

The main features of medical quality are as follows:

- The method has to be effective. Its efficacy has to be controlled in situ (safety) and off-situ immediately after and a long time later (follow-up).
- The method has to be controllable and accurate; a definite dose parameter has to be involved.
- The method must be reproducible with certain limits on the same patient or on other patients.
- The method must be safe (low risk/benefit ratio) (Note that the safest method is the non-acting one. This is an inherent contradiction with the efficacy, and this requires the risk/benefit ratio calculation).
- The method has to be cost-effective (the cost/benefit ration is calculated, again with the contradictory tendencies with the efficacy).

It is a well-known fact – right from the beginning of human medicine – that, in order to maintain homeostasis, we have definite lower and upper limits of the interactions and conditions of life. The body homeostasis is stable within a certain interval of the parameters; the level of any interactions is determined and measured by its harm. However, harm is a relative notion: safety and the harmless categories are not identical. The "no action" treatment can be safe, but harmful as well, because the uncontrolled disease does the harm, which we can stop by action. The changes that are acceptable have the direction of returning to normal, healthy homeostasis, or if this is not possible any more, then as near to it as possible. The Hippocrates phrase, "Nil nocere" also has to be understood in this way, otherwise the meaning is "Do nothing."

There has been demand for gauging, measuring medical actions quantitatively for as long as medicine has existed; however, personal differences and conditional factors modified it adapting to the actualities. Also, statistical conditions and data management were not sufficient for handling the collected data in a satisfactory way in order to assure that the observations are objective enough. The dose was fitted to the immediate actualities, and the result measured afterwards, with the facility of prompt or later modification of the given medical action.

Nevertheless, the rapidly developing available medications and medical methods, as well as information exchange in society, and the carefully built social network in modern, post-industrial nations does not mean we have to accept arbitrary dosing and accidental success.

The dose has to be a well-characterized quantity, fitted to other quantities and measurable within usual conditions. The dose is predestinated to measure the actual action and to compare it with the usual helpful action. The dose has to have accepted physical quantities (fits to the International system of units (SI); retraceable to the seven basic quantities: length (m), mass (kg), time (s), electric current (A), thermodynamic temperature (K), amount of substance (mol), luminous intensity (cd). The SI-derived units (e.g. force, frequency, etc.) are also acceptable, and the constants of nature could be used, (e.g. Plank constant, Boltzmann constant, light velocity, etc.). All the non-SI units have to be reduced to SI-ones by condition-independent factors, eg. "inch," (2.54 cm), "foot" (30.48 cm), "yard" (0.9144 m).

The general challenge of dosimetry in oncology is in its objectivity. Oncology, like medicine in general, deals with persons, the medical problems are personalized, depend sharply on the patient and her/his general conditions.

The dose has to have a quantitative measurability; we must be able to measure the given/applied quantity objectively. This demands that the dose is size (e.g. mass, volume, surface, etc.) proportional; the double of the original size has to have the double of the applied dose.

The conventional treatments have very definite dosing: surgery uses volume (which is resected) (cm^3), radiotherapy uses the absorbed energy per absorbing mass, Gray (Gy=J/kg), chemotherapies use the mass of the active medicament per unit-surface of the patient (mg/m^2). The dose in chemotherapy is mainly a safety formulation, establishing the acceptable tolerance limit by dose-escalation studies. Because of toxicity considerations, this dose is entirely independent from the size of the tumor, it depends only on the weight of the patient. This impersonalized dose in chemotherapy allows the hypothesis-based statistical probes in biostatistics, creating the necessary normal distribution for the evaluation.

Hyperthermia uses the temperature as the dose and as the safety limit as well. Unfortunately the temperature-dose does not satisfy an important requirement of the dose: the extensive behavior, it does not depend on any size parameters. Using the time dependence of hyperthermia the time and temperature was used together, making doses how long was the actual temperature valid. This simply creates a unit (temperature multiplied by time, [Ks], which has no physical relevance. Later CEM43°CT$_{90}$ was introduced, which – using the Arrhenius relationship – measures the cumulative equivalent minutes at 43°C where the temperature exceeds 43°C at 90% of the locations during the treatment (it is called thermal isoeffect dose at 90% of the area.) Unfortunately, it is such a complicated construction, with a very complex way of measurement, that in practice it is not viable. The problem is well shown in the case of whole-body hyperthermia, in which it is very easy to measure this dose (basically, the body and the tumor inside have homogenous temperature), but the results are very different from the same dose provided by local/regional treatments. It is even more interesting, that a lower CEM43°CT$_{90}$ dose, applied by local/regional treatment provides better results than more of this dose in whole-body treatment. Therefore, we can claim that this dose unit does not satisfy the basic requirements for the dose concept in general.

2.5.5 What We Expect?

The definite expectation with application of hyperthermia is the same as the overall accepted paradigm in oncology: kill the tumor cells.

Destroying the tumor or at least diminishing the number of malignant cells has criteria: find them selectively, without considerable damage to healthy tissue. The primary way to destroy a cell is necrosis [557]. This effect is a massive, unconditional stimuli accompanied by severe hypoxia and liberation of toxins. In fact no internal energy is used from the biosystem to reach this state. Cells swell in their integrity and in parts (organelles) also. The cell disintegration is complete. Disruption of organelles, DNA breakdown and lyses of plasma membrane occur. The process stimulates inflammation and neutrophil infiltration to degrade dead cells. The process is toxic.

The other, smoother and more natural way is apoptosis (programmed cell death) [558]. This process includes pathologic and physiologic stimuli using internal energy sources (ATP) to perform the process. Contrary to necrosis it affects usually scattered individual cells, causing death of isolated cells. Contrary also to necrosis the cells contract, shrink in this process, chromatin condenses and apoptotic bodies are formed. DNA laddering occurs, the DNA is fragmented to base-pair units. The cell membrane is not lysed, only becomes blebby. Apoptosis is generally not inflammatory, and no neutrophil infiltration occurs; the apoptotic bodies are phagocytized and intact. The final state could be a secondary necrosis or more likely a phagocytosis ends the process.

Necrosis is more abrupt than apoptosis, which takes more time to be performed. This enhanced time from initialization to completion works in apoptosis like a "memory," which accomplishes a definite program by the given stimuli.

Our expectation from hyperthermia is cell killing by thermo-necrosis as well as by thermally stimulated apoptosis.

The most trivial thermal necrosis is thermal injury, the burn, occurring on the skin over 48°C, causing considerable intra-vascular hemolysis. This drastic injury procedure is used in tumor therapies also, like a special thermal ablation by RF [18] or by laser [559] techniques.

Non-trivial necrosis can be shown by severe cellular membrane damage [560] and changes to the membrane proteins [561] or other heat damage [264].

The cellular lyses and liberation of toxins of course could create limits for the distortion process. However, apoptotic cell death or any systemic immune action would be more natural and free from toxic complications. Thermally induced apoptosis [562] and the activation of natural killer cells [563] are both possible to solve this task.

For a successful treatment it is of course mandatory while treating the (in many cases deep-seated) area where the tumor is located, and performing an action on the tumor cells, that the healthy neighborhood is not damaged. On the whole we have strong expectations for safety, rigorous control, and for dose-characterized treatment guidelines.

The principal requests are to:

- unify the methods and the quantification, create an acceptable dose concept to measure and compare the treatments;
- develop a process that is reproducible enough and simple to follow.

On the basis of these considerations, the expectations for a well-conducted hyperthermia in summary are:

- an effective deep-heat targeting;
- a definite selectivity to choose the cancer cells in the targeted range;
- cell killing preferably by apoptosis or activation of other immune reactions;
- a safe and reproducible, dose-characterized treatment.

2.5.6 Possible Solution: Oncothermia

A change in the hyperthermia paradigm looks necessary. Oncothermia is a possibility.

Oncothermia concentrates on the useful energy absorption and regards the temperature as a consequence of the energy absorption but not as an aim and controlling parameter. As we will show in Section 3.2.2, the energy has to be directed for chemical reactions and for structural rearrangements. The temperature in this meaning is wasted energy on the average energy-absorption of the entire treated environment, irrespective of its use. Measuring the average energy (temperature) would be such an approach in our case, as if we were to guess the height of rocks falling down to a lower glade from the measured temperature of the soil at the bottom. Of course the temperature of the ground has a connection to the potential energy, but other significant effects (like the hurl of the rocks on their arrival, the sound energy, the heat conduction of the soil, etc.) obscure the principally important height parameter.

These principal requests have to be realized by an adequate technical solution, which could solve the main challenges:

- deliver the heat to deep targets in the body;
- have adequate and measurable feedback from the running treatment;
- select the malignant cells to treat;
- apply effective cell killing and effective control of it;
- reduce the risk and be safe for the patient, lower the possible side effects;
- be safe for the treating personnel and for the environment;
- be a relatively simple procedure and have acceptable treatment complications;
- have an attractive cost/benefit ratio.

Oncothermia is devoted to solving the main challenges noted above and giving an adequate answer to some special questions and unsolved problems in the field. To explain its ability we have to summarize the basic biophysical and biotechnical effects and understand the complex phenomenon of the heating procedure.

Chapter 3
Thermo-Biophysics

3.1 Factors of Physiology Heating

There are some basic physiological factors connected to the heating phenomena. Two of them are essential for heating: the metabolic rate, which generates additional heat to the external energy intake, and the heat sinks (mainly the blood flow) cooling effectively the locally heated volume. The absorbed energy from outside energy sources is measured by the specific absorption rate [SAR, W/kg]. The SAR increases the temperature but due to the cooling of the physiologically regulated blood stream this heating mechanism is very complex and the temperature is definitely lower than in a regular phantom without a blood stream, even if the phantom material fits well to the targeted real tissues.

The metabolic rate (M) determines the liberated heat in the unit volume of the tissue. There 60% of the daily energy expenditure is used by the metabolism [basal metabolic rate (BMR)], in humans [564].

Among normal conditions the average daily energy expenditure in the case of most living species ranges from 1.5–2-times BMR [565]. The metabolic rate and the body temperature are definitely connected having Arrhenius-like behavior with 0.6–0.8 eV activation energy and a mass-dependent pre-exponential factor [566]. However, the constrained temperature rise rapidly increases the metabolic rate [MR(T)] exponentially, see Fig. 3.1 [221]. Seven degrees increase in the temperature

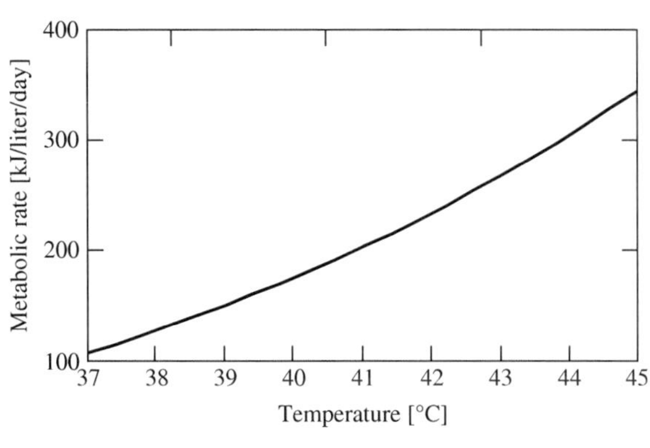

Fig. 3.1 The metabolic activity is temperature-dependent (1 kJ/liter/day ≈ 11.6 mW/liter)

A. Szasz et al., *Oncothermia: Principles and Practices*,
DOI 10.1007/978-90-481-9498-8_3, © Springer Science+Business Media B.V. 2011

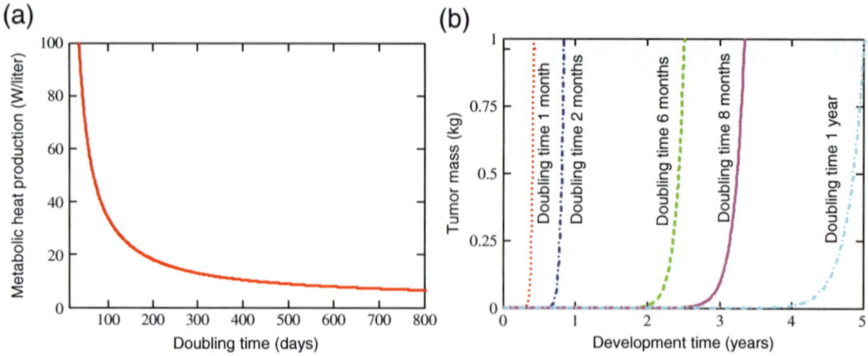

Fig. 3.2 Tumor development with time: metabolic heat production (**a**) and the tumor mass (**b**)

doubles the metabolic rate. (This approximately doubles the ATP consumption as well.) This effect significantly modifies the heat distribution. It is a positive feedback mechanism by the well-focused hyperthermia.

As few as 6°C increase in temperature will increase metabolism by 1.8-times.

The source of the difference between the regular normal body temperature and the environmental one lies in the metabolism. The daily energy expenditure on the thermal effect of food is approx. 8% [564].

The tumor in any case, fundamentally has a higher metabolism rate, which depends on the tumor growth rate, see Fig. 3.2 [567, 219].

Detailed analysis was given as early as 1959 on heat regulation [568], which showed there was a sudden increase in the conductance of healthy tissues by internal temperature change at a turning point of 36.7°C. This heat-conducting capability must not be ignored during the heating of the tissues (see Section 2.2).

Detailed reviews and discussions on the latest results on tumor blood flow affecting the applied temperature have recently been published [253–255]. The change of the blood flow on heating is one of the most important selection factors for the targeted temperature increase in a tumor compared to its healthy environment. The quantitative analysis [569] shows the blood flow in characteristic tissues (see Fig. 3.3), varying with temperature. The deviation (selection) of the tumor blood flow starts just above 38°C (see Fig. 3.4).

Further selection differences could be observed in the variations of specific heat and the structural differences of the tissues.

Because of the variation of the blood flow, the necessary energy in a mass unit (SAR [W/kg]) changes with the actual temperature (see Fig. 3.5).

Figure 3.5 shows the well-emphasized change caused by blood perfusion [569]. The results certainly show different slopes for SAR to maintain the given temperature constant in the tumor tissue. Naturally, the energy request is less over the threshold.

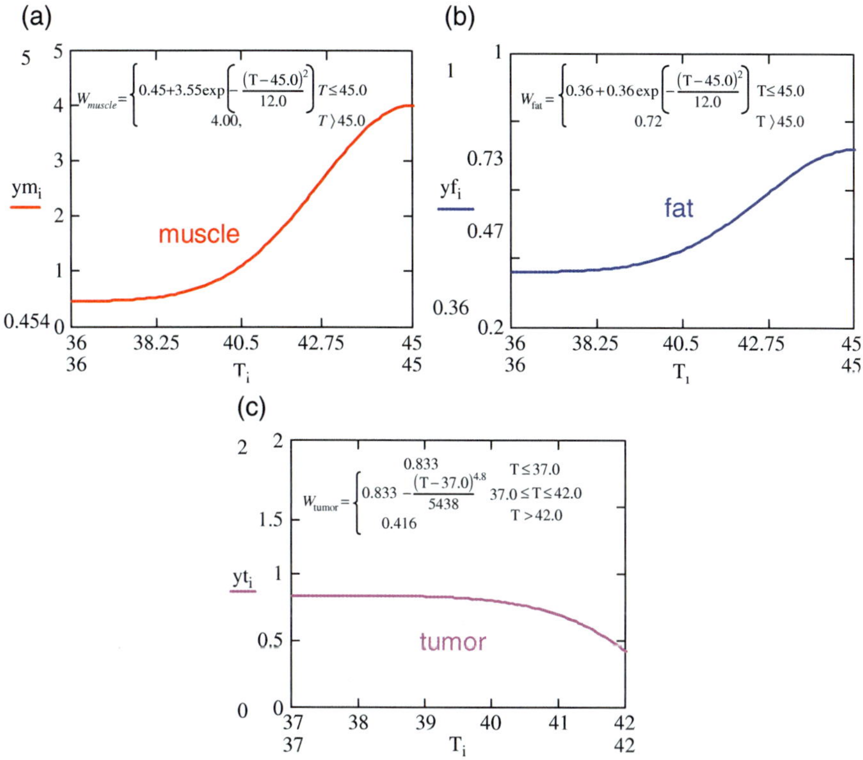

Fig. 3.3 Quantitative changes of the blood flow with temperature increase in (**a**) muscle; (**b**) adipose tissue; (**c**) tumor lesion

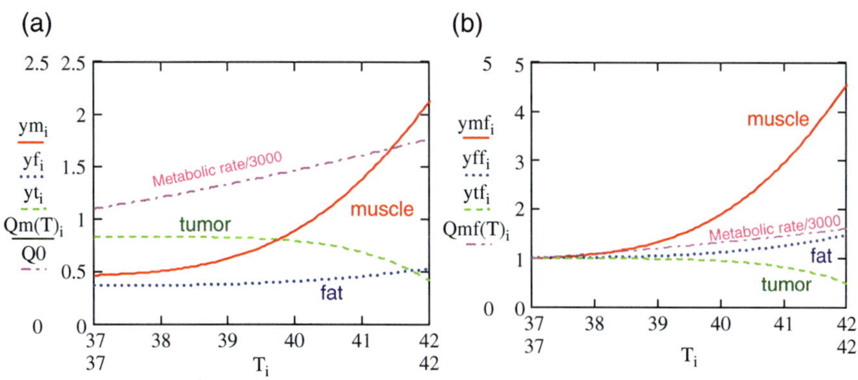

Fig. 3.4 Comparison of blood flow in different tissues (**a**) absolute values, (**b**) relative values

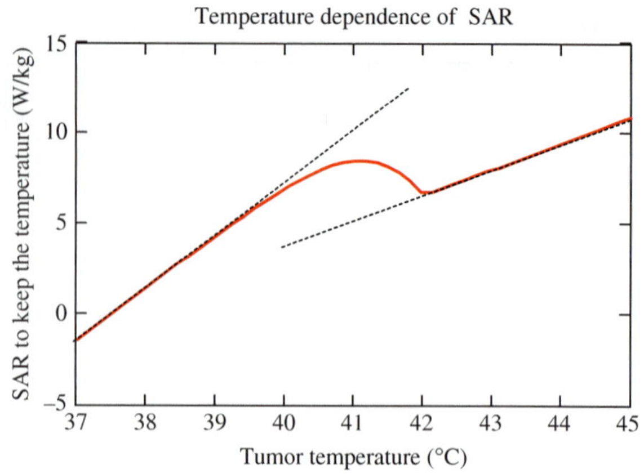

Fig. 3.5 Variation of the requested SAR to maintain the given temperature in the tissue

3.2 Biothermodynamics

Energy, heat, and temperature are connected but not identical terms in physics and in biophysics. Energy is a general term, defined by the ability/capacity to work (or produce heat). Energy is a property of a system, but the work which it could do is not a system property, it is not a system's character, it is a process. If a process spontaneously produces work, that is *exergonic* (*ergon* = work [Greek]), but if it exploits work we call it *endergonic*. The latter could occur in coupled (cascade) reactions, when a particular reaction liberates energy which is absorbed to perform the other (coupled) reaction.

The popular terms *heat dose*, *temperature gain*, *thermal dose*, *energy intake*, *energy dose* may have different definitions by the various individuals who use them. Our first task is to clarify the differences between these terms. In popular literature they are often used as synonyms.

The transformation of energy forms to each other is usually not reversible. Someone can easily transform the potential energy of a falling ball into heat: the energy, which the ball had in relation to the height from which it fell, will heat up the ball and the environment when it is in equilibrium again on the floor (the same happens if water falls). However, by heating up the ball and its environment, the ball will never levitate again using this heat energy alone. (Energetically, the heat energy would be large enough to provide the actually required potential energy by losing its temperature, spontaneously cooling down the object, while using this energy to lift up it.) The energy conversion is directional, the process can not be reversed with the same conditions.

Let us perform a simple experiment to show the principle: push some rocks down from a plateau terrace (beetling height) into the deep, (idea of Alberts et al. [613]). The falling rocks rapidly grow their kinetic energy, and at the end they lose it entirely, at the end all the fallen pieces are at rest at the bottom. The kinetic energy was transformed to break the pieces, to make noise and to heat themselves and the environment (see Fig. 3.6a). On the other hand, we can construct a device, which

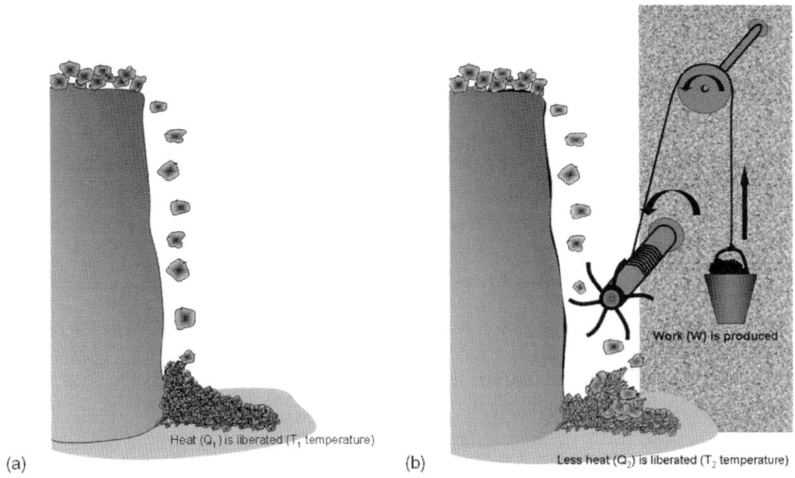

Fig. 3.6 The illustrations show the conversion of potential energy. The falling rocks lose their energy at the bottom, and convert it to heat (temperature). If we use the energy before the rocks reach the bottom, the energy which is used for the lifting of the mass, will be missing from the converted energy, and the heat (temperature) will be lower there. The kinetic energy is not converted entirely to heat, but a part of it makes work to lift up the bucket of brash. Consequently a smaller amount of energy is transformed to heat, the measured temperature will be less ($T2 < T1$), however the full energy remains the same: $Q1 = Q2 + W$

uses a part of the kinetic energy of the falling rocks to make other movements or work (see Fig. 3.6b). (Note this method is frequently used for water-falls to energize various industries (e.g. mills) and even large power stations are constructed to use this energy source.) Do not forget however, the full energy is conserved, the energy used for work by a rotating wheel will be missing from the other kinds of energy: the rocks will not be broken as frequently and or into such small pieces, the noise will be less, the liberated heat (and the temperature of the heated environment) will be less than before the constrained work of the wheel.

The "diverted" kind of energy could be used again for mechanical work and in this meaning it is reversible. However, the losses through friction and the resistance of air etc. could take a part of this energy also into the "not available for work" energy status. The energy in this status of course is not lost as such, only lost for use again for mechanical work.

"The use of thermodynamics in biology has a long history rich in confusion." [570]. The main complication is the fact that life can not be studied isolated from its environment, and so the energetically open system could lead to numerous uncertainties, leading sometimes to mystification as well.

3.2.1 Energy, Heat, and Temperature

What is temperature? Temperature is the average kinetic energy of the given set of particles (usually this is such a large number as 10^{24} particles in the range). Simple

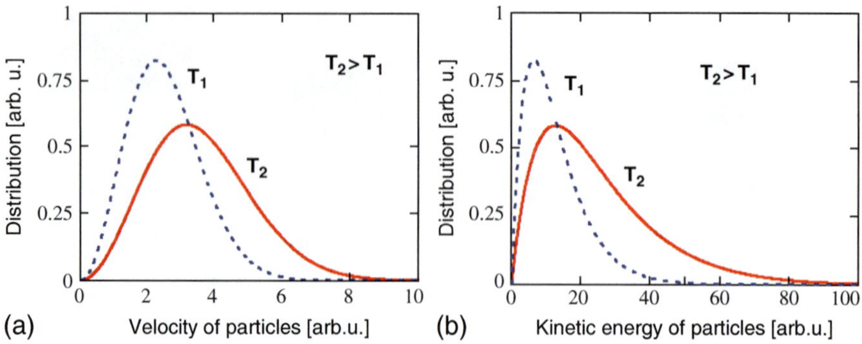

Fig. 3.7 The Maxwell distribution of the (**a**) speed (absolute value of the velocity) and (**b**) energy of non-interacting particles in equilibrium

Fig. 3.8 The average of the energy, and the energy which is obtained by most of the particles (maximum of the curve) are different (asymmetry of the distribution)

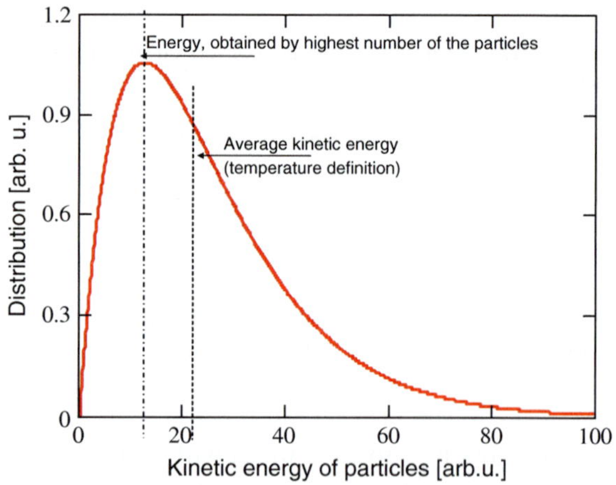

speaking, temperature measures the molecular/particle activity (moving energy) of the matter. The internal mechanical energy is distributed for thermal motion of the particles in the involved systems, having a definite energy distribution (Boltzmann statistics, [571]). The actual distribution of the particles covers a wide range of energies (Maxwell distribution, [572]; see Fig. 3.7). In general, the average energy differs from that which obeys the maximal number of particles (see Fig. 3.8).

Temperature, in this sense, is a hypothetical value of the average energy, and it could be that no particle in the given set has such a definite energy as we defined by the temperature. (The average is a funny calculus: for example, the average number of children in households is 2.3, but of course, it is impossible to find a family who has this number of children.) When temperature is different between the parts of the system the difference forces have to be equalized; so the heat energy (not the temperature!) starts to flow to reach equilibrium, the same average all over, so the process equalizes the temperature. Temperature characterizes the system itself, and all the parts have the same temperature in equilibrium. But it is a relative value. The "hot" and "cold" character has a relative meaning. Because of the relative measurability, historically different scales (Celsius, [°C]; Rankine/Reomur, [°R]; Fahrenheit [°F]) were introduced to measure the temperature. The absolute temperature could be introduced by the approaching of the absolute minimum (zero) of the "particle activities." The absolute scale of temperature was introduced by Lord Kelvin, and for scientific purposes we have to measure the temperature in "Kelvin" [K] ($x\,[°C] \approx 273.2 + x\,[K]$).

The average energy of the particles (in a non-interacting ideal gas system) at room temperature (20°C = 293K) is ~ 25 meV ($\sim 4 \cdot 10^{-21}$J per particle or ~ 2.4 kJ/mol). The average energy of the various particles at normal human body temperature (ideal thermodynamic model) is ~ 2.5 kJ/mol). This relatively large energy is embedded and blocked in the actual system. (It is so large, that if it could be liberated within 1 s, the obtained power would be 2.5 kW/mol.) This average thermal energy limits the internal bonds and interactions, because any lower energy bond will be destroyed by this thermal background. This internal energy could make abrupt changes by such chemical reactions, whose activation energy is smaller (or equal) than the actual thermal average energy. The weakest bonds in life are the hydrogen bridges, having 18 kJ/mol in ice [573] and ranging 3–30 kJ/mol in various compounds in living objects [574].

The heat is definitely not temperature. Heat energy behaves differently than the temperature does. The heat (even in equilibrium) depends on the volume (or mass, or chemical potential, etc.) of the systems. The addictive sum of these partial heat energies will be the heat of the entire system (which however could be characterized with unique, homogeneous temperature), (see Fig. 3.9).

To mix the terms *temperature heat dose* and *energy dose* is incorrect and misleading. We may list numerous examples on the difference of temperature, heat, and energy. A very trivial example to differentiate between heat and temperature is their mass-dependence: heat is the observation of the cooling ("losing" the heat) from the same temperature. This process is very different in a bath-tub or in a glass of water taken from the tub which is at the same initial temperature at the start of the observation. Obviously the water in the glass cools down much quicker than the

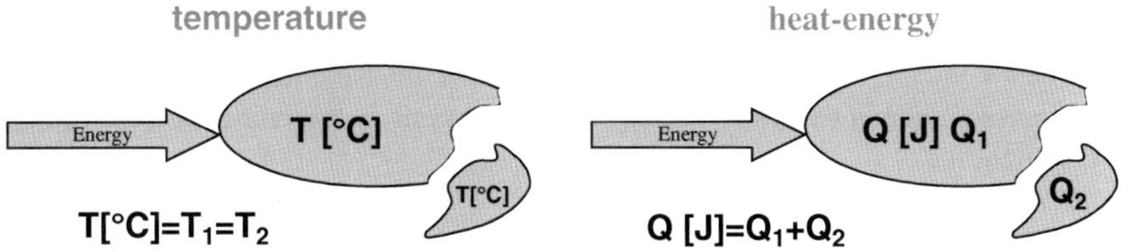

Fig. 3.9 The temperature is intensive, it is equal in the entire object, does not depend on its volume or mass. The system always spontaneously equalizes the intensives in such a way. The extensive parameters (like the mass, the volume, the particle number, etc.) are additive values from the subsystems, the total character is the sum of those over all the subunits

water in the bath-tub, even though their temperature was initially equal. Another popular example could be firewood. Firewood contains a large amount of chemically bonded energy; however its temperature before inflaming corresponds to its environment. Its energy (later heat) content is not proportional to its temperature; these are not proportional even during the period it burns away when the energy is liberated.

Conditions to measure the invested work with temperature have to be discussed and applied very carefully. We use the temperature and heat as synonyms deducted from a simple approach (see Appendix 1) where the temperature (T) and the heat (Q) which generates it are proportional. They could be transformed to each other by a simple constant multiplicator (M):

$$\Delta Q = M\Delta T \quad (M = mc) \tag{3.1}$$

where Δ denotes the change of the given (Q and T) physical quantities, m is the mass and c is a specific heat, (a definite material characteristic giving the amount of heat transferred per unit mass), of the heated material. This is the origin of the misuse of equivalence of the two physical quantities. The interexchange of these parameters is oversimplified and misleading in most of applications. It could be successful only in a very special case: in homogeneous material with steady-state (stationer) conditions and without any other energy-consuming reactions (closed system; e.g. no chemical changes, no phase changes, no heat loss, no energy wasted to the environment, no internal flows, etc.).

To clarify the challenge; go back to the falling rocks in Fig. 3.6. The kinetic energy will be equal to the heat energy down on the ground, if no energy loss occurs

- the rocks fall down in a vacuum, (no loss from the resistance of air);
- the rocks are not broken (no loss of the breaking energy);
- they do not bounce back (no elastic loss in the rock);
- no sound is generated (no loss from the pressure-delivery through the air).

Under these conditions the kinetic energy is transformed directly to heat on the ground. But this does not mean, that the temperature is the parameter that characterizes the kinetic energy! Another set of conditions (entire heat isolation) would have to be met for that:

- no heat conducted away from the area, where we measure the temperature;
- no masses take away heat energy from the system (no wind or other mass conduction);
- no heat radiation to the environment.

These conditions are unrealistic in general, only very special laboratory conditions could approach these requirements. To measure the kinetic energy of falling rocks by the temperature which they had generated down on the ground, is trivial nonsense.

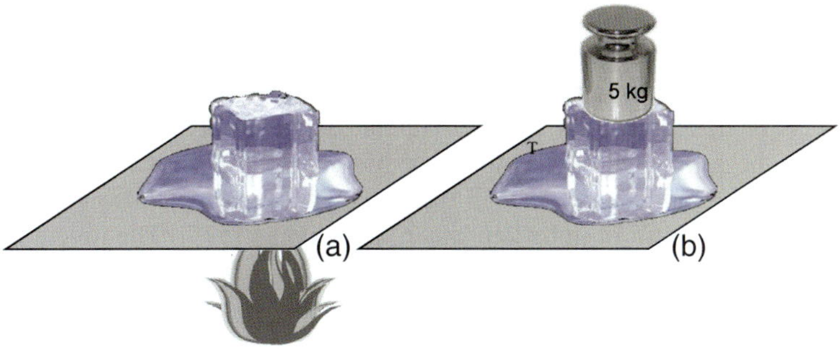

Fig. 3.10 Melting an ice cube with direct heat (*thermal*) or by pressure (*nonthermal*)

If the rocks would have phase transitions during the procedure (lets say they are not rocks but pieces of ice, of which a part melted), the picture would be untraceable even in laboratory conditions. The cold ice-cube follows the rule of Eq. (3.1), till it reaches zero centigrade. Then (from this point) all the energy- (heat-) input will be taken for the distortion job: the ice will be melted, the hydrogen bridges (making the ice solid) are broken, and the ice became liquid (see Fig. 3.10a). During this melting procedure the temperature is unchanged: the phase transition is occurring at a definite, zero-centigrade temperature. Finishing the melting, the temperature grows again by Eq. (3.1), till the next phase transition (the evaporation at 100 centigrade) starts. The *M* multiplicator changes by phases, although the mass is constant, (the specific heat will differ), but the linear characteristics are definitely valid in one phase, if we avoid any heat loss during the process. The melting of the ice is edifying in another way: the unchanged temperature during the transition shows a non-temperature-dependent (NTD) behavior of this particular process (however do not forget, we need a definite temperature to reach this situation).

So, we are doing the melting at an unchanged temperature, but are we able to melt the ice also without heat? The answer is surprising: yes! The ice under pressure melts without any heat needing to be applied (see Fig. 3.10b.)! Hence the process of melting the ice at zero centigrade could be nonthermal in both meanings – no temperature and/or no heat changes are requested to derive the liquid from the solid form of the water.

There is a definite balance between the entropy and energy defining the actual state of the bioreactions (see Appendix 1).

3.2.2 Energy of the Chemical Bonds and Reactions

The rules of spontaneous processes drive the thermodynamic actions in closed, undisturbed systems. However, if the system had connections with the environment, through that it could have had exchange of various thermodynamic parameters (e.g. energy, mass, pressure, temperature, etc.), and it is not so simple to follow the process or make any predictions on its development. Living objects are typical open systems, their guiding differ from the spontaneous processes in the closed ones. Life is not in equilibrium, it is in permanent development, permanent change and energy intake. This is the reason, why a Nobel-laureate, A. Szent-Gyorgyi, formulated the situation in the following way: in terms of life-energy it is important to note that not only does the monkey go through the jungle, but the jungle also goes through the monkey, in the form of nutrition, water, and oxygen [575]. The jungle became a part of the monkey and in this manner all the living objects there are interconnected, we are not able to discuss an energy cycle of a species without considering the energy cycle of the other lives there. The unique, ultimate energy source of the life is sunlight.

The actual source of energy may be different, depending on its kind: the same energy from nutrition and nuclear radiation could not be used in the same manner, unless both released their energy to heat. Usually the energy could be stored in four basic interaction mechanisms (strong [nuclear], weak [nuclear], electromagnetic, gravitational). Definitely: the entire chemical, biological, and all other biologically relevant energies are covered by the electromagnetic interactions. According to our present knowledge, no other natural forces have a role in biology, only electromagnetism, the nuclear and gravitational interactions are excluded from the bioenergies.

In the frame of the properties of electromagnetic interactions, materials could be characterized by chemical interactions (bonds) of the particles constructing the given object. The strength of the chemical bonds in a vacuum is usually much greater than the same in aqueous solution (see Table 3.1). Van der Waals bonds are weak, they could easily be destroyed by thermal energy. Ionic and covalent bonds are much stronger to be destroyed near room temperature by thermal way. However, the hydrogen-bridge bond is not as strong. It is only less than double of the average thermal energy at room temperature. The hydrogen-bond is most likely to be broken triggering some thermally enhanced chemical reactions.

Spontaneous reactions could be instant after the reagents are connected, Fig. 3.11a. Then the energy difference is released promptly. However, there is another, locally blocked reaction also, when the reagents can not react even after they are mixed. (An initial push is requested to start the process of falling rocks in Fig. 3.6, however, sometimes this can be spontaneous if a rock is placed on a dip.)

Table 3.1 The approximate strengths of common chemical bonds in vacuum and water

	Approx. bond strength (kJ/mol)	
Chemical bond	In vacuum	In water
van der Waals bonds	0.4	0.4
Hydrogen bridge	17	4
Ionic bond	340	13
Covalent bond	380	380

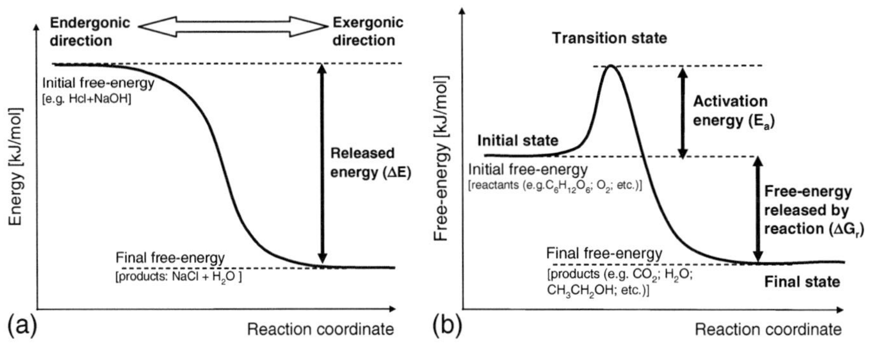

Fig. 3.11 The spontaneous reaction and the barrier (transition state) blocked reaction

However, from static initial conditions, without pushing the rocks down, (investing "ignition" or "activation" energy), the procedure does not start. For the reaction of many chemical mixtures an "ignition-like" energy intake (like in the internal combustion engines e.g. in our car) is mandatory (Fig. 3.11b). In consequence in this case a definite transition state exists at the beginning of the reaction. These types of processes have a basic role in all living objects. Life cannot be without activation energy and the transition state; otherwise the nutrition reacts immediately and an unlimited explosion-like energy liberation takes place. A trivial example: the wood does not catch fire without the initial flame. However, afterwards the wood produces much more energy then was required for the kindling.

The probability of the given reaction (or movement) is naturally the opposite (proportional with the reciprocal value) to the energy diagram in Fig. 3.11, (see Fig. 3.12).

It is possible to construct machines (e.g. the steam engine), which can transform heat to mechanical movements (we can lift up the bucket of brash with a steam-engine-powered wheel also). With this energy conversion potential energy

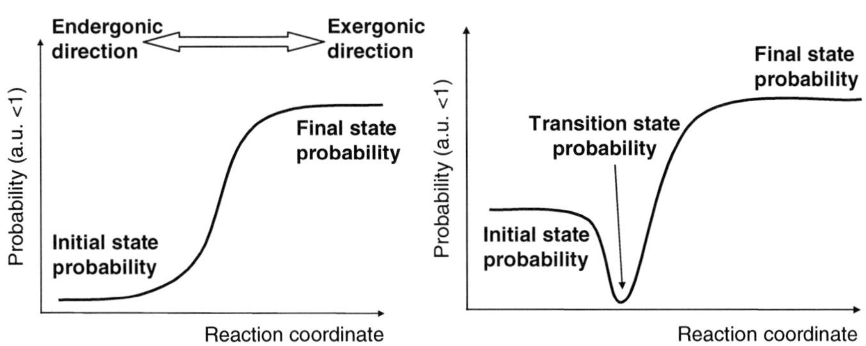

Fig. 3.12 A sketch of the probability function of spontaneous and unspontaneous reactions

is "pumped in" to the system. The only problem is that much more energy has to be used in total to reach the same original position of the rock brash again, from where they were pushed down, the efficacy of such a procedure could never be 100%. This negative statement is the consequence of a proven law of physics, so such machinery, that go against this rule, are usually rejected by patent offices without investigation of the details ("second kind of perpetum mobile"). The mechanical energy could be converted to heat in full, but the opposite path is impossible in full. The maximum efficacy has a principal limit, which is the percentage of the temperature change (caused by the used heat-consumption) to the environmental temperature (Carnot machine [576]).

To put it simply, heat and energy are, in principle, the same categories: heat is a special form of the wide category of energies; but their conversion to each other shows essential differences. The (nonheat) energy could always be transformed into heat with 100% efficacy; but heat cannot be transformed to energy in the same way, only with considerable losses. Why is this? Because the energy has definite forms, able to work directly. Heat (in thermal equilibrium) is a randomly distributed form of the energy in a system. In general, its components contain a product of so-called intensive-extensive pairs of the thermodynamic parameters. The intensive properties characterize the system (or subsystem) independently of their mass or volume; while the extensive properties vary directly by the mass or volume. Extensive properties of the part of a system characterize quantitatively the actual subsystem, while it has an equal intensive property for all the subsystems, irrespective their size.

In medicine a trivial intensive property is the body temperature, which characterizes the system (homeostasis), while an extensive property is for example the volume of blood, which naturally depends on the size of the body-part under investigation.

Heat energy is easily transferable in radiative, conductive, or convective forms. The only point, which complicates this transfer, is its direction. Spontaneously the heat always transfers by the direction determined by the difference of the temperatures: the colder object in spontaneous processes is always warming-up and the hotter is cooling-down and never the opposite. Reactions that run opposite to this (e.g. fridge, air-conditioner, etc.), need extra energy investment, and can not be spontaneous.

The (bio-)chemistry could be interpreted by the Gibb's free energy as a driving force. The reacting species (reagents) lower their common free energy by the reaction, and seek to reach the available minimum of the G (see Fig. 3.13).

A comparison of these pictures is summarized in Table 3.2.

There is an important factor of the rate of changes, which is the concentration change by a definite time interval:

$$\text{Rate of change} = \frac{[\text{concentration}]}{\text{time of observation}} \tag{3.2}$$

Because the changes could be very rapid, so the time interval has to be small for precise description. For this, in more rigorous formulation the rate of change is the

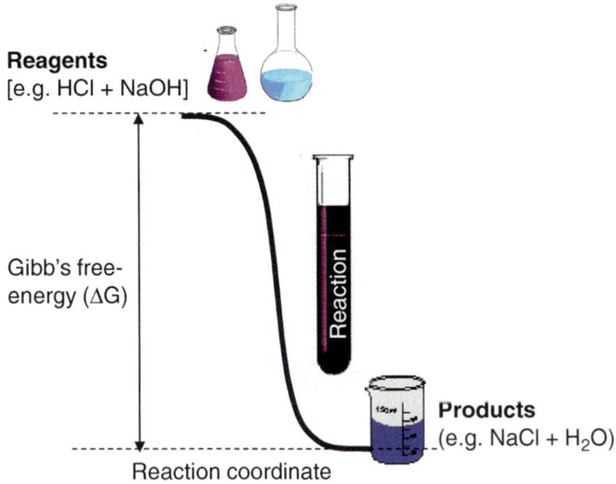

Fig. 3.13 The reaction is going to realize the minimum of the Gibb's free energy

Table 3.2 Similarities of various interactions

Interaction	Driving force	Intensity
Mechanical (flow)	Reduce the potential energy	Mass-flow
Electric	Surmount potential difference (voltage)	Charge flow (current)
Chemical	Minimize the Gibb's free energy	Reaction rate

time derivative of the concentration and the reaction velocity (v_r) is a normalizing factor (f_{norm}) to keep the reaction equation valid. So:

$$v_r = f_{norm} \frac{d[c]}{dt} \tag{3.3}$$

An example of the water reaction:

$$\begin{aligned} &2H_2(g) + O_2(g) = 2H_2O(l) \\ &v_r = -\frac{d[H_2]}{dt} = -\frac{1}{2}\frac{d[O_2]}{dt} = \frac{d[H_2O]}{dt} \end{aligned} \tag{3.4}$$

The systems need an initial activation to start the process like the starting push on the falling rocks (see Fig. 3.6) or like a switch in an electric circuit, or like a simple activation energy ("ignition") for some chemical reaction, (see Fig. 3.14), etc. mechanically the same activation could be constructed by a barrier blocking the water transfer from up to down. An appropriate tube, between the two reservoirs is not able to lead the water down, but an appropriate "suck" could give the initializing activation and then the water goes over the barrier alone from that moment.

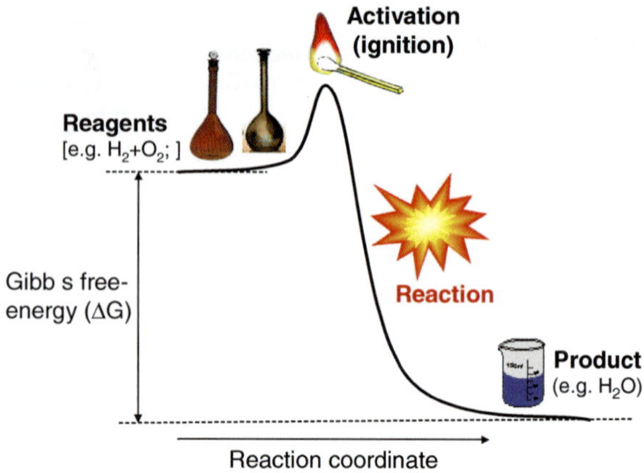

Fig. 3.14 Reaction of a mixture of H2+O2 gases to form water via explosion

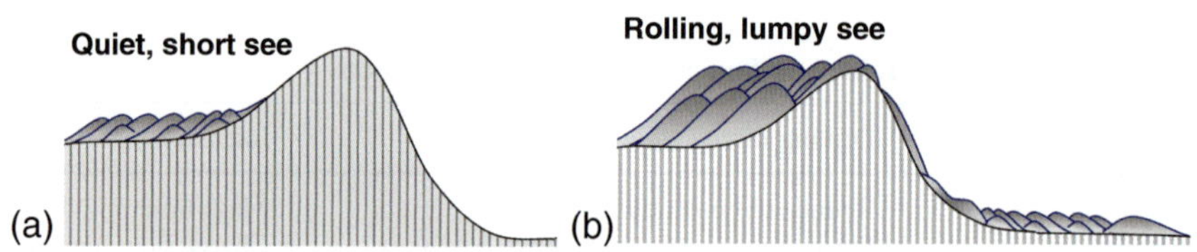

Fig. 3.15 Water can sweep over a barrier should the height of the waves exceed the height of the barrier

Water could also go over the barrier with by waves sweeping (e.g. see Figs. 3.15a, 3.15b). The process depends on the ratio of the height of actual water waves to the height of the barrier.

Similarly we may model the barrier's role by moving balls. If the balls do not have enough kinetic energy to go over, they repeatedly roll back, see Fig. 3.16a. However, with satisfactorily high energy the ball goes over the barrier, see Fig. 3.16b.

The average kinetic energy of the balls corresponds to the energy distribution among the particles in a system. The temperature is the energy average, so its value compared to the value of the energy barrier (activation energy) decides the rate of the particular reaction. Naturally, the reactions need minimal energy to start, and the energy from the temperature is additive, a fluctuation-promoted process to go over

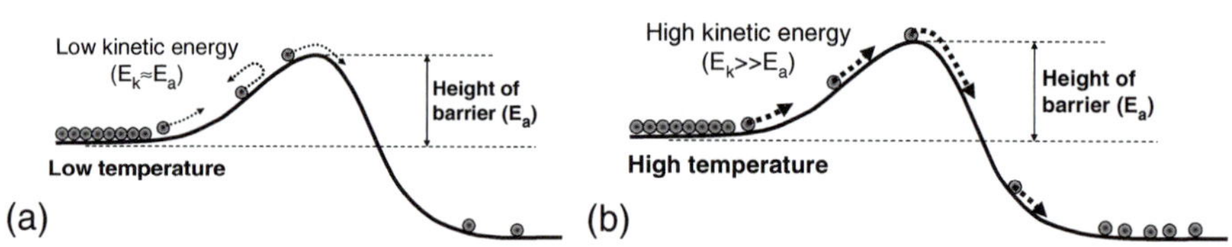

Fig. 3.16 The value of the kinetic energy determines the overcoming on the barrier

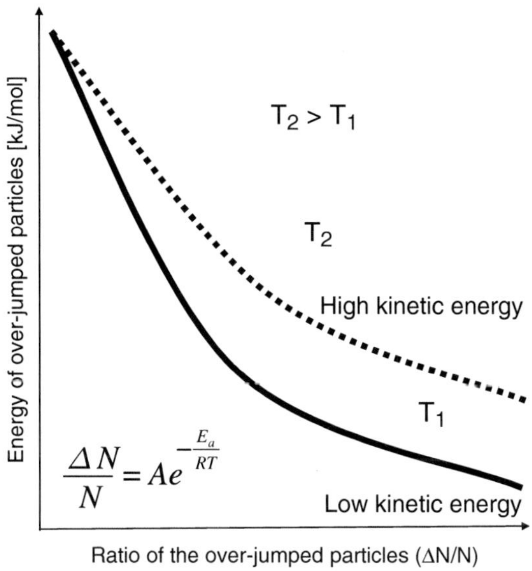

Fig. 3.17 Higher temperature makes the probability to react larger

the barrier. The activation energy normally can not be reached by the ambient temperature, because than the process became spontaneous. Higher temperature makes the reaction more likely, see Fig. 3.17. The kinetics of the processes depends on the overall conditions. We may calculate the probability of the particular reaction (reaction rate). A non-interacting (ideal gas) system of particles in thermal equilibrium at T temperature has a large variety of particle velocities and consequently in their kinetic energy. Boltzmann described [577] the probability (p) of the actual energy (E) of a particle in an equilibrium system.

$$p = Qe^{-\frac{E}{kT}} \tag{3.5}$$

where k is the Boltzmann constant ($k \approx 1.38 \cdot 10^{-23}$J/K) and Q is the normalizing factor. In fact the probability is determined with the ratio of the actual E energy of the particle to the average energy (kT) in the system. Equation (3.5) is valid for a single reaction. If we calculate by mol-value, than instead of k we use R; (universal gas constant, $R = k{\cdot}6{\cdot}10^{23} \approx 8.3$ J/K/mol). This probability distribution is the basis of simple chemical kinetics, and determines the Arrhenius equation [578, 579]:

$$D = Ae^{-\frac{E_a}{RT}} \tag{3.6}$$

where E_a is the activation energy of the given reaction; D is the rate constant of the given reaction in T temperature, (when the initial energy is suitably large, exceeds the barrier); A is the pre-exponential factor.

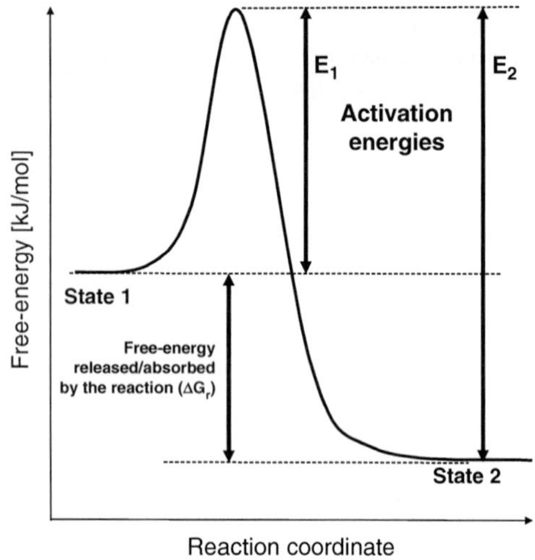

Fig. 3.18 The reaction energy-schematics between "State 1" and "State 2"

The *D* rate constant is characteristic for the actual reaction rate. Having two states *State 1* and *State 2*, with $[C_1]$, $[C_2]$ and E_1, E_2 concentrations and activation energies. If they are the two sides of a reaction equation, (see Fig. 3.18), then:

$$\begin{array}{ccc} \{\text{site 1}\} & \rightleftarrows & \{\text{site 2}\} \\ D_1 = qe^{-\frac{E_1}{RT}} & & D_2 = qe^{-\frac{E_2}{RT}} \end{array} \tag{3.7}$$

In the equilibrium stage the rate of conversion {State 1} → {State 2} is equally balanced by the reverse {State 1} ← {State 2} reaction, so:

$$D_1[C_1] = D_2[C_2] \tag{3.8}$$

So the reaction rate in equilibrium is:

$$D = \frac{[C_1]}{[C_2]} = e^{-\frac{E_2-E_1}{RT}} = e^{-\frac{\Delta G_r}{RT}} \tag{3.9}$$

The E_a activation energy in Eq. (3.6) is actually an energy barrier, keeping the particular chemical state in its actual position, (e.g. holds atoms together in a molecule), avoiding a jump to a lower energy bond, (e.g. free the atoms from the actual bond), and reacts spontaneously (see Fig. 3.19).

The rather complex phenomenon could contain various independent (or not strongly dependent) thermodynamic processes. These are present at the same time, commonly determining the actual state. Considering their thermal behavior a mixed Arrhenius picture could be applied, with two different processes having their own

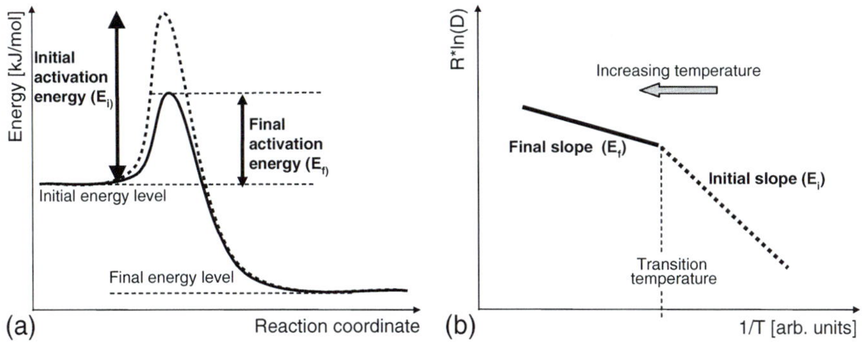

Fig. 3.19 The activation energy of a reaction/chemical bond, (**a**) and the corresponding Arrhenius plot (**b**)

activation energy. These two activation energies (two thermal processes) could fit to the measurements [580]. An example of this picture is shown in Fig. 3.20.

A more complex description could be made on the assumption of the cascade chemical reactions when many species are present but actually not involved in the reaction, while additionally their amount could be time-dependent. Convex Arrhenius plots are obtained from these reactions, interpreted by [581] on the basis of the early works of Tolman [582].

The definite change of the activation energy is plausible in all the phase transitions [583], when the material is transformed from one to the other state. The initial and final stages have different chemical bonds, which characterizes the phase itself. At least in this point when the chemical transformation occurs the Arrhenius line (the *1/T*-function) will change its slope. All the above considerations

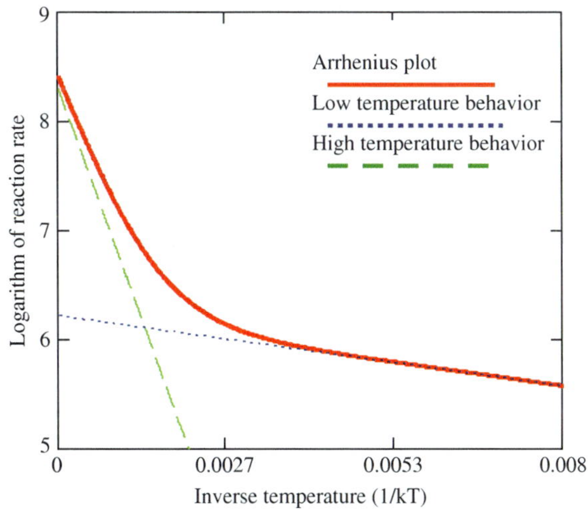

Fig. 3.20 Double Arrhenius fit for a mixture of the reactions. (Case of $c1N = 500$, $c2N = 4,000, W1 = 80$ and $W2 = 1,600$)

Table 3.3 The breakpoints of fitted Arrhenius parameters in bio-objects

	Temperature range (°C)			
Sample	min.	max.	E_a (kJ/mol)	ln(*A*) [-ln(*s*)]
Porcine epidermal histology	44	50	416	148.83
	50	58	665	241.76
Joint capsule shrinkage	44	60	34	12.90
	60	70	234	74.30

are formulated for a thermal equilibrium state, where the Boltzmann distribution exists [584]. Various reactions do not have this simple (constant E_a) Boltzmann distribution, having phase transitions (precipitations, phase transitions, aggregations, nucleation, growth, etc.). The phase-transition-like changes could be observed by breaking points (see Table 3.3; 585]). These kinetic processes of course determine not only the direct chemical changes but simple restructuring (e.g. protein folding) also [586].

In practical applications we apply the formula for a mol-quantity and use the logarithm of Eq. (3.6):

$$\ln(D) = -\left(\frac{E_a}{R}\right)\frac{1}{T} + \ln(A) \tag{3.10}$$

Equation (3.10) is suitable to determine the activation energy of a particular process, if we measure the (*1/T*) dependence of the logarithm of the chemical reaction rate.

The value of pre-exponential factor *A* depends on the distribution of velocity (kinetic energy), on the collision frequency, and on the reaction cross-section. It depends on the temperature by its square-root.

The Arrhenius fits to the experimental plots for some reactions are listed in Table 3.4. The (E_a/RT) ratio is always larger than 1, keeping the reaction in metastable position till the ignition is taken. The minimal energy of the barrier naturally has to be higher than the activation energy.

The Arrhenius picture is useful for various heating-induced changes in tissues and in physiological phenomena, (see Table 3.5; [585]). The measured parameters in these biologically quite high temperatures are remarkably higher than those of the simple chemical processes.

Quantum-mechanical considerations were used for transition-state theory [587–589]; which has good perspective for use in inter-disciplinary applications [590]. In this approach the barrier could be passed by a special quantum-mechanical effect [591] (tunneling), and so the rate could be higher at the same activation energy, (see Fig. 3.21). The larger slope of the calculation from the transition-state theory is a consequence of the transition state (a complex, metastable compound) assumed for the overcoming of the energy barrier.

Table 3.4 Arrhenius parameters of some reactions

Reaction	E_a (kJ/mol)	ln(A) (liter/mol/s)	Temperature range (°C)		(E_a/RT)	
			Min.	Max.	Max.	Min.
$H + HCI \rightarrow H_2 + CI$	14.6	23.85	73	227	8.8	3.5
$H + HBr \rightarrow H_2 + Br$	15.5	25.42	727	1427	1.9	1.1
$O + O_3 \rightarrow 2O_2$	20	23.21	0	627	8.8	2.7
$F_2 + ClO_2 \rightarrow F + FClO_2$	33.5	16.37	46	26	17.8	16.3
$CH_3 + C_6H_6 \rightarrow CH_4 + C_6H_5$	38.5	17.04	183	327	10.2	7.7
$H + D_2 \rightarrow HD + H$	39.3	24.61	27	477	15.8	6.3
$CO + O_2 \rightarrow O + CO_2$	213.4	21.97	2127	2727	10.7	8.6

Table 3.5 The fitted Arrhenius parameters of various tissues and physiological processes

Sample	Temperature range (°C)		E_a (kJ/mol)	ln(A) [$-\ln(s)$]
	min.	max.		
Epidermal necrosis	44	70	627	226.78
Porcine epidermis	48	57	339	123.45
Porcine cornea	60	95	106	35.27
Chordae tendinae shrinkage	60	95	357	122.30
Rat skin collagen birefringence loss	40	60	306	104.09
Rabbit muscle birefringence loss	53	74	128	47.19
Rett syndrome birefringence loss	50	60	370	129.59
Rett syndrome calorimetry	57	60	521	183.80
Rett syndrome in acetic acid calorimetry	35	37	1310	503.30
Kangaroo tendon shrinkage	58	62	589	206.03
Lens capsule calorimetry	40	65	860	316.80

Generally in most of the chemical reactions the Gibb's free energy is the driving thermodynamic potential, (see Appendix 1) and the Arrhenius-like formula derived from the transition-state theory. Note in this case a temperature-independent exponent appeared and the pre-exponential factor has been expanded by the structural part. The rate constant in transition-state theory (D_{tst}) is considerably modified by any specialties or combinations of the reactions. In biosystems one of the most important categories of these combined reactions is the diffusion-controlled reaction category. The modification of the reaction rate is calculated and checked experimentally [592].

The thermal equilibrium, describes well the temperature dependence [Eq. (3.10)] in the case when the activation energy is constant. However, it has been shown (e.g. statistical rate theory, [593]) that the Arrhenius equation does not adequately describe reaction kinetics in non-equilibrium systems.

Characteristic non-Arrhenius behaviors could be observed in complex systems [594]. These reflect multi-step reaction mechanisms or radical change of the

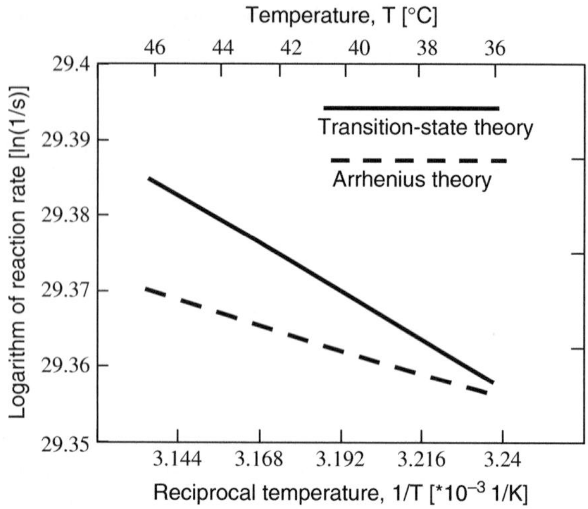

Fig. 3.21 Comparison of the classical and quantum-mechanical calculations under the same energy conditions. (The activation energies are identical.)

mechanism during the process (e.g. a phase transition occurs). Such a multi-step process for example is the decomposition of diacetylene [595], where two contributing transitions present (D_1 and D_2 rate constants have $A_1 \approx 10^{9.15}$ and $A_2 \approx 10^{14}$ pre-exponential factors, and $E_1 \approx 218$ (J/mol), $E_2 \approx 168$ (J/mol) activation energies, respectively). Living objects regularly consume the energy in multi-step processes.

These phase-changing phenomena have been well formulated by the so-called Avrami equation description pioneered by Kolmogorov [596], Johnson, Mehl [597], and Avrami [598], and modified by others [599, 600]. According to the proposal made by F.W. Cope, the Avrami equation could serve as a mathematical model of different biological processes [601, 602]. Experimental data which were collected by Cope [603, 604], and by others [605, 606], show a definite universality of the Avrami equation to describe real processes. The meaning of universality in this case is the possibility to study different processes without knowing the exact structure and dynamics of the given system. The situation is similar to the description of the critical phenomena [607], where the physical laws near to the phase-transition temperature are connected to very general function categories.

These phase transitions are symmetry-breakers, and the following structure construction is mostly based in a self-organized way. Self-organizing behavior of the materials is a well known and widely investigated topic in science [608] and especially in biology [609] (see Appendix 3). The phase transition changes the activation energy, and could promote or detain the given process.

The task to effectively go over the energy barrier could be solved by the increased temperature (gain in the reaction rate) or by a lowered energy barrier, without the increased temperature. The first is the so-called thermal, the second the non-thermal solution. A typical solution for a non-thermal process is catalysts suppressing the activation energy below the threshold.

The driving forces of thermodynamics in general are the gradients (inhomogeneity) of any intensive (e.g. pressure, temperature, electric field, magnetic filed, chemical potential, etc.) parameters. If there is inequality in the intensives in the given volume, then the spontaneous direction is to seek to smear and cause this to vanish, to establish again equal values for the intensives all over the given volume, to give homogeneity to the system (see Appendix 2).

What happens if the intensives start to equalize? A current(s) of extensive parameters (e.g. mass, particle, heat, volume, entropy, charge, dipole, etc.) is(are) generated, and flow till the gradient of the intensive exists (e.g. see Table 3.2).

The various intensives could strongly interact, and one could make better equilibrium conditions than the other. In this case the spontaneous process to make equality in the space of one intensive could create inhomogeneity for another one, which will be homogenized in a following spontaneous step only. Such a situation for example is when the temperature rises in a geographic area, until the point is reached where the up-to-now equal air pressure increases in the heated area, and then later a wind begins to equalize this created difference. Also the wall above a central heating radiator becomes dirty quicker than the other parts of the wall, because the locally heated air starts to move, and a mass-current (air with dust) starts to flow from that area upwards.

As we have seen in the above examples, the interacting intensives generate interacting flows of extensives also. The temperature increase in an area starts a heat flow. This generates a pressure gradient, which generates a mass (air) flow. This air flow (wind) causes water precipitation from the air, which makes clouds and again a mass flow (rain), etc. The interactions of these currents could be discussed by the Onsager relation [610], which describes the cross-currents by their linear combinations. Onsager established a theorem, that the interactions are symmetric: the degrees of the cross-effects are equal; the linear combinations have symmetrical values. Generally the cross coefficient of the interactions is one order of magnitude less than the direct currents of the intensive–extensive pairs (see Appendix 2).

Living objects are open systems, but there exist some environmental, conditional, sets in the meaning of life. Probably the solar system could be regarded as energetically closed for life on Earth. Solar energy is the exceptional and only energy source of life. Disregarding this "energy pump," the Earth alone, (for biological energies) could be considered as closed. From the physical point of view, the living objects lower the overall average environmental energy, and increase the overall average entropy. However microscopically it fluctuates, the tendency is valid only in macroscopic average. These overall processes are identical with all the spontaneous processes in nature, where the driving force is "equalization," diminishing the existing differences, and decreasing the gradients. In this sense, life follows the basic thermodynamic laws: the living process continuously "burns" the incoming "nutrition." The only energy pump, which does the opposite, is the incoming sun energy, which makes the differences, creates original gradients which later divide other differences by spontaneous processes. The life process tries to diminish the working energy of sunlight to the unusable energy form, by increasing the overall entropy (decreasing the overall free energy) in the world. Naturally: without a permanent energy pump life could not be continued.

The living process lowers the electron energy, which is caused by the oxidation of the outgoing final "products." The gradual loss of electron energy of the "nutrition"

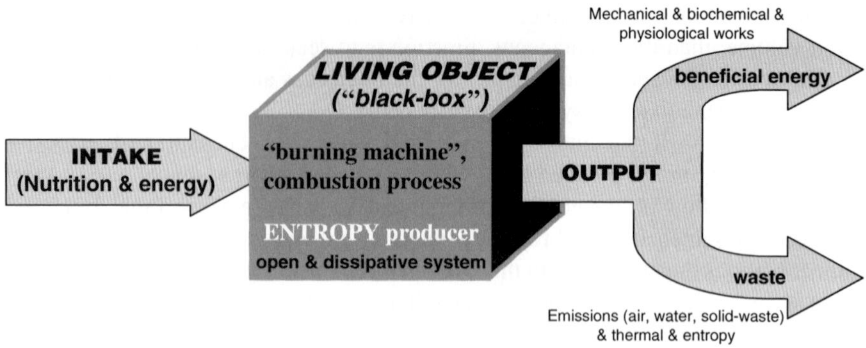

Fig. 3.22 The main functions of the "living machine"

molecules is the energy to sustain life. Simply speaking, the living process is only an entropy producer, an open, dissipative system, see Fig. 3.22.

We take a definite amount of energy from food (we could measure it in kJ [or with older units kcal]). For example, an energy table of some different vegetables/fruits is summarized in Table 3.6 [611].

The energy for this huge average equilibrium is the result of our nutrition intake and the thermodynamic exchange with the environment (see Table 3.7 [612]).

Naturally, we may gain energy from a radiative heat source (e.g. from the Sun, or from a heat stove, or from hot water, etc.) measured also in kJ. However, the two kinds of energies are not equivalent: we must energize ourselves by eating, drinking and breathing, and it is not satisfactory to have the energy from heat exchange only, even if the heat energy is larger than the energy we get from nutrition. Even the energy from the sun is useless if it turns only to heat, we are not able to use this kind of energy for further processes. The difference is obvious: the heat itself is a disoriented, distributed energy, without making gradients no useful utilization is possible.

The kinetics of life processes could be described by the Arrhenius equation in dynamic equilibrium (in homeostasis). These stationary states are governed by the

Table 3.6 Energy values per 100 g vegetables/fruits

	Apple	Avocado	Cactusfig	Cucumber	Grapes	Kiwi	Lime	Litchi	Mango	Melon
Energy (kJ)	218	670	151	52	310	202	189	269	279	153

Table 3.7 The energy liberated from the main nutrition components

	Glucose	Fat	Protein	Typical human
Energy [kJ/g]	16	39	18	
Oxygen request [l/g]	0.75	2	1	
Energy by 1 l oxygen [kJ/l]	21	20	18	19

principle of minimum entropy production (minimal "loss" of the potentially available energy) [576]. The comparison of the actual activation energy to the average energy (kT) in the case of chemical reactions under isotherm and isobar conditions could be replaced by the change of the Gibb's free energy (ΔG) divided by the kT. The Arrhenius exponent will have two terms, one enthalpy-dependent, and another entropy-dependent part. The slope depends on the temperature while the intercept depends on both the kinetic properties and the structure of the system. In some experimental evaluations the increase of the rate constant by a 10°C increase of the temperature is determined. This increase is characterized by an enhancement ratio (denoted by Q_{10}) from the initial to the final reaction rates between the 10°C temperature-states [579]:

$$\ln(Q_{10}) \cong \frac{10E_a}{RT^2} \tag{3.11}$$

The living systems are open; their interactions with the environment are mandatory. During this interaction they have an intake of nutrition and energy, in general they incorporate/create compounds with high electron energy, and by special surface-controlled processes. The liberated energy is used for the needs of life (synthesis of macromolecules, internal transport, mechanical motion, etc.) and a definite large part of it is wasted by various modes (heat loss, evaporation, liquid- and solid-waste, etc.).

Because of its openness, living systems are far from static equilibrium, but in normal conditions it has a stationery state, a steady-state development which is kept constant in time under definite conditions. This is called homeostasis, a stationary state of life. In this special state the living object has the lowest available entropy, governed by the least dissipation principle [576]. The living system loses the minimal amounts of free energy in its dynamic stability; in energetic standpoint it is the most economic state available. Its consequence the special structural and functional arrangement characterizes living objects.

Bio-oxidation cannot be a process liberating large amounts of energy in one step. If the liberated energy would be too large, the well-balanced chemical equilibrium could not tolerate it, the fluctuations could overwhelm the balance. To avoid this explosion-like fluctuation numerous coupled chemical reactions use a step-by-step oxidation procedure, gradually losing the highly energized chemical bond and producing the final low-energy compound, the waste. This process could be described by a ladder-like reaction scheme (see Fig. 3.23). The total free energy is released by small energy blocks, and the various steps are separated by various activation energies, corresponding to the actual chemical reaction.

Chemically the process is very simple: the energy delivery is via high-energy chemical compounds or by an externally radiated energy source (e.g. sunlight); and the liberated energy loss governs the essential life processes and is wasted in various forms of losses. The high energy in chemical compounds means high energy of electronic states (bonds). In principle nothing else occurs, the high-energy electrons simply lose their energy gradually along a path of multiple coupled reactions.

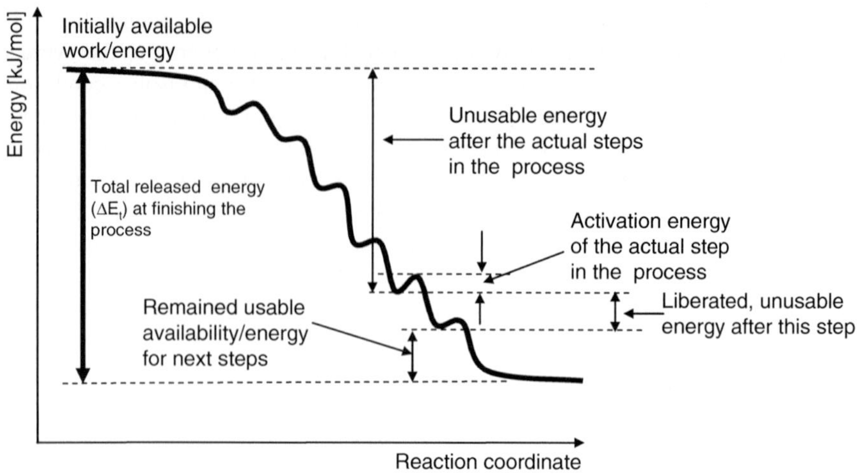

Fig. 3.23 Ladder-like reaction. It divides the sudden one-step energy liberation to many small steps, providing the energy on a continuous basis

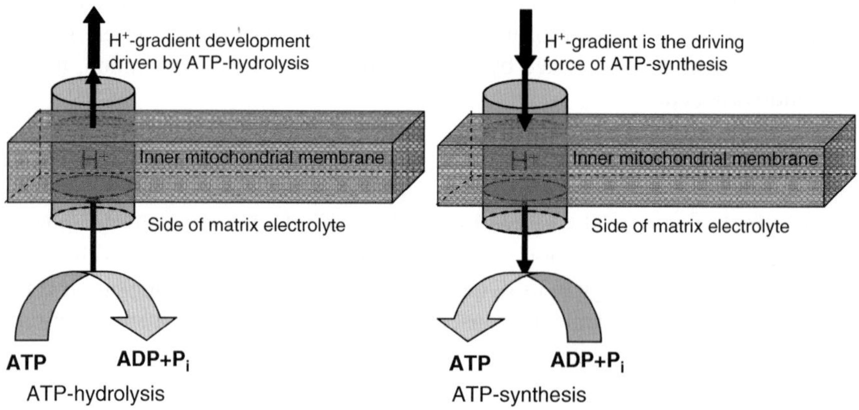

Fig. 3.24 Creation of the proton gradient with ATP energy in eukaryotes

Every living organism however, irrespective of their source of energy, uses an overall "currency" for the energizing: this is ATP (adenosine-triphosphate). ATP is the dominantly used and convertible free-energy donor in all basic bioprocesses. Three adenosine-phosphates are the players: mono- (AMP), di- (ADP), and tri- (ATP) phosphates. ATP energizes to create the trans-membrane proton gradient, which is predominantly produced in the mitochondria in eukaryotes (see Fig. 3.24).

The fundamental energy exchange is the ADP $\leftrightarrow$ ATP reaction in all living objects, producing and storing convertible energy for most of the important trans-membrane processes. ATP drives the endergonic ($\Delta G > 0$) reactions, which without ATP could not spontaneously occur. ATP is transported (or produced in situ) where the energy is requested. This is a mandatory process of the living energy-exchange:

the reactions in living systems are blocked by the activation energy, to avoid spontaneous explosive-like liberation. The role of ATP is to aid in overcoming this barrier, and liberate the free energy from the actual reaction.

The ATP $\leftrightarrow$ ADP hydrolysis/synthesis reactions give definite free-energy exchange at pH=7 conditions:

$$\mathrm{ATP} + \mathrm{H_2O} \Leftrightarrow \mathrm{ADP} + P_i + H^+ \quad \{\Delta G = -30.5[\mathrm{kJ/mol}]\} \tag{3.12}$$

$$\mathrm{ATP} + \mathrm{H_2O} \Leftrightarrow \mathrm{AMP} + PP_i + H^+ \quad \{\Delta G = -32.2[\mathrm{kJ/mol}]\} \tag{3.13}$$

where *Pi* and *PPi* are the orto- and pyro-phosphates:

$$P_i = PO_4^{3-} \quad \text{and} \quad PP_i = P_2O_7^{4} \tag{3.14}$$

It is of course not the case that all of the possible reactions have such an activation energy that could be surmounted by the ATP $\rightarrow$ ADP liberated one. Most of the conversion is made in the cellular "power-plants," in mitochondria, which are on average 1,000–2,000 units in a mammalian cell, but of course it is very different tissue-by-tissue, ranging from a single unit to several million [613, 614]. The overall energy-source for the chemical processes are the ATP $\rightarrow$ ADP energy conversion, which drives the actual microprocesses. The human body consumes an extremely large amount of ATP even at rest: this amount is about 40 kg/day, and in need the consumption could go up to about half-kg/min! (Of course, there is permanent regeneration from the ADP.) The transferable phosphate group between the molecules gives huge flexibility to ATP-assisted reactions and produces an appropriate volume of energy-equivalent ATP in the actual conditions, e.g.:

$$\frac{\left\{\begin{array}{ll} \mathrm{ATP} + \mathrm{H_2O} \Leftrightarrow \mathrm{ADP} + P_i + H^+ & \{\Delta G = -30.5[\mathrm{kJ/mol}]\} \\ \mathrm{Glycerol} + P_i \Leftrightarrow \mathrm{Glycerol-tri-phosphate} & \{\Delta G = +9.2[\mathrm{kJ/mol}]\} \end{array}\right.}{\mathrm{ATP} + \mathrm{Glycerol} \Leftrightarrow \mathrm{ADP} + \mathrm{Glycerol-tri-phosphate}\ \{\Delta G = -21.3[\mathrm{kJ/mol}]\}} \tag{3.15}$$

Because of its central role in life-bioenergetics and its general convertibility the production of ATP is one of the main functions of metabolism. One of the basic reactions to produce ATP, to "charge" the energy-machinery and produce the energetic "currency" is glucose metabolism. Glucose is a relatively simple organic compound, but complex enough to have various pathways for its decomposition. First glycolysis produces lactic acid ($CH_3CHOHCOOH$) and at the end two ATPs and a remarkable energy is liberated:

$$\begin{aligned} &C_6H_{12}O_6 + 2\mathrm{ATP} + 2\mathrm{ADP} + 2\mathrm{NAD}^+ \rightarrow \\ &\rightarrow 2\mathrm{NADH} + 4\mathrm{ATP} + 2CH_3CHOHCOOH + \mathrm{Energy}\ \{196.6[\mathrm{kJ/mol}]\} \end{aligned} \tag{3.16}$$

The NADH is recycled in this reaction:

$$2(3\mathrm{ADP} + \mathrm{NADH}) \rightarrow 2(\mathrm{NAD}^+ + 3\mathrm{ATP}) \tag{3.17}$$

win 6ATPs in this part of the oxidization. The most energetic for producing ATP is Krebs' cycle, (e.g. [615, 616]):

$$2(CH_3CHOHCOOH + 6O_2 + 15ADP) \rightarrow 2(3CO_2 + 3H_2O + 15ATP) \quad (3.18)$$

Consequently the overall basis, which became the basic life process for all living objects is:

$$C_6H_{12}O_6 + 6O_2 \underset{\xrightarrow{\text{oxydative way}}}{\xleftarrow{\text{photosynthesis}}} 6CO_2 + 6H_2O + \text{Energy } \{2881 \text{ [kJ/mol]}\} \quad (3.19)$$

The process is used in both reversible directions:

for plants it is a product, the building up of living mass; (photosynthesis by absorbed light energy),
for objects living by chemo-oxidation (prokaryotic and eukaryotic cellular structures) the oxidative path produces energy for ATP generation.

The most optimal process in eukaryotes cells is:

$$C_6H_{12}O_6 + 6O_2 + 36P_i + 36ADP \xrightarrow{\text{oxydative way}} 6CO_2 + 6H_2O + 36ATP \quad (3.20)$$

while the eukaryotes produce different amounts of ATPs:

$$\eta_{ATP}^{\text{eukaryotes}} = \frac{36 \cdot 30.5}{2881} \cong 38.1\% \quad (3.21)$$

Under actual biological conditions it could be even higher [617]. This efficacy is remarkably high, notably better than most of the non-biological energy conversions (e.g. the efficacy of the gasoline engine ranges from 10–20%); and even far more than from any thermodynamic machinery. From the thermodynamic point of view, if the human organism were a heat engine then the efficiency could be calculated by the basic thermodynamic equations. According to the ideal Carnot process, the maximal efficacy (η) of a thermal engine depends on the temperature difference of the device (T_d) and the environment (T_e) as:

$$\eta = \left[\frac{\text{(useful ⟨beneficial⟩ work)}}{\text{(overall invested energy)}}\right] = \left[\frac{\text{(work in – use)}}{\text{(energy consumed)}}\right] = \left(\frac{T_e - T_d}{T_d}\right) \quad (3.22)$$

Consequently, if the human being would be governed by the thermal rules then the maximal efficacy of the human body would be:

$$\eta = \left(\frac{T_{body} - T_{environment}}{T_{body}}\right) \cong \frac{310K - 295K}{310K} \cong 0.03 \quad [3\%] \quad (3.23)$$

This is very low compared with the above ATP production efficacy. Principally ideal thermodynamic machinery, (Carnot machine) have to have a temperature

150°C above room temperature to be as effective as 40%. Consequently it is trivial, that life is not a thermodynamically governed system, but governed by high-efficacy chemistry.

The loss in the biochemical path however is also huge. Glucose oxidation liberates 2881 kJ/mol but the ATP produced, represents only 1159 kJ/mol. The "missing" value (1,722 kJ/mol) is wasted by the non-used chemical species and by the heat, which makes the body temperature independent from its environment.

Because of the proton gradient through the mitochondria membrane a facilitated diffusion of H^+ ions exists. The protons re-enter the mitochondria matrix without taking part in ATP synthesis (proton leak). The responsible proton channel (thermogenin, [618]), is a 33-kDa protein primarily found in brown adipose tissue; it is responsible for non-shivering thermogenesis.

The temperature is definitely the factor to measure the non-utilized energy, which is unusable for the next steps.

The real processes of living objects are chemical, cutting, rearranging, and building up various chemical compounds in large amounts. Living objects are much more chemical than thermodynamic "machines." The construction of biological objects, i.e. synthetic biology [619, 620], uses this trivial fact well.

The cellular ability to perform reactions fuelled by ATP-hydrolysis depends on the relative concentration of ATP and its various products. Accordingly, the energy state of the cell is measured by the energy charge (E_{charge}) [621]:

$$E_{\mathrm{charge}} \equiv \frac{[\mathrm{ATP}] + \frac{1}{2}[\mathrm{ADP}]}{[\mathrm{ATP}] + [\mathrm{ADP}] + [\mathrm{AMP}]}; \quad \{0 \leq E_{\mathrm{charge}} \leq 1\} \tag{3.24}$$

which is in most of the healthy cells $E_{\mathrm{charge}} \approx 0.9$, so?

$$11[\mathrm{ATP}] \approx (9[\mathrm{AMP}] - [\mathrm{ADP}]) \tag{3.25}$$

If $E_{\mathrm{charge}} \approx 0$, than $\sim$100% [AMP], when $E_{\mathrm{charge}} \approx 1$, than $\sim$ 100% ATP exists.

Again, this energy storage can not be replaced by heat or any temperature-related processes, this is a definite chemical and not thermodynamic relation.

Because of the exponential dependence [Eq. (3.6)] of the free energy the reaction rate of the ATP-promoted process is drastically enhanced. In numbers: given a reaction $R1 \to R2$, (which has activation energy $E_{R1\to R_2}$) on human body temperature ($T = 36$°C), the reaction rate is:

$$k_{R1\to R2} = \frac{[R1]}{[R2]} = e^{-\frac{E_{R1\to R2}}{RT}} \tag{3.26}$$

and with the help of ATP the free energy in the exponent will be modified by the ATP $\to$ ADP energy liberation:

$$k_{\mathrm{promoted}} = \left\{\frac{[R1]}{[R2]}\right\}_{\mathrm{promoted}} = e^{-\frac{E_{R1\to R2}+\Delta G(\mathrm{ATP})}{RT}} = k_{R1\to R_2} e^{-\frac{-30500}{8.3\cdot 309}} = 1.46\cdot 10^5 \cdot k_{R1\to R2} \tag{3.27}$$

So the increase in the ATP-promotion is almost 150-thousand! Note the temperature was the same; no temperature change was requested for this enormous enhancement!

Multi-ATP processes have their amount of ATPs in the exponent, so a process with involvement of n molecules of ATP will increase the actual reaction rate by $\sim 1.5{\cdot}10^{5n}$. In principle any of the non-spontaneous (endergonic) processes could be converted into favorable exogenous processes by the hydrolysis of a sufficient number of coupled ATP-molecules. ATP provides energy for many cellular functions, like membrane ion pumps, protein synthesis, nerve signal conduction, muscular contraction, etc. [564].

In most living objects Q_{10} [see Eq. (3.11)] is more than two, close to three [622, 623]. From here the most common activation energy could be calculated:

$$E_a \cong \frac{RT^2}{10}\ln(Q_{10}) \approx 50 \div 82\frac{\text{kJ}}{\text{mol}} \tag{3.28}$$

which is far above the average energy represented by the body temperature (2.5 kJ/mol), and the energy of 2–3 ATPs are necessary to activate a general process.

The ATP synthesis is temperature-dependent, and also follows the Arrhenius law [624]. Extrapolation of the data obtained for rainbow trout in [625], which was re-evaluated in [624], showed that the ATP production more than doubles from 36 to 46°C. Also the ATP synthesis could be promoted by physical training [626].

Naturally the (potentially usable) energy is not expended only on the ATP synthesis: some of the energy exchange and processes are non-ATP dependent (e.g. diffusion, coagulation, etc.), but most of the decisional life processes are energized by ATP.

Sometimes the ATP energy is not involved directly in a chemical process. It could drive so-called stochastic resonance phenomena. Simple speaking it is like a vibration for a granulated system, when the entire volume of the given mass could be less, the system become denser, more ordered. The vibration could give a chance for the various microenvironments to leave the local energy minima and occupy a lower energy state. This process makes possible to move the given structure from a side minimum (like the ladder steps in Fig. 3.23) to deeper minima. The process has to have sufficient energy avoid the frozen-in (blocked) state in the side minima. This has an especially important role in forming the proper protein structures. Proteins have a complex structure, having different levels. Three structural levels exist, subsequently embedded in each other:

The primary structure (the polypeptide chain) is a long chain with a backbone of amino acids connected by peptide bonds in a definite (characteristic) sequence. The amino-acid units are variations of 20 different structures, all organized around a central carbon atom; and their chain length ranges between 50 and 3,000 amino-acid units in a protein.

The primary chains form a secondary structure, they are curled-up into alpha-helixes or braided into beta-sheets. Both the arrangements use hydrogen bonds to fix the actual structure.

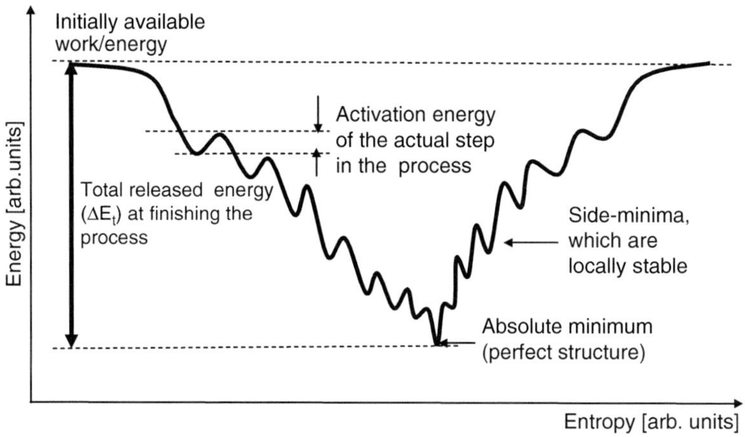

Fig. 3.25 The energy-funnel landscape for protein folding

The secondary structures form three-dimensional architectures, called the tertiary structure, of the protein. This structure appears irregular compared to the very regular secondary one. The tertiary structure depends on the secondary one.

The process to find the proper structure is not simple at all [627, 586]. The primary protein structure has to find its optimal (means lowest available free energy) structure. The "Levinthal paradox" [628], estimated an unbelievably long (in order of the magnitude of the age of the universe) time to find the proper folding by an energy-minimizing procedure for a simple protein. However, the energy-landscape theory of protein folding solves this problem: the protein folding is governed by a funnel-like energy landscape (see Fig. 3.25), which orients the process to have the proper structure.

There are three strong conditional factors to construct the funnel:

- The folding is performed in aqueous solution, which like a cramp forces the structural selection by the hydrophobic and hydrophilic bonds; the water constrains the hydrogen bonds and so repels the hydrophobic parts of the giant protein molecule and seeks to contact with their hydrophilic parts. The water matrix like a "cramp" pushes the proteins toward their native structure.
- The parts of the protein chain are also seeking to bond by hydrogen bridges, providing a driving force for folding.
- The thermodynamic fluctuations are large enough to jump over the actual local activation-energy barriers leaving the side-minima, and seeking to achieve the absolute minimum through subsequent small steps.

To make the folding problem simpler, it could be that the final state is only near the absolute free-energy minimal, so it is frozen in a side-minimum, the optimal state is not actually realized. It is possible that not all the proteins do the folding in a relatively short time, and some have additional energetic "help" from specialized chaperone proteins (stress proteins) to find their correct way.

Table 3.8 Energy consumption of some important processes. These are not accessible from the heat form of energy

Process/state	Approximate energy (kJ/mol)
ATP hydrolysis	30.5
Glycolysis (lactic acid formation)	196.6
Photosynthetic glucose production	2,881
Denatured protein to native state	960
Thermal energy at room temperature	2.4
Thermal energy at body temperature	2.5
Change of membrane permeability	50

Some approximate characteristic energies of the living processes are collected in Table 3.8. Obviously, animals have to have food to energize these reactions; to cover the reaction energies from heat energy is impossible. (In the case of plants this energy comes from the sun and for the non-organic compounds from the environment, simply heating the plant with adequate energy is not enough to keep it alive.) The heat developed by the processes is "waste energy". This will be equally distributed in the system. It is in fact had been lost for further use; can not be transformed into definite functions of the living objects.

3.2.3 Energy Sources and Driving Forces

The energy for life comes from (sun)light or from chemical compounds (nutrients). The latter anyway, originally drew their high energy demand from the Sun's energy also. The solar radiation on the surface of Earth is $\sim$150 W/m^2 [629]. Only a small fraction of this solar power can be used by plants, through photosynthesis, to obtain compounds from carbon dioxide and water. The photosynthetic power on the Earth's surface is all together 10^{12} W, and "only" 0.06% of the total arriving energy [630]. There are various calculations concerning the metabolic rate and energy consumption and about the effect of temperature on the metabolic theory in ecology [631].

The electron energy of the various compounds is the source alone. Nothing else, only the electron-jumps liberate energy for use, mainly from organic carbon compounds. The average energy content per unit mass of the organic carbon compounds is approximately equal for most living species and their constituents, about 42 kJ/$g_{(\mathrm{Carbon})}$ [632]. (An extreme deviation could range from its half to its double.)

However, in aqueous solutions the electrons are not transferable as simply as in a metal. The charge transfer is based on the H^+ ions – protons – and other ions in the electrolytes. To make the effective gradients for the driving force, membranes divide the various electrolytes and the microscopic energy processes are basically transmembrane electrochemical procedures by a proton (H^+) concentration gradient (pH-gradient) through the membrane (see Fig. 3.26). The proton-concentration gradient is used for active membrane transports, for energy production and storage, and for

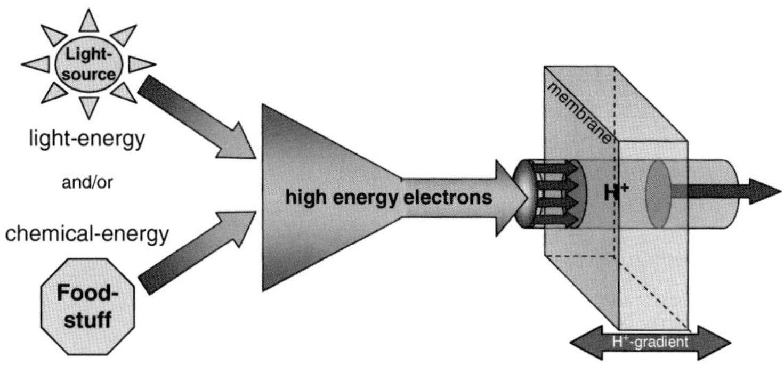

Fig. 3.26 The basic microscopic process: the trans-membrane proton gradient

some mechanical (flagellar) motions in primitive life forms (bacteria). The original energy source divides the living organisms into two systems, those which use the energy from oxidation of food stuffs, and other systems, which use absorbed light as an energy source.

The process and the basic energizing paths are similar in plants and animals, only the resources are different, however the initial and final products of the two mechanisms are connected (see Fig. 3.27).

A question naturally arises: what is the general driving force of the spontaneous processes in general? On the basis of the previous examples, the answer is trivial: the driving force of spontaneous processes is always the seeking of equilibrium by the systems: a decrease in free energy and increase in entropy. The intensive parameters measure the equilibrium, like the temperature and the pressure. In simple cases their equality in a system characterizes the equilibrium. However, various and very complex equilibriums exist, where the intensive parameters interact. We can construct such functions (called thermodynamic potentials, see Appendix 1) which describe the actual interactions of the intensives, their extremes will fix the equilibrium in the given conditions (like enthalpy or Gibb's free energy).

Heat put into a system (ΔQ) plus work done on a system (ΔW) is equal to the increase in internal energy of the system (ΔU), (first law of thermodynamics [633]):

$$\Delta U = \Delta Q + \Delta W \tag{3.29}$$

Sign Δ means a small change. (This is the same equation as (3.1). The only difference is the additional outside non-heat work ΔW.) During this small change the system does not alter (quasi-static approach). In a more rigorous description a differential calculus has to be used. Equation (3.31) shows that the internal energy U is determined by the energy exchange. This exchange has various forms, all the available interactions (denote their number by n) have to be calculated. These terms can be easily defined by the pair products of intensive (Y_i) and extensive (X_i) parameters:

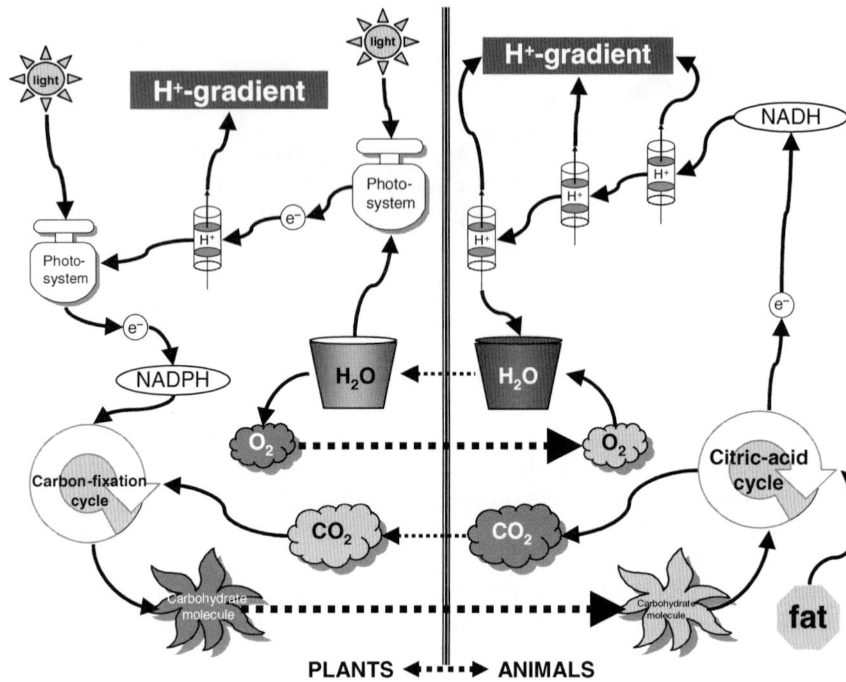

Fig. 3.27 The energy scheme and interconnections for plants and animals. The *dark-shaded elements* are final, while the *light-shaded elements* are initial compounds. The *non-shaded steps* are the main energetic conversions of the internal combustion

$$\Delta U = \sum_{i=1}^{n} Y_i \Delta X_i \tag{3.30}$$

For example some of the terms are:

$$\Delta U = T\Delta S - p\Delta V + \Phi\Delta e + (E \cdot \Delta P) + (H \cdot \Delta M) + \sum_{j=1}^{k} \mu_j \Delta N_j + \alpha \Delta s + f \Delta \ell + \cdots \tag{3.31}$$

where $T, S, p, V, \Phi, e, \boldsymbol{E}, \boldsymbol{P}, \boldsymbol{H}, \boldsymbol{M}, \alpha, s, f$, and ℓ are the absolute temperature, entropy, pressure, volume, electric potential, electric charge, electric field, electric polarization, magnetic field, magnetization, surface tension, surface area, linear force, length, respectively; while N_j and μ_j are the number/mass and chemical potential of the various (k) particles (molecules, ions, clusters, etc.) in the system. Many other pair interactions (all the energetic terms) may be included in this energy balance. All the pairs have special biological meanings: $T\Delta S$ is the absorbed heat, $p\Delta V$ is the work of pressure (volume changes), $\Phi\Delta e$ is the work of the moving electric charges, ($\boldsymbol{E}\cdot\Delta\boldsymbol{P}$) and ($\boldsymbol{H}\cdot\Delta\boldsymbol{M}$) are the work of the electric and magnetic fields, $\mu_j \Delta N_j$ are the terms of the chemical reaction energies of various ($j = 1, 2, 3, \ldots, k$) species, $\alpha\Delta s$ can be the energy of the surface (membrane) changes, $f\Delta\ell$ could be the work of the muscle fibers, etc. Only one of the terms, namely, the heat energy has temperature, all the other terms are other kinds of energy consumptions. In reality all the thermo-processes are time-dependent. The gradients of intensives (deterministic or fluctuations) are the driving forces of the changes, forcing the flow (current) of extensives to equalize the actual differences (see Appendix 2)

Certainly, if there are no any structural, chemical etc. interactions in the system, only the heat is absorbed, than (3.1) and (3.31) has no ΔW term, and the equation becomes very simple:

$$\Delta U = mc\Delta T = \Delta Q \quad \Rightarrow \quad \Delta T = \frac{1}{mc}\Delta Q \tag{3.32}$$

or

$$T = T_o + \left(\frac{\Delta Q}{mc}\right) = T_o + \frac{\Delta Q}{V\rho c} \tag{3.33}$$

where T_0 is the original body temperature [°C], c is the specific heat [J/kg/K] showing how much energy is required to heat up 1 kg of tissue by 1 K. (ρ, V, and m are the density [kg/m^3], the mass [kg] and the volume [m^3] of the heated tissue, respectively.) ΔQ is the energy delivered [J] into the heated tissue. This picture could be the basis of the misleading interchange of the heat dose and temperature change; in (3.33) they are proportional. Do not forget that (3.33) is only valid if there is no interaction other than the heat absorption. Of course, this is not the case at all in hyperthermia, where our definite goal is to change the structure and the biochemical constituents. In the study of (3.33) it is also important, that the time (dynamic effects) is not included in these static considerations. However, if the energy balance is time-dependent (the heat delivery and heat sink have power-like relations, [J/s]), then the temperature also becomes time-dependent.

3.2.4 Energy and Structure

The driving forces in all respects are the laws of thermodynamics: i.e. lower the energy and increase the entropy. We are convinced it is the final cause of the development and diversity as well as the actual biomachineries.

Modern physiology is essentially an inter-disciplinary subject; it applies numerous principles and discoveries from other science fields and synthesizes the macroscopic interactions (like electromagnetic fields and potentials) with microscopic (like cellular, subcellular, molecular and sub-molecular) effects. The electronic structure approach of solid-state physics (e.g. Szent-Gyorgyi, [634, 635]), superconductivity (e.g. Cope, [636]), electromagnetism (e.g. Liboff, [637, 638]), thermodynamics (e.g. Schrodinger, [639], Katchalsky & Curran [576]), etc. are all parts of the physiology, and make it really complex as the phenomena of life itself is. The living organism develops itself, rearranges, reorganizes the incoming chemicals and builds up its own structure, consequently lowering the entropy.

Various modern approaches have been developed in the last decades in relation to this complexity, like self-organization ([640–642, 627]), fractal physiology ([643–646]), and bioscaling ([647–649]). Oncothermia uses these new approaches to achieve the best curative performance.

Anyway, there are numerous debates about life and its physiological description. However, it is commonly agreed: life is rather more a chemical machine than a thermodynamic one. This is the key factor of the oncothermia approach.

As we saw in Eq. (3.31), the structural changes always cause certain energetic changes in the system by the product of the temperature and entropy change ($\Delta U_{\text{structure}} = T\Delta S$).

The living systems have controlled energy combustion; otherwise the chemicals undergo sudden and coincident reactions producing explosion-like impulses instead of continuous energy support for the system. The chemicals for energy liberation are transported to the reaction locations by various methods, and so the transport properties define the energy balances. The metabolic activity has a scaling behavior in all the ranges of living matter from the subcellular to the entire organism. The chemical reactions and the transport of reagents and the signal transductions are rather unified in all the living cells, so their scale-free networks [650] are not surprising. All the reactions are surface-controlled, so we expect an exponent for scaling by the mass 2/3. [The mass of the living object is volume-dependent [scaling by Eq. (3.3)], while the surface is scaled only by Eq. (3.2)]. However, the mass-dependent scaling of metabolism goes with $^3\!/_4$ [651], as if life would be four dimensional [652] (see Appendix 3). The bioscaling depends on the energy supply of the system (see Appendix 4)

3.2.5 Energetics of Malignant Cells

From an energetic point of view the important fact is that malignant cells undergo frequent and permanent cellular division. The energy consumption for this intensive division is definitely higher than the energy request of healthy cells in homeostasis. A high intensity mass-production of ATP is necessary to fulfill this strong energy demand. There are two ways to produce ATP: the oxidative and the fermentative path.

The similarities of oncogene activity and anti-apoptotic functions in cancer and in various healthy processes (like growth and reparation) are one of the most challenging facts in the present research. This is because the apoptotic processes as well as the oxidative ATP production which are suppressed in numerous growth and reparative processes degrade (at least temporarily) the function of mitochondria. This is the reason for the renaissance of Wartburg's theory, tumor metabolism and its mitochondrial connection is undergoing intensive investigation [653–655]. According to the main idea of Warburg, the primary cause of cancer is non-oxidative glucose metabolism. Because the oxidative metabolism is the task of the mitochondria, the missing oxidative metabolism is a dysfunction of the mitochondria. According to Warburg, the mutation of the genome is a consequence of fermentative metabolism: the hypoxia causes malignant transformation.

There are, however, some important facts to be taken into consideration. The early stage of the development of life on our globe was fermentative and also the early stages of the development of new life after fertilization also have a basically non-oxidative period till the circulatory systems have been developed. The division speed of these stages is similar to the development rate of the cancer. The first development stage always uses large amounts of energy, which is mostly produced under anaerobic conditions. The oxidative ATP production will dominate only in later developmental stages, when the circulatory systems are established. The division speed of the cells in the oxidative stage is slower.

Reparation is a special process aiming at the re-establishment of the original morphology and proliferative homeostasis. The basic processes involved are:

- Stem-cell differentiation and division are controlled by definitive conditions.
- Dedifferentiation of some normal cells at the wound or damage could be possible by unregulated oncogenes, which are down-regulated after the division, and the cell redifferenciated in the normal way. (This is a dangerous form of division, probably used only when the previous two do not work properly.).
- What could be the fault when the first two mechanisms do not work? Either the number of stem cells is not enough or the redifferentiation process is not working or at least not intensively enough. The redifferentiation could be promoted by the injury currents [656], used for wound healing [657].
- The organism itself has such natural local processes, which use the large energy flux of the fermentative metabolism in normal functioning as well.

In the case of any damage some cells produce a functional state which repairs the actual dysfunction. The special growth- or repairing-phase genes are activated to produce such cells, which repair the damaged tissue [658, 659]. Growth and reparative factors are released [660, 661]. Proto-oncogenes become active in the area [64], collecting stem cells to the wound [662], which repair via their differentiation [663]. After the process all the activation genes are down-regulated or tumor suppressors are activated [664], and the normal homeostasis becomes re-established. However, the reparation mechanism could be blocked or limited. An obvious reason could be the permanent irritation by a mechanical, chemical, or physical (e.g. ionization) factor. However there is a more sophisticated reparation block also: the injury current does not become blocked, the current becomes permanent. The simplest reason for this error is the grouping of the new-born cells, having no possibility to neutralize themselves to the potential level of the normal surroundings. (The new-born cells are charged more negatively than the cells in the normal host tissue.) The reparation has an extra high energy demand, so the massive ATP production of the fermentative path is preferred in this stage.

In the cancerous state the repairing conditions are not blocked after finishing the reparation itself. Permanent reparation demand depletes the available stem cells [665, 666], and emphasis will be placed on the third way to repair: the proto-oncogenes are activated, and malignant transformation could occur inducing clinical cancer [667]. The malignant transformation of the wound causes secretion of proto-oncogenes [668–670], and intensive capture and stimulation of stem cells from other places [671, 672], the cancer cells produce repairing [673, 674] molecules increasing the stem-cell concentration in the cancer tissue [675].

In normal, healthy cells both the oxidative and fermentative metabolism is present. (One of the basic questions of sports medicine is the control of the optimal ratio by exercise.) The oxidative metabolism has high efficacy (Szent-Gyorgyi–Krebs cycle, produces 36–38 ATP) but due to its complex, multi-step reaction chain, it has low energy flux (low intensity), while the fermentative path (mostly finished in lactate) has low efficacy (produces only 2 ATP in a step), but due to its simplicity (only a few steps) it has a huge energy flux. The first is used in homeostasis, to keep the cell in the given normal state, while the second is used under extreme conditions. Examples of such extreme conditions are embryonic development, hypoxic state, permanent strong stress, tissue repair, or a defective cell state.

The higher metabolic rate could be routinely measured by positron emission tomography (PET) [676]. The reaction rate of the simple fermentative reaction could be 100-times quicker (approximated from the positron annihilation data, [677]) than the oxidative path. However, the oxidative path has at least 18-times higher efficacy due to the ATP amount at the end. The result is funny: the simple, primitive fermentative process produces at the end 6-times more ATP than its high efficacy, but too complicated oxidative counterpart. This is why the extreme situations constrain the anaerobic metabolism, the balance is broken out of a normal homeostasis situation.

Of course its glucose demand is also 6-times higher, which intensifies the glucose transport. The end-product (lactate) also has to be transported away, and the increase in blood lactate concentration after an oral glucose load or after intensive exercise in normal subjects is well known. The extra lactate transported by the blood stream to the liver, and the Cori cycle [678] reproduces the glucose for the next metabolic run.

The oxidative metabolism is performed by mitochondria in the cell. The mitochondrion has its own DNA, and it replicates itself when the host cell makes its own replication. These divisions are synchronized. According to Warburg's theory, cancer is primarily caused by mitochondrial dysfunction. However, the challenge could be formulated by questions:

- Do all the mitochondria in the host cell work improperly, or is it enough if one of them does not function?
- Is the dysfunction inheritable (then all the mitochondria carry it)?
- After the division of the cancer cell the mitochondrial dysfunction remains. This means it is at least inheritable in a malignant cycle. How does the DNA of the host cell change the DNA of the mitochondria?

Answering these questions is easier if we assume a mechanism, which does not descend (at least at its start) from the DNA but is constructed by the overall conditions in the intra-cellular electrolyte of the host cell. In this case the heredity would be found in the cytoplasm and not in the DNA, consequently the conditions of the cytoplasm are inherited. This could be formulated also by the altered metabolism: the microenvironment of the tumor selects special metabolic paths [679]. The conditions for the interactions with the environment are decisional in this process. The same is valid in complete organisms as well: all living systems are energetically open. This means intensive and permanent interaction with the environment, a definite energy and material flow (together with the flow of other thermodynamic parameters like entropy), goes through these living objects. Nobel laureate Albert Szent-Gyorgyi formulated this well [680]: instead of the monkey's migration through the forest, he centered his attention on how the forest migrates (flows) through the monkey.

The host cell and its mitochondrion are energetically different: the mitochondrion function needs pyruvate supply and metabolizes only in the oxidative way, while the host cell can produce ATP only in the anaerobic way. This is energy-supply symbiosis. Mitochondria are shielded against the direct oxygen flux by the host cell. Its proper function supposes then an effective proton transporter which transports the hydrogen farer from the mitochondria to a more oxygen-rich part of the cell. This process is driven by the active proton pumps energized by the ATP (produced by mitochondrion) as well. The proton alone does not exist in aqua-solution, it associates with a water molecule and could be transported only by a slow diffusion to the proton pumps.

What is a quick, effective proton transporter with small energy dissipation? This is simply ordered water [681], which can transport the proton with high speed.

The monomer water molecule has a simple tetrahedral structure, which is slightly asymmetric by the two proton-occupied positions and two lone pairs. However, the simple water has a rather complex structure in bulky conditions [682, 683]. The stochastic proton migration in hydrogen bonds makes the bulky water collective [681]. The solid water (ice) has hydrogen-bridge connections all over the volume [684]. The entire bridge-bonded ice turns to water by a first kind phase transition, but only a fraction of the bridges broke, about every seventh, approximated from the evaporation and melting heat ratio ($W_e \approx 2256$ kJ/kg and $W_m \approx 334$ kJ/kg). A remarkably high number of hydrogen bridges exist even at the boiling point of the water [685]. In fact the water is always a mixture of two phases [686, 687]: disordered, highly dynamic, mainly monomer and ordered, clustered ones. (Even the clusters could have various structures, including an entirely closed clathrate with icosahedral symmetry [688]). The statistical, stochastic transformations of the phases make the water so complex. The simple bifurcative phenomenon (the proton charge oscillates between the two possible positions), of the hydrogen bridges is shown in Fig. 3.28. The charge transfer of such oscillating bonds could be varied by the dwelling in the different states. (In general the oscillation of chemical bonding could be multistate.) The generalized solution of the bifurcative phenomena in living materials was worked out earlier by us [689].

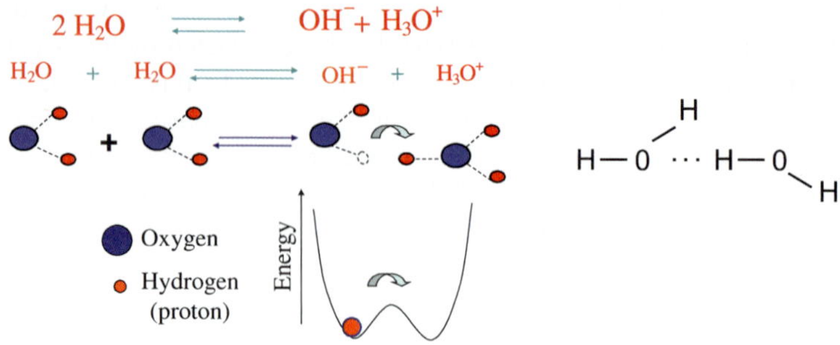

Fig. 3.28 Bifurcation conditions of protons in a hydrogen bridge (points in schematics symbolize the hydrogen bridge)

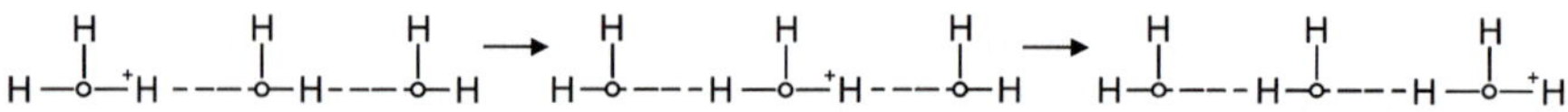

Fig. 3.29 The Grotthuss mechanism of proton jumping (three subsequent steps of the process are shown)

The hydrogen ion can be transported by the hydrogen bridges. The high speed and low dissipation of the transport propagation is based on the Grotthuss mechanism [690, 691], where the proton tunnels (jumps) from one water cluster to the other bridged by hydrogen bonds [see Fig. 3.29, Eq. (3.34)]. The lifetime of H_3O^+ (hydronium ion) is rather small ($\sim 3 \cdot 10^{-12}$s) so the speed of proton transport by the Grotthuss mechanism is approximately ten times higher than the one by diffusion.

$$H_3O^+ + H_2O \leftrightarrow H_2O + H_3O^+ \tag{3.34}$$

The Grotthuss mechanism is in fact the propagation of the ionization of a water molecule. The dissociation and recombination steps alter during the "travelling." Recombination-dissipation is a quantum-mechanical process, in principle free of dissipation [692]. However, it has temperature dependence, and also the vector potential is able to modify the quantum states of the water [693, 694], which could modify the chain processes.

The ordered water is a good conductor for hydrogen ions (by hopping which we described), and bad conductor for other ions. The water ordering selects between the ionic flows and prefers the proton against all the other reaction products. The effect of the outside electric field could conduct the hydroxyl (OH^-) and hydronium (H_3O^+) ions by the same Grotthuss mechanism (see Fig. 3.30).

The tetrahedron of a single water molecule is nonregular (its edges differ by the actual proton occupancy of the corners), so the migration of the hydrogen ion by this chain jumping looks like a rotation of the given water molecule: the occupied corners become vacant and the empty positions filled up (see Fig. 3.31).

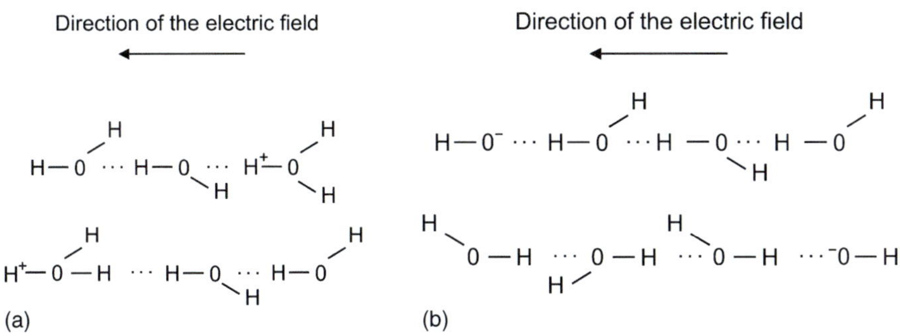

Fig. 3.30 Grotthuss' mechanism of hydronium (**a**) and hydroxyl (**b**) ion conduction by an outside electric field

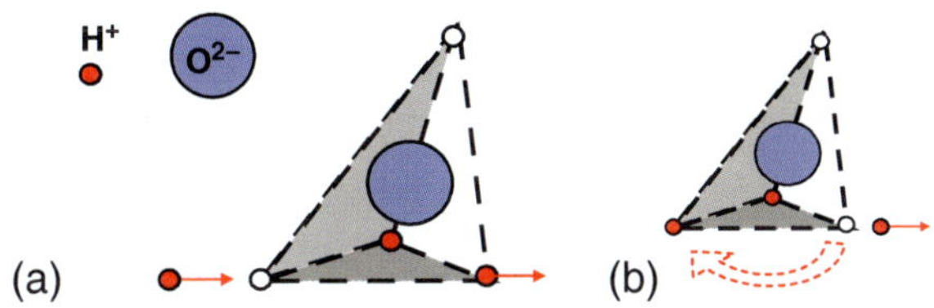

Fig. 3.31 Apparent rotation of a water molecule by proton conduction. (**a**) The arriving proton occupies an empty space, and one of the protons travels further. (**b**) By changing structural distances the tetrahedron looks like swiveling by the newly arrived proton

Living matter uses the hydrogen bridges not only in water, but in most of its important structures of life. They could bond the amino acids, the nucleotides and appear in many other bonds also (see as example Fig. 3.32).

The normal proteins in aqueous solutions have ordered water on their surfaces. (The dry protein has definitely different properties compared to its wet counterpart.) This water-"coat" is a good conductor for protons to promote the forming of secondary and ternary structures of the protein.

The living system is not an ordered solid, it is an aqueous solution, where the co-operativity is not easily introduced, (contrary to crystals [695]). However in the living state the water is mostly well ordered, nearly crystalline (semicrystalline, [696]). This relative order turns the "dilute salted water" into a system having entirely new mechanical, chemical, physical, etc. behaviors as the normal aqueous solutions. Indeed, the important role in living systems of so-called ordered water was pointed out in the middle of the 1960s, and later it was proven [697]. At first ordered water was suggested to be as much as 50% of the total amount of the water in living bodies [698]. The systematic investigations have shown more ordered water [699, 700] than was expected before. Probably the ordered water bound to the membrane is oriented (ordered) by the membrane potential, whose decrease probably decreases the order of the connected water, so increases the electric permeability of the water [701], and so decreases the cell–cell adhesion and could be the cause of cell division or even of proliferation [701].

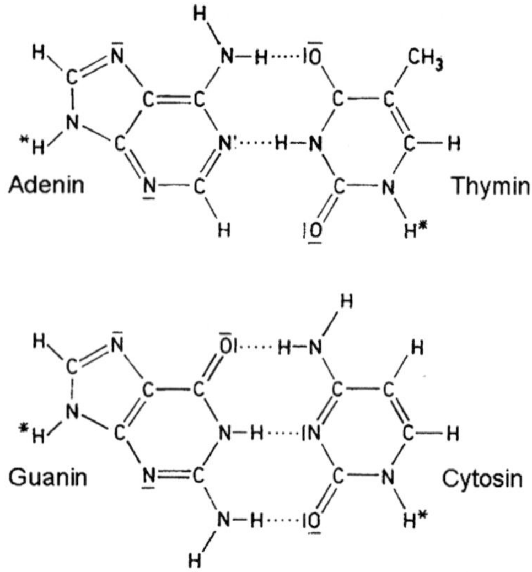

Fig. 3.32 Some hydrogen bonds with oxygen

In consequence of the malignant changes the metabolism gradually favors the fermentation path, where the host cell carries out the ATP production instead of the blocked mitochondria. The end-products of both the metabolic processes are ions in the aqua-based electrolyte. The oxidative cycle products dissociate like $6CO_2 + 6H_2O \leftrightarrow 12H^+ + 6CO_3^{2-}$ while the lactate produced by fermentation dissociates: $2CH_3CHOHCOOH \leftrightarrow 2CH_3CHOHCOO^- + 2H^+$. Assuming the equal proton production (by more intensive fermentation energy flux) the main difference is in the negative ions. The complex lactate-ion concentration grows rapidly, and increases its osmotic pressure. To reestablish the normal, the dissolvent (monomer water) has to be increased as well, seeking to solvent by non-ordered water. Indeed, it is measured in various malignancies that the water changed to a disordered form [702–704], so in these cases the ordered water concentration in cancerous cells is smaller than in their healthy counterpart. Consequently the hydrogen ionic transmitter became weak, the removal of the hydrogen ions became less active. This decreases the intra-cellular pH and the proton gradient in mitochondria, which directly worsens the efficacy of ATP production. To compensate for the lowered proton gradient, the membrane potential of mitochondria grows. This lowers the permeability of the membrane, and decreases the mitochondrial permeability transition (MPT), which has a crucial role in apoptosis [705, 706]. (The high mitochondrial membrane potential and low K-channel expression have been observed in cancerous processes, [707]). These processes lead to apoptosis resistance, and for the cell energizing the ATP production of the host cell (fermentation) became supported. The free-ion concentration increases in the cytoplasm, and so the HSP chaperone stress proteins start to be produced. This process needs more ATP as well as it is an anti-apoptotic agent, so the process could lead to a complete blocking of apoptosis. Rearranging (disordering) the water structure needs energy [708]. It is similar to the way in which the ice is melted with latent heat from a zero centigrade solid to liquid with unchanged temperature conditions. This drastic change (phase transition) modifies the physical properties (for example the dielectric constant) of the material without changing the composition (only the microscopic ordering) of the medium itself.

The decisional role of the two metabolic pathways (the oxidative and the fermentative) was studied by Szent-Gyorgyi [701], from an etiology approach, and using other formulation. His interpretation describes the cellular states by two different stages. The alpha-state of the cell is the fermentative status. This was general in the early development of the life, when free oxygen was not available. The aggressive electron acceptor was not present [709]. In this stage only simple, primitive life forms could exist, The main task was to maintain life by their unlimited multiplication. This state was only reproduction-oriented, they were not able to develop complex structures and complicated work division. All living objects in the alpha-state are autonomic, they are competing with each other, and cooperative communication does not exist between them. With the later presence of free oxygen the beta-state of life was developed. The oxygen made it possible to exchange a higher value of electric charges, the unsaturated protein allowed more complex interactions, and started the diversity of life. The cells in this state are cooperative, the task since the multiplication-only phase became more complex, including optimal energy consumption, and diversity for optimal adjustment to life. This is the phase, which integrated the mitochondria for oxidative ATP production, and so produces energy in high efficacy.

The historical development of life from the alpha- to beta-stages had been generalized [701], introducing the same states for the actual stage of the cells in developed complex living systems. (Further we use the same notations but we will use Greek letters α and β for those states.)

The highly organized living objects mainly are built up from cells in the β-state. Their cell division became controlled. This control was mandatory, because the division needs autonomic actions, the cooperative inter-cellular forces slack, a part of the structure has to be dissolved and rearranged, so the cell in a division state is again in a non-differentiating state, similar to the α one.

The α-state is the basic status of life. In this the highest available entropy is accompanied by the lowest available free energy. All complex living systems could easily be transformed into this basic state when they became instable. Then by the simple physical constrains (seeking low free energy and high entropy) the cells try (at least partly) to realize the α-state again. Again the system (or a part of it) contains cells with high autonomy and proliferation rate. By simple comparison Szent-Gyorgyi's states and Warburg's metabolic pathways are common: the α- and β-states correspond to the fermentative and oxidative metabolism, respectively. In other words the α-state prefers the host-cell ATP production (anaerobic) path while when a perfect mitochondrial function works that is the β-state. These states are mixed (the cell works in both forms of metabolic activity) and it is only a question of quantity of each category. About 70% of the cells are in the β-state in normal homeostasis. The balance could be formulated by the cell status of co-operability

($\alpha \leftrightarrow \beta$); or formulated by metabolic paths (fermentation $\leftrightarrow$ oxidation) or could be formulated with acting parts of metabolism: (host cell $\leftrightarrow$ mitochondrion). The meaning of all the formulations is equal: the actual energetic state is described. Note the interesting relation between the energy flux and co-operability. The high energy flux makes the cells less cooperative and more primitive, while the low energy flux makes the cells not only cooperative but also sophisticated, highly effective in energy production and in environmental adaptation as well. (It has interesting similarity with the organizing of societies also [710], but this is outside of our present topic.)

Differences in the metabolic processes of vertebrates and invertebrates (terrestrial, pelagic, and benthic) are well mirrored in the scaling exponent [711]. The benthic invertebrates ($n = 215$) have the lowest average scaling exponent ($p_{\text{mean}} = 0.63$, [near to $2/3$], $\text{CI}_{\text{mean}} = 0.18$), which metabolizes basically in the anaerobic way [712], while all the studied animals ($n = 496$) have ($p_{\text{mean}} = 0.74$, [near to $3/4$], $\text{CI}_{\text{mean}} = 0.18$) [711]. Also the scaling of the metabolic activity is different in mitochondrial or non-mitochondrial metabolism. The mitochondrial metabolism is always aerobic, its scaling exponent is nearly $p = 3/4$ [713, 649], while the non-mitochondrial respiration scaling is near to $2/3$ [714].

A question arises: what mechanism controls the balance of β-and α-states in highly developed living objects? It is probable the electromagnetic behavior of the electrolytes in living systems gives the answer [715]. The cooperative cells mostly run on oxidative metabolism, and their division is controlled by the cells in their neighborhood. There are two basic reasons for normal cellular division, it could be a regular division maintaining homeostasis of the given tissue, replacing elder cells with young daughter cells, or it could be a forced, constrained division (like in wound healing, reparations, embryonic development, constrained tissue-specific cell production, etc.). The questions: what is the process that starts the division, and what finishes it?

It is easy to start the division. There have to be unusual conditions (extreme needs), which desperately increases the energy requirement. This could be such a mechanism as was described above: the changing concentration of one or more components needs more dissolvent, which is provided by the order–disorder transition of the intra-cellular aqueous electrolyte as well as the osmotic water flow through the cellular membrane. The concentration misbalance can be created by outside stimuli (like injury currents) or by inside enrichment of a component due to aging or due to metabolic misbalance. The order–disorder water transition does not only change the hydrogen-ion diffusion, but also changes the dielectric constant of the medium [715]. The more disordered liquid increases the dielectric constant (in simple words, the ability of electric isolation has increased). This is directly connected with the promoted charge division and the suppressed polymerization activity at the sub-cellular level, creating positive feedback to the fermentation processes. The balance is broken, and turned to the phase where the α-state is dominant. It is not necessarily a malignant transition. This happens with any regular cell division as well. This is the "motherhood" of the cell, making it possible for it to "deliver" the daughter cells. The "individualism" of the mother cell is explainable by the extreme high energy demand of the division process. When the daughter cells appear they must accept the previous order. Their "infancy" is normal, as "babyhood" is normal after deliveries. The only not normal case would be if the "babyhood" of a newly born human became 10-years long... Consequently the process could go wrong, if after finishing the division, the daughter cells do not find the path back to co-operability, the β-state again. When this is not the case, and the cells become stuck in the α-state, their proliferation becomes uncontrolled. This unfortunate case, however, is not a simple process originating from one single defect. It is a disturbance of a complex controlling mechanism [701], which well correlates anyway with the single "renegade cell" concept [716], showing a long process to produce "a renegade cell" as the ancestor of the billion-cell group called cancer. How complex damage has to be happen by cancer developing, is shown well by epidemiological research. It is shown, that at least five different mutations have to be coincidently present to be malignant [29].

Again we are back to the main question: what is the mechanism to re-establish the β-state after the division of the cell. We think, the down-regulation of the energy flux has the same active elements as the up-regulation had at the start of the division. The clue is again the order–disorder transformation in the aqueous solution. As we stated, at the start of the division a huge amount of energy has to be ready to supply the process, a large number of proteins and other cellular elements (lipids, enzymes, etc.) have to be produced, and they all need ATP desperately. In α-state the conditions are ready for that. When the division is over, and the two new daughter cells appear, the energy consumption drastically lowers to a double of the original mother cell. The doubled cytoplasm and all the cellular elements had enough dissolvent capacity even in the ordered water case. The hydrogen-bridge proton bifurcation can be reorganized, no opposite environmental driving force. The sudden doubling of the cellular elements serving like the cooling down of a liquid to a solid, going through a phase transition (disorder–order transition), just the same as (only in the opposite direction) when the division started. This again (like in the liquid-phase transitions) lowers the free energy, and in all (together with the environment, where the extra heat is radiated) increases the entropy. Note, the entropy apparently decreases (information build up) at the local cellular level, the overall conditions have to be considered for a full picture.

As we have shown, the metabolic pathways could drastically modify the development of the cell, and it could be the primary source of the malignant deviations. The balance of the oxidative and fermentative metabolism tunes the cellular ability to behave collectively or constrict autonomy, be individual. These conditions of course well depend on the energy (and signaling) exchange of the cell with its actual environment. The intra-cellular transport properties also have to be different on changing the metabolic pathway. The intensive energy flux of the fermentative metabolism increases the liberated heat in the cell, and so the temperature gradient between the extra- and intra-cellular compartments. The growing temperature difference could reach a critical threshold, when the heat flow turns from a conductive to a convective one [717]. (This phenomenon works like the well-known Benard instability, [718].) The convective path promotes the ionic flows through the cellular membrane, increasing the glucose permeability and so supports the fermentation path of metabolism together with the changes of the intra-cellular circulations [719, 720]. This complex change could down-regulate mitochondrial oxidative metabolism.

The divisional processes probably increases the intra-cellular flows and all the pathway activities both in regular and malignant cell division. Probably order–disorder transition of the aqueous solution also has a role in the changes [721].

However, finishing the division, the daughter cells separated, a higher surface suddenly appears and the separated volumes limit the intra-cellular flows and change the order of structure as well. It decreases the gradient through the membrane. This regulates the heat flow through the cellular membrane and changes the energy exchange from the convective to the conductive one again [717]. The conductive heat exchange does not support the intensive diffusion of the large-molecule glucose, so the oxidative path became necessary and regular. The two daughter cells have less than half the energy consumption (each) than was requested by the mother cell, which was because the mother cell was large (doubled its volume) and was intensively producing various elements to complete the daughter cells. Instead of the division conditions where the high energy request increased the energy demand and thus preferred the high-energy flux fermentative metabolism, the normal homeostatic conditions will dominate again.

The metabolic rate does not depend on the cultured mass (has no scaling) in cell cultures [722]. It shows well: the scaling is a behavior of the cooperative, collective structures, and does not appear in cases, when the nutrition is available practically infinitely due to the passage of the culture. This raises the question of the autonomy of the cancer cells. Probably, at the start of the malignancy the situation corresponds to the infinite availability of nutrition for the "renegade" cell. However, with the growing number of "individuals" nutrition starts to become limited. At this stage some cooperative features lead to the death of weak or internal members of the "colony." (Study the development of ant colonies which also support this type of organization [723].) This was formulated theoretically [724] and experimentally by the linear growth [725, 726] explained by the similarities with molecular beam epitaxy (MBE) [727]. The proliferation was observed to be highest on free surfaces, and the size of these determines the proliferative activity. The average radius (L) of the cancer colony is linear with time. The colony develops well-known fractal structures, where the edges promote the division of the cells, while the wells allow the cells to survive. The deviation [$w(t, l)$] of the radius is a self-similar function of the time (t) and length of the arc of the circle with the average radius (l), like:

$$w(t, l) = \begin{cases} t^{\beta} \\ l^{\alpha} \end{cases} \tag{3.35}$$

where $\alpha \approx 0.9$ and $\beta \approx 0.36$ for all types of measured tumors.

The bioscaling makes it possible to explain the ontogenetic growth (see Appendix 5).

3.2.6 "Non-Thermal" Effects – The Thermodynamic Approach

As far as we saw above, the energy exchange with the environment changes the internal energy of the thermodynamic system, so in this meaning all the energy kinds give a thermal effect, irrespective of its temperature dependence. The temperature dependence is not equivalent to the thermal behavior, as we saw when the ice was melted by simple pressure without any temperature changes. The relevant literature, however, uses in most cases "nonthermal" and "non-temperature dependent" like identical categories. This is false, like it was discussed (the false equivalence of "heat" and "temperature"). To state that the temperature-independent processes are nonthermal leads to many misunderstandings in the interpretations. The environmental energy change is thermal per definition. However, it could be that the energy exchange does not change the temperature (the average energy pool), and is "only" used for one definite interaction.

The NTD effects could also be connected to the temperature. Usually a definite average energy (temperature) is necessary to make a process, which is actually NTD. Cell killing needs energy. In this process the overall energy of the system decreases from a well-ordered state to a disordered one. To break the chemical bonds and make the structural rearrangement energy is invested. In this way the absorbed energy (at least a part of it) will not increase the temperature, the breaking of the chemical bond made directly by the energy. Energy must be pumped into the system for the transition energy to move from the ordered state to the disordered one.

The NTD effects are usually referred to in the interactions with the electromagnetic fields. This distinguishing factor in most cases depends on the applied power having not enough energy to increase the temperature, but the effect of the applied radiation (field) is measurable [728]. The distinction was theoretically described also [729] by shift of concentration at both sides of the membranes, and tries to determine the threshold when below the electromagnetic energy absorption regarded as nonthermal. Others formulate this situation as a "subtle" thermal effect [730, 731]. From the thermodynamic point of view all of the kinds of energy are a term in the energy balance [see Eqs. (3.31) and (3.35)], irrespective of whether they change the temperature or not. The only point is the addition to the internal energy of the system. The temperature change in the internal energy means uni-directional "smeared" energy incorporation by the system, involving a distribution of the energy to all the parts, particles involved in the system. However, the directional energy intake acts only in a special way in parts of the system. The parts are selected by particular interactions (like the electric field acts on the charges, the quantum/chemical effects act on special selection rules, etc.) In this case the energy does the job by that purpose-made interaction and it could be that the excess energy is distributed to the other parts of the whole set of the parts of system. When the energy transfer is correctly targeting the goal, then all the pumped-in energy is devoted to performing the desired job and not "waste" the energy by distributing to all the involved parts. Initiating a change in specific chemical bonds could be done by a particular chemical effect that targets the energy-intake only to that reaction, but this could also be done by pumping in energy to the whole system involving all of its parts, thus reaching the desired energy-intake for the originally desired reaction.

A good example is simple washing. The greasy material could be washed by high-temperature hot water without any chemicals. However, applying washing powder allows the same job to be performed at a lower temperature. The energy for the reaction (dissolving the grease-spot in the solvent water) is given without wasting much energy to pump it into substrate heating. The chemical reaction made the selective energy deposition, without any unnecessary unselective energy intake. Here the initializer of the full process was of course the temperature; according to the Arrhenius law the higher temperature promotes the reactions. However, we could also wash a temperature-sensitive textile (wool) so we use a low temperature and highly effective chemicals to target the dirt selectively.

The same can be done in such highly complex and extremely well-organized systems like living objects: we choose the actual interaction of the desired target, which is specifically characteristic only for that set inside the complex organization. This is the idea behind chemotherapy: find a specific reaction which selectively reacts only with cancer cells and destroy/block/paralyze/change only them. Here there is no intention to distribute the reaction to all cells. When we apply electromagnetism, our goal is the same: find the specific electromagnetic reaction which selects our targets, without wasting too much from the field energy to interact with other parts of the system. Like the chemical components in chemotherapy, we change the properties of the bioelectromagnetic interactions (field, intensity, frequency, phase, pulsing, etc.), choosing the appropriate electromagnetic reactions. Anyway, when we are clever (and lucky) enough we can work with greater clarity and more precisely in the bioelectromagnetic way then in the chemical one. The chemistry is basically the electromagnetic interaction between the reactants, changing their electrons, without any changes in their atomic nuclei. The biology is simple, no nuclear reaction has to be taken into account. This means from the four natural forces (weak nuclear, strong nuclear, electromagnetic, and gravitational) only the electromagnetic force has relevance, irrespective of whether it is transmitted in a chemical, biological, or physical way. The physical (purely electromagnetic) way is of course the "cleanest," but it is also very general, interacting with every material on its way to the effect, and this is the complication.

Clever bioelectromagnetic selection chooses only the actual bioelectromagnetic reactions to modify, and then the job is solved. Even if we have the properly selected interaction and the effect is dedicated only to the desired job, the electromagnetic qualities (fields, potentials, charges, currents, etc.) have to reach the target. In the case when it is directly focused on the reaction in question, without interacting with other, unwanted species (like in cellular or many in vitro experiments), the situation is simple. However, when the target is a part inside an object, surrounded by many other non-targeted parts, the challenge is to pass through these without interaction,

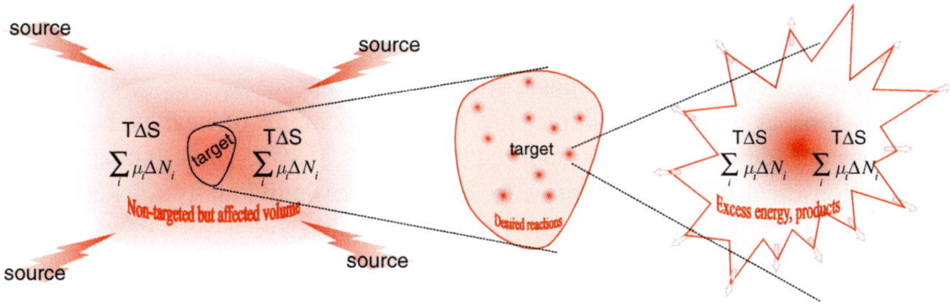

Fig. 3.33 The reactants (electromagnetic or chemical species) aiming for the target through unselected volumes. On their path to the target effects to the structure (and temperature) $[T\Delta S]$ and chemical reactions $[\mu_i N_i]$ could occur in the non-targeted volumes. Finally, the desired reactions are forced in the reached target. However, excess or produced reagents (heat or chemical reactants) produce a further secondary effect as they transport out of the targeted volume

and perform the active reaction only in the target. This is the challenge, which is decisional in categorizing the temperature effects. Should there be considerable interaction with non-targeted parts also, and this depletes the energy irrespective of the target then the temperature arises. Should this only be a low-level "energy loss" then we will only have a "subtle" change in temperature.

This targeting problem is not only a problem for bioelectromagnetic interactions. To follow a path to a target object inside a complex involute system without interaction with non-targeted (desirably untouched materials) parts is a challenge also in chemotherapies, radiotherapies, and all the targeted modes of selection (see Fig. 3.33). This "wasted" part that affects non-targeted objects is a common category in medicine: side effects. In this meaning the temperature increase is a side effect. Temperature is clearly a side effect when the absorbed energy does not reach the target, but its excess production by target reactions themselves is also unwanted. (It is like the ADP $\leftrightarrow$ ATP reactions, when the reaction energy supports the actual cellular reactions. The heat production is normal, maintaining homeostatic conditions for the reactions [body temperature]. However, the produced excess heat (e.g. tumors, inflammations, fever, etc.) could improve the reaction rate in a positive way (it is desirable to accelerate the reaction rate), or in a negative only waste of energy, which we could measure on average, by the temperature).

Heat is a kind of energy, which generally could be characterized by the SAR (integrative approach). Microscopically, this energy is depleted by various mechanisms in the actual heating applications. Heat as a physical quantity is an extensive thermodynamic parameter: the heat energy is proportional to the mass/volume/part of the targeted material. In most real cases we pump heat (energy) into the targeted system to change its chemical bonds and/or reactions. Hyperthermia in principle uses this energy to destroy the malignant cells/tissue, and to reach the definite aim of the treatment.

The temperature certainly characterizes this process differently: it is an intensive (average) thermodynamic parameter; it characterizes the actual state irrespective of its mass/volume/part. It is a parameter for the control of the local equilibrium (homeostatic state), does not depend on the mass or volume, and it has the same value in every subunit. (Note the basic difference from heat.)

Let us look at a simple example: the body temperature of healthy humans is fairly constant (deviation is less than ±1°C, approx. 0.3%), while human energy consumption (heat equivalent) varies in a wide range individually (deviation can be more than 100%). Even the same individual could have a very hectic energy intake depending on complex conditions (deviation can be more than 50%), with no change in body temperature. Without a notable change in the temperature we are able to pump energy (heat), mass, volume, entropy (information), etc. into the body. These processes are characterized by extensive thermodynamic parameters (heat, mass, volume, etc.) identifying the processes quantitatively. Intensive parameters (like temperature, pressure, chemical potential etc.) distinguish the thermodynamic state, describing the actual quality (momentary equilibrium) of the targeted system. Gradients of these intensive parameters are the driving forces of the flow of the extensive ones. The flow of the extensives changes the equilibrium, so redefines the intensives in the system. Let us study a simple steam engine: the applied energy (heat) heats up the system, raises its temperature. However, the engine starts to work only when the water temperature has reached a definite value. From this moment the temperature remains stable, the engine starts to work, and the pumped-in heat is converted to useful mechanical energies, which is the aim of the engine application. If we stop taking away the mechanical energy (block the engine), but the energy intake remains as it was, then the excess energy starts to raise the temperature. The active use of the energy and the rise in temperature of the steam-boiler in these conditions are contradictory.

Most of the debates about "non-thermal" effects are connected with the electromagnetic interactions, which we discuss in Section 3.3.3.

3.3 Bioelectrodynamics

The bioelectrodynamics phenomena are a complex interaction of the external fields and the biosystem. This relatively complicated functionality could be modeled with similar, non-thermal systems, (e.g. mechanical, electrical, etc.), and make the similarity obviously self-explanatory. Running water in a watercourse has a fall between the top and the bottom, allowing the water to run. The driving force is the height-represented potential energy, and the intensity (current) of the flow is the volume of the water at a given time period (see Fig. 3.34).

A similar picture could be drawn for a simple electric circuit: the driving force here instead of being in proportion to height, is the battery potential energy, the voltage (see Fig. 3.35). The water flow is replaced here with the charge flow (electric current).

The case where a creek is divided into two prongs (see Fig. 3.36a) could also be trivially modeled with electrical circuits (see Fig. 3.36b).

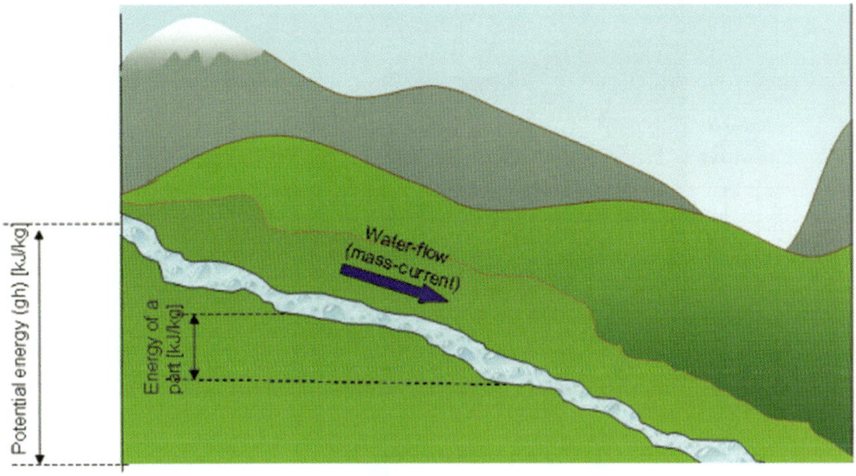

Fig. 3.34 The flow of a watercourse with a driving force of potential energy in relation to height (gravitation)

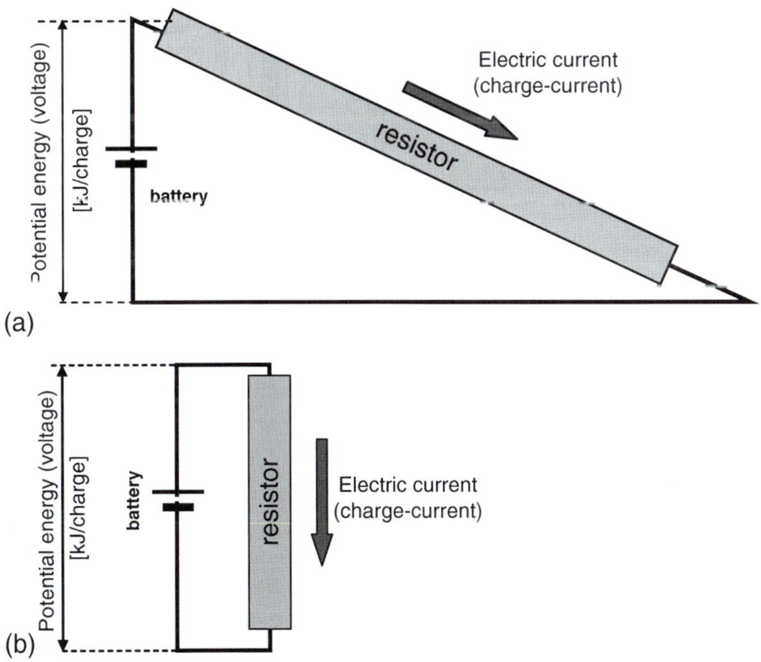

Fig. 3.35 The electric equivalent of the water flow shown in Fig. 3.34 [Drawn in similar geometry (**a**) and in the conventional manner (**b**)]

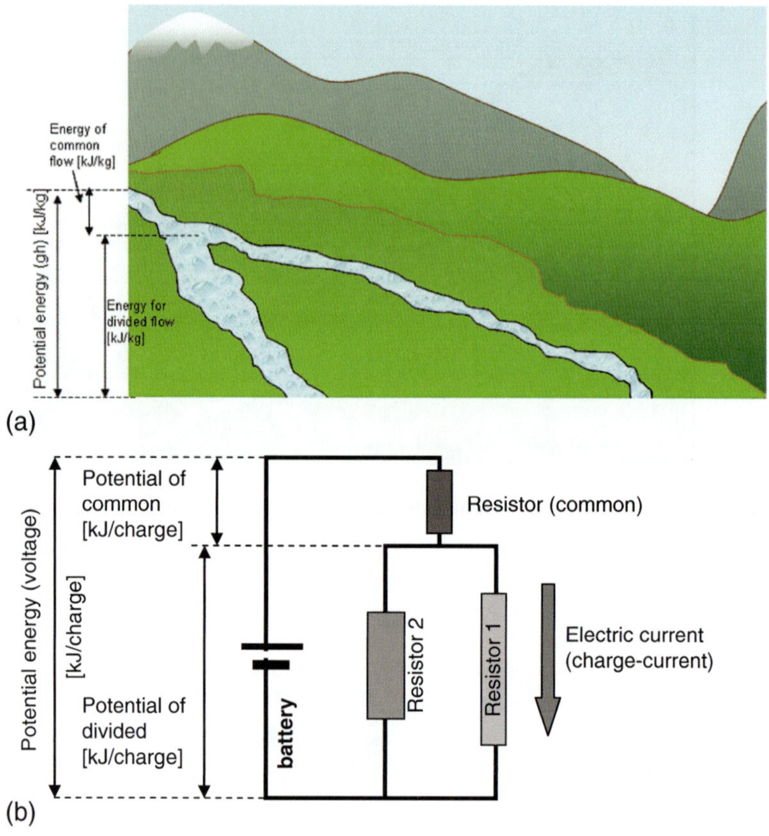

Fig. 3.36 Similarity of the water-flow (**a**) and current-flow (**b**) arrangements

3.3.1 Basic Interactions

Primary sensing in humans is rather complex. There are the basic sensing effects connected to mechanical forces, light effects, heat conduction as well as available recognition of some liquids and gases (taste and smell). All the phenomena around us must be transformed to one of the well-known senses to be recognized. Also the recognition has definite (far from linear) physiological rules, but its conscience appreciation is always adapted so as to be linear: we carry out interpolations and extrapolations of our senses on a linear basis. The electromagnetic phenomenon hurts both rules: we have no direct sensing of the electromagnetic fields, a transforming action is necessary to recognize its presence; and the mechanical action of the electromagnetic forces are not linear with distance.

This nonlinearity (the force between charges depends on the inverse square of the distance) caused many complications for the first modern scientific investigators, i.e. Coulomb and Ampere. Maxwell solved this problem by linearization of the topic, transferred the non-linear space dependence in the newly introduced physical quantities to the electric and magnetic fields. The force became, however, a linear function of the field. This new construction (the fields) had no direct human sensing facility; the fields could be activated (detected) by materials and/or charges, currents. Actually the electromagnetic fields from a thermodynamic point of view have the role of intensives, forcing extensive currents (charge flow) in the biomaterial (see Appendix 6).

3.3.2 The Bioimpedance

The electric circuit contains an energy source (e.g. battery, mains network, generator, etc.) which pushes through the charges on the material, which is connected. The load (where the charges are passing through) could have various properties. The simplest load, is when the material allows the charges through without any rearrangement in the material itself, so the current does not have any frequency dependence, the only change which it causes is the energy delivery so the increase of the temperature of the load. Frequency dependence appears if the charges could be accumulated by a special arrangement (condenser effect) or the charge flow builds up a field to foreshow the movements of the next coming charges, inhibit their free movements. (inductivity effect). Both situations could be constructed with simple materials: a condenser made from isolated conductors, storing the charges by their attractive forces and the induction coil by coiled wire, building up a magnetic field, which suppresses the free current-flow.

There is other, more variable modification of the applied current in the case when the structure of the material, which contains the charges, changes during the loading. The material (it does not matter which phase, solid, liquid, or gas) is modified, restructured by the current (or field) through the material. The energy transfer in this case depends on the structural changes of the conductive material. The most spectacular feature of the complicated energy transfer in this case is the frequency-dependent energy intake. Because the material properties are changing in a short period of time (frequency dependence) the resistivity of it could drastically change by the applied frequency (called frequency dispersion). The resistivity caused by this phenomenon is called impedance. Biomatter is of course more complicated than a simple resistance, it has impedance. Bioimpedance is a very complex value, because the biomaterial is complexly structured and has no homogeneously unified conformation. The conductivity is mainly made in electrolytes, which are "encapsulated" by various membranes. The membranes are lipid compounds having good electric isolation, construction capacitance in the biosystem. So the bioimpedance behaves like a (nonperfect) capacitor. Nonperfect means, the isolating dielectric material between the electrodes is a non-perfect isolator, it has conductivity as well (see Fig. 3.37).

The complex structure of various electrolytes and their isolating layers (membranes and other wall structures like vessels), make the phenomena inhomogeneous

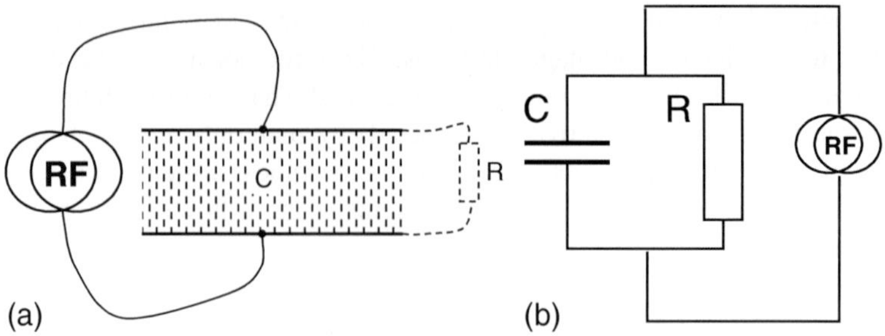

Fig. 3.37 Condenser with non-perfect dielectric material: (**a**) sketch, (**b**) schematics

and complicated. The currents flowing through the system could be real moving "free" charge, an energy-bag, but also could be sliding through the system by the transmission of the forces between neighbors through the full target. This constructs frequency-dependent resistivity of the currents, which is called impedance (see Appendix 7).

Measuring the contact bioimpedance (below the usual radiation [<100 MHz] regime) we have to distinguish two different conditions:

- when we are in direct contact with the measured electrolytes of the living systems;
- when the measurement is made outside of the system having no direct contact with the electrolytes.

Both methods could apply a pair of electrodes or multiple (mostly quadruple) ones. The first category is applied in laboratory experiments in silicon or in vitro, and sometimes invasively in vivo as well. The non-invasive measurements are in the second category, when the electrodes are capacitive (rarely inductively) coupled to the measurable system, having various isolating layers between the electrodes and the measured objects. The largest differences, however, between the methods are in their targets:

- they are simple biomaterial (biologically uniform object, like cell- or tissue-cultures in vitro, prepared tissue-part in vivo, etc.), having only two [extra- and intra-cellular] electrolytes together with their separating membranes, or
- they are definitely complex systems, multielectrolytes with multiple separation layers (including the skin in most cases) are involved in the investigated object. [The EIT (Electro-Impedance Tomography), and in most cases the BIA (Bio-Impedance Analysis) are characteristically in this category.].

The methods operate via differences in the frequency dispersion of various compartments in the target. Depending on the frequency, there are different mechanisms in the systems for dispersions, denoted by the first letters of the Greek

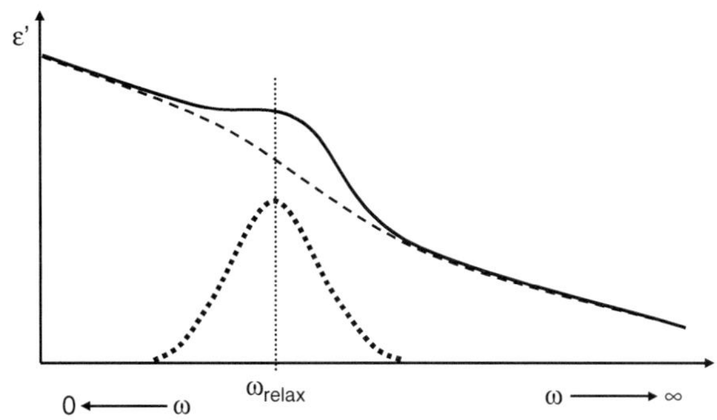

Fig. 3.38 An additional dispersion peak creates a bump in the overall curve of the dielectric constant

alphabet (α, β, γ). The α-dispersion [732] starts in the very low frequency region (as low as mHz) up to the kHz region. It is generated by various effects like a near-membrane counter-ion interaction, active cell-membrane changes, voltage-gated ionic channels, ionic diffusion, and dielectric losses. The β-dispersion [733] (Maxwell–Wagner effect) is between 0.1 and 100 MHz (characteristically membrane capacities of cell and intra-cellular organelles, molecular responses, bound-water response, etc.) while the γ-dispersion is in the range of 0.1–100 GHz [734], originating from the dipolar mechanisms in polar electrolyte, small molecules in aqueous solution, or the water relaxation itself. The bound water to the membrane has the upper part of the β-dispersion, denoted by δ [735]. Generally the dielectric permittivity decreases rigorously monotonically by the frequency (see Fig. A.7.1. and (A.7.4)), however the various dispersions (like additional peaks, see Fig. 3.38.) produce smaller or larger humps on the generally smooth curve.

The β-dispersion is involved in the most complex biointeractions, emphasizing not only the properties of the membranes, but how they fit to the electrolytes on both sides. The treatments also have to be chosen in the frequency range of the β-dispersion, expecting most of the changes in the complex system [736]. The well-chosen frequency of investigation could depend on the number of cells in the target, their structure (individual and collective), their connections (junctions, adherent connections), and the properties of the cellular membrane. Notably, as the electrolyte and membrane properties differ between malignant and healthy tissue [737–739], this provides the possibility of diagnosis but also could be a selective factor in active treatments. Practically the low frequency limit is chosen as 20 kHz, so as to not disturb the other sensitive bioelectric measurements (like ECG, EEG, EMG, etc.) and to avoid any muscle or nerve stimuli [740]. The above-described three-component electric circuit is generally used for modeling the β-dispersion [741].

The main selection factor for tumor heating by RF-electric current is their complex impedance. This could be measured by a complex physical value, by the bioimpedance. The impedance is a resistivity-like parameter that depends on the frequency (dispersion) and also can be used to characterize the biomatter by a simple electrical measurement technique [742]. By the impedance distribution we have a full picture of the tissue's specific electric field distribution as well. As early as 1940 both the whole-body electrolyte status [743] and the local changes (ECG) [744] were studied by the method. Nowadays the bioimpedance is a widely investigated [745, 746] and extensively applied diagnostic method to select tissues on their impedance differences for numerous diseases [747–754]. Guidelines and comprehensive reviews have been published about the bioelectric impedance analysis method [755] and clinical applications [756] summarizing the available results for lung cancer, prostate cancer, renal diseases, hemo-dialysis, peritoneal dialysis, cirrhosis, malignant melanomas, breast cancer, HIV, body electrolytes, obesity check, malnutrition, gastrointestinal disorders, and cancer of mixed origin. The method has been approved for a long time by the FDA for breast diagnostics [757].

In a simple theoretical investigation [758] an elliptical "tumor" was introduced into an otherwise homogeneous body. The appropriate Green's function is able to identify the changes in conductivity between the tumor and the surrounding region. Precise diagnostics have been established by careful calculation of electric impedance of the human thorax also [759], and the 3-D electrical impedance tomography is intensively studied [760].

The increase of the current density in the tumor could be visualized by the RF-CDI (radiofrequency current density image), which is an MRI-conducted and well-controlled measurement of real processes [761–764].

The question naturally arises: why does the impedance difference between the tumor and its healthy counterpart exist?

The tumor has intensive metabolism. This process increases the ion transport and the ion concentration in the neighborhood of the malignant cell, and produces higher ionic concentration in the extra-cellular electrolyte in its vicinity compared to its healthy counterpart. The increased ionic concentration means higher conductivity [765, 742] or in other words lower impedance. It could be used to distinguish between healthy and malignant situations [766]. In consequence the RF-current (with specially chosen frequency and modulation) will self-selectively flow toward the malignant cells. The higher current density in the tumor is one of the characteristics of oncothermia [767] as well (see Section 4.1.10). The selectively absorbed energy increases the temperature in the tumor quicker than in the healthy environment. The increasing temperature creates a positive feedback by its natural increasing effect on the conductivity [768], which increases the conductivity of that volume.

Also the increase in blood perfusion through the increasing temperature will lower the impedance (increase the conductance) [237, 241], which is an additional further, positive feedback selectivity.

The prompt necrotic effect trivially changes the impedance in the extra-cellular matrix. During the capacitive coupling of oncothermia most of the RF-current flows in the extra-cellular electrolyte, the cells are electronically capsulated (shielded) by their membrane by more than one-million V/m field strength. When the membrane is damaged, the till isolated electrolyte would be "liberated" and becomes a part of the conductive process. The conductivity drastically changes [768] by this process (Before this point, the trend of the impedance was opposite: it increases because the cells swell and the conductive extra-cellular volume shrinks [768]). The cellular volume and the forming edema are also measurable in situ [769]. The destructive phase of hyperthermia is well measurable, the necrotic and even the mitochondrial damage, as well as the cellular histolysis is observable [768]. The process of the observed characteristic cellular response includes: cellular swelling, progressive membrane damage, cellular shrinkage and subsequent progressive histolysis. During the measurements, the histological changes [770] and the induced coagulate necrosis in human xenografts [771] are reflected. A special self-similar structure, the hierarchical circuit is also assumed [772], which well agrees with the fractal physiology approaches and the connected dynamism (noises) [773].

Some measured electric impedance changes are:

- In advanced cases the blood perfusion increases by neo-angiogenesis [774]. This extra visualization lowers the impedance (increases the conductivity), which is again a selectivity factor.
- The malignant tissue water content is higher than their healthy counterpart. The significantly larger permittivity and conductivity in tumor tissue in vitro is explained on this basis [775].
- Also the proliferating cells control their cell volume by their water content, in the malignant growth [776], and this effect increases the conductivity in the given tissue, too.
- The malignancy has a special fractal structure, which could be identified by impedance measurements on Erlich solid tumors [777]. This structure (due to its definite percolative self-similarity) is a better conductor [778] than the non-fractal healthy tissue.
- The malignant "autonomy" blocks the cellular connections. This develops a characteristic isolation of the malignant cells from each other [779, 780] lowers the impedance as well. With this mechanism the extra-cellular pathway is free between the cells, definitely increasing the conductivity of the extra-cellular electrolyte.
- Furthermore, a decrease of epithelial barrier function (tight-junction permeability changes) has a role in the development of colon tumors, which could be measured by electric impedance measurement [781].
- The impedance measures selectively: differentiates between the cancerous and healthy tissue, and is able to distinguish the extra- and intra-cellular electrolyte as well.

It is clinically proven, that:

- The cancerous and healthy tissue of hepatic tumors is significantly different [782].
- The VX-2 carcinoma can also be measured [783]: rabbit liver at low frequency in vivo had a conductivity 6–7.5-times higher, permittivity 2–5-times lower than in healthy liver (600%), for the 10-MHz region (200%).
- Well applicable is the impedance method for breast-cancer biopsies [784] and for breast-cancer diagnostics and prophylactics [785, 786].

- Advanced lung cancer could be followed by impedance observations [787, 788]
- It has been successfully tried in skin-cancer diagnosis [789].
- Commercially available devices for impedance tomography for mammography use the method [790].
- Impedance spectroscopy is effective in early tumor stages [791] and single-cell characteristics [792] too. The method is cell selective [793].
- It is temperature-dependent and the hyperthermia can be monitored [794]. Arrhenius activation energy is also measured by impedance [795].
- Numerous important physiologic parameters, like apoptosis [796], and ischemia [797–799], can also be followed by monitoring the bioimpedance.
- The special membrane effects (rectification on membranes, [800]), are also factors of the cellular selectivity of cancer by RF electric current.
- Impedance measurement is useful for the control of other treatment modalities. It adequately measures the distortion made by irradiation [801], as well as, the drug effect can also be controlled [802]. Common practices, like following wound healing, are also objectively traceable [803]. The bioimpedance vector pattern could distinguish between cancer patients without disease versus locally advanced or disseminated disease [788].
- The method could be used for the control of tissue cultures [804]; causing significant impedance change by thickness and homogeneity of the culture growth. Combining optical and electrical impedance techniques for quantitative measurement of confluence in MDCK-I cell cultures [805]. An interesting tissue and cell-culture characterization method has been established, the ECIS Method, masterminded and chaired by a Nobel-laureate [806]; and it is widely used by the Pharma-industry. Also the electric impedance could be sensitive for drug reactions [802]. The dielectric loss changed by chemo-infiltration. This process also could be a helping factor in vivo, when the chemo-enriched tumor has more ions, increasing the selectivity by oncothermia.
- In some hyperthermia cases the impedance measurement becomes the control for treatment quality. It is widely applied for the RF-ablation/interstitial technique, without any control of the temperature [807, 808].
- Nowadays, the largest commercial application of the impedance method is in cosmetics. It can selectively and systemically measure the electrolytes and fatty tissues [809] in the whole body [810, 811]. The volume of extra-cellular electrolyte can be determined [812].
- Magnetic resonance electro-impedance tomography (MREIT) shows spectacular data on the impedance selectivity. The structural (MRI) and MREIT images are shown as corresponding [813], and so the theoretical calculations and selectivity assumptions are proven [814–816, 762].
- One of the best selections could be attained in the breast [742]. Breast tissue is in any case widely investigated and reviewed [765]. The impedance tomography that is used for mammography purposes became a commercial tool to allow breast control with a cheap and harmless procedure [742].
- Other comparative studies were provided for malignant tissues, however the results are not identical; the measurements very much depend on the conditions.

(This is a trivial consequence of many factors, which I have listed above.) However, all the studies measured lower impedance in the tumor than in their healthy counterpart in all of the tissues and staging of the tumor, in vitro and in vivo as well. The smallest impedance difference was measured for kidney (4–6%) [817], but other measurements had basically different results for renal tumors [818]: 1.43±0.39 S/m, (while 0.73±0.77 S/m for renal healthy ($p < 0.05$) [96% increase]. The largest impedance difference was measured in breast: up to 40-times lower in the breast tumor than in the healthy environment [819]. It could also be possible to distinguish between living and necrotic malignancy by impedance tomography [782]; both the conductivity and permittivity are higher in malignant liver, but the frequency dependence of necrotic tissue differs.

- Significant separation of malignant and benign tumors was successfully applied to the skin [820]. This application was extended by temporal resolution of the skin impedance measurement in the frequency domain method [821]. In any case, noise analyses of electric currents become a new tool in the field. Cell motility was probed by noise analysis of thickness shear mode resonators [822], and adsorption and fluctuations of giant liposomes were studied by electrochemical impedance measurements [823]. Electrical impedance scanning was applied for lymph-node evaluation in children [824] and has also been successfully introduced.
- The method could be applied for prognosis. Phase-angle investigations of electric impedance [825], give a good indicator of cellular health and integrity. In this manner the phase angle is a prognostic indicator; having low phase angle were living significantly shorter ($p = 0.003$), independent from the tumor stage. (Below 5.17 the median survival was 10.1 months, at angle 5.17–6.19 22.6 months, and above 6.19 was 25.6 months.)
- The pre-malignant check by non-invasive (virtual) biopsy using the electro-impedance method is also an option. For example in case of modular basal cell carcinoma, significant impedance change was observed [826]). Virtual non-invasive biopsy is applied for Barrett's esophagus [827], and for other non-invasive tests reducing the need for real biopsies [828].
- Numerous surfaces and other inhomogeneities make the system complicated to describe. However, the penetration depth of the electromagnetic fields into the biomaterial has many common (see Appendix 8).

3.3.3 "Non-Thermal Effects" – The Electrodynamic Approach

Remembering Section 3.2.6, the NTD effect does not mean it is nonthermal. Most of these effects are thermally induced, or at least competing with the thermal noise in the object. All the factors which change the internal energy of the system are thermal, but not all internal energy change has an influence on the temperature. For example: a certain structure at a certain temperature could be rearranged (for example mechanically) to another structure, without any temperature change. This is, however, a change in the entropy and so a change in the internal energy.

The diffusion constrained by the concentration gradients (inhomogeneities) could change the internal energy of the system by NTD conditions. When the system rearranges its structure or chemical bonds (e.g. in a cell [829]), and also the electrochemical reactions are NTD [830]. However, the fixed temperature, the actual conditions, and the environmental energy exchange are part of the thermal dependence. It would not be correct to say that the melting of ice at zero temperature to water of the same temperature is nonthermal. This process needs a large energy input, which could be provided by different sources like heat or mechanical energy, as is shown in Fig. 3.10. The melting process is thermal, but NTD. The only point is: the observed effects are the consequence only of the increased average energy of the system, or it has effects that are directly modifying the processes, structures, actions in the system, with the terms of intensive–extensive pairs in internal energy [see Eqs. (3.31) and (3.35)], independent of the overall temperature. Even the change of behaviors of monkeys was measured and published in *Nature* using electromagnetic radiation with as low a frequency as 7 Hz, and as low a field strength as 7 V/m [831].

The temperature-dependent effects are definitely macro-scale processes, and of course these are irrelevant for the microscopic changes, however the electromagnetic interactions are microeffects. Again and again we have to emphasize: the NTD effects certainly could be part of thermodynamics (of course the non-equilibrium one) and by far it does not mean that these are nonthermal or "athermal," they only have no macroscopically measurable temperature change at all.

Logically we expect changes in many terms of the internal energy, regarding the fact that biological objects are more chemical than thermodynamic systems. Its chemical energy exchange, however, can be described in the frame of non-equilibrium thermodynamics, considering non-linear (phase-transition like) effects and definite chemical reactions. It is trivial, that living matter needs nutrition from the environment, and it would be by far not enough to have equivalent energy simply by heat exchange. It was clearly formulated by Adey in 1993 [832]: "There is increasing evidence that these events relate to quantum states and resonant responses in biomolecular systems, and not to equilibrium thermodynamics associated with thermal energy exchanges and tissue heating."

From the discovery of electromagnetism, the electric and magnetic field was one of the most controversial topics in biology. Electromagnetism in biology is difficult not only because of the complexity of the biosystems, but because of problems of measurement and realistic documentation. The effect of external electromagnetic fields on bioprocesses is the topic of a few "miracle" experiments and of course the target of many unscientific approaches too. The "traditional" controversial "battle" involved bioelectromagnetism, the effect of electromagnetic fields on living objects. A new field of discussions that has given rise to much controversy is connected with some hypothetical effects and possible environmental dangers. Debates on "electrosmog" [833], energy medicine" [834], "new biophysical field" [835], "force-free actions" [836], "scalar-wave effects" [837], "subtle energies" [838] are all creating upheaval around "silent" bioelectromagnetics, having many serious opposing opinions from the physical [839, 840] and even from the purely mathematical point of view [841, 842].

The newest publication [843] emphases: "... EMF safety limits must be changed from the current thermal standard, based on energy, to one based on biological responses that occur long before the threshold for thermal changes." This is one of the clues. The thermal (meaning temperature-dependent) changes of course are important and build up an effect on the biological tissues, when the electric energy selectively acts and produces changes much below the thermal limits. The field effect could cover numerous "non-thermal" (meaning NTD) changes in the cells by extra-cellular stimuli [844]; including pearl-chain deformation, change of orientation, movement or rotation of the cells or its compartments, shape deformation, destruction of the cell or its parts, fusion of cells, vesicle ejection from the cell, blip formation on the cell membrane, protoplasmic streaming, intra-cellular flows, and all are definitely declared as NTD [845]. When the order strengthens through the electromagnetic interactions, it is trivially NTD; because if the higher energy absorption increases the temperature, that has to disorient the structures and not vice versa. However, there are definite observations on structure reorganization by ordering through applied electromagnetic fields [846, 847].

The membrane potential and membrane transport are in any case the center of the NTD effects, because of the β-dispersion. Studies have investigated electric field-driven effects on proton flow [848], on the Na/K ion pump [849], and on the changes of messengers [850]. The trans-membrane potential also could be changed by an external electric field. The Schwan equation of electric field effect on the trans-membrane potential [851] based on the dispersion relation does not contain temperature, only the field acts. Its validity is experimentally proven [852]. The static picture could be modified by the correlation between the amplitude of plasma-membrane fluctuation (fluctuates the C_{memb}) and the applied electromagnetic effects [853].

Cell destruction by ultrasound in the frame of NTD was studied [854, 855], and there was a controversial debate on the non-thermal effect of the 430-MHz treatment in clinical practice in Australia [856, 176]. However, a spectacular experiment was presented, where the same temperature was reached by conventional and microwave heating, and the reaction was significantly different [857]. It is of course not easy to separate the temperature-dependent and NTD effects [858], but the NTD effect no doubt exists, and other laboratory experiments have supported it [859–861]. A large series was published about NTD electric-field mechanisms on DNA transfection ([862–866]). The effect of electromagnetic fields on DNA was studied as a charge distribution effect [867], and tried to explain the molecular mechanism of ion pumps influenced by external electric fields [868, 869], which later led to a controversial debate [870, 871]. Anyway, a wide range of publications have reviewed bioelectromagnetic effects in morphogenesis [872], trying to explain many of the electromagnetic connections in the development of the human embryo as well as in the development of cancer.

Many observations were made on various electromagnetic effects without mentionable temperature changes. Results showed how the electric field promotes cellular fusion at low [873] and high [874] frequencies, a field-strength-dependent hemolytic effect which was observed under RF exposure [875], activation of ion channels at the cellular membrane [876], a membrane-mediated Ca^{2+} signaling affect on the immune system [877], and trans-membrane Ca^{2+} induced by alternating current (AC) (low-frequency electromagnetic fields [LFEMF]) [878]. Anyway, the biological effects of LFEMF have raised significant interest and debate in the past. Numerous reviews [879, 880] and articles report on the response of biological matter to LFEMF [881–888]. However, these observations proved controversial [840] and the explanations describing these observations were often found to be inconsistent.

Electrically controlled cell properties and its effects on drugs was investigated [889]; furthermore the NTD effect are obvious in the antibiotic action under a 10-MHz RF current [890], as well as in germination [891]. This latest is even more sticking in the experiments with vector potentials [892], which effect definitely has no temperature increase in the system.

The absorbed bound water on the membrane has an important role in relation to SAR and electric field distribution [893]. At low frequencies the membrane impedance is extremely high, so the transversal-incident electric field is huge. The large β-dispersion in the bounded water on the outer membrane (δ-dispersion [735]), was measured at high frequency (2.45 GHz microwave) [894]. This has considerable SAR, which could destroy the water structure. The disordered water is one of the important factors for cellular isolation [715], and so could lead to the autonomy of the cells, supporting possible malignant transformation. The bounded water, however, could give δ-dispersion starting much lower than the GHz region [895], and could be divided into subcategories in the range of 1 MHz–10 GHz [896]. A unified model for impedance, and induced trans-membrane potential was elaborated [897]. This model uses electronic circuit schematics where all the elements (resistors and capacitors) are calculated as nonperfect components, representing them with complex properties. The numerical evaluation of the model confirms well the characteristics of the various measured data around 10 MHz.

The electromagnetic field directly affects large, complex molecules like proteins and DNA [867]. It was shown much earlier [898] that the exposed surface charge-density of the protein exposed by water remains constant irrespective of the charge or the size of the area. This leads to the conclusion, that the extra area enlarging the water-exposed surface by unit charge is inversely proportional to the charge density [867]. The charge density is increased by the gradient of the permittivity, which is large in the case of molecules with high dielectric permittivity. This important effect shows well the NTD reaction of large molecules to an external electric field. It is clearly shown however, that an 80-μT magnetic field and 43°C heating (both for 20-min duration), cause equal gene expression in HeLa cells, (magnetic field is NTD), but their combined application has synergy [899].

Development of thermotolerance is also partly NTD. There are a set of chaperone proteins (stress proteins), which are called for historic reasons heat-shock proteins (HSPs). HSPs [300], could be produced by various stresses, including chemical [900, 901], mechanical [902], electrical [903], and not only as a heat response. These relatively small proteins however, need energy to be formed, and naturally use the energy resources of the cell. Of course the HSPs represent one of the central problems of hyperthermia theory and practice [904]. The energy for HSP creation could be sourced from a previously accumulated and stored energy or could originate from external sources. The presence of a large amount of HSPs limits the success of hyperthermia [288].

The literature relating to HSPs is diverse and not clear in all aspects. The main consensus covers the different categories of HSPs. There are HSPs that recognize a stress and block aggregates forming in the cell. These activate immediately after the stress. These HSPs have an apoptotic effect also. Other, much later activated HSPs repair the consequences of the stress. The HSP is able to repair the membrane proteins as well, but they are not only involved in molecular reparations, but take part in global control as well, controlling the fluidity and the micro-domain structure of the membrane. They are able to repair or at least to compensate for the changes made by the actual stress, and in that way switch off the production of new HSP molecules. If massive stress associates too much HSP to the membrane and they change the membrane function considerably, than spontaneous apoptosis occurs. The HSPs could penetrate through the membrane, and interfere with the immune system in the extracellular electrolyte. The membrane localization of HSP70 [905] promotes apoptosis [293], and has a very important role (more than chaperone) in the membrane "fluid' to keep it functional [906]. In terms of tumor-specific membrane localization of HSP70 [907], it is mainly localized in the cholesterol-rich microdomains of the membrane [908] and well activates the NK-cells in immune response [909, 563]. Overexpression of HSP70 reduces the dysfunction caused by ischemia-reperfusion in myocardial ischemia. It could have an important role in ischemia reperfusion. A broad-band (0.2–20 MHz) electromagnetic field increased HSP70 expression and so gave some degree of ischemic protection [910]. To produce the same increase of HSP70 expression by temperature, the perturbation would have been 14 orders of magnitude greater [911]! This is the great advantage of the NTD effect. The role of extra-cellular HSP is a topic of increasing interest in relation to the overall immune reactions of biosystems [912, 913].

The application of the NTD electric field has been the focus of a wide range of research. For example, the electric field is recognized as an active factor for endocytosis [914], the electric field could regulate and control biosynthesis [915], and it was extendedly studied in articular cartilage research [916–918].

Electromagnetic NTD treatment of inflammation [919, 920], or pain management by magnetic fields [921], has become more and more accepted in the medical community.

3.3.4 "Non-Thermal Effects" – Approach of Electric Currents

One of the important and controversial phenomena in biophysics is the actual biocurrents. These are spontaneous biological charge transfers having an important role, which are hypothesized and supported by some observations [922, 923].

The concept of "biologically closed electric circuits" (BCEC) was developed by B. Nordenstrom, Karolinska Institute, Sweden [924, 925], alongside the introduction of a treatment method [926, 927]. These types of methods apply electric field to generate currents without an increase in temperature (using less than 5-W power). It has been found to be effective against cancer [928–930], by using galvanic (DC) current applications. (see Appendix 9).

Control of cells by electrical manipulations without any temperature increase is well proven [931], so the NTD effects are generally accepted when the field is static, or the current is constant [932, 933]. Electrochemotherapy (ECT) has been applied for a long time in cancer treatment [934, 935]. The first commercial unit of the galvanic application for tumor treatment was presented to the medical community as a precedent to oncothermia in 1992 [936]. The results were amazing, and it was soon well accepted in Japan and China [937–939] with results reported in several peer-reviewed journals [940–944]. However, the method was invasive, and a non-invasive safe method was demanded.

Special biocurrents, called injury currents, exist [945]. These physiological currents are induced by wounds/injuries, and their typical value is around 100 $\mu A/cm^2$. The physiological potential gradient drop is approx. 100 mV/cm and extended over a few mm distance from the wound [931]. It works as "electrotaxis" in a way similar to chemotaxis. Chemotaxis increases the density of ligands by the gradient of the available chemicals, while electrotaxis increases the density of the membrane receptors by the potential gradient (electric field). This very weak power (approx. 0.01 mW/g) definitely does not cause any temperature increase; it is a typical NTD method [946]. It is measured using high-tech methods at the wound-healing process [947–949]. The electric field in the tissue is oriented to the wound area, and the current has a closed loop (similar to Nordenstrom's BCEC [924]) through the surface of the epithelium, with the current running from inside to the surface in the wound itself. This electrically controls the woundhealing, and persistently acts until the wound no longer exists. It regulates the orientation and the frequency of cell division [950], and directs the cell migration to heal [951]. The edematic processes also have an important role in this mechanism (see Appendix 10).

The injury current has a role in cancer development. The malignant cells are more negative on their surface than their healthy counterparts, and their membrane potential is markedly lower [952, 953].

There are considerations that the malignant tissue has a certain potential gradient to its healthy neighborhood [924, 934], which acts to promote and direct cancer-cell migration [954]. There are telling arguments in relation to the cancerous process being a wound repair [955]. The biosystem falsely recognizes the tumor as a wound, and stimulates its environment to heal the irregularity (meaning to produce cells to heal). This wound-healing mechanism is actively supported by the actual injury currents due to the potential gradients.

The dielectric polarization depends on the gradients of the permittivity and of the natural logarithm of conductivity [Eq. (A.7.15)]. Because of to the weak dependence of conductivity, let us consider the permittivity gradient only:

$$\bar{E}.\Delta = \rho \tag{3.36}$$

where Δ is the gradient of permittivity [$\Delta = \mathrm{grad}(\varepsilon)$], E is the electric field, and ρ is the density of charge in the target. For the easiest geometry let us study a skin cancer (surface tumor), having a disk shape with full cylindrical symmetry (see Fig. 3.39). We know its higher permittivity and negative charge compared to its healthy neighbors, so the assumption shown in Fig. 3.40, could be a realistic distribution of the physical parameters.

The geometry involves the maximal value of the space charge, the $\boldsymbol{E}$ and $\Delta = \mathrm{grad}(\varepsilon)$ are parallel enough to calculate the product of their absolute values. The permittivity gradient is opposite to the field-strength vector, because of the negative charge of cancer. To compensate the negative space charge, the field constrain

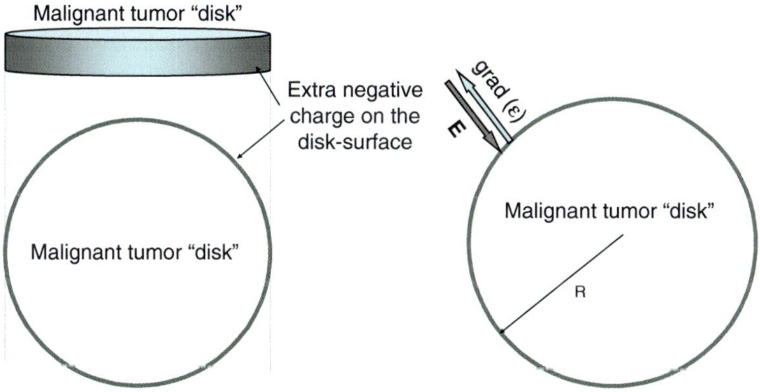

Fig. 3.39 Schematic disk shape of the surface tumor

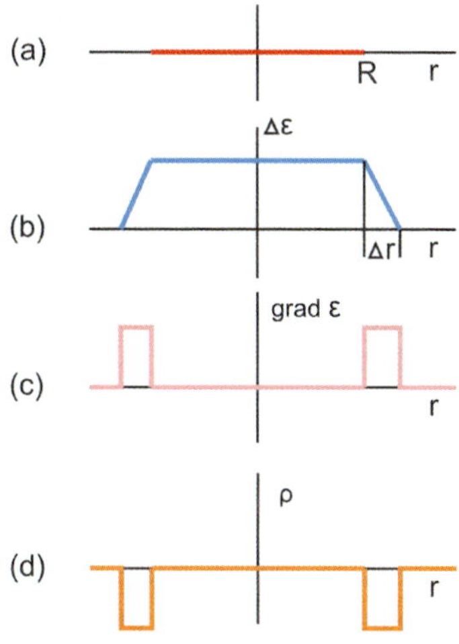

Fig. 3.40 Distribution of important physical parameters at the disk-shaped tumor

electric current to the disk of cancer; initiates an injury current between the cancerous and healthy parts thought their interface. This current could dedifferentiate the healthy cells to multi-potent ones, which become autonomic and redifferentiate to cancerous cells. This mechanism creates the "pre-cancerous cells" measured by Loewenstein [780]. The challenge is the injury current is not able to compensate for the space charge. Because of the high metabolic rate the cancerous cells are in permanent division and produce the negative space charge. Therefore, the natural mechanisms of the biosystem, are not able to block the cancerous development after a definite size, so an artificial intervention to deliver positive charges to compensate could be helpful. For this the necessary electric-field strength has to be a parallel vector with the gradient of the permittivity.

However, the effect of body electrolytes on the electrode surface in the anyway chemically reactive biomaterial, cause the next problem: due to the developing Warburg impedance [956] (Appendix 11). The DC current is quickly compensated by the double layer on the applied electrodes. The boundary between the target and the electrode affects the impedance by building up a polarized layer. These are important phenomena of the electrodes, even if they are indifferent. (Indifferent electrode: the electrode does not dissolve in the concerning electrolyte. This is the case for the well-chosen invasive electrode or the non-invasive surface-touching electrode as well.)

Special NTD current processes are supposed by superconductive currents in biosystems. This typical low-temperature phenomenon was observed at room temperature in biocompounds [957–960], as well as in various bioprocesses [961–964]. A possible explanation of the superconductivity in life processes has been suggested [689] based on the electronically driven frustration in the chain of proteins. The collectivity and the fine balance of the living dynamic equilibrium can be explained by these ideas, similar to the high-temperature superconductivity in metastable materials [965–967], and fullerenes [968]. The thinking behind the interconnection of metastability and superconductivity could be established in a rigorous theoretical way [969]; and also based on philosophical reasons [970].

3.3.5 Membrane Effects

Numerous chemical materials are transported through the membranes. A part of the transported species are ionized compounds (parts of small or large molecules) or simple elementary ions (like H^+, Na^+, or Ca^{2+}), so the mass transport is accompanied by charge transport (electric current) as well. The full transport has a diffusion and drift (forced by electric field of the membrane potential) transport current-density. The membrane permeability also changes, (see Appendix 12), allowing a higher rate of chemical reactions in the system.

Therefore, either the change of the volume or the permeability changes the membrane potential directly. These calculations show well the character of the temperature dependence: the phase transition (as a condition) is thermally initialized (the transition temperature or the other transition conditions must be reached), but the process itself is NDT under those conditions. Furthermore, the phase transition could be the order–disorder transition of the water or other species in the electrolyte, which is basically a structural change, and could be caused by direct electromagnetic effects instead of the temperature.

A periodic external field could change the ion-transport properties through the membrane by an NTD process [971, 972]. NTD membrane disruption is in any case one of the targeted aims [973–975], as well as the pretty high field, aggressive electroporation technique, creating "pores" on the membrane, [976] which is also a typical NTD method.

3.3.6 Stochastic Processes

Living systems are open, dynamic structures, performing random stationary stochastic self-organizing processes [977]. The self-organizing procedure is defined by the spatial-temporal-fractal structure, which is self-similar both in space and time [978]. A special noise (called pink noise, temporal fractal noise) – as a fingerprint of the self-organization [979], – is a typical and general behavior of the living biomaterial [646]. The biosystem is based on cyclic symmetry and has an infinite degree of freedom arranged on self-organizing principles. On this basis a new approach to the living state has been developed: fractal physiology [644, 645]. Instead of deterministic actions stochastic processes exist in the living systems (see Appendix 13), so their description always have random, unpredictable elements.

This power spectrum characterizes the so-called pink (1/f-, or Flicker-) noise. In general, a stationary self-similar stochastic process follows the pink noise if its power spectral density function is proportional to 1/f, like Eq. (A.13.21). Because of the self-similarity and stationary stochastic processes of the biosystems, all of those are a priori pink-noise generators [179, 773], with definite autocorrelations (see Appendix 14).

This remarkable result shows that the autocorrelation function of a pink noise of stationary random processes is similar both as a function of time-shift τ and frequency f. So the autocorrelation of living effects is inversely proportional to the time-shift, characterizing the interdependence of the process events. This is a clear fingerprint of the self-organizing structure of the living processes, which could be proven in generalized cases also [179]. It had been shown [179] that the energy equation under self-similar conditions is enough to generalize pink noise in a given system described by any kind and number of variables. These conditions make possible to understand the common pink-noise behavior of living objects, where self-similarity and stationary random stochastic processes together with their self-organizing dynamism are ready presented.

One of the origins of the stochastic (probability) behavior of living matter is the intrinsic bifurcation in all the levels of living organization. The basic bifurcation mechanism could be introduced by a simple nonlinearity of the potential wells of

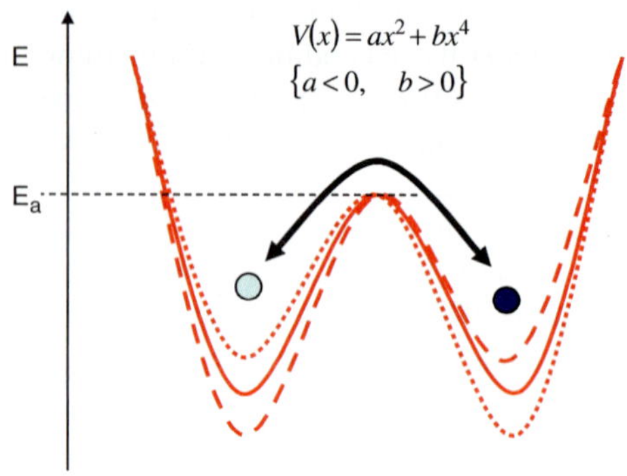

Fig. 3.41 Simple potential structures. Double-well potential, describes a metastable situation. A particle with higher energy than *E*a (*gray dots*) has both possible minima with equal probability, but below this barrier energy one of the wells will have a higher occupation probability. Double-well bifurcation (the standard harmonic force is modified with a positive [opposite] anharmonic term) is the most common in living objects. The *dashed* and *dotted curves* are possible noise-induced changes in the symmetric [$V(x)$] arrangement

chemical reactions [980, 981] showing nonlinear behavior by a double-well potential (non-harmonic potential, chaotic arrangement), see Fig. 3.41. The noise in the system makes the potential wells slightly changing, which modifies the optimal energy situation and constrains the bifurcation. Simply speaking, the bifurcative behavior is similar to exposition of a double-possibility event (like betting by throwing a coin). A multi-possibility (multifurcation) event (like roulette) could be described as a special arrangement of connected double wells (see Fig. 3.42).

The entire living system is organized by an embedded multi-furcation structure starting with the structure of water and finishing with the arrangement of the whole organism and follows a bifurcative mechanism of hydrogen-bridge bifurcation and proton/hydronium-hydroxyl conduction [982–984]. This transport is arranged by equal multi-well potential in water, and produces soliton waves through the system

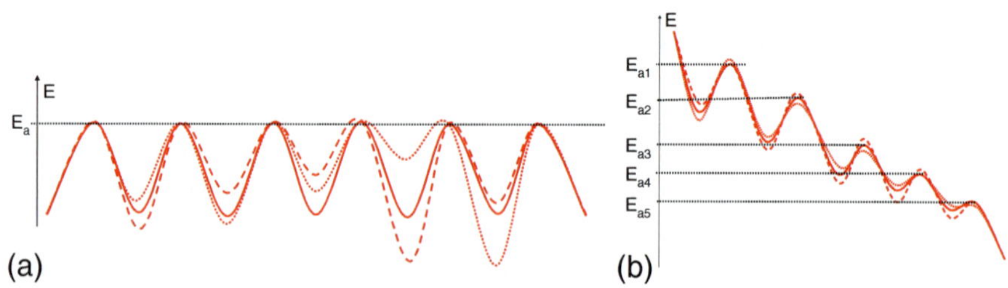

Fig. 3.42 Multi-well potentials. (**a**) Equal multi-well potential (solid line, *roulette rule*), where one well will be occupied while all others are empty. The state probably occupied has energy under *E*a while others have very little probability, decreasing to zero when the actual state is occupied with probability one. Unequal multi-well (dotted and dashed lines) causes unequal probability distribution. (**b**) The cascade multi-well potentials make step-by-step reactions possible, the wells affect the reaction (state occupation) time

[985]. The basic bifurcation of the equal multi-well solution is the hydrogen bridge (double-well potential for a migrating proton). Special proton bifurcation (hydrogen bridges) is observable in protein bending and it even forms the structural backbone of the DNA helix, connecting the nucleotides.

Electron transfer by various reagents is an essential life mechanism at the sub-molecular level. The metabolic processes power this reaction, which is realized in a charge flow making the final oxidation through various cascade steps by different reagents (oxidation). The step-by-step oxidation process transfers definite electric charge in each step forming a special charge transfer mechanism. Collectivity appears in the conduction chain of information between individual protein molecules.

A stable compact protein molecule is nonconducting, inert. Taking part in living processes however, the protein state has been destabilized by mutual interactions. Interactions ionize the large molecules by missing or excess charges using exchange with their neighbors. The molecules try to reach their energy minimum (inert, stable state) again, seeking to fill up their shells to an electronically closed, stable situation. The non-saturated (or "oversaturated") macromolecules seek to reach their stable, saturated form. In the living system, the local (seeking molecular stability) and global (establishing collective mode stability) energy requirements compete with each other resulting in a bifurcation at the protein molecular level. The momentarily optimal short-range order cannot be frozen in, because its neighborhood becomes unstable by actually stabilizing the given molecule. However, the non-saturated proteins fit well to long-range requirements due to their average state, which is able to transfer the electrons over a long distance. This process can occur only under aqueous conditions, explaining why the dry and wet proteins behave so differently [698]. The transport is a long-range process, the system seeks to satisfy this, but in the short-range it is not optimal in energy, so the molecules are seeking the saturated situation. This local energy condition contradicts the global energy request, forming the intrinsic bifurcation of the system. A "stability-bag" (a so-called "soliton") will be formed, sliding through the material [986, 987]. (The soliton concept is valid in muscle contraction as well [988].) The molecules in the neighborhood of the one that saturates itself will be unstable due to the missing charge. It is in fact a CDW (Charge Density Wave) based on Peierls–Fröhlich instability [989].

The same collective bifurcation-based sliding wave is fixed inherently in the topology of the protein connections. The protein arrangements have no proper space filling [990, 991], their translational repetition can not fill the space without holes and/or overlaps. Their microarrangements varied mainly by five-fold of non-crystalline symmetries [992]. This topology has a double role:

The bifurcation is completely established by this, because the five-fold symmetry in the clusters are the most space-filling ones in short range (lowest energy in the cluster [993]) [994], but they are not able to fulfill the long-range energy minimalization [681], to fill translationally the space (fill it like crystalline materials do).

The incompleteness of the short-range order with the long-range minimal energy protects the living material from being trapped in a low-energy crystalline state, the minimalization of the free energy for the entire system can not be effective. This is a ready situation for geometrical bifurcation (frustration) [995, 965], which are topologically "frustrated."

Bifurcation exists at every level of living organization, including membranes [646], cells [575], and even the organism [996]). The system itself also has bifurcative behavior [689, 965], which is very similar to the high-temperature superconductivity phenomenon [969]. This construction is the basis of the cyclic structure [773]. Because of its self-similarity, every biostructure satisfies the cyclic symmetry criterion.

We have shown [773] that the white-noise-excited linear system with infinite freedom and cyclic symmetry emits pink noise. It works like a special filter creating 1/f noise from the non-correlated white-noise spectrum, which was measured [997]. The infinite freedom of the biomatter is trivial, so the biosystems are pink-noise filters if the excitations are uncorrelated. These conditions give additional support to common pink-noise behavior of living objects in their interactions with environmental (natural and artificial) radiation. The colored noises could have a role in the phase transition on membranes of neurons [998], indicating the role of colored noises not only in normal material transport properties but in information exchange as well.

Environmental electromagnetic exposition (*electrosmog*) is a rapidly growing research fields [999]. Electrosmog in most cases has no correlation in time, so general it is usually a white-noise spectrum. The well functioning living matter filters and correlates the exposed spectrum and turns it into pink-noise radiation.

This remarkable biomatter effect could be important for two reasons:

1. Most of the environmental excitations are uncorrelated, white noise, which is transformed to the pink one by biosystems. This is in general a negentropy production, so the biomatter gains the negentropy in its vicinity as well.
2. It is also well established, that the pink noise can characterize the dynamism of the living structures [1000–1002]. The original pink-noise source has an additional pink-noise gain, which has its origin in the environmental interactions.

3.3.7 Noises and Signals

There is a thermal noise in any of the existing matter. This is the consequence of the actual temperature of the object. Namely, the average thermal energy (see before) is considerable, having the energy of a thermal background photon. In the past, theoretical approximations compared thermal-noise fluctuations of the cell-membrane field strengths to the field strengths induced by LFEMF at the membrane [1003]. The final conclusion of these comparisons was that LFEMF-induced changes were several orders of magnitude lower than thermal-noise-induced fluctuations. Therefore, the authors concluded that thermal noise limits the electromagnetic influences and no biological impact can be expected from LFEMF.

The effective field strength of thermal noise was first calculated by Weaver and Astumian [1004]. The Weaver and Astumian model (W-A model) assumed changes in the field strength is a result of fluctuations of space charges on both sides of the cellular membrane, and further showed a thermal noise limit at low frequencies. Kaune [1005] revisited the W-A model and showed that the field strengths typical of thermal noise converge to zero at low frequencies, therefore, the W-A model does not describe this region appropriately. However, thermal noise in Kaune's model [1005] is assumed to be synchronized (coherent) over the entire cell membrane. This assumption is called the coherence condition. Unfortunately, thermal noise is unlikely to be coherent over a large structure such as a cell, therefore, the calculation that followed is limited to a highly unlikely special case. On the basis of the coherence condition, Kaune set all noise generators to be equipotential by assuming parallel connectivity and the equivalent electrical circuit. As the coherence condition does not hold in the general case, the equipotential assumption also does not hold in the general case.

We generalized the problem and developed a solution [1006]. Our results proved when there is only zero-mode currents present, the limit does not exist. However, at non-zero currents the thermal noise does limit the efficacy of electromagnetic effects in low frequencies [1006]. The zero mode is the action by central symmetry for all individual cells instead of the translational symmetry of the usually applied external field effects (see Fig. 3.43). This is one of the factors making oncothermia active by zero-mode to overcome the thermal-limit problems (see Section 4.1.4).

However, the signal between the cells, the "social signal" [1007] shall propagate between the cells without loss of information. Therefore, the communication assumes an effective filtering. Counting the billions of cells expected to be organized by social signals, the communications system shall be organized in such a way that it shall have significant noise suppression in the certainly noisy environment.

To explain the noiseless communication, we suppose that the role of the individual cells is the same within one aim-oriented cell group (part of an organ). The interactions have to be characterized by some simple modes. This situation can be constructed with cyclical interaction. We suppose that the social inclination coded in the DNS by evolution is able to create a form of motion in which only the noiseless modes are performed. This solves two problems at the same time. The first one is the noiseless communication. The other one is the issue of an alarm signal in a network of higher organization to the organ specialized for carrying out the control (e.g. brain). If the organ works normally then we have noiseless modes. In the living organism the mental state of good state of health belongs to this physiological condition. If a disorder occurs then these modes become noisy and change. Apparently, the physico-chemical processes coming off in the individual cells are identical, however, this is not true for the accompanying communication signals. This change gives information to the central organizer that there is a problem. (This speculation could respond to Schrodinger's enigma: what is the connection between the physiological and consciousness processes [1008].)

The described communication means as well that the individual DNS molecules can not be assigned as a controller, whereas they control in parallel the operation of the organ. This condition provides for the astonishingly high security at the

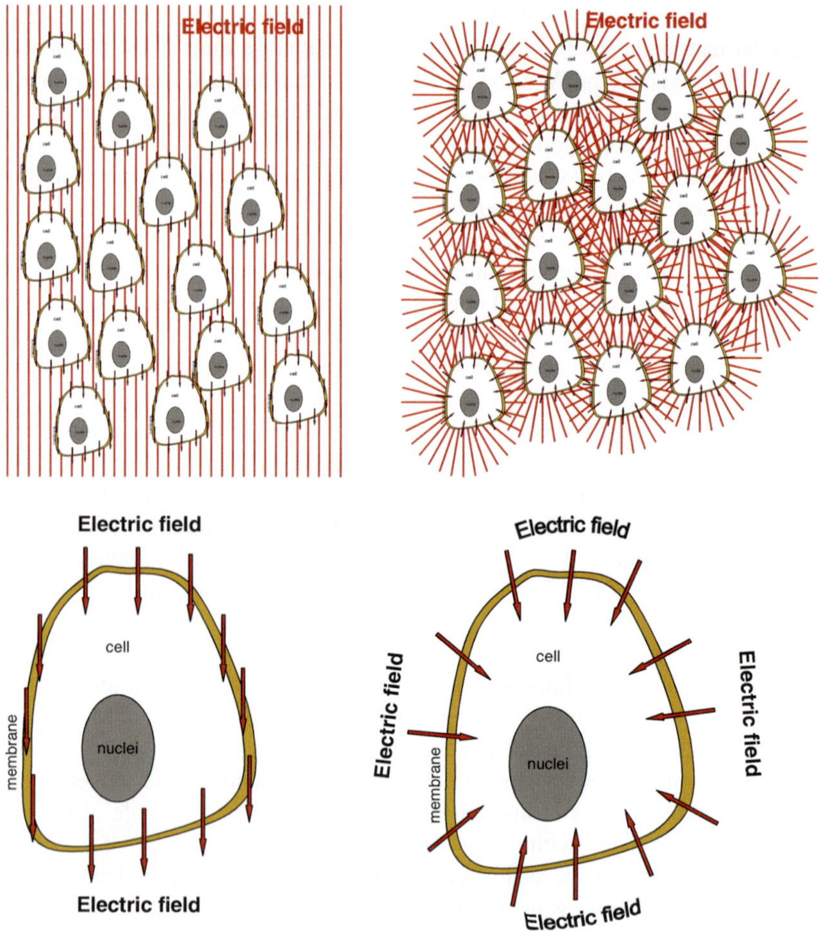

Fig. 3.43 Translational symmetry (**a**) and central symmetry (**b**) of the applied field. Only the central symmetry has zero-mode interaction, the translational one has a noise limit by the thermal energy

operation of the organ. Essentially, this is the Neumann conception [1009] about the realization of a system of high reliability from subsystems of low reliability. The described communication permits that the organ operates in the case of damaged parts as well. Therefore, the DNS has a double role within the collectivized cell: it is a parallel controller and a polymerization pattern for the generation of the new cell.

The above-supposed cyclical interaction symmetry is important in the biological interactions because it allows effective noise suppression. Interaction of cyclical symmetry can be established among the parts of a biological system without a cyclical symmetry in their arrangement in the configuration field (see Appendix 15). This can be achieved by a refined communication network among the parts.

Important experiments were performed [1010] by the patch-clamp (cellular microelectrode) method. The applied individual electrode stimulation was typically a current of central symmetry (zero order). Naturally the resulting (sum) of the currents is zero (the current pumped in is equal to that that goes out and vice versa), but the geometry of stimuli was zero order, so the noise suppression was effective. The membrane potential could be changed (even depolarized) by this stimuli, and which is important, experimentally showed the same effect in the neighbor cells without the direct stimuli. The signal was effectively transmitted. On the basis of the experiments an equivalent electric circuit of a single cell can be constructed as shown in Fig. 3.44. The electric circuit of the connected cells in zero-order info exchange is shown in Fig. 3.45.

The zero order works like an electropermeabilization effect. In the case of any contact with neighboring cells (it could also be a foreign cell as in any part of the immune reactions) the zero-order communication creates a noiseless signal.

However, the malignant cell does not connect to this communication network, and such an electropermeabilization effect does not exist [1010]. This could happen when the potential drop on R_S is low. Indeed, it could be derived from the equivalent electric circuit. The resistance of the membrane of the cancerous cell is higher by several orders of magnitude [1010] than its healthy counterpart. The probable reason for this is that the cholesterol content of the cell membrane is higher [1011] and

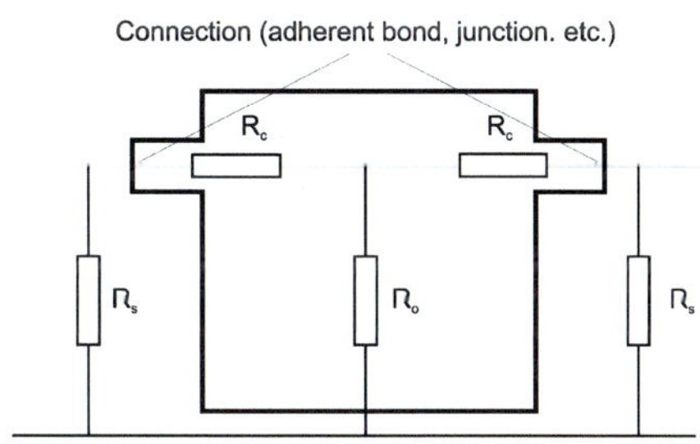

Fig. 3.44 The electric circuit of a single cell. The junctions connect the neighbors, R0 - non-junctional membrane resistance, RC - junctional membrane resistance, RS - non-junctional membrane resistance at the junction

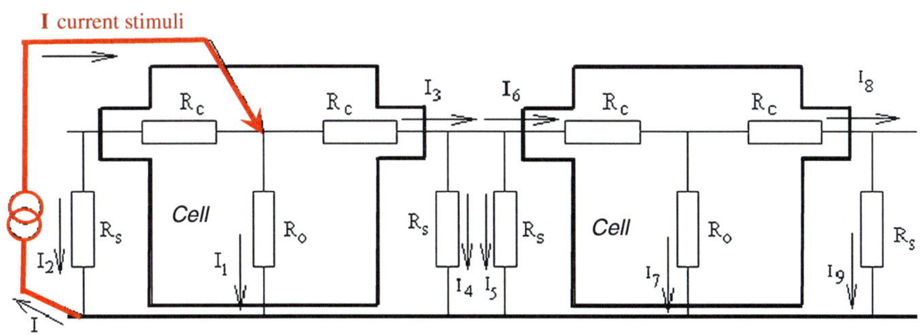

Fig. 3.45 The inter-cellular communication with zero-order current between healthy cells. The cell at the *left-hand side* is stimulated by current *I*, and it is transferred to the next cells by connections (junctions and adherent bonding)

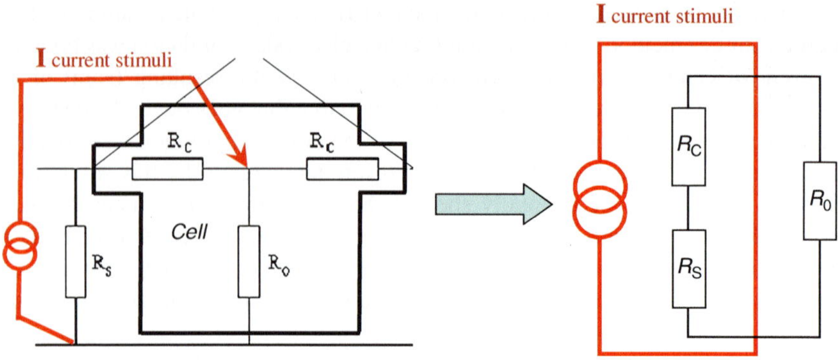

Fig. 3.46 Equivalent circuit of the stimuli on the malignant cell

works as an isolator. Consequently, the membrane of the cancerous cell is noisier than that of the healthy one. We may suppose as well that the distribution of cholesterol is not uniform, therefore the individual branch resistances in the equivalent circuit of Fig. 3.44 are not the same. As a consequence of this the cyclic symmetry breaks up, that is, the conduction matrix will no longer be cyclic. The result of this is: the individual symmetric component modes will be coupled. Therefore, these are symmetrical component currents of non-zero index with a thermal noise limit producing noise potential in the zero sequence networks. Hereby, the field strength component of zero modes will have also a thermal noise limit. The signal-to-noise ratio (SNR) falls off, and possibly the communication with the cell becomes impossible. Cancer appears in this meaning as a communication inability of cells. The measured ratio [1010] $R_S/(R_S + R_C) < 0.002$. However, because of their large membrane resistance it is clear from the equivalent circuit (see Fig. 3.46) that R_O and R_C must be large, so R_S is small, so the potential drop on it is indeed small.

The organization of the cells builds up the multi-cellular living object. This means that groups of cells carry out together certain functions. It is easy to assume that the cells are organized by means of a special network. Obviously, this network shall protrude into the interior of the cell, and has to have adequate connection points outside the cell wall. These connections are made by adherent connections in a chain of trans-membrane proteins connected to the cytoskeleton in the cell interior [1012, 1013]. It was shown [1014], that in accordance with the Froehlich theory this network develops through polymerization. The cytoskeleton micro-filament structure drastically changes by electric field [1015]. The extent of the micro-filament reorganization by electric field was measured on the human hepatoma cell line (Hep3B) as NTD [1015]:

$$F_{\%}(E) = 31.6 \cdot e^{0.22E} \quad \text{and} \quad \ln[F_{\%}(t)] = -0.031 - \frac{270}{t} \tag{3.37}$$

where $F_{\%}(E)$ and $F_{\%}(t)$ are the fraction of cells in % changed by external electric field E [V/cm] and by time t [s], respectively. The fit of measured curves is pretty impressive, $R_E^2 = 0.97$ and $R_t^2 = 0.99$.

The cytoskeleton network is connected to its surroundings through ordered water in the matrix part to be found on the external side of the cell, or occasionally – as e.g. in the case of epithelial cells – through a special coupling unit (the so-called "inter-cellular filaments"), a well organized protein-chain network.

We emphasized the failure of communication in the cancerous state. As we described, we need a good coupling which is cyclic at the same time for perfect communication. The communication is performed through electric signals. Therefore, the inability for communication means – in general – an electric isolation. Unfortunately, this is a normal process. Namely, when the cell divides then it has to insulate itself from the communication, otherwise the division could not take place. When the cytoskeleton network carries the signal then disintegration shall occur or the electric resistance has to change significantly. It is demonstrated by way of experiment that in the case of cancerous cells the resistance among the cells increases. We assume that the (normal → cancerous) biological transformation is a communication phase transformation which is similar to the stochastic resonance (see Section 3.3.8). This appears on the level of the individual cell, a ($\beta \rightarrow \alpha$) change of cellular state [575]. As we mentioned in Section 3.2.5, ordered water is necessary to generate an effective proton-conducting mechanism, which does not exist in disordered water. Furthermore, the ordered water does not weaken as much the electric field as the disordered, because of its low dielectric permittivity. However, the disordered water on the cell membranes isolates the cells not only by their proton exchange, but by their electrostatic forces with high permittivity. Also when the membrane has more and more isolating compounds (cholesterol) and/or the adherent connections and junctions are broken or blocked, the communication between the cells becomes noisy, the communication signal becomes weak compared to the actual noise. The noisy communication might cause cancer as well. This might happen also if the arrangement of cells in relation to each other does not make possible the interaction resulting in cyclic symmetry.

The topological construction is an important factor for the cellular organization [1016], irrespective of whether it is alive or not. The cellular structure for some topological reasons develops preferring special coordination arrangements [1017], and could arrange a self-organized collectivity [1018, 1019]. It was discovered that the division tendency is very low in a cell population small in number [1020]. For the start of a significant cell division a critical cell density is necessary. This was later observed on a self-synchronization of chemical oscillators [1021]. The topological importance was assumed in living cellular cultures also [1022], declaring that not the cell density but the position (coordination number) of cells related to each other determines what is favorable or not favorable from the point of view of division. This hypothesis was later justified experimentally [1023]. The conclusion can be drawn: if the symmetry is cyclic then the cell divides, otherwise it does not. However, this is only a geometric situation, a topologic request. This is valid in all the cellular structures irrespective of its existing forms, nonliving or living, normal or malignant. The topological order is valid for all the cellular structures of cancer also, however the communication makes important modifications. The living cellular structures are energetically open, they need transport of the energy sources in and transport of the waste out. Without direct cellular communication (no "social signal") this organized transport would be missing. The malignant transformation breaks this organized transport and seeks to build up new one, according the new situation. However, basic differences exist: the healthy construction is driven by the collective signal seeking to optimize the energy use for highest efficacy, the malignant one is only driven by the topology and physics of the mass number of individual, competing cells, irrespective of the efficacy of the energy conversion. This collectivism produces a difference in the geometric arrangement, not only in the cell–cell correlations, but the autonomic behavior forms cells individually as well. The topology in this meaning has important diagnostic (pathologic) meaning: the more the pattern of the tumor (mesh and form of the cells) resembles the healthy cells the less malignant the tumor is.

3.3.8 Resonances

One of the most important NTD processes is the so-called "window" effect [1024]. Optimum values both of the frequency and amplitude was observed interacting with the cellular membrane [1025]. This effect has resonant character. The measured frequency dependence very widely depends on the experimental conditions, and could be in synergy with chemical effects [1026]. The "window" was measured in multiple power ranges [1027], depending on the applied power (amplitude of the signal at the same impedance load), with such a small energy which categorized these experiments definitely as NTD. (They used max. 5 μW/g energy). The active Na^+ flux pumping was observed as maximum between 0.1 and 10 MHz [876], whose "window" effect could be well explained by the active transport system model in the membrane [1028]. The "window" to increase DNA concentration in the specimen was measured at 10 Hz between 0.03–0.06 V/m and 4–5 μA/cm^2 electric field and current density, respectively.

The NTD biological effects of low-level, non-stationary magnetic fields have been observed [1029] and adopted [999]. Recently, summaries have been published demonstrating a lot of experimental evidences from the field of ionic cyclotron resonance (ICR), [638] supporting widely the existence of the NTD phenomena [1030]. At the same time, we can find publications that could not characterize this phenomenon [1031] even by applying precise, cell-level measurements.

For the explanation of the ICR resonance there are several theoretical recommendations [1032–1035, 637]. Among these, for the explanation of this phenomenon, the most effective theory, until now, has been given [1036, 637] on the basis of assuming that the ions passing the membrane have a special trajectory specified by the ionic velocity and outer magnetic field.

For the development of the adequate resonance frequency we shall assume a long impact time at ionic cyclotron resonance (in order for the trajectories to form and endure for a long time), molecular excitation at parametric resonance, and superconduction at coherent resonance. We have carried out [1037] an exact mathematical analysis in the frame of the Drude–Lorentz model. Relying upon these findings, we showed [1037] that without any special assumption a resonance effect occurs in the ionic velocity on the $\Omega = \frac{q}{m}B_0$ cyclotron resonance, which is present also in the fundamental harmonics of the stationary and periodic ionic velocity component. The stabilized ion velocity has a time-independent and a periodic part. That is easy to understand physically, as the steady magnetic field tries to force the ion to a cyclotron orbit. The upper harmonics of the cyclotron frequency also appear, since the non-stationary magnetic field deforms this orbit.

The remarkable result [1037] is that these harmonic currents show resonance-wise behavior too. If it is valid for the biological systems, then the independent measurement of resonance is not suitable for the demonstration of the appearing oscillating members, as they measured for example the ion efflux [638], which is a time average. A similar statement can be made about the Ca^{2+} experiment as well [1038], where a very small decay for Ca^{++}-efflux (0.026 s) was shown [1038].

A decrease in the extent of mobility for ion transport through the membrane is acceptable because the steady magnetic field forces the ion to a longer trajectory. Besides, we may see that the field strength $\boldsymbol{E}$ and the adjoin velocity are not parallel, therefore, anisotropy takes place in the mobility of the ion and, accordingly, in the electric conductance. Consequently, the magnetic field generates anisotropy, which is the well-known Hall effect. Thus, from the analysis of the non-resonant term we may see [1037] that the switching off of the magnetic field means the increase of the mobility to the normal value, in other words, the magnetic field reduces the ion mobility, and the resonant conditions "only" restore the ordinary zero magnetic field conditions. With regard to this, the observation is not surprising and can be readily explained: The erythrocytes [1039] of human blood have been examined in zero magnetic conditions also screened by the geomagnetic field. They doubled the Ca^{2+} and Na^+ influxes and K^+ effluxes, while their hemolysis increased [1039]. This was interpreted by the authors as an ageing process. The significant changes of human blood serum measurable by the exclusion of the geomagnetic field make diagnostic differentiation [1040] between healthy persons and sick persons suffering from asthma-bronchial conditions possible.

In the case when the value of mobility is lower than the ordinary ion mobility, our result shows [1037] that the mobility can be described as a function of field components, which has also been measured [1041], and explained by the assumption of ionic parametric resonance.

At the cyclotron resonance measurements the ratio of the alternative magnetic field to the constant one is small in general [1042, 1034]. Namely, if we measure the ionic mobility in the case when these two magnetic fields are equal under the resonant conditions, and compare it with the ionic mobility of a simple case of zero magnetic fields, then, by tuning the frequency to the resonance, only the mobility will change to the trivial, zero-field values. In the resonance state, by decreasing the

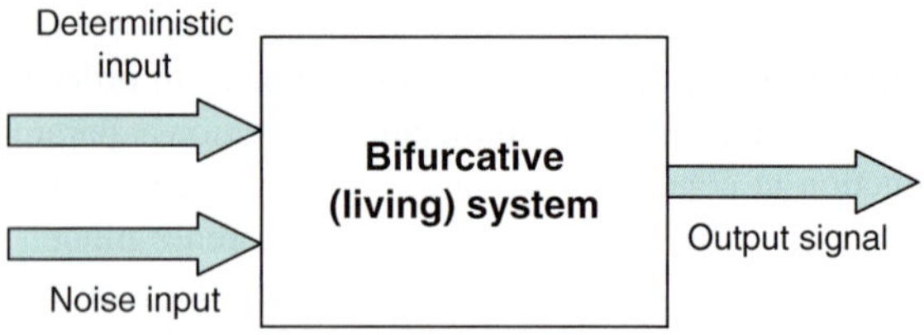

Fig. 3.47 The system with stochastic resonance

value of ε, the mobility will decrease as well. Under magnetic conditions and in the non-resonance state the leaking ionic currents flowing through the membrane will fall off, as their mobility decreases as well. Therefore, the increment in resonance is merely apparent, and serves for the re-establishment of the decreased ionic mobility caused by the magnetic field, or rather, in the resonance state and at a suitable ε the mobility may also increase. A good example for that is the application of different magnetic assuaging of pain, which might be effective by the usage of an adequate magnetic manipulation [1043, 1044], or even by the exclusion of the magnetic field [1045]. The time dependence of the efflux measurements does not come because the measurement itself integrates out the variation by time (e.g. [1046, 1047]).

The other important category is the stochastic resonance [1048] of the biosystems. In general the mixture of deterministic signals and noise could produce stochastic resonance as output (which is also noise) in a non-linear system (see Fig. 3.47). The output noise could be characterized by its distribution function, by its autocorrelation function, or by its power spectrum. In a simple description we use the last two.

The simple output by a non-linear transmission without deterministic input has definite decay at the high-frequency end, but we may observe definite resonance-like peaks on top of the output noise when an additional deterministic input is applied. This is the consequence of the correlation of the noise with the deterministic signal. The deterministic signal will be transformed on this way to a noisy one corresponding with the entropy-law.

The amount of resonance is described by the SNR which is the signal over the nose compared to the noise value in the actual frequency. This phenomenon is the stochastic resonance (see Appendix 16).

Its consequences for living systems are tremendous, mainly affecting the catalytic processes (the so-called "catalytic wheel," [1049]). This model describes a cyclic catalytic reaction having a two-conformation state of the enzyme governing the speed of the actual process. This classical model (Michaelis–Menten Enzyme model, MME) well describes the enzymatic processes in steady state [1050] (see Fig. 3.48):

- Enzyme in conformal state E1 connected to S substrate state and forming the E1S complex: E1+S→E1S.
- The complex transforms to the P product, while the enzyme transforms to the E2 state: E1S→E2P.

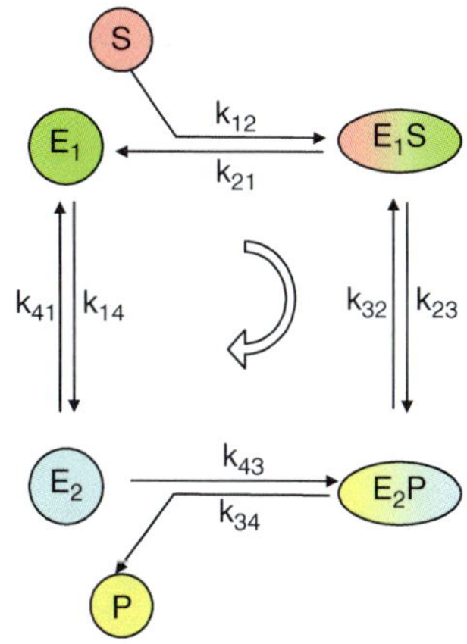

Fig. 3.48 The enzymatic "wheel"

- The product and the enzyme separate: E2P→E2+P.
- The enzyme is transformed back to E2→E1.

The overall direction (direction of the "rotation") is determined by the free enthalpy (G). When G decreases, then the "direction is S→P (like in Fig. 3.48), when it increases the process is the opposite. The energy supply of the "rotating engine" is the ATP → ADP transition. By stochastic concept the free energy can be obtained from the inherent fluctuations and outside electric noises [1051].

From the point of view of NTD processes the important point is, that the above rotation engine could be modified by an external electric field! This is the so-called electro-conformational coupling (ECC, [1052, 1053]), which activates to overcome the barrier by oscillatory activation [1054]. Let us apply the MME model to enzymes acting at the membrane transport (promoting ion pumps like Na^+/K^+). These processes are influenced by the huge membrane potential (order of magnitude 10^7 V/m) acting on some biopolymer species that are strongly polar molecules in consequence of having huge dielectric permittivity. The change of polarization state in the various conforming states of the enzyme is a realistic assumption, so we expect an energy $\Delta W = (\boldsymbol{E} \cdot \boldsymbol{P_1}) - (\boldsymbol{E} \cdot \boldsymbol{P_2}) = (\boldsymbol{E} \cdot \Delta \boldsymbol{P})$; where $\boldsymbol{E}$ is the membrane electric field vector, $\boldsymbol{P_1}$ and $\boldsymbol{P_1}$ are the polarization vectors in the E_1 and E_2 conformational state of the enzyme. In consequence ΔW is able to modify the energetics of the catalytic wheel, and the outside field ($\boldsymbol{E_{out}}$) could do the same by $\Delta W_{out} = (\boldsymbol{E_{out}} \cdot \Delta \boldsymbol{P})$. Therefore, there is no question: there is a direct coupling between the outside electric field and the enzymatic processes at the membrane. Because of the relatively small field interaction compared to the normal "wheel" process, this could be notable when it could be amplified. This amplification is expected by stochastic resonance. In a simple model the wheel is energized by ATP hydrolysis with 10^{-16}–10^{-17} W, while the molecular scattering due to the thermal effects provides 10^{-8} W [1055]. When we assume the ATP hydrolysis as a periodic process we can apply the stochastic resonance conditions. This is a direct stimulation of the given reaction. Also the

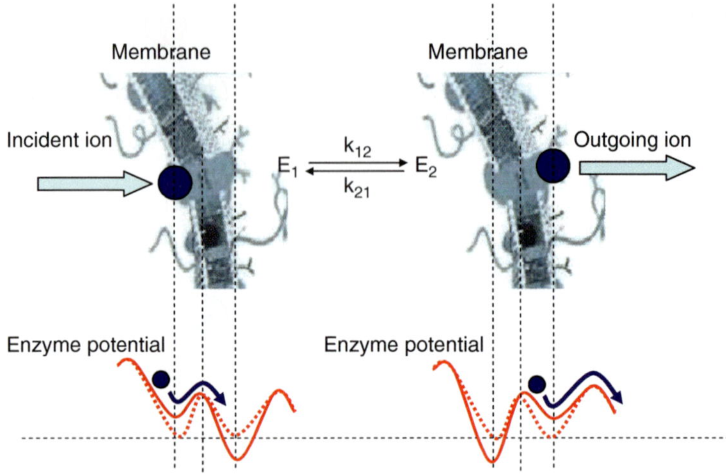

Fig. 3.49 Stochastic excitation of enzyme at the membrane

same could be produced by a periodic external field, when we act through the EEC effect and support the stochastic resonance, however the fluctuation-driven directional flow described by ECC needs more effort to clear-up the ion-pump processes in detail [1056].

The other consequence of the stochastic resonance near the membrane is transport promotion by a weak but periodic signal (Brownian engine, ratchet, [1057, 1058]). The translational symmetry can be broken in one direction by the periodic signal superimposed on the double-well symmetric enzyme potential (see Fig. 3.49). The Brownian ratchet is the multiple minima energy situation (as shown in Fig. 3.42) It could cause a sliding stability bag (see Fig. 3.50). These models was proven by experiments as well as by ECC coupling of K^+, Na^+, Rb^+ through membranes [1059, 1060, 1056].

The cellular machinery is based on various and numerous catalytic reactions. These work well till the catalyst is not "poisoned" (participates in a reaction where this catalytic activity is blocked), and of course till the time when the reagents are available to catalyze. The full process could be successfully described by stochastic resonance (see Appendix 17).

3.3.9 Modulation–Demodulation

There is a huge debate about the modulation possibility of living material. A special workshop was organized to discuss the question [1061], but there was no definitive outcome. Two definitely contrary opinions clash: one has positive modulation–demodulation theory (e.g. [1062]) experiments in humans (e.g. [1063]), others

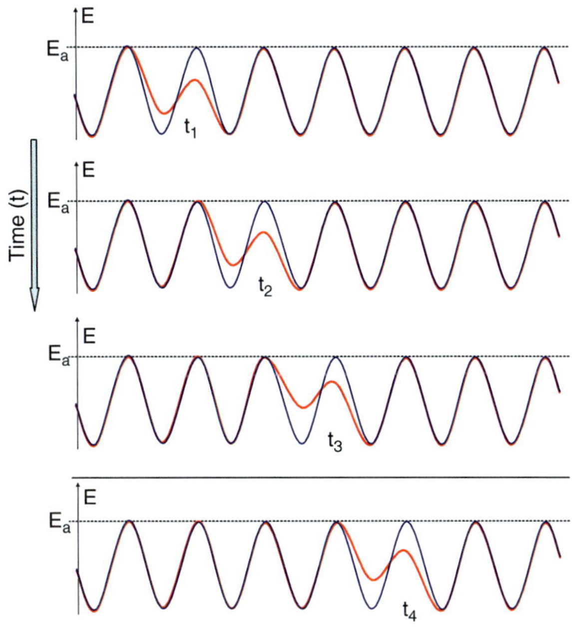

Fig. 3.50 The "siding" bistability by the enzymatic processes

referring to their experiments (e.g. [1064]), insist: there is no evidence of demodulation in living objects. The question is crucial for many applied methods in human medicine (e.g. resonant methods [1065]), but also sensitive for the energy and information transmission industry, as well as for "green" environmental politics in relation to "electrosmog" pollution.

The modulation–demodulation problem is an important topic in radio broadcasting, and could be applied to hyperthermia using electromagnetic radiation. (This is the oncothermia solution.) In various telecommunication and broadcast systems (radio, TV, GSM, etc.) the carrier frequency is high, having clear info transmitting. The same carrier frequency does not mean at all the same transmitted info. Our popular-music radio broadcast could be on the same carrier frequency as an absolutely different info channel in far-away territories. The same frequency does not mean the same info, which is the direct logical consequence of the large variety of the sounds, which we have from the radio (audio range, up to 20 KHz) with its rigidly fixed carrier frequency onto which we tuned our gadget. The carrier frequency (F) has three "free parameters" fixing this radiation: its amplitude (A), frequency [$f = \omega/(2\pi)$], and phase (φ): $F = A\sin(\omega t + \varphi)$. These parameters are ready for modulation: amplitude, frequency, and phase modulations are possible. They look like ...(see Fig. 3.51).

The modulated signal is very complex, and after it is received demodulation (mining the info, detach the carrier) is necessary. Definitely the easiest is the amplitude modulation–demodulation pair, because the modulation is only the change of

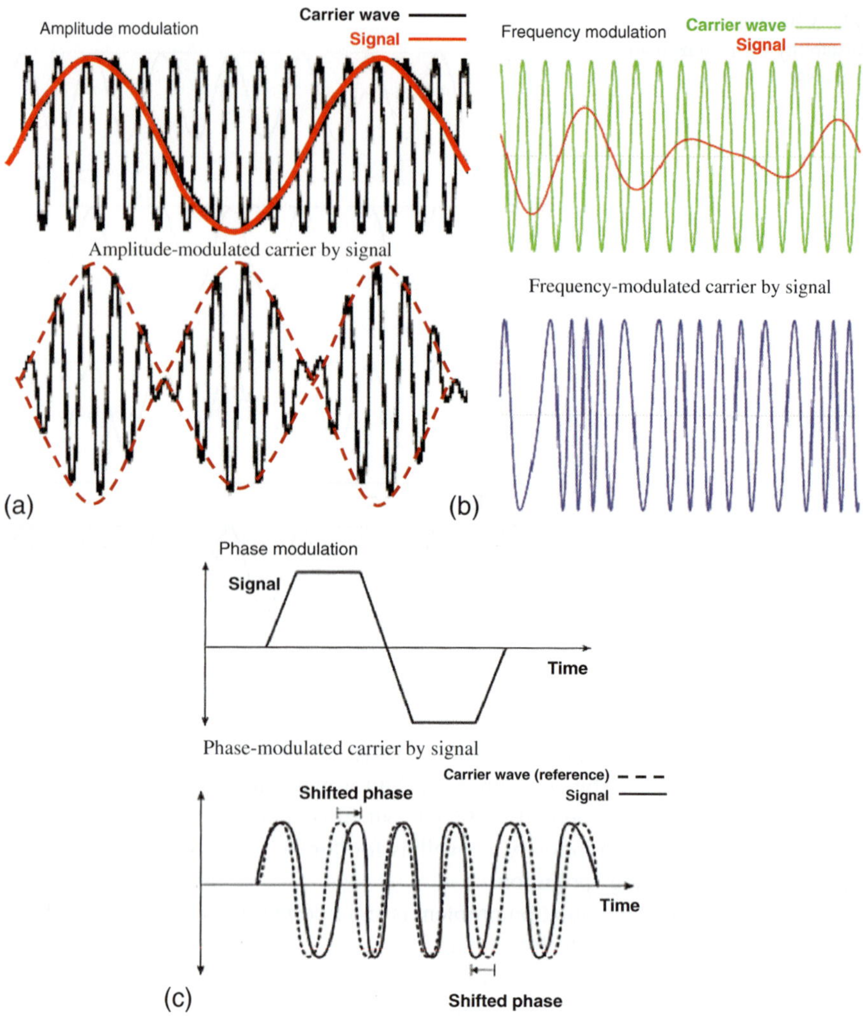

Fig. 3.51 Modulations: amplitude [AM], (**a**); frequency [FM], (**b**); and phase [PM], (**c**)

the "strength" of the signal by the info it has to carry, and the demodulation is a simple rectification, cutting the symmetric signal (see Fig. 3.52).

The rectification needs asymmetry and non-linearity. One solution for the demodulation problem could be stochastic resonance (see Appendix 18). We showed that the amplitude-modulated signal could excite stochastic resonance.

In conclusion: all small amplitude modulations of the carrier frequencies (if the modulation is chosen on the stochastic resonance frequency) could cause a definite resonant effect in all two-state Markovian situations (e.g. enzymatic processes,

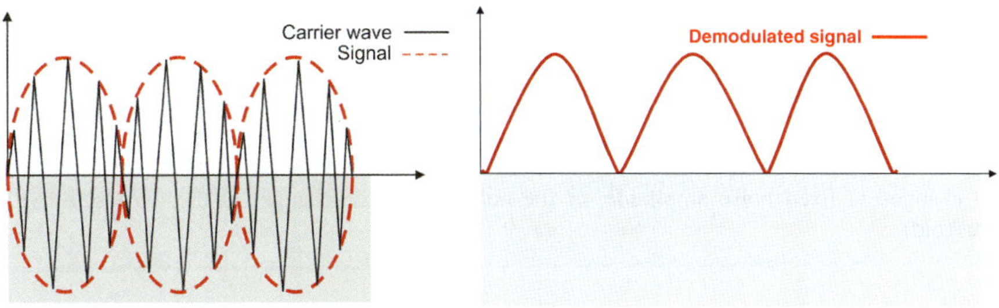

Fig. 3.52 Demodulation of an amplitude-modulated signal. The process involves "cutting" the negative part (rectification)

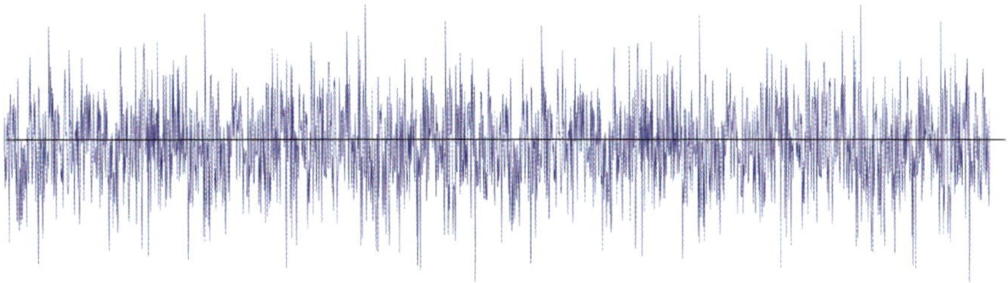

Fig. 3.53 The noisy environment (*white noise*)

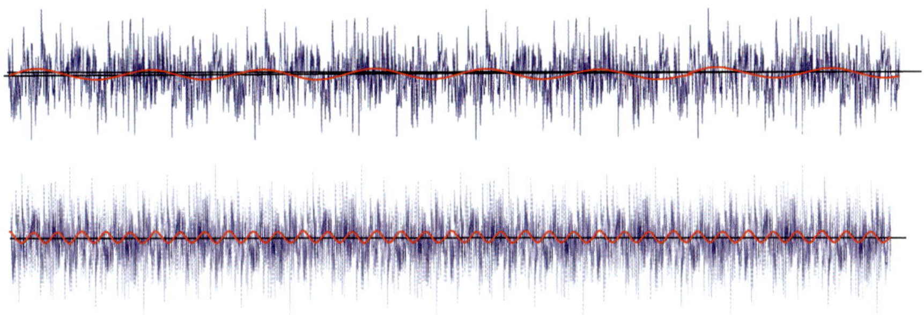

Fig. 3.54 A small deterministic signal is mixed (modulated or simply added) with the noise. Note, the signal strength is much less than that of the noise

voltage-gated ionic channels, etc.). Because of the very high number of such possible reactions in a living organism, these microscopic effects have a macroscopic resultant.

The sensitivity threshold application gives a simple explanation of the demodulation-like effect by stochastic processes. The white noise (uncorrelated, normal distribution around zero level) has no regularity (see Fig. 3.53). Adding (modulating) a low or high frequency, but low level deterministic signal apparently

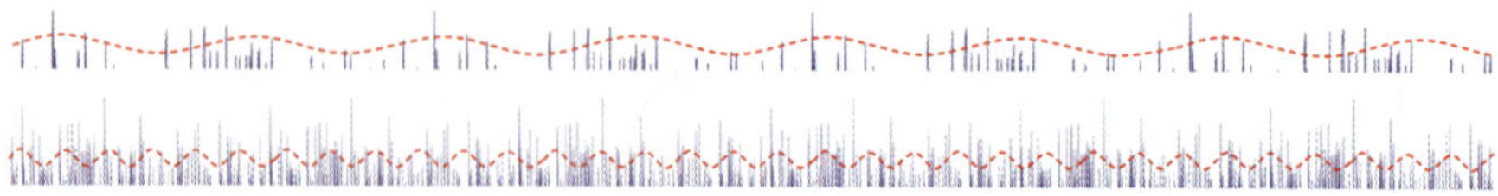

Fig. 3.55 Demodulation-like process by the well-chosen threshold-noise relation. (Or the threshold changed at fixed noise amplitude, or the noise amplitude changes at the fixed sensitivity threshold)

does not change the noise (see Fig. 3.54). However, if we have a sensitivity threshold cutting the mass of the amplitude of the wave and show only the high amplitudes over the threshold, the amplitude modulation (or simple addition) is observable, and the deterministic signal could be reconstructed above the threshold (see Fig. 3.55). The threshold cut is not a real modulation, but its effect is identical. The effect has a minimum where the over-threshold signal is recognizable, and another one, when the threshold is so low that practically the complete noise is going over, and of course it has in between a maximum.

Of course it is possible to change the threshold at fixed noise amplitude, but in practice the threshold is fixed in living objects. In this case the noise amplitude can be tuned up to reach the threshold at an appropriate level. Of course a mixture of deterministic waves also can be recognized in this way.

Research shows the alternating field effect on enzyme activity [1066] and signal transduction [1067]. Recently, research on amplitude-modulated RF in human medicine became very active, and [1068] also clinical trials show its progress [1069, 1070]. Research showed how an AC electric field inhibited metastatic spread of a solid lung tumor [1071].

Direct application of a low frequency current is also possible, without any demodulation demand. The AC-field-induced ponderomotoric forces (typical NTD) were studied by Schwan [1072] as early as 1982 and the redistribution of the membrane receptors is also possible by action of AC [1073]. Numerous applications of AC are effectively applied for cancer treatments [1074, 1075]. There are various applications of alternating field from power-line frequencies (50–60 HZ) [1076, 1077], supplied by theoretical considerations [1078].

The success of AC applications does not hinder the modulation–demodulation problem, because applying a carrier frequency is beneficial to target the chosen part of the object (deep tissue, membrane effects, etc.) There are different aspects to the modulation–demodulation phenomena. One argues on the classical electromagnetic basis: demodulation needs macroscopic nonlinearity, which can not be measured in biological systems; while others insist that the microscopic (cellular and subcellular) nonlinearity makes the rectification necessary for demodulation and the nonlinearity is mainly connected to the non-equilibrium thermodynamic behavior [1079, 1080]. We showed the stochastic resonance solution of the demodulation, which is out of the frame of classical considerations. The nonlinearity of biomaterial from an electromagnetic point of view is an important addition to the demodulation process (Appendix 19).

3.3.10 Special Field Effects of Biosystems

The general properties of biosystems can not be described by conventional simple electrodynamics; these structures are too complex to be simplified. Together with the impact of the active fields, there are many controversial debates on the possible effect of the zero fields when only potentials exist. Two giant milestones inflamed the discussions: Nikola Tesla, with his special bioelectromagnetic views [1081] and the Aharonov–Bohm effect [1082] with the "field-free" shifting of quantum-mechanical interference.

Living matter is a highly self-organized hierarchical structure. It is in nonequilibrium and its processes are nonstationer [642]. Their subsystems are multiply connected with various physical, chemical, and physiological processes and the interacting signals change in a wide range. The simplest biological systems show various processes on different time scales in vivo, which are connected by bioscaling. No two identical living objects exist, living matter is variable, changeable, mutable [646]. It basically differs from the lifeless [1083]. While the thermal and quantum fluctuations in the lifeless are becoming negligible by the growing size of the system; the living object has a high number of homologous phase states randomly transformed and altered into each other. These mutate with time and are in dynamic equilibrium among identical environmental conditions. In contrast, the permanent and immanent change makes the living object possible for adaptation, for mutation, for natural selection. This dynamism appears in the change of the confirmation state of proteins optimizing the enzymatic reactions of life. Because of these fluctuations, the living matter is more "noisy," and because of its self-similar [1084, 1085], and self-organized [609] behavior its power spectrum shows pink noise (1/f noise) [644, 645].

As we have shown before [1086], the symmetry drastically changes through the effect of an axial vector, and Onsager's symmetry is replaced by the Casimir's antisymmetric relation [1087]. This effect is nothing else except the rearrangement of fluctuation-noise distribution by the changing of the coupling of interconnected processes. This effect is independent of the presence of fields, only the action of the vector potential ($\boldsymbol{A}$) is necessary. The vector-potential $\boldsymbol{A}$ is a phase-shifter of the de Broglie waves [1088]. In a zero-field ("field-free") case the microscopic (quantum level) $\boldsymbol{A}$ could vary freely. This is a direct potential effect without any change of the actual energy state, it only rearranges its fluctuation distribution.

Resulting from the above considerations, the presence of any axial vectors (e.g. magnetic field $\boldsymbol{B}$ or vector-potential $\boldsymbol{A}$) could destroy the symmetry of the coupling of transport properties in living organisms [1086]. Consequently any axial vector changes the coupling between the transport processes, and so effectively affect the noise spectra and the interconnection of the various homologous phases of the actual living state. This special interaction behavior could give a clue to explaining the special respiration change by magnetic field [1089] or a proposed tumor-genesis theory by magnetic-field interactions [1090, 1091]. The action of axial vectors on the biosystem could affect its self-organizing ability, with the direct consequence of the autonomy of individual cellular organizing. The pink-noise fluctuation (and the connected large-scale maximal entropy) is broken by the effect of the axial vectors, modifying the transport properties and the symmetry of the interactions. This effect could modify the critical state and so the correlation length of the interactions [1092], and could create stress-like effects in the organism.

To demonstrate the above considerations we use a simple example: gambling. The casino has a definite income (macroscopically it is calculable, well defined.) The gambler goes to the casino to use his/her "luck," to win on the fluctuations. The casino's income is independent of the microscopic fluctuations over a longer time, the profit is entirely independent of who, when and how many wins in a single process. (We assumed an entirely correct, equal probability game, and systemic calculations breaking this equality are excluded. Of course this example oversimplifies the interacting life processes, because in the casino no long range and no multiple interactions exist between the gamblers.)

We are able to formulate our results on the basis of thermodynamics as well. The highest deficiency of information (highest entropy) is achieved by the noise, which has Gaussian distribution [1086] (Gaussian noise). Because the effective power density of pink noise is constant in all the characteristic scales, the Gaussian pink noise then has maximal entropy in all the scales. The living system has special fractal dynamism [1093], in consequence of its self-similar stochastic behavior it fluctuates by pink noise [179, 773].

The maximal entropy of Gaussian pink noise allows an important conclusion: the noise of the living state has maximal entropy (stable dynamic equilibrium) in all of the characteristic scales. By applying the Fokker–Plank equation [1094] we showed [1086] that the entropy fluctuation is connected to the coefficients of the Langevin equation too. In this case the elements of the cyclic matrix will determine the change of the entropy. The applied field-free potential could change the configurational entropy and the noise spectra of the living matter. This agrees well with the meaning of the minimum value of the volume integral of vector-potential squared ($A^2_{\min}$), which is connected to the topological structures of the matter [1095].

There is a possible bioeffect of the electromagnetic potentials without the presence of any electromagnetic fields [1096]. The potentials are actually effective electrodynamic parameters to describe and modify the living systems [1097]. The effect is expected on quantum level. It is based on the change of interactions of stochastic processes in living objects.

The practical benefit of the results is evident. It is not only a great possibility to work out new bioeffects, but it has industrial application possibilities also. The amplitude of field-free potential does not decrease, because it does not induce an Eddy current by the induction law that would dissipate its energy. Consequently, an effective communication method can be achieved by applying low energies. Extensive research has been carried out in this field. As an example, on behalf of Honeywell Inc. several patents were filed and granted on a communication system of this type [1098].

Chapter 4
Oncothermia – A New Kind of Oncologic Hyperthermia

4.1 Oncothermia Characteristics

4.1.1 Electrochemotherapy (ECT)

The very first oncothermia application was electrochemotherapy (ECT) which we showed basically works on the effect of charges pumped into the target by an external electrode. This charge causes special cellular distortion in the target. Its selection was regulated by the invasive insertion of the electrodes. The target tissue in ECT is a part of the closed electric circuit, so it could be directly controlled by the circuit parameters! Neither magnetic nor antenna radiation applications have such possibilities; in those cases the target is independent from the generating electromagnetic source. So in our concept the conduction (as clear as possible) has a central role in treatment control. This was the starting point for electrohyperthermia, the root of the complex oncothermia method.

After the European start [1099, 1100, 924] the method become very popular in China and other Far-Eastern countries. A few-thousand patients were treated with remarkable results [1101–1105].

The invasive manner and the method and its relatively small target area (a circle of ca. 3 cm around each inserted electrode) limited its application and gave other disadvantages such as possible infections, possible support of metastases by constrained blood flow from the wound, and initializing possible inflammation or ulcerous processes. There is a way to apply the electrodes noninvasively directly touching the skin surface over the target volume. This touching electrode method in principle eliminates the above disadvantages and avoids the complications. However, the effect is not the same, because of the variable layered skin structure and the uncontrolled pathway of the applied current. The challenges include the adipose layers in the skin structure, which isolate the direct current (DC) from the deeper parts, the conductance of the outermost surface due to sweating, and outside contamination, which conducts the current in non-controlled surface directions.

A. Szasz et al., *Oncothermia: Principles and Practices*,
DOI 10.1007/978-90-481-9498-8_4, © Springer Science+Business Media B.V. 2011

4.1.2 Concept of Oncothermia

To surmount the surface problem of the ECT method, a radiofrequency (RF) current has to be applied, which goes through the mismatched resistors taking their impedance into account. The complex impedance depends on the permittivity as well (not only conductivity) and also could be modified by a well-chosen frequency, because of the dispersion relation of the layer components. These layers could be modeled by resistors and capacitors. The capacitive impedance inversely depends on the applied frequency, lowering the impedance. Consequently not DC, but possible RF has to be applied. However, if the frequency is too large then the radiative component (as in a radiobroadcast) becomes more dominant, and the desired conductivity becomes less emphasized. However, the penetration depth to the body of the high frequency is low, we have to apply RF with not too high frequency.

Certainly it is necessary to keep one of the main advantages of the ECT method: its direct electric control on the treated target by its active involvement in the actual electric circuit. The tumor in these methods represents impedance in the circuit, directly influencing the source of the energy. The radiative or magnetic methods target the tumor as an independent energy absorbent. In ETC the inserted needles point to the target, constraining a current through it from needle to needle. While the non-invasive RF-current flows through the full volume, it is not located to a point-like needle source. The RF-current passes through the surface and selects a lower impedance (higher conductivity) path. The impedance drives the current densities in the tissue, the current flows of course along the "easiest path," the current density is higher where the resistivity (impedance) is smaller. This is represented as a trivial solution in water flows: the water always chooses the easiest path to follow. This everyday observation can be shown rigorously by a basic physical law in electronics as well (see Appendix 20).

To choose the right frequency we have to consider various components of the effects and practical applicability. From these considerations the most important factors are:

- the request is to treat deeply reaching the body cross section, so the frequency for effective penetration depth must not to be higher than 25 MHz;
- to be in the beta-dispersion regime to have an effect on the cellular membrane (i.e. at around 10 MHz);
- have the possibility of low frequency modulation (in a range up to 20 KHz), to use the resonance effects (the carrier frequency must be at least 100-times higher than the modulation, for accurate info-transmitting, so the carrier frequency must be not less than 2 MHz);
- be in a safe region, above the level for nervous excitations (more than 10 KHz) and below the level of microwave radiation (1 GHz).

Together all of the considerations above, it is practical to use a free frequency, which does not requires shielding. The free frequencies are 13.56, 27.12 MHz, or 40.78 MHz in the requested regime. In comparison all of these it was a trivial chance to choose for carrier frequency the lowest possible, 13.56 MHz.

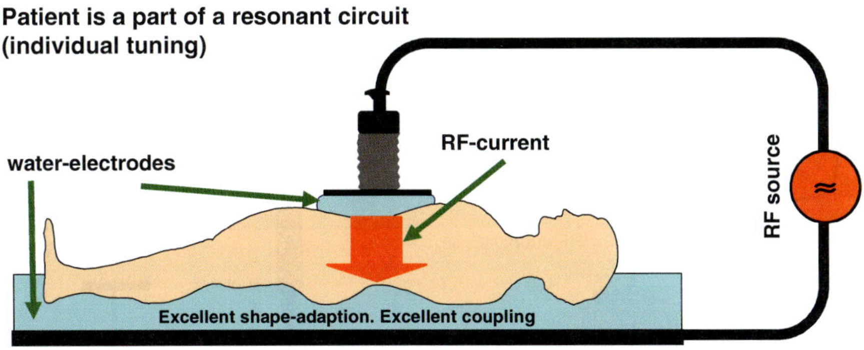

Fig. 4.1 Patient is a part of the closed radiofrequency current circuit making oncothermia well controllable

Oncothermia is a capacitive-coupled energy transfer, forming the capacitor by the target. The capacitive-coupling technique is a relatively old technical solution [1106–1110, 357]. In capacitive coupling. RF-current flows through the patient from one electrode to the other. Electrodes are flat metals, both under a water pillow: one is in the bolus; one is under the water mattress. Water is a transmitter of the RF-current, making a good fit of the human body to the flat metals possible. Both water electrodes (the water bed and the water bolus) are parts of the highly sophisticated electric circuit and not only a matter of convenience. The RF-energy flows in a controlled way in the constrained directions, the current delivers the energy to the malignancy. Both electrodes are active, current flows through them in all the frequency periods (see Fig. 4.1).

Its main advantage is the large efficacy and less deep hot spots due to the pure electric field action. However, its disadvantage is unfortunately remarkable: surface (adipose) burn is more frequent than in the radiative technical solution. For its further development, oncothermia applies a complex approach of physics and physiology.

The carrier frequency is 13.56 MHz, fixed like the frequency of your favorite radio station. The carrier frequency is only the basis, carrying the music and speech that you hear. Our 13.56 MHz is also only a carrier, which has information to transfer; the fractal modulation (time fractal fluctuation) helps to optimally select between the malignant and healthy cells.

The RF-electrode has to be arranged like the invasive one, using the electric field under controlled conditions. This is the reason why we chose the plan-parallel electrode arrangement (simple condenser). The patient in this situation simply appears as a dielectric media in a large condenser (see Fig. 4.2), regarding the body part as an electric component of the circuit.

Roughly speaking, the tumor is more disordered than its healthy counterparts, and so we are able to "pump-in" work to order it by the external electric field (see the sketch in Fig. 4.3).

The current changes its direction very rapidly (flows in one direction only ∼40 ns). It is remarkably important that the energy flows rigorously in the same

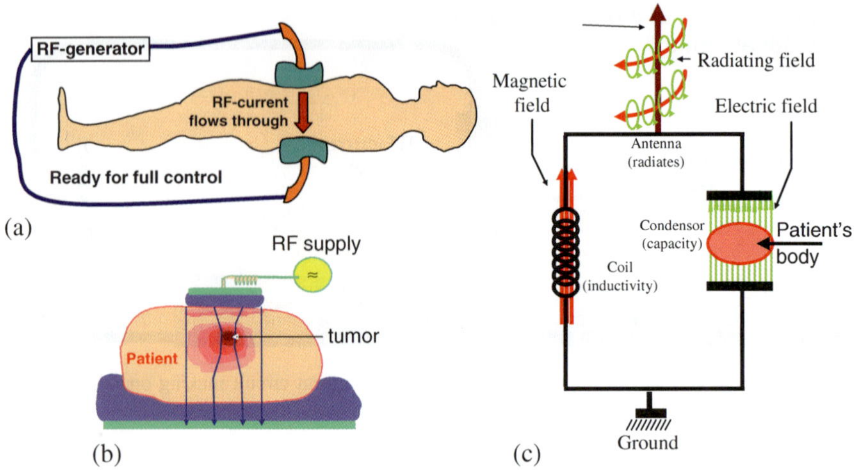

Fig. 4.2 The part of the body of the patient is directly involved in the electric current flow: (**a**) positions of the electrodes; (**b**) cross-section shows the RF-current; (**c**) the condenser part of a circuit is used for the arrangement

Fig. 4.3 The target has two dielectric materials: the tumor (highly disordered) and the healthy counterpart (ordered) (It is a rough explanatory model only.)

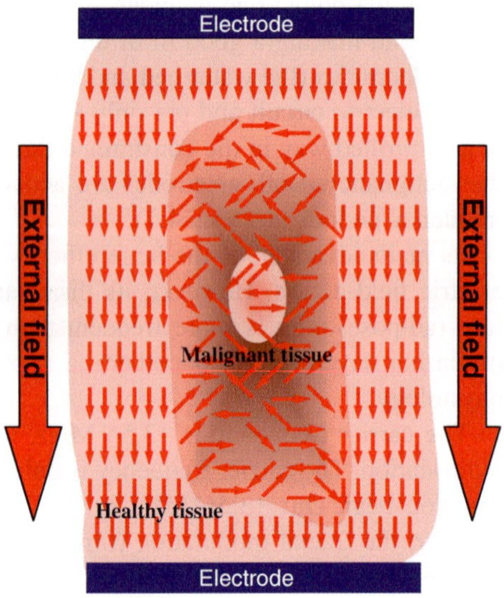

direction independently of the rapid changes of the actual polarity of the electrodes (see Fig. 4.4). The ***S***-vector is the Poynting vector [Eq. (A.6.12)], which describes the energy flow, measured in W/m^2. The energy flows always from the circumflex of the electrode to its center (see Fig. 4.5). This is one of the crucial behaviors of the well-constructed electrode system in the oncothermia solution.

The treated area and the tumor itself are nonhomogeneous, creating charge inhomogeneities during the flow of the RF-current (see Appendix 21).

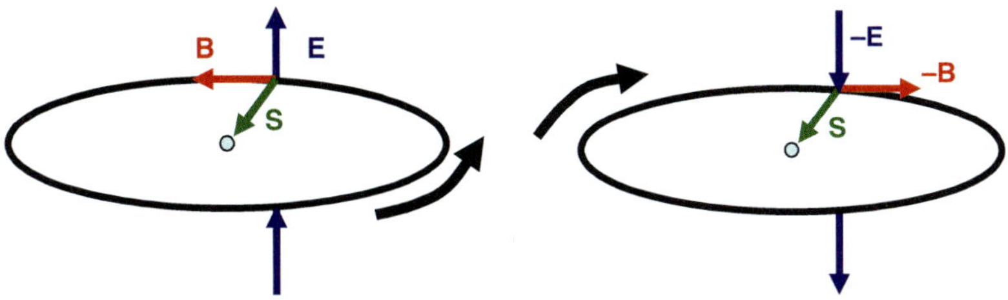

Fig. 4.4 Important behavior of the Poynting vector: it is an axial vector, and always points to the center, the energy flow is definite, irrespective of the direction of the current

Fig. 4.5 Physical model of dielectric heating. The Poynting vector (direction of energy flow) is permanently directed to the center of the electrode, irrespective of the applied frequency

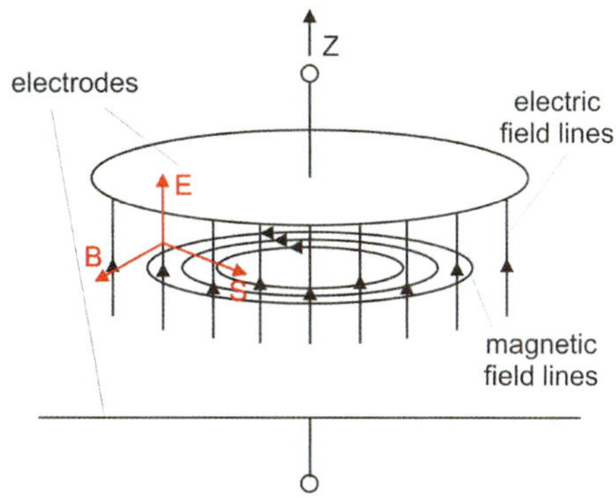

The characteristic time of the applied frequencies [~0.1 μs (=10 MHz)] is much less than the average folding time of a protein, which is 100 ms (=10 Hz) [586], so the charge exchange will not be able to cause the same effect as in the invasive case. The pH decreases by the increasing of the temperature. This effect definitely higher in the invasive ECT case. The effect of the change of temperature alone is not enough to make the pH so low, which denaturizes the protein: the pH decreases by 0.5, when the temperature rises from 25 to 60°C. The energy depletion effectuated dominantly by the RF electric field (capacitive arrangement) does not operate on the same basic principles as the DC, but offers other advantages:

- Extra selectivity factors can be included by electric conduction and permittivity (complex impedance).
- The conduction has deep penetration compared to the galvanic action, which has an approx. 3-cm diameter spherical volume around the DC-acting needle [940].
- The penetration depth of the electric field is at least double the penetration of energy absorption, and depends sharply on the conductance of the tissue (see Fig. 4.6).
- The energy penetration of the conductive RF solution is limited by the subcutaneous capillary bed, which tries to balance the local homeostasis. The capillary blood perfusion cools the skin in a physiologically controlled way. The

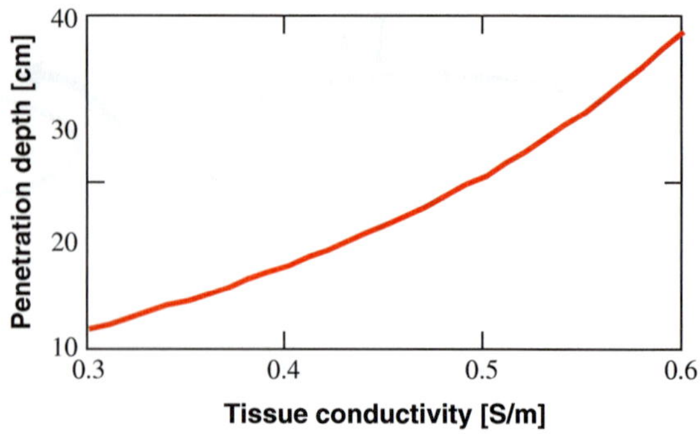

Fig. 4.6 Change of penetration depth vs. tissue conductivity at 13.56 MHz

blood stream functions in deeper layers as a heat sink, rapidly delivering the heat away from the targeted area.

The physical/electrodynamic parameters guide the heat absorption, which may be selective in a way other than the desired one (e.g. bone treatment could be problematic). Hot spots/layers/areas could be generated in unwanted locations involving risks for the healthy tissue. Conduction of the tissue could be modified by many physical and physiological factors, such as:

- The local RF-current density increases by the decreased cross-section of flow.
- The reflected waves are additively amplified at the internal tissue boundaries.
- Tissues are nonhomogeneous, their complex impedance may change drastically by location.
- The body cavities could serve as hollow-space resonators caused by standing waves and their local field maxima.

The deep-heat targeting and control of non-invasive transfer of energy constitute a major share of the problems of hyperthermic oncology. Adequate measurability and reproducibility should form the fundamental basis for the treatment. Additionally, the possibility for control is an indispensable requirement in the process of the treatment of patients. For that very reason the theoretical discussion and strict biophysical examination of the actually used, mainly experimentally introduced, control parameters and dose concepts cannot be left out of serious investigation.

The definition of the function of hyperthermia seems to be of vital importance in oncology. There are two approaches in the literature. The classical formulation concentrates on the temperature and tries to reach as high a temperature as possible. The temperature concept of course is supported by the equilibrium processes, (Arrhenius equation), and by numerous lethal processes over the human systemic physiological (42°C) temperature. This approach dominates also the hyperthermia in oncology, and all the controls and guidelines concentrate on this issue. The other approach places the energy absorption in a central position, irrespective of how high a temperature could be reached with the actual energy intake. This energy concept mainly concentrates on the possible micro- and macrochanges in the target, considering the overall temperature increase as a consequence and not as the goal of the actual process.

Oncothermia requests technically two definite effects: selectivity and cell killing. [1111]:

1. Apply a mechanism that is self-selective (the focusing in this case would be automatic).
2. The internal energy distribution should be applied in a way that does not cause an average heating only, but definitively works in the places where the energy can be applied in the most optimal way.

The first point was approached by a general mechanism of the tumor: the malignant cells have autonomy (renegades as Weinberg says), they are in permanent competition with the others for nutrition and for life conditions. The healthy cells are generally collective, their control is made by "social signals," and no real competition is introduced only a labor division is active. This means, that the active ionic exchange near the malignant cells (in most cases) is more intensive than in their healthy counterpart. This allows the introduced current to find the optimal path, following the path of greatest conduction. So the current self-selectively moves to the malignant cells. The applied current starts on one electrode and ends on the other, changing its direction 13.56 million times in a second. The electric field reaches all the depths involved in the treatments (between the electrodes) but the temperature development could be observed only at limited depths (about 20–25 cm, depending on the type of tissue). The growing phase shift by depth between the current and potential increase the reactive part of the applied power. It has been shown [1177] that the distortion effect depends 25% on the temperature and 75% on the electric field. This makes the method extremely effective at depths in the body. Technically (and simple speaking) we are simply introducing current through the tissue, which will find the malignant cells automatically using the optimal path. We performed in vitro and in vivo experiments, and observed the effect at work.

The second point is more sophisticated. The application of energy at a particular location may increase the temperature of the target but could also perform other actions. Like in the case of ionizing radiation: the absorbed energy increases the temperature but that is not the desired effect. The effect expected is to damage the DNA, destroy the chemical bonds, and rearrange the tumor structure. A simple example: if we had a dirty dish after dinner, we could wash it only with very hot water, but a clever housewife uses detergent to reduce the water temperature, and performs the job at the place where it is needed – on the surface of the dish, thereby not wasting energy on non-important volumes. To simply raise the temperature on its own in the tissue, because of heat diffusion the selection task cannot be fulfilled. The energy has to be distributed not generally into the target but specifically to the place where we want to achieve the distortion (as ionizing radiation does). The target of oncothermia is the cellular membrane! If we keep the current in the extra-cellular matrix (ECM) then the energy heats up only this electrolyte, and a heat flow begins from the extra- to intra-cellular regions through the membrane. This heat flow is accompanied by different ionic flows and water transport, changes to the Hodgkin–Huxley equilibrium, the membrane becomes more transparent, and finally it would be destroyed. (Anyway a transparent membrane could also be helpful to kill the malignant cells, because a large concentration of the intra-cellular HSP could be expressed extracellularly, which has a direct effect on the apoptosis and stimulation of the systemic immune reactions.)

The philosophy of oncothermia and in consequence the applied concept definitely differs from the other oncological hyperthermia considerations. The conventional tumor therapies have a concept to apply the largest tolerable dose (mg/m^2), (J/kg), measured in volume/mass-dependent doses/values. Their efficacy is measured by off-situ diagnostics (e.g. MRI, CT, US, etc.). Their safety is measured by a toxicity limit, established by dose-escalation studies and measured in the same doses as the therapy concept describes ([mg/m^2], [J/kg]).

Conventional temperature-controlled hyperthermia measures everything in temperature terms: the concept is to apply the largest tolerable temperature (°C), the efficacy is measured by the reached in situ temperature (°C), and the safety limit is measured by hot spots (in situ temperature) (°C). The latter is a very important factor, because the only toxicity effect caused by the hyperthermia is burning (hot spots). Temperature measurement at depth is necessary to avoid such (often very dangerous) internal burning. This makes the quality guidelines of hyperthermia very complicated. Measure the temperature invasively or approximate it by large devices like MRI.

The temperature measurement is definitely a safety issue and not a treatment one. The required optimal temperature may not be reached due to several factors including large blood flow as in the kidney or brain; air-cooling (breathing) as in the lung; liquid-cooling (as in the urinary tracks). The opposite is also true, if the required temperature could be reached, but the patient's tolerance limits the power, then the treatment is again down-regulated, the prescribed temperature is again not reached. Also the same limit is in action when hot spots form outside the tumor, so that the energy intake must be limited to avoid the burn. In this case it is also problematic to reach the temperature guidelines. The above challenges make the temperature controlled hyperthermia unsettled.

The temperature anyway has problems. The temperature is not a dose! (It does not change by the volume/mass.) Its measurement also is not easy, it is practically impossible at depth (MRI $\leftrightarrow$ phantom). Some controversial studies are explained by the missing reference point (Rotterdam Group: "Reference point is needed!").

Oncothermia turned to the well-known gold standards, using the energy-dose concept in their protocol. The energy is controlled with a concept to apply the largest tolerable energy dose (J/kg). The efficacy is measured by the absorbed energy (J/kg) and the safety limit is by energy transfer through the skin (J/m^2). Control of the latter makes a safe and complication-free oncothermia process possible.

The presently applied radiative hyperthermia devices, operating at a one order of magnitude higher frequency than oncothermia, are in fact also capacitive-coupled, because the applicators are definitely in near-field arrangements. However, these are by far not optimally coupled and their frequency is also too high to be able to provide the desired effects. In oncothermia no artificial focusing is needed for selectivity, and no isotherms in space and time have to be controlled. Both effects are solved in oncothermia with a directed electric field. It represents a well-designed capacitive coupling on the 13.56-MHz free frequency [1112]. Oncothermia is controlled by the changes of the impedance, and by the absorbed energy, which both are accurately measured. In this meaning oncothermia is very similar to RF-ablation hyperthermia, where the temperature is not measured, the effect is controlled by the measured impedance of the tissue. The power ranges from 30 to 150 W, which is adequate for heating up the tumor over 42°C by well-controlled focusing. (If you touch a working 12-W halogen lamp you can feel its burning efficacy. Less than 20 W is enough to heat up a 5-cm diameter tumor from 36 to 44°C at 3 min! The only point is, how we concentrate the energy on the target making energy-density high. This is the question of the focusing.)

4.1.3 Pennes Equation Revised

First we need to decide on the objective of hyperthermia in oncology. The definite objective is to eliminate the malignancy. This needs of course clear measurable qualitative goals and a dose quantification. Two main qualities are obviously in consideration: the achieved temperature (measured grades) and the absorbed heat (measured in joules). Herein, we directly propose to use the absorbed energy (heat) to measure the hyperthermia dose. We disagree that the actually reached temperature should characterize the process. According to our position the temperature could only be a possible tool to reach the objective, but must not be a goal. In simple theoretical formulation: if the absorbed energy for the destructive purpose increases only the temperature than no energy is consumed to destroy the malignancy, to break the actual chemical bonds. A part of the energy has to be expended for the bond-breaking. This energy will not increase the average energy (temperature) of the tissue, it will be missing from the overall temperature control of the system. Another wording of the argument: if the destruction has a demand of energy then this energy will be absent from the internal energy characterized by the temperature. Therefore, the temperature alone is not enough for a description of the processes taking place in hyperthermia. This is the reason why hyperthermia characterization extends the temperature measurements by its time-average also [1113]. The temperature characteristics and their dynamism was theoretically discussed by Pennes [1114] in 1948. He studied only the temperature change. His formulation [see above, Eq. (A.22.3)], describes a non-equilibrium heat flow when only the temperature is the driving force. The Pennes equation has derivatives by time (t) and place (x), describing correctly the temperature-dependent part of the heat flow in the tissue excited by external energy (see Appendix 22). The equation is applicable only in the case when no other phase transitions or energy sinks exist, so the distortion phenomenon is excluded.

However, hyperthermia is devoted to destroying the cells, it is not only a temperature process. Besides the temperature development during hyperthermia three things can happen with the thermally excited cells:

- they return to their original healthy state (reversible change, this is not an expected effect);
- they become lifeless by apoptosis [768];
- they become lifeless by necrosis [771].

All the above effects could be temperature-dependent, but energy consuming also. The individual energy intake of the above processes will definitely modify the average energy distribution (temperature). Unfortunately, this type of energy sink is missing from the Pennes bioheat equation [1114]. Pennes created this equation for an examination of the blood stream of the human forearm at rest, excluding all the changes which could use energy for something other than the temperature. (He emphasized for the description of the equilibrium, fixing the forearm in a resting state.) For this purpose Pennes' equation is correct and usable, since the internal energy depends exclusively on the temperature, for which numerous model calculations have been provided [1115–1118]. To clarify the time-dependent transient problems some solutions were published [1119, 1120].

The solution of the partial differential equation (bioheat equation) (A.22.3) should be reasonably given numerically because of the following arguments:

- The exact geometry of tissues is unknown.
- The equation parameters are not available in their exact form, the values change by tissues, tumors, and individuals. The tumor is nonhomogenic, the parameters widely vary in the tumor tissue.
- A certain part of the equation parameters – for example perfusion rate, metabolic thermal power, electric thermal power – can be expressed as a function of temperature, therefore the equation may be nonlinear as well.
- The treating electric field and for this reason the electric thermal power can be expressed as a function of position, and because of the skin effect it depends also on the temperature.
- The parameters of the transitional zone between the tumors and healthy tissues are unknown; therefore the exact definition of boundary conditions is almost impossible.
- The unanimity conditions of the bioheat equation (A.22.3) are the initial data and boundary values.

Some papers were published devoted to modifying Pennes-like equations [1121–1124], but none of them took into account the energy required for the distortion of the actual arrangements. In the only temperature investigations (studying the average energy), the change in chemical bonds has no role at all, and the energy consumption required for this special purpose was not included. However, in the case of oncological hyperthermia we have to describe the goal of the treatment: the cellular distortion. It is clear that if the energy could be used only for the chemical changes (distortion of molecules and restructuring arrangements) then the temperature would remain constant, no energy consumption occurs for the average energy distribution. We have to construct a model resembling better the reality if we suppose that the internal energy is the state function of not only the temperature T but of other parameters too, which could describe the energy consumption of the structural and chemical changes. The non-thermal parameters have to be measured in vivo through their measurable consequences (e.g. impedance of the tissue, dielectric constant of the tissue, heat, and electric conduction of the tissue, etc.). We are going to refer to these parameters as internal variables of a given cellular composition.

Pennes' equation could be generalized [1125] considering the cellular destruction and structural rearrangements [1126], which are both important in the effect of hyperthermia. On the basis of the generalized Pennes equitation we introduced the energy dose which contains a clinically observable term, the memory effect, namely the effect of the irreversible changes depending on the time, which is characteristically longer (or at least comparable) than the treatment time. We showed that neglecting the distortion energy, (memory effect) the newly introduced energy dose and the Separeto–Dewey empirical dose are identical. This is a control of the new dose calculation and at the same time shows the reality of the rigorous thermodynamic basis of the empirical dose as well. By studying the differences between the energy- and empirical doses, we established that they are near to each other if the energy intake is large enough to neglect the energy of the distortion, or the distortion process is so immediate that its time is negligible compared to the treatment time.

Considering the definitive task of hyperthermia to destroy the malignant cells, the cell disruption and the energy expended on this is mandatory in the process. In this regard, our present calculation is important to clarify the quality assurance and all the quality guidelines of oncological hyperthermia.

Because of the expected cellular distortions this last term (as many of the hyperthermia experiments show [769]) depends explicitly on the time. Consequently, in every case when the electric power is not constant, the source term of internal energy balance depends on time, there is no way to construct any stationary (equilibrium) conditions, so again, the process cannot be determined by the temperature alone. Because of the definite time-dependence, the equations became nonhomogenic; the process cannot be controlled by the temperature alone.

In the case where no cellular distortion and/or other cellular/tissue changes take place, and only the general temperature rises without other changes, the temperature characterization is definitely correct. However, this process does not cure, simply heats without the expected hyperthermia effects.

The problem of the widely and incorrectly applied Pennes bioheat equation is not the fault of the original publication. That was correct regarding the analysis of the tissue's arterial blood temperatures in the resting human forearm. In that case no cellular/tissue changes are expected, the stationer (homeostatic) case trivially exists, and even the author emphasizes this equilibrium by the investigated "resting" cases.

Do we have any reason, why hyperthermia should use an intensive thermodynamic parameter (temperature) regarding the quasi-equilibrium of the treatment? No, of course, we do not. It is only a historical "bad habit." This bad characterization is mostly responsible for the contradictory results, for the loss of comparability of the results, and for the blaming of physics/physiology!

According to our view, hyperthermia (it is definitely a heat-dose treatment) is a temperature-dependent but not temperature-determined process. The temperature concept is not bad as long as the physiological factors (blood flow/vascularization, metabolism, chaperone-protein production, dissemination, apoptotic action, etc.) are included, and the tissue can be regarded as homogeneous, semi-isolated from the surroundings. Unfortunately, these conditions are not common, so sometimes the temperature gives statistically significant, sometimes random results. This is the reason why some trials check the patients before, dividing them into the "heatable" and "not-heatable" groups [1127], randomizing for the trial only the previous group. For this group the anyway scientifically incorrect and assumed equivalence of the heat dose to the temperature is an acceptable approach. On the other hand, this preselection excludes a large number of patients from receiving hyperthermia; however, this treatment could be a help for them as well. The exclusion was made on an insufficient characterization of the method.

A relevant characterization of oncological hyperthermia for quality guidelines has to begin in the aim: to destroy the malignant cells! This demand contains some more precise requests: act selectively on the malignant cells, block further proliferation, and stop the dissemination of tumor cells. Distortion could be promptly direct (the cells become necrotic during the treatment) or indirect (tune killing conditions; the cells become necrotic or apoptotic after the treatment). The demands do not actually contain any temperature request. The desired temperature could be a tool to manage the process; but the goal never should be replaced by any actually applied tools.

The appropriate characterization has to be a definite adverb of degree: characterizes the process, determines the actual state, measurable and repeatable. Usually we have to measure the state of the actual process and not the status of the tool; so the distortion process itself is the measurable target.

The prompt effects during treatment seem to be easier to measure, it is really a quantitative factor, but to measure the conditions for the effective distortions afterwards, needs assumptions. Changes supposed to lead to some postponed actions (ischemia/hypoxia, stress factors, acidosis, etc.) are also measurable in situ. With the temperature concept (using it like an aim, and not like a tool) the assumption is very simple: the suitably high temperature does all the jobs, the prompt and the postponed effects can be expected, the requested thermal dose is reached in the target. As we see from Eq. (3.31), the temperature-dependent term ($T\Delta S$) in the internal energy (ΔU) is only one of many others. The main factor of the real desired action: to have cellular distortion, to have chemical reactions. If the biosystem undergoes chemical reactions, the non-temperature parts of the internal energy become important [576].

In spite of its inadequate character, the temperature had gradually become the base of the quality assurance and treatment control. I think it is time to change, and choose a heat-dose (energy-dependent) characterization. There are various possibilities to choose from to study such a parameter, but I think, the physiologically and physically well-studied ionic environment offers the best option. This environment well depends on the metabolic rate, on the chemical reactions, and on the structural changes as well. (Sure it is also temperature-dependent by these extensives.)

Some special considerations were devoted to modifying the Pennes equation [1128, 1129, 1123, 1124], but no one had considered the energy for the distortion of the actual arrangements. In the only temperature investigations (studying the average energy) the change of chemical bonds has no role at all, and the energy consumption required for special purposes is not included at all. However, in the case of oncological hyperthermia we have to describe the goal of the treatment: the cellular distortion. Definitely, if the energy was used only for the chemical changes (distortion of the molecules and restructuring the arrangements) then the temperature would remain constant, no energy consumption occurs for average energy distribution. We can construct a model resembling the reality better if we suppose that the internal energy is the state function of not only the temperature T but of other parameters too, which could describe the energy consumption of the structural and chemical changes. The NTD parameters have to be measured in vivo through their measurable consequences (e.g. impedance of the tissue, dielectric constant of the tissue, heat and electric conduction of the tissue, etc.).

These above challenges led to a revision of the Pennes equation (see Appendix 23).

The difference between the results of the two approximations grows in relation to the relative energy portion of the cellular distortion in the complete energy intake. If we have a definite high temperature then of course we pump in much greater energy than the distortion requests, the temperature is in fact only the "non-directly-used" part of the energy intake. The separation of patients into "heatable" and "non-heatable" in reality means the condition that the energy expenditure for the distortion has to be negligible ("heatable patient"), so the temperature could therefore be a control parameter instead of the correct energy control.

A simple example could show the difference in a very peculiar way: An average human has a 2,100 kcal/day diet. This energy (together with the environmental energy sources like sunshine), supplies all the daily activities, movements as well as the breathing and heartbeat and all the physiological changes, including approx. 10^{11} healthy cellular divisions and producing approx. 60 mol ATP ($\sim 3.6 \cdot 10^{25}$ molecules) from ADP during a single day. All these enormous changes are supplied with only the daily energy intake, which is equivalent to approx. 100-W continuous energy absorption (2,100 kcal$\approx$8,640 kJ$\approx$100 J/s). This all happens in the full system of the human body (about 70 kg), so the energy supply from nutrition on average is $\sim$1.4 W/kg. This relatively small amount of energy produces immense changes in the structure, works intensively in the "chemical factory" of life. If we were clever enough to input the energy as accurately to the cell as in the natural nutritional supply, then a really small amount of energy would be necessary for any desired changes. Even if we use the applied power to heat up the tissue only (not calculating the blood flow and other heat-conduction-modifying factors) as small as ~7.5 W/kg heating for 60 min would be enough to heat up the tissue from 36ºC to over 42ºC. Having realistic heat losses (e.g. blood flow, heat conduction, etc. all together ~3 W/K/kg) ~20 W/kg for 60 min would be enough for the same 36ºC → 42ºC heating. However, this is more than an order of magnitude higher than the request for the chemical modifications! Why is this so? Because we give to every molecule the same energy irrespective of whether we wish to use it for change or not. The crucial point is: the necessity to find appropriate effects acting directly and selectively on the chemical machinery, so we would be able to use much less (and consequently much safer and probably much more effective) energy in the method then in overall heating. This is the objective of oncothermia!

4.1.4 Thermal Limit Problem

As we showed in Section 3.3.7 no thermal limit exists at zero-mode noise. It is difficult to induce a pure zero-mode field on a single cell. The external fields have translational symmetry, which is limited by thermal noise.

Artificially, we can produce pure zero-mode field indirectly by changing the ECM homogeneously: changing its composition and thus inducing ionic currents, or by heating the ECM and thus producing thermo-diffusion on central symmetry. The surrounding ionic or thermal gradient through the cellular membrane will have zero-order noise, unlimited in the thermal case Such zero-order noise is produced by oncothermia. The primary ECM heating makes the temperature gradient centrally symmetric (acts around the cell in the same direction). This is achieved by capacitive-coupled electromagnetic field application within a certain frequency range [171, 334]. Provided that the frequency is low enough, the fields do not readily penetrate the cell membrane and thus the bulk of the energy is deposited within the ECM. This leads to thermal gradients (currents) from the ECM towards the inside of the cell. This thermal current also carries ions through leading to thermo-diffusion, thus creating a zero-mode electric current, which in turn induces a zero-mode electric field in the cell membrane. Therefore, even small fields with zero-mode components could elicit biological effects.

Note that the zero-mode calculations primarily considered situations in which the thermal energy noise was significantly higher than the external electromagnetic effect in the low energy range (LEMF conditions) [1130]. Although these methods are true in general, there could also be frequency and noise-mode spectrums, where the signal-to-noise ratio (SNR) is clearly dominant, resulting in an effective low-energy external signal [1131, 1025]. In the oncothermia case the above-described membrane effects do not favor any direction in space, only the structural arrangement of the cells has a role. Demodulation of the time-fractal fluctuation modulated carrier is direction independent, while additionally beta-dispersion and the forming of hot spots in the membrane do not request any directional conditions. This (together with the amplification mechanisms of the stochastic resonance phenomena) makes it possible to act with a low level signal despite the noisy thermal environment.

4.1.5 Energy Transfer Through the Body Surface

Transferring the energy deep into the body is a general problem of hyperthermia. This process is necessarily accompanied by a massive surface energy absorption, which could absorb the useful energy transforming it into extensive heating and therefore increasing the risk of burn. This is a real challenge. The possible burn of adipose tissue by capacitive coupling is well known, due to the selective electric field-energy absorption of the adipose tissue. Calculations show, that the ratio of the adipose tissue (P_a) and muscle tissue (P_m) absorbed power is selectively different, and their ratio (P_a/P_m) is large. The ratio of the absorbed power of the adipose and muscular tissues is:

$$\frac{P_a}{P_m} = \frac{\sigma_a}{\sigma_m}\frac{\left|\varepsilon_m^*\right|^2}{\left|\varepsilon_a^*\right|^2} \tag{4.1}$$

where σ_a, ε_a^* and σ_m, ε_m^* are the conductivities and complex dielectric constants of adipose and muscular tissues, respectively. In consequence of the relatively small (σ_a/σ_m) and dominantly large $(|\varepsilon_m^*|^2/|\varepsilon_a^*|^2)$ ratio, a relatively large absorption of the adipose tissue occurs [1132, 1133], [1134]. Normally all values have frequency dispersion. We consider only the values at 13.56 MHz. The muscular data spend on the orientation (parallel or perpendicular to the field) of the fibres. The values are: $\sigma_m \cong 0.9$ (S/m) [1135], $\sigma_m \cong 0.6$ (S/m) [1136], $\sigma_a \simeq 0.02$ (S/m) [1135], $\sigma_a \cong 0.3$ (S/m) [1137], $\varepsilon_m \cong 100{-}120$ [1135], $\varepsilon_m \cong 160$ [1136], $\varepsilon_a \cong 10$ [1135], $\varepsilon_a \cong 30$ [1137], $(\sigma_a/\sigma_m) \cong 0.02{-}0.5$, $(|\varepsilon_m^*|^2/|\varepsilon_a^*|^2) \cong 12{-}260$). The absorption ratio with these data could be normal but can be as high as 130 as well. Realistic calculations [1132] show a power ratio of about 5. This could be a real burning problem. However, if the field is not perpendicular to the boundary of adipose muscle tissue, then the problem is controllable [1133], Also the vertical potential gradient can be well controlled by technical arrangements: controlling the blood perfusion of the skin tissues. If the blood perfusion is adequate in the skin, (the blood data are the same, a little bit higher in its values than the muscle one [1138]), then the dielectric constant increases [1137], the relative absorption dominance is suppressed in this way, and the heat- and electrical-conductance also grows [1137], helping to prevent the problem of burning. (The temperature dependence of the blood perfusion is selectively high [225]. A 20-fold increase of the blood flow was also observed in the skin [1139]) To avoid harm well-designed cooling is applied, too. In the water bolus and the applicator material itself nearly no energy is absorbed because of the use of deionized water and non-absorbing materials. Definitely the problem cannot be eliminated by any of the known techniques, but could be well controlled and minimized. In our applications (for a long time a few tens of patients have been treated daily with our installed devices) less than 3% burning problems were reported.

The other (and massively problematic) issue of skin burn is the intensive tangential currents in the near-surface area. The origin of this phenomenon is the inhomogeneous distribution of the electric field, (edge effects and distribution inequalities of the RF potential, dielectric frames, etc.), on the electrodes. This could be compensated by a special resonant circuit, (a further development of [1140]), which corrects the potential inhomogeneities at the electrode surface, while also proper development of the geometry of the electrode offers a solution to this problem.

Technically two points are important for the characterization of the full process:

1. Keep the phenomenon in the RF-current region, avoid the radiation losses and make accurate impedance control possible.
2. Keep the sign of the gradient of energy transfer positive, (no hot spots in healthy tissue), make accurate energy control possible and avoid mistreatments at depth.

The problem could also occur at bone or in other isolating layers in the treated area. The current must flow through the skull bone; otherwise the circuit is not established (see Fig. 4.7).

The situation in the case of other bones (like ribs, femur, etc.) is basically different. In these cases the RF-current has a way to go "around" the bone, and shortcut the isolating part (see Fig. 4.8).

The crucial point of oncothermia is its conductive approach, reducing the radiative part of the electromagnetic interactions to as small as possible. This means the treated target is a definite impedance load of the circuit. However of course, various arrangements (mainly the electrode structure) could modify the conductive situation to radiative. In this case we are using the near-field approach, because the distance of the target is much smaller than the wavelength of the applied RF. The wavelength of 13.56 MHz RF in vacuum is ~22.1 m, while in the body it is ~2 m. Applying frequency an order of magnitude higher, (in the range of the FM-radiobroadcasts) then the wavelength in the human body will be about 20 cm, which easily modifies the near-field approach to a far one (when the wavelength is in the range of the target–source distance). In general the near-field treatment uses significantly lower

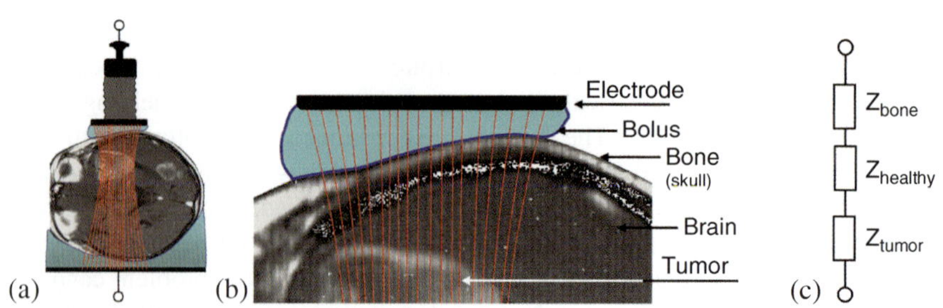

Fig. 4.7 The current flow in the case of the skull (**a**), the details (**b**), and its equivalent electric circuit with the relevant impedances (**c**)

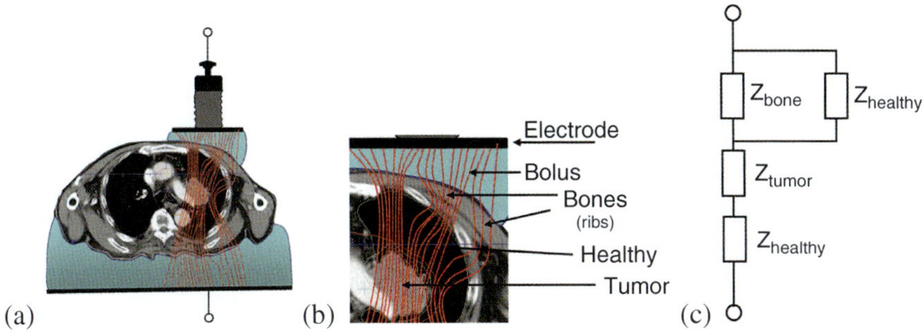

Fig. 4.8 The current flow in the case of abdomen treatment (**a**), the details (**b**), and its equivalent electric circuit with the relevant impedances (**c**)

frequencies, and allows interference-focusing. The SAR values depend on the situation (far- or near-field) [1141], their empirical ratio in the few tens of MHz range (<40 MHz) was observed [1142]:

$$\gamma_{SAR} = \frac{SAR_{near}}{SAR_{far}} = \frac{1}{\left[1 + \left(\frac{5}{\lambda_v}\right)^2\right]\left[1 + \left(\frac{2.2}{\lambda_h}\right)^2\right]} \tag{4.2}$$

where λ_v and λ_h are the wavelength in air at the target place, the indexes denote the vertical and horizontal wave positions. Using these data, the actual values of γ_{SAR} are 0.8 and 0.2 for 13.56 MHz and 135.6 MHz, respectively. The far-field effect is well dominant at high frequencies.

4.1.6 Penetration Depth

In the case of oncothermia the penetration depth has a different meaning than we defined for radiative of magnetic RF-heating, where the energy absorption was independent from the electric circuit itself. In conductive oncothermia the RF-current flows through the whole volume between the applied electrodes. (In the radiative solution a single radiative antenna is the energy source, here the flowing current between the electrodes generates the energy.) In the oncothermia case the area of the cross-section that the RF-current flows through changes by depth, and it decreases the current density (current through a unit area). The energy deposition of the current in a unit volume, however, depends on the current density, which makes the energy absorption nonuniform in relation to depth. However, it is not a penetration depth in the same meaning as the radiative case.

However, in the case of the strict RF-current energy transfer, the Joule heat and the delivered direct energy will dominate the process; the current density determines the actual energy delivery. The current goes through the entire body, it starts out in an electrode and finishes in the other one, and no current loss occurs (see Fig. 4.9). The absorbed energy in all identical layers is identical if the medium is homogeneous,

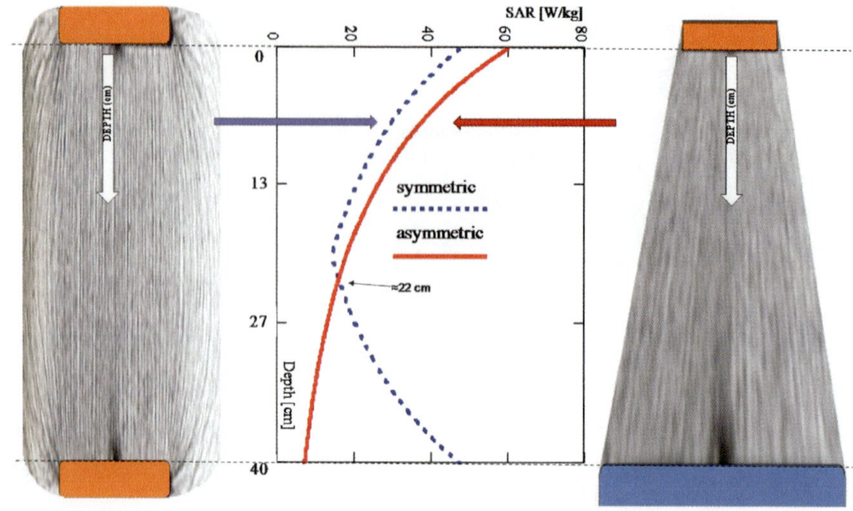

Fig. 4.9 The energy distribution of the conductive solution by depth of the absorption

only the volume of the layers changes by depth. This is crucial for the control and the quality-reproducibility of the treatment [526].

4.1.7 Arrangement of Electrodes

The most important part of oncothermia systems are the applicators, the electrodes. Their construction and materials definitely decide the success of the treatment and the safety of the processes. Most of the know-how is connected to these applicators, which were developed through the combination of long experience in biophysics and high-tech applications. The materials, shapes, arrangements are carefully selected and set after long experimental work. The dielectric constants, isolations, water circulation, and the bubble-free solutions are all production secrets to provide the best service and highest safety.

The symmetric and asymmetric electrode arrangements differ in their SAR distribution (the symmetric has equaled heating at both ends, while asymmetric has higher energy absorption until a depth of 22 cm). The equal point of the absorbed energies could be defined here as penetration depth, which is, in general, over 10% higher than the radiative solution. In the case of the not-well-controlled conductive solution the radiative and conductive techniques are mixed, the penetration depth cannot be definitely determined.

The conductive solution has a self-focusing automatism. The area which has lower impedance (simply higher conductivity), will have more currents flowing through it, and more energy will be absorbed in that volume. The absorption has similar mechanisms for both the conductive solutions – symmetric and asymmetric –

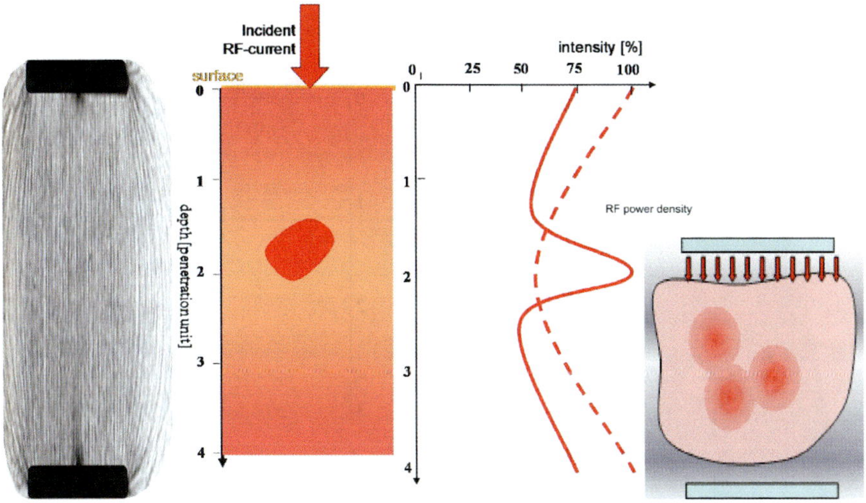

Fig. 4.10 The self-focusing mechanism in the symmetric conductive arrangement

as shown in Figs. 4.10 and 4.11, respectively. The asymmetric arrangement has the same conditions; in the equal thick layers an identical amount of energy is absorbed. The only difference is the continuous widening of the treated volume. This lowers the SAR steadily. However, the absorbed energy is higher in the asymmetric case than in the symmetric one over half of its depth, because the higher conductance of the lower part will request higher current at the same power. So the efficacy of the tumor-destruction is definitely larger in asymmetric case.

To evaluate the self-focusing efficacy we may perform some estimative calculations. The actual schematics (built in by discrete elements instead of the real continuously distributed electronic parts) can be constructed by dividing the treated volume into three layers: the healthy layers above and below the tumor-containing layer. The boundary of the division has to be chosen by the size of the tumors, and the dividing line has to be equipotential. This is not necessarily a flat plane, its curvature is not relevant in the calculation). The three hypothetic layers and their schematics are shown in Fig. 4.12.

Because of the conductive solution, in the homogeneous case the RF-current heats up the surface to the highest temperature. The deeper tissues have no possibility to be heated up higher, except where their conductivity focuses the RF-current. This focusing could be the abdominal acitis, pleural liquid, etc., which anyway must be drained before any hyperthermia treatments. The blood vessels are isolating enough to avoid heating of the blood, which anyway intensively circulates and cools down immediately any overheated blood volumes. In consequence of these, controlling the surface temperature alone is enough to control the safety in full. (The surface temperature is anyway a "sensitive point," due to the adipose tissue layer in the near-surface region of the body). The surface control is simple to solve

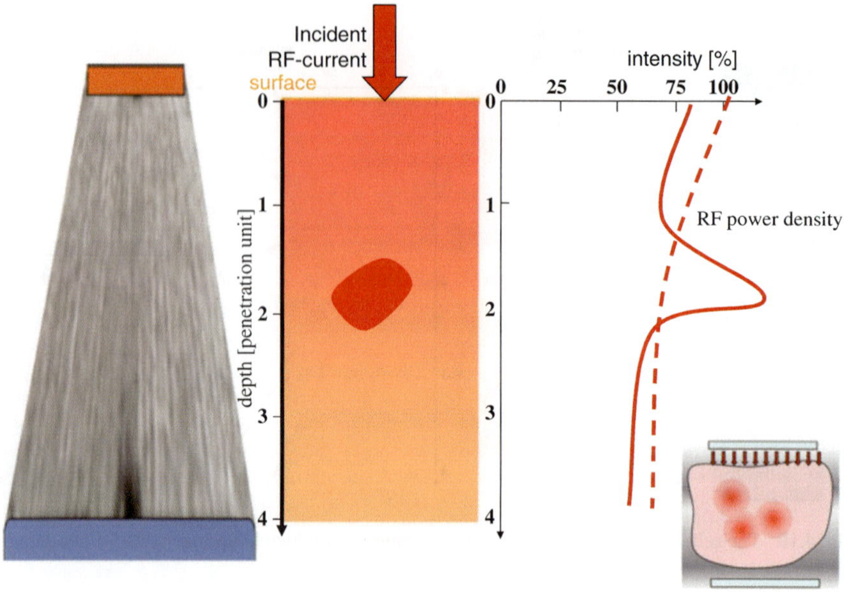

Fig. 4.11 The self-focusing mechanism in the symmetric conductive arrangement

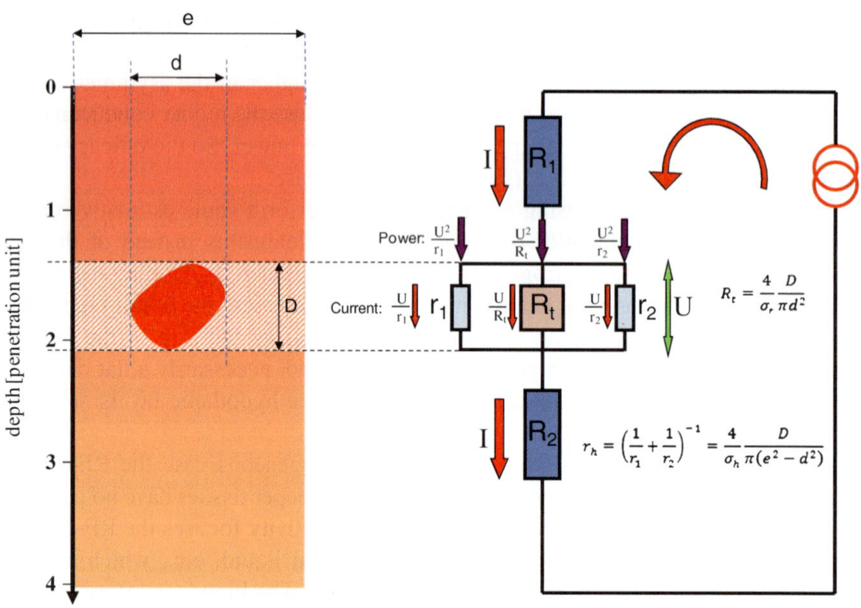

Fig. 4.12 Schematics of the layer division and the corresponding circuit built by discrete elements (The *red arrows* denote the currents, the *purple* are the actual power depositions)

by the patient's usual thermal sensing, which is physiology determined and its "sensors" are located in the near-surface area. Therefore, the safety of oncothermia is controlled by the patients. (To treat patients with no or limited thermal sensing is contraindicated, or at least needs extra care and temperature measurement under the electrodes on the surface.)

However, in the development phase of the method, the temperature was carefully checked in the tumor for clear safety reasons. The invasively placed thermosensors were handled by interventional radiologists under image guidance (MRI, CT, or ultrasound). One of the CT-guided human temperature validations is shown in Fig. 4.13.

The applied thermometer was a remote-controlled four-channel extremely thin thermocouple (see Fig. 4.14), developed for such purposes, or a fluoro-optical system was successfully (see Fig. 4.15) used in laboratory experiments as well.

The developed intra-tumoral temperature of course differs from the temperature, which could be reached by the healthy tissues in the same volume. The standard phantom for the healthy material could be agar-agar infiltrated by isotonic water, and heated by the same power. The temperature reached by the phantom is of course different than we can reach in the same volume of tissue. The phantom temperature is called "equivalent" temperature. The meaning of this temperature is the possible

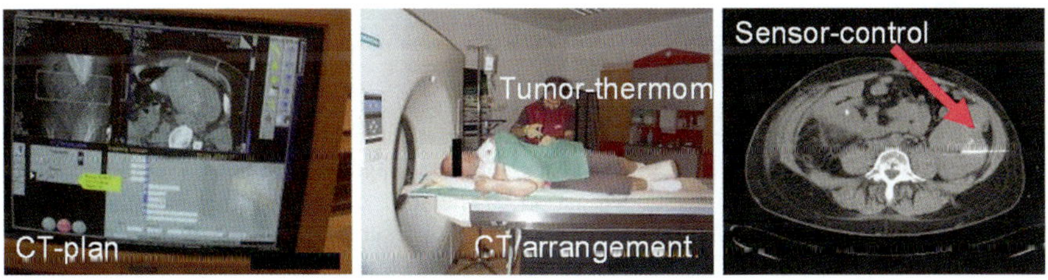

Fig. 4.13 CT-guided sensor placement {Institute: University Witten-Herdecke, Institute of Microtherapy, Bochum, Germany, Investigators: Prof. Dr. D. Gronemeyer, and Dr. H. Sahinbas}

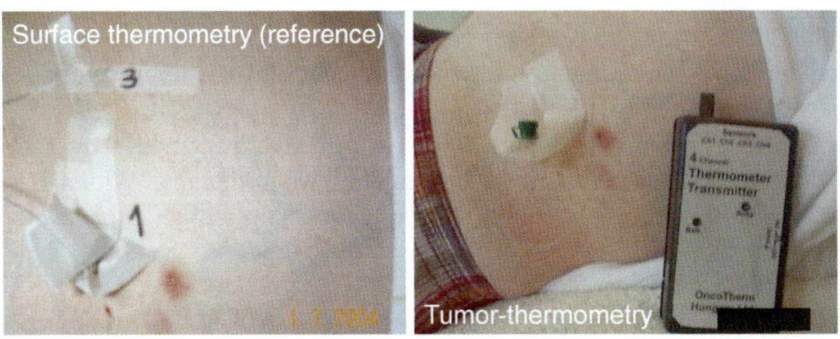

Fig. 4.14 Remote-control operated temperature measurement on the body surface and inserted into the patient. {Institute: University Witten-Herdecke, Institute of Microtherapy, Bochum, Germany, Investigators: Prof. Dr. D. Gronemeyer, and Dr. H. Sahinbas}

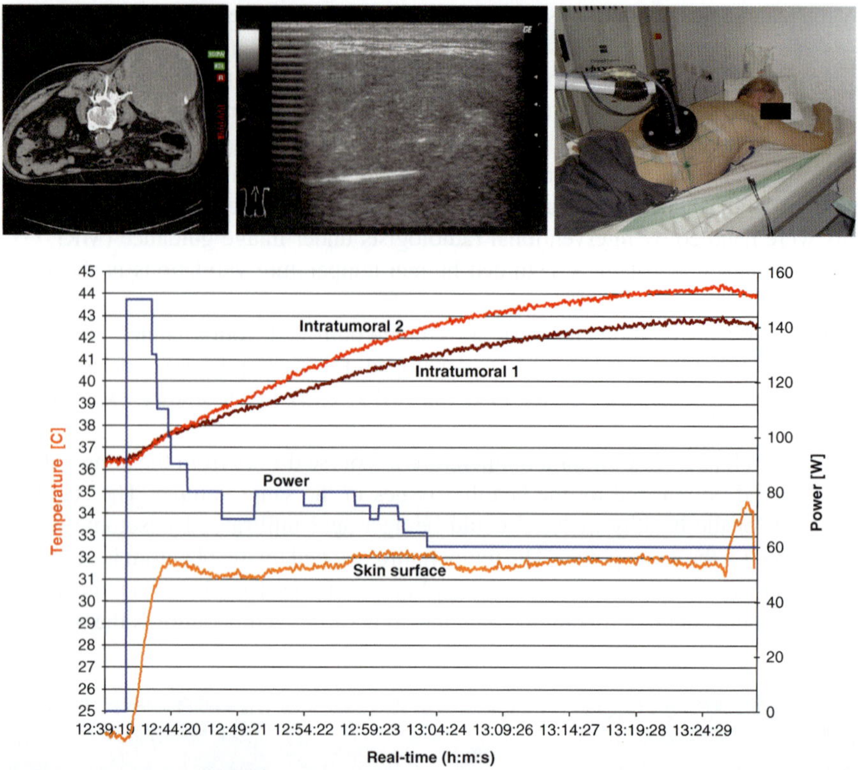

Fig. 4.15 Temperature measurement of a huge sarcoma. Patient: male, 87 years; Tumor: soft tissue sarcoma on the right side of the back, Primer diagnosis: 12/07 CT-guided biopsy, Histology: Malignant fibrotic hystiocytoma G3, Therapy: curative, Radio-Thermo-Therapy (Double-modality), Intra-tumoral in situ temperature measurement first Oncothermia, afterwards radiotherapy, Dose 22 Gy, 6 Fractions. {Praxis at Klinikum Nord Nuernberg, Germany, Investigator: Prof. Dr. H. Renner}

achievement when no chemical or physiological changes are counted. The equivalent temperature is the maximal which is taken by the target and physically this means that all the provided energy is converted to temperature.

A temperature measurement in a gynecological case. The temperature development is well controlled (see Fig. 4.16).

The non-invasive measurements were performed transcranially by a Toshiba 0.062 T open MRI system, in the Institute of Microtherapy, Witten-Herdecke University (see Fig. 4.17). The results show pretty well the temperature development in the tumorous lesion.

4.1.8 Far from Equilibrium

The consequence of the selective and direct chemical process targeting approach is the non-equilibrium energy distribution. (Sure, the energy has to be distributed

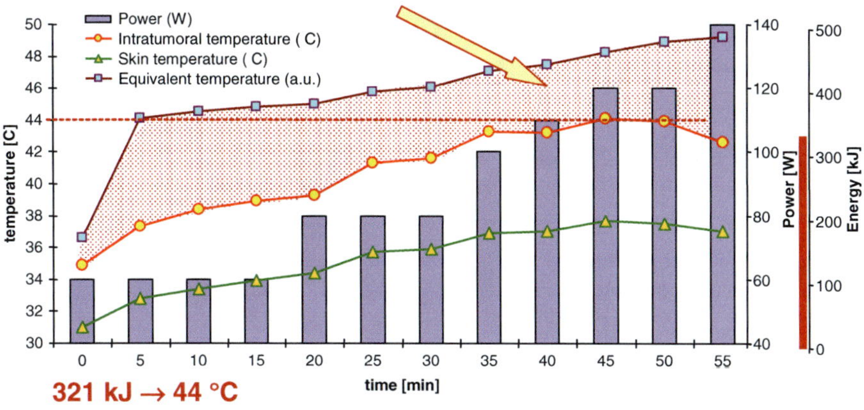

Fig. 4.16 Invasive temperature control in a breast cancer case. The gap between the equivalent temperature and the real temperature is the "missing" energy expended to break the chemical bonds. This gap includes the wasted energy also, which is minimized by oncothermia. {Institute: University Witten-Herdecke, Institute of Microtherapy, Bochum, Germany, Investigators: Prof. Dr. D. Gronemeyer, and Dr. H. Sahinbas}

nonequally and it has to support special processes and not affect all the processes in the living system.) In the well-focused case, the system shifts far away from the thermal (thermodynamic) equilibrium due to the thermal gradients between the target volume and its neighborhood.

The specialty of oncothermia is to deliver energy which is distributed nonhomogenously in the target. Consequently oncothermia causes field and temperature gradients at microscopic scales, having constrained temperature gradients together with gradients of other thermodynamic intensives (e.g. pressure, chemical potential, electric field, surface tension, etc.) The constructed gradients are the driving forces for internal flows and changes. The constrained flows are governed in the target electrolytes through their membrane-separators by Onsager's relations (see Appendix 2). In consequence oncothermia modifies the membrane ionic exchanges together with various other microscopic (e.g. cellular junctions, adherent connections, signal pathways, etc.) structures and processes. These could change the homeostatic equilibrium in the selected volume and so could be manifested macroscopically (e.g. by blood and lymph flow, by injury currents, etc.) as well. The energetics of these processes has to be approached by non-equilibrium thermodynamic considerations rather than thermostatic equilibrium. While in classical hyperthermia according to Arrhenius (Boltzmann) formulations the equalized temperature was expected to perform the job, oncothermia centers on permanent microflows accompanied by macroscopic heat flow due to the local heating. In oncothermia the various currents (ionic, displacement, soliton waves, mass flow, etc.) and their interactions (cross-currents, electro-osmosis, etc.) are present and produce certain changes to promote the selective cell destruction. These are microscopically definitely thermal processes, absolutely governed by the temperature

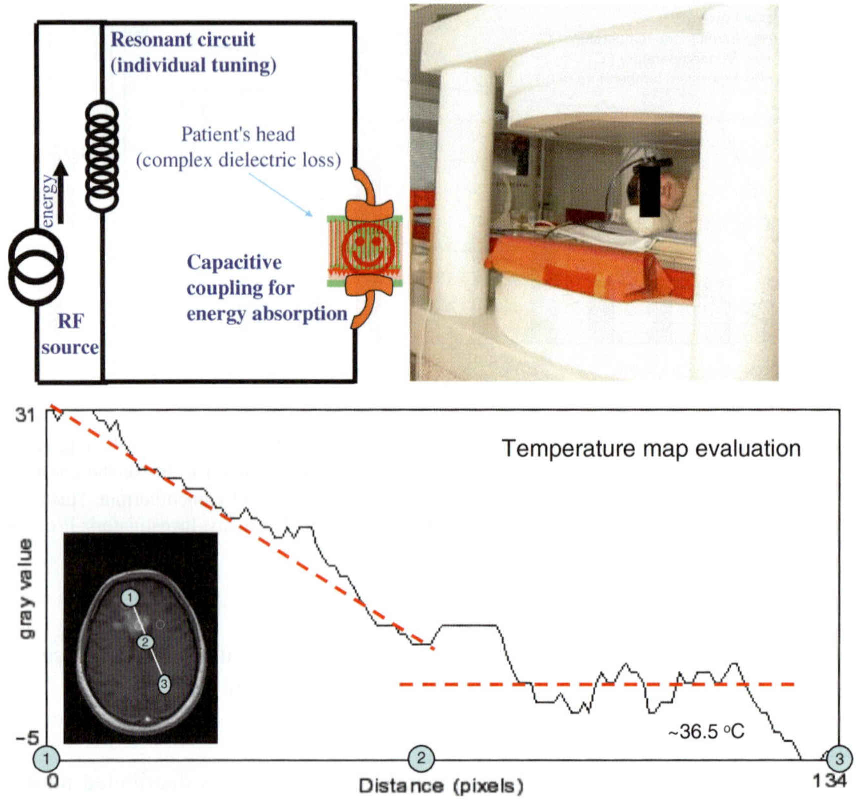

Fig. 4.17 Non-invasive temperature measurement in brain with an open MRI. {Institute: University Witten-Herdecke, Institute of Microtherapy, Bochum, Germany, Investigators: Prof. Dr. D. Gronemeyer, and Dr. H. Sahinbas}

gradients in the usual way of non-equilibrium thermodynamics. At the microscopic level we cannot speak of any NTD, while the *nonthermal* and *athermal* categories are also surely a wrong characterization of this situation. However, macroscopically the temperature change is not necessarily expressed; even the most drastic microscopic temperature gradients could be hidden in a macroscopic average. The macroscopic temperature change is observable only for a longer time.

The equilibrium thermal process always could be characterized by Arrhenius conditions (see Appendix 1), when the reaction rate depends on the exponential function of the inverse temperature. This simple relation is the consequence of the ratio of the activation energy (E_a) to the thermal one (kT), with increasing thermal energy the reaction (going over E_a) is more likely. The growing thermal energy means growth of the white-noise component also, which could certainly destroy the deterministic signals, and decreases the SNR.

To discuss hyperthermia in oncology on the basis of a linear Arrhenius analysis has numerous challenges:

- the kink of the Arrhenius graph depends on the applied chemotherapy [1143, 336];
- the kink of the Arrhenius graph depends on the prehistory and dynamics of treatment [1144–1148];
- the Arrhenius graph gives different time doses for the different points of the target (because of its non-homogeneous structure).

The above challenges could be explained by a double reaction kinetics as is shown by (3.10), Fig. 3.20. Reaction kinetics could be globally summarized by the reaction of cellular death: the energetic perturbation activates, excites the cell whose "excited" state could turn to "death" or back to "alive," Fig. 4.18. The final state could also be split into two basic cellular death processes: apoptosis and necrosis (see Fig. 4.19). These two death processes are always presented together in hyperthermia [768].

An Arrhenius plot of the general two-state case was shown in Fig. 3.20. The change of the Arrhenius plot by the number of states and by the activation energies is shown in Fig. 4.20.

4.1.9 Energy Intake and Temperature

The penetration depth in the RF-radiation process sharply and inversely depends on the wavelength [1149] and could be modified by the wave-source symmetry

Fig. 4.18 The global action mechanism of cell killing on a symbolic level. All the transitions have multiple reactions, which are characterized by an overall symbolic transition, characterizing the group

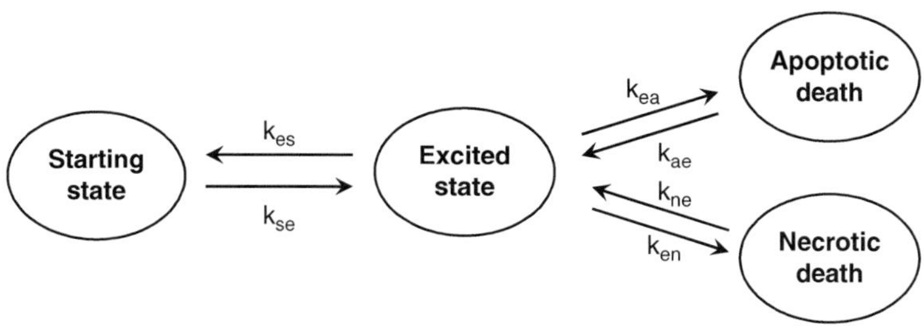

Fig. 4.19 The cell death has two distinguishable categories: apoptosis and necrosis. All the transitions have multiple reactions, which are characterized by an overall symbolic transition, characterizing the group

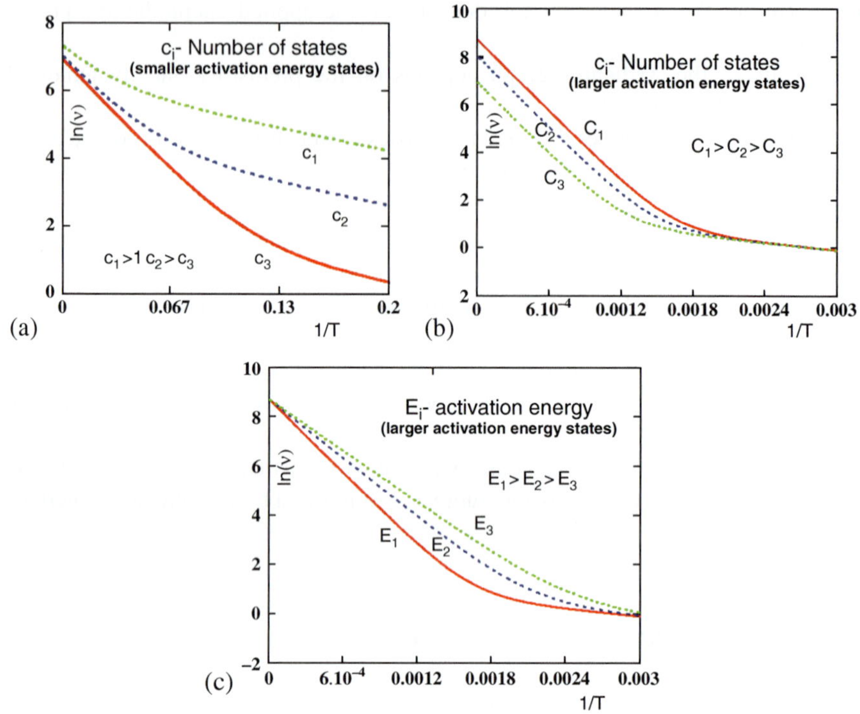

Fig. 4.20 The decreasing number of states of smaller activation energy process (**a**) decreases cell death in the low-temperature range; the decrease of the number of states of larger activation energy process (**b**) decreases cell death in the high-temperature range; while when the activation energy of the higher state shifted down, (**c**) cellular death decreases in almost all the studied interval

[1150]. Also, the absolute value of the absorbed energy in a definite depth is linearly proportional to the incident energy. The planar waves at the applied relatively low frequency generated by capacitive-coupled antennas penetrate into the body by 14–20 cm [1151, 1152]. (The penetration is defined by the depth where the energy intensity is about 36% of the incidence beam.) However, in the case of the strict RF-current energy transfer, the Joule heat and the delivered direct energy will dominate the process, the current density determines the actual energy delivery. This is crucial for the control and quality-reproducibility of the treatment [526].

Selection of the energy absorption defines the efficacy of the actual treatment. It could be shown [1153] that microscopically the extra-cellular matrix and the cellular membrane absorb the main energy at the chosen frequency. According to the calculations [1154, 1114, 1117] a relatively small amount of energy could heat up average-sized tumors to the appropriate temperature, if it targets the tumor accurately enough [1125]. To calculate very roughly the temperature gain, overestimating the heat losses, we can start with the basic equation of the elementary thermodynamics: (Eq. (3.1):

$$\Delta Q = mc\Delta T \tag{4.3}$$

where ΔQ is the absorbed energy, m is the heated mass, c is its specific heat, and ΔT is the temperature gain. Assume a large tumor, with a mass of $m = 3$ kg ($c = 4{,}183$ Ws/kg/K, the density of the tumor is 1.1 kg/dm^3) and the heating conditions are: power is $P = 150$ W, standing wave ratio SWR = 1.06, tumor absorption factor (ratio of the absorbed power from the initial) $a = 30\%$, and the constant energy loss of the tumor is 5 kW/m^3 (which is a huge overestimation, because the blood-flow power loss is about 3 kW/m^3, and the metabolic extra-power is not considered), and with initial body temperature, derived from (4.3):

$$\Delta T = \frac{\left(\frac{2aP}{\mathrm{SWR}+1} - \frac{m}{\rho} W_{\mathrm{loss}}\right)\Delta t}{mc} \tag{4.4}$$

after $\Delta t = 60$ min treatment time the tumor temperature is 45°C. Refinement of this estimation can be found in many publications [1147] dealing with the topic. Consequently, relatively low energy could be enough to reach the desired treatment conditions.

The real challenge is: to focus the energy on the target, to select the malignant cells (as microscopically as possible), to distinguish them from their healthy counterparts. Two definite selection mechanisms force the delivery of energy to the malignant target: the blood-perfusion regulation, and the bioimpedance. The specific absorption rate (SAR) is completely proportional to the temperature increase if nothing else (no other kind of energy is involved) happens in the heated object, only the average energy increases. (See Eq. (3.31) in the case where only the first term is nonzero, all the others vanish.) Because of the physiological changes (first of all due to blood-stream variation) the linear dependence breaks (see Fig. 3.5). However, when the above shown necrotic and apoptotic cell death could occur, certain energy is taken to those special processes, so a part of the energy will be taken for this distortion. This means that larger SAR is necessary to keep the given temperature (see Fig. 4.21).

The variation of the requested SAR to keep a given temperature naturally means: at constant SAR the temperature increase will not be constant. This alone would be enough to regard the temperature as false for dosing.

Changes of the blood flow are summarized in Table 4.1. The change affects not only the temperature itself, but affects the complementary applied chemo- and/or radiotherapy efficacy as well. The blood perfusion is mandatory for the drug delivery, but their oxygen content defines the efficacy of the radiotherapy. In consequence, both the frequently applied conventional complementary therapies are by far not independent of this perfusion issue (see Table 4.2). This specialty of the blood flow has to be considered for complementary chemo- and radio-potentiating (or resensitizing).

The vascular change of the heating-up process was tested on healthy Beagle dogs. The temperature was measured subcutaneously. A typical example of its obtained pattern is shown in Fig. 4.22 together with a scintigraphy check. Oncothermia

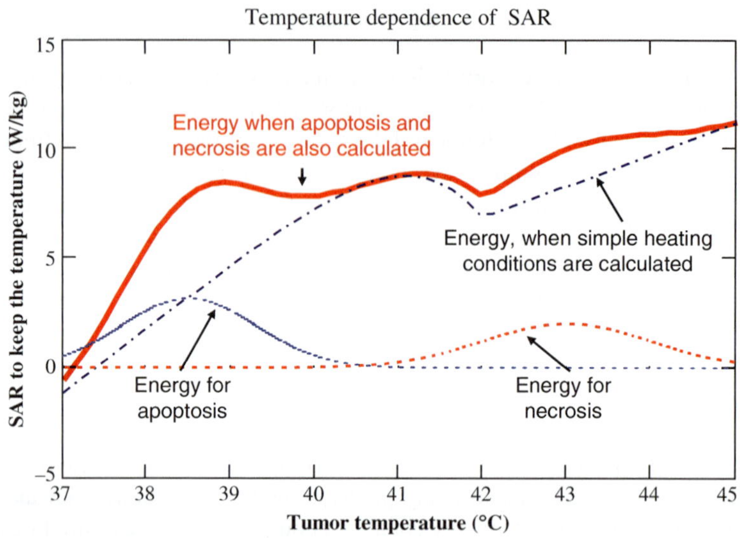

Fig. 4.21 The SAR which is necessary to keep the temperature shown in the abscissa. The simple heating (Fig. 3.5) is shown for comparison. The energy request of apoptosis and necrosis has different activation energies, so their energy-absorption peak is located at different temperatures

Table 4.1 Blood-flow changes and its consequences below and above the actual threshold for healthy and tumorous tissues

	Under threshold		Over threshold	
Temperature dependences	Healthy	Tumorous	Healthy	Tumorous
Blood flow	⇧	⇧	⇧⇧	⇩
Vascular permeability	⇧	⇧	⇧⇧	⇩

Table 4.2 The potentiating differences of oncothermia under- and over-threshold conditions

	Under threshold		Over threshold	
Potentiating dependences	Healthy	Tumorous	Healthy	Tumorous
pO2 (radio-efficacy)	⇧	⇧	⇧⇧	⇩
Drug delivery (chemo-efficacy)	⇧	⇧	⇧⇧	⇩

increased the temperature over 43°C and also promoted the uptake of the bone-specific radiopharmacon by 17%.

The focus could be measured by ^{99m}Tc-labeled radiopharmacon signaling, measured by Micro-CT and SPECT combination (see Fig. 4.23). In case of simple ^{99m}Tc-HAS the injected dose: 60 MBq; 5 min post treatment; increasing activity in the tumor compared to control: 12%. The ^{99m}Tc-liposoma experiment was

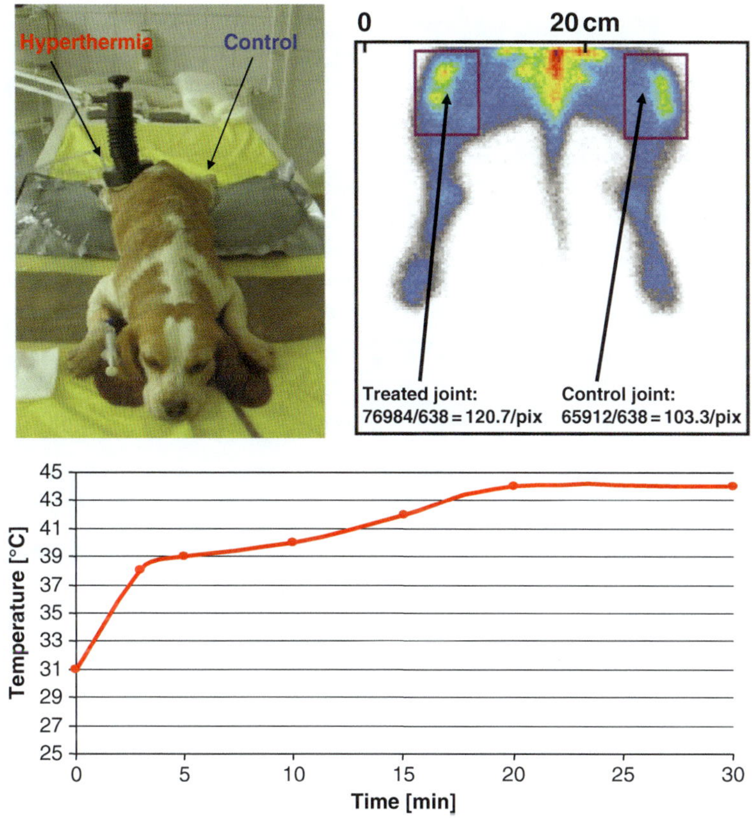

Fig. 4.22 The temperature development in a healthy animal treated with 20 W for 30 min. Scintigraphy of the joint was performed at the end of oncothermia, 1 h after the intra-venous injection of the bone-specific radiopharmacon labeled by 400 MBq 99 mTC-MDP. {Institute: Department of Pharmacology and Toxicology, Faculty of Veterinary Science, St. Istvan University, Budapest, Hungary, Investigators: Dr. G. Andocs}

performed by 20-MBq injected dose; 30-min post treatment; increasing activity in the tumor compared to control: 38%. The selective radiopharmacon accumulation of the treated tumor by oncothermia is proven also in the case of the liposome-jacketed doxorubicin (^{99m}Tc-Caelyx, injected dose: 20 MBq; 60-min post treatment; increasing activity in the tumor compared to control: 20%.)

4.1.10 Macroscopic Focusing on the Tumor

Both the conductivity and permittivity (so the complex dielectric constant) differs between the healthy and tumor tissues at the applied 13.56 MHz frequency. The difference originates simply from the structural and functional deviations of the malignant tissue from its healthy counterpart. These changes affect the impedance

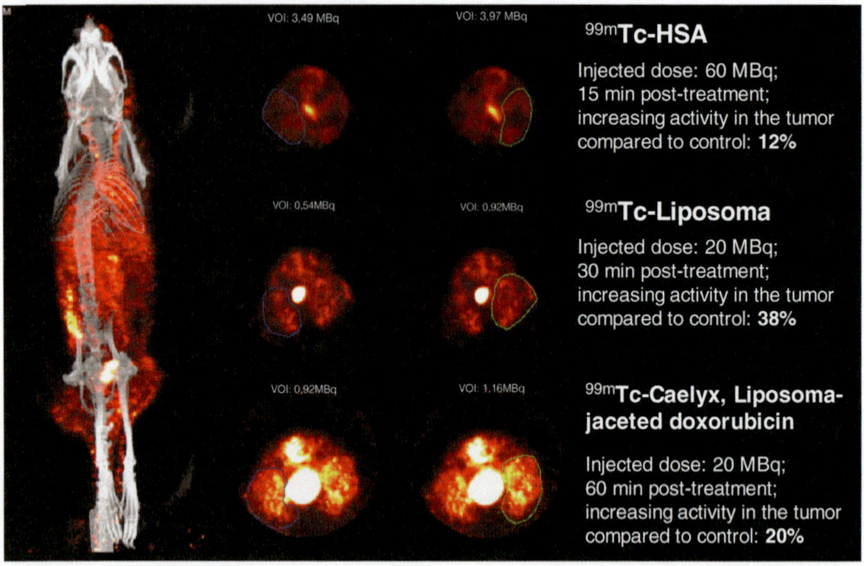

Fig. 4.23 Radiopharmacon: ^{99m}Tc-signaling, measured by MicroCT-SPECT {Institute: Department of Pharmacology and Toxicology, Faculty of Veterinary Science, St. Istvan University, Budapest, Hungary, Investigators: Dr. G. Andocs}

of the tumor, which makes it possible to selectively distinguish it from the healthy environment by the RF-current. The most important factors of the impedance change are:

- Tumors have mostly anaerobic (fermentative) metabolism, which request high transport volume of exchanges of necessary species (primarily glucose in and lactic acid out [71]). (The high glucose metabolism could be sensitively detected by Positron Emission Tomography (PET), which provides one of the most sensitive diagnostic images of the tumor. The higher molecular and ionic transport definitely changes the extra-cellular electrolyte in the vicinity of the malignant cell, causing higher ionic concentrations, which lower their impedance.
- In more advanced stages (when the angiogenesis is active) the blood perfusion in the area is more than the usual healthy situation [1155], which is again an impedance-lowering factor. It is characteristically enhanced by the development of the tumor.
- The water state in the extra-cellular liquid around the tumor is more disordered [697] (less hydrogen bridges exist), which decreases the proton (hydronium ion) mobility, but at the same time increases the mobility of all the other ions, which decreases the impedance.
- Most tumors have lower pH [1156] (mainly due to the lactic acid) than their environment, which is again a factor that decreases the impedance.

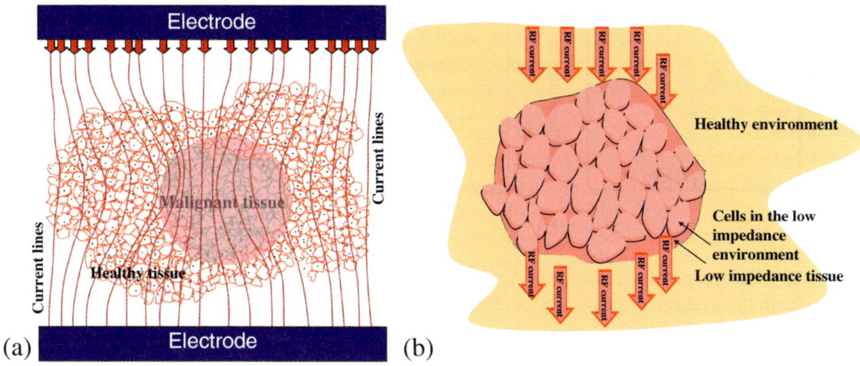

Fig. 4.24 The current-density distribution near the tumor. The tumor focuses the RF-current to higher conductivity (**a**), the current flows dominantly in the extra-cellular matrix, due to the isolation of the membrane (**b**)

- Most of the tumors metabolic rate is higher than their healthy counterpart (at least 15% over [1157, 1158]), which selectively increases the temperature of the tumor. The higher temperature decreases the impedance of the tissue [794].

As we have shown, the impedance effect settles the current flow in the tumor (see Fig. 4.24), which indeed could be measured by a Current Density Image in a special MRI arrangement.

This selective factor (which according to a wide range of literature ranges from 20 to 4,000% difference between the healthy and malignant tissues) is applied by the oncothermia technique. Impedance selectivity is applied in some of the tomography devices (for example: Siemens oncological impedance-tomography device [790]. A special index has also been introduced for impedance selection [766]. The results are convincing: the mean value of the indexes of normal tissues (various human tissues were investigated and bone was excluded) is 0.68 and for cancerous tissues is 1.8. This selectivity factor leads to the self-focusing of the energy to the tumor (see Appendix 24).

A sample of the results is shown below. The parameters which we used are: forwarded (original) power $P = 200$ W, cooling power $P_c = 45$ W, cooling penetration depth $L_H = 3$ cm, electrode radius $R = 6.75$ cm, electrode distance $L = 30$ cm, starting depth of the tumor $z_1 = 10$ cm, final depth of the tumor $z_2 = 15$ cm, tumor radius $r = 2.5$ cm, original body temperature $T_o = 36.5°C$, metabolic rate in healthy tissue $M = 2$ W/dm^3, metabolic rate in the tumor tissue $M_t = 2.5$ W/dm^3, blood-flow rate in healthy tissue $w_b = 2,127$ g/s/dm^3, blood-flow rate in the tumor $w_t = 0.5$ g/s/dm^3. The results for this parameter set are shown in Fig. 4.25.

A further macroscopic (focusing) selective heating factor of oncothermia is the self-supporting positive feedback from the growing temperature. By the temperature enhancement the impedance is lowered, [794] increasing the selective differences

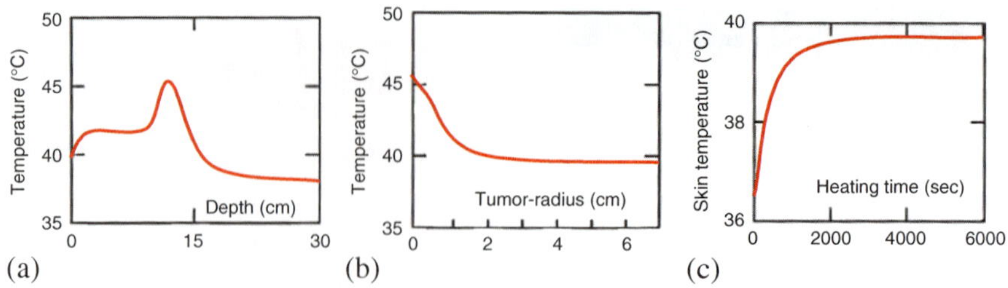

Fig. 4.25 The temperature change in stationer state by depth (**a**), by the radius of the tumor from its center in the average direction (**b**), and the outermost skin by elapsed time (**c**)

between the malignant and healthy areas. The 36→43°C heated tumor increases its relative conductivity at 10 MHz normalized to 10 kHz (10 MHz/10 kHz) by 15% compared to the unheated tumor [768]. The measured gain of the selectivity is 2%/°C [769].

The auto-selective focus of oncothermia was measured in vivo also. The experimental animal was a SCID mouse with human medulloblastoma tumor. Location of the tumor was in the femoral region, both sides of the animal was symmetrically inoculated for control purposes. Radiopharmaceutical: ^{99m}Tc labeled-liposome (experimental product of OSSKI). Injected dose: 35 MBq/0.1 ml micro-SPECT – Single Photon Emission Computed Tomography (internal radiation by pharmaceuticals labeled by radioactive isotopes). SCID mouse xenograft model, investigating to effect of oncothermia to the radiopharmaceutical (^{99m}Tc-labelled liposome). Injected dose: 35 MBq/0.1 ml. The tumor on the right femoral region was treated with oncothermia right after the radiopharmaceutical injection for 40 min, the left-side tumor was the control. The animal was anaesthetized during the treatment. (Ketamin+Xylazin, i.p.) The slide-by-slide analysis of the scan shows well the localization of the oncothermia effect in the target (see Fig. 4.26).

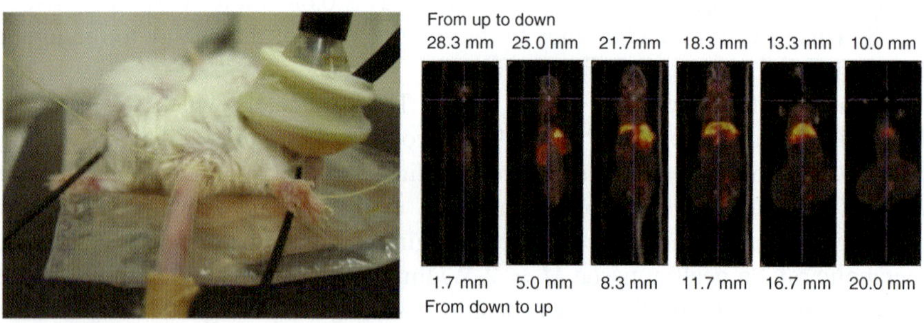

Fig. 4.26 Single-Photon Emission Computed Tomography (SPECT) of a SCID mouse (slides are shown in order from the bottom up). The focusing (radiopharmacon) is well demonstrated at a depth around 11 mm. The same enrichment of radiopharmacon could not be seen in the control tumor on the *left-hand side* {Institute: Department of Pharmacology and Toxicology, Faculty of Veterinary Science, St. Istvan University, Budapest, Hungary, Investigators: Dr. G. Andocs}

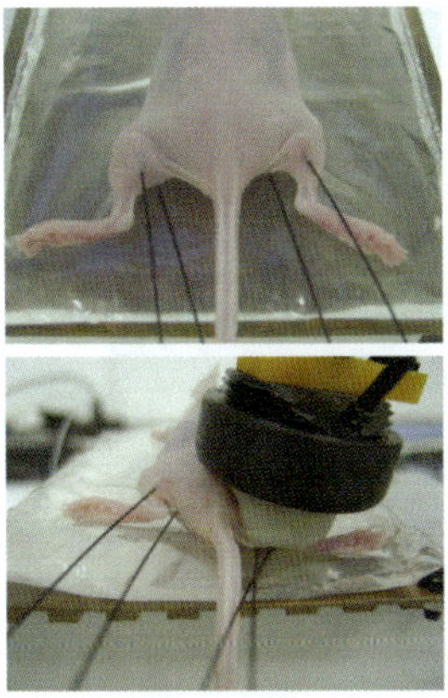

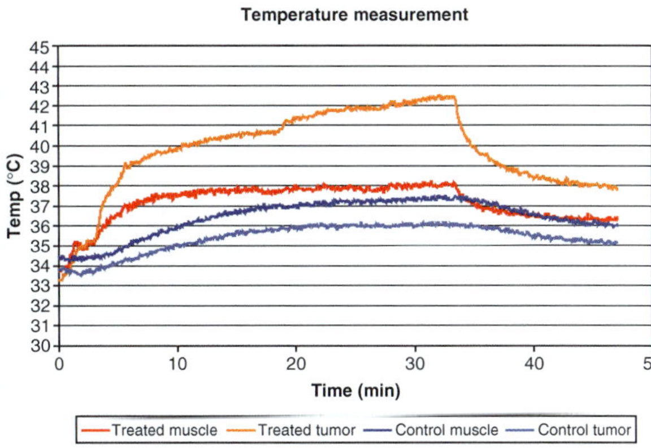

(a) (b)

Fig. 4.27 Selectivity in vivo experiments: (**a**) HepG2 tumor xenograft nude mice with temperature measurement probes in the tumors: (**b**) comparative energy absorption (temperature) measurement {Institute: Department of Pharmacology and Toxicology, Faculty of Veterinary Science, St. Istvan University, Budapest, Hungary, Investigators: Dr. G. Andocs}

The focusing was measured again by temperature distribution in the tumor and the surrounding healthy muscle (see Fig. 4.27). The appropriate focus had been observed in these animal models.

The proper focus is morphologically well demonstrated in the brain treatment of mice too, where the sharp border of the cancer area is untouched, see Fig. 4.28.

A significant difference in application of the classical and oncothermia methods lies in this macroscopic focusing mechanism. The artificial focusing (radiative and magnetic approaches) directs the energy absorption into the area where the malignancy is located, trying homogeneously to heat up the volume. It is in most cases cannot be homogeneous at all. In the case of multi-centered tumor-lesions the heated volume includes both the healthy and malignant areas, see Fig. 4.29a. One of the great advantages of oncothermia that the actual differences in complex impedance and conductivity determines the area of the heat absorption irrespective on its size or multiplicity, (see Fig. 4.29b).

4.1.11 Heating the Extra-Cellular Electrolyte

The selection process on the macroscopic level is completed with microscopic selectivity, governed by the cellular and sub-cellular processes induced by the oncothermia treatment.

From a dielectric point of view, the simplest approach is to study the cellular structure in its three rather different parts (see Fig. 4.30). As we showed above, the membrane has large impedance (isolation) for the current flow, which of course decreases linearly by the increasing frequency.

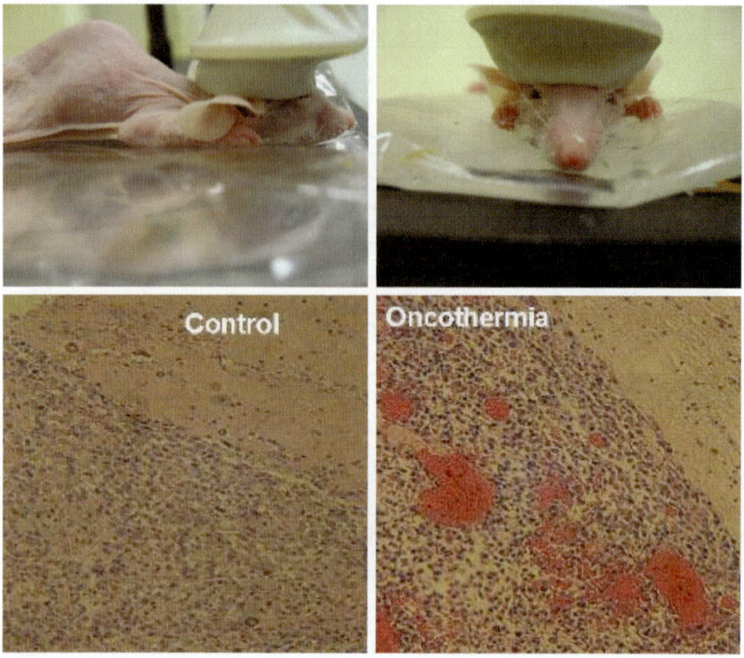

Fig. 4.28 Selectivity in vivo experiments (fixed sample): The definite borderline of the GL261 murine glioma (brain) tumor in nude mice shows the tissue selectivity. {Institute: Department of Pharmacology and Toxicology, Faculty of Veterinary Science, St. Istvan University, Budapest, Hungary, Investigators: Dr. G. Andocs}

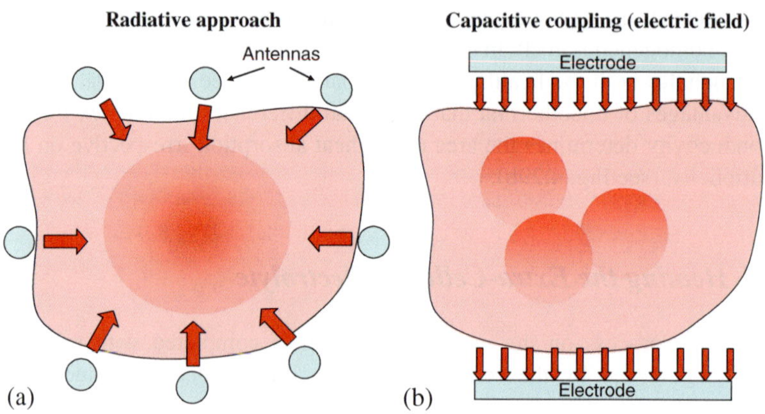

Fig. 4.29 Focusing arrangement by artificial focusing (radiative and magnetic approaches) (**a**) and arrangement in oncothermia (self-selective techniques) (**b**)

Fig. 4.30 Cellular structure by its dielectric properties

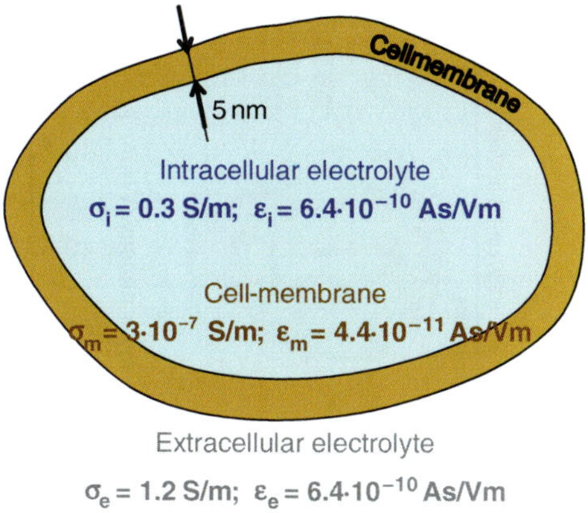

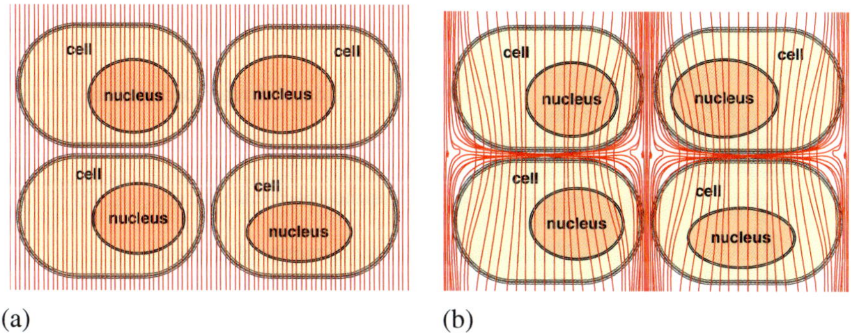

Fig. 4.31 The current density distribution is frequency-dependent. In the high frequency range (>20 MHz) the distribution is homogeneous, irrespective of the various components in the tissues (**a**), while at low frequency (<20 MHz) the RF-current dominantly flows in the extra-cellular "channels."

The applied 13.56 MHz of oncothermia partly allows the RF-current through the membrane, but the current mainly flows in the extra-cellular electrolytes, while at higher frequencies (over 20 MHz) there is no distinguished pathway observed (see Fig. 4.31).

The frequency-dispersion energy absorption of the characteristic cellular structures is calculated [1153] (see Fig. 4.32), and the absorption in the intra-cellular electrolyte has a "shoulder in the region of 10 MHz, and much lower energy intake takes place there than either in the extra-cellular liquid or in the membrane itself. The membrane absorption of the energy is considerable and dominating at the frequency range lower than the GHz-region.

The membrane-bound water has special beta-dispersion energy absorption in the RF-range [893]. The electric field is highly enhanced in the membrane bound water in the oncothermia frequency range, which produces an additional effect in the oncothermia action (see Section 4.1).

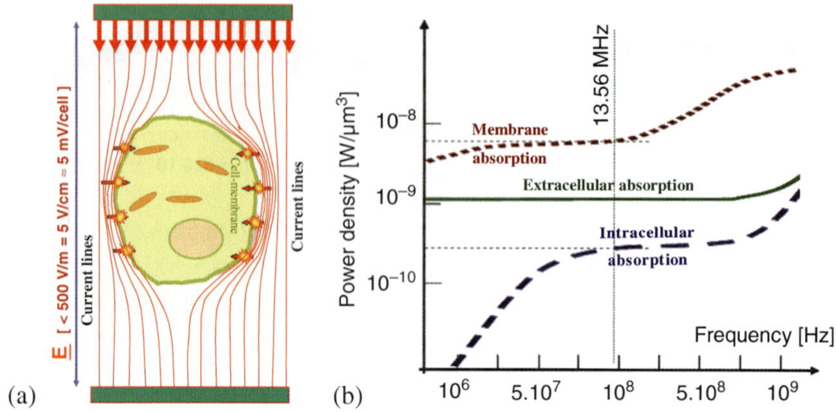

Fig. 4.32 Frequency-dependence of the absorbed energy for the cellular parts [1155]

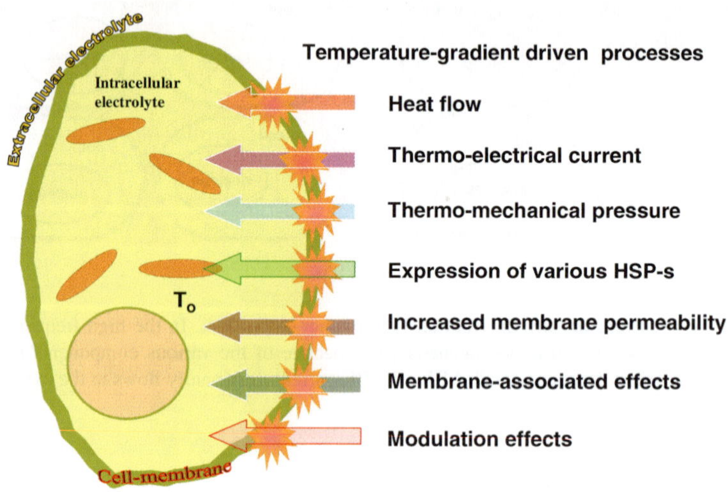

Fig. 4.33 Numerous factors of membrane effects in action

We summarized these microscopic effects in Fig. 4.33, which form the essence of oncothermia. In the following sections of this chapter detailed descriptions of these processes are given.

4.1.12 Temperature Gradient and Heat Flow on the Membrane

When the frequency is low enough to allow selective energy absorption, then the temperature gradient has an essential role in membrane destruction. In this case the

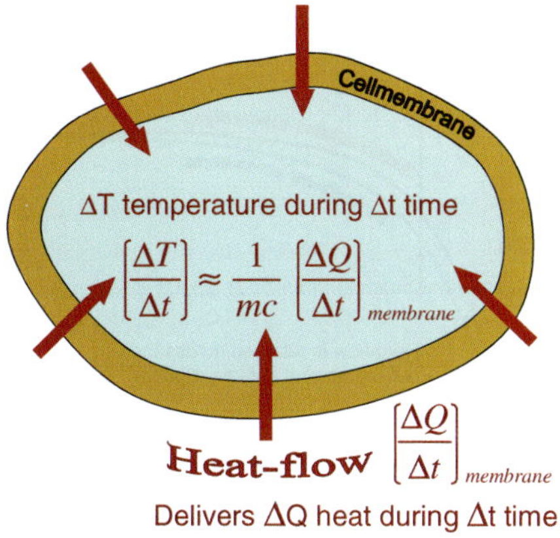

Fig. 4.34 Central symmetry of heat flow (direction is to the center of the cell, having no unidirectional constraint) makes the excitation noise zero order

current flows mainly in the extra-cellular electrolyte and creates a temperature gradient between the inner and outer side of the cellular membrane. This temperature gradient has an essential role in the oncothermia processes (see Appendix 25).

However, the rapid membrane equalizing process means only rapid heat transfer, but does not mean infinite energy transfer, it does not realize prompt thermal equilibrium between the extra- and intra-cellular electrolytes, see Fig. 4.34:

$$\left(\frac{\Delta T_{\mathrm{i}}}{\Delta t}\right) \approx \frac{1}{m_{\mathrm{i}}c_{\mathrm{i}}}\left(\frac{\Delta Q_{\mathrm{e}} - \Delta Q_{\mathrm{memb}}}{\Delta t}\right) \cong \frac{1}{m_{\mathrm{i}}c_{\mathrm{i}}}\left(P - \frac{\Delta Q_{\mathrm{memb}}}{\Delta t}\right) \tag{4.5}$$

where the indexes i, e, memb, denote the intra-cellular, extra-cellular, and membrane parameters, respectively; P is the RF-power.

The extra-cellular electrolyte is permanently heated by the current flow constrained by the outside RF-source. The extracellular heats up the intra-cellular liquid, but it takes time to pump into the cell such an amount of energy, which heats up the cytoplasm to the level of the surroundings (see Fig. 4.35a). The temperature difference between the electrolytes on the two sides of the membrane changes by the elapsed time (see Fig. 4.35b), and disappears in equilibrium. This disappearance leads oncothermia to equilibrium heating, which is identical to other classical hyperthermia processes. The full oncothermia process is tuned for 60 min to reach equilibrium in an average deep-seated solid tumor, however in the case of the intensively cooled lung (by breathing) and liver (by blood flow) 90 min is characteristic.

A simple example: the heater in the room has a metallic wall with extremely rapid heat flow through this wall, however it takes a pretty long time to heat up the whole room, because the rapid heat flow does not mean unlimited energy intake. The heating surface area, and the heat capacity of the heated volume together with the temperature gradient and the rapid heat exchange determine the time to reach the equilibrium temperature, which could be pretty long even with the prompt heat

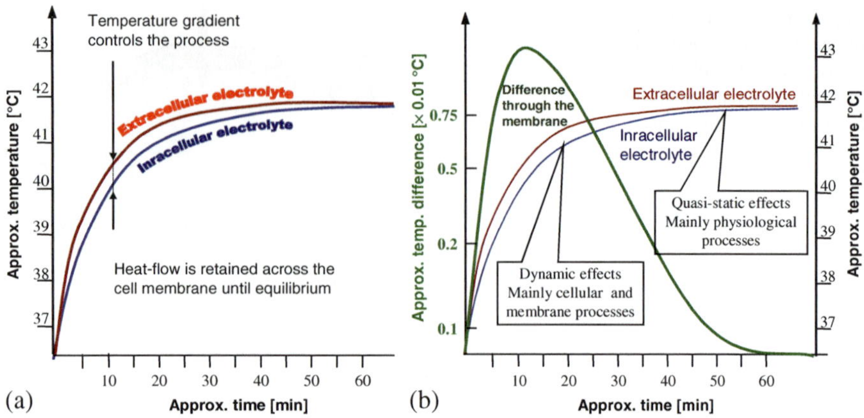

Fig. 4.35 The temperature gradient develops with time between the inner and outer side of the membrane. Approximately 60 min is necessary to reach the equilibrium, (**a**). (However, its timing depends on the possibility of energy transport into the extra-cellular volume.) The gradient between the membrane sides develops to a maximum at a third of the process, and goes back to zero by the end of it (**b**)

exchange of the system, when the quantity of the full energy (heat-energy flow) is limited by other conditions.

Despite the identical temperature characterization, we observed distinct differences between the treatments of oncothermia and classical hyperthermia, (suspension culture HL-60 human leukemia cell line, [1159]). The reason – as we described above – is the constrained non-equilibrium thermodynamic transport in oncothermia, which destabilizes the cell-membrane, increases its permeability and could make its bubbling and distortion [1160, 1161]. Due to the different energy-transfer mechanisms the efficacy of oncothermia is higher than its temperature-equivalent hyperthermia counterpart (see Fig. 4.36).

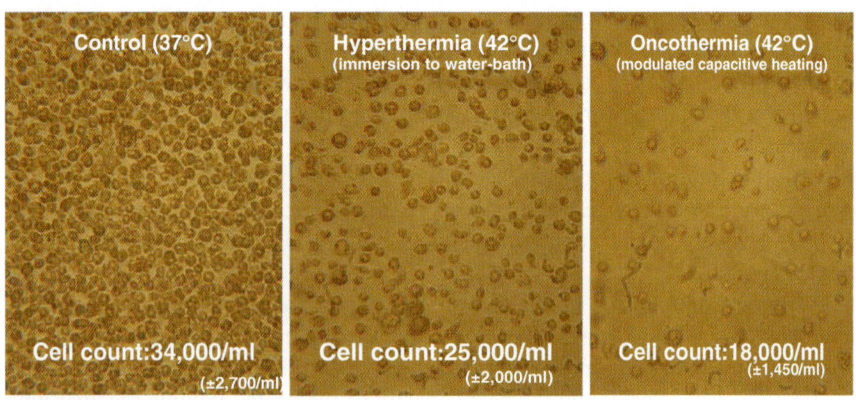

Fig. 4.36 Comparison of hyperthermia and oncothermia in vitro (suspension culture) at the same temperature in a HL-60 leukemia cell line [1162, 1159]

Selectivity on the cellular level could be demonstrated by co-culture experiments. Malignant cells are cultured together with healthy fibroblasts, and studied. The aggressively malignant A431 cell line was drastically reduced in the culture by oncothermia, while the healthy fibroblast matrix remained intact (see Fig. 4.37). In the case of the less aggressive HaCaT cell line the selection was not so marked (see Fig. 4.38).

The metabolic activity also changes through the oncothermia, depending on the malignancy of the cell line. In the measurements cells were plated into slide flasks (100,000 cells/ml) and incubated for 24 h at 37°C. Oncothermia was performed, and the cellular metabolic activity was measured using the MTT assay [1163] and quantified at 630 nm (E630). The metabolic activity (as a factor of the selectivity of oncothermia) determines well the ability of the cellular killing (see Fig. 4.39).

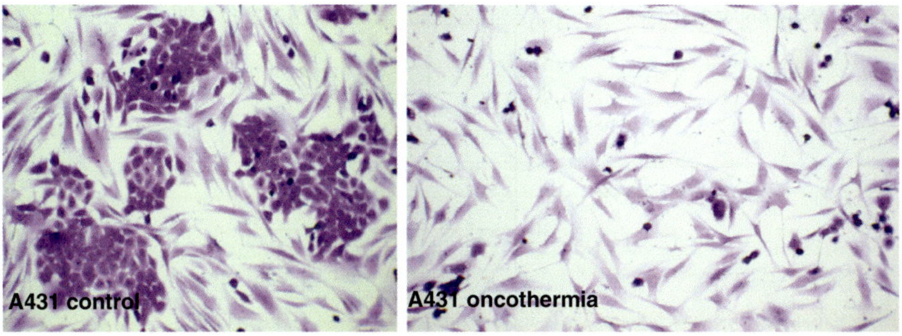

Fig. 4.37 Selectivity in vitro experiments: only the aggressively malignant A431 cells are destroyed in a co-culture with non-malignant fibroblasts (100,000/ml, incubated 24 h at 37°C, stained with crystal violet after fixing) [767]

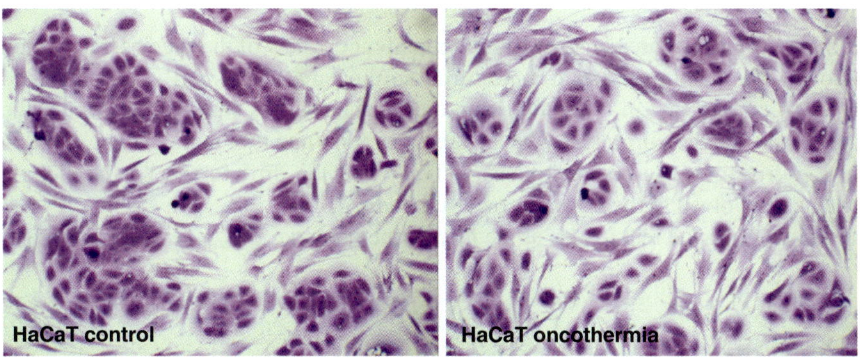

Fig. 4.38 Selectivity in vitro experiments: less effect could be observed on the non-aggressively malignant HaCaT cells in a co-culture with connective healthy fibroblasts (100,000/ml, incubated 24 h at 37°C, stained with crystal violet after fixing) [767]

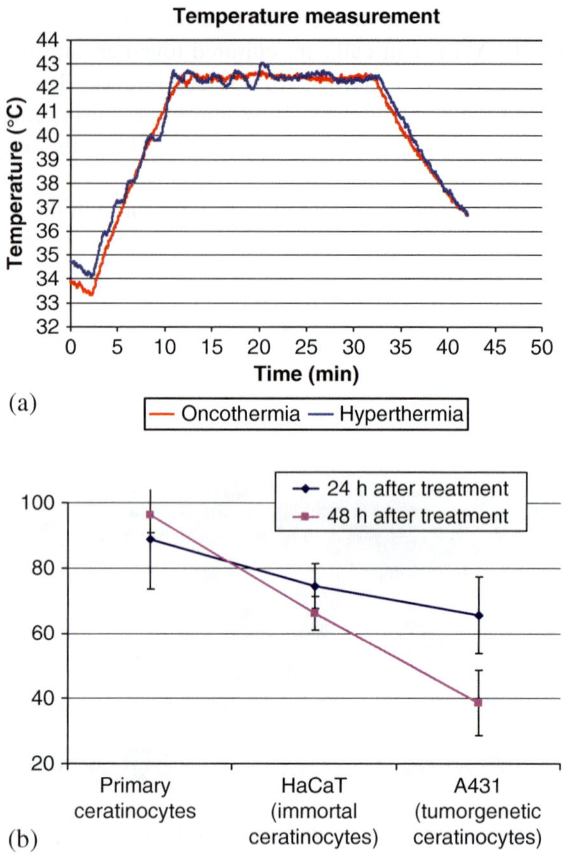

Fig. 4.39 Co-culture experiments with a mixture of healthy human fibroblasts and other cell lines. The temperature pattern of the control and the active specimens were identically managed, (**a**) but the treatment was different. The selected cell killing is well correlated to the cellular metabolic activity of the cell line (**b**). Data represent the mean value with standard error of mean (±SEM) of 4–6 separate experiments assayed in triplicate, but some experiments were repeated up to 12 times to obtain reliable data

4.1.13 Changes of the Membrane Potential

Multiple transports are active on the cell membrane. Many of the species transported through the membrane are in ionic form, so their mass transport is accompanied by charge transport (in other words by electric current) as well. Consequently the mass- and charge-current density of the transports has two components: the diffusion driven by the concentration gradients and the drift forced by the electric potential involved in the transport (see Appendix 26).

Substituting the real values of the membrane potential $\psi_{memb} = -70.2$ mV, and

$$C_{Na}^{(i)} = 145,\;\; C_{K}^{(i)} = 5,\;\; C_{Cl}^{(i)} = 125,\;\; C_{Na}^{(e)} = 15,\;\; C_{K}^{(e)} = 150,\;\; C_{Cl}^{(e)} = 9,$$
$$\omega_K : \omega_{Na} : \omega_{Cl} = 1 : 0.04 : 0.45,\quad T = 36°C \tag{4.6}$$

the slope of its temperature-dependence from (A.26.8) equals −0.23 mV/K. We may conclude that the cellular membrane will not be affected to too great an extent by the

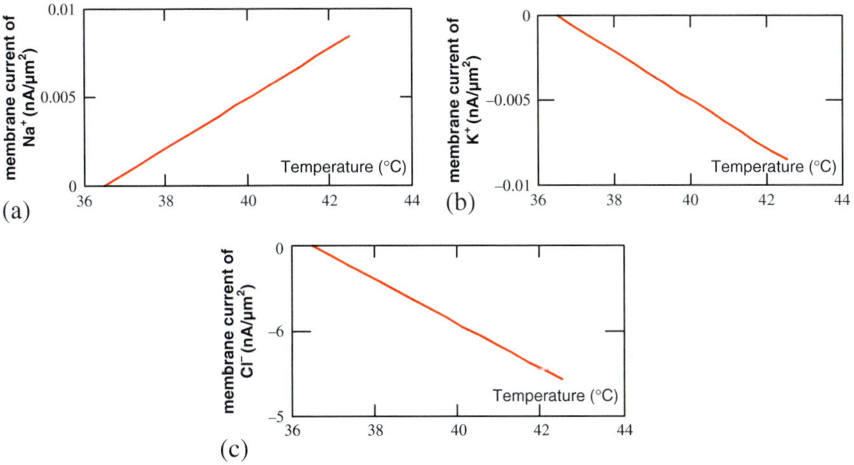

Fig. 4.40 Temperature dependence of the ionic currents through the membrane Na^+ (**a**), K^+ (**b**), and Cl^- (**c**)

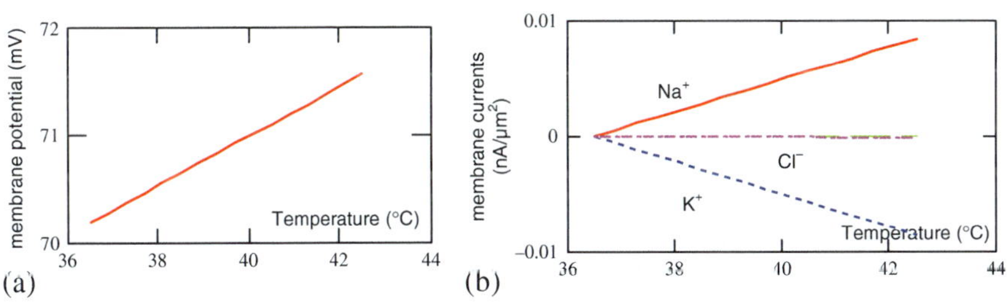

Fig. 4.41 Temperature dependence of the membrane potential (**a**) and the summary of the main ionic currents through the membrane (**b**)

temperature alone. The various currents (change from the $T = 36°C$ homeostatic value) are shown in Fig. 4.40.

Consequently, the membrane potential increases, and the cell is hyperpolarized under increasing temperature (see Fig. 4.41).

4.1.14 Membrane Damage by Constrained Ion Currents

The large temperature gradient through the membrane is one of the main driving forces of the oncothermia effects. One of the most important microscopic processes is the intensive heat transfer through the cellular membrane, which intensifies the ionic transports [334], and (in positive feedback) changes the ionic motility and conductivity, while it also could influence the membrane potential, the membrane permeability, in consequence changing the injury currents.

According to Onsager's non-equilibrium thermodynamic approach, all the various transport processes are tightly coupled, so in our present case, heat transport is accompanied by electric current and mass transport in the membrane. The membrane itself is anyway a good RF-conductor due to its high dielectric constant (permittivity). The Onsager relation for heat and electric transport coupling takes the form

$$j_q = -L_{qq}\frac{\Delta T}{T^2} + L_{qe}\frac{\Delta \phi}{T}, \quad j_e = -L_{qe}\frac{\Delta T}{T^2} + L_{ee}\frac{\Delta \phi}{T} \tag{4.7}$$

where ϕ is the electric potential, and the L values are the transport coefficients. The non-coupled observed values are one order of magnitude larger than the coupled one [335], so we use the $L_{qe}/L_{qq} \approx 10^{-1}$ approximation. Hence: $j_e \approx j_q \cdot L_{qe}/L_{qq} \approx 1.5 \cdot 10^{-1}$ A/m^2. The usual data [200] are $j_{\mathrm{Na}} = 1.26 \cdot 10^{-2}$ A/m^2 and $j_{\mathrm{K}} = 4.53 \cdot 10^{-2}$A/m^2 for Na^+ and K^+ ions, respectively. In consequence, the forced current density is significantly larger than the natural one. This forced current is mainly Na^+ influx. It depolarizes and therefore destabilizes the membrane, and stimulates the Na^+/K^+ pump activity which results in ATP transformations and further heating at the membrane. The forced current depending on its actual direction can change the membrane potential to a depolarized or over-polarized state, see Fig. 4.42. It could be hypothesized that the injury current as a constrained current, depolarizes the cell membrane. The depolarization of the cell membrane is connected to the proliferation [953], therefore could induce cell division filling up the wound and healing the injury. The membrane potential essentially changes by the effects (see Appendix 26).

The extra-cellular spaces between the malignant cells are not equal, having the same wide channels and narrow (even blocked through touching cells) flow paths. Hence other membrane damage is induced by the RF-current. The current density varies by the width of the conductive channel, while also the absorbed energy alters through these structural factors. However, this is a mechanism, heating up the near or touching points of the cells: the current density and the locally absorbed power become large in the narrow paths between the neighboring cells (see Fig. 4.43).

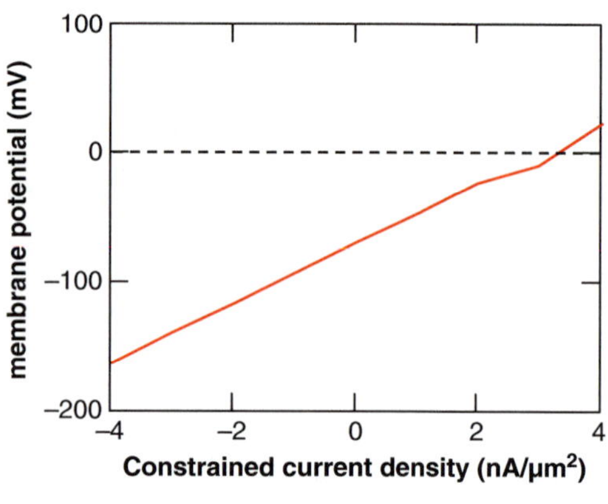

Fig. 4.42 The constrained current flow affects the membrane potential. A high enough flow could repolarize the membrane

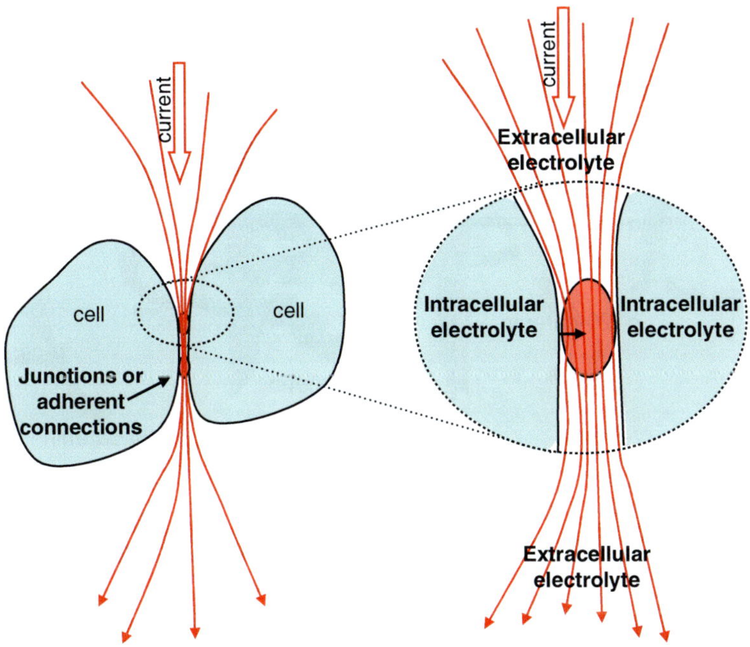

Fig. 4.43 The hot spot mechanism hot spot for the cell membrane

This phenomenon is used by the DC treatment also, but the RF-heating amplifies it considerably. In the RF-case, the applied current (and power) is considerably higher and the high dielectric constant of the large molecules in the narrow path of the field concentrate the field strength and create hot spots extremely effectively using the current density in the nearby membranes.

The membrane is loaded by the developing pressure inside the cell (see Appendix 27). This could change the membrane permeability promoting the transport of such large molecules as chaperone proteins. The appearance of HSP70 on the membrane could promote systemic immune reactions in the body.

4.1.15 Effect on Cell–Cell Connections

Oncothermia effectively acts on the adherent connections between cells. The malignant cells are autonomic, they lost all their "social connections" with the tissue. These cells are competitive unlike the healthy collective units. They lost their junction connections (tight- and gap-junctions) and also their adherent connections (cadherins) are broken (see Fig. 4.44).

Their re-establishment could be a strategy of the treatment due to

- rebuilding possible social signals for apoptosis;
- reduce the free dissemination of the malignant cells;

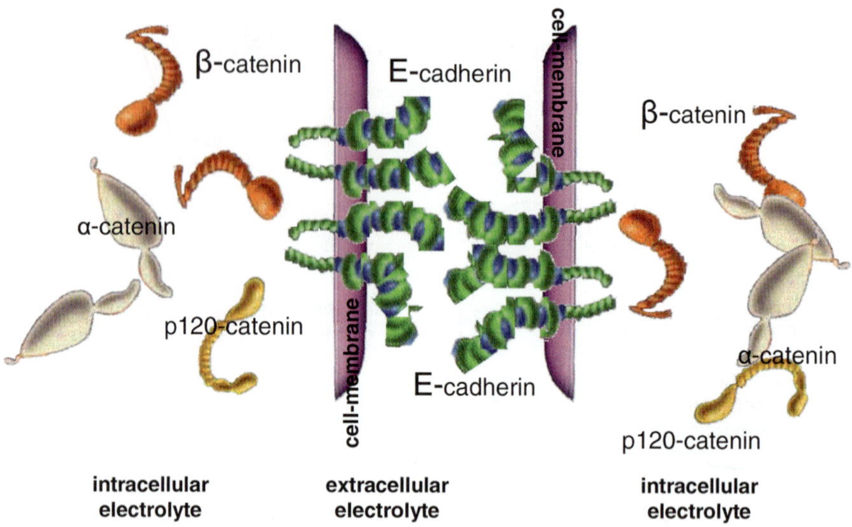

Fig. 4.44 The autonomic cancer cells break all their adherent connections with their "competition" cells in the neighborhood

- make a new effective pathway for the constrained current to destroy the membrane (creating new hot spot actions);
- start redistribution of the proteins in the cytoplasm.

Various trans-membrane proteins exist including adherins [1164]. Their density is not homogeneous in the cell membrane, their concentration at the touching membrane-points is considerably higher than in other places [1164]. The malignant cells have lower expression of adherins on their membrane than their healthy counterpart. When it is not bonded, the extra-cellular part of the adherins moves randomly. When the end-dimers of these adherent proteins move close to each other in the presence of the Ca^{2+} ion, a weak attractive force appears, and this is amplified by the interactions with the other dimers from the proteins, constructing an adherent bond which is a complex bonded state. Forty percent of the cadherins are bonded at 5 s [1164]. Important, the first dimer contact does not suppose the intracellular bond by catenins. These connections of course could be broken also, and the bonding equilibrium will characterize the homeostasis. Their associative and dissociative kinetics are described by Bell's theory [1165].This mechanism dynamically develops the adherent connections (see Appendix 28).

We saw that (see Fig. 4.31) the electric current density is concentrated in the extra-cellular electrolyte, as well as that due to the beta-dispersion the SAR is high in the membrane (see Fig. 4.32).

Perpendicularly to the cell membrane an inhomogeneous force-field is developed (see Fig. 4.45). This field $\boldsymbol{E}$ creates a dielectrophoretic, cataphoretic force ($\boldsymbol{F}$) according to (4.8) (see Fig. 4.46) to the molecules having $\boldsymbol{p}$ dipole moment:

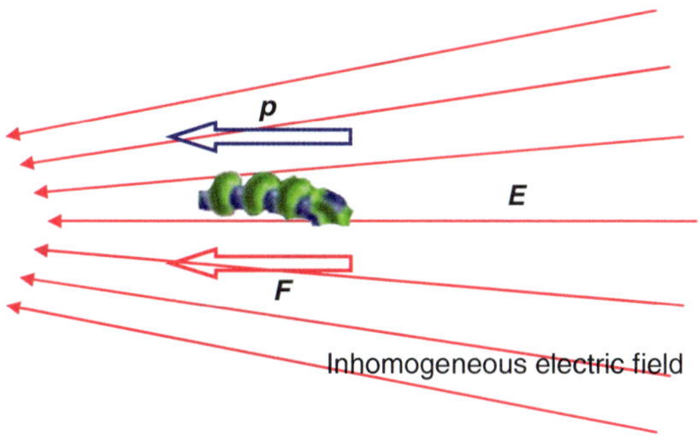

Fig. 4.45 The proteins have high dielectric constant, their dipole moment is considerable. The inhomogeneous electric field turns them into the direction of the field gradient

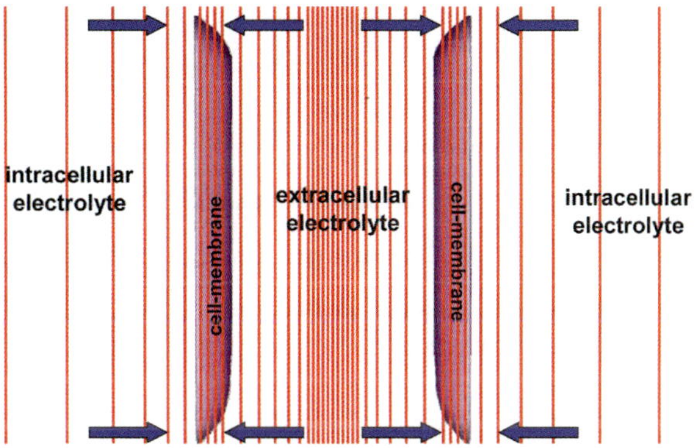

Fig. 4.46 Current gradient creates forces perpendicular to the membranes

$$\underline{F} = (\underline{p} \cdot \text{grad})\underline{E} \tag{4.8}$$

The cataphoretic forces orient well the membrane microdomains [906] (dimers [1166]) of such large molecules like E-cadherin on the membrane surface, and they could bond again (see Fig. 4.47), connecting the dimers to each other. The gradient inside the cell orients the beta-catenins and other anchor proteins (e.g. p120 catenin) to connect the actin or filament network of the cell.

The field has three effects (see Fig. 4.48):

- Promotes the bond of the non-connected adherins (decreases the k_D dissociation reaction constant).
- Increases the activation energy (E_0 barrier and decreases the k_D dissociation reaction constant).
- Creates attractive forces between the nearby membranes (decreases the f force of the single bond).

The re-established connections, the field lines are conducted by the adherins which have extremely high (sometimes a few thousand) relative permittivity (see

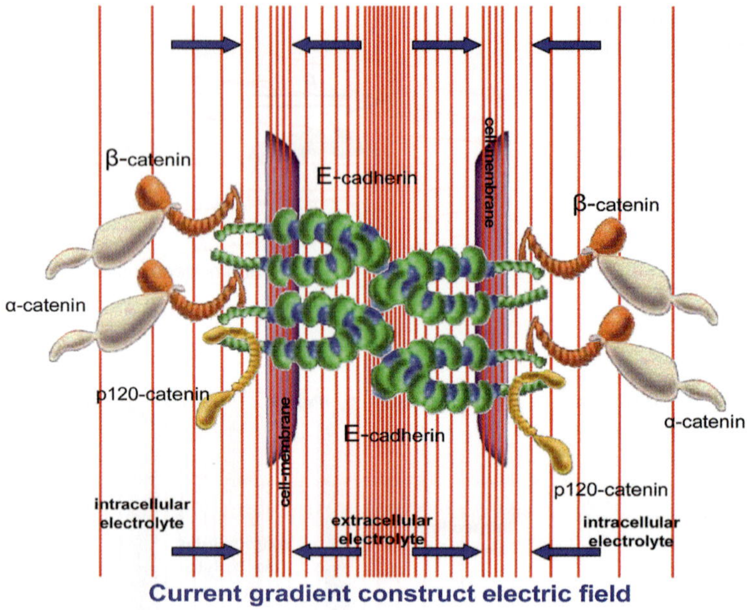

Fig. 4.47 The oriented bonds of Fig. 4.44 could find their connective way, re-establishing the broken adherent connections

Fig. 4.48 The bonds are newly formed by the promotion of the constrained forces

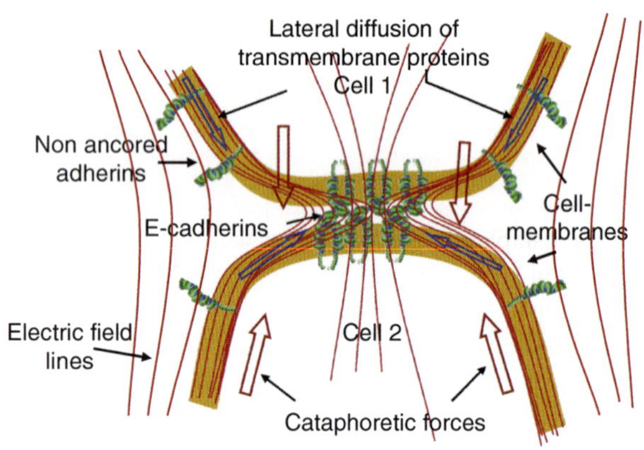

Fig. 4.49). This creates "hot spots" again on the membrane, and promotes the membrane damage.

The effect of promoting adherent connections is directly opposite to the dissociation forced by the growing temperature, so it is a clear difference of oncothermia from the classical heat therapies.

We studied the change of adherent connections under carefully arranged experimental conditions (see Appendix 31). We observed the differences in the HepG2

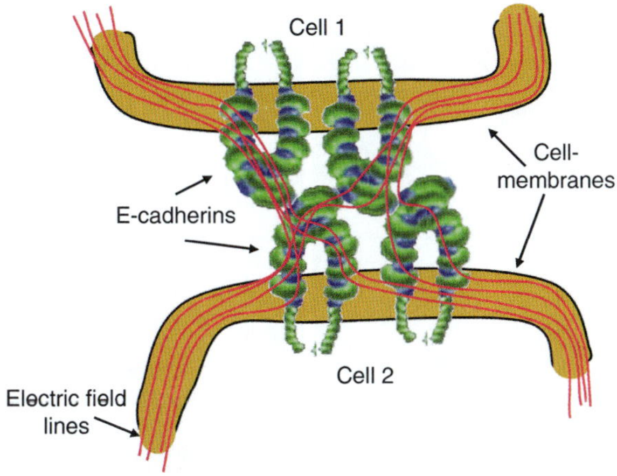

Fig. 4.49 The re-established bonds are good conductors of the electric current, "channelizing" the electric effects between the cells

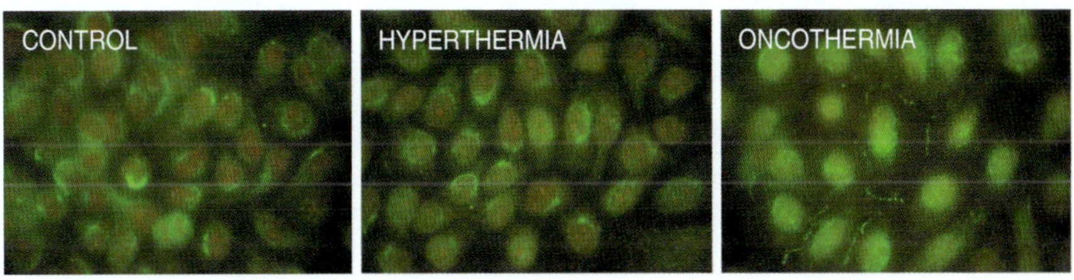

Fig. 4.50 Differences in β-catenin between hyperthermia and oncothermia at the same temperature (HepG2 human hepatocellular carcinoma cell line, treated 30 min, incubated 1 h afterwards, and fixed in cold methanol, pattern is taken by confocal microscopy)

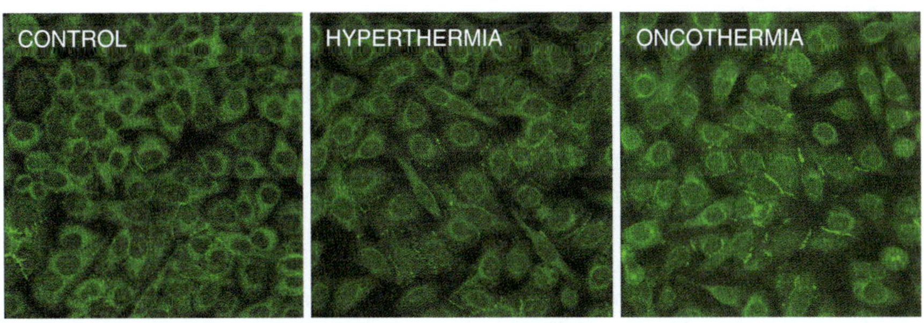

Fig. 4.51 Differences in p120-catenin between hyperthermia and oncothermia at the same temperature (HepG2 human hepatocellular carcinoma cell line, treated 30 min, incubated 1 h afterwards, and fixed in cold methanol, pattern is taken by confocal microscopy)

(human hepatic carcinoma) cell line in both the β-catenin and p120-catenin detection. The confocal laser-microscope images are shown (see Figs. 4.50–4.52) after the 1-h incubation period of a single treatment for 30 min at the same temperature (42°C) [1167].

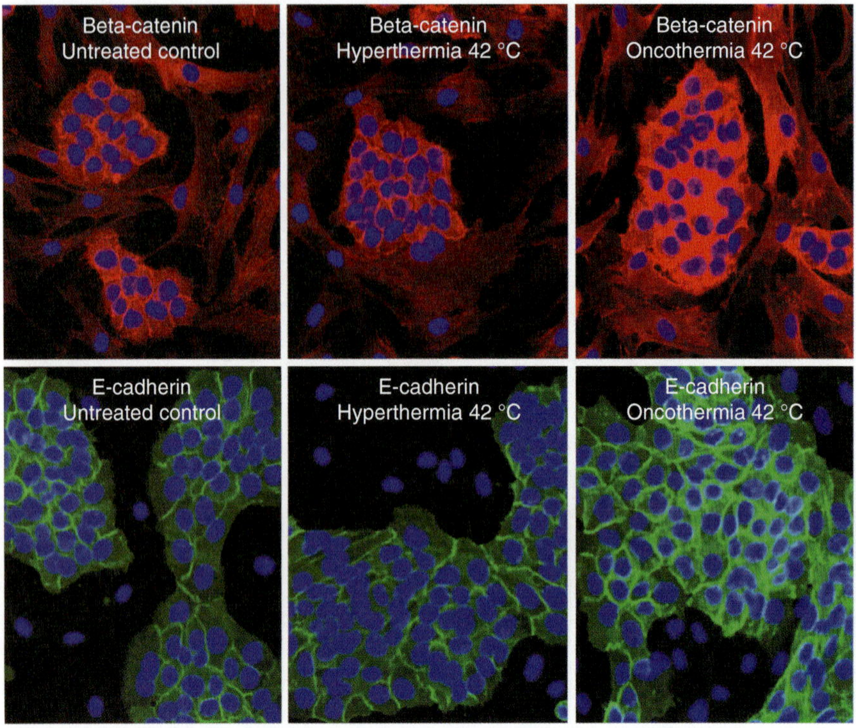

Fig. 4.52 β-catenin (detected by Zymed mouse anti-beta catenin) and E cadherin (detected with BD mouse anti-e-cadherin) in a comparison between untreated and hyperthermia as well as oncothermia-treated samples. (Immuno-fluorescent microscopic images, *red*: β-catenin, *blue*: cell nuclei. The confocal microscope setup was strictly kept identical for the images taken with the Nikon-BioRad confocal microscope)

The other novelty and one of the effective factors of the oncothermia treatment is the modulation of the carrier frequency. This helps select and destroy the malignant cells. Certainly, the same carrier frequency does not mean the same effect. The modulation carries the information like the broadcast of a radio station on a given carrier frequency.

The 13.56-MHz RF frequency carries energy to heat but its effect is indefinite in its target. It heats up everything along the path of the conduction, and the heat flows into the neighborhood by heat diffusion, seeking to equalize the temperature all over the body. This "smearing" effect is well supported by the active blood flow, which promotes the equalizing process by physiologic control as well. These conditions request additional and effective selection factors to the above-described impedance selection; otherwise the possible contrast of the malignancy disappears. The situation is even stricter when we calculate the disseminated cells, which request an accurate selection at the cellular level. This was solved by the modulation. The modulated carrier frequency performs a selective energy delivery, and makes the cell killing optimal in a mostly apoptotic way. Generally, a thermal limit of electric effects was expected on the cellular membrane [1004]. It was questioned [1005], and later it was rigorously shown, that the effect of well-modulated signals is not limited to this issue [1006]. Special modulation [179] is applied in the oncothermia process. This is devoted to enhancing the selection for membrane distortion on an accurate cellular level, and to enhance the collective (apoptotic) control of cell death [773]. The modulation is structured in connection with the highly self-organized hierarchic order [1168, 609, 642] and bioscaling [647] of the living matter [646], having a special pink-noise power spectrum [609, 645]. This power spectrum is the basis of oncothermia modulation (see Appendix 29), making possible the precise selection by re-establishing the adherent connections at the cellular membrane and destroying the malignant cells. Its biophysical effect could be described by the spectrum which we apply.

This generates high-amplitude voltage peaks (on a well-designed time domain basis) to make the dielectrophoretic (cataphoretic) effects and all the voltage-dependent factors effective. The pink noise is the self-organizing fluctuation, the operating noise of the homeostasis. The essence of this is the random physiology effects that are collectively controlled by their spectrum deviation which is constant in homeostasis. This externally constrained fluctuation promotes the collectivity, constrains to equalize the deviations of the random events. The collectivity, however, is expressed by the cellular communications in living organisms, so the adherent connections and junctions are controlled by this modulation method. However, the deviation control is the original pink noise having a zero-frequency center, so the demodulation of the modulated signal is essential. This is performed by the above-discussed stochastic resonance mechanism, which necessarily has a window for action. So a too low or too large signal would be ineffective. This is an important factor of course for signal control and tuning, as well as for the electrode arrangement and fine details of the coupling in oncothermia. The well-tuned oncothermia system has a reaction-promoting behavior as catalysts do. To damage and destroy the membrane a temperature over 42°C is required to exceed the energy barrier of the membrane stability. In oncothermia this barrier is suppressed, a lower temperature (lower thermal energy) is then required to carry out the requested action, see Fig. 4.53. The energy difference is provided by the well-organized fluctuation of the potential, which is the pink-noise-modulated electric field.

A remarkable change could be observed on the HepG2 cell line through dynamic development of β-catenin with time after the treatment, see Fig. 4.54. This considerable change after 24 h of treatment is significantly different to hyperthermia at the same temperature, and supports the other observations on the non-temperature-dependent processes [1169]. The sudden regrouping of the beta-catenin and its enrichment at the cell nuclei could be an indicator of apoptosis [1170].

We studied the extra-cellular adherent connections (E-cadherin) and their intra-cellular β-catenin counterpart in the same A431 cell culture (co-culture of squamous carcinoma growing in normal human skin fibroblast cells (100,000/ml) exposed to oncothermia, incubated for 24 h at 37°C. The measurements clearly showed (see Fig. 4.52) the real differences despite the identical temperature pattern [1171].

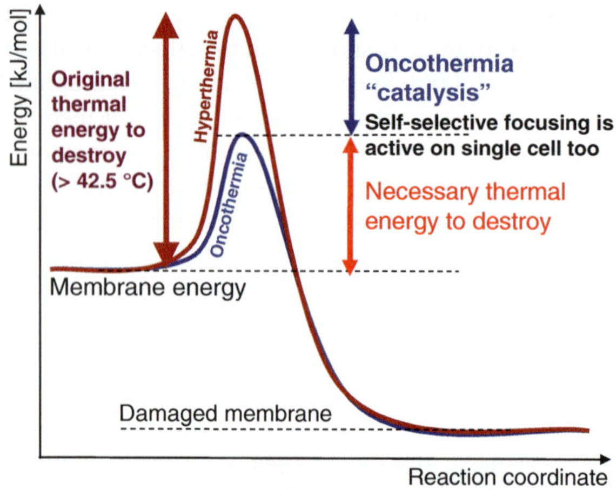

Fig. 4.53 The proper modulation helps to focus, to destroy the membrane and provide a better contour for the tumor (a previously inoperable tumor became operable)

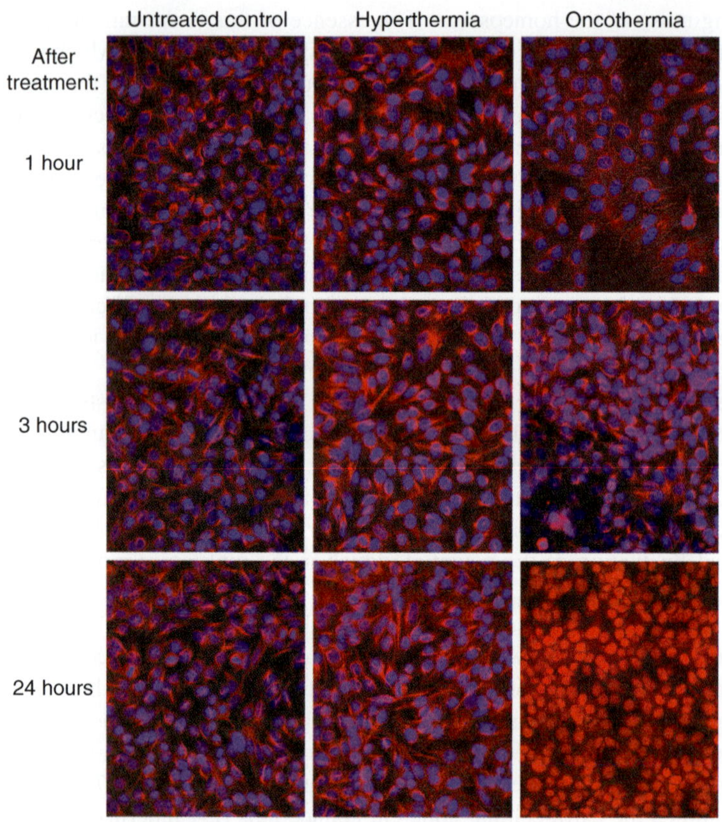

Fig. 4.54 Development of β-catenin with time elapsed after treatment in a comparison between untreated and hyperthermia as well as oncothermia-treated samples. Sampling: 1, 3, and 24 h after the treatment; (Immuno-fluorescent microscopic images, *red*: β-catenin, *blue*: cell nuclei) [1169, 1170]

The in vivo experiments (see Appendix 32) show nuclear relocalization of beta-catenin as well, on an HT29 cell line in nude mice, see Fig. 4.55 [1172].

The experiments indicate the gain of the social signals [1007], expecting to promote apoptosis [1173, 580]. The beta-catenin relocalization into the cell

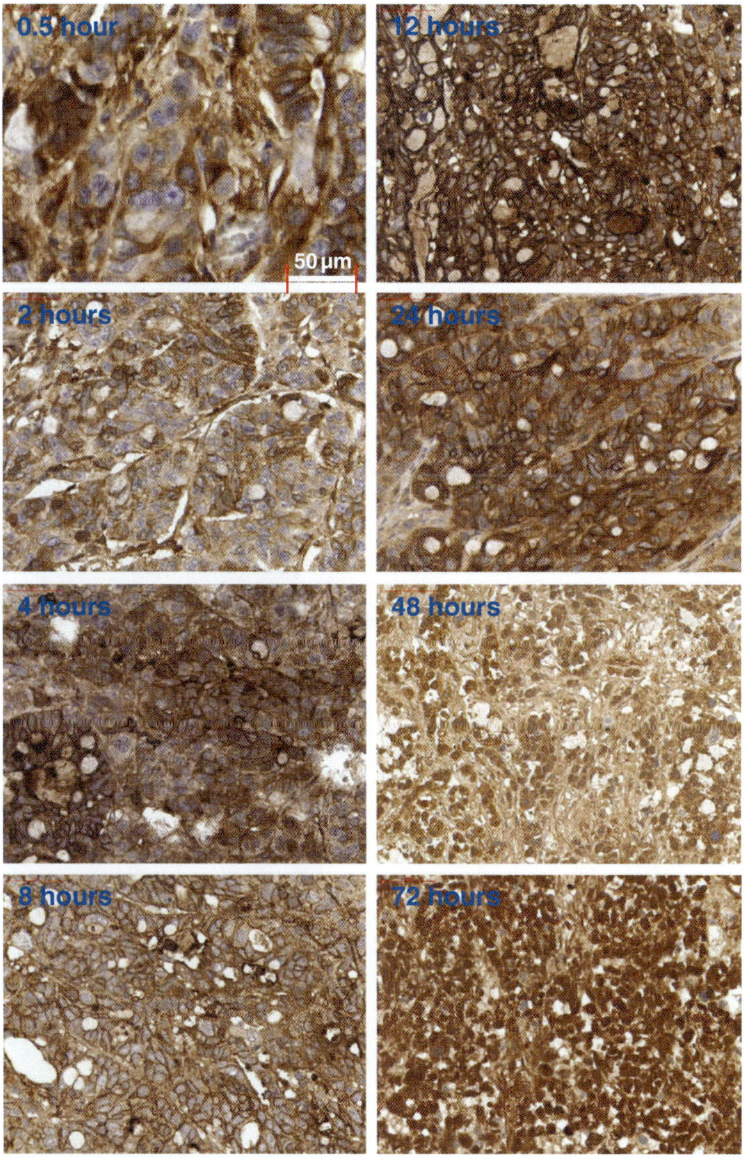

Fig. 4.55 The beta-catenin relocalization is observed in vivo (HT29 xenograft nude-mice model) also. The difference from the in vitro case is the time of the effect, in vivo the relocalization needs 72 h after the treatment [1172]

nuclei is a probable sign of this. The indication of apoptosis was directly also measured in vivo. The model system was HT29 xenograft on nude mice. We detected the double strains of DNA (DAPI staining, see Fig. 4.56) and measured the enzymatic labeled strain-breaks of DNA (TUNEL-FICT, see Fig. 4.57) also. According to the proofs, apoptosis is highly likely in oncothermia, while at identical temperature in classical hyperthermia necrosis is preferred. This observation shows also a major difference between conventional hyperthermia and oncothermia.

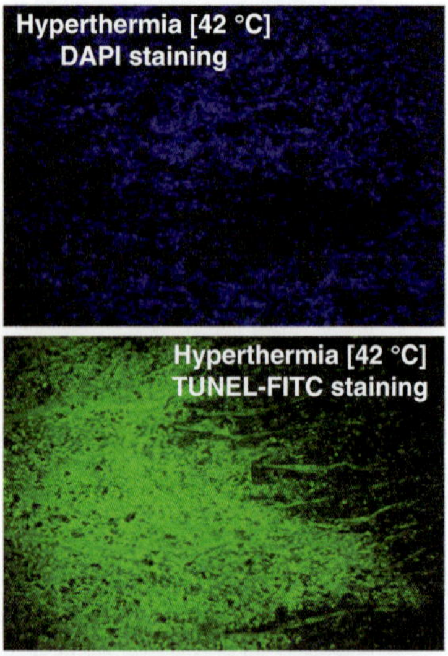

Fig. 4.56 DAPI staining (stains the double strains of DNA only) and TUNEL-FITC staining (enzymatic label of the strain-break of the DNA) of specimen treated by hyperthermia at 42°C. {Institute: Department of Pharmacology and Toxicology, Faculty of Veterinary Science, St. Istvan University, Budapest, Hungary, Investigators: Dr. G. Andocs}

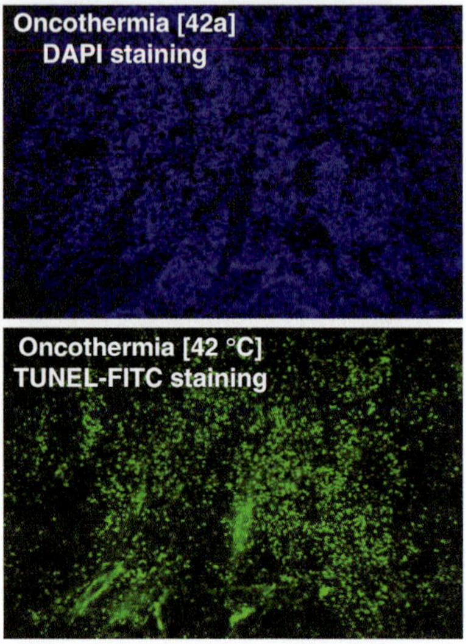

Fig. 4.57 DAPI and TUNEL-FITC staining of specimen treated by oncothermia at 42°C

The mechanism of the selective membrane damage by the pink-noise-modulated signal is probably connected to the difference of the membrane state of malignant cells from their healthy counterparts. The membrane of the malignant cells is more rigid [1174]. The adherent connections and junctions could be formed only when the reactive ligands are close to each other. In consequence the cellular connections have a geometric requirement to be re-established (see Fig. 4.47). The fermentative path of metabolism of malignant cells develops a strongly negative glycocalix shell. This work counter to build a proper geometric order, blocks bonding access the appropriate ligands. However, the extremely highly fluctuating cataphoretic forces from the pink-noise modulation (see Fig. 4.48) compensate for the repulsion, and make possible the adherent connections (see Fig. 4.49). This is helped by thermal conditions of the tissue according to the Arrhenius law, but the thermal background has to be small enough so as not to destroy the starting attractive forces between the ligands. This makes the situation sensitive for the appropriate balance of the thermal and electromagnetic forces, for which the pink-noise modulation is the clue. Anyway, in most cases classical hyperthermia is not able to produce such a thermal shock, which denaturizes the proteins [1175]; therefore a stable oncothermia mechanism is necessary to be sure of the cell killing.

The membrane contains stress-sensor microdomains (rafts) operating as a trigger of the cellular processes [906]. The cataphoretic forces generated by oncothermia induce lateral movements and sensed by the rafts in the membrane, HSF1 begins to activate and ultimately modulates the actual HSP (mainly HSP70) level. The touching points of the cells bridging the electric current, creates a large SAR at these points (hot spot). These have enough instability to make the membrane permeable for HSP70. The extra cellular (or outer membrane bonded) HSP70 is an immune modulator, and has the potential to activate the innate and adaptive immune system [906].

This shows that assuming linear dependence of the local and total fields is correct. Consequently the two stress components: the field [Eqs. (A.30.1) and (A.30.2)] and the heat [Eq. (A.30.3)] are acting differently in the total arrangement. A part of the stress is proportional linearly with the applied RF-potential while the other part is squared to the electrode potential (see Appendix 30).

An important xenograft study was carried out on SCID mice to prove the efficacy of oncothermia as monotherapy in the case when no other therapies are effective enough. Human pancreatic adenocarcinoma cell line (BxPC-3) was grown as a monolayer culture in minimum essential medium (a-MEM) supplemented with 10% fatal bovine serum (FBS), 0.1-mM non-essential amino acid, 1.0-mM sodium pyruvate, 100 U/ml penicillin, 100 mg/ml streptomycin, 0.25 mg/ml amphotericin B, and 2 mM L-alanyl-l-glutamine at 37°C in an atmosphere of 5% CO_2 in air. The tumor cells were routinely subcultured twice weekly by trypsin-EDTA treatment. The cells were harvested from subconfluent logarithmically growing culture by treatment with trypsin-EDTA and counted for tumor inoculation. An acclimation period of at least 7 days was allowed between animal receipt and commencement of tumor inoculation. When the female CD-1 mice were 7 weeks of age (~25 g), each mouse was sub-cutaneously injected at the right flank with 5.5 × 106 BxPC-3 human pancreatic adenocarcinoma cells in 0.1 ml of PBS to induce tumor growth. The following treatment (or control) conditions were evaluated for this experiment. Group 1: Untreated Control ($n = 10$); Group 2: oncothermia ($n = 10$). After the tumor size reached an approximate volume of 50 mm^3, treatments are initiated. Each group contained ten tumor-bearing mice [1176]. On Day 72, mice were sacrificed and their tumors excised and weighed. There was a big difference observed in the total weight of excised tumors between control and treated groups. In summary, there were a total of ten treatment doses administered to each mouse (six doses in cycle 1 and four doses in cycle 2) using oncothermia (see Fig. 4.58). The mice responded well to the treatment during cycle 1 and no adverse side effects were observed. Tumor growth inhibition was found to be significant when tumor volumes were compared between the control and treated groups on Day 22 when cycle 1 was concluded ($p < 0.0005$) and 40% growth inhibition was achieved. There was no significant mean body weight change observed between the control and treated mice during the entire duration of the study. The tumor growth inhibition was 66% after the second cycle observed based on excised tumor weights.

In other in vivo studies the experimental animals were female nude BALB/c (nu/nu) mice. Animals were provided from the Division of Animal Experiments of the National Research Institute for Radiobiology (Budapest, Hungary). Animals were maintained in a sterile environment, sterilized food and water was provided ad libitum. All animals were kept on a daily 12-h light/12-h dark cycle. The animals were 6–8 weeks old and weighed 22–25 g at the time of tumor induction. All studies were approved by the local animal experimentation committee and were carried out in compliance with national guidelines for the care and use of experimental animals.

For easy and accurate comparison most of the experiments were performed with animals developing two symmetric tumors. Each of them was sub-cutaneously

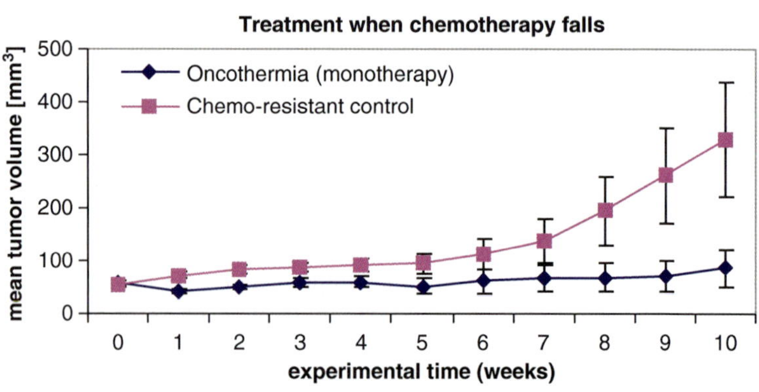

Fig. 4.58 The tumor inhibition by oncothermia on human pancreatic adenocarcinoma cell line (BxPC-3) in female CD-1 mice (10–10 animals are involved) (Institute Lorus Therapeutics Inc. Toronto Canada, Investigator: Dr. Y. Lee and Dr. R.C. Peralta)

injected at the femoral region on both sides with 6×10^6 cells in 0.1 ml of medium to induce tumor growth on both sides, symmetrically. On average, animals were used for experiment 18 days after the tumor inoculation, when the tumor volume reached approximately 0.5–0.8 cm^3. For the experiment we only used animals that developed their tumors symmetrically and at approximately the same size. No animals were paralyzed by the growing tumors.

The evaluation of the results was made by morphological comparison, which was verified and validated for applicability (see Appendix 34).

The aim of the further investigations was to compare oncothermia with its conventional hyperthermia counterpart at the same temperature. We excluded the differences due to the variation of the temperature in the same manner as we did in vitro.) Xenograft models were performed by a standard cell-line culture and its standard preparation. The cells were maintained in DMEM+GlutaMax, high glucose (4.5 g/l) medium (GIBCO, Invitrogen), supplemented with 10% heat-inactivated fatal calf serum (FCS) (GIBCO, Invitrogen), and gentamycin (10 μg/ml). Cells were grown in 75-cm^2 cell-culture flasks (BD, Falcon) incubated at 37°C with 5% CO_2 in humidified air. Cells were harvested with 0.25% Trypsin + EDTA (GIBCO, Invitrogen) of sub-confluent monolayers, were washed once in DMEM serum-free medium, and counted. Cells were resuspended in serum-free medium to achieve the desired cell concentration.

We compared the area change of the dead part of the control and treated tumor originating from the same animal. The differences are significant (see Fig. 4.59).

A more sophisticated study was done also with a xenograft tumor model of the HT-29 human colo-rectal carcinoma cell line [1177]. We performed a low-temperature oncothermia experiment, where the bolus of the upper electrode was cooled down. The intensive cooling kept the tumor near the physiological temperature (38°C) while the oncothermia field was identical with the heating conditions.

The experimental animals were female nude BALB/c (nu/nu) mice provided by the Division of Animal Experiments and Experimental Animal House of the National Research Institute for Radiobiology and Radiohygiene (Budapest, Hungary). All animals were maintained in a sterile environment, sterilized food and water was provided ad libitum. All animals were kept on a daily 12-h light/12-h dark cycle. The animals were 6–8 weeks old and weighed 22–25 g at the time of tumor induction. All animal studies were approved by the local animal experimentation committee and were carried out in compliance with national guidelines for the care and use of experimental animals. Animals were used for experiment 18 days after the tumor inoculation, when the tumor volume reached approximately 0.5–0.8 cm^3. Each mouse was sub-cutaneously injected at the femoral region on both sides with 6×10^6 cells in 0.1 ml of serum-free medium to induce tumor growth on both sides, symmetrically. Treatments were systematically made only on the right tumor of the animals, while the left was kept for individual control. Four experimental groups were formed; each had seven animals (14 tumors in pairs).

Group 1 was the untreated control group. The animals in this group did not receive any kind of treatment. Although each participant of all the groups had their

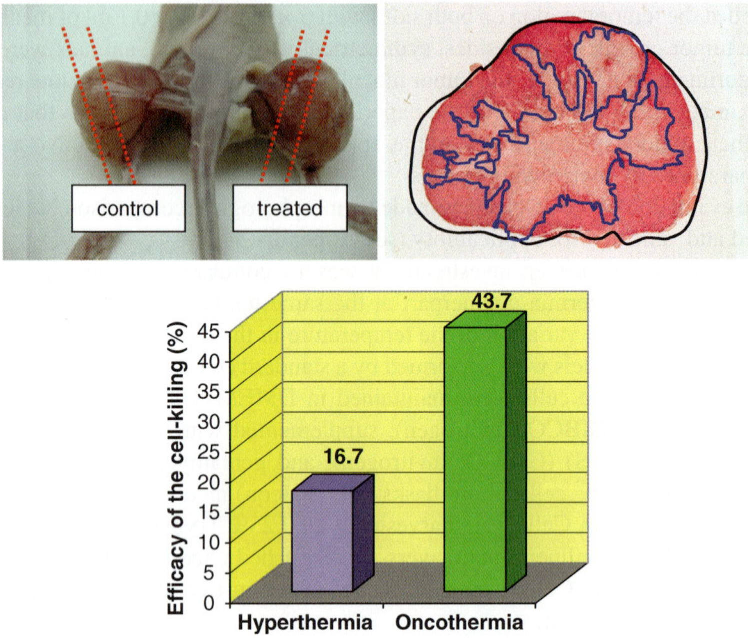

Fig. 4.59 The macroevaluation of the efficacy of oncothermia in comparison to hyperthermia in HT29 tumor xenograft. Change of the areas of dead and vivid parts as a percentage of the untreated control for the same experimental animal (data average of three animals each). Similar experiments were carried out with the same results for A431 human epidermoid carcinoma xenograft model and GL261 murine glioblastoma model {Institute: Department of Pharmacology and Toxicology, Faculty of Veterinary Science, St. Istvan University, Budapest, Hungary, Investigators: Dr. G. Andocs}

own control tumors, to which we could compare the effect of one single treatment, it was necessary to have one control group, which received absolutely no treatment in order to let us compare the proportion of the living/dead tissues in the tumors developing independently on the two sides of the animal.

Group 2 was treated by classical hyperthermia reaching and keeping 42°C in the left-side tumor, while the other was kept untreated. The temperature development was registered in real-time (see Fig. 4.60).

Group 3 was devoted to the same 42°C temperature but heated by oncothermia treatment (see Fig. 4.61). The identical temperature history was carefully managed within the experimental accuracy (see Fig. 4.62).

Group 4 was treated identically with Group 3 animals, only the tumor was kept cooled down near the physiologic temperature. We have calculated also the approximately 1°C systemic base-line increase by the heating of the tumor; so 38°C was chosen for this treatment (see Fig. 4.63).

The heating process was performed for all the randomly chosen animals, and the four processes were characterized by the average of seven animals with 14 tumors.

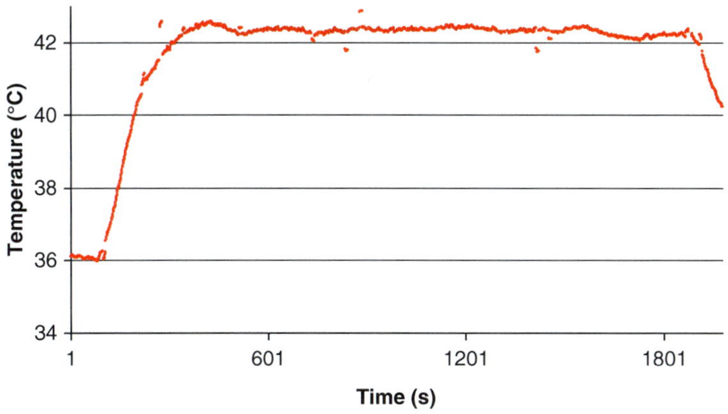

Fig. 4.60 Temperature development in the bulk tumor by hyperthermia (classical heating). (More than 2,000 points, all the points are average of seven measurements)

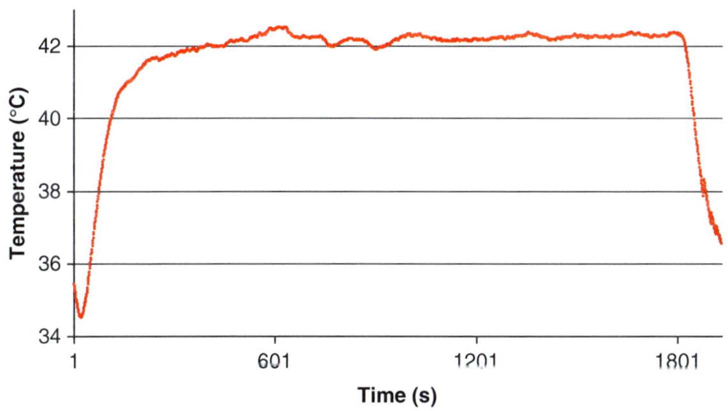

Fig. 4.61 Temperature development in the bulk tumor by oncothermia (modulated electric field heating). (More than 2,000 points, all the points are average of seven measurements)

The treatments were carefully managed and registered (see Fig. 4.64).

The four groups were evaluated and compared by their efficacy in this way. The results are collected in Table 4.3.

All samples were evaluated in an identical morphological way and with the same magnifications (see Fig. 4.65) the morphology indicates well the actual differences, especially at higher magnifications.

The dead-cell part was compared to the full area of the tumor section. The results are shown in Table 4.4.

The statistical evaluation shows a good (identical) correspondence between the tumors inoculated on two sides of the control animals, while the difference is definitely significant in all the treated groups (see Table 4.5).

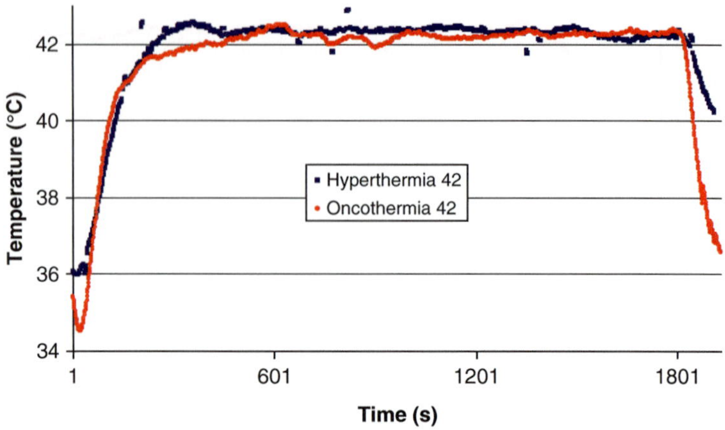

Fig. 4.62 Temperature development in the bulk tumor by hyperthermia and oncothermia. (More than 2,000 points, all the points are average of seven measurements)

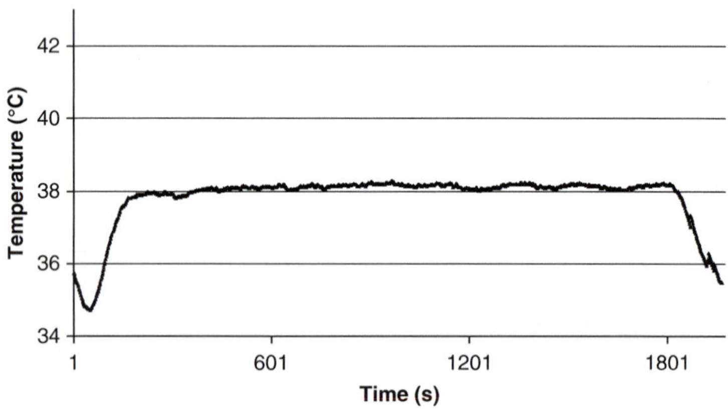

Fig. 4.63 Temperature development in the bulk tumor by oncothermia kept cooled at 38°C with the same applied electric field energy as for heating to 42°C. (More than 2,000 points, all the points are average of seven measurements)

The average temperature graphs of the 7–7 animals from hyperthermia at 42ºC, oncothermia at 42°C, and oncothermia at 38°C are shown in Figs. 4.66–4.68, respectively.

The average of the temperature in control tumors for the seven animals is slightly increased (see Fig. 4.69) by the temperature gain in the whole body of the animals (see Fig. 4.70), but the average of the treatment temperature was maintained according to the protocol (see Fig. 4.71).

The temperature differences (average) between the treated and untreated tumors in the same animal are shown in the case of hyperthermia 42°C, oncothermia 42°C, and oncothermia 38°C, in Figs. 4.72–4.74, respectively. The difference in the cases

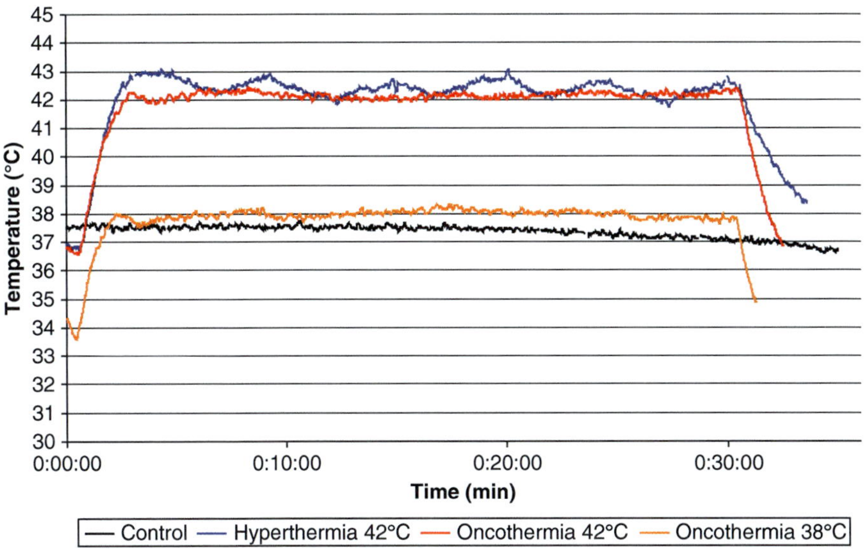

Fig. 4.64 An example of the temperature in various treatment processes

Table 4.3 The actual treatment parameters in the four groups of treatments (all values are averaged from seven animals)

Name	Protocol	Temp. (°C)	Temp. accuracy (±°C)	Field (V/m)	Field accuracy (±°V/m)	Dead-volume (%)
Group 1.	Normothermia	37.5	0.8	0	–	6.1
Group 2.	Hyperthermia	42	0.9	0	–	17.9
Group 3.	Oncothermia	42	0.3	32	5	57.1
Group 4.	Cooled oncothermia	38	0.5	32	5	45.9

of the high-temperature treatments (42°C) was stable around 4°C, while in the 38°C case it was less than 1°C.

The control tumors had two definite different groups: the untreated tumor of the treated animals (symmetric tumor growth, all together 21 control tumors) and both the tumors of the untreated (intact) animals (14 control tumors). The temperature differences of the different control groups and their average is shown in Fig. 4.75.

The above results show clear differences between the temperature-dependent and field-dependent killing effects by hyperthermia in the HT29 xenograft model in nude mice. The cellular death inside the tumors is a normal process due to the missing blood supply in the center of the tumors. (The two centigrade temperature increase in the case of the control of treated animals works against the active effect observations; greater differences in the treatment results are expected if the control tumor remains at constant temperature.) Both control groups (i.e. control in

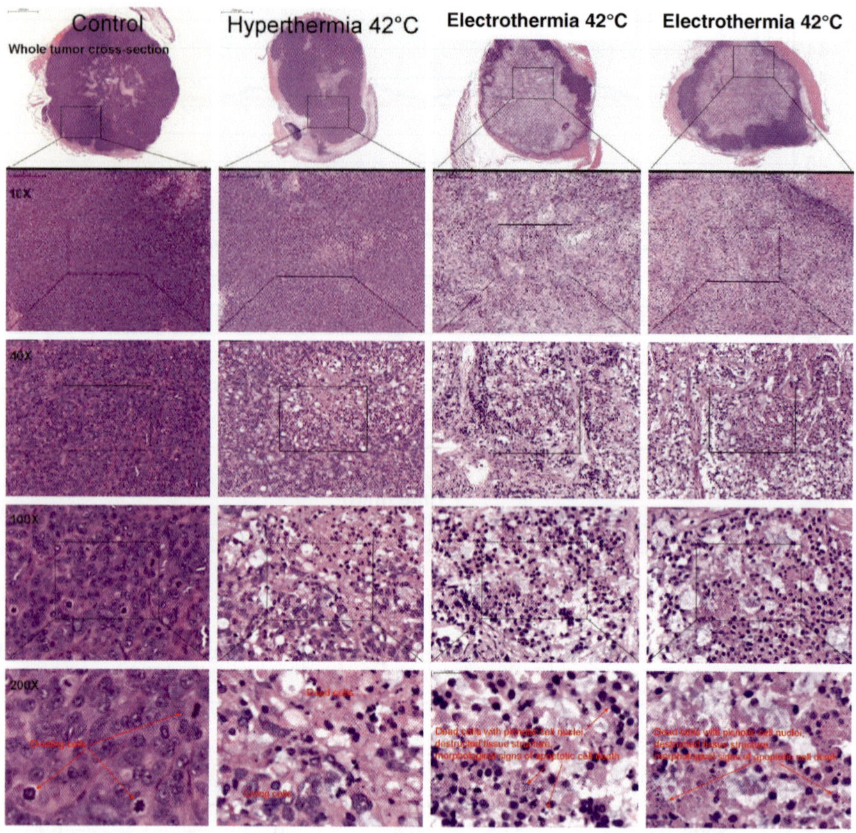

Fig. 4.65 The morphologic differences of the samples [1177]

Table 4.4 The distribution of the dead parts in the tumor (seven tumors in each category)

Dead-cell ratio in the tumor (%)	Group 1 (control)		Group 2		Group 3 (oncothermia)		Group 4 (oncothermia)	
	Left	Right	Left (control)	Right (active)	Left (control)	Right (active)	Left (control)	Right (active)
Mean	6.1	5.2	9.2	17.9	4.8	57.1	5.2	45.9
Minimum	1.0	1.2	1.7	4.5	1.8	43.6	1.0	20.0
Maximum	13.5	11.2	31.2	48.7	11.3	77.7	15.6	69.7
Confidence (95%)	4.5	3.9	6.8	13.3	3.6	42.3	3.8	34.0

treated and tumors in untreated animals) naturally showed small difference in their measured dead-cell ratio so the control arm is stable.

However, hyperthermia multiplies the normal cell death by both the applied effects: temperature (TER) and the applied field (FER) increase the natural cell

Table 4.5 The Wilcoxon matched-pairs signed-rank evaluation of the dead parts in the tumors in different groups (seven tumors are in each category)

Wilcoxon matched-pairs signed-rank of dead-area ratio	Group 1 (control)	Group 2 (hyperthermia) (42°C)	Group 3 (oncothermia) (42°C)	Group 4 (oncothermia) (38°C)
p	>0.93	<0.02	<0.02	<0.02

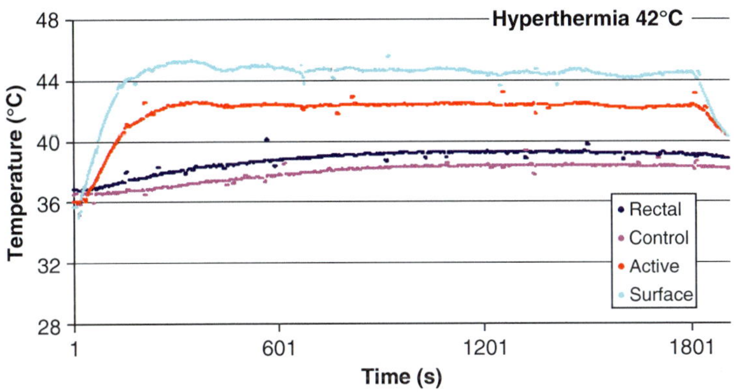

Fig. 4.66 Temperature-dependence average of seven hyperthermia 42°C experiments

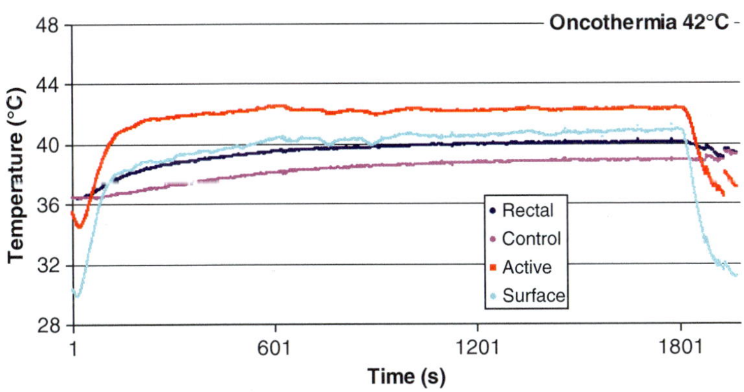

Fig. 4.67 Temperature-dependence average of seven oncothermia 42°C experiments

death by a factor of about 3. The temperature effect was expected to make such a drastic change [1178], but the effect of the electric field was only theoretically predicted [334, 1006]. The evaluation of the processes shows remarkable enhancement of the destroyed tumor volume. The thermal enhancement ratio (TER) of the killing process (introduced by Overgaard [1179]) without field application is high, 2.93, but the enhancement ratio with the field application (FER) at constant temperature is higher (3.19) (see Table 4.6). The high efficacy could be controlled by

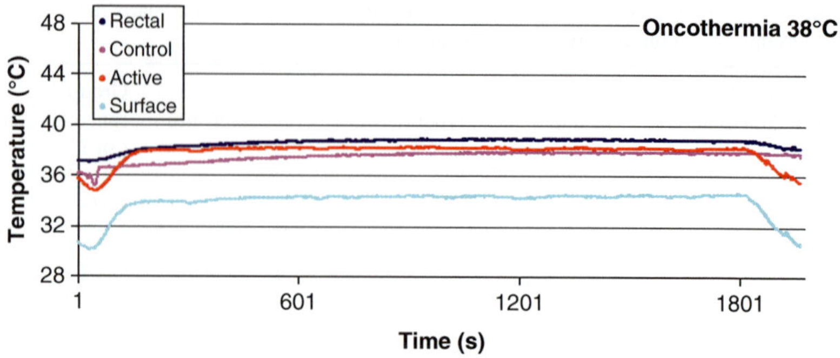

Fig. 4.68 Temperature-dependence average of seven oncothermia 38°C experiments

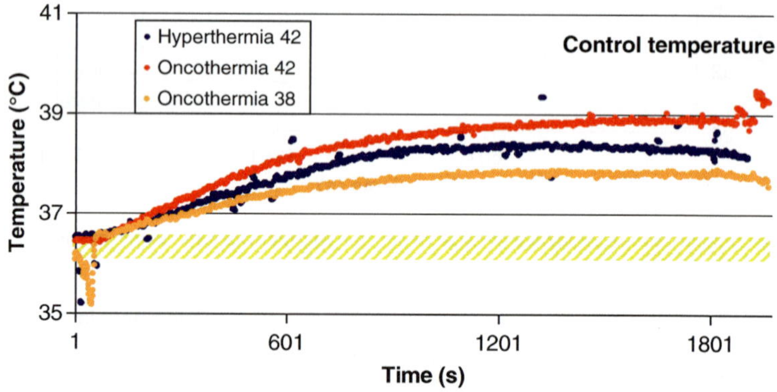

Fig. 4.69 Development of the temperature of the control tumors in average of seven animals. (The *shaded pattern* shows the data of the untreated animals)

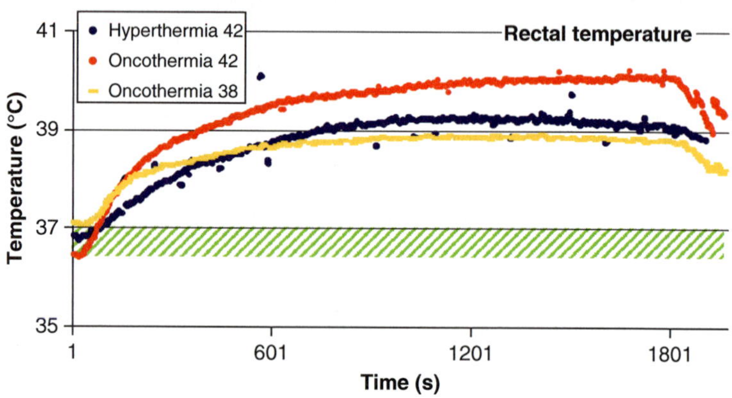

Fig. 4.70 Development of the average rectal temperature in the treated seven animals (The *shaded pattern* shows the data of the untreated animals)

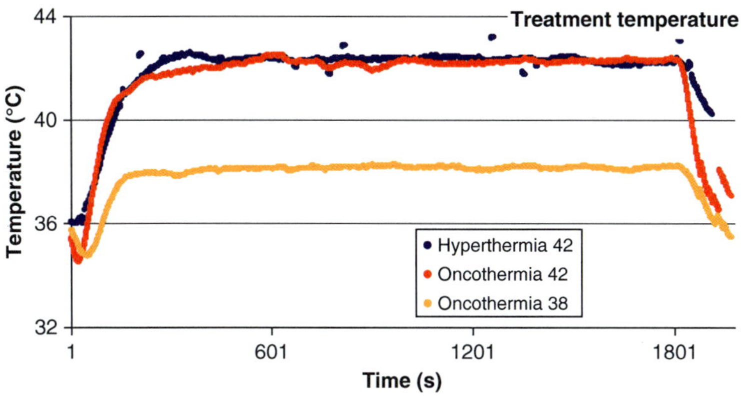

Fig. 4.71 Strictly maintained temperature protocol on the treated tumors verified by the temperature graph of the tumor core temperature average

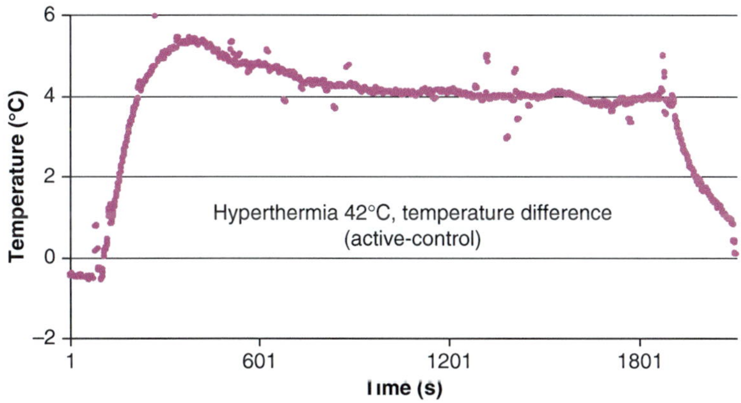

Fig. 4.72 Temperature difference between the hyperthermia 42°C treated and untreated tumors in the same animals. (Every point is the average of seven measurements)

the cooled oncothermia experiment, excluding the temperature effect (Group 4, see Table 4.7).

The cell killing of the temperature and the electric field is well observable from the data. The effects are internally controlled by introducing the cooled down oncothermia, showing the same additive gain of the temperature and the field under different conditions. The identity of the results is significant. The data are shown in Fig. 4.76.

The many applications of hyperthermia in oncology [22, 152] and in radio-oncology [21] provide reasonable input to a discussion of the method, however emphasizing the temperature effect alone is misleading [526]. The question arises: why was the effect of the electric field not realized before, despite the huge number of hyperthermia applications in oncology using capacitive-coupled (electric-field

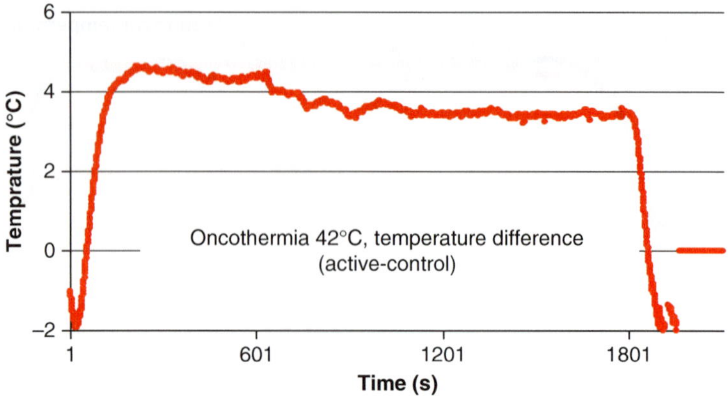

Fig. 4.73 Temperature difference between the oncothermia 42°C treated and untreated tumors in the same animals. (Every point is the average of seven measurements)

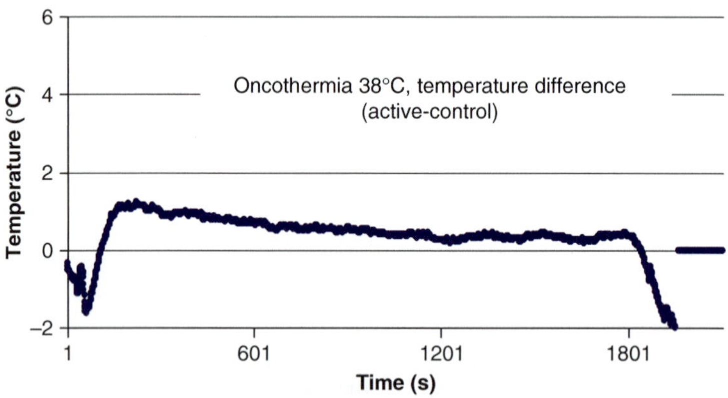

Fig. 4.74 Temperature difference between the oncothermia 38°C treated and untreated tumors in the same animals. (Every point is the average of seven measurements)

dominated) devices. The reason for this is the controlling parameter of hyperthermia which is solely the temperature. The change in this paradigm was urged on a theoretical basis [1180, 1125] supported by many clinical observations also [1181, 171], but no step forward was made. The other reason for ignoring additional effects of the electric field is the many unsolved questions in bioelectromagnetics, including the effect of the frequency. Our electric field approach used fractal physiology achievements in the dynamics of living matter [179, 773, 1372].

The electric field effect is probably initialized by temperature. The RF-current creates a microscopic temperature gradient between the extra and intra-cellular electrolytes through the cell membrane, and could destroy the membrane by various parallel actions [334]: directed heat flow, modified ionic exchanges, increased internal pressure of the cell, expression of membrane HSP, and enhanced energy

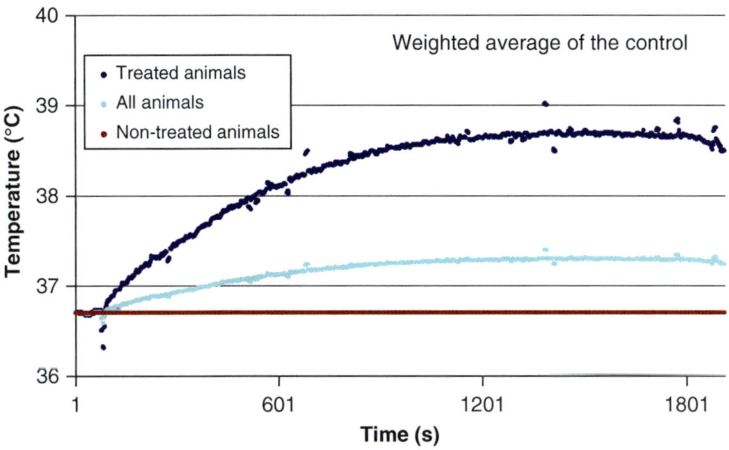

Fig. 4.75 Temperature development of the different control groups. The treated group is the average of 21 tumors, while the untreated is 14 tumors

Table 4.6 The efficacy of the processes shows definite addition between the heat and the field effects (all values are averaged from seven animals)

Process			Efficacy			
				Absolute	Relative	
Change	Temperature (°C)	Field (V/m)	Enhancement	Gain	TER	FER
Temperature changes (no field)	38 → 42	0	6.1 → 17.9	11.8	2.93	
Field changes at constant temperature	42	0 → 32	17.9 → 57.1	39.2		3.19
Combined (field + temperature)	38 → 42	0 → 32	6.1 → 57.1	51	9.36	

Table 4.7 The control comparison shows well that the temperature effects show a gain in cell killing of below 12%, while the field applied alone has more than 39%. These values remain the same under any constant conditions: the temperature adds the same under any field and the field addition is valid at all temperatures (all values are averaged from seven animals)

Control process			Control changes			
				Absolute	Difference	
Fixed parameter	Temperature (°C)	Field (V/m)	Enhancement	Gain	Absolute	Relative
Normothermia	38	0	6.1 → 6.1 (fluctuates)	0	0.18	0.03
Field is constant	38 → 42	32	45.9 → 57.1	11.2	0.60	0.05
Temperature is constant	38	0 → 32	6.1 → 45.9	39.8	0.60	0.02

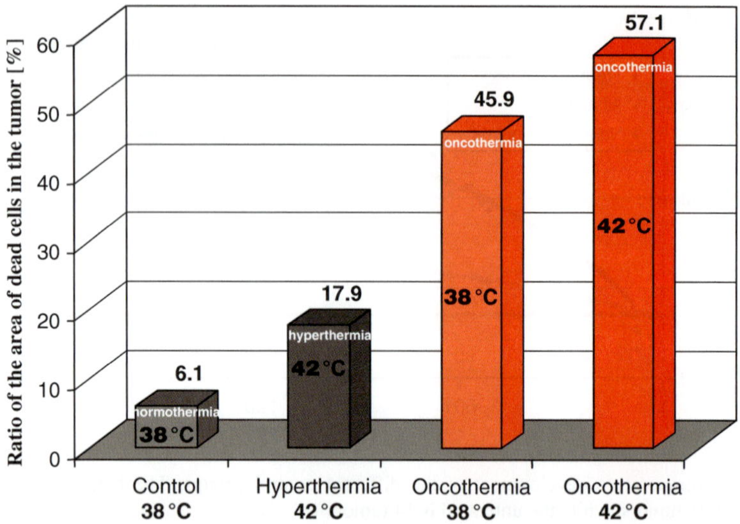

Fig. 4.76 Ratio of the dead cells in the studied tumors, evaluated by morphological methods. The only temperature addition is 11.8%, in field-free and 11.2% in the presence of field. The field effect is 39.8% at 38°C while 39.2% at 42°C

absorption in the membrane. These effects are effective together with the zero-order noise structure [1006].

These thermally induced field effects could cause different mechanisms of cell killing than the simple, equilibrium-based thermal one. The electric effects could be different from the mainly necrotic cell destruction by hyperthermia over 42.5°C. The difference could be observed in the various experiments that have been documented [1182, 767, 1161]. The research in this direction is in progress.

There is a remarkable synergy between the heat-induced cell death in classical oncological hyperthermia and the electric-field-driven cell destruction in oncothermia applications for the HT29 xenograft model in nude mice. The morphological evaluation made it possible to measure the largest cross-sectional areas only, which formed the basis of the quantitative evaluation. It became clear that both factors (the temperature-characterized heating and the field-characterized cell destruction) have definite effects in cell killing in tumors, however the effect of the electric field is more robust. The thermal and the thermally induced non-thermal (electric) effects are well measurable and show more than three-times higher efficacy in advantage of the electric field.

A comparison of hyperthermia and oncothermia combined both methods with Mitomycin-c (MMC) single-dose chemotherapy in vivo at tissue and cellular level. HT29 human colo-rectal carcinoma cell line-derived xenograft tumor model in nude mouse: for hyperthermia (42°C) + 3 mg/kg MMC i.p. (30 min before the treatment); and for oncothermia (42°C) + 3 mg/kg MMC i.p. (30 min before the treatment). For

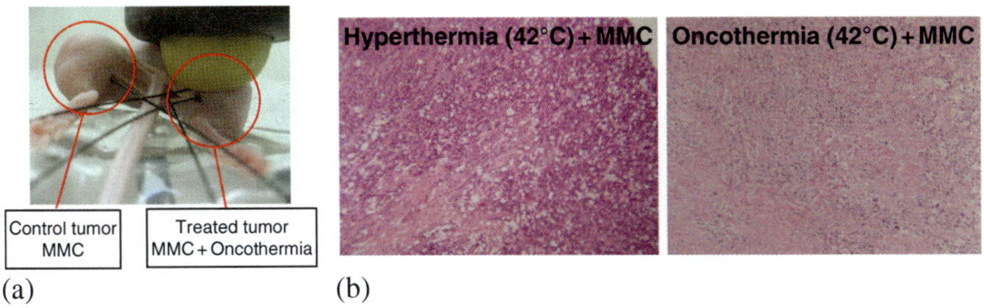

Fig. 4.77 Investigating the difference of the effects of i.p. administered Mitomycin-C. (**a**) The experimental arrangement, (**b**) Hemalaun-eosin-stained microscopic images of tumor samples {Institute: Department of Pharmacology and Toxicology, Faculty of Veterinary Science, St. Istvan University, Budapest, Hungary, Investigators: Dr. G. Andocs}

Fig. 4.78 Experimental comparison of hyperthermia with Mitomycin-C and oncothermia with the same dose Mitomycin-C, cell killing is relative to the control tumor in the same animal. (Values are from two–two animals/four–four tumors)

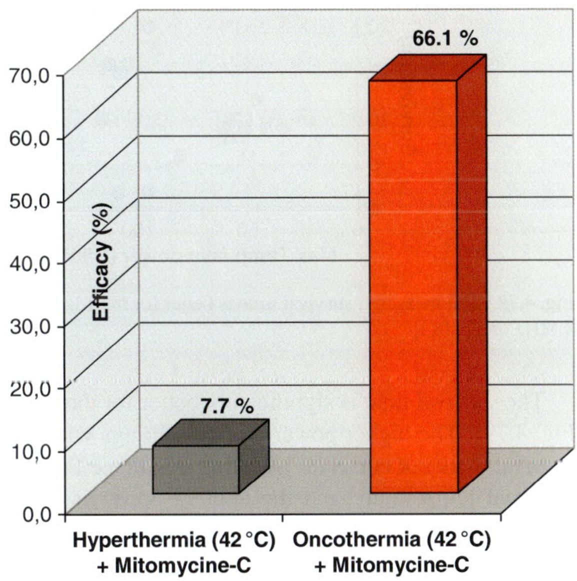

identification histological examinations were used, shown in Fig. 4.77. The relative difference (both compared to its control averages) is shown in Fig. 4.78.

4.1.16 Oncotherm Comparison

In a prospective post-operative intra-thoracic chemothermotherapy study two kinds of capacitive devices with 8 and 13.56 MHz frequencies were used [1183]. Analyzing the data we may derive some conclusions through the comparison of the results obtained from the two frequencies.

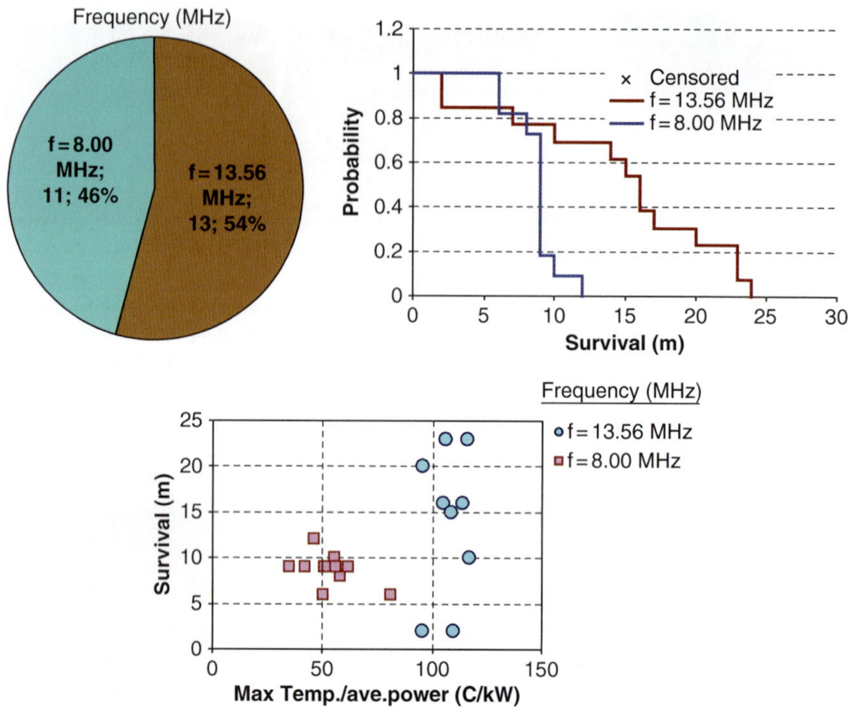

Fig. 4.79 The measured survival time is better for the 13.56-MHz capacitive coupling than for the 8 MHz one [1183]

The survival time is significantly better for the higher frequency application (see Fig. 4.79). The higher power naturally brings with it a higher temperature, but the survival does not correlate with this trend (see Fig. 4.80). The ratio of the reached maximal temperature and the applied power is important as it shows the actual focusing possibility. In this comparison heating at 13.56 MHz is better. Both the additive and synergetic approach of the combined dose (heat and chemo) show advantages for the 13.56-MHz heating frequency.

4.2 Oncothermia Treatment Guidelines

The classical form of hyperthermia has some inherent risks:

- Classical hyperthermia uses in situ MRI to reduce the possibility of hot spots. Most of the installations have no such expensive facility, so they measure the temperature by invasive point measurements or by semi-invasive catheters in a lumen close to the heated lesion. However, the point measurements are inaccurate and do not provide a detailed picture of the hot spots in the radiated volume. Their inaccuracy is mainly from the very limited detection area and the non-negligible heat absorption and conduction of the measuring sensor itself.

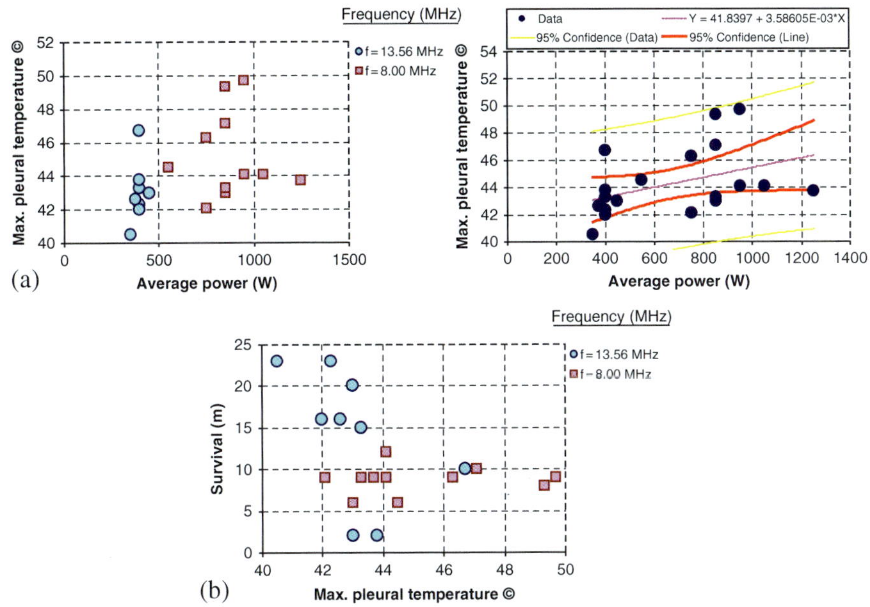

Fig. 4.80 The higher power causes the temperature to rise (**a**) but the survival does not necessarily follow it (**b**)

- The applied 500–2000 W power is alone a real danger if it is misfocused or the surface cooling is not intensive enough.
- Accurately focused heat delivery to the localized target is of course soon smeared by the intensive heat diffusion in the area. Consequently, the idea of local heating in the limited target volume becomes illusory.
- The natural movements of the living body (mainly because of breathing) move the organs, and these movements are not followed by the focusing. The theoretically assumed full accuracy of the focusing (which could never happen due to the non-ball-like shapes of the tumor) would also not be enough to avoid overheating of the healthy environment.
- The high-intensity RF heating is a danger to the environment and the operating personnel. In consequence special electromagnetic shielding is necessary, and the operation is performed outside of the chamber.
- The classical RF heating, due to its construction, can treat only the trunk below the armpits, so the head and neck area, brain, or the upper chest is excluded from the treatment.
- The classical RF heating is insufficient in the case of such intensively naturally cooled organs like lung (breathing) and liver (blood perfusion).
- The classical RF hyperthermia is contraindicated for brain lesions, due to the danger of edema and increased intra-cranial pressure. Only invasive (interstitial) versions of classical hyperthermia can be applied in brain [1184, 380].

These risk factors request accurate temperature measurement which is mandatory for focusing and hot spot control as well. The treatment from a medical point of view is to deliver heat exactly to the target, and not to the healthy neighborhood. The temperature in the target could be as high as possible; the limit exists only for the healthy surroundings. When hot spots or heated healthy tissue becomes significant, down-regulation is necessary, irrespective of the measured value in the target itself. On the other hand, if there is no dangerous heating in the healthy environment, an increase of the temperature in the malignant volume is desired, irrespective of what temperature is reached in the volume. If the temperature is higher, the tumor destruction is greater. This means, the temperature measurement in classical hyperthermia is necessary first of all from the safety and focusing point of view and not for real control of the treatments.

Oncothermia eliminates the disadvantages and risks in classical hyperthermia treatments. The developed safe and versatile oncothermia apparatus is devoted to solving the problems noted above. The method itself is popularly known under the designation "oncothermia" in remembrance of the company's name. Its safety principles and its complete solution definitely differ from the classical way of heating.

Oncothermia treatment is simple and easily applicable. The self-selective mechanism ensures targeting of the malignant cells by the treatment, and the electric field effect is safe and does not request high power which has a possible danger of hot spots or other unwanted burning. The main differences and characteristics (supported by extended medical references) of oncothermia are as follows:

- The main idea behind oncothermia is its connection to the electric field effects, the temperature is a condition only (as we discussed above in detail). The efficacy of electric field distortion is significantly higher than that of classical hyperthermia at the same temperature. The electric field has no effect on conduction, diffusion, convection as the heat energy has, so it strictly remains there where it is necessary, irrespective of the heat-conduction behavior of the environment.
- Oncothermia needs lower temperature for effective cell killing than the HRF conventional hyperthermia. These results are consequences of the applied modulated electric field, with active tumor killing even at low temperatures. The lower temperature makes the treatment safer with higher benefit.
- Oncothermia is cellular-selective in a self-selection manner. This allows safe and accurate concentration of the energy on the malignant cells.
- Oncothermia uses blood-flow physiology to improve the tumor selection at the macroscopic level. The constrained blood flow boosts drug penetration and the chemometabolism in combined application with any chemotherapy, as well as boosting the efficacy of the complementary applied radiotherapy by oxygenization of the tissue. These effects make oncothermia complementary to the conventional therapies.

4.2.1 Treatment Planning

The simplicity of oncothermia means in fact there is no request for such complicated procedures as in classical hyperthermia. Important for patient safety: no temperature measurements are necessary (no risk of inflammation, ulcers, dissemination increased by the wound, etc.).

4.2.2 Treatment Consensus

Oncothermia is a personalized, non-toxic treatment. It supports the natural processes (apoptosis, immune reactions, conditional effects, etc.). Oncotherm never promises miracles. Twenty-years of activity have proved the solid scientific and medical basis and given oncothermia a definite legacy. Oncothermia efficacy is focused on patient-centered values: on the survival time and quality-of-life (QoL).

Oncothermia treatment has a consensus-based protocol, which fixes the proposals of the treatment. However, the treatment is not an automatism, the physician must decide about the actual treatment on a personalized basis. The main points proposed are as follows:

- It is curative treatment also in "hopeless," high-stage, advanced cases.
- Inclusion criteria includes pediatric cases (from the age of the possibility of full communication), and any elderly ages, only for solid tumors, primary or metastatic.
- It could be applied over 3rd-line treatment and complementary with all the presently known tumor therapies, except those which are directly thermo- or electro-destructive.
- Exceptional care is necessary in the case of large volume liquids in the treatment area (ascites, pleural liquid, cerebral edema, etc.), which have to be drained first. It is requested that the patient empties her/his urinary bladder and colon before treatment in the pelvic area, and an empty stomach is also desired when treating the chest or close to the stomach.
- Also great care (and a ready duty emergency package including defibrillator, electric-shock etc.) should be taken in cases where the patient has a pacemaker or other built-in electronics, (muscle, sense support like hearing, visual aid, etc.) supporting or functional bone replacements, implants, stents, stitches, etc. in the treated local volume. The physician has to recognize the high risk of treatment of the eye, brain, and spinal cord.
- Inflammation, fever, viral, or bacterial infections must be evaluated before any heat treatment. Use in inflammatory areas low power, apply mostly the electric field. Also extra care is necessary for patients who are sensitive to

electromagnetic fields, or have epilepsy. Careful fixing of the applicators is necessary in the case of tremor or other unexpected movements of the patients during treatment.

- The sudden inundation by tumor degradation (tumor-lysis) could cause dangerous toxicity, as well as at the same time increases the risk of dangerous internal bleeding. Consequently pause the treatment or considerably down-regulate its application frequency when the risk of rapid tumor degradation is consists.
- Exclusion criteria: non-solid tumors, non-communicative patients, unconscious patient, baby patient, organ-transplanted area, open wound, internal bleeding, heavy inflammation. There is no information on pregnant patients, so we exclude this case as well.
- No local or systemic anesthesia, analgesia, or heavy sedation should be applied.
- The patient should not wear/have any metallic or magnetic belongings (jewelry, piercings, memory cards, coins, etc.) during the treatment.
- Oncothermia should be applied as palliative care where necessary.
- Oncothermia is a combination therapy, it is applied complementarily when conventional cancer therapies fail, due to resistances, organ failures, blood-count insufficiencies, and other blocking factors. It could be applied to resensitize earlier treatment or substitute them in combinations.
- Apply it immediately after the chemotherapy and/or radiotherapy.
- In radiotherapy treatment in hypoxic cases apply moderate oncothermia first (boosting).
- Treatment time is 45–90 min (average is 60 min). A shorter time shows good blood-perfusion increase but the expected curative oncothermia effects are less than optimal; while a longer treatment time could produce heat resistance and may down-regulate the effect of oncothermia. However, use a shorter time (30 min) in tumors close to the surface, when the electric and heat sensing causes an inconvenience for the patient. Use more than 60 min in the case of lung (due to the cooling by breathing), liver, spleen, kidney etc. (due to the robust blood-flow cooling). The treatment of the central nervous system needs special timing: start with 20–30 min, and gradually increase it to normal within three to four treatment durations.
- Treatment frequency 2–3-times a week (sometimes a low dose everyday for blood perfusion). One treatment or less in a week could be also feasible in the case of a danger of tumor disintegration, rapid tumor degradation, danger of in situ internal bleeding, or other special conditions.
- Treatment number 4–12/cycles (average is 6); however more numerous treatments are also possible when necessary. There is no practical limit from a medical point of view.
- The treatment cycle follows the combination (average is 2). In some cases the oncothermia works like a treatment for any chronic diseases (like dialysis): repeating the cycles from time to time to control the tumor growth.
- Apply step-up heating, with gradually increased power (follow the adaptability of the patient). The various sizes of electrodes have different recommended starting default values and the protocols recommend step-by-step heating.

- Allow time (slow adaption during two to three treatments) to adapt the modulation (in the case of sensitive organs like the brain, spinal cord, etc.).
- A patient's feedback is an important part of the personalization; it has to be used to the greatest extent possible.
- The best performance can be reached in a no-stress situation, in a convenient environment, with relaxing music or other entertainment being applied.
- Nothing else except the oncothermia device should be within touchable area of the patient.
- The medical personnel have to be trained, certified to use the device.
- It is not recommended that pregnant staff operate the device.

The treatment process is simple and comfortable. Its steps are as follows:

- The physician explains the therapy, informs the patient of the possible protocol and discusses the treatment frequency (usually 2–3-times a week).
- The patient signs a written consent that he/she has been informed, and accepts the therapy.
- The patient removes all metallic and magnetic materials from her/his body.
- The patient comfortably lays on the treatment bed. The patient must be between the upper and counter electrodes. The optimal placement of the applicator is as horizontal as possible. Such an arrangement gives the most effective heating power. Note that in many cases only a low power is required for the treatment (for example when treating a brain tumor) which can be regulated by the output power control or by placing the electrodes in a non-horizontal arrangement.
- The medical staff places the applicator in its proper place, over the treated tumor.
- Treatment starts, the patient feels comfortable heat.
- In case of any discomfort the medical staff on duty has to regulate the power to a comfortable level.
- Communication from the patient in relation to his/her feeling of comfort is very important. The treatment must be optimally comfortable. The thermal dose has to be chosen to fit the best to each patient. Like in chemotherapy, a higher dose doesn't mean higher efficacy. The treatment efficacy is fitted to the patient's comfort. (Most patients sleep during the treatment.)
- The treatment has to be performed in harmony with other oncotherapies, and has to be checked regularly (with image diagnostics about a month after the treatment session the result can be seen.)

Oncothermia is a new reliable technique [1185]. It is easy to use and cheap, fitting optimally with both the patient's possibilities and the social insurance budget. Its requested space is one third of classical hyperthermia applications, its price is one sixth of classical hyperthermia applications, its human operating resource is only a trained nurse instead of at least two experienced graduate specialists in classical hyperthermia, its electric energy consumption is less than one fifth of classical hyperthermia, its treatment capacity is seven patients/8 h, which is much higher than classical hyperthermia allows.

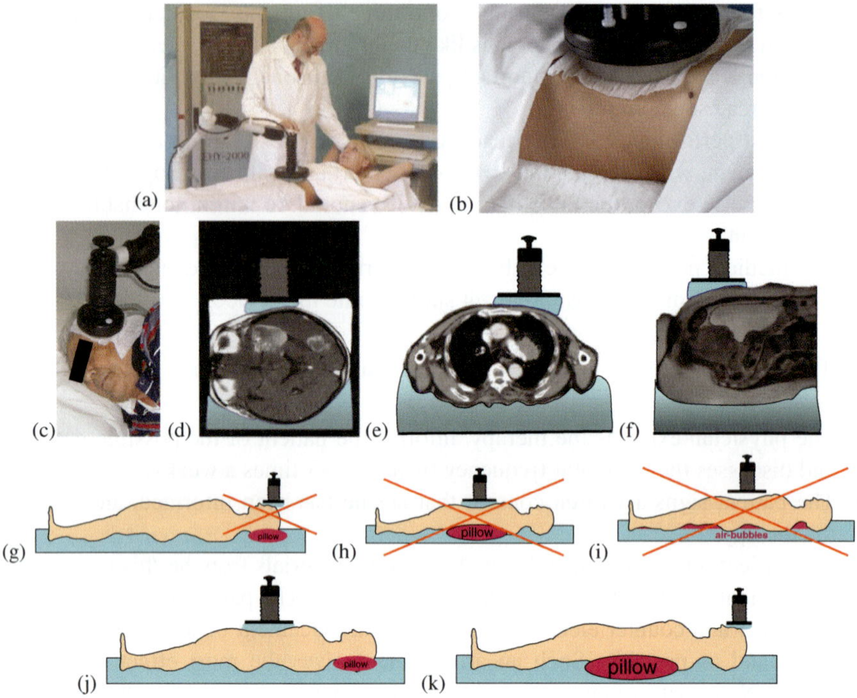

Fig. 4.81 The patient lies on a water bed (**a**), touched by the upper bolus electrode (**b**), which could be applied to the head (**c**, **d**), and any other organs (**e**, **f**). Some incorrect (**g**–**i**) and a couple of correct (**j**–**k**) positions of the patient. When the RF-current flow is disturbed [e.g. by air bubbles, (**i**)], oncothermia does not work properly

The clue to proper treatment is a free flow of the RF-current through the treated volume, between the electrodes to allow self-selective focusing and to make effective treatments (see Fig. 4.81). Blocking of current can cause burn and/or electric shock! Pillow or air bubbles can be placed only if it does not block or deviate the current flow! Direct your attention to the careful arrangement of the patients to eliminate possible air bubbles in the electrodes.

The most frequently applied bimodal treatments are oncothermia combined with chemotherapy or radiotherapy. While in trimodality it is combined with radiochemotherapy. The most applicable multimodality is additional supportive therapies (like vitamins, trace elements, minerals, immune supporters, high-calorie special nutrition, etc.). The therapy has specific uses depending on how it is combined. The main driving force of the combined therapies is the conventional therapy as a basis, and oncothermia is applied in harmony with those.

Special care is necessary if the patient has hair in the treatment area [e.g. pubic hair, or hair on the head], because the hair electrically isolates, and so the burning and mistreatment is very likely. Please shave before treatment if necessary, or at least keep very tight control of the treatment, using less power for a longer time. If you are not able to shave, please use ultrasound/ECG gel on the hair for better contact, or at least please wet the hair with infusion solution. Please ask the patient about his/her cavity (bladder, stomach, pleural cavity, etc.) sensing. Stop the treatment immediately if anything unusual occurs near the cavities, and continue it after the hair has been removed.

4.3 Complementary Applications

4.3.1 Complementary to Radiotherapy

It has to be fitted to the blood perfusion and neo-vascularization of the actually treated tumor. Oncothermia is applied before or after the ionizing radiation, depending on the original blood-perfusion status of the tumor.

In the case of low blood perfusion, the aim is to increase oxygenization, to support the ionizing efficacy of the radiotherapy. A mild temperature increase is expected to increase the blood flow and so help the radiotherapy through extra oxygenation of the area This requires a low-dose oncothermia before each radiotherapy (see Fig. 4.82). The proper electrode application in this case is shown in Fig. 4.83.

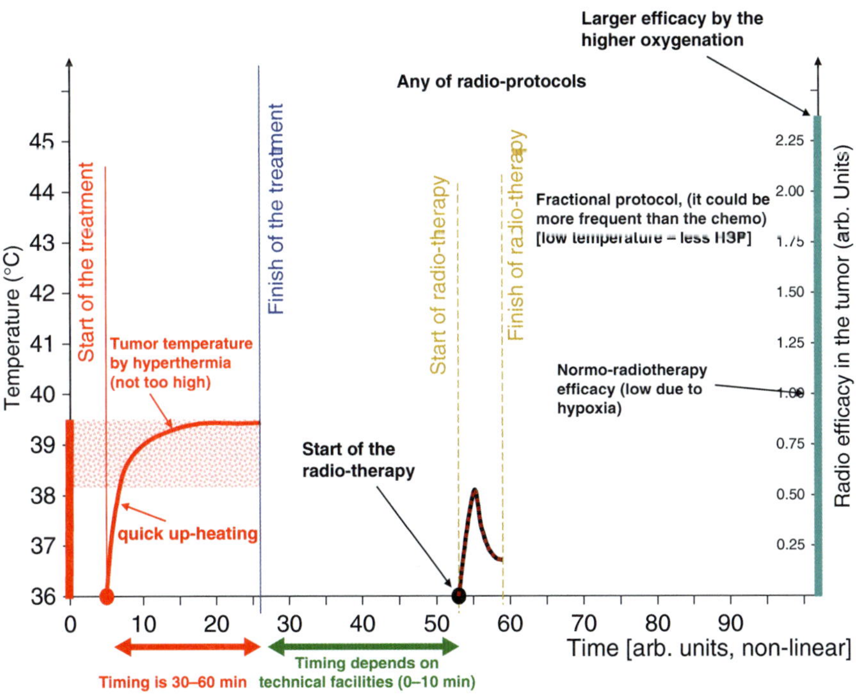

Fig. 4.82 Oncothermia has to be applied before radiotherapy for oxygenization of low blood-perfunded tumors. A low dose treatment has to be applied just before every radiation

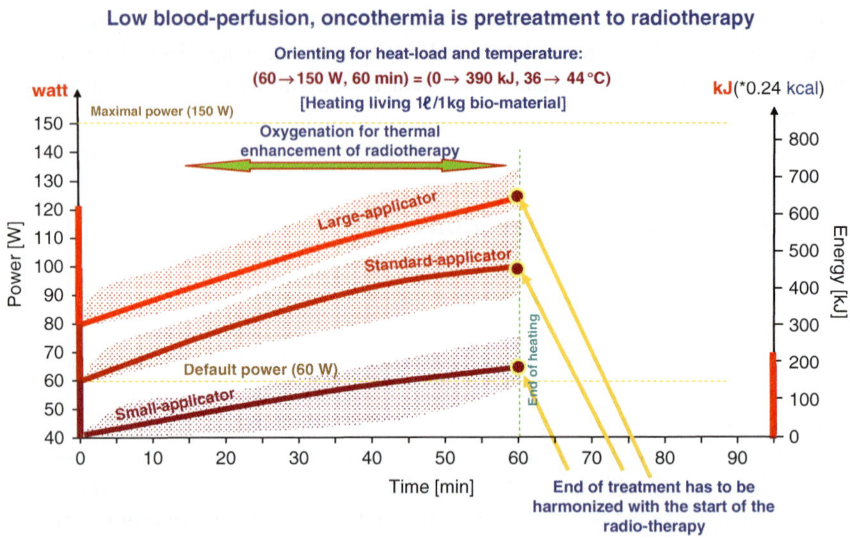

Fig. 4.83 Application of various electrode sizes in low blood-flow cases before radiation therapy

In the case of high blood perfusion the radiotherapy efficacy is expected to be high, its primary application is desirable. In cases of high blood perfusion we expect high radio-efficacy, so the sequence of the combined therapy must be started with the radiotherapy. Oncothermia has to be applied afterwards with the highest tolerable dose to achieve the maximal result! (See Fig. 4.84).

The proper electrode application is shown in Fig. 4.85.

Oncothermia has to be applied parallel or before in the case to complete the chemotherapy, where the chemo-perfusion and the chemometabolism have to be promoted. However, oncothermia (or generally hyperthermia) could down-regulate the blood flow when the temperature is high. This does not help the effect of the primary therapy. In these cases oncothermia has to be applied afterwards when the chemo-infiltration is highest in the tumor, and localize the high chemo-concentration with physiological down regulation of the blood-flow by high temperature, as we had discussed in Section 2.2.2.

Protocols for a high blood-perfusion tumor treated with fractioned radiotherapy and complementary oncothermia are shown for everyday and not everyday applications (see Fig. 4.86) The protocol for low blood-perfusion cases treated with fractional radiotherapy complementary with oncothermia is shown in Fig. 4.87.

4.3.2 Complementary to Chemotherapy

Oncothermia has to be fitted to the chemometabolism and pharmaco-kinetics of the actually applied drug. It has to be applied before or concomitantly with the

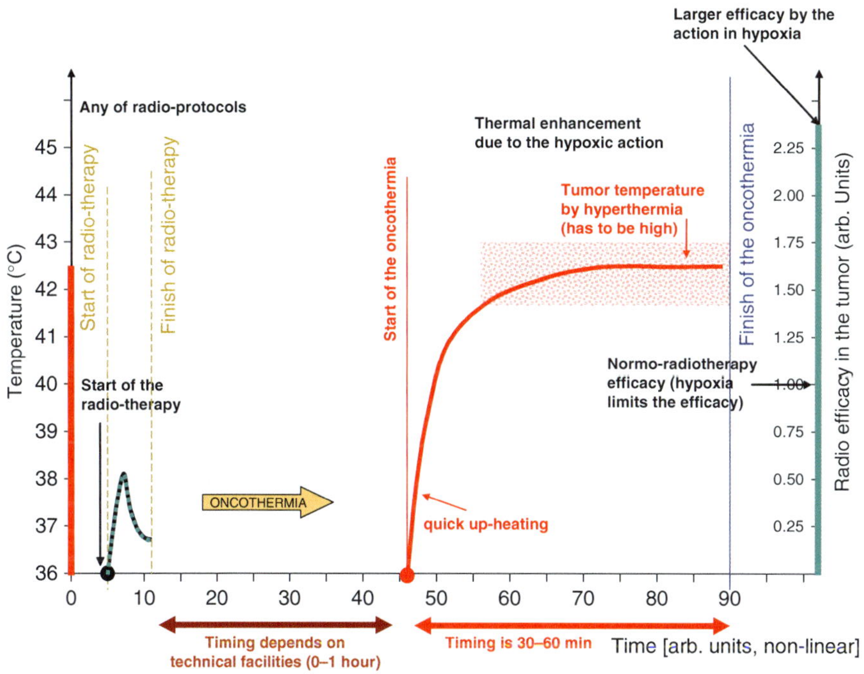

Fig. 4.84 Oncothermia follows the complementary radiotherapy with the maximal tolerable dose

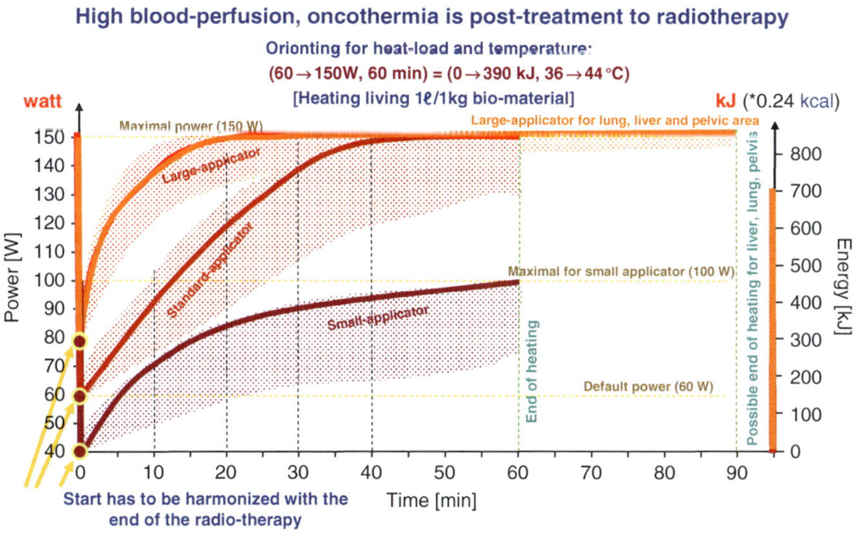

Fig. 4.85 The electrode powering in the high blood-perfusion case with radiotherapy complementary to oncothermia

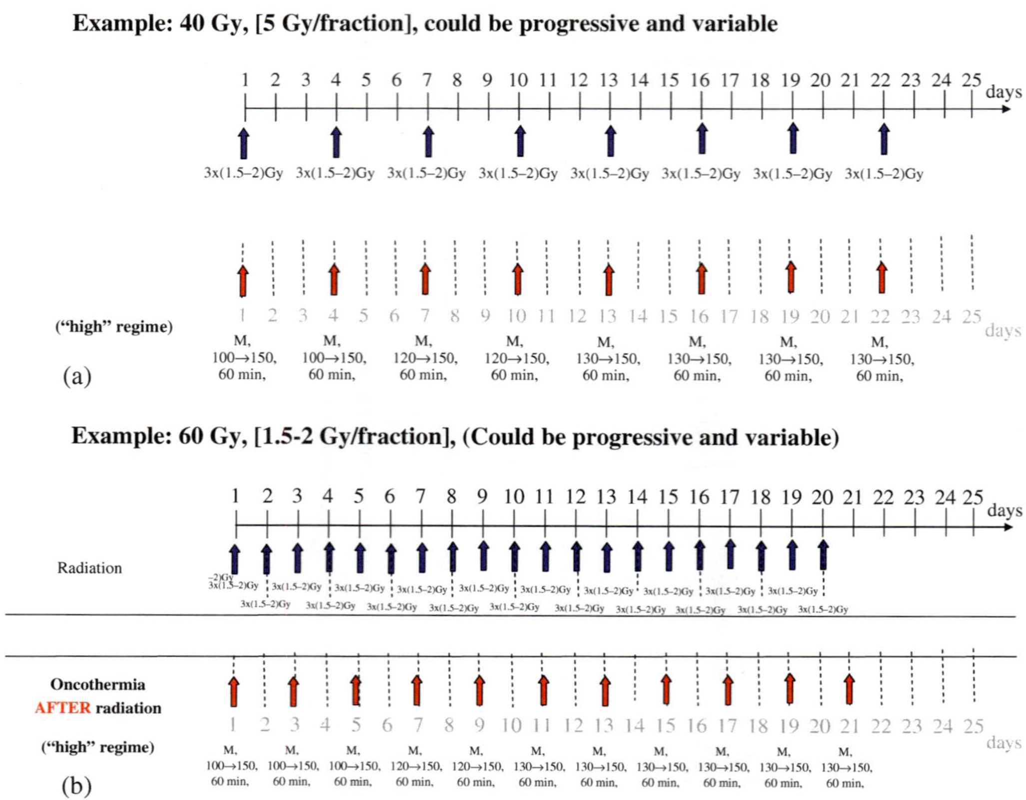

Fig. 4.86 Protocol for high-dose fractions, applied every third day (**a**) and low dose fraction applied everyday (**b**). "*M*" denotes the appropriate "Medium" size (20 cm diameter) of the bolus electrode

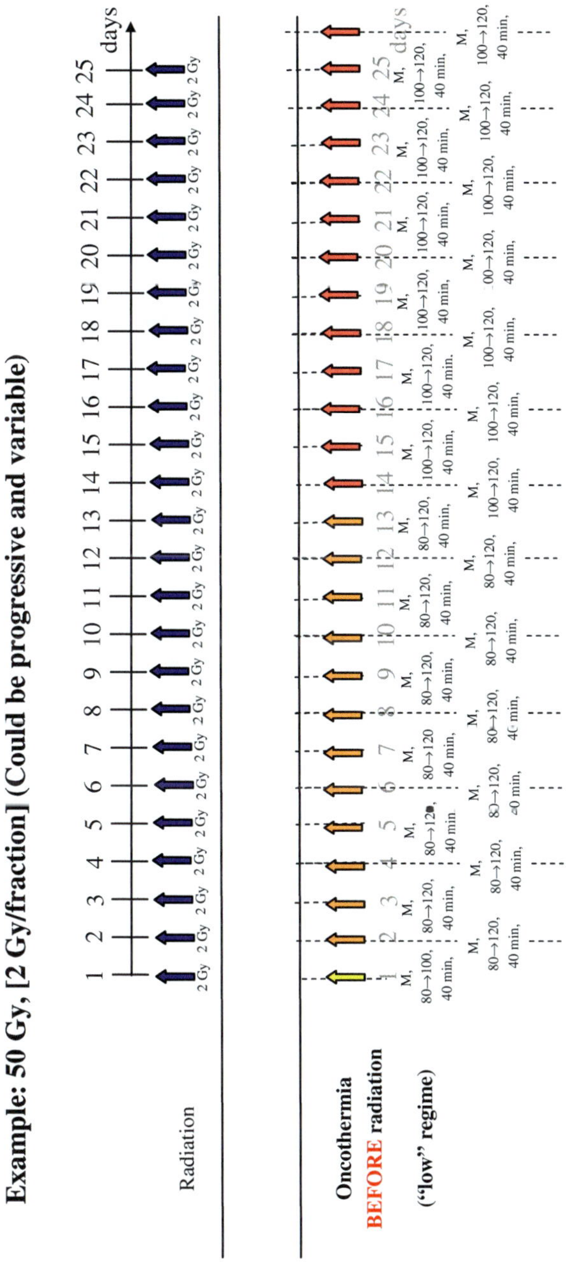

Fig. 4.87 Protocol for low-dose fractional radiotherapy applied complementary with oncothermia

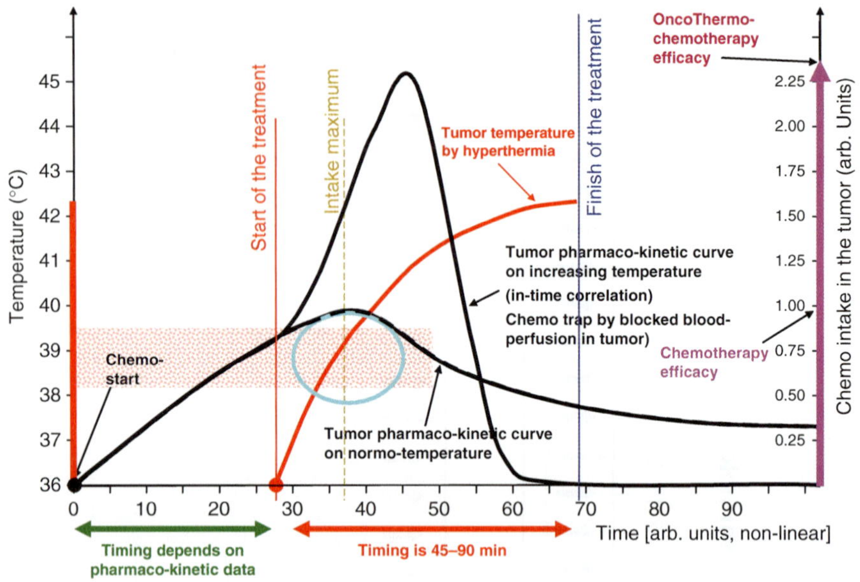

Fig. 4.88 The oncothermia synergy with chemotherapy

chemo treatment. Administering the chemotherapy after the hyperthermia could decrease the chemo-intake of the tumor, because hyperthermia suppresses the neoangiogenetic blood flow.

The best performance of the combination can be achieved, if the oncothermia is performed at the time when the given drug has the highest chemo-dose in the tumor (see Fig. 4.88). The kinetic data of the actual drug are usually provided by the drug producer.

Oncothermia can be applied as triple- or quadruple-modality (radio-chemo-thermo therapy and additional to surgery (adjuvant or neoadjuvant), and also some supportive therapies (vitamins, enzymes, etc.) can be given alongside.

The preliminary mild temperature increase (one of the advantages of oncothermia), could be followed by definite high-energy treatment in a chemo-combination, to keep the chemo agents in the tumor, and to wash them out from the surrounding healthy volume (see Section 2.2.2).

In the case of well blood-perfunded tumors (not too large and/or diffuse) oncothermia could be applied afterwards (after the chemo treatment not more than 1 h). Here the treatment protocol has to prefer the quick increase of the temperature, a dynamic increase of the energy. Treatment time is 60 min, but the near-equilibrium has to be reached within 20 min.

In the case of badly perfunded tumors (large solid tumors with a necrotic inner zone), the chemo treatment is preferably applied in parallel with oncothermia. In this case the treatment has to start slowly, and adjusted to the kinetics of the drug. The starting slow energy intake increases the tumor perfusion at first, promoting

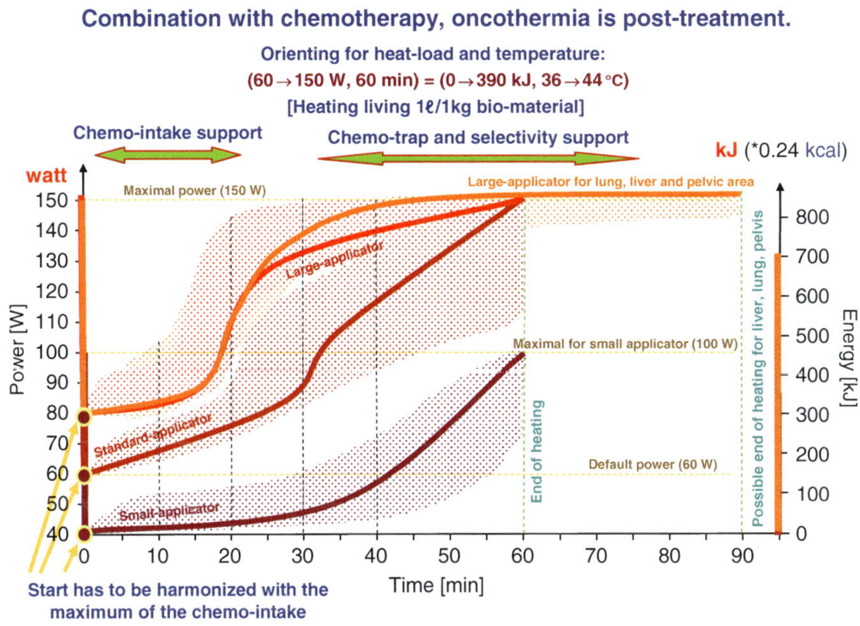

Fig. 4.89 The proposed electrode protocols for complementary chemotherapy applications

blocked perfusional channels and perfunding the tumor by the drug. Later (through gradual increase of the power) we need to have maximal power after 40–60 min, and the total treatment time should be 60–90 min.

The proper electrode application is shown in Fig. 4.89.

An example protocol of the chemotherapy complementarily applied with oncothermia is shown in Figs. 4.90 and 4.91.

To choose the gradual step-up or step-down heating we have to evaluate also the various kinetic processes.

The step-down heating could start with high energy (temperature) treatment, which promotes the chemical reactions, and so lowers the activation energy [1186]. The step-down heating uses the hysteresis process in the physiology of the neoangiogenesis also. As stated, the neoangiogenetic vessels shrink at high temperature. To reach the high energy necessary for the shrink (down-regulation of the blood flow), immediate temperature growth is necessary. When the shrinking occurs, the blood flow lowers considerably. This drastically decreases the cooling effect of the blood stream. In consequence of this low blood supply, much less energy is adequate to maintain the same temperature than was necessary when the blood intensively cooled the area. The process is directly connected with the temperature expectations, and the actual immediate real-time changes in the tumor status. Its real advantage is the relatively low energy supply after overcoming the relatively high activation energy. Choosing step-down heating is a good option for temperature-oriented

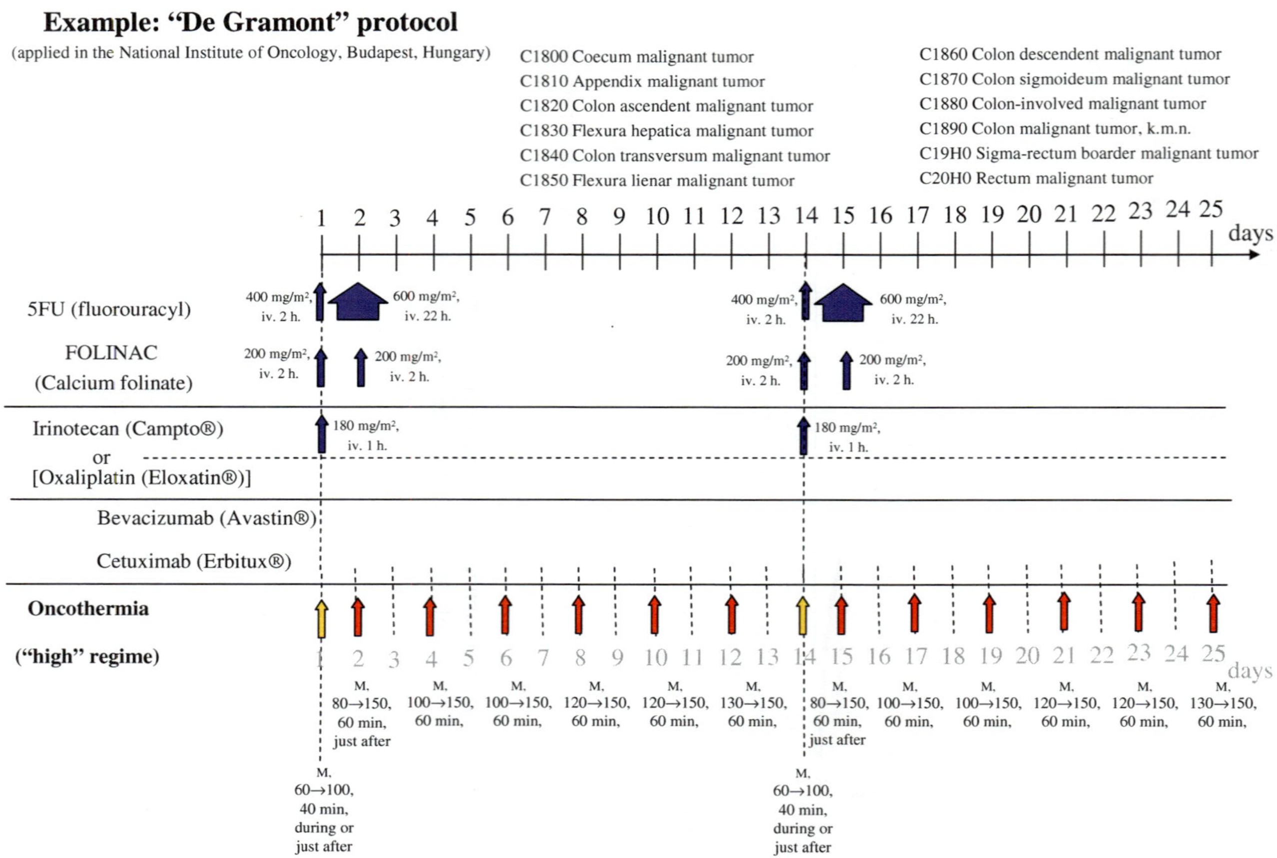

Fig. 4.90 Oncothermia applied complementarily to the "deGramont" chemoprotocol for various kinds of malignancies (Protocol of Prof. Dr. I. Lang, Hungarian National Institute of Oncology)

Week	1	2	3	4	5	6	7	8	9
Days	**1**	**8**	**15**	**22**	**29**	**36**	**43**	**50**	**57**
ChTx	C	C	C	C	C	C			
Oncothermia	**O O O**	**O O O**	**O O**	**O O**	**O O**	**O O**			
US-Sono	UU		U				U		

ChTx=Chemotherapy; US-Sono=Ultrasound

Fig. 4.91 Oncothermia applied complementarily to Avastin-Campto-LV-5FU chemo combination

hyperthermia approaches. Complementary application of step-down heating with chemotherapy needs careful consideration of the chemometabolism and its kinetics. To block the blood flow before the chemo has reached its maximum intake suppresses the chemo-efficacy and so decreases the advantage of the common application.

The step-up heating has a different philosophy. In this case the chemo-intake is insured in the beginning, gradually enriching the tumor with the drug. Reaching the blood-flow reduction region the tumor traps the chemo, which is washed out at the same time from the healthy neighborhood by the blood flow. This is an extra selection to destroy the tumor. Another advantage of the step-up heating process is it develops HSP in both the tumor and healthy cells, but the gradual temperature growths induce relatively much more HSP change in healthy than in malignant cells. This makes healthy cells more resistant to the treatment than their malignant counterpart. Another advantage of the step-up process could be observed by the better tolerance of high energy intake when it is gradually introduced and does not become suddenly high. In the oncothermia process gradual step-up heating has to be applied. With this not only the above advantages could be used, but the non-equilibrium heating will be longer and so the temperature equilibrium where the gradient driving force on the cellular membrane vanishes, occurs later, giving more room for the special oncothermia effects.

Oncothermia can be applied as triple- or quadruple-modality (radio-chemo-thermo therapy and additional to surgery (adjuvant of neoadjuvant) and also some supportive therapies (vitamins, enzymes, etc.) might be given alongside. An example protocol of such an application is shown in Fig. 4.92.

Week	1	2	3	4	5	6	7	
Days	**1**	**8**	**15**	**22**	**29**	**36**	**43**	
Radiation therapy	**RRRRR**	**RRRRR**	**RRRRR**	**RRRRR**	**RRRRR**	**RRR**	**RRRRR**	**RRRR**
Chemo-therapy	**C**				**C**			
Oncothermia	**O O**	**O O**	**O O**	**O O**	**O O**	**O O O**	**O O O**	

Fig. 4.92 Example of trimodal therapy involving oncothermia

The surface temperature is controlled by the patient's senses. The surface of the body senses the heat (receptors are in the subcutis), so this process is an important part of the personalization.

The best performance can be reached in a no-stress situation, when the patient's ability to sense is undisturbed, and the other physiologic processes are normal. This is the reason, why a simple quiet environment and relaxing, stress-free entertainment (music, video, reading, etc.) is so favorable during the treatment.

4.3.3 Clinical Toxicity, Safety

The hot-pain starts at approx. 44°C and rapidly grows above this threshold [564]. The systemic body temperature does not change while the surface of the heated area reacts rapidly to the ambient temperature, [1187] by the vasodilatation, the capillary network is an effective heat-exchanger [1188]. The regulation is very effective at the peripheries. In consequence the blood flow in the peripheries changes drastically the skin heat- as well as electric-conductance.

Oncothermia does not cause deep hot spots. The capacitive coupling has a well-known inverse penetration function, [1149, 1151], and inverse square energy absorption from the surface down to deep regions [221]. Consequently, oncothermia has the largest energy deposition on the surface (this is gained by the low dielectric constant near-surface adipose tissue) and deeper it is decreasing. The monotonically decreased energy intake is modified only by the area of the malignancies, where the higher metabolic activity (as described above) concentrates the current density [1189, 1190 762, 763, 813]. Extreme hot spots at depth can be only possible by this mechanism, which increases the temperature only inside the tumor. The surface, adipose burn is physiologically controlled by the patients. The temperature sensors (heat sensation) are located on the sub-cutaneous area [564], sensing well the dangerous overheating. The oncothermia protocol strictly fixes the considerations of the patients signals, excludes the patients with low temperature sensitivity in the treated area, or unconscious patients or simply those that are not able to communicate. Furthermore, the surface energy-density current remains 1 W/cm^2, which is the possible hurting limit. Also important a surface measuring device could be attached to the electrode being sure of the harmless process. Consequently no MRI and no invasive temperature control is necessary to be sure of proper safe treatment. Furthermore, all valid standards of electromagnetic radiation [1191–1194] are carefully checked at 13.56-MHz operating frequency so no health risk exists. For comparison:

- electric field: standard: $E < 800$ V/m; while in oncothermia: $E < 100$ V/m;
- magnetic field: $H < 0.3$ A/m, while in oncothermia: $H < 0.03$ A/m.

No shielding is necessary.

The adverse/side effects of oncothermia treatments are minor. A minor number of erythema (<8%), and rarely sub-cutaneous fibrosis may occur; no other toxicity was observed except the usual toxic reactions of the complementarily applied conventional treatments (radio- and/or chemotherapies). Patients reported (subjective) the decrease of adverse effects of parallel conventional therapies, i.e. decrease of pain and other subjective symptoms. Most patients reported improvement of their general well-being. During and after the treatment a slight increase of the systemic temperature could be measured by simple checks. It could cause extra perspiration and temporary fatigue in the patients. (The majority of patients sleep during the treatment.) Increase of the body temperature is the consequence of the high energy intake, heating the circulation blood. Approximately 30–60 min after the treatment the normal body temperature is automatically re-established.

To prepare the patient for treatment she/he has to be free of any metallic pieces (necklaces, rings, other jewelry, watches, pipes, coins, phones, hairpins, pens, etc.) or equipment like earphones, hearing aids, music devices (walkman, mp3 player, etc.) and/or any wire-connected instruments, also credit cards and/or any other magnetically sensitive products (diskettes, tapes, etc.) should be kept away from the treatment.

4.4 Oncothermia Case Reports

In order to provide a comprehensive insight into the indications under which patients are being treated with oncothermia we would like to present a few representative cases. Some of the cases are very new; some of them were treated over a 5-year period, used for follow-up cases and other possibilities. These old treatments are new results for the follow-up, which is the critical issue for the feasibility of future prospective trials.

We would like to demonstrate the efficacy of the oncothermia method in some peculiar cases. These are selected to show the very advanced cases as well as such localizations that are contraindicated by conventional hyperthermia, but where oncothermia is able to handle the case (brain, eye, etc.) We have to add an important safety note however: these cases require great expertise of the oncothermia applications, special care with extreme attention at the treatment always counting the inherent risk of causing harm. The patients senses in these cases give crucial control and have to be regarded as the main feedback mechanism during treatment.

4.4.1 Near-Eye Treatments

The near-eye application needs special attention and control, because the eye could be easily damaged by heat. The direct RF-radiation can cause temporary or permanent blindness. However, as oncothermia does not use high temperatures, the treatment is possible with care. The case-report examples for near-eye localizations are shown in Fig. 4.93 [1195].

An eye involving a non-operable non-Hodgkin-lymphoma case is shown in Fig. 4.94 [1196]. The local success of oncothermia is obvious despite the eye

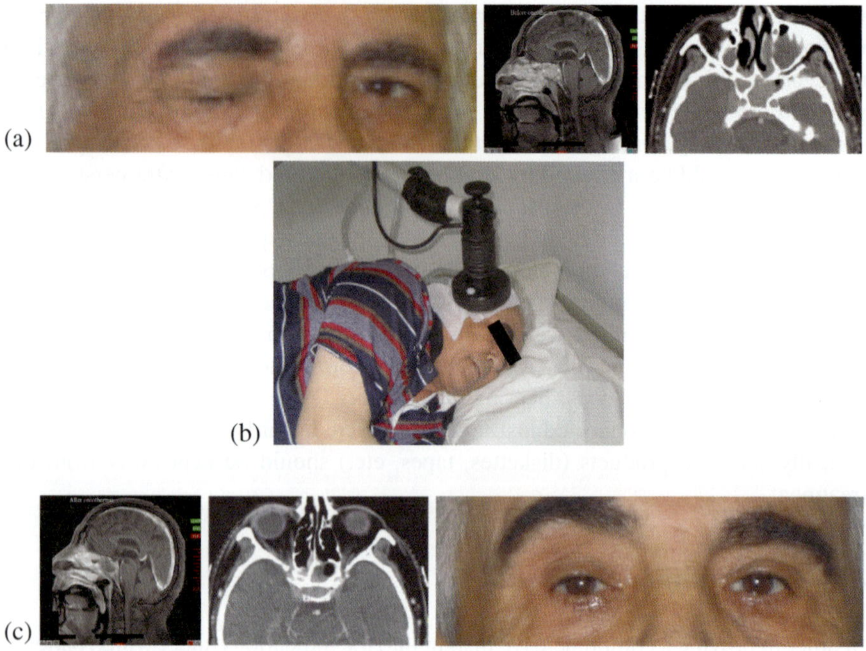

Fig. 4.93 Case of inoperable tumor of sinus sphenoidalis. Investigator: Prof. H. Renner; Praxis at Klinikum Nord, Nürnberg, Germany. Patient: 67 year old, male; Histology: squamous epithelium carcinoma; Development: complete right opthalmoplegy (**a**). Therapy: radiotherapy: 54 Gy, fractional + oncothermia (simple regular approach, electrode diameter is 10 cm; 6 sessions, 60 min/each, immediately after radiotherapy) (**b**). Result: partial remission (PR), spectacular success shows the safety of the method (**c**)

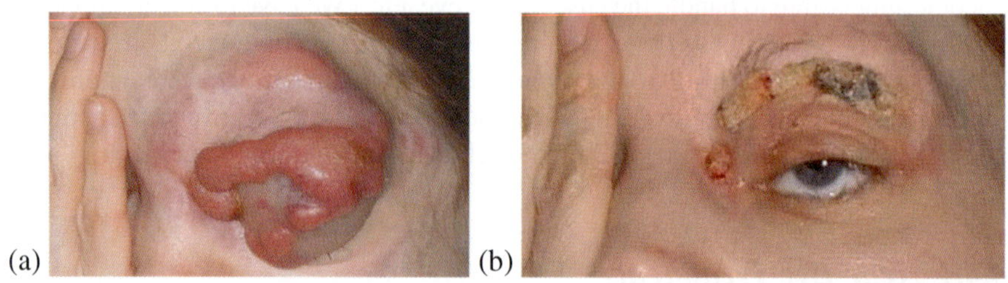

Fig. 4.94 Case of inoperable advanced non-Hodgkin-lymphoma Investigator: Prof. A. Herzog; Institute: Fachklinik Dr. Herzog, Nidda, (Bad Salzhausen), Germany. Patient: 38 year old, female; Stage: WHO IV; (**a**). Treatment: chemotherapy (Bendamustin) + oncothermia. Result: complete remission, (CR), success shows the safety of the method (**b**)

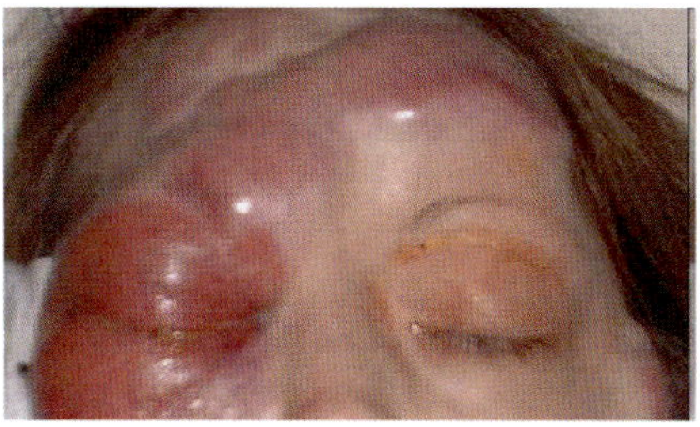

Fig. 4.95 Development of the same case as Fig. 4.94. Definitely no relapse was occur in the area which was treated by oncothermia before

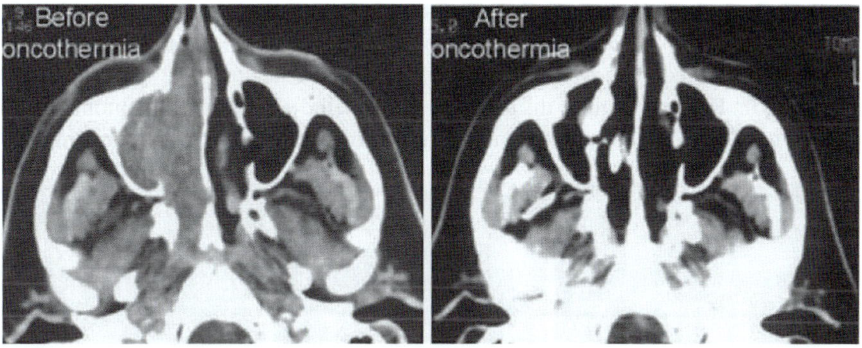

Fig. 4.96 Case of metastasis, infiltration in rectus internus and rectus internus inferior; inoperable. Primary tumor: colon adenocarcinoma. Investigators: Prof. D. Grönemeyer, & Dr. H. Sahinbas. Institute of Microtherapy, University Witten-Herdecke, Bochum, Germany. Patient: 46 y, male; (**a**). Therapies: local chemo perfusion and concomitantly oncothermia. (12×60 min/each). Result: complete remission (CR); (**b**)

involvement. However, the development of the disease was intensive, but the area which was treated before had no relapse, see Fig. 4.95.

Inoperable sinus metastasis of colon adenocarcinoma is successfully treated with oncothermia (electrode 10 cm diameter) (Fig. 4.96) [1197], the large tumor mass was eliminated.

The near-eye treatments show the unique capability of oncothermia: the field acts. The higher temperature of the regular hyperthermia is contraindicated in these cases.

4.4.2 Brain Cases

The brain is a sensitive organ. It is mostly contraindicated to treat with external conventional radiation hyperthermia, because the elevated temperature in the brain could cause harm and increase the intra-cranial pressure by developing edema. Oncothermia, due to its field effects, could treat this organ with high efficacy and safety.

A 52-year-old female patient [1198] presented with a pre-history of raised light-headedness, headache, and fatigue of 2 months' duration. Neurologic examination revealed central facial paralysis and homonymous hemianopia. Laboratory evaluation was without significant pathologic findings. MRI showed a lesion of 6×5×4 cm in the left frontal region, and a mass in the left parieto-occipital region measuring 4×3×3 cm (see Fig. 4.97a). Surgical intervention was not possible. She underwent stereotactic biopsy of the left frontal mass, where the histopathology was reported as GBM, WHO grade IV. Treatment was started with fractionated RT (54 Gy total doses: 1.8 Gy × 5 d/wk for 6 weeks). Her complaints were reduced though neurologic signs did not completely resolve. It was decided to start an adjuvant therapy with local hyperthermia and temozolomide being administered concomitantly (TMZ) (100 mg/m^2/d × 21 days, 1 week rest, for six cycles). Local hyperthermia was continued. The MRI evaluation showed a near complete remission (see Fig. 4.97b). Her complaints disappeared and all the neurologic symptoms completely resolved. She has not required any further treatment from the time the

(a)
(b)

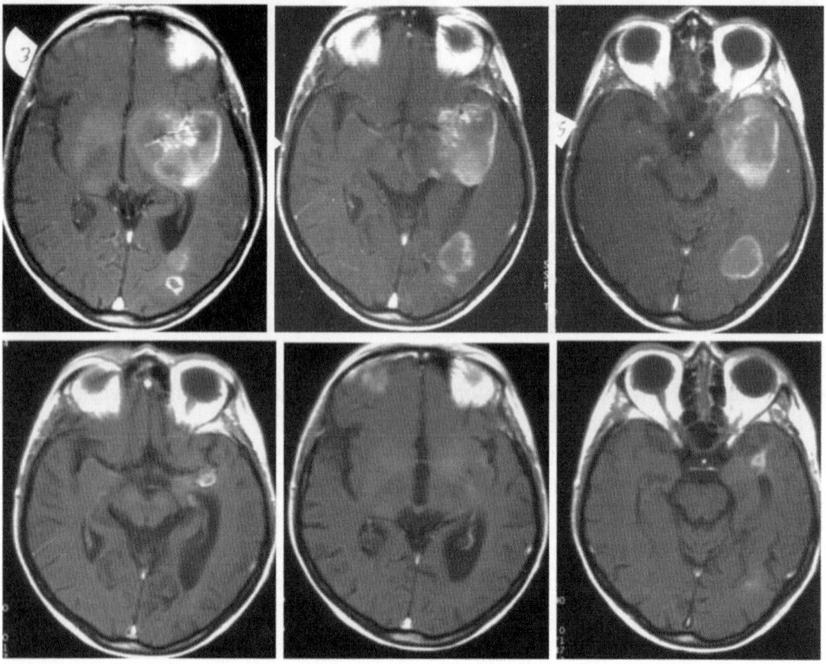

Fig. 4.97 Case of glioblastoma multiform (GBM), WHO IV. Investigators: Prof. D. Grönemeyer, & Dr. H. Sahinbas. Institute of Microtherapy, University Witten-Herdecke, Bochum, Germany. Patient: 52 year old, female. Lesions: 6×5×4 cm left frontal and 4×3×3 cm left parietooccipital region, (**a**). Treatments: trimodal therapy; radiotherapy fractionated (54 Gy total dose: 1.8 Gy×5 d/wk for 6 weeks) and temozolomide (TMZ) (four cycles). Concomitantly with temozolomide oncothermia was applied. Power ranged between 40–150 watts and the average equivalent temperature in the tumors was above 40°C more than 90% of the treatment time. Oncothermia was performed in two/three sessions per week (total: 25 sessions). After the 1st diagnosis patient was alive more than 24 months. Post treatment, imaging with MRI-T1, (**b**)

adjuvant therapy was finished to date and appears for regular check-ups. She is in good health and has an active QoL.

A 47-year-old male patient [1198] with a pre-history of Anaplastic Astrocytoma. Laboratory evaluation was without pathologic findings. MRI showed a lesion of 8×6×5 cm in the temporal region. Staging classified the tumor as anaplastic astrocytoma grade WHO III. The patient underwent a partial resection. RT (60 Gy total dose) was started with partial overlapping of oncothermia. The neurologic signs being observed before were not completely resolved. A second oncothermia cycle was started in combination with Temodal (Temozolomide). The MRI evaluation showed a complete remission (CT) (see Fig. 4.98). His complaints disappeared and all the neurologic symptoms completely resolved. He has not required any further treatment from the time the adjuvant therapy was finished to date and appears for regular check-ups. He is in normal health and has a good QoL.

A 30-year-old female patient [1198] with Anaplastic Astrocytoma. MRI showed a lesion of 5×4×3 cm in the left central brain region. It is staged as grade WHO III. A metastatic lesion in the contra-lateral central region of 2×1 cm plus edema was observed. The patient underwent a partial resection. Unfortunately, shortly after that procedure the tumor relapsed and RT was started in combination with seed implantation. Oncothermia monotherapy was applied in combination with systemic

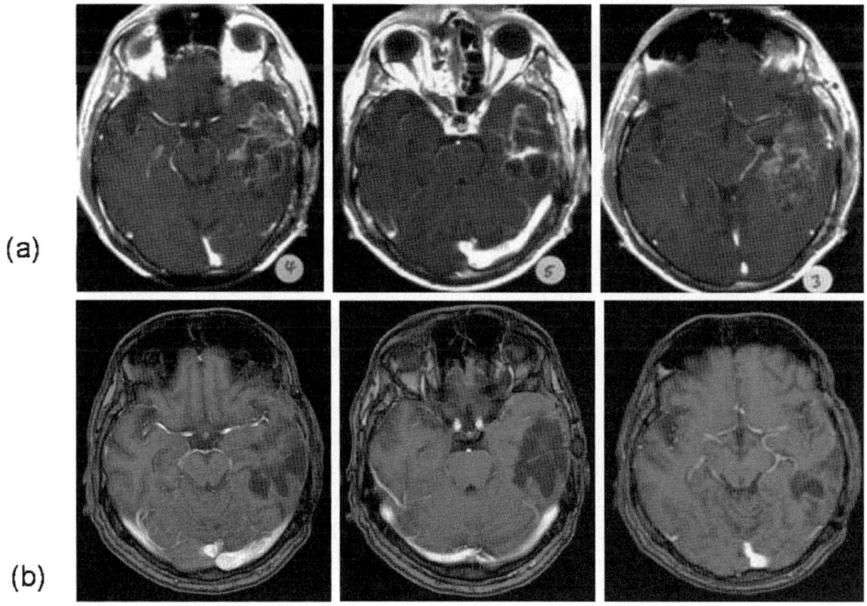

Fig. 4.98 Case of anaplastic astrocytoma, WHO III. Investigators: Prof. D. Grönemeyer, & Dr. H. Sahinbas. Institute of Microtherapy, University Witten-Herdecke, Bochum, Germany. Patient: 47 year old, male. 6×8×5 cm lesion in temporal region, (**a**). Treatments partial resection and radiotherapy + oncothermia; 2–3 weekly. Followed by chemotherapy (Temozolomide) and oncothermia was continued. Result: complete remission; CR (**b**)

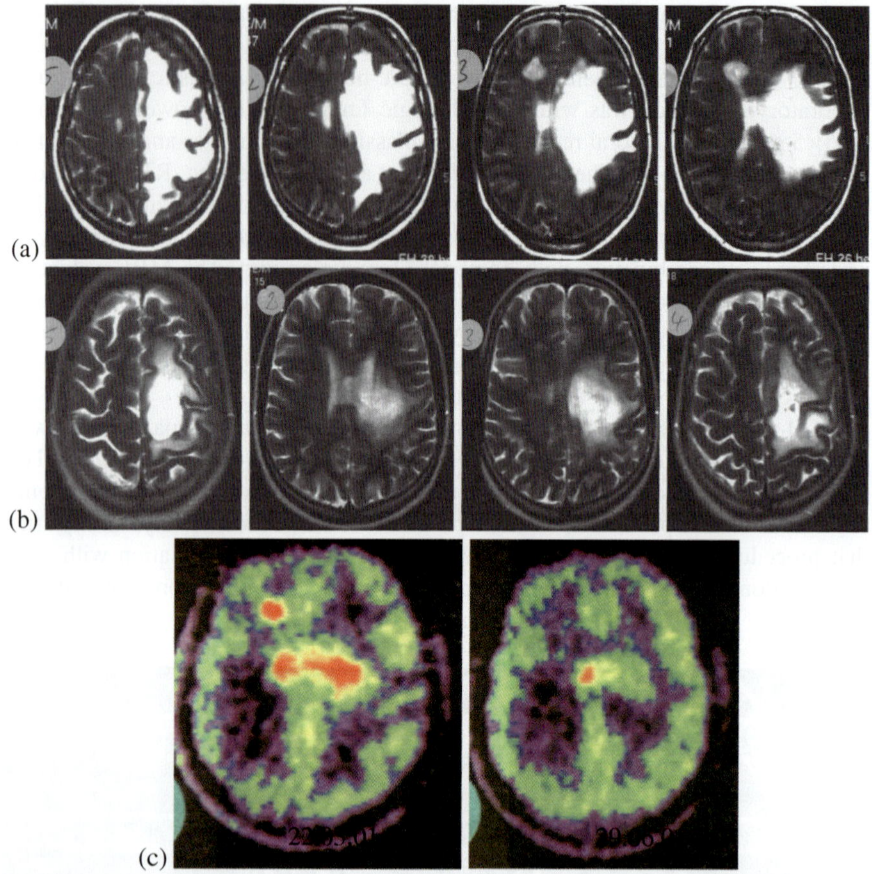

Fig. 4.99 Case of anaplastic astrocytoma WHO III/IV. Investigators: Prof. D. Grönemeyer, & Dr. H. Sahinbas. Institute of Microtherapy, University Witten-Herdecke, Bochum, Germany. Patient: 30 year old, female. Stage was WHO I. at the first diagnosis, multi treatments. Bifrontal and central recidivs, (**a**). Therapies: radio-therapy: seed-implantation (25 Gy)+ 25 Gy external radiation. After it oncothermia + systemic chemotherapy (temozolomide) during 21 days / 1 week; (**b**). PET images before and after oncothermia treatment (**c**)

chemotherapy using Temodal for 21 days at 100 mg/m^2 with 1 week free interval. The MRI images after loco-regional oncothermia showed a partial remission (see Fig. 4.99). Her complaints disappeared and the neurologic symptoms decreased.

Another astrocytoma case is shown in Fig. 4.100 [1197]. The 38-year-old male patient received oncothermia as monotherapy after Fortecotin (4 mg) and fractional radiation (40 Gy [2 Gy/d]). Oncothermia did not induce perifocal edema, which in conventional hyperthermia operating only with temperature would be the case.

Figure 4.101 shows a patient at an advanced stage [1196], with a bad QoL that was treated by oncothermia complementary to ACNU (3×50 mg/5 w). The QoL was drastically improved.

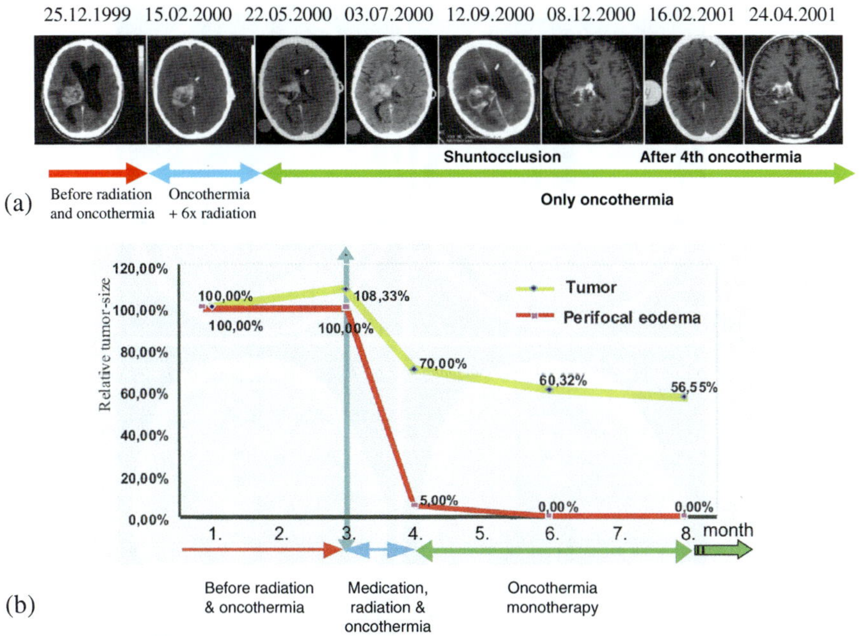

Fig. 4.100 Case of non-operable anaplastic astrocytoma WHO III.; hydrocephalus occlusus, neurofibromatosis. Investigators: Prof. D. Grönemeyer, & Dr. H. Sahinbas. Institute of Microtherapy, University Witten-Herdecke, Bochum, Germany. Patient: 38 year old, male. Treatment radiation: 40 Gy (2 Gy/d); chemotherapy: Fortecotin 4 mg + oncothermia. Next therapy: oncothermia as monotherapy, (**a**). The important factor, that the perifocal edema considerable decreased (**b**). Follow-up: 10 months

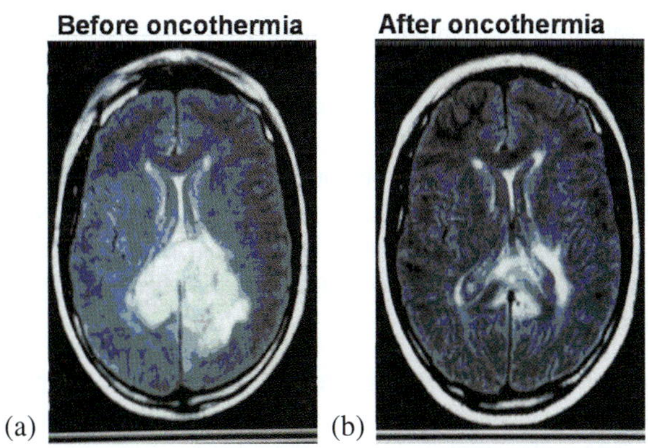

Fig. 4.101 Case of glioblastoma multiform, WHO IV. Investigator: Prof. A. Herzog; Institute: Fachklinik Dr. Herzog, Nidda, (Bad Salzhausen), Germany. Patient: 64 year old, male. Prior to treatment: unable to walk, aphasia; (**a**) Treatment: ACNU 3×50 mg every 5 weeks + oncothermia; Results: partial remission, (PR) (**b**). After 3 cycles of treatment patient walks again, speaks fluently

(a)

(b)

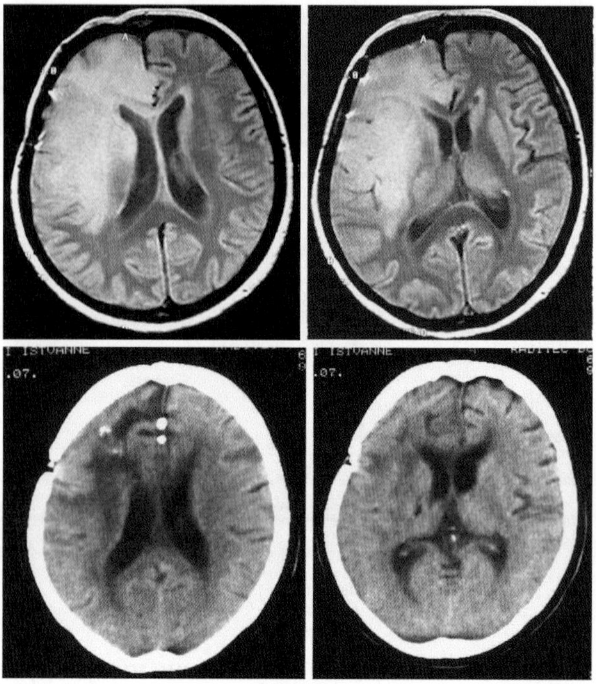

Fig. 4.102 Case of anaplastic astrocytoma; WHO III. Investigator: Dr. A.Varkonyi; HTT-Med Polyclinic, Budapest, Hungary. Patient: 45 year old, female. Before oncothermia (**a**). Treatment: oncothermia as monotherapy. Result: after 2nd session of oncothermia complete remission (**b**)

A 45-year-old female patient suffering anaplastic astrocytoma [1199] refused the gold standard therapies and was treated by oncothermia as monotherapy (see Fig. 4.102). A year later a complete remission (CR) was observed.

A patient (49-year-old female) was heavily treated by various standard treatments, which failed. Afterwards oncothermia as monotherapy was applied (see Fig. 4.103 [1198]). The treatment was successful.

A Phase I safety study was performed [1200] in the Neurology Clinic of Regensburg University (Regensburg, Germany). (Institute: Neurology Clinic, Regensburg University, Phase I prospective clinical trial. Investigators: Prof. Dr. U. Bogdahn and PD. Dr. P. Hau). Two typical cases for heavily pre-treated advanced tumors could be cited as examples:

1. Case of anaplastic oligoastrocytoma, WHO III. Investigators: Prof. U. Bogdahn & Prof. P. Hau. Department of Neurology, University of Regensburg. Patient: 48 year old, male. 2 resections (partial), 3 relapses. Treatments: radiotherapy, after this PCV (Procarbazine, CCNU and Vincristine); and after Temozolomide. After these Nimustin (ACNU) + Oncothermia. Three cycles of ACNU, with three cycles of oncothermia five times a week 60 min each (60 sessions). Karnofsky score at start: 70, at end: 60. Best performance: stable disease (SD).

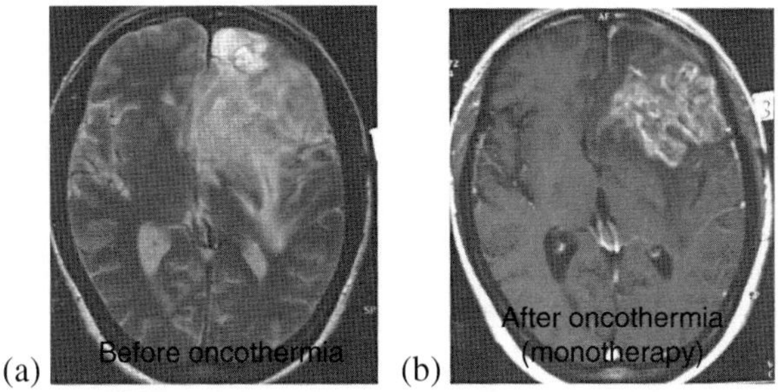

Fig. 4.103 Case of anaplastic astrocytoma WHO II. Investigators: Prof. D. Grönemeyer, & Dr. H. Sahinbas. Institute of Microtherapy, University Witten-Herdecke, Bochum, Germany. Patient: 49 year old, female. Treatments: surgery: two times within 10 months; after surgeries radiation: 60 Gy; 2nd. 30 Gy; after this chemotherapy: 12 cycles of Temozolomide. Recurrence (**a**). After these oncothermia was applied as monotherapy; 8× sessions, 60 min/each. Result: histological tumor-free, only necrotic tissue (**b**)

2. Case of glioblastoma multiform, WHO IV. Investigators: Prof. U. Bogdahn & Prof. P. Hau. Department of Neurology, University of Regensburg. Patient: 49 year old, male, 2 resections (partial), 3 relapses. Treatments: radiotherapy, after this chemotherapy Temozolomide/PEG-Dox; 2nd: Epothilone; 3rd: Temozolomide/int.; 4th: Nimustin (ACNU) + Oncothermia. Two cycles of ACNU, with three cycles of oncothermia five times a week 60 min each (55 sessions). Karnofsky score at start: 80, at end: 80. Best performance: stable disease (SD).

During the Phase I safety study an inadvertent treatment selection was observed Fig. 4.104. The oncothermia complementarily with ACNU treated localization (primary lesion) was reacting well, but a distant metastasis (relapse) was not recognized at the beginning of the treatment session, so was not treated. At the end the non-treated relapsed lesion was growing further, while the treated one was drastically shrinking. The ACNU systemic chemotherapy was of course active for both the lesions.

An advanced pediatric ependymoma case with low QoL (see Fig. 4.105) [1201]. Oncothermia was applied as monotherapy, because all the conventional facilities were no longer available. The treatment was successful, a spectacular improvement was observed.

4.4.3 Gynecology Cases

The pelvic gynecology cases are classical applications for hyperthermia treatments. Oncothermia is also applicable in this important field of treatments. The treatments obviously start with gold standards, and when those are unsuccessful, oncothermia begins to be applied. In consequence most of the cases are advanced, heavily pretreated. In most cases drastic improvement of QoL was observed.

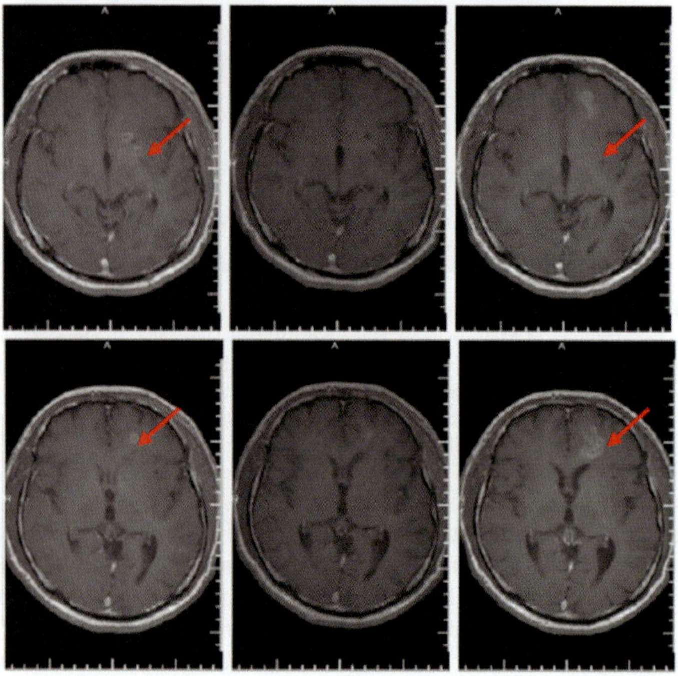

Fig. 4.104 Case of inoperable anaplastic astrocytoma, WHO III. Investigators: Prof. U. Bogdahn & Prof. P. Hau. Department of Neurology, University of Regensburg. Patient: 59 year old, male. Two relapses, Therapies: Temozolomide (/13-cis-RA); 2nd: radiotherapy, 3rd: Nimustin (ACNU) + Oncothermia. Two cycles of ACNU, with three cycles of oncothermia four times a week 60 min each (48 sessions). Karnofsky score at start: 90, at end: 90. Best performance: stable disease (SD). The oncothermia treated primary lesion shrinks, (**a**), while the relapsed, only ACNU treated lesion was in progression (**b**)

One of the most common lesions is the cervix of uterus. Three case reports are shown for this in Figs. 4.106–4.108.

A typical case of polypoid tumors is shown in Fig. 4.109.

4.4.4 Gastrointestinal Cases

The full gastrointestinal track can be successfully treated by oncothermia.

One of the treatments of the salivary glands and the tongue-base is shown in Figs. 4.110 and 4.111. The case was investigated by Prof. H. Renner, in Praxis at Klinikum Nord, Nürnberg, Germany. The patient was 46 years old, male, with tongue-base cancer. It was inoperable, histological squamous cell carcinoma having metastases: cervical lymph nodes. Tumor classification: cT3 cN2b M0 G2 R2. The applied treatment was a trimodal protocol: radiotherapy: 61 Gy, PT+LA + interstitial PT 19.2 Gy, Chemotherapy: Cisplatin + 5-FU, Oncothermia: 60 min, diam. 10 cm, (14×, 2/week). Result: CR March 2003, with follow-up till recidiv involved cervical lymph nodes.

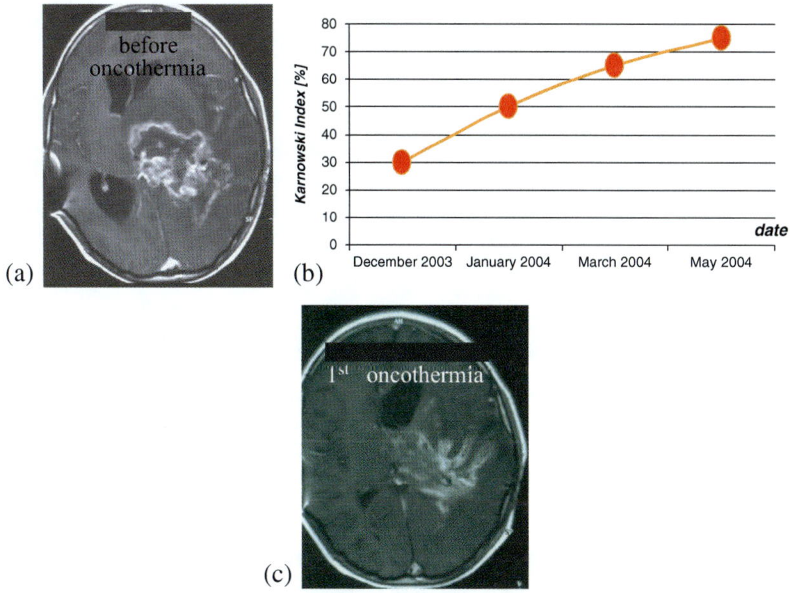

Fig. 4.105 Case of advanced, progressive pediatric ependymoma, bithalamic; stage: WHO III. Investigators: Prof. D. Grönemeyer, & Dr. H. Sahinbas. Institute of Microtherapy, University Witten-Herdecke, Bochum, Germany. Patient 10 years old, male. Conventional treatments (surgery, chemotherapy, radiotherapy) fail (**a**). Karnofsky score at start of oncothermia 30 (**b**). Oncothermia was applied as monotherapy. (60 min, 2–3 times weekly). During half a year the quality of life was drastically improved, the tumor-progression was blocked, partial remission was achieved (**c**)

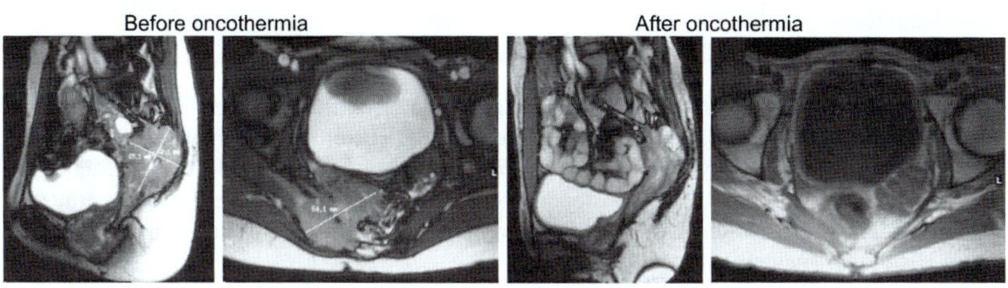

Fig. 4.106 Case of cervix uterus, cervix conisation P5, HPV positive. Investigator: Piko B, Institute: Department of Clinical Oncology, Hospital K. Pandy, Gyula, Hungary. Patient: 28 year old, female. Treatments: total uterus exstirpation with bilateral adnexectomy. Observation: planocellular undifferentiated cancer with vascular tumor cell invasion, pT1 N0 M0. 18 month later strong pain; incontinence. Locally recurrent tumor 6.5×5.5 cm (**a**). CT-guided biopsy: locoregional recurrence of the original carcinoma epidermoid non-cornescens. Treatments: combined radio-chemo-thermo therapy (trimodality). Radiation: pelvic, paraaortic and bilateral parailiacal irregular large field technique: 20×2 Gy. Chemotherapy: Carboplatin, Ftorafur plus oncothermia once a week. After this further radio-chemo-thermo therapy: pelvic and parailiacal CT-planned radiotherapy with craniocaudally adjusted fields: 8×1.8 Gy, Carboplatin/Ftorafur chemotherapy + oncothermia. Results: no more incontinentia, no more pain. MRI: considerable regression (**b**). Further improvement: the patient is able to work again

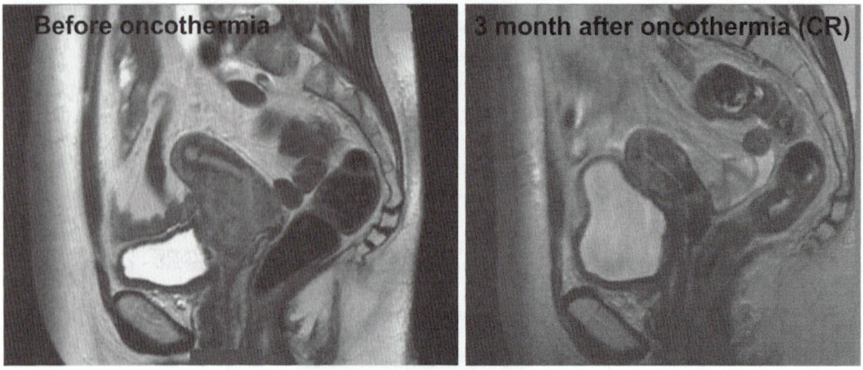

Fig. 4.107 Case of carcinoma of cervix uterus; cT4 cN0 M0 G3. Investigator: Prof. H. Renner; Praxis at Klinikum Nord, Nürnberg, Germany. Patient: 61 year old, female. Histology: squamous cell carcinoma (**a**). Therapy: bimodal, radiotherapy: 50.4 Gy; (5×1.8 Gy/weeks); oncothermia: 6 sessions. Control: 3 months later, hysterectomy (Wertheim). Result: pathologically complete remission ypT0ypN0 (**b**)

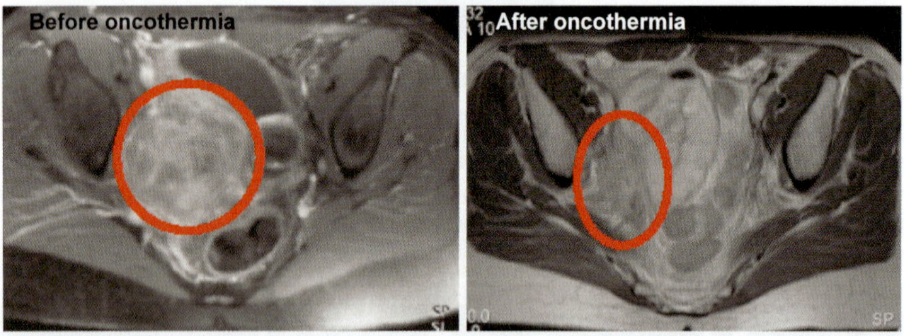

Fig. 4.108 Case of carcinoma of cervix uterus. Investigator: Prof. H. Renner; Praxis at Klinikum Nord, Nürnberg, Germany. Patient: 38 year old, female. Histology: squamous cell carcinoma; local recurrence (**a**). Therapy: local curative trimodal therapy (radio-chemo-thermo). Radio therapy: 45 Gy preoperative and 16 Gy postoperative. Chemotherapy: 2 + 1 cycles Carboplatin/5-FU Oncothermia: 12 + 4 sessions Duration: 5 + 2 weeks 9 month pause, Result: complete remission 100% Histology – negative (**b**)

Another tongue-base treatment (Investigator: Dr. Brockmann W-P. Institute OncoLight Hamburg, Germany, Patient: 57 years old, female, Diagnosis: Tongue-base cancer, Therapy (1): surgery: resection; radiotherapy: postop. 60 Gy. Result (1): recurrence – refusal of glossektomie. Therapy (2): radiotherapy; 36 Gy (2 × 1.2 Gy/day); chemotherapy: Carboplatin 10 × 60 mg; Oncothermia: 2–3/week, 10 treatments, 60 min each. Result (2): complete remission Follow-up: no relapse for a year.

The main emphases of oncothermia applications are in the most common and aggressive gastrointestinal tumors, like esophagus, stomach, pancreas, colon, rectum and their metastatic situations.

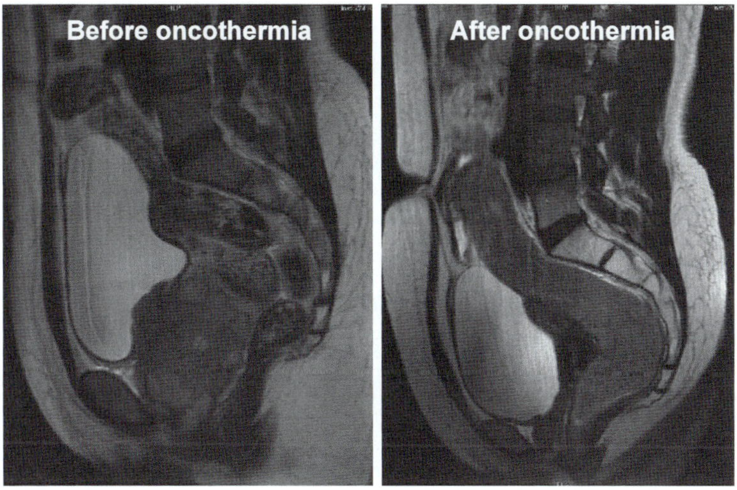

Fig. 4.109 Case of removal of polyploidy tumor on the right wall of the vagina. Investigator: Dr. A. Csejtey, Markusovsky Hospital, Szombathely, Hungary. Patient: 43 year old, female. Histology: planocellular cancer. Huge local/regional recurrence infiltrating the urinary bladder (**a**). Treatment: Radio-thermo-therapy (bimodality) pelvic irradiation: 17×15 cm AP-PA opposed fields, 15 MV photons, daily 2 Gy fractions, total 30 Gy; after it oncothermia (2× weekly pelvic) combined with 6 and 15 MV photon radiotherapy of the same pelvic region from 12×10 and 12×6 cm fields with 4 field box technique, daily 2 Gy fractions, total 26 Gy. Result: considerable regression (**b**), she was able to work again

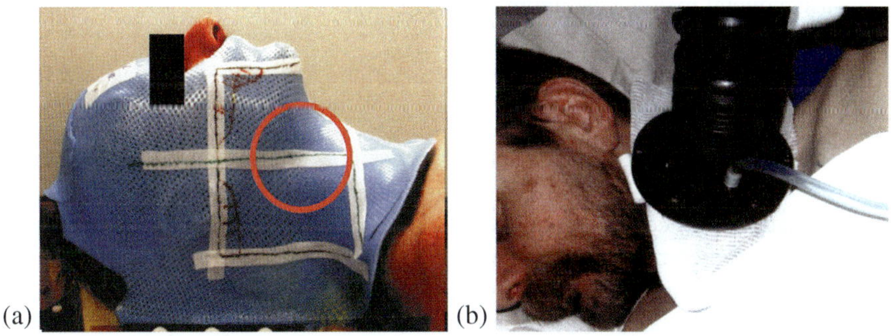

Fig. 4.110 Case showing practice: treat tongue-base cancer. Investigator: Prof. H. Renner; Praxis at Klinikum Nord, Nürnberg, Germany. Bimodal therapy, oncothermia with radiotherapy. Fixing for radiotherapy (**a**) and for the oncothermia (**b**)

Typical esophagus cases are shown in Figs. 4.112–4.114.

Primary and metastatic liver tumors are frequently and successfully treated by oncothermia. Some examples for advanced liver metastases from primary:

- colon transversum carcinoma (see Fig. 4.115),

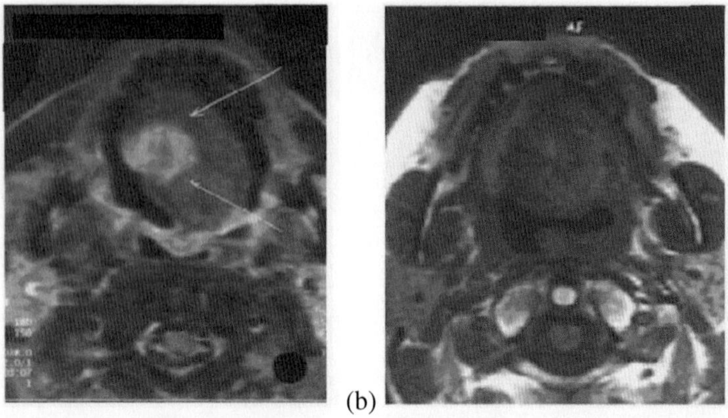

(a) (b)

Fig. 4.111 Case of tongue-base carcinoma. Investigator: Dr. W-P. Brockmann Institute OncoLight, Hamburg, Germany. Patient: 57 year old, female. Treatments: surgical resenction and postoperative radiation 60 Gy. After it recurrence (**a**). Refusal of glossectomy. Trimodal therapy is applied: radiotherapy 36 Gy, chemotherapy: Carboplatine, and oncothermia 2–3 times a week, 10 sessions, 60 min/each. Result is complete remission (CR) (**b**). Follow up 7 months, no relapse

- colo-rectal carcinoma, second-line treatment with oncothermia (see Fig. 4.116)
- rectum carcinoma (see Fig. 4.117),
- uroepithelium carcinoma (see Fig. 4.118),
- sigma carcinoma (see Fig. 4.119),
- mamma carcinoma (see Fig. 4.120),
- invasive ductal mamma carcinoma see (Fig. 4.121) [1202].

Important, long-term observations were made for some hepatic metastases from primary:

- Breast (male); see Fig. 4.122. Metastasis was not observed for 5+ years. When it was diagnosed, regular oncothermia decreased it in the first year of metastasis. Termination of oncothermia stopped the regression. Renewed oncothermia gave regression and complete remission at the end of the second year. After termination of oncothermia relapsed metastasis was observed.
- Colo-rectal metastases. Metastasis was not observed for 1 year (see Fig. 4.123). After its detection, a relatively rarely applied regular oncothermia led to no change in the liver metastasis in the first year. Tumor progression was observed after termination of oncothermia. Reapplied oncothermia gave regression, but after termination of oncothermia progression of the metastasis was observed again. Other colo-rectal metastasis was not observed for 1+ year. After its detection regular oncothermia achieved a stable condition, the tumor did not grow further. This situation persisted for 2 years. After termination of oncothermia tumor progression in the liver was observed again (see Fig. 4.124).

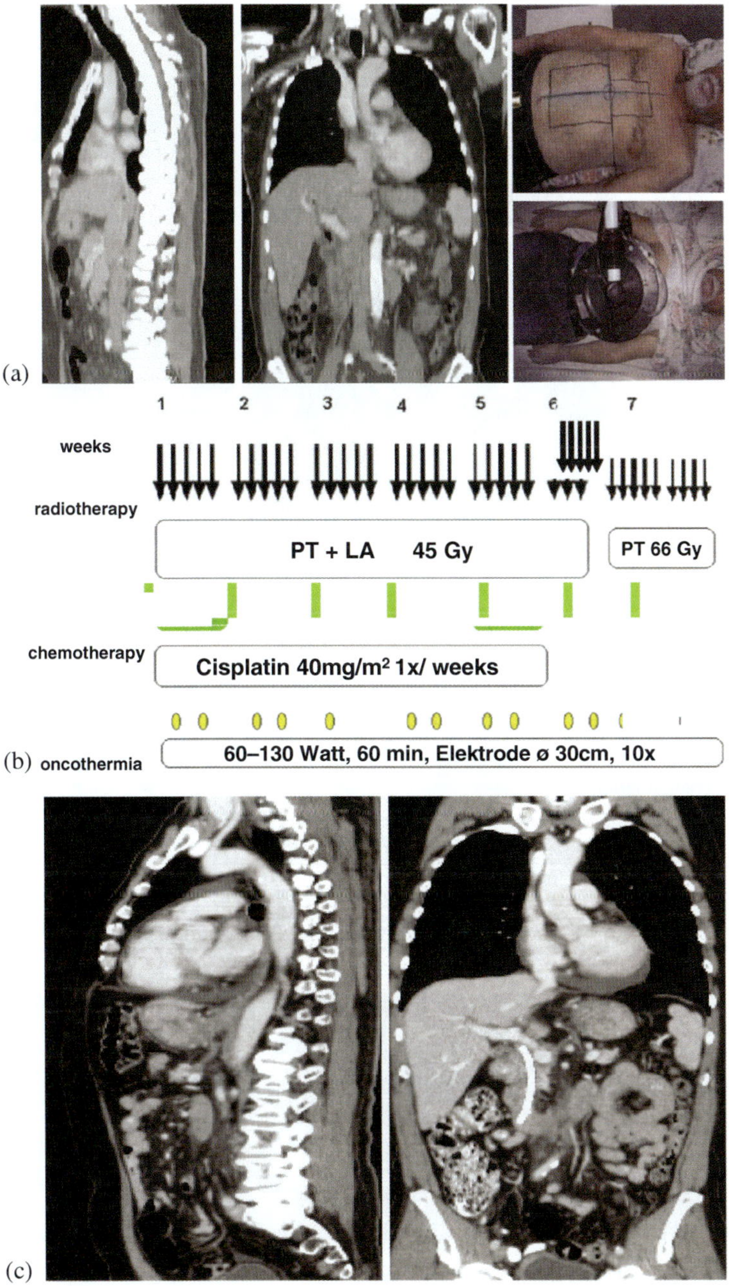

Fig. 4.112 Case of inoperable esophagus carcinoma (Squamous cell G3) in the middle of the esophagus. Investigator: Prof. H. Renner; Praxis at Klinikum Nord, Nürnberg, Germany. Patient: 50 year old, male. Metastases: in mediastinum & celiac ganglia. Tumor-classification: cT2 cN1 M1a G3 R2 (**a**). Treatment: trimodal protocol (**b**). Result: complete remission (CR). Follow-up: after 12 month tumor-free, (**c**)

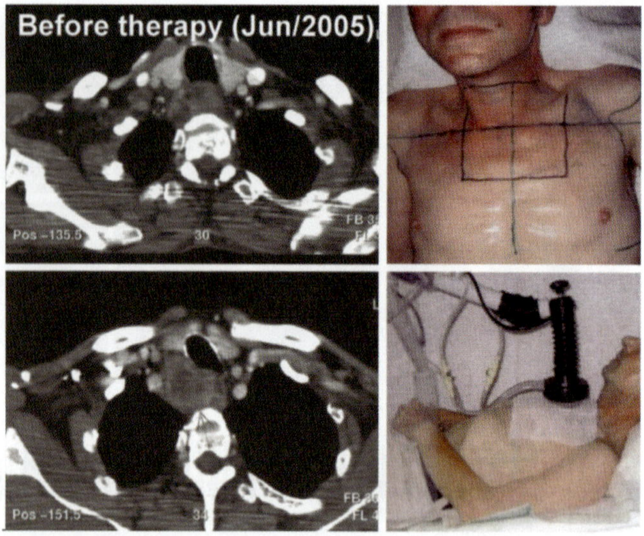

(a)

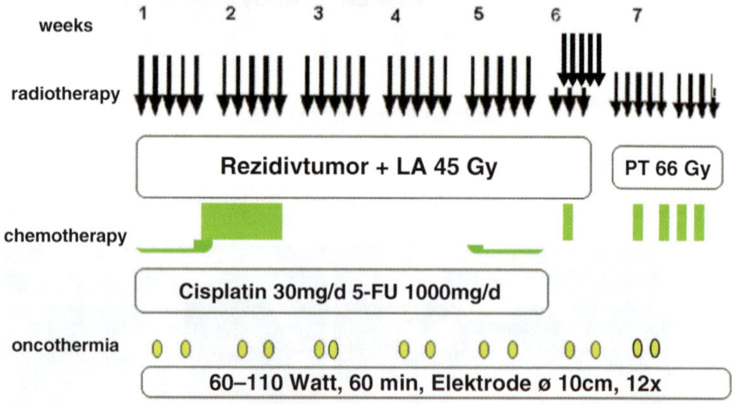

(b)

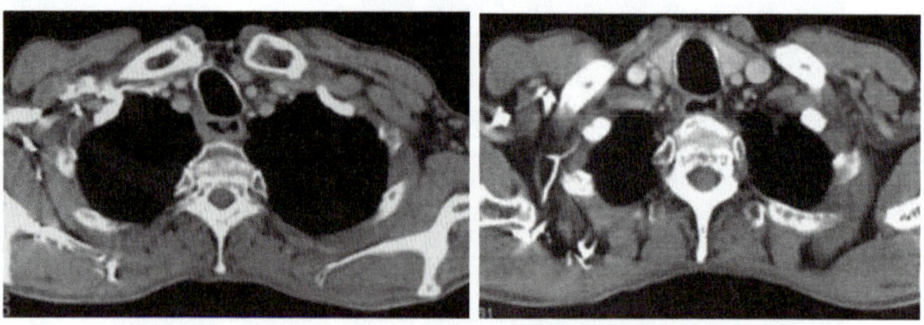

(c)

Fig. 4.113 Case of inoperable esophagus carcinoma (Squamous cell G3), upper esophagus. Investigator: Prof. H. Renner; Praxis at Klinikum Nord, Nürnberg, Germany. Patient: 49 year old, male. After resection inoperable recidiv. Metastases: involved multilocal lymph nodes (**a**). Treatment: trimodal protocol. Result: complete remission (CR). Follow-up: after 30 month tumor-free, (**b**)

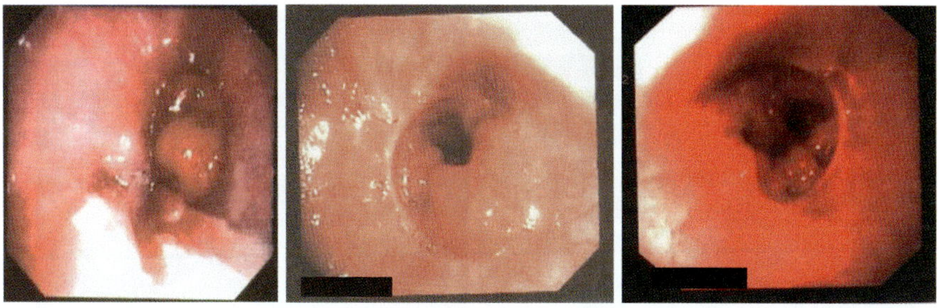

Fig. 4.114 Case of inoperable esophagus carcinoma. Investigators: Prof. D. Grönemeyer, & Dr. H. Sahinbas. Institute of Microtherapy, University Witten-Herdecke, Bochum, Germany. Patient: 46 year old, male. Therapies: surgery followed by multiple chemotherapy and afterwards radiotherapy (50 Gy). Recidiv, anastomose, block of food-passage (**a**). Oncothermia applied as monotherapy. Result: complete remission (CR), free-food passage; (**b**)

(a) (b)

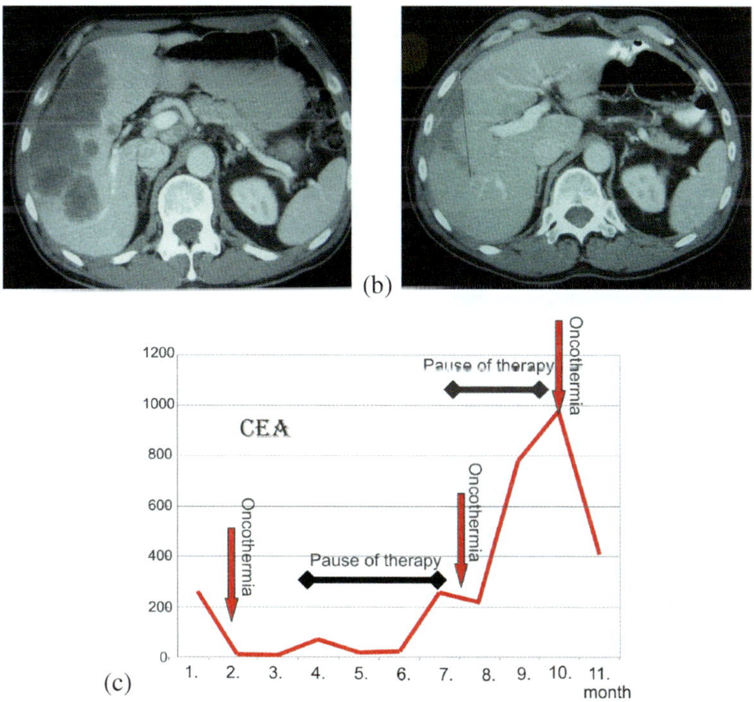

(c)

Fig. 4.115 Case of hepatic metastasis of colon transversum carcinoma. Investigator: Prof. H. Kirchner, Medical Department III. (Hematology & Oncology), Hospital Hannover-Siloah, Hannover, Germany. Patient: 61 year old, male. Tumor classification: pT4, pN2, M1 (Liver). Treatments: surgery hemicolektomie (**a**). Therapy: Oxalyplatine, Leukovorine, 5-FU + oncothermia on liver. Result: good partial remission (PR). Therapy-pause consequently progressive disease (PD). Erbitux, Campto + oncothermia on liver. Result (3): good partial remission (PR) tumor (**b**) and tumor marker regression, became normal (**c**)

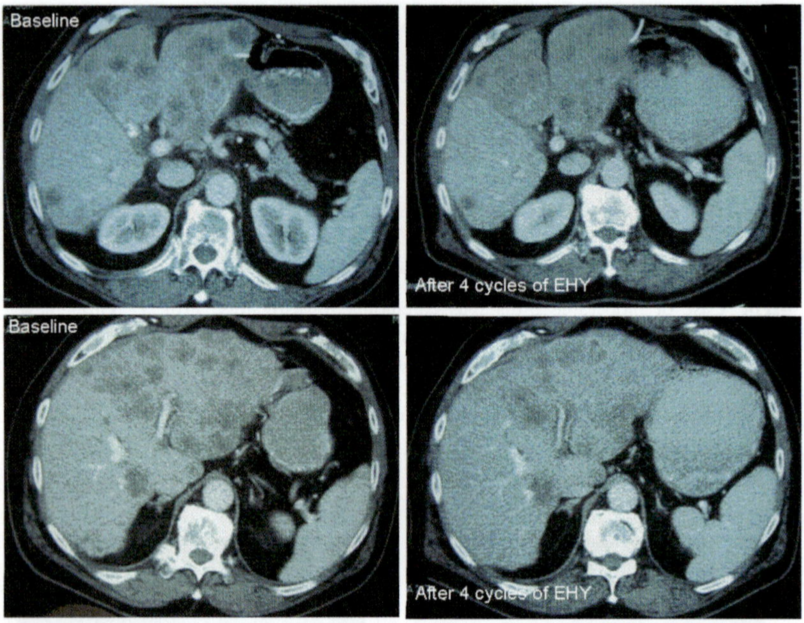

Fig. 4.116 Case of liver metastasis of colo-rectal carcinoma. Investigator: Prof. H. Kirchner. Medical Department III. (Hematology & Oncology), Hospital Hannover-Siloah, Hannover, Germany. Therapy: Oxalyplatine + 5-FU + folic-acid (**a**), progression. Further therapy: Irinotecan and Capecitabine + oncothermia on liver. Result: good partial remission (PR) of the tumor (**b**) and tumor marker regression after 4 cycles of oncothermia (**c**). Follow up: 26 months

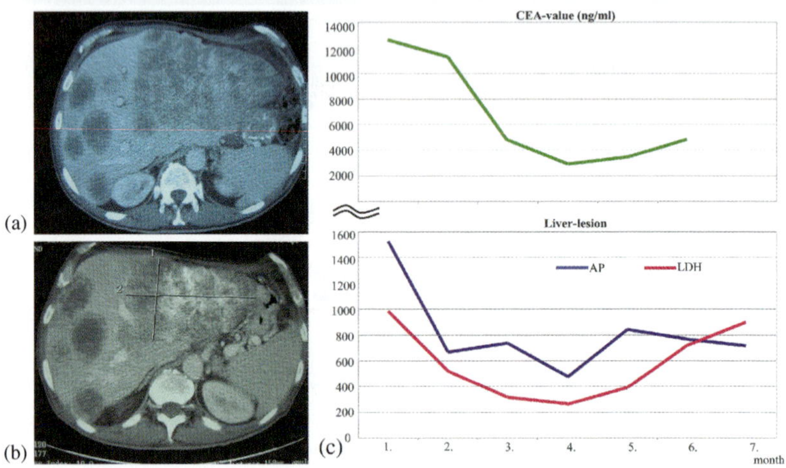

Fig. 4.117 Case of hepatic metastasis of rectum carcinoma. Investigator: Prof. H. Kirchner. Medical Department III. (Hematology & Oncology), Hospital Hannover-Siloah, Hannover, Germany. Patient: 56 year old, male. Tumor classification: pT3c, pN2, M1 G2 (Liver). Therapies: surgery: OP anterior rectum resection, followed by Campto, Leukovorine, 5-FU (**a**) + oncothermia on liver Result: no change, (NC) (**b**), but good improvement of the tumor-marker (**c**) and good QoL

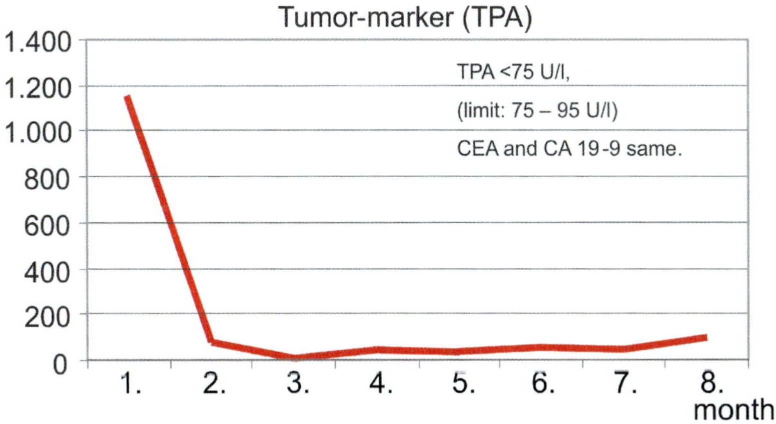

Fig. 4.118 Case of hepatic metastasis of uroepithelium carcinoma. Investigator: Prof. H. Kirchner. Medical Department III. (Hematology & Oncology), Hospital Hannover-Siloah, Hannover, Germany. Patient: 69 year old, male. Surgery: nephroureterectomy, Tumor-classification: pT2 N0 Mx R0 G2. Metastasis: multiple hepatic. Therapy: Gemcitabine/Cisplatin (3 sessions) + oncothermia (Liver) followed by PNP Gemzar/Carboplatin (4 sessions). Result: good partial remission (PR) of the tumor (>1cm) and tumor marker regression

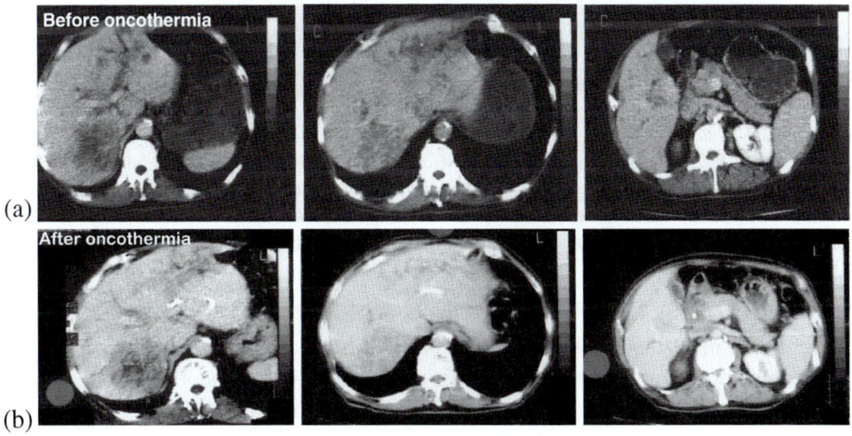

Fig. 4.119 Case of hepatic metastasis of sigmoid carcinoma. Investigators: Prof. D. Grönemeyer, & Dr. H. Sahinbas. Institute of Microtherapy, University Witten-Herdecke, Bochum, Germany. Patient: 61 year old, male. Metastasis: hepatic. Tumor-classification: pT3 N1 M0; Size: 4×5 cm; Therapy (for metastasis): oncothermia. Result: partial remission (PR) tumor and tumor marker regression

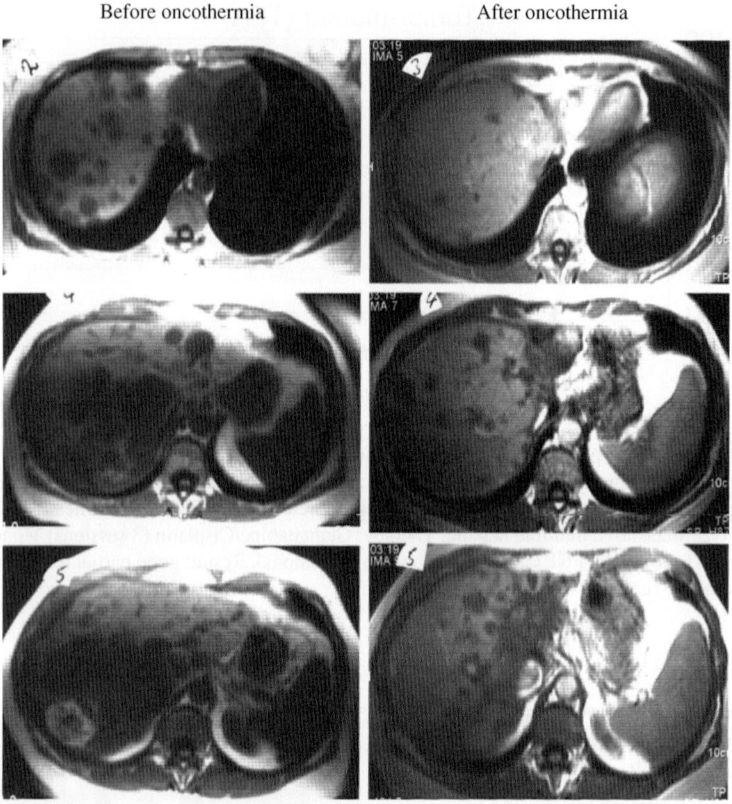

Fig. 4.120 Case of hepatic metastases of mammary carcinoma. Investigators: Prof. D. Grönemeyer, & Dr. H. Sahinbas. Institute of Microtherapy, University Witten-Herdecke, Bochum, Germany. Patient: 49 year old, female. Hepatic metastases two years after the primary diagnosis, followed by the cerebral metastases after 6 years. Treatments: Radiotherapy 30 Gy + chemotherapy: Xeloda + oncothermia; 60 min, 2–3 weekly. Result (on liver): partial remission; (PR)

- Stomach metastasis in liver was not observed for a few months. After its detection chemotherapy kept the metastasis unchanged. A few oncothermia treatments were concomitantly applied afterwards, and the stable condition was continued. By intensifying the oncothermia treatments (more frequent and later higher doses) the liver metastasis started to shrink, and a good partial remission was achieved. A year after termination of oncothermia progression of the liver metastasis was detected again, see Fig. 4.125.

These investigations show the possibility of handling advanced metastases as chronic diseases by regular long-term oncothermia with an extended, high number of regular treatments. Termination or break of the regular treatment causes tumor progression in these cases.

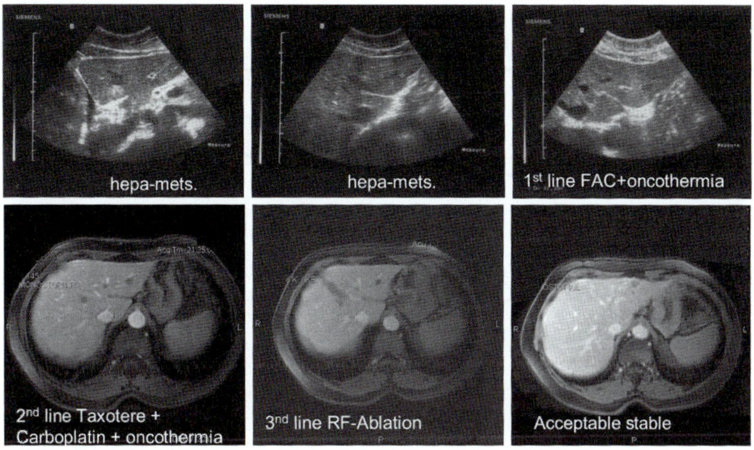

Fig. 4.121 Case of hepatic metastases of invasive ductal mammary carcinoma. Investigator: Prof. Dr. I. Lang; National Institute of Oncology, Budapest, Hungary. Patient: 49 year old, female. Therapies: Quadrantectomy and ABD. Result: complete resection (R0). Follow up: lymphatic tumor cell invasion. Treatment: adjuvant treatment: 3x CMF → Radiotherapy → 3x CMF; Result: complete remission (CR). Follow up: for the next 6 years symptom- and complaint-free. Afterwards: hepatic metastases. Therapy: FAC + oncothermia → SD; followed by Taxotere – Carboplatin + oncothermia, followed by RF-Ablation. New liver metastatis appeared. Treatment: surgical metastasectomy + intra-arterial chemotherapy (FAM). Result: patient has been in good general state with acceptable hepatic function

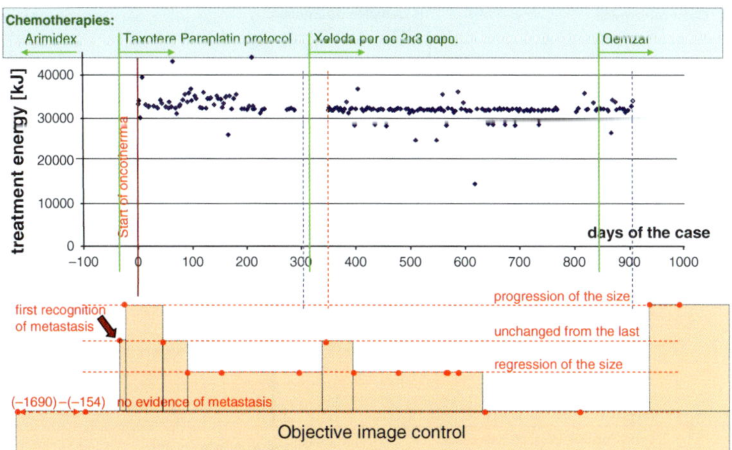

Fig. 4.122 Case of hepatic metastases of breast cancer (male). Investigator: Dr. A. Csejtey. Markusovsky Hospital, Szombathely, Hungary. Patient: 46 year old, male. Therapy for primary: Arimidex. Metastases appear more than five years later in liver, rapid progression. Therapy for liver: Taxotere + oncothermia; followed by Xeloda + oncothermia; and Gemzar + oncothermia. (Notes: green points indicate the chemotherapies, the blue points are the individual oncothermia treatments showing its treatment energy in kJ, the red points show the objective image control by MRI.)

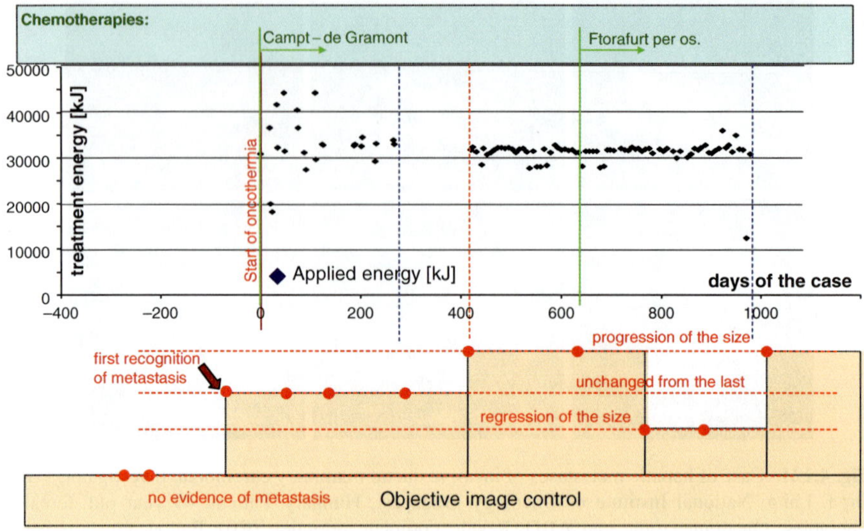

Fig. 4.123 Case of hepatic metastases of primary colorectal tumor. Investigator: Dr. A. Csejtey. Markusovsky Hospital, Szombathely, Hungary. Patient: 62 year old, male. One year after the treatment of the primary tumor, metastases appeared in the liver. Therapy: Campto [deGramont] + oncothermia; followed by Ftorafurt per os + oncothermia. (Notes: green points indicate the chemotherapies, the blue points are the individual oncothermia treatments showing its treatment energy in kJ, the red points show the objective image control by MRI.)

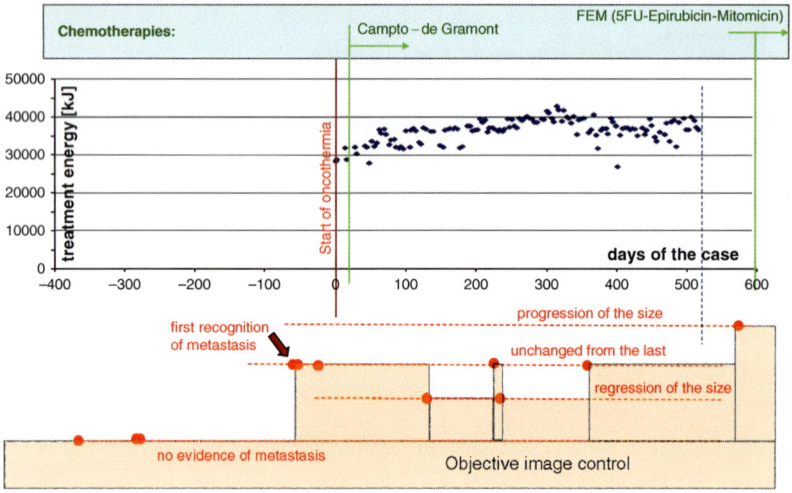

Fig. 4.124 Case of hepatic metastases of primary colorectal tumor. Investigator: Dr. A. Csejtey. Markusovsky Hospital, Szombathely, Hungary, Patient: 53 year old, male. 13 months after the treatment of the primary tumor, metastases appeared in the liver. Therapy for liver: Campto [deGramont] + oncothermia; FEM + oncothermia. (Notes: green points indicate the chemotherapies, the blue points are the individual oncothermia treatments showing its treatment energy in kJ, the red points show the objective image control by MRI.)

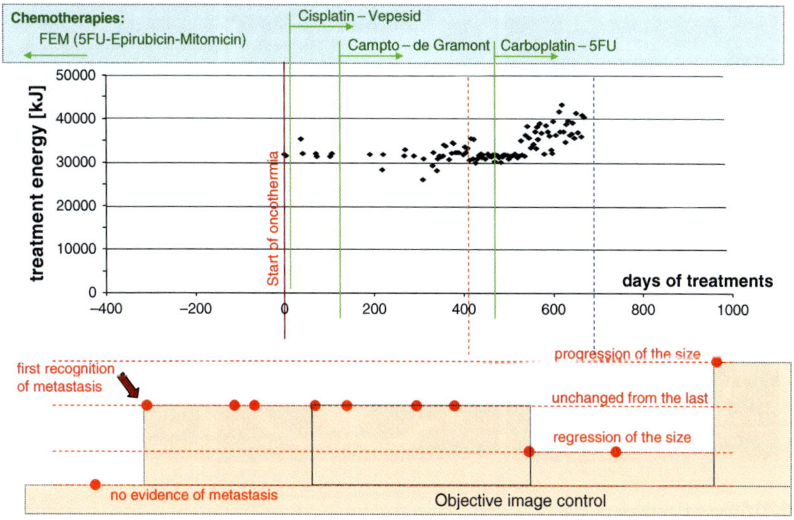

Fig. 4.125 Case of hepatic metastases of stomach tumor as primary. Investigator: Dr. A. Csejtey. Markusovsky Hospital, Szombathely, Hungary, Patient: 48 year old, male. Therapy for primary: FEM (5FU, Epirubicine, Mitomycin-C). Follow-up: no evidence of metastasis (400 days), but afterwards metastases appear in liver. Therapy for metastasis: Cisplatin + Vepesid + oncothermia; followed by Campto + oncothermia; and after Carboplatin + 5-FU + oncothermia. (Notes: green points indicate the chemotherapies, the blue points are the individual oncothermia treatments showing its treatment energy in kJ, the red points show the objective image control by MRI.)

Other gastrointestinal cases are the Jejunely Carcinoid, Fig. 4.126 and the Intrahepatic bile-duct carcinoma, Figs. 4.127 and 4.128.

A case of rectum carcinoma treated by oncothermia is shown in Fig. 4.129.

4.4.5 Pulmonary Cases

One of the most common cancers in humans is the lung tumor, which is dominantly non-small-cell lung cancer (NSCLC). Oncothermia is applicable with great success in this important field. Some case examples are shown in Figs. 4.130–4.138. The parallel brain metastasis is also well controllable by oncothermia (Fig. 4.139).

Metastatic lung tumors are also treatable by oncothermia. A case of a breast primary is shown in Fig. 4.140.

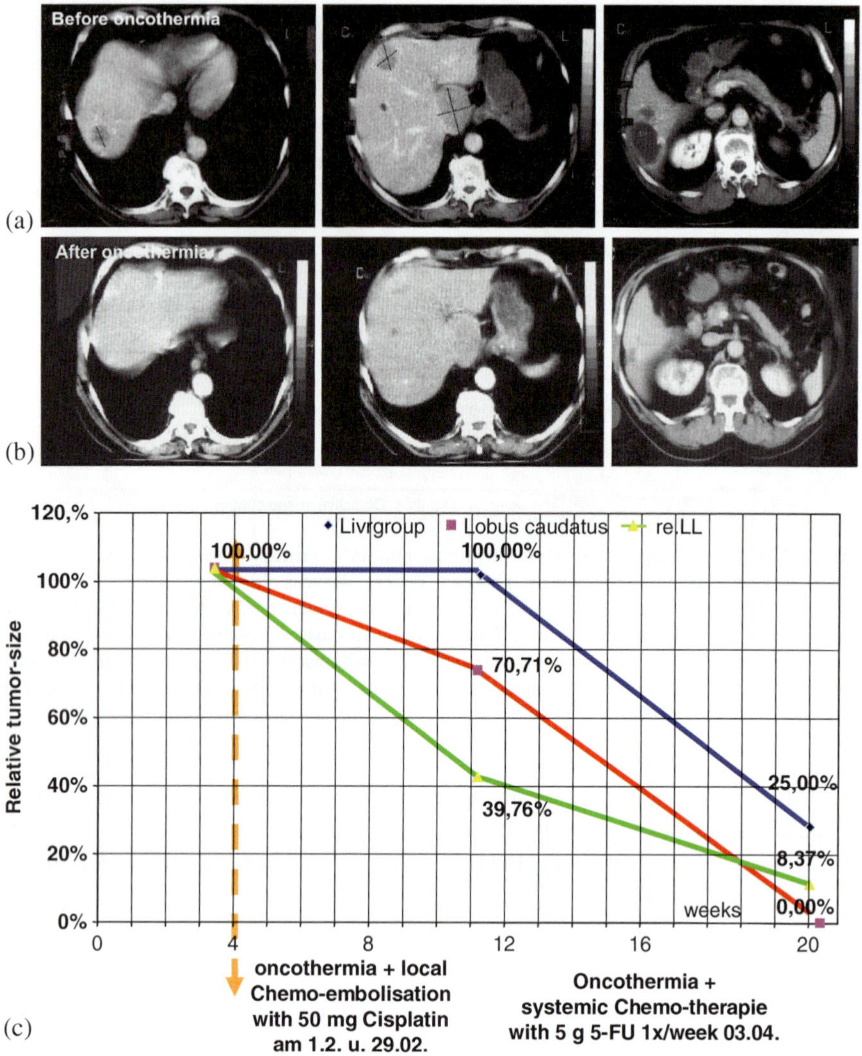

Fig. 4.126 Case of hepatic metastases of jejunely tumor. Investigators: Prof. D. Grönemeyer, & Dr. H. Sahinbas. Institute of Microtherapy, University Witten-Herdecke, Bochum, Germany. Patient: 63 year old, male. Metastasis: hepatic (**a**). Therapies: Chemo-embolisation (Cisplatin) combined with oncothermia; followed by i.v. 5FU (5g/week), combined with oncothermia. Result: partial remission (PR) of the tumor (**b**) and tumor marker regression (**c**)

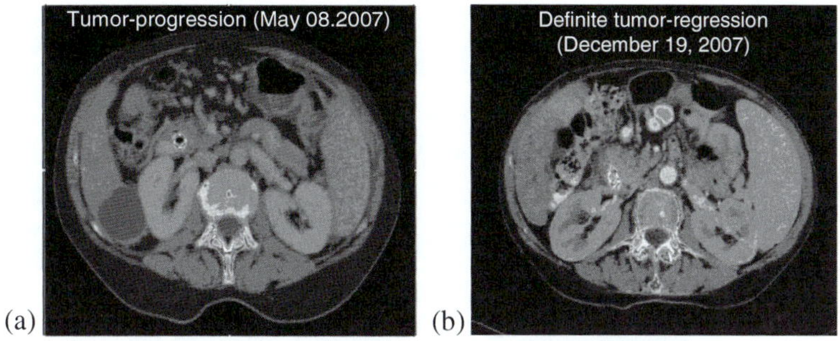

Fig. 4.127 Case of inoperable intrahepatic bile-duct carcinoma. Investigator: Dr. A. Csejtey. Markusovsky Hospital, Szombathely, Hungary. Patient: 57 year old, female. Inoperable lesion (**a**). Prognosis: overall median survival 6 months. Therapy: oncothermia applied as monotherapy with concomitant supportive vitamins only. (Due to the patient's status, no any other therapies was possible.) Result (after 6 months treatments): complete remission (CR) (**b**)

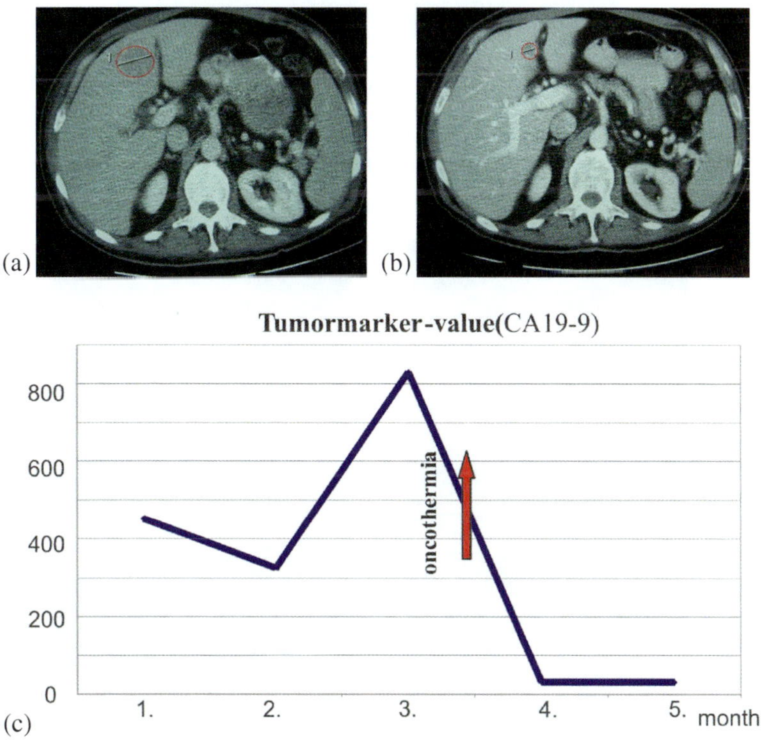

Fig. 4.128 Case of Cholangio-cellular carcinoma. Investigator: Prof. H. Kirchner. Medical Department III. (Hematology & Oncology), Hospital Hannover-Siloah, Hannover, Germany. Severe Cholangitis; (**a**). Therapy: 4-week antibiotic no result. Chemotherapy: Gemcitabine/ Oxalyplatine. Result: progress; Therapy: Erbitux, Capecitabine + oncothermia on liver. Result: good partial remission (PR) tumor (**b**) and tumor marker regression, good QoL; (**c**)

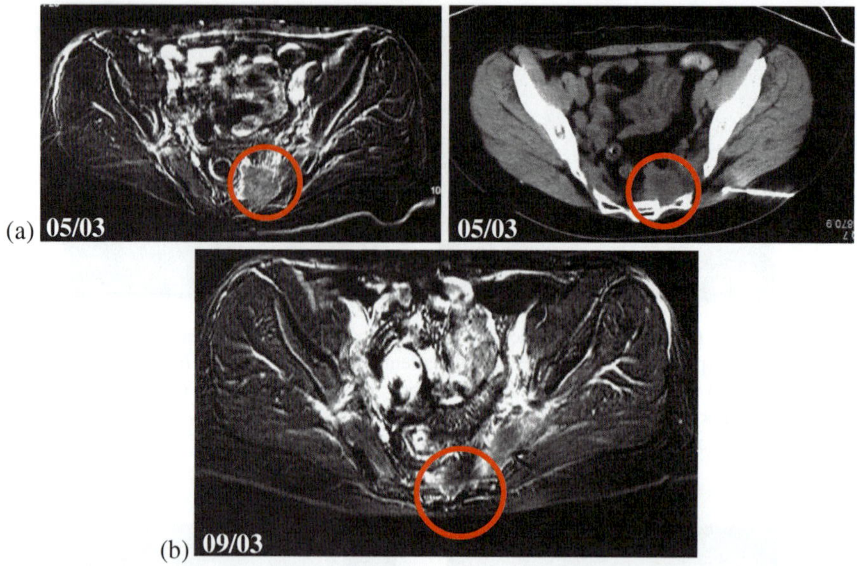

Fig. 4.129 Case of inoperable rectum carcinoma. Investigator: Prof. H. Renner; Praxis at Klinikum Nord, Nürnberg, Germany. Patient: 36 year old, female, Histology: Neuroendocrine carcinoma; Tumor-classification: rcT4 cN0 M0 G4 R2, (**a**). Treatment: trimodal protocol; radiotherapy 45 Gy, + Chemotherapy, Cisplatin, + oncothermia, 60 min, (14×, 3/week, electrode 20cm). Result: partial remission (PR) (**b**)

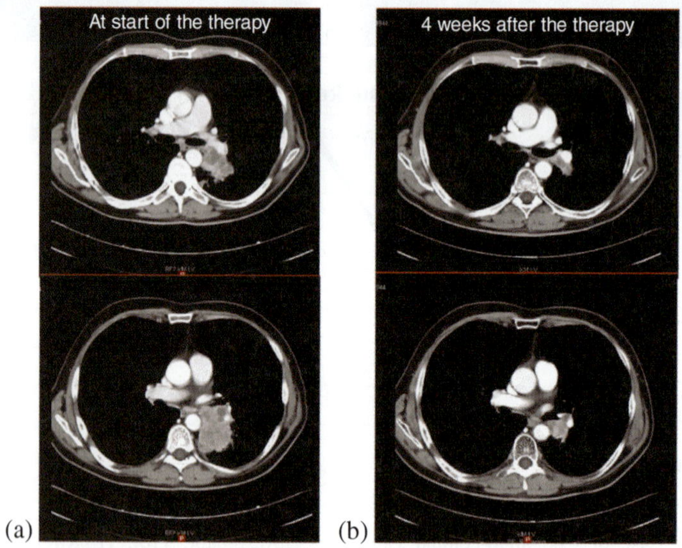

Fig. 4.130 Case of bronchial carcinoma linked to main-bronchus. Investigator: Prof. H. Renner; Praxis at Klinikum Nord, Nürnberg, Germany. Patient: 61 year old, male. Tumor classification: cT4 cN0 M0 Gx R2. Histology: Squamous cell carcinoma, (**a**). Comorbities: multiple sclerosis, diabetes (type II.). Therapies: radiotherapy: 72 Gy, Chemotherapy: Cisplatin, 1xweeks, 4 weeks, + oncothermia: 8 sessions. Result: partial remission (PR)(**b**)

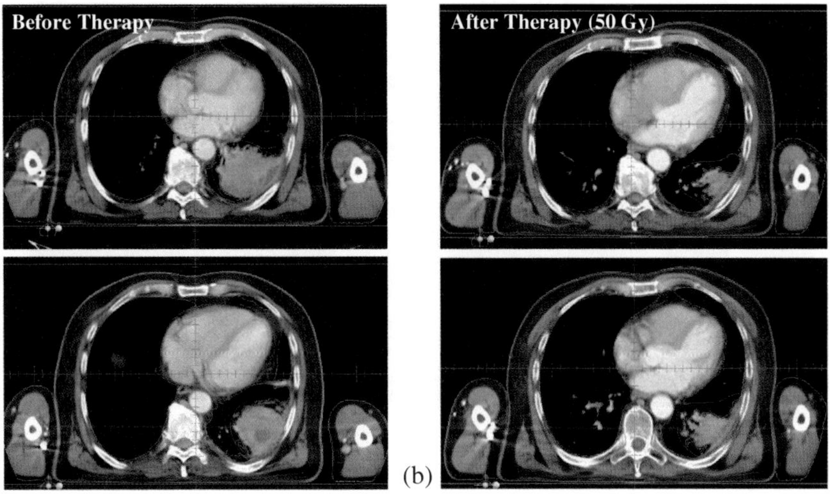

Fig. 4.131 Case of bronchial carcinoma left-lower-lobe. Investigator: Prof. H. Renner; Praxis at Klinikum Nord, Nürnberg, Germany. Patient: 68 year old, male. Tumor-classification: cT3 cN2 M0 G3 R2. Histology: squamous cell carcinoma; (**a**). Therapy: radiotherapy 60 Gy, chemotherapy Cisplatin 1×/week 5×, + oncothermia (10 sessions). Results: after 8 weeks of treatments (**b**). Surgery: lung-lower-lobe-resection, R0. Histology: no tumor was observable in the resected tissue

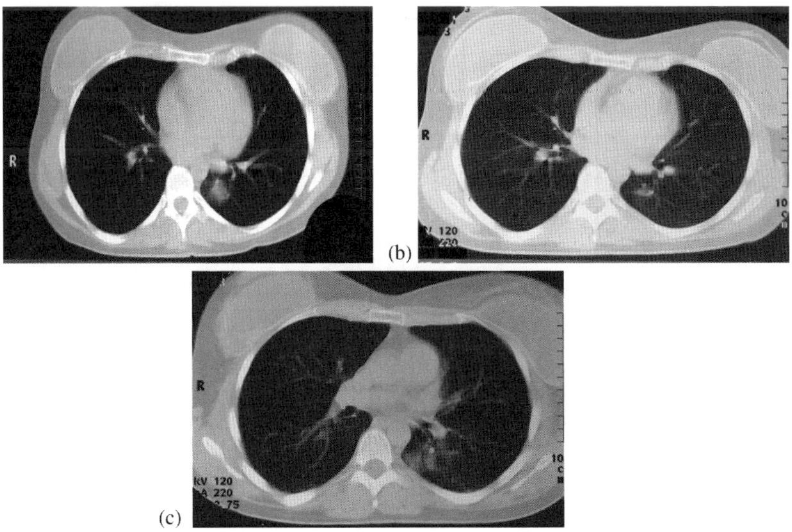

Fig. 4.132 Case of lung metastasis of primary carcinoma of uterus cervix. Investigator: Prof. H. Renner; Praxis at Klinikum Nord, Nürnberg, Germany. Patient: 38 year old, female. Histology: squamous cell carcinoma. Lung-metastasis is diagnosed 19 months later than primary (**a**). Therapy: palliative radio-thermotherapy: 55 Gy, + oncothermia: 12 sessions, duration: 6 weeks. Result: after 40 Gy: partial remission (PR, 80%), (**b**). Result (after 4 weeks): complete remission, (CR, 100%) (**c**)

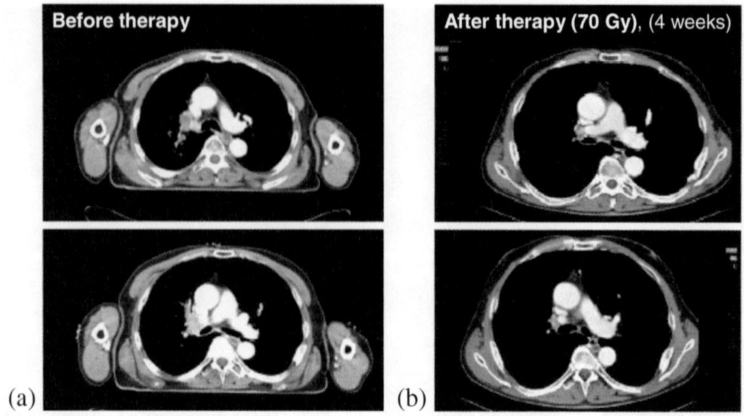

Fig. 4.133 Case of bronchial carcinoma right upper-lobe. Investigator: Prof. H. Renner; Praxis at Klinikum Nord, Nürnberg, Germany. Patient 58 year old, male. Tumor-classification: cT3 cN0 M0 Gx R2. Histology: squamous cell carcinoma. (**a**). Therapy: radiotherapy 70 Gy; Mitomycin C 1×/week 3×; + oncothermia (10 sessions); Result: partial remission (PR), (**b**)

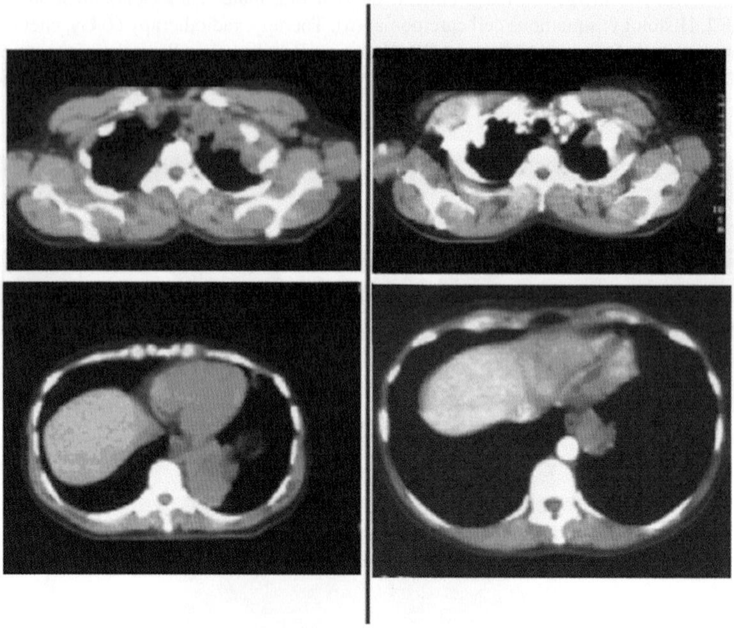

Fig. 4.134 Case of lung metastasis of soft-tissue sarcoma. Investigator: Dr. W-P. Brockmann Institute OncoLight, Hamburg, Germany. Patient: 38 year old, female. Diagnosis: adenoid cystic carcinoma of the parotid gland with two extended soft-tissue lung-metastasis. (**a**) Therapy: surgery, resection, radiotherapy, postoperative fast neutrons. Result: relapses small cell lung carcinoma + uterine cervix. Therapy: radiotherapy on lung metastases, 18 MV, 50.4 Gy (5× 1.8 Gy/week) + oncothermia: complementary 2–3/week, 13 sessions, 60 min each. Result: good partial remission in both the lung nodules (**b**). Follow-up: further reduction, ended by complete remission (CR) and no obvious metastasis was observed for three years

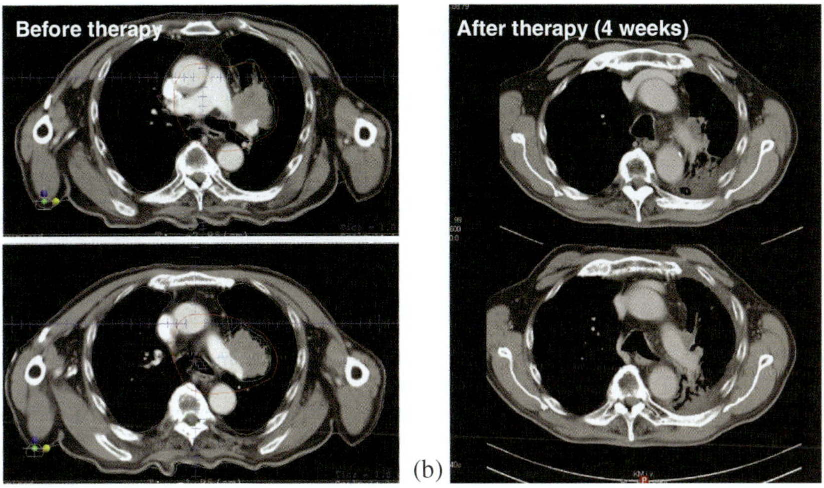

(a) (b)

Fig. 4.135 Case of non-small cell lung cancer (NSCLC) left-lobe. Investigator: Prof. H. Renner; Praxis at Klinikum Nord, Nürnberg, Germany. Patient 61 year old, male. Histology: squamous cell carcinoma; classification: cT3 cN0 M0 G3 R2; (**a**). Therapy: radiotherapy 72 Gy; + oncothermia (16 sessions). Result: partial remission, (**b**). Experience: not evaluate too early, and not apply too high dose of radiation

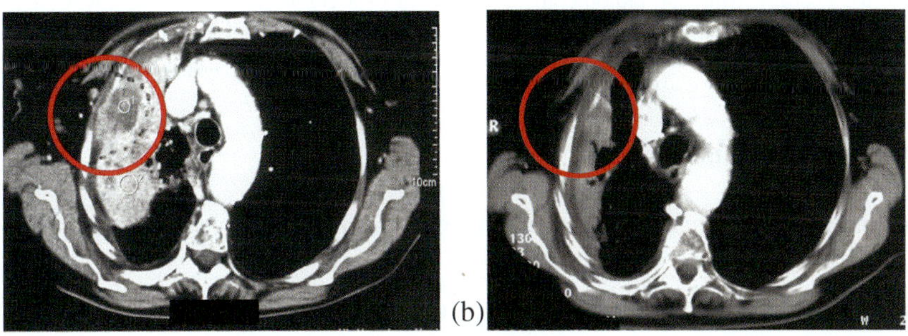

(a) (b)

Fig. 4.136 Case of non-small cell lung cancer (NSCLC) periphery, inoperable. Investigator: Prof. H. Renner; Praxis at Klinikum Nord, Nürnberg, Germany. Patient: 79 year old, female. Histology: squamous-cell carcinoma; classification: cT3 cN0 M0 G2 R2 (**a**). Treatment: bimodal palliative protocol; radiotherapy, 54 Gy, + Chemotherapy, Cisplatin + oncothermia, (9×, 1–2/week [6 weeks], 60 min/session, electrode 30cm). Result: partial remission (PR), (**b**)

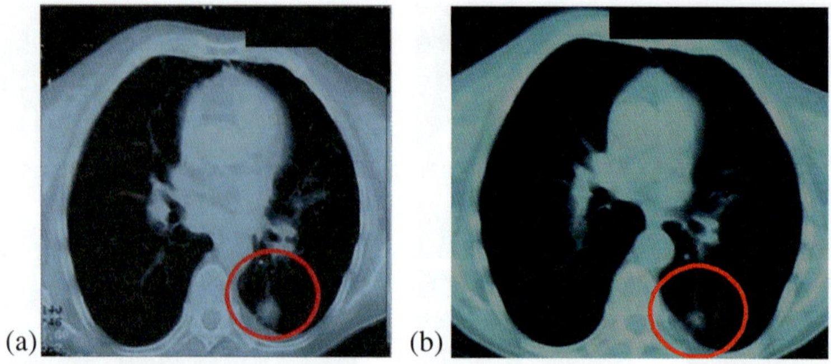

Fig. 4.137 Case of non-small cell lung cancer (NSCLC). Investigator: Prof. A. Herzog; Institute: Fachklinik Dr. Herzog, Nidda, (Bad Salzhausen), Germany. Patient: 66 year old, female. Symptoms: cough, shortness of breath (**a**). Treatment: oncothermia + Irressa (Gefidinib). Results: partial remission (**b**), disappearance of symptoms, improved, good QoL

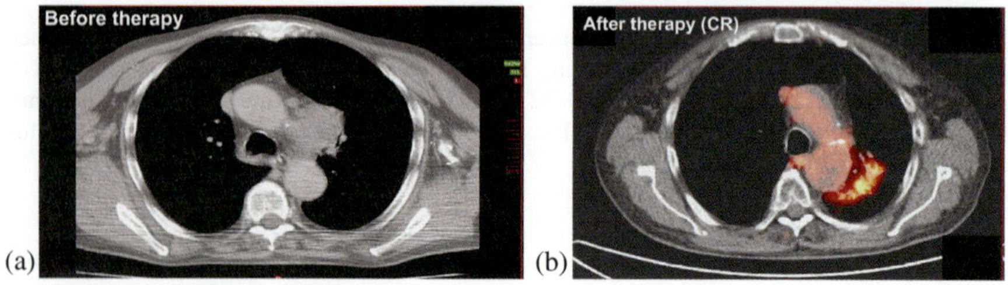

Fig. 4.138 Case of non-small cell lung cancer (NSCLC). Investigator: Prof. H. Renner; Praxis at Klinikum Nord, Nürnberg, Germany. Patient 76 year old, male. Tumor-classification cT3 cN2 G3 (**a**). Therapy: trimodal; radiotherapy, 50.4 Gy; (5×1.8 Gy/weeks), chemotherapy, Cisplatin + oncothermia (12 sessions). [Control: CT: 1× in every 3 months; PET-CT: 2× in every 9 months.] Result: complete remission after 3 months (**b**)

4.4.6 Other Cases

We show a urethra carcinoma case having metastasis in hilus, see Fig. 4.141.

4.5 Evaluation of Oncothermia Studies

The clinical trials of oncothermia are dominantly retrospective. Prospective trials are in progress. To develop randomized clinical trials involves a challenge for patients. Patients do not agree to be in the control arm under any circumstances. In most cases they are registered for oncothermia because other conventional gold standards have failed. This could in any case involve progression, resistance, organ-overload (kidney, liver, etc.), relapses, sometimes psycho-resistance, etc.

The advanced cases under the conditions described above, emphasize not only the complexity of the individual situation of patients, but also underlines the fact that oncothermia is applied as a facility for the "no other treatment is possible" often

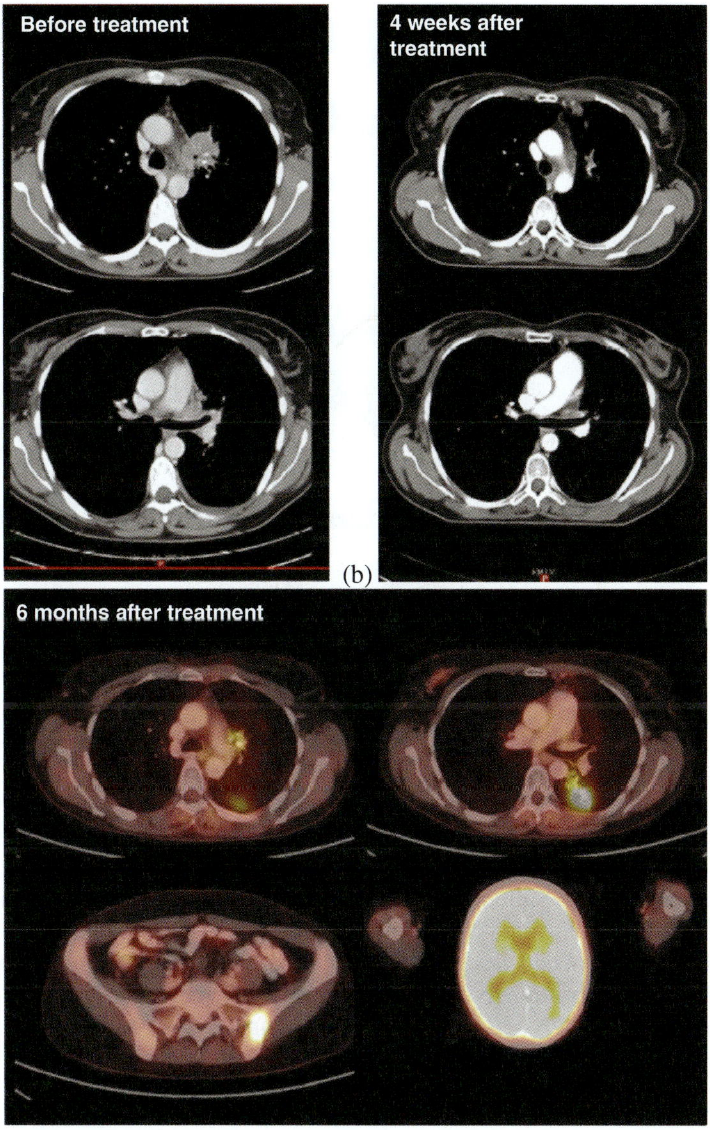

(a) (b) (c)

Fig. 4.139 Case of bronchial-carcinoma left-lobe with multiple brain metastases. Investigator: Prof. H. Renner; Praxis at Klinikum Nord, Nürnberg, Germany. Patient: 47 year old, female. Histology: adeno-carcinoma; classification: cT3 cN2 M1 G3 R2, (**a**). Therapy: palliative brain-radiation, connected palliative chemotherapy for 6 month, ineffective. Local-tumor therapy: 72 Gy, without chemo-therapy + oncothermia (23 sessions). Result: complete remission (**b**). Follow-up: 6 months, free of disease (**c**)

(a)

(b)

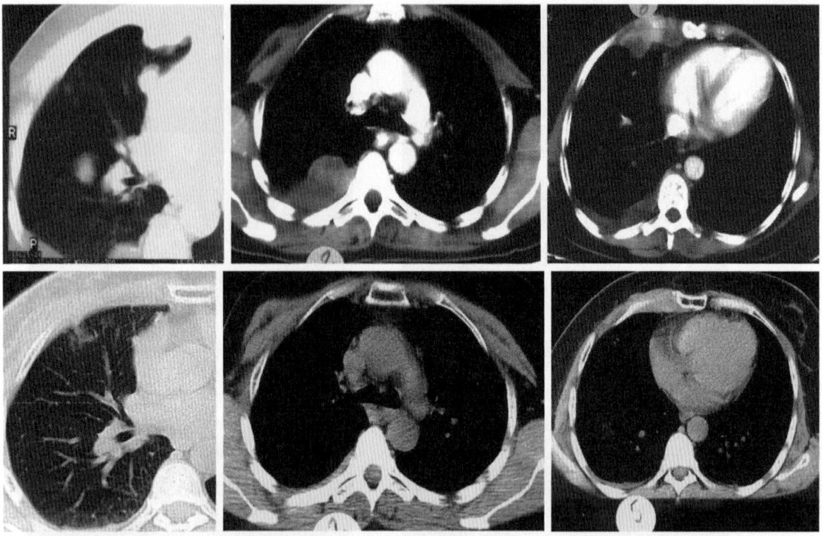

Fig. 4.140 Case of breast cancer with metastases (thoracic-wall and intrapulmonal). Investigators: Prof. D. Grönemeyer, & Dr. H. Sahinbas. Institute of Microtherapy, University Witten-Herdecke, Bochum, Germany. Patient: 48 year old, female. Therapy: chemotherapy, Tamoxifen. Progression (**a**). Complementary oncothermia (16 sessions). Result: complete remission (**b**). For preventive purposes two more oncothermia cycles were performed

hopeless cases providing an over 3rd line treatment approach. This high-line treatment process is in general palliation (the first goal is to provide acceptable QoL), which is an important factor for oncothermia as well. However, oncothermia even in these advanced situations has curative value, and provides curative therapy at 3rd-line or over. The professional literature shows well the rarity of evidence-based clinical trials for these high-line treatments. Other evidences have to be shown when randomized controlled trials are not possible [1203]. The challenge of evaluation appears specifically strongly in the case of patients in advanced stages, in inoperable (or partly resected), relapsed or patients resistant to gold standards, and this is a challenge facing oncothermia as well.

4.5.1 Evaluation Conditions

We have to give special attention to the evaluation of the clinical results of oncothermia. The complications mean definite challenges for objective evaluation. The main challenges are:

- Oncothermia is applied in higher (usually third and subsequent) treatment lines, boosting or resensitizing the effect of the conventional therapies. Oncothermia is mostly applied in cases where the conventional therapies fall away. The

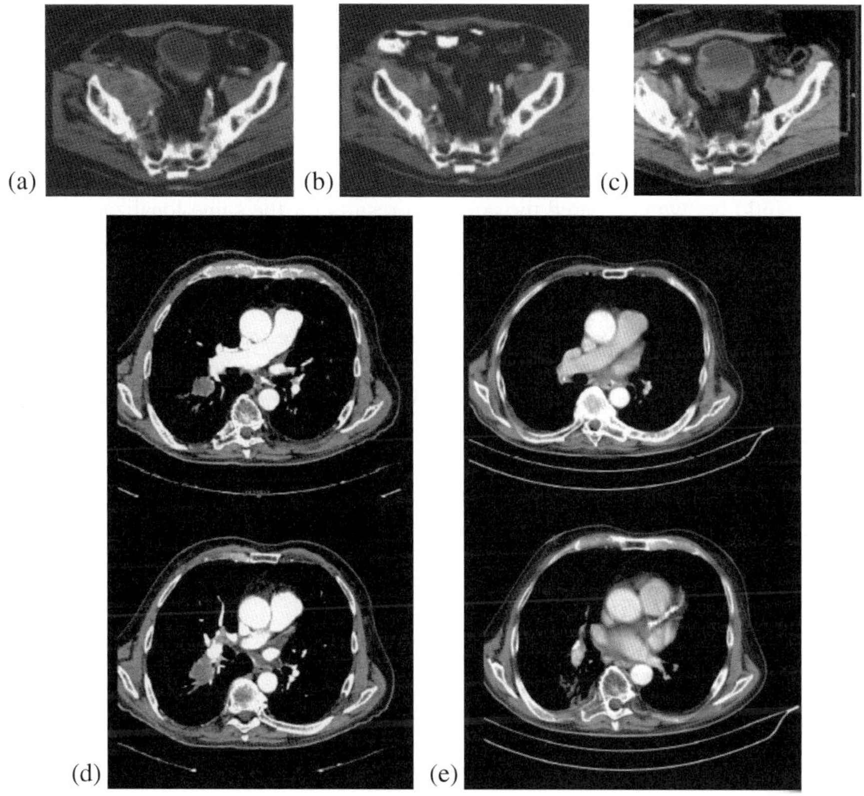

Fig. 4.141 Case of ureter-carcinoma. Investigator: Prof. H. Renner; Praxis at Klinikum Nord, Nürnberg, Germany. Patient: 80 year old, male. Histology: transitional cell carcinoma. Therapy: surgery. Local-recurrence, (**a**). Therapy: palliative radio-chemo thermo therapy, radiotherapy, 54 Gy, chemotherapy, 1 cycle of Carboplatin + oncothermia (12 sessions, duration 6 weeks). Result: partial remission (PR) (**b**), followed by a complete remission (CR) a year later (**c**). Follow-up: metastasis hilus right (one year after the CR) (**d**); Therapy: palliative radio-thermo-therapy; radiotherapy 51 Gy, (5×1,7 Gyw/eek), + oncothermia (17 sessions; duration: 6 weeks). Result: complete remission (CR) again; (long follow-up, 9 months) (**e**)

most probable reasons include when the cases are inoperable, radio-resistant, chemo-resistant, have low blood counts, liver failure, kidney failure, and sometimes psycho-resistance alone or in combination. Because of these conditions, oncothermia is applied in a higher line of the therapies. This sequence of treatments is mostly determined by individual decisions of the physicians [115], usually without the help of any evidence-based statistical approvals.

- Usually it is applied in palliative care; many patients are in the terminal phase. This type of patient care has very limited statistical evidence-based trials; the medical decision-making processes are usually well tailored to the individual patients [1204, 1205].

- Only a few controlled randomized clinical trials are available for oncothermia. The results have to be concluded from observational studies and from historical and database comparisons. Mostly used are the USA- and EU-databases (SEER [47], Eurocare [53]). Because of the unsolved problematic between the hypothesis check confidence of evidence-based medicine and observational studies [111, 112], this data comparison is acceptable. To make the result as objective as it could be, we compared the collected results for the same localizations and same protocols from various clinics. The common significant difference from the databases could be accepted as evidence.
- In the case of long survivals, we have to consider, that oncothermia forms only a small fraction of the overall survival. The patients are treated with oncothermia in their definite late stages, after prognosis of "no curative help" by continuation of gold standards alone. The long overall survival generally is not the result of the oncothermia, because patients start oncothermia in their advanced stage only. The survival effect of oncothermia in the very late, hopeless cases could be negligible, irrespective of its real efficacy. An opportunity to measure the efficacy is the first-year survival rate (%), when the patients with the most aggressive kind of the given cancer do not survive the second year after the diagnosis. If they start the oncothermia in the first year, the survival-rate result could form an objective sign of the survival efficacy.
- Special cases are treated on Intend-to-Treat population, where in many cases patient chooses the treatment. This makes the selection not objective enough, but the advanced, dominantly metastatic patient's spectrum compensates of this lack of selection. We have an automatic selection of advanced (often terminal) cases.
- Because of the generally low QoL of most of the patients, and a combination of supportive therapies, this again weakens the measurability of oncothermia alone. However, as oncothermia is not proposed as monotherapy, the combination objectively measures the benefit, if the supportive therapies alone would not be successful at all.
- The patients are treated with oncothermia in their very advanced, metastatic states. The local oncothermia treatment of course concentrates on definite localizations (primary or metastatic), which again lowers the full measurability of oncothermia in the development of the cancer. This is the main reason, why oncothermia measures first of all the overall survival rates, which are good objective parameters of the treatment efficacy in general.
- The QoL of the patients is an important characteristic of oncothermia. In general a trial effect exist in controlled randomized studies [120]. It could be a negative outcome, as in a strict competitive market opinions are not independent and objective [127], and also a conflict of interest could cause considerable bias [128, 129]. These are not characteristic in oncothermia applications.

In summary: there are some recognized negative biases in the retrospective data:

- Reimbursement bias: The treatment is paid for by private insurance or by the patient. This is an additional factor to the inclusion criteria in selecting patients eligible for treatment.

- Voluntary bias: Treatment is on a voluntary basis (intention to treat [ITT] population), which selects patients on the basis of their own decisions.
- Historical bias: Controls are from the historical arm or from the large databases. This makes the comparison statistically not exact.
- Clinical bias: The protocols are identical, but the selection of patients and practice of cure (mainly the supportive therapies) could differ.
- Personalization bias: Oncothermia is a personalized treatment, no definite overall dose or other overall valid parameters could be applied.

There are however some positive biases which could underestimate the results:

- Late-stage bias: Oncothermia is applied when conventional treatments fall aside. Results of this kind are not comparable with the regular curative processes.
- High-line bias: Oncothermia is applied in high lines of the treatment process, in which states evidence-based trials rarely have been performed. The falling-away of conventional therapies takes oncothermia out of the evidence-based line, and makes it palliative in most cases.
- Private clinic bias: Private clinics dominate in terms of number, taking care of patients in advanced or even in the last stage. Oncothermia is applied mostly by these medical units. Private clinics usually lack the high-tech set of diagnostic and treatment equipments of non-private centers. To achieve and control the same results is sometimes difficult.
- Lack of "trial-bias": No extra trial attention is taken on the patients, who makes the results more objective.

On the basis of the above bias structure the characteristics of the clinical studies could be described as – single arm, open label, observational for (ITT) population, dominantly for patients in late/advanced stages, where the conventional methods have failed. Mostly the survival rate was the studied endpoint. The inclusion criteria was inoperable and in progression after chemo- and/or radiotherapy. Exclusions were only the well known, above-described contraindications of oncothermia.

The temperature during the oncothermia treatment is calculated from the absorbed energy, which is provided, and it is displayed in real-time. This parameter allows a "classical" orientation of the physician in terms of the safety of the actual process. However, in the tumor considerably higher temperature than the average has to be considered. The temperature calculation is based on solid measured data of absorbed energy (not on forwarded alone). The approach is the so-called "equivalent-temperature" idea. The equivalent temperature is the equivalent distortion ability of the cells with static overheating. Oncothermia heats up the tissue by a dynamic, gradient method, using the field effects at the cellular membranes. So the distortion is more efficient by oncothermia than by simple heating. To reach the same efficacy as oncothermia does, considerably higher temperature (equivalent temperature) has to be managed than the actual temperature in the oncothermia process. However, this approach of "equivalent temperature" is widely applied for temperature measurements by MRI as well. The MRI signals (T1 and T2 relaxation times) are temperature sensitive, but without calibration to a realistic system it is inapplicable. The calibration of the *T*1 and *T*2 time-shifts are made on phantoms, which are static (no physiology effects are included), and the MRI measures the phantom-equivalent, static temperature as well.

Note, the patient's ability to sense is the best safety alarm for any unwanted, unexpected events. Do not ignore it, react immediately. Oncothermia is a highly effective power transfer, its mismanagement by ignoring the patients sensing (or suppressing it by any analgesic medication) could be unsafe.

4.5.2 Evaluation Methods

The challenge of oncothermia is its use when conventional treatments are unsuccessful. In consequence its effect could be active only in a small (last) fraction of the overall survival. Patients with long overall survival have already benefitted from a long period of conventional treatments, without the application of oncothermia, and the last stage oncothermia application may not be observable in terms of life-elongation, even if the oncothermia was effective.

To make an objective evaluation we have developed methods to obtain the evidences from the available information pool and determine the objective evidences. These methods highlight the objective information and their parallel results make the obtained data evidence-based. It is a complex challenge, having five basic approaches:

1. *Fast course case comparison*. Use the survival of the rapid, fast course cases (most advanced, drastic quickly developing cases) as a comparison with large databases. (Only the survival is considered as a relevant parameter, the clinical outcomes (responses) are not studied as evidences.)
2. *Comparison of clinics*. Compare the retrospective data of the independent clinics, using the same protocol for the same cohort.
3. *Quality-of-life comparison*. Collect the QoL data and the adverse effects limiting the application of oncothermia.
4. *Create a quasi-control arm*. Patients having no benefit from oncothermia could form a quasi-arm for control. The "no benefit" category could be defined when the patient survival is short from the time of the first oncothermia treatment.
5. *Parametric evaluation*. Use the available latest statistical knowledge to find the relevant parameters of the survivals and use the best fit of the parametric distribution for evaluation.

The details of the approaches are complex.

Fast course case comparison. Compare the first-year survival rates (percentages of the surviving cases in the first year), to the large national and international databases. The comparison is realistically unfavorable for oncothermia, because it is generally applied in late stages, but the databases consider all the available cases of the given disease. Consequently if oncothermia shows any benefit in this comparison, it is a strong probability that it is reliable evidence. However this is only indicative, qualitative information, no quantitative conclusion could be made. The overall survival rate of the patients treated by oncothermia is mostly dominated by the non-oncothermia therapies, as oncothermia began only when the expectations from the gold standard therapies were weak. In consequence, to make any conclusion about the overall survival time is very complicated, it could even not be done without definite extra information and solid extra facts. However, the first-year survival ratio shows the most aggressive cases only, so the applied oncothermia could have a considerable effect in these aggressive cases. When the first-year survival is significantly increased by oncothermia, this should give a certain indication of the success of the oncothermia. The evaluation was made by regular descriptive biostatistics and a log-rank survival test. Comparison to large studies and databases (SEER [47] and Eurocare-3 [53]) as well as local historical data was made.

Comparison of clinics. The survival rates can be collected from various clinics practicing oncothermia. These clinics when applying the same oncothermia protocol in the same patient groups (cohorts), are ready to compare. The data could be analyzed by their coherence. When clinics using oncothermia have statistically significant congruence with each other that could be regarded as evidence. When these data are comparable within a statistically acceptable level (confidence interval) the objectivity of the data would be statistically proven. The weighted average of the results of these clinics could be compared to the large national and international databases too. The possible difference between the database data and the statistical average of independent oncothermia applications from independent resources could provide evidence-based data to support or disprove the oncothermia benefit. Both the first-year survival and the median survival could be chosen for comparison.

Quality-of-life comparison. The collection of data of QoL could show two different aspects of oncothermia: QoL during the curative process and also in the palliative actions. Because of late-stage applications, oncothermia is a palliative approach in most cases. In this stage the QoL has extreme importance. Also the connected adverse and side effects have to be evaluated for correct evidences. This is mandatory for a treatment like oncothermia, where an evidence-based clinical study is not a real option for the high-line curative and/or palliative actions.

Create a quasi-control arm. Because of the single-arm data collection, the information of the oncothermia effect is hidden in the available set of data. A realistic control would be in a prospective, randomized trial, but as was discussed above, it has certain complications. The other possibility (less strong evidence) is to have historical control from the same clinic/hospital. However, the locations are highly specialized for very advanced patients, and their oncothermia treatment is overall routine, having no cohort patient group without oncothermia treatment. This method has an old history starting with Pauling's proposal [1206]. However, Pauling's mathematical construction has no realistic application in our case, where the patients have successive sequences of treatments, and oncothermia is applied only at the end. The method of propensity scores [1207, 1208] has also numerous problems.

Because of the very advanced and heavily pre-treated cohort it is complicated (or impossible) to choose correctly the confounding variables and the proper cohort for control. The treated patients are in a hopeless, mostly untreatable state, which has to be treated by oncothermia as the last chance. The approach is the sequenced trial [1209, 1210], which is inherently applied in most cases, however the proper documentation of the data is not always satisfactory enough to derive strong conclusions from the dataset.

It is appropriate in our case for the choice of control group to select patients in whom the oncothermia was ineffective. This selection considers the method oncothermia has no harm for the patients (no adherent effects to cause tumor progress by the treatment alone), so the results have two categories only: effective or ineffective in terms of the applied treatment. The patients are involved in the treatment in the stage after which the gold standards have failed, so if they die soon after the first oncothermia, we may assume, that oncothermia was ineffective in their case. The inefficiency in no-benefit is that oncothermia is not able to change their stage and should they die shortly after therapy begins, they would have received only a few oncothermia treatments. The choice of this group as a control does not mean the group is identical with the group of patients having short overall survival. The patients overall survival is dominantly determined by the elapsed time and the type of pretreatments before starting the oncothermia.

Parametric evaluation. This method evaluates the available data by parametric statistical methods. The right information is hidden in the overall survival due to the single-arm data, but it could be mined from that. The basis of the mining is the fact that oncothermia is only a fraction of the applied treatments, and the patient is included when the gold standard treatments fail.

Evaluating clinical trials we determined the empirical distribution function of survival probability. The description of survival curves [1211, 1212, 104, 115] [Kaplan–Meier (KM), log-rank test (Cox-Mantel)], could be approached by fitting the parametric Weibull (Avrami) curves [1213–1216], on the actual probability function. Using the Weibull distribution function to approach the survival curve parametrically is theoretically and practically established for clinical applications [1217]. Fitting the distributions on the real survival curves divides the distribution in to two subgroups of the patients in the cohort. The originally homogeneous cohort is divided by this fit into subgroups of eligible and not eligible oncothermia effects for patients. (Anyway this approach used temperature development criteria, and the patients who were not "heatable," were declared as resistant to hyperthermia and were excluded from the trial (exclusion criteria) [1127].

In this way we study a split of the original cohort distribution into two different groups [1217] having Weibull approach [1217]. Patients in the first group are for whom the treatment had no or minor influence, while the cases where the treatment was effective are in the second group. The "inclusion criteria" for the patients to oncothermia treatment is when they are no longer eligible for the "gold standards." These criteria could be checked by studying the elapsed time to the first oncothermia from the first diagnosis. The time from the first diagnosis to the first oncothermia has to be a cohort (when the inclusion of the patients to oncothermia had identical criteria) consequently it has to be characterized by a single-Weibull parametric formulation.

The fit of parametric curves splits both the survival plots (the overall and the oncothermia) into two subgroups (see Appendix 35). The split of the Weibull distribution can be used as a statistical formulation of the two arms. The quasi-arm is determined by the short-living subgroup surviving oncothermia treatments, very similarly (but parametrically expressed) to the quasi-control arm method. The responding and non-responding patients can be measured by their ratio as well. The two arms in overall survival, however, will slightly differ from those of the oncothermia survival. The latter shows clearly the eligibility of oncothermia, while the split in overall survival has numerous interacting factors due to taking into account the long time of the previous treatments. The patients ratio, however, has to be identical in the two survivals, so the best fit ratio obtained from oncothermia is fixed in the overall survival as well. In this way the control arm in the overall survival could be obtained from the best fit, and the evaluation of oncothermia would be possible. The detailed exact description, validation and verification of the method is presented in Appendix 35.

4.6 General Overview on a Large Patient's Pool

Numerous clinical studies have been performed on a huge number of patients. Presently (2009) there are more than 150 active oncothermia devices working intensively all over the word, providing more than 100,000 treatments in 2009. This huge number of treatments indicates the usefulness of oncothermia, and the unharmed patients (no reported safety problems occurred) show the extreme safety of the method.

A remarkable amount of retrospective clinical studies are available to indicate the oncothermia effect in humans. It is commonly used for such complex and very frequent tumors like lung, liver, pancreas, brain, gastrointestinal, gynecological, etc. Only a few prospective evidence-based clinical trials have been performed till now on oncothermia. The reasons for the predominance of observable trials are:

- Oncothermia is applied over the second/third line of treatments (in far more advanced cases). No evidence-based trials exists in this treatment line for pharmaceutical products also.
- The evidence-based studies are too expensive for the companies involved.
- Most of the users run a private clinic, with no interest in performing such studies.

Studies on a huge number of patients in any case show amazingly good results in all the registered localizations. One of the largest coherent databases ($n = 1180$) was retrospectively collected by Hungarian hospitals (HTT, PTF) [1218], and presented

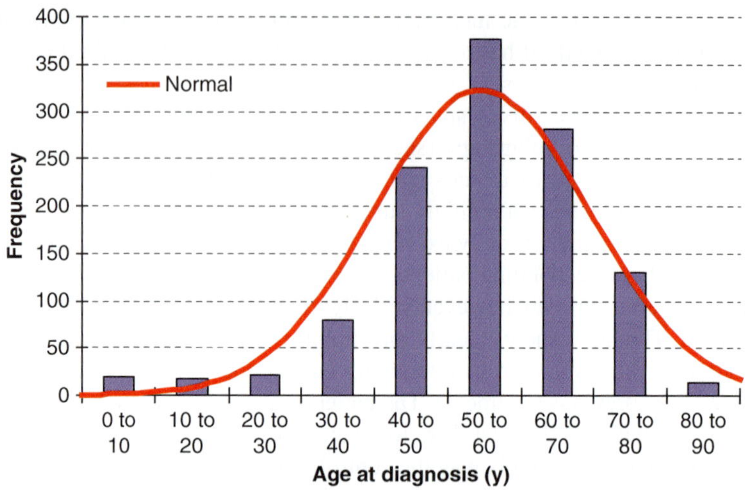

Fig. 4.142 Distribution of patients in the study

at conferences [1199, 1262]. The study time was from 1997 to 2003 (recruiting 1997–2002). Generally heavily pre-treated patients, with various localizations, mostly in late stages, were studied. Some localizations (non-small-cell lung cancer [$n=258$], pancreas [$n=99$], stomach [$n=68$] colon [$n=114$], rectum [$n=92$], breast [$n=103$], cervix uteri [$n=38$], kidney [$n=39$], brain [$n=24$], and ovary [$n=27$]) are large enough to derive some sort of conclusions. The endpoint was the survival time as a primary check of the efficacy of a curative method in such a lethal disease. The date of death (or of being alive) was checked in the Hungarian National Death Register, so actual and accurate data were collected. The final check of the deaths was December of 2003.

The age distribution of patients follows well the normal distribution, (see Fig. 4.142) average/median ages of the patients were 54.1/56 (1–87) years, the standard error of mean is 0.41, the standard deviation 14.2, and kurtosis/skewness is 1.88/–0.96. No outliers were present.

The male/female ratio is 612/568, their median ages are 57/54, (mean 55.2/52.9), respectively. 928 patients were treated in HTT (HTT-Med Polyclinic, Budapest, Hungary), while 252 were treated in the PFY (Peterfy Hospital, Budapest, Hungary) center. The groups of patients not less than 10 is shown in Table 4.8.

Patients were heavily pretreated (2.62 pretreatments for one patient on average); Fig. 4.143 shows the advanced cases in the study. (The pretreatment combinations are also shown).

The number of metastases also was high (see Fig. 4.144), about a quarter of the patients had no metastasis, and also about a quarter had more than one. The topographical ICD (ICD-02) codes of metastases per patients are shown in Fig. 4.145.

Table 4.8 Main localizations investigated in the present study (localizations having less than ten cases are not listed)

Disease	ICD	Number of cases (pts.)
Head and Neck	C00–C14	64
Esophagus	C15	12
Stomach	C16	68
Colon	C18	114
Rectosigmoid junction	C19	12
Rectum	C20–C21	92
Liver	C22	25
Other biliary	C24	14
Pancreas	C25	99
Larynx	C32	10
Lung and bronchus	C34	258
Skin	C43—c44	32
Soft tissue	C49	16
Breast	C50	103
Cervix	C53	38
Ovary	C56	27
Prostate	C61	18
Kidney	C64+C65	39
Urinary bladder	C67	18
Brain	C71–C72	29

The average/median oncothermia treatment time was 68.8/60.0 min (30–180), the average/median equivalent temperature 50.9/52 C (37.4–59.9), the applied average/median treatment number 8/6 (1–69). Patients tolerate the treatment very well, the pain relief was obviously presented, the subjective and objective QoL was increased. No serious toxicity was observed.

The overall survival [median 25.20 months, (0.87–299.6) and the mean: 35.24 months, (std.err.:1.06)], and the survival from the first oncothermia treatment [median 7.67 months, (0.03–75.3) and the mean: 14.29 months, (std.err.:0.45)].

Survival times are of course dependent on the metastases. Generally, patients with metastases have much worse life expectancy than those without. The difference (considering all the studied localizations) is statistically massively significant (for overall survival $p<0.00004$, and for survival from the first oncothermia treatment $p<0.000002$; see Fig. 4.146)

The pre-treatment efficacy in the surviving fraction of the oncothermia-treated patients has also been checked. While surgery ($p<0.002$) and radiotherapy ($p<0.004$) have significant effect, chemotherapy ($p>0.17$) was not significant in the overall survival rate and none of those significantly changed the survival from the first oncothermia.

The average oncothermia treatment parameters (number of treatments, average treatment time, and the equivalent temperature of the treatment) were also studied, having no significant change on survivals (except the naturally different treatment number) (see Figs. 4.147, 4.148).

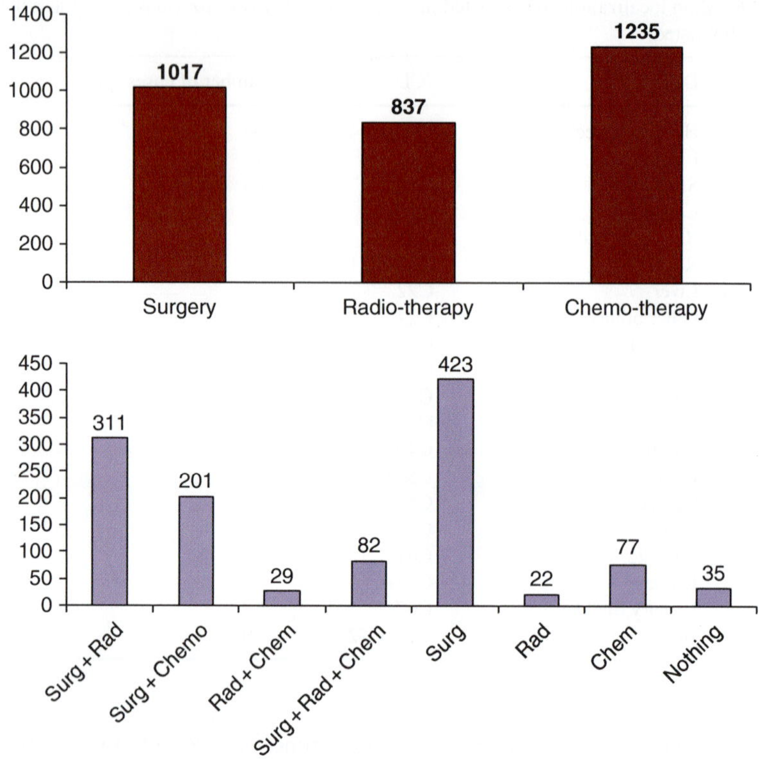

Fig. 4.143 Pretreatment statistics of the 1,180 patients. The *right panel* shows the therapy combinations

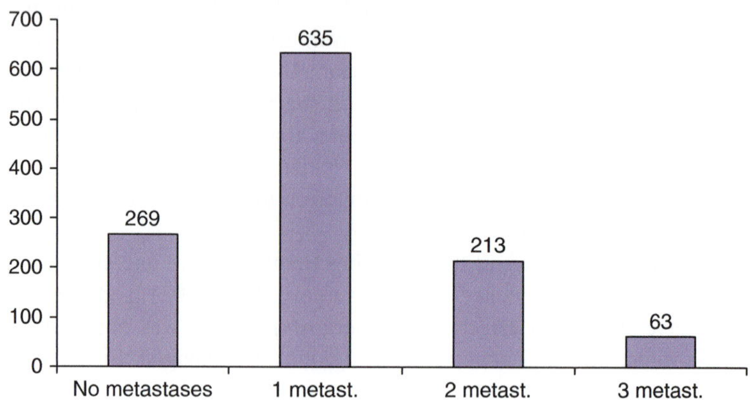

Fig. 4.144 Distribution of metastases

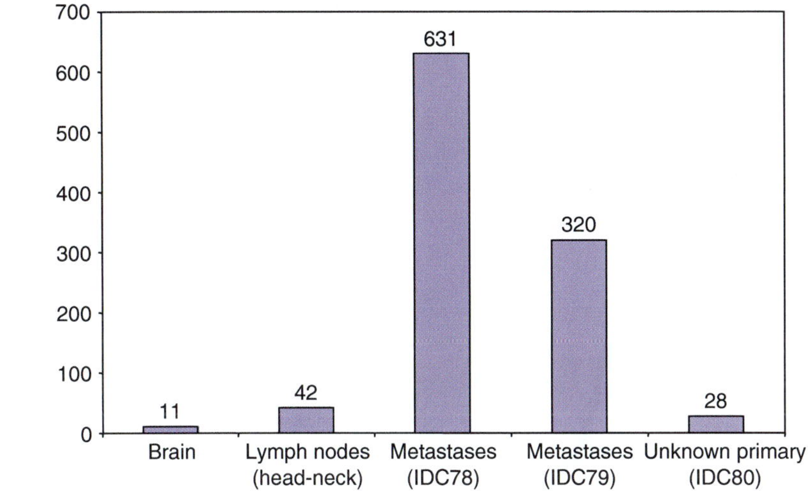

Fig. 4.145 The ICD codes of metastases

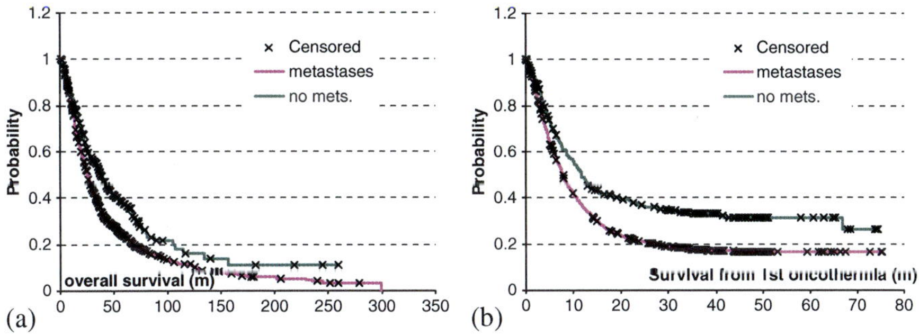

Fig. 4.146 The survival dependence of the metastatic diseases: (**a**) overall survival from the first diagnosis of the disease, (**b**) survival from the first oncothermia treatment

Average/median time from the 1st diagnosis to the first oncothermia treatment was 20.9/10.6 (0–265.7) months, which compared to the average/median overall survival [35.2/25.2 (0.9–300)], shows, that the patients were treated only in their second-half of their survival time, [median of the ratio of elapsed time to overall survival is 57.14 (0.0–99.8)] and the confidence intervals show hectic practice to include hyperthermia in the applied treatment protocol. However, the survival time of the patients from the first oncothermia treatment is generally longer, when the elapsed time to first oncothermia is smaller.

Age dependence of the results was also considered. The elderly (>68) and the young (<18) groups were independently studied and the differences were measured (see Figs. 4.149, 4.150).

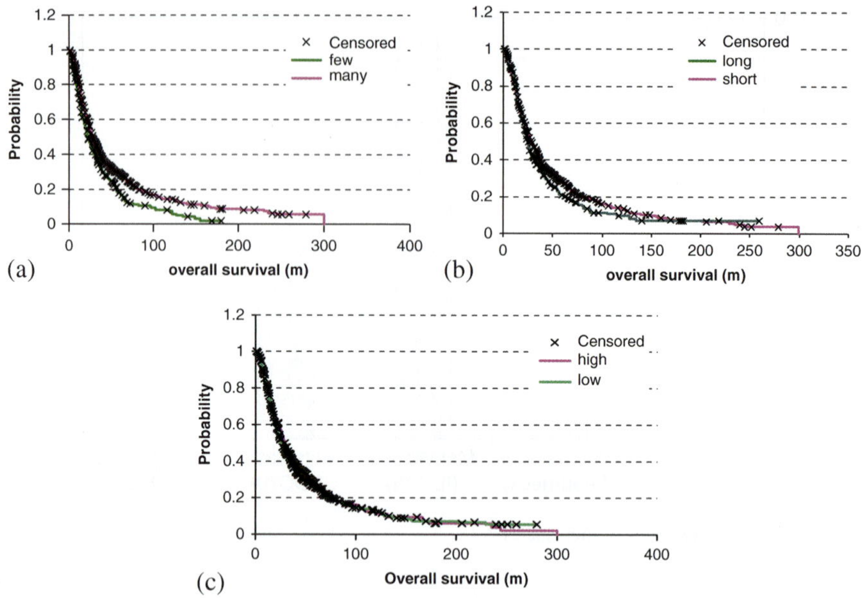

Fig. 4.147 Effect of the average treatment parameters on the overall survival: (**a**) number of treatments ($p<0.00015$), (**b**) treatment time ($p>0.14$), and equivalent temperature ($p>0.42$). The two–two quantities are divided by the median value (below or above) of the given parameter

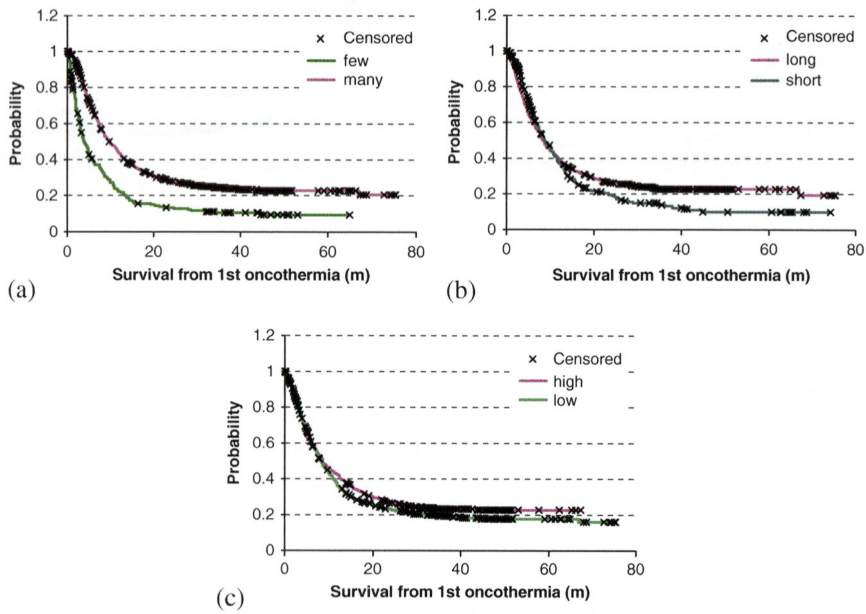

Fig. 4.148 Effect of the average treatment parameters on the survival form the first oncothermia treatment: (**a**) number of treatments ($p<10–17$), (**b**) treatment time ($p<0.17$), and equivalent temperature ($p<0.17$). The two–two quantities are divided by the median value (below or above) of the given parameter

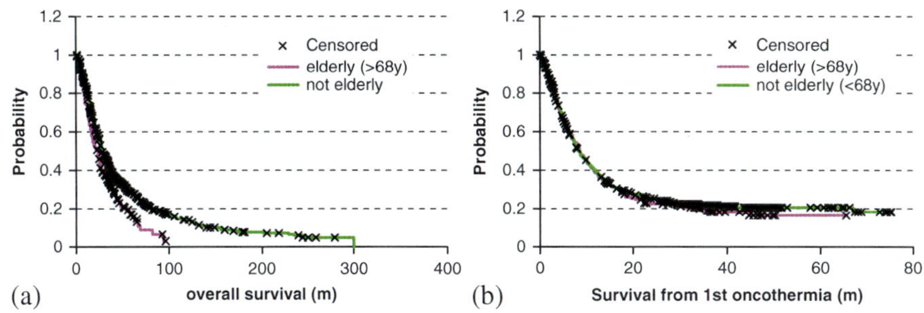

Fig. 4.149 The age dependence of the results from the gerontology point of view: (**a**) overall survival ($p < 0.00016$) and (**b**) survival from the first oncothermia ($p > 0.63$)

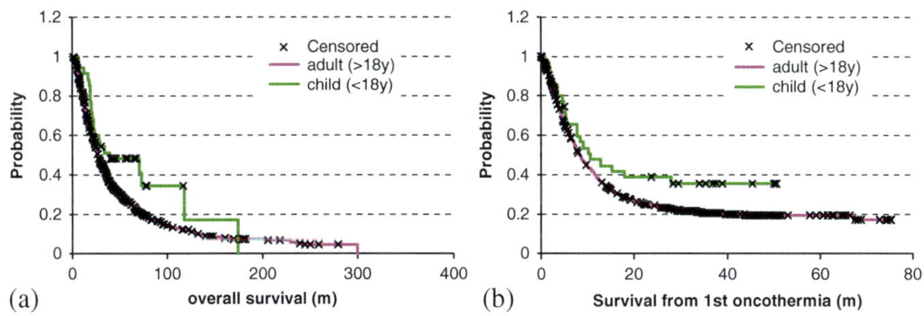

Fig. 4.150 The age dependence of the results from the pediatric point of view: (**a**) overall survival ($p < 0.038$) and (**b**) survival from the first oncothermia ($p = 0.09$)

No serious toxicity/burn was reported during the full study. The low forwarded energy was well focused on the actual tissue; there was not enough energy loss to cause surface burn.

Patients reported subjective improvements in their QoL.

Results show the general behaviors of the oncothermia treatments and are not of course satisfactory enough to derive any final conclusions on the actual cancer cure. The huge number of patients is adequately normally distributed to make some general arguments about the oncothermia method. The cohort studies in detail will be shown in the next parts of this series.

However, in general we may state: the heavily pre-treated, advanced cases of the patients had in median survival after the first oncothermia almost one third of their overall survival, which in the case of such a group (beyond the limit of the traditional treatments) is a remarkable result. It is a fact that patients with one or multiple metastases have less survival than their non-metastatic counterparts, and of course this difference is larger in the last third of their survival. In terms of the pre-treatment efficacy we are not able to make any remarks, because it is very much localization- (and specific protocol) dependent.

To evaluate the treatment parameters, naturally the larger number of treatments is connected to the longer survival. However, the data shows the after oncothermia

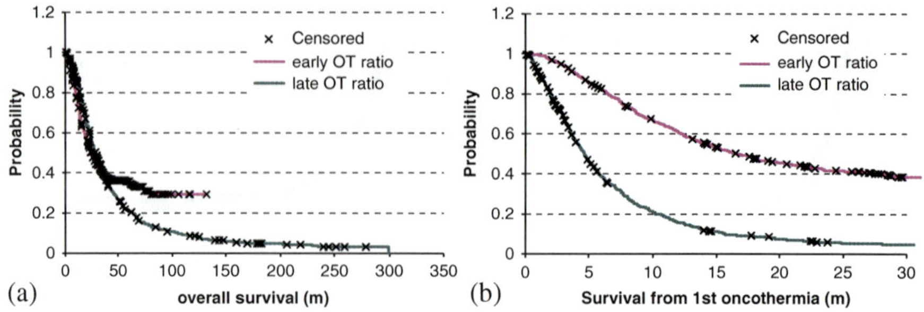

Fig. 4.151 The ratio of the elapsed time from the first diagnosis till the first oncothermia treatment to the survival from the first oncothermia shows a difference. The earlier oncothermia is significantly better in both the survivals [(**a**) overall survival ($p = 0.07$), (**b**) oncothermia survival ($p < 10–74$)] (The two–two quantities are divided by the median value (below or above) of the given parameter)

survival is significantly longer ($p < 10^{-5}$) in the higher treatment numbers than in the fewer ones. This is natural: the longer survival has more possibilities to treat. However the longer survival in most of the cases was measured with a long follow-up period after finishing oncothermia.

However, a question remains: does the time of the start of oncothermia have a role? The analysis of the ratio of the elapsed time to the survival from the first oncothermia shows the difference between the early- and late-beginning oncothermia (see Fig. 4.151).

Also there could be significant influence on the results dependent on the experience of the treating staff. Data analysis shows there has been increasing beneficial effect from the ongoing oncothermia training and increasing experience of the users since the very beginning of its application. In comparison between the early experience (first half of the study time) and the late experience (second half of the study time) there is almost a significant difference ($p = 0.052$) in the oncothermia survival, but no significance could be observed in the elapsed time to the first oncothermia and in the overall survival, see Fig. 4.152.

One of the further advantages of oncothermia is its effect on the QoL and in suppressing the side effects of the complementarily applied conventional methods. The QoL was measured on anecdotal, subjective reports only. The dominant opinion (>70%) is the pain reduction and improved well-being leads to a better QoL.

4.7 Brain Studies

4.7.1 Brain Safety Study (Phase I)

Glioms are one of the most common primary brain tumors. Despite surgery and radiotherapy (RT) with or without adjuvant chemotherapy, malignant glioma

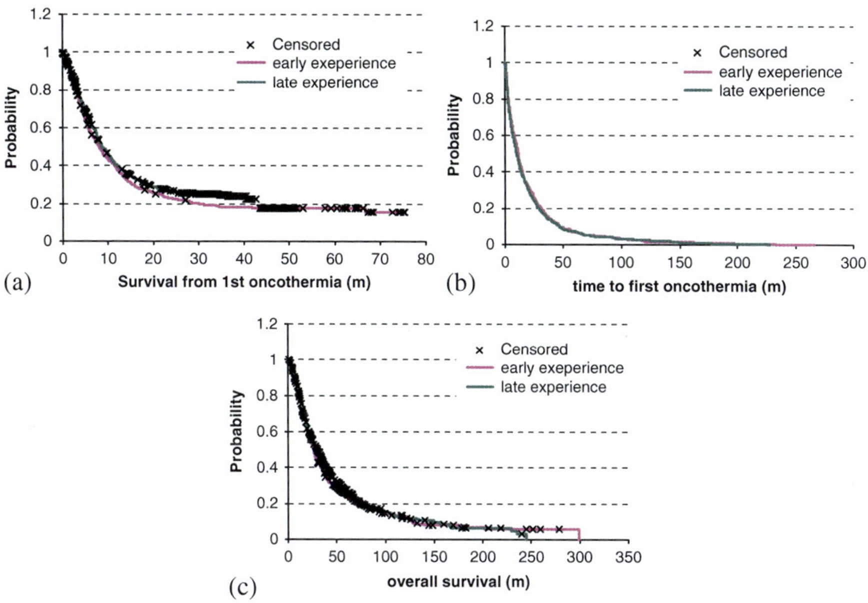

Fig. 4.152 Differences by the experience of the treatment procedure during the trial, dividing the results into two groups: obtained in the first or in the second half of the study time. (**a**) Survival from the first oncothermia ($p = 0.052$), (**b**) elapsed time to the first oncothermia ($p > 0.44$), and (**c**) overall survival ($p > 0.29$)

remains an almost uniformly fatal disease characterized by a rapid and devastating clinical course. Oncologic hyperthermia (oncothermia or modulated electro-hyperthermia) applied either alone or in combination with chemo- and/or radiotherapy is a new modality of brain glioma (BG) treatments.

A monocentric prospective single-arm Phase I/II study ($n = 15$) was performed by the Neurology Clinic of Regensburg University, Germany (Investigators: Prof. U. Bogdahn and Prof. P. Hau) [1200, 1219, 1220]. The main inclusion criteria were recurrent high-grade glioma WHO Grade III or IV, age 18 to 70, and Karnofsky Performance Score (KPS) ≥70. Primary endpoints were dose-limiting toxicities (DLT) and maximum tolerated dose (MTD) with the combined regimen. Patient characteristics are shown in Table 4.9. Groups of 3–4 patients were treated 2–5-times a week in a dose-escalation scheme with oncothermia (see Table 4.10.). Alkylation chemotherapy (ACNU, Nimustin) was administered in a dose of 90 mg/m^2 on day 1 of 42 days for up to 6 cycles or until tumor progression (PD) or DLT occurred. All together 15 patients with high-grade glioms were included. Relevant toxicities were local pain and increased focal neurological signs or intra-cranial pressure. No DLT occurred in the trial. In some patients, the administration of Mannitol during oncothermia or long-term use of corticosteroids was necessary to resolve symptoms. Summary of adverse effects are collected in Table 4.11. (The observed adverse effects are originated from the combination of ACNU with oncothermia.)

Table 4.9 Characteristics of the patients involved in the trial

Number of EHT/week	Code	Age at inclusion	Gender	Histology	Grade	No. of relapses	Extent last resection	No. of resections	First-line chemotherapy	Second-line chemotherapy	Third-line chemotherapy	KPS before OT	KPS at end of study	Cycles of OT (completed)	Sessions of OT	Cycles of ACNU	Best response (neuro-radiologist)
2/w	RE-001-1	52	m	GBM	IV	2	R0	2	RTX+ TMZ	Glivec/Litalir		80	50	2	16	5	PD
	RE-002-1	58	m	GBM	IV	1	R0	1	RTX+ TMZ/PEG-Dox			80	uk	3	28	4	SD
	RE-003-1	59	f	GBM	IV	1	R0	1	RTX+ TMZ/PEG-Dox			90	50	3	20	3	SD
	RE-004-1	66	m	GBM	IV	1	R0	2	RTX+ TMZ/PEG-Dox			70	60	2	16	2	PD
3/w	RE-005-2	53	m	GBM	IV	1	R0	1	RTX+ TMZ/PEG-Dox			100	70	2	20	2	SD
	RE-006-2	58	f	GBM	IV	1	R1	1	RTX+ TMZ			90	50	1	11	1	PD
	RE-007-2	45	m	GBM	IV	3	R1	2	RTX+ TMZ	TMZ int.		70	uk	1	12	1	PD
	RE-008-2	51	f	GBM	IV	1	R0	2	RTX+ TMZ/PEG-Dox			80	50	1	13	2	PD

Table 4.9 (continued)

Number of EHT/week	Code	Age at inclusion	Gender	Histology	Grade	No. of relapses	Extent last resection	No. of resections	First-line chemotherapy	Second-line chemotherapy	Third-line chemotherapy	KPS before OT	KPS at end of study	Cycles of OT (completed)	Sessions of OT	Cycles of ACNU	Best response (neuro-radiologist)
4/w	RE-009-3	57	m	GBM	IV	2	R0	1	RTX+ TMZ	Epothilone		50	40	5	77	5	SD
	RE-0010-3	56	f	GBM	IV	3	R0	2	RTX+ TMZ	TMZ int.	PCV	80	80	1	16	1	PR
	RE-0011-3	64	m	GBM	IV	2	R1	2	RTX+ TMZ	Epothilone		80	30	2	31	2	PR
	RE-0012-3	59	m	AA	III	2	Biopsy	0	TMZ/ 13-cis-RA	Radio Tx		90	90	3	48	2	SD
5/w	RE-0012-4	48	m	AOA	III	3	R1	2	RTX	PVC	TMZ	70	60	3	60	3	SD
	RE-0014-4	49	m	GBM	IV	3	R1	2	RTX+ TMZ/ PEG-Dox	Epothilone	TMZ int.	80	80	3	55	2	SD
	RE-0014-5	73	m	GBM	IV	1	R1	1	RTX+ TMZ/PEG-Dox			100	100	1	23	1	PD

OT – Oncothermia; **GBM** – Glioblastoma multiforme; **TMZ** – Termozalonide; **AA** – Anaplatic astroytoma; **PCV** – Procarbacine; **AOA** – Anaplastic oligoastocytoma; **KPS** – Kasnofsky Perfomance Score; **RO** – Complete resection; **PD** – Progressive disease; **R1** – Partial resection; **SD** – Stable disease; **PR** – Ratial

Table 4.10 Dose-escalation in the safety study

Group	Number of patients	Chemotherapy (single dose of a 6-week cycle)	Oncothermia (4 of 6-week cycle)
1	3 (6)	ACNU 90 mg/m2	Oncothermia 2×/week
2	3 (6)	ACNU 90 mg/m2	Oncothermia 3×/week
3	3 (6)	ACNU 90 mg/m2	Oncothermia 4×/week
4	3 (6)	ACNU 90 mg/m2	Oncothermia 5×/week

Table 4.11 Collection of adverse effects of the dose-escalating trial. Toxicity evaluated according to NCI CTC criteria (Grade I–IV), [version 3.0]. Grade IV was occurring in 1.3% of patients

NCI CTC	I	II	III	IV	all (%)
Local pain	3	9	–	–	12(80)
Hemiparesis	0	7	5	–	12(80)
Headache	0	3	6	–	9(60)
Fatigue	0	6	3	–	9(60)
Thrombopenia	2	3	2	1	8(50)
Leukopenia	1	5	2	–	8(50)
Vomiting	2	6	–	–	8(50)
Confusion	1	4	3	–	8(50)
Speech impairment	0	0	5	1	6(40)
Seating	3	3	–	–	6(40)
Nausea	3	3	–	–	6(40)
Dizziness	4	2	0	0	6(40)

4.7.2 Brain Efficacy Study (Phase II)

The study is an open-label, single-arm, monocentric, sequential Phase II study which was finished in March 2006 [1198]. The involved patients were analyzed according to ITT. Recruiting time was 56 months. The primary endpoints of the study were the overall survival time (OST) and the survival time from the first oncothermia treatment (TST). The applied test was KM log-rank. Inclusion criteria were: (1) Inoperable or sub-totally resected or recurrent BG, (2) progression after radio- and/or chemotherapy, (3) KPS $\geq$ 30%.

Oncothermia was applied with power ranging between 40 and 150 watts and the calculated average equivalent temperature in the tumors was above 40°C more than 90% of the treatment time. The applied bolus fixed the skin surface at 20°C. The targeted area was self-selectively treated by the well-covering electrode system. Oncothermia was performed in two/three sessions per week. Treatment time and power range per session started with 40 W in 20 min, and (step-by-step) gradually and linearly this was raised up to 60 min, 150 W over 2 weeks.

Distribution of the BG patients by their WHO-grade showed mostly advanced cases: diffuse astrocytoma, (DA): 8, (5.7%); anaplastic astrocytoma, (AA): 40, (28.6%); GBM: 92, (65.7%); (see Fig. 4.153). Most of the patients had failed to respond to the applied traditional therapies.

Fig. 4.153 Distribution of the 140 patients involved in the study

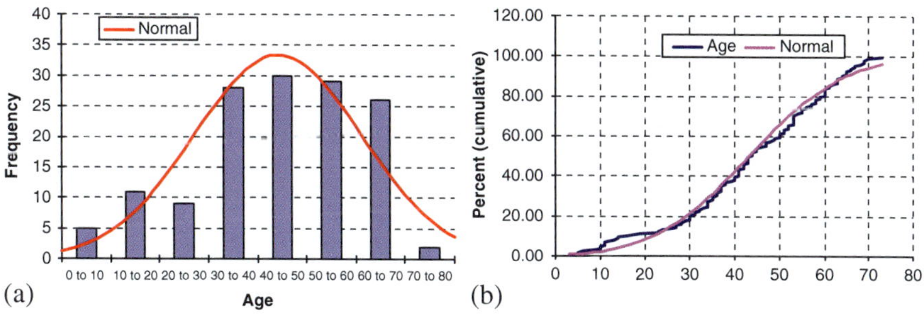

Fig. 4.154 Age distribution of BG patients ($n = 140$). (**a**) Distribution by 10 year categories, (**b**) probit function (inverse cumulative distribution function)

The age distribution shows near to normal ($p < 0.001$ by Chi-square test for discrete variables), and no outliers ($p < 0.05$) were present. The median age was 43.5 years (3–73), the mean age was 43.2 years (Std.err=1.42), 15 (10.7%) patients were below 18 years, and 8 (5.7%) were over 68 years. The gender distribution was 50/90 female/male. The epidemiologically shown [1221] more frequent BG in the elderly population (in Japan BG-incident is 2.40/100000/y over 70 years, while under it is only 1.42/100000/y [1222]) did not appear in our case. A slight increase from the normal distribution could be observed in the range of 50–70 year ages (see Fig. 4.154).

Pretreatments were applied in 364 cases (~2.6/patient), and its distribution by the main categories is shown in Fig. 4.155. The chemotherapies were: in 117 cases (84%), the radiation was in 129 (92%), and there were 117 cases of surgery (84%). (Two patients had no pretreatments for individual reasons.) As a mean, 69% of all patients had all three therapy modalities.

Oncothermia was applied adjuvant in most cases. The distribution of adjuvant treatments is shown in Fig. 4.156. The chemotherapies (in most cases TMZ) were present in 102 (73%) cases and radiation in five (3.6%) cases, while supportive therapy was present in 105 (75%) cases. Characterization of the applied supportive therapy is shown in Table 4.12. This therapy was started together with oncothermia, and was applied for 3 months. Application of oncothermia as mono-therapy (2 cases (7%)) and only combined with supportive therapies (27 cases, (19%)) were applied if no other modality was possible.

The oncothermia treatment cycles were on average 1.8 (1–9) while the treatment number average was 21.5 (2–108). The median oncothermia treatment number was

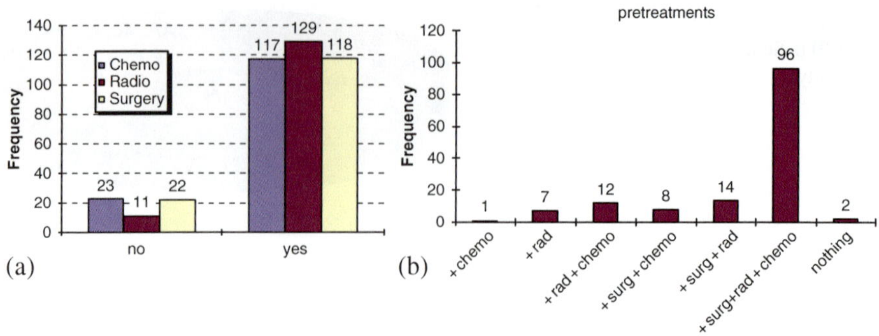

Fig. 4.155 The pre-treatment distribution kind (**a**) and by its combinations (**b**)

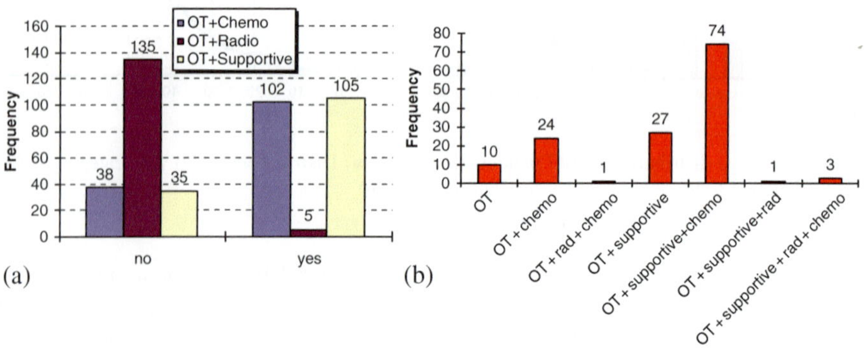

Fig. 4.156 The adjuvant/neoadjuvant treatment distribution with oncothermia by kind (**a**) and by their combinations (**b**). (OT = oncothermia)

Table 4.12 The applied supportive therapy

Supportive drug	Dose
Boswelia carterii (Weihrauch)	6 g/d, (3×/day)
Mistletoe (Mistel, Lectinol)	15 ng, (3×/week), subcutan
Selenium	300 μg/d

15. The applied dose of oncothermia was regarded as low if it did not exceed the 8-times 60-min load, (dose-threshold, DT). Such a low dose was provided for 28 patients. The median time of total duration of the oncothermia treatment period was 1.7 m (1 d–36.4 m) on average 3.3 m (Std.err=0.4).

The median time elapsed to first oncothermia was 10.8 m (0.2–181) 21.7 m (std.err.=2.5) on average. The median follow-up time after the last oncothermia was 3.4 m (1 day–49.1 months) on average 6.6 m (Std.err=0.8).

No toxicity or other problems were observed during the treatment, only 10–15 times did we observe headache, there was no increased edema, but all the points were clinically controllable. In most cases the edema decreased and the intra-cranial pressure was also decreasing. No surgical or other intervention was necessary during or after the oncothermia treatments for any of the patients. All the patients tolerated the treatment very well, and subjectively they reported better QoL, but this was not objectively evaluated.

The MST of overall survival and TST for all of the patients were 19.8 m (1.4–190) and 6.7 m (0.3–50), respectively. The average (mean) survival time (AST) of overall survival and TST were 31.7 (std.err=3.0) and 10.0 (std.err=0.9), respectively. The corresponding KM plots are shown in Fig. 4.157. The same survivals categorized by their WHO-grade are shown in Table 4.13. and Fig. 4.158.

The dose analysis shows (see Fig. 4.159) the relative dependence to DT is not significant for overall survival ($p = 0.129$) and significant for TST ($p < 0.01$).

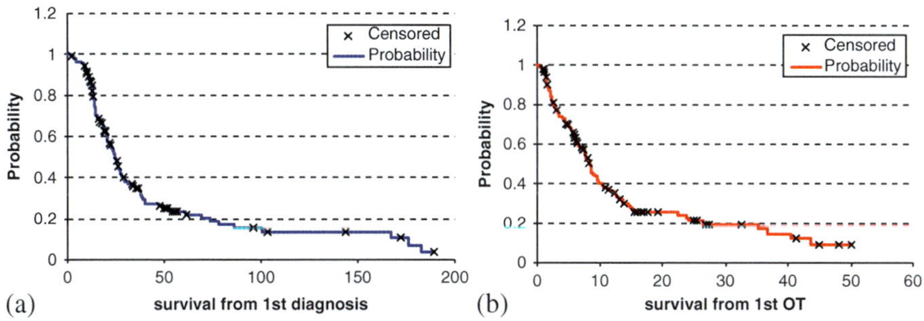

Fig. 4.157 (**a**) Overall survival and (**b**) TST KM survival plots for all of the treated patients

Table 4.13 Median and mean data of the survivals

WHO grade	#Pts.	MST OST (m)	(min.-max.) (m)	MST TST (m)	(min.-max.) (m)	AST OST (m)	(Std. err.) (m)	AST TST (m)	(STD. err.) (m)
DA	8	59.2	22–190	11.6	1.1–41	73.6	18.8	15.6	5.2
AA	40	25.8	3.6–183	9.1	1.4–50	43.3	7.0	13.4	2.0
GBM	92	16.0	1.4–176	6.1	0.3–48	23.0	2.5	8.0	0.9

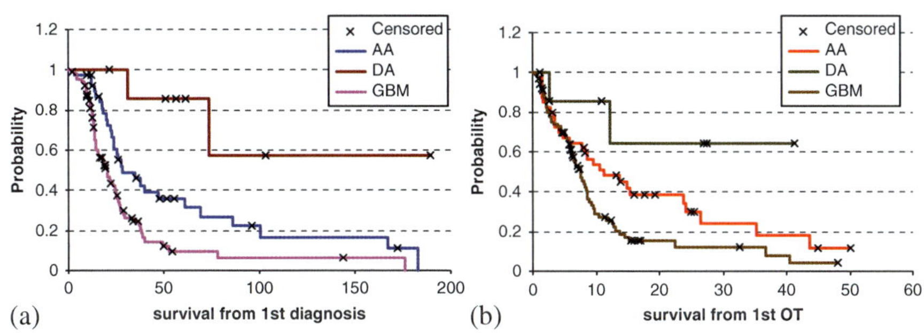

Fig. 4.158 (**a**) Overall survival and (**b**) TST KM-survivals for patients with DA, AA, and GBM

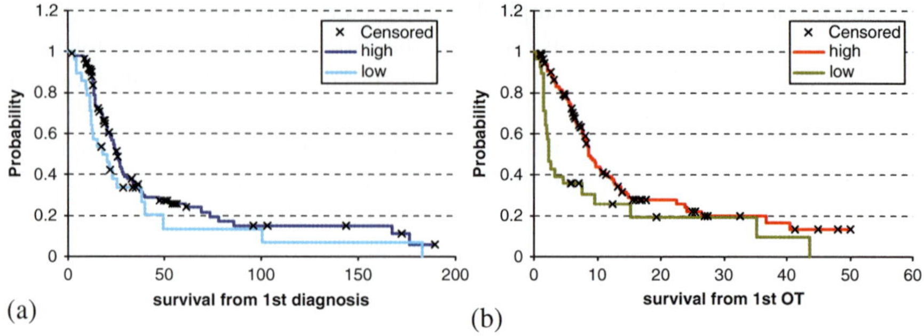

Fig. 4.159 (**a**) Dose dependence to DT for overall survival ($p = 0.129$) and (**b**) for TST ($p = 0.003$) survivals (KM survival plots)

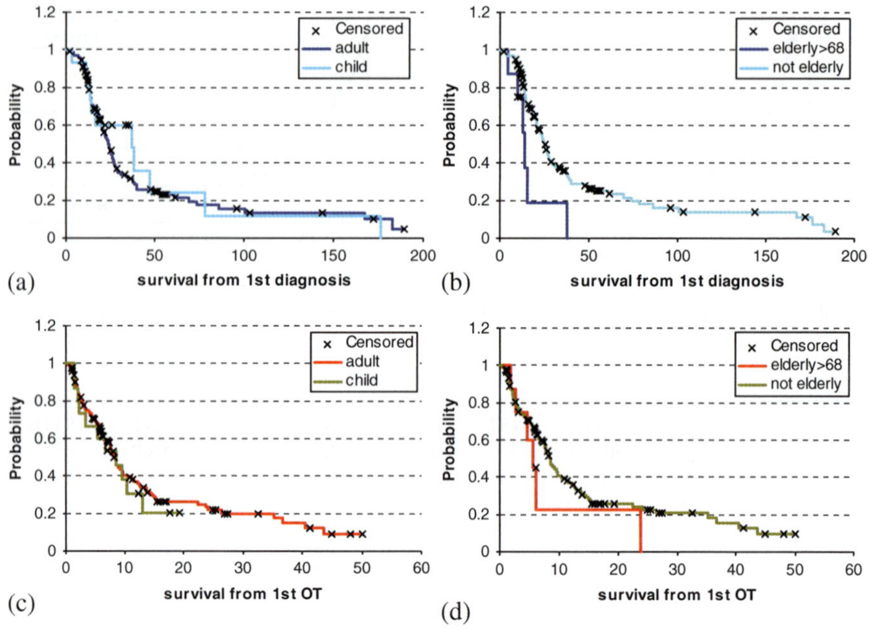

Fig. 4.160 Comparison of adults with young (<18 years) and elderly (>68 years) patients. (**a**) Overall survival for youngsters ($p = 0.65$), (**b**) overall survival for elderly ($p = 0.016$), (**c**) TST for youngsters ($p = 0.66$), (**d**) TST for elderly ($p = 0.24$)

The age categories by young (<18 year, $n = 15$) adult (between 18 and 68 years) and elderly (>68 years, $n = 10$), show no remarkable differences (see Fig. 4.160). The significance in the overall survival (at KM plot, Fig. 4.160b) for elderly patients could be for natural reasons, and this assumption is supported by the no-difference results in the TST KM-plot (see Fig. 4.160d).

No serious side effects were observed (see Table 4.14). Patients tolerated the treatments well during the whole treatment period. Most of the patients were well

Table 4.14 The observed side effects during the study

Side effects	Rel. val. (%)
1. Short term (<2 h) asthenia after the treatment	9
2. Local redness (rubor) of the skin	8
3. Complications	15
Subcutan fibrosis of fat tissue	1
Skin burn (diam. <1.5 cm) grade I–II	2
Headache and vomiting (<2 h)	12

relaxed, some even fell asleep during the treatment. Patients reported better QoL, but this information was not objectively measured.

The expected MST for BG patients is over all 11.3 m, which is well short of the actual 19.8 m (an increase of 75.2%). According to the RTOG classifications [1223], we divided the patients into two groups: age under- and over-50 years. The obtained patient's distribution is shown in the pie diagram in Fig. 4.161.

The overall survival and of TST in general definitely with high significance differs according to these categories, as the KM-plots show, see Fig. 4.162.

By categories the MST of overall survival was (except one category) systematically significantly well higher (see Table 4.15 than the expected ones for corresponding stage BG patients: the under and over 50-years patients median

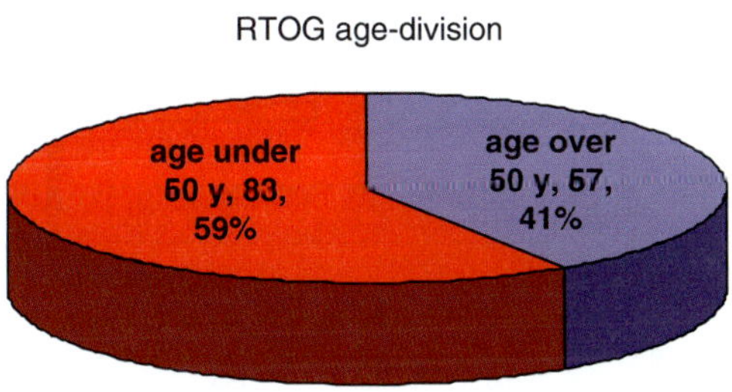

Fig. 4.161 Distribution of patients using a 50-years-old age threshold

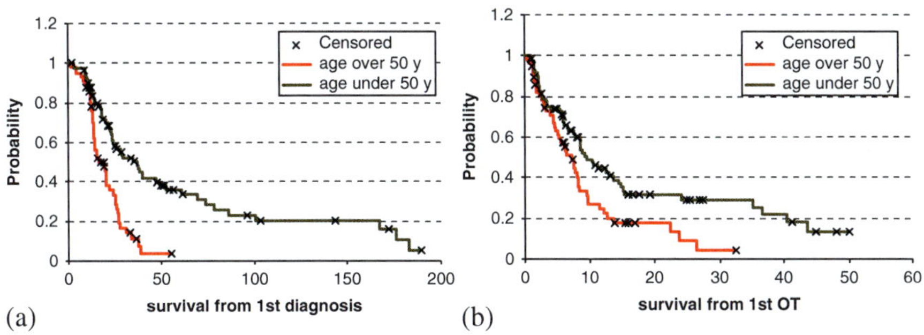

Fig. 4.162 KM plots using a 50-years-old age threshold: for (**a**) overall survival ($p < 0.0003$) and (**b**) TST ($p < 0.009$)

Table 4.15 The main statistical characters by the RTOG division

WHO grade	Patients no. (n)	MST OST (m)	(Min.-max.) (m)	AST OST (m)	(Std.err.) (m)	MST RTOG (m)
DA+AA (<50 year)	36	37.7	3.6–190	56.7	8.5	49.4
DA+AA (>50 year)	12	18.4	9.9–56	23.3	3.8	21.7
GBM (<50 year)	47	19.0	2.4–176	28.7	4.7	13.7
GBM (>50 year)	45	14.4	1.4–39	17.1	1.3	9.7

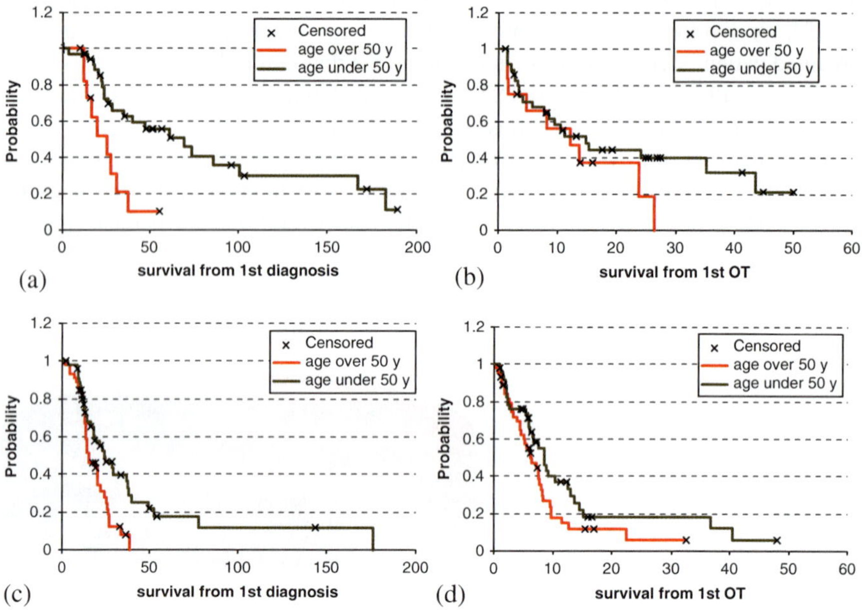

Fig. 4.163 The KM-plots of patients by RTOG categories. For AA+DA (**a**) overall survival ($p<0.003$) and (**b**) TST ($p=0.22$), as well as for GBM (**c**) overall survival ($p<0.009$) and (**d**) TST ($p<0.08$)

gains are −28.5, 24.0% for DA+AA and 39.4, 46.4% for GBM, respectively. The KM-plots show well the significant differences (see Fig. 4.163). The gain is obviously large except the DA+AA patients under 50 years of age. The reason for this discrepancy is not known. The next trial will have to decide on this issue as well.

The results could be well compared to the available SEER [47] data. Comparison of the overall survival of our retrospective 140 patients and SEER retrospective

28,970 patients is shown in Table 4.16, selected by the grade categories, The gain of the MST overall survival in various categories is 38,6, 146, and 57.0% for DA, AA, and GBM patients, respectively.

The parametric evaluation shows the definite benefit of oncothermia. The overall survival shows significant curative benefit, the slope of survival modified from 1.8 to 0.85 (gain: 53%) and from 1.95 to 1.42 (gain: 27.5%), for AA and GBM, respectively; see Fig. 4.164.

Table 4.16 Comparison of the data of SEER and our present study

WHO grade	Patient number (n) (present)	MST OST (present) (m)	MST OST (min.–max.) (m)	Patient number (n) (SEER)	MST OST (SEER) (m)
DA	8	59.2	22–190	2749	42.7
AA	40	25.8	3.6–183	3273	10.5
GBM	92	16.0	1.4–176	5801	10.2

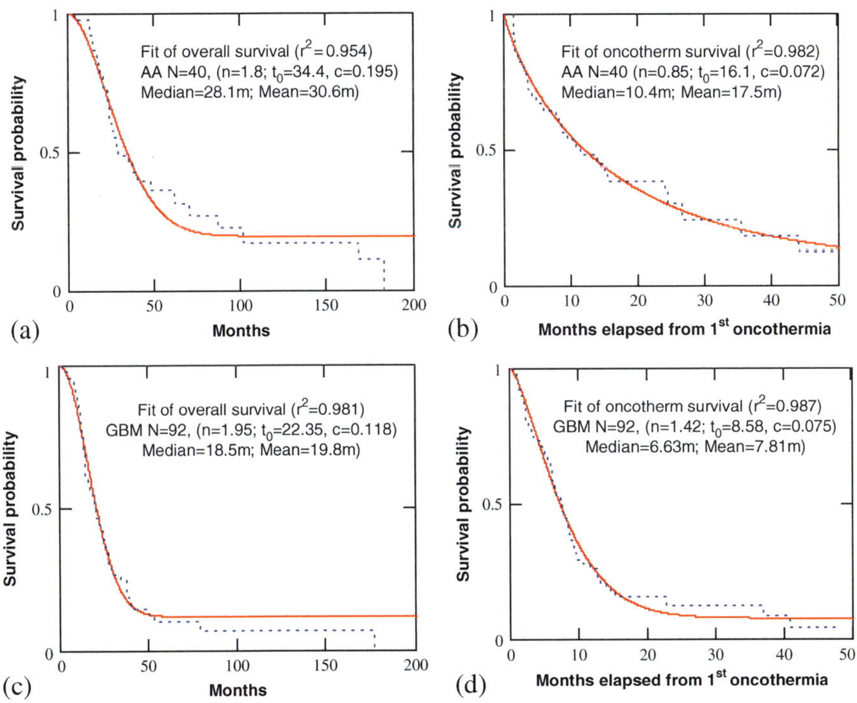

Fig. 4.164 The accurate Weibull fit (*solid lines*) to the survival plots [AA in (**a**) and (**b**), GBM in (**c**) and (**d**), while the overall survivals respectively are on (**a**) and (**c**), and the survivals from the first oncothermia treatment are on (**b**) and (**d**)]

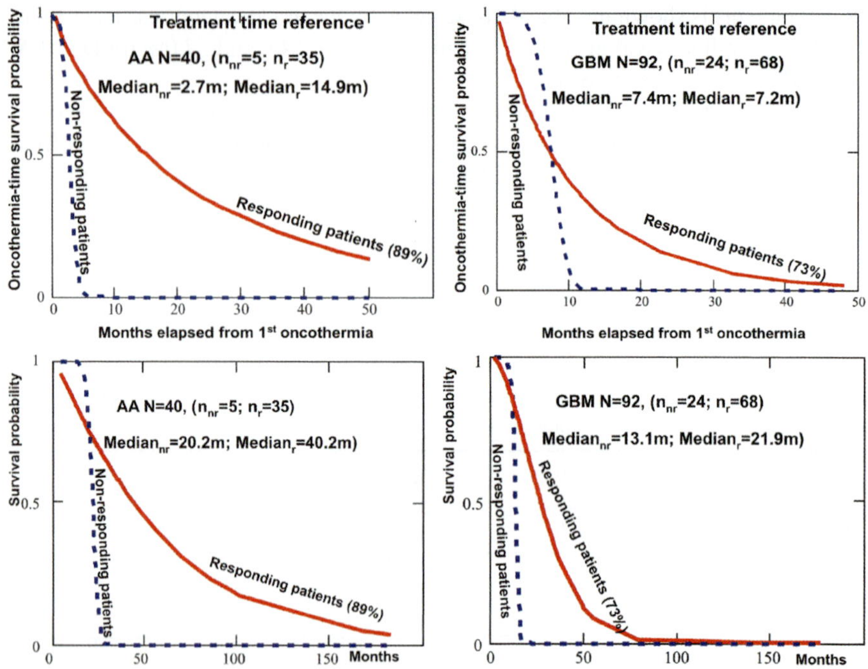

Fig. 4.165 The parametric fit (method is described in Appendix 35) of responders of Anaplastic astrocytoma cases (**a** and **c**) and glioblastoma multiform (**b** and **d**). The treatment reference (choosing the non-responding patients as the control arm) (**a** and **b**) and the overall survival (**c** and **d**) are shown independently. Both fits for AA and GBM show highly significant benefit of oncothermia for those patients who respond (The subscripts nr and r denote the non-responding and responding patient's data respectively)

This general gain could be studied further, by separating the responders and nonresponders. The parametric fit of the survival plots shows high response for treatment (responders are 89 and 73% for AA and GBM, respectively), see Fig. 4.165. The relatively large significant difference between the non-responding-arm and the actively responding one is the consequence of the inclusion criteria. When the patients are included in their last phase of treatments, than usually only a short time remains of their lifetime. Oncothermia in its active long treatment period shows a considerable elongation of the lifetime. However, it could also be explained from the side of efficacy. The oncothermia was effective for the patients who were in the active arm of the oncothermia study, while the control arm collects the patients for whom the oncothermia (and also other concomitant treatments) was not effective at all. Their percentage is rather large, therefore showing a more complicated picture then appeared from only the probability comparison.

In a recent publication [1224], the 1- and 2-year survivals with TMZ were 58% and 31%, respectively. Compare these results with ours, the increase is also remarkable; 71.7% (66/92) and 30.4% (28/92) for 1 and 2 year survivals, respectively.

(The 2 year survival for GBM by RTOG study (no TMZ application) is only 17%, [1223].) The most recent TMZ randomized clinical trial for GBM [1225], summarizing the results of 573 patients from 85 cooperating centers shows a gain of MST from 12.1 m (without TMZ) to 14.6 m (with TMZ). Former TMZ results [1226] were similar, having MST in only the RT group ($n = 24$) 11.2 m, RT+CT (not TMZ) group ($n = 32$) 12.7 m, and for RT+TMZ group ($n = 23$) 14.9 m. The 2-year survival in the new study [1224] increased from 10.4% (without TMZ) to 26.5% (with TMZ). Our present results were even better than the presently published best TMZ applications.

Long-term survival in GBM is very rare, ca. over 3 years being only 1.8% in a 279-patient trial [1227]. In our case from 92 GBM patients 13 (14.1%) had longer overall survival than 3 years, which is a remarkable gain.

Studying the MRI images we have some indicative hints to suppose an extended apoptosis initialized by the oncothermia. This could be in good correspondence with some theoretical considerations [1228], as well as with some experimental facts [1229–1231]. More intensive investigations on this line are in progress.

The results of this study well indicate the feasibility and the benefit of the oncothermia treatment, so the present study was sequentially continued with further observations. The data were published elsewhere [1232], and also extended using additional data from other clinics [1233] (bicentral trial).

The QoL of any of the oncothermia-treated patients did not worsen. According to their subjective reports, the QoL was considerably increased in most cases. (No objective evaluation has been carried out yet.)

4.7.3 Hungarian Brain Glioma Study

Some of the first preliminary oncothermia results of primary brain glioms ($n = 27$), were published in Hungary [1234, 1199]. (HTT-Med Polyclinic; Budapest, Hungary. Investigator: A. Varkonyi.) In this study the stages were not distinguished. Overall survival median was 23.6 m, while the overall average was 46.7 m. Survival times from first oncothermia were: median 6.4 m, average 14.8 m; see Fig. 4.166. The parametric evaluation of responders shows 43% only (see Fig. 4.167).

4.7.4 Small Prospective, Double-Arm Brain Glioma Study

A small prospective double-arm (control arm $n = 36$, active arm $n = 9$) study for advanced primary brain tumors (glioms WHO IV) was carried out in Nurnberg (Praxis at Klinikum Nord, Nürnberg, Germany, Investigator: Prof. Dr. H. Renner). Trimodal therapy was applied: radiotherapy (50–60 Gy), chemotherapy (Temodal), and oncothermia (6–12×; 60 min) [1235]. The median survival was measured on the control arm as 9 m, while in the active oncothermia arm it was 15 m (see Fig. 4.168).

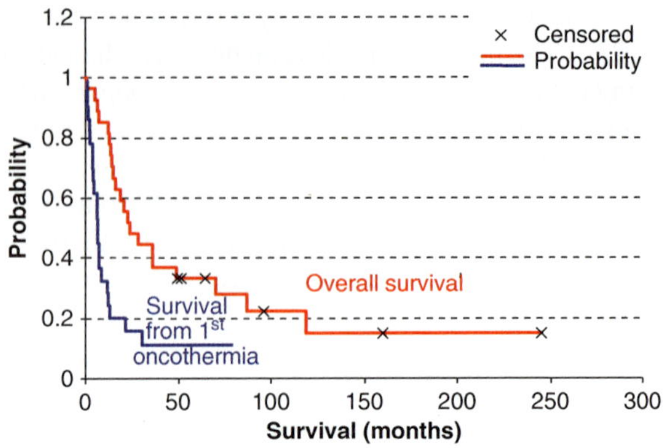

Fig. 4.166 The overall survival (median 23.6 m) and the survival from the first oncothermia (median 6.4 m) are shown in the Kaplan–Meier plot

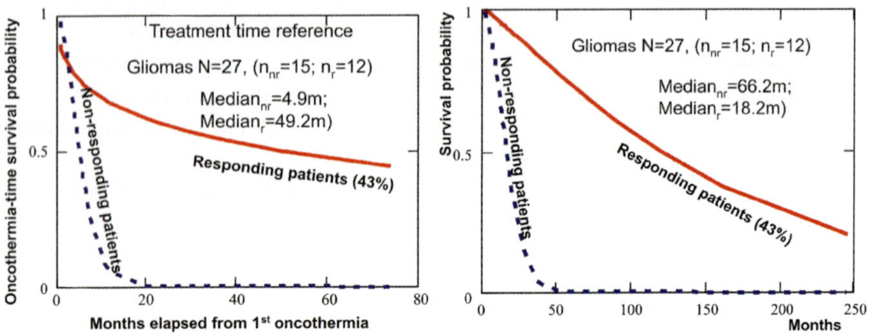

Fig. 4.167 Responders are well distinguishable from the nonresponders, but only less than half (43%) of the patients were responding in this very first study

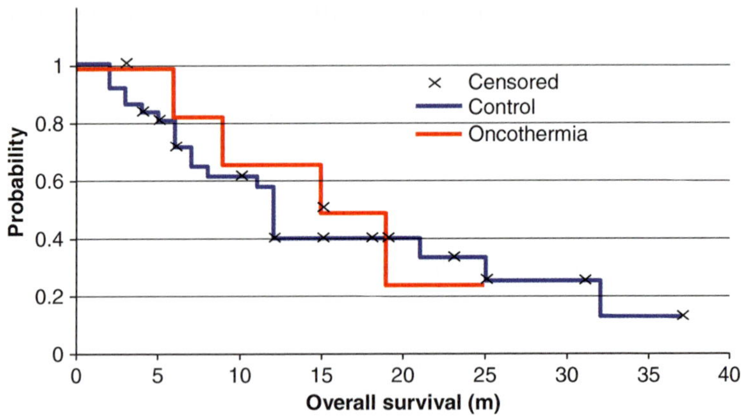

Fig. 4.168 The double-arm study shows remarkable improvement for advanced glioblastoma multiform patients (WHO IV), but the change is not significant, due to the low number of patients ($n = 9$ in the active arm)

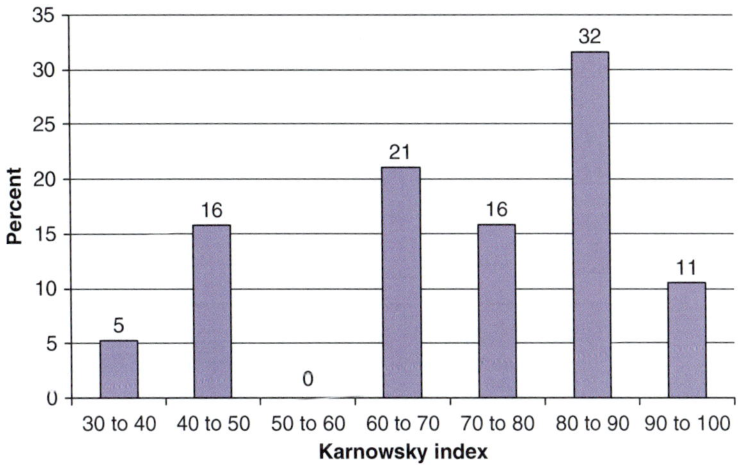

Fig. 4.169 Distribution of KPS for the patients involved in the study (percentages)

4.7.5 Study of Brain Gliomas with Local Clinical Responses

The same ACNU combination (60 mg/m^2), which was used for the safety (Phase I; dose escalation) trial was used in Phase II as well for recurrent glioblastoma ($n = 19$). (Clinic St. Georg, Bad Aibling, Germany, Investigator: Dr. F. Douwes, [1236]). The QoL of patients was measured by standard KPS (index) (see Fig. 4.169). The obtained median of overall survival was 21.8 m (average survival 25.5 m); while the median survival from first oncothermia was 8.8 m (average survival 13.5 m) (Fig. 4.170). The local clinical response had no complete remission, and 58% of the patients were in progressive disease (PD) (looks like not responding to oncothermia, see Fig. 4.171). However, the responding patient's ratio was 59% (falling 41%) calculated by the parametric selection (see Fig. 4.172). This difference could be caused by the relatively short follow-up in the trial.

The local clinical response categorizing by KPS, shows an interesting distribution, the local response was more dominant in the relatively low KPS, see Fig. 4.173. This fact can be explained by the number of treatments, showing better results when the treatments are more frequent, see Fig. 4.174.

4.7.6 Brain Glioma Study with Relapses

Treatments of brain tumors actively use oncothermia, because of its non-invasive trans-cranial effect and its safety. The first ASCO presentation of primary brain tumors treated by oncothermia was made in 2003 [1237]. Results show the feasibility and efficacy of oncothermia: median survivals were measured as 106 m ($n = 9$) and 20 m ($n = 27$) for AA and GBM, respectively; see Fig. 4.175.

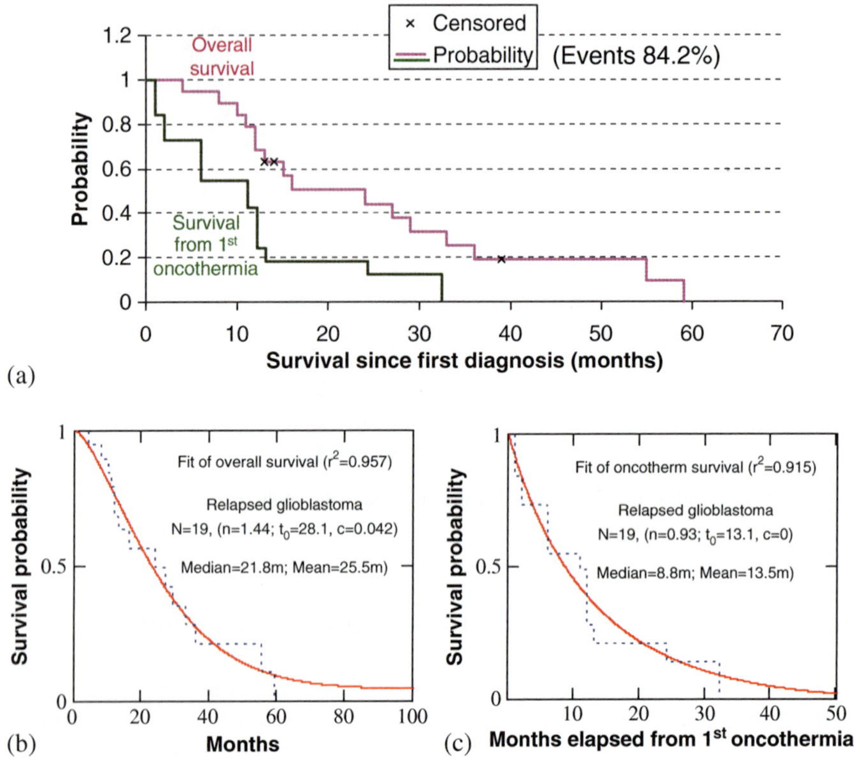

Fig. 4.170 Kaplan–Meier survival plot for recurrent glioblastoma multiform. The median of overall survival is 21.8 m. The parametric fit shows 35.8% gain by oncothermia

Fig. 4.171 Local response was falling from 58% of the patients

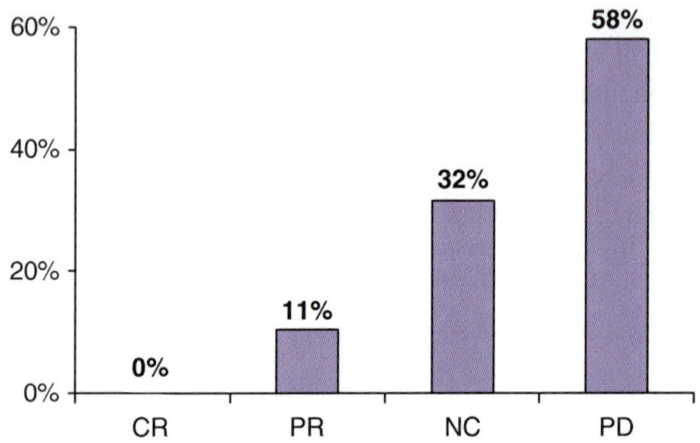

4.7.7 Bicentral Brain Glioma Study

Further evaluation was carried out by common pool of the patient databases where the same treatment protocol was applied [1233]. The local complete response (CR)

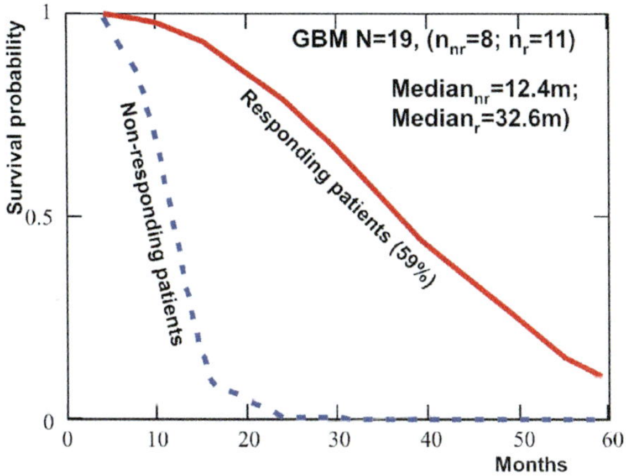

Fig. 4.172 The parametric selection significantly divides the pool of the cohort into two subgroups: responding (subscript "r") and non-responding (subscript "nr") patients

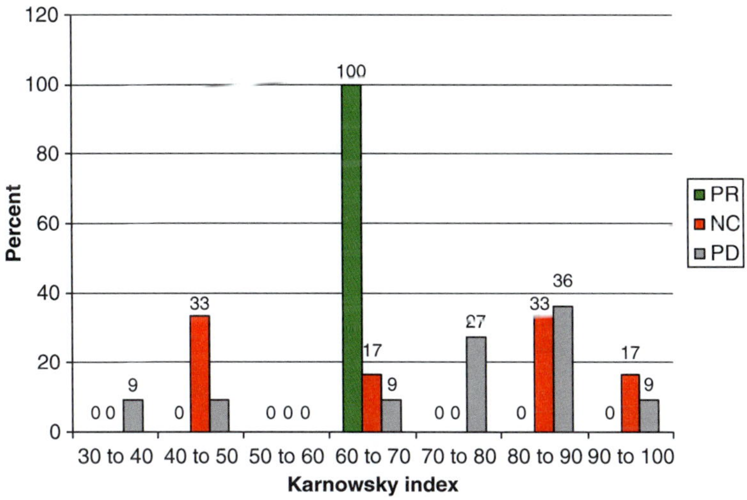

Fig. 4.173 The local clinical response (PR or NC) was in case of 60≤KPS≤70

was more than 16% for AA ($n = 53$) but CR was not obtained for GBM ($n = 126$). Contrary to this the PD local response was pointed for GBM (64%), see Fig. 4.176. The KM survival plot (see Fig. 4.177) and the details of survivals were similar to before: the medians of overall survival, of the first relapse progression time after the conventional therapies and of the oncothermia treatment time are 103, 10.6 m and 20.6, 7.6 m for AA and GBM, respectively.

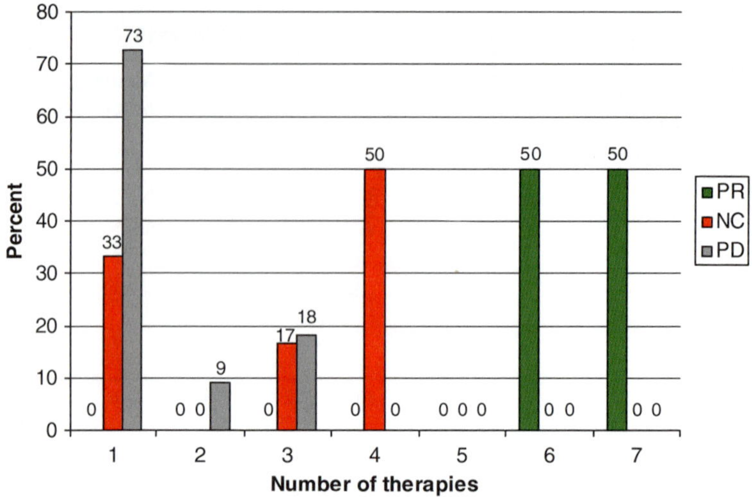

Fig. 4.174 The number of treatments significantly drives the local clinical response

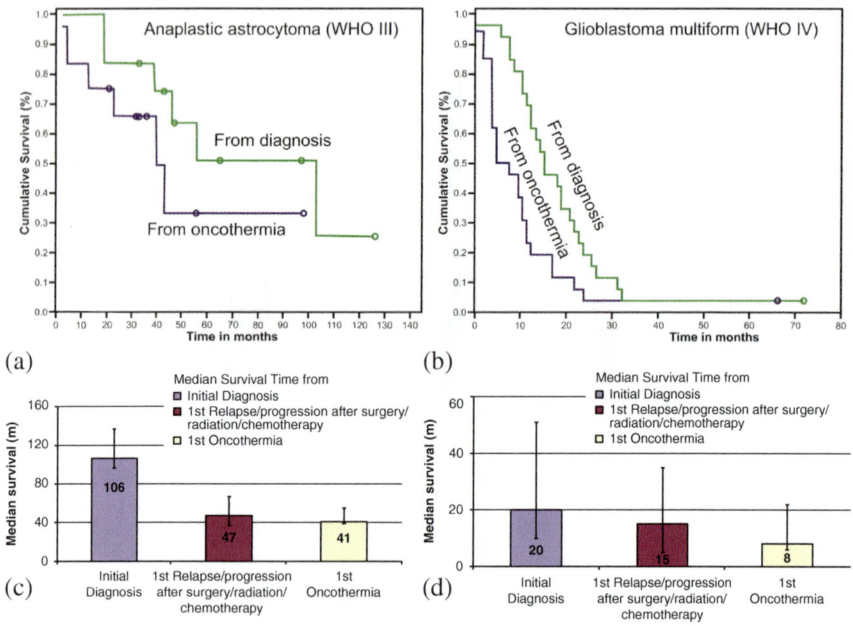

Fig. 4.175 The cumulative survival (Kaplan–Meier plot) of overall survival and survival of oncothermia is shown for AA and GBM, on panels (**a**) and (**b**), respectively. The medians of overall survival, of the first relapse progression time after the conventional therapies and of the oncothermia treatment time are 106, 47, 41 m and 20, 15, 8 m for AA and GBM, respectively. (Biomed Clinic, Bad Bergzabern, Germany. Investigator: Dr. D. Hager)

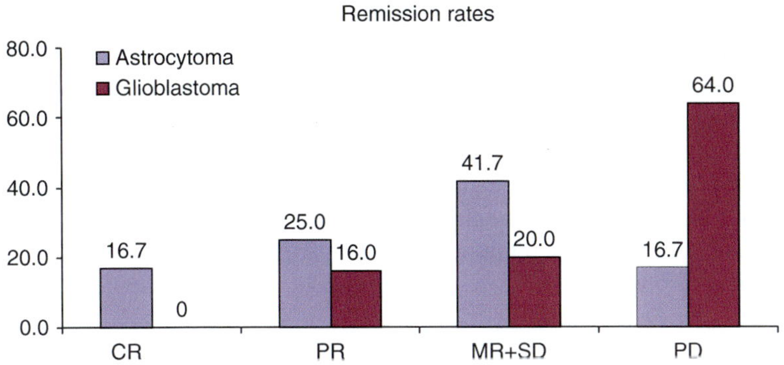

Fig. 4.176 Local responses for AA ($n = 53$) and GBM ($n = 126$) in the study [1233]

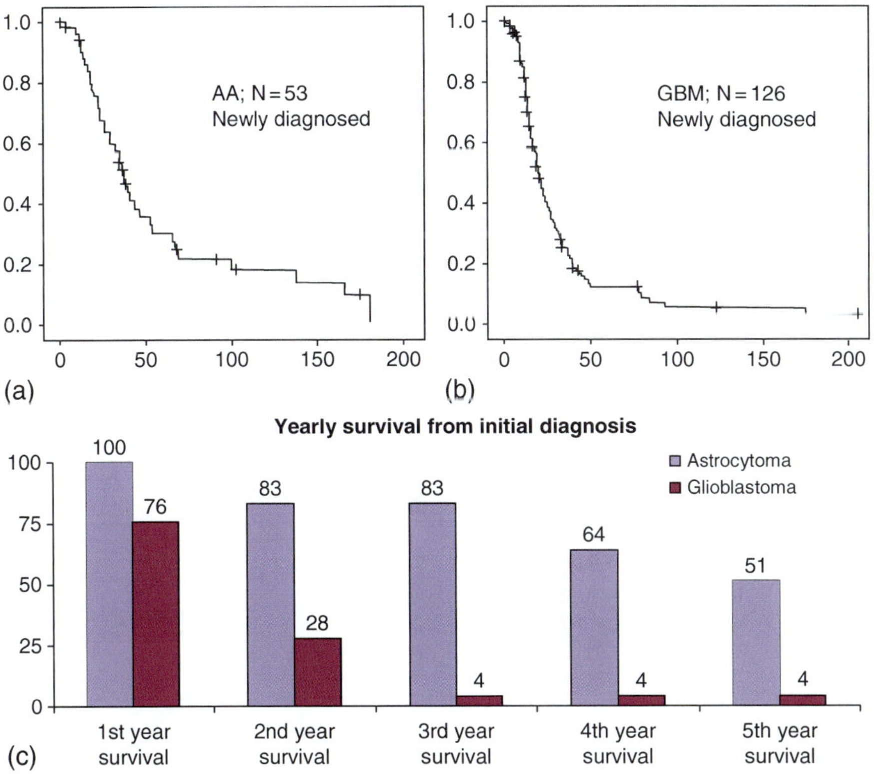

Fig. 4.177 The cumulative survival (Kaplan–Meier plot) of overall survival and survival of oncothermia is shown for AA ($n = 53$) and GBM ($n = 126$), on panels (**a**) and (**b**), respectively. The yearly survival rate is shown in panel (**c**)

4.7.8 Oncothermia for Heavily Pretreated and Relapsed Brain Gliomas

Another Phase II trial ($n = 12$) was carried out on very advanced glioblastoma cases [1238]. All patients were heavily pretreated (chemotherapy [temozolomide-based], radiotherapy) and relapsed. The median duration of response in this was 10 m (4–32), and the survival from the first oncothermia treatment was 9 m, with a 25% survival rate. One complete remission and two partial remissions were achieved in local clinical response.

4.7.9 Study of Metastatic Brain Tumors

Various metastatic brain tumors were treated by oncothermia [1199] ($n = 15$). Despite the various primary origins of the tumors, the parametric evaluation of overall survival shows significant curative benefit: the slope of survival modifying from 1.3 to 0.65 (gain: 50%); see Fig. 4.178.

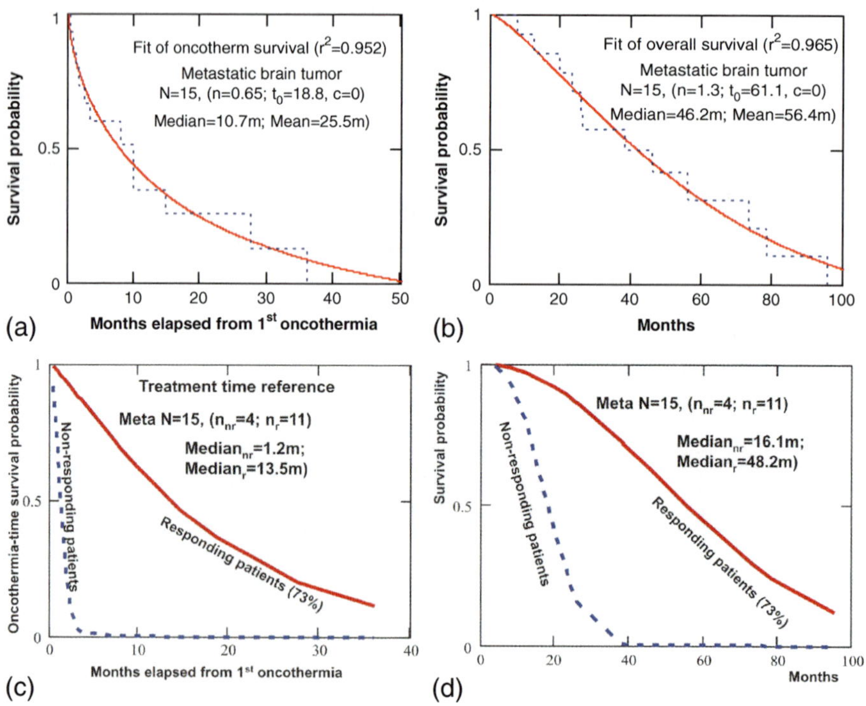

Fig. 4.178 Results of metastatic brain tumors ($n = 15$). The ratio of responding patients was 73% (The subscripts nr and r denote the non-responding and responding patient's data, respectively)

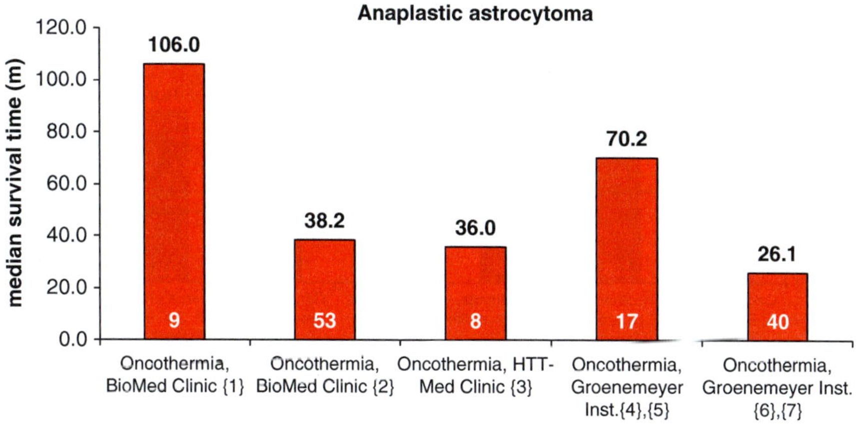

Fig. 4.179 Results of median survival time for advanced anaplastic astrocytoma in different, independent clinics using the same oncothermia protocol. ({1}=[1237]; {2}=[1233]; {3}=[1234]; {4}=[1239]; {5}=[1240]; {6}=[1232]; {7}=[1232].) (The number of patients involved in the study is shown at the bottom of the columns)

4.7.10 Comparison of Oncothermia Brain Studies

A few other open-label, single-arm, monocentric, retrospective, ITT frame oncothermia studies have been published at professional conferences [1239–1243].

Comparison of the median survivals for anaplastic astrocytoma and for GBM is shown in Figs. 4.179, 4.180, respectively.

According to the RTOG classifications [1223], we divided the patients into two groups: age under- and over-50 years. In this division oncothermia is also better, (see Fig. 4.181). The method shows successful applications in pediatric cases as well [1245].

The results are pretty coherently above the statistical values of the large databases SEER [47] and the gold-standard radiotherapy (RT) and RT+PCV [1244]. The results of oncothermia show advantages in comparison with the recent publications on Temozolomide [1224, 1225], too.

The first-year survival rates compared to SEER [47] and EUROCARE [53] databases as well as to the recent chemotherapy of Temozolomide shows also significant advantages (more than 25% increase) of oncothermia (see Fig. 4.182).

No serious side effects were observed [1232]. Patients tolerated the treatments well during the whole treatment period. Most of the patients were well relaxed, some even fell asleep during the treatment. In all studies patients reported better QoL after the treatments, but this information was not objectively studied.

Results indicate well the feasibility and the benefit of the oncothermia showing a valid treatment potential and safe application. Oncothermia is a possible way of escaping from the present impasse situation to treat brain glioms successfully. The question posed by the JAMA Editorial "Where to go from here?" [1246] could be answered using oncothermia.

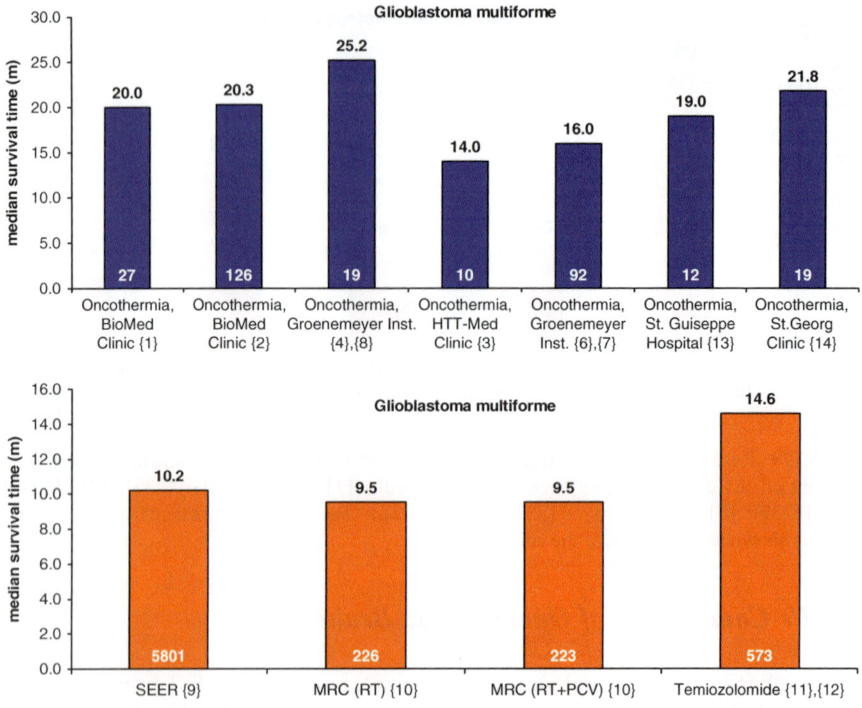

Fig. 4.180 Results of median survival time for advanced glioblastoma multiform in different, independent clinics using the same oncothermia protocol. Data from large database (SEER USA) and from other treatment results are shown for comparison. ({1}=[1237]; {2}=[1233]; {3}=[1234]; {4}=[1239]; {6}=[1232]; {7}=[1232]; {8}=[1243]; {9}=[47]; {10}=[1244]; {11}=[1224]; {12}=[1225]; {13}=[1238]; {14}=[1236] (The number of patients involved in the study is shown at the bottom of the columns)

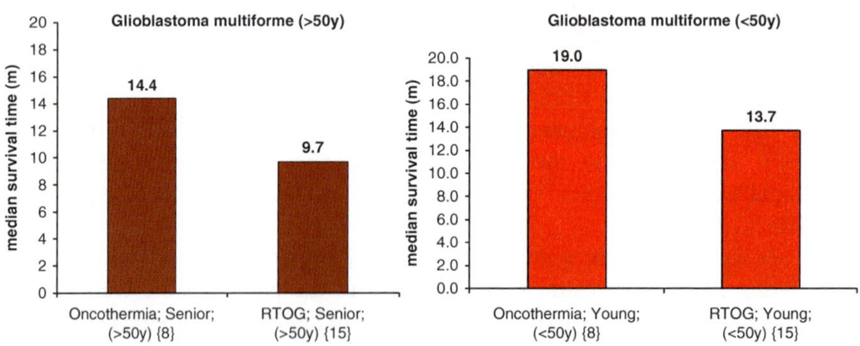

Fig. 4.181 Data from the Radiation Therapy Oncology Group (RTOG) compared to the data of treatments made by oncothermia in age groups under and above 50 years. ({8}=[1243]; {15}=[1223])

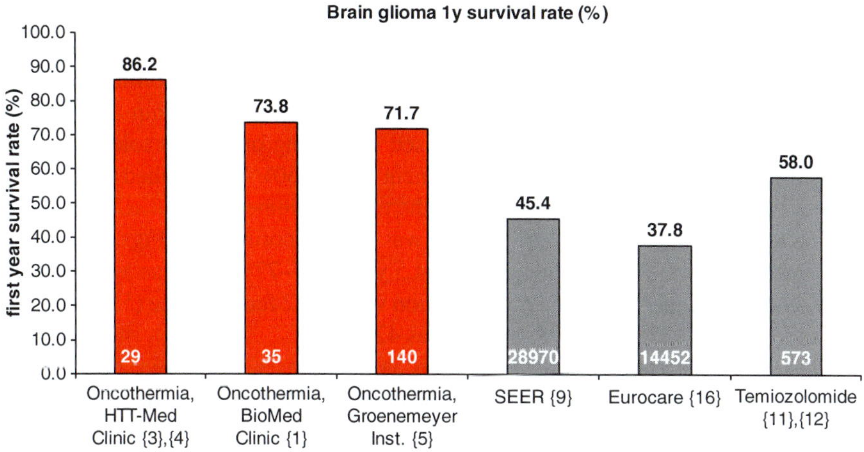

Fig. 4.182 First-year survival ratio (%) for advanced brain glioms treated in different, independent clinics using the same oncothermia protocol. Data from large databases (SEER USA; Eurocare, EU) and from recent temozolomide treatment results are shown for comparison. ({1}=[1237]; {3}=[1234]; {4}=[1239]; {5}=[1240]; {9}=[47]; {11}=[1224]; {12}=[1225]; {16}=[53]) (The number of patients involved in the study is shown at the bottom of the columns)

4.8 Pancreas Studies

4.8.1 Pancreas Efficacy Study I

The location of the study was the Peterfy Hospital in Budapest. The investigator was Dr. Magyar T. This Phase II study was designed to be observable, open label, single arm. The involved patients were analyzed according to an ITT schedule. Recruiting time was from April 1997 to August 2002, all together 64 months. Patients ($n = 26$) were dominantly in late/advanced stages, at which point the traditional therapies had become unsuccessful. The primary check of the efficacy of a curative method in such a lethal kind of disease is the survival time. The primary endpoints of the present study therefore were the overall survival including the oncothermia treatment time and the survival time from the first oncothermia treatment (oncothermia treatment survival time). The date of death (or of being alive) were checked by the Hungarian National Death Register, so actual and accurate data were collected. The latest check of deaths was made on 31st December, 2003.

Inclusion criteria were: (1) Inoperable or sub-totally resected or recurrent primary pancreas tumor, (2) progression after surgery and/or chemotherapy, (3) KPS $\geq$ 30%, and the inclusion was irrespective of the localization of the lesion in the pancreas. Most of the patients failed to respond to any of the applied conventional therapies. Exclusion criteria were only the well-known contraindications of the oncothermia method (metallic implants or replacements in the treated area, missing heat-sense in the treated area, pacemaker or other field-sensitive implants in the patient).

The calculated average equivalent temperature in the tumors was above 43°C for more than 90% of the treatment time. The targeted area was treated by the properly covering applicator system. Oncothermia was performed in two/three sessions per week. Treatment time and power range per session were 60 min and 150 W. The power was gradually and linearly raised depending on patient tolerance. The applied average energy was 300 kJ/treatment (250–450). The applied applicators were 3.1 dm^2 and 7.1 dm^2, depending on the tumor volume.

The age distribution of $n = 26$ patients was near to normal (see Figs. 4.183); no outlier was present. The median age was 64.5 years (37–77), the mean age was 62.5 years (Std.err= 1.99). The gender distribution was 14/12 female/male (53.8/46.2%). The ratio of the elderly (>68 y) patients was 42.3%.

Most of the patients (23, 88.5%) had distant metastases. They were heavily pretreated, everybody received at least one chemotherapy and most of them underwent surgery (see Fig. 4.184).

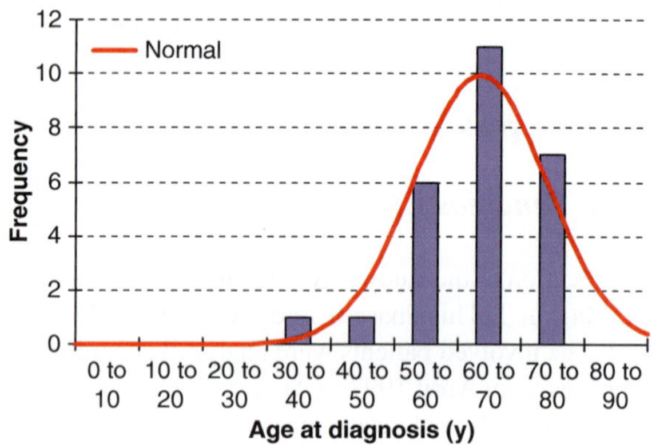

Fig. 4.183 Age distribution at diagnosis

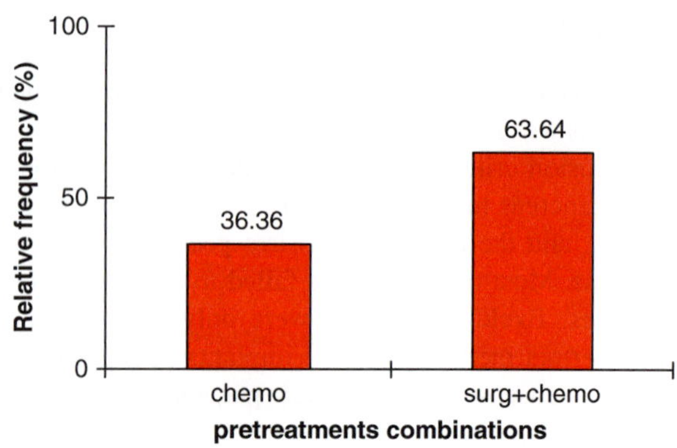

Fig. 4.184 The pre-treatment distribution in the patient population

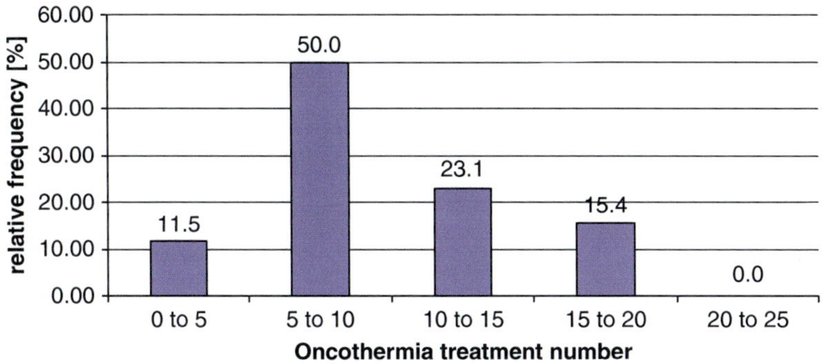

Fig. 4.185 Oncothermia treatment number is dominantly in the 5–10 interval

The actual staging was made at the first diagnosis: 23, [88.5%] were in advanced [WHO III or IV] stages, and at the first oncothermia treatment 100% were in advanced stages, 19 (73.1%) were in the worst stage.

The median of the elapsed time from the first diagnosis to the first oncothermia was 4.1 m (0.8–75), while its mean was 8.6 m (st.err.3.0). The elapsed time ratio to the overall survival was more than 35% [median 37.3%, (5.9–86), mean 44.3 (st.err.4.2)].

The oncothermia treatment was provided 2–3-times a week, the treatment number was on average 9.0 (st.err.0.86) and its median 6 (3–16) (see Fig. 4.185).

The KM plots of the overall survival [median 12.0 m, (2.3–115.5); mean 17.5 m, (st.err.4.4)] and the survival from the first oncothermia treatment [median 6.3 m, (0.7–40.4); mean 8.9 m, (st.err.1.9)] are shown in Fig. 4.186. For elderly patients the survival plots were not different ($p=0.41$ and $p=0.61$, for overall survival and oncothermia survival, respectively).

The parametric decomposition of the survival plots gives for responding patients 58%, the decomposed distributions are not significantly different.

The survival was significantly different for patients without or with metastases in their overall survival ($p=0.039$), but was not significant in terms of their oncothermia survival ($p=0.20$), see Fig. 4.187.

We studied the effect of the experience of the treating medical personnel on the data before and after the median time of the study. No observable effect was registered: overall survival ($p=0.86$) and oncothermia survival ($p=0.69$). The elapsed time to the first oncothermia from the first diagnosis is lower in the late experience, (medians are 5.17 m [0.8–75.1] and 3.27 m [0.9–35.3], in the early and late experience period, respectively) but the difference is not significant either ($p=0.29$).

4.8.2 Pancreas Efficacy Study II (HTT)

The trial (Phase II) was identically designed to the previously described one. It was performed at the day-clinic HTT-MED (HTT) [Investigator: Dr. A. Dani]. The

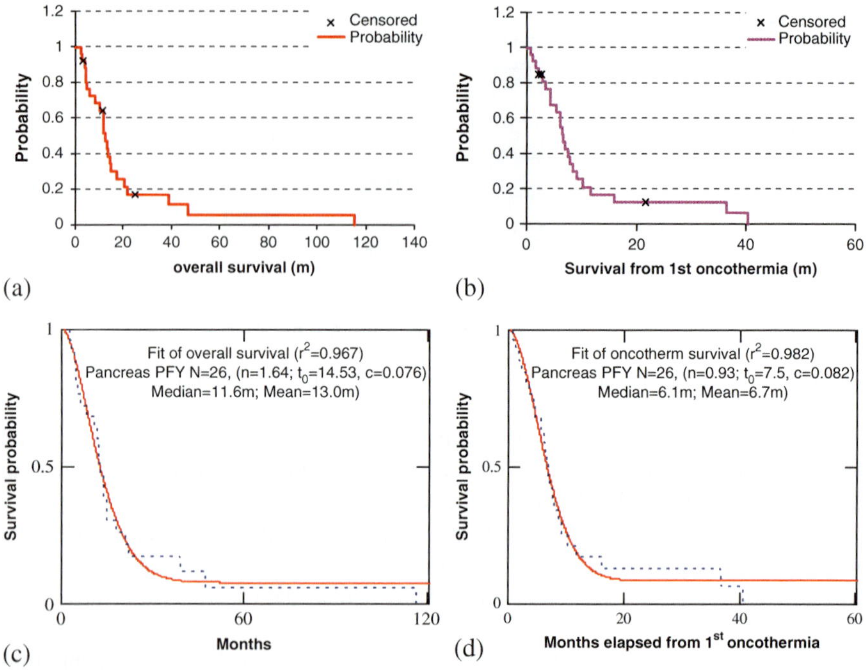

Fig. 4.186 The overall survival (**a**) and oncothermia survival (**b**) Kaplan–Meier plots. The Weibull parametric fit and its results are shown on panel (**c**) and (**d**)

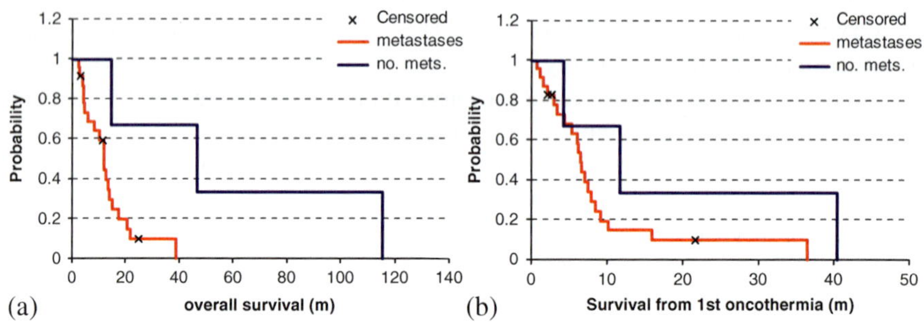

Fig. 4.187 Overall survival (**a**) and oncothermia survival (**b**) depend on the preliminary surgery

age distribution of $n = 73$ patients was near to normal (see Fig. 4.188); no outlier was present. The median age was 58 years (24–79), the mean age was 59.1 years (Std.err= 1.3). The gender distribution was 33/40 female/male (45.2/54.8%). The ratio of the elderly (>68 year) patients was 26.0%.

The actual staging was made at the first diagnosis [45, 61.6% were in advanced (WHO III or IV stages)] and at the first oncothermia treatment they were in a more advanced status.

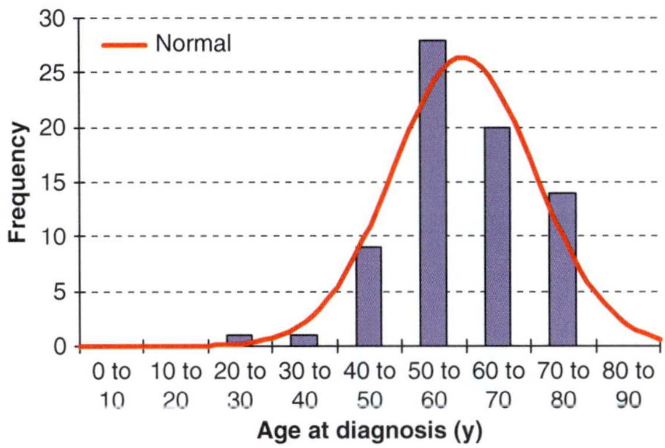

Fig. 4.188 Age distribution of patients in HTT trial Most of the patients (54, 74.0%) had distant metastases, (one, two, and three metastases were observed for 43 (58.9%), 10 (13.7%), and 1 (1.4%) patients, respectively). They were heavily pretreated, most (93.4%) had undergone surgery and subsequent radiation and/or chemotherapies, see Figs. 4.189 and 4.190).

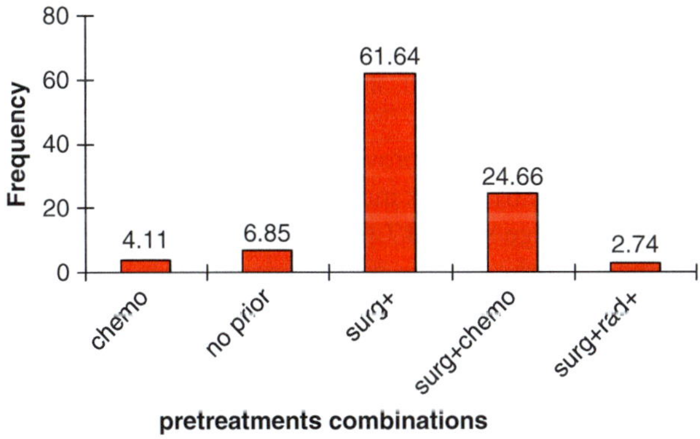

Fig. 4.189 Pre-treatment combinations

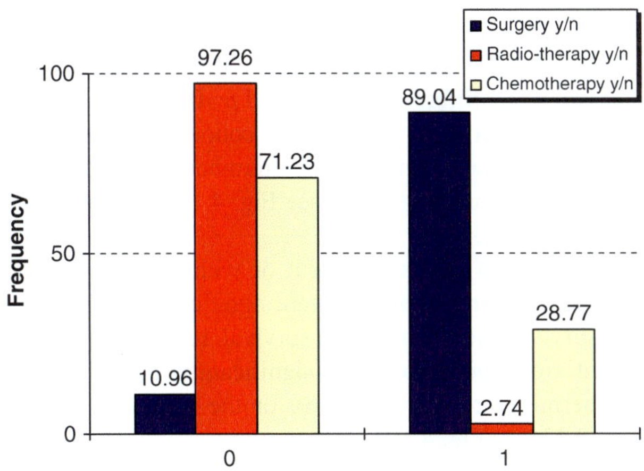

Fig. 4.190 Pre-treatment distribution

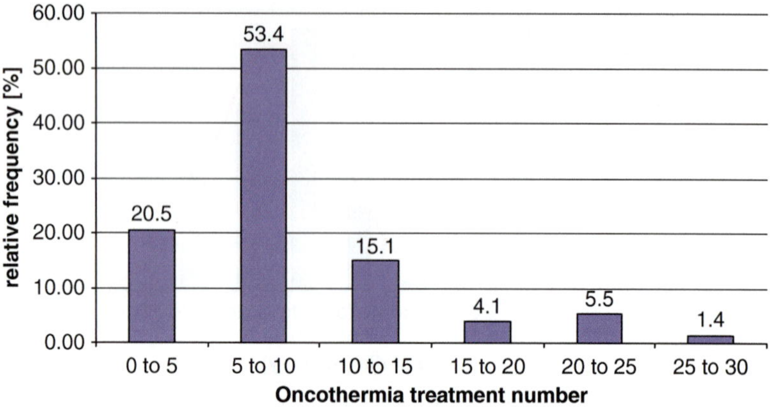

Fig. 4.191 Number of oncothermia treatments

The median of the elapsed time from the first diagnosis to the first oncothermia was 3.3 m (0.3–85.7), while its mean was 6.6 m (st.err.1.3). The median of the elapsed time ratio to the overall survival was 37.1% (5.1–96.0), mean 41.2 (st.err.3.4).

The oncothermia treatment was provided twice a week, the treatment number was on average 8.0 (st.err.0.6) and its median 6 (3–26) (see Fig. 4.191). The equivalent temperature on average was 50.7 (sd.err.0.6), median 51 (43–59). (Note that the equivalent temperature is not the real temperature. It is the calculated value from the actual energy absorption and the impedance, meaning the actual destruction rate, which is as high as it would be in a purely temperature-oriented case.) The applied treatment time on average was 67.2 min, (st.err.1.8) and its median was 60 (45–120).

The KM plots of the overall survival (overall survival) [median 12.7 m, (1.2–94.5); mean 19.2 m, (st.err.2.1)] and the survival from the first oncothermia treatment (oncothermia survival) [median 4.7 m, (0.3–49.2); mean 12.6 m, (st.err.1.7)] are shown in Fig. 4.192. For elderly patients neither the overall survival nor the oncothermia survival were different ($p \sim 0.23$ and $p \sim 0.42$, respectively).

The parametric decomposition divides the cohort significantly into two subgroups (responders and nonresponders). The number of the responders was only slightly more than nonresponders (58%) (see Fig. 4.193); corresponding well with the results in the PFY study.

The differences between patients without or with metastases in terms of their overall survival and oncothermia survival were significantly different ($p = 0.016$ and 0.004 for overall survival and oncothermia survival, respectively), see Fig. 4.194.

The number of treatments does not significantly influence overall survival ($p = 0.24$), oncothermia survival ($p = 0.16$), or the follow-up time after the last oncothermia ($p = 0.23$, see Fig. 4.195.

Interestingly, the in any case necessary surgical pretreatment wasn't significantly important for the longer survival of these patients, neither for overall survival

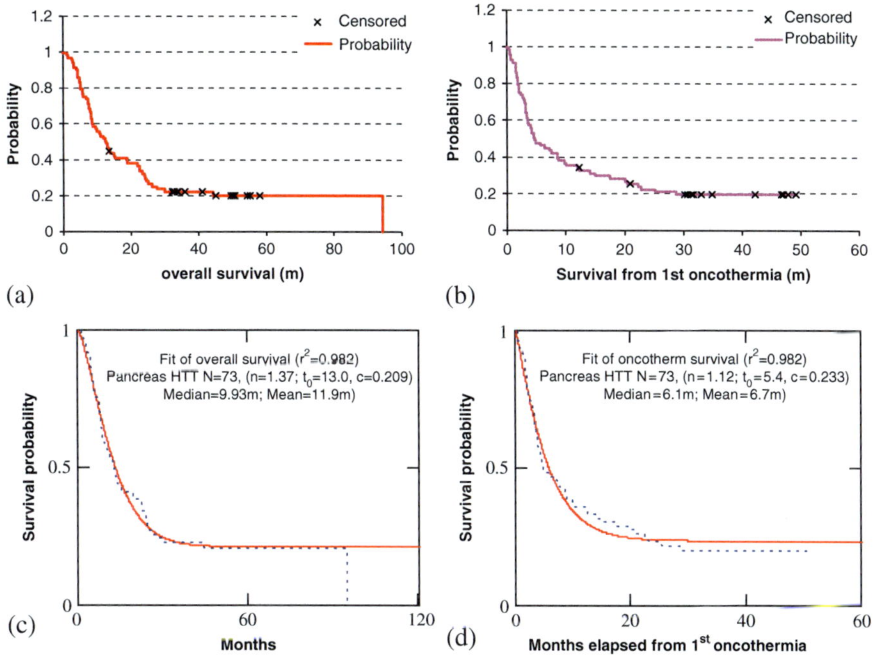

Fig. 4.192 Overall survival (**a**) and oncothermia survival (**b**) of HTT study, and their parametric fits, (**c**) and (**d**)

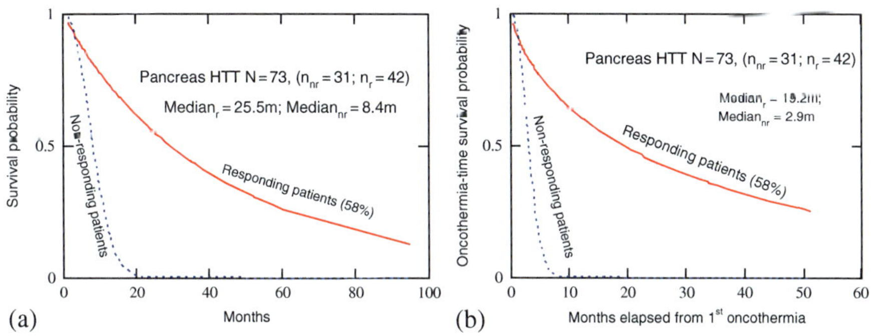

Fig. 4.193 Decomposition curves from parametric fits: overall survival (**a**) and oncothermia survival (**b**)

($p = 0.84$) nor oncothermia survival ($p = 0.87$) (see Fig. 4.196). This was probably because the surgery in these patients was not complete, the tumor was only partially resected, or the surgery was only for palliation.

We studied the effect of the experience of the treating medical personnel on the data before and after the median time of the study. There is some difference (not significant) in the overall survival, oncothermia survival, and elapsed time to the first oncothermia ($p = 0.15$, 0.077, and 0.52, respectively) (see Fig. 4.197).

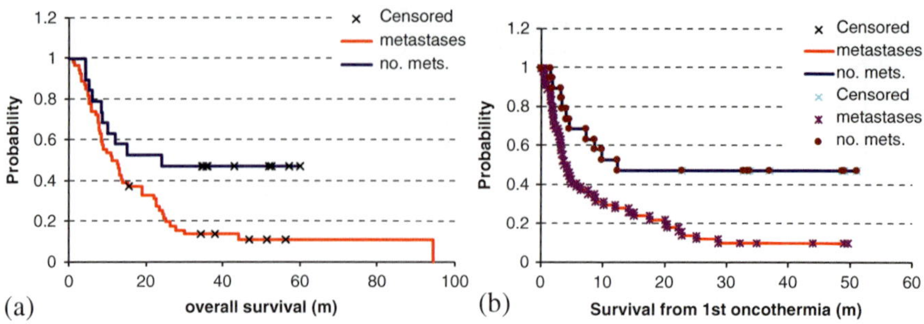

Fig. 4.194 Dependence of the survival on the number of metastases [(**a**) overall and (**b**) oncothermia survival, respectively]

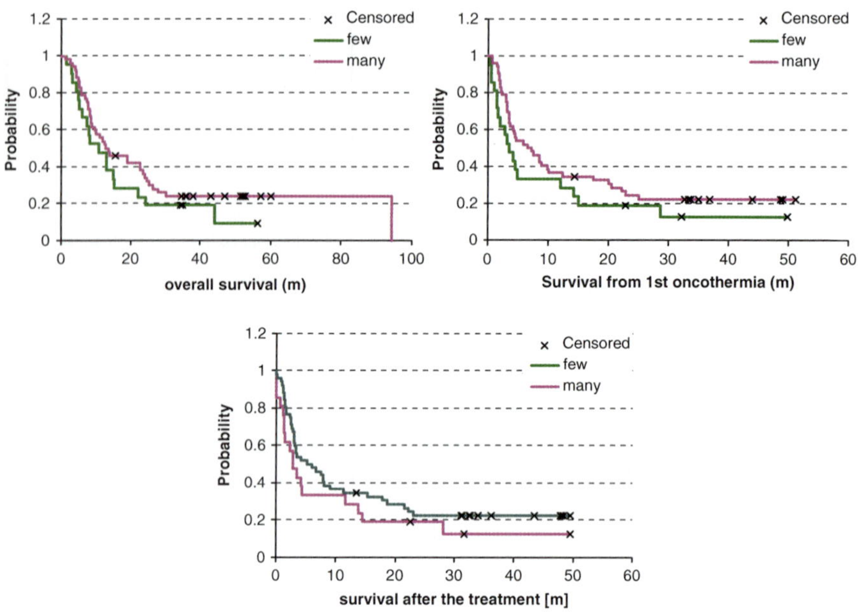

Fig. 4.195 The typical survival times do not depend significantly on the number of treatments (few = below median, many = above median number of treatments)

4.8.3 Additional Historical Control to HTT Pancreas Study

As an additional check a historical control ($n = 34$) Was given as a comparison to the HTT data. The median overall survival of the control was 6.5 m (1–31), and the mean survival was 8.7 m (St.err.1.29), while for the compared HTT ($n = 73$), median 12.7 m (1.2–94.5), mean 19.6 (std, err.2.1). This control arm is equivalent to the published data from the Hungarian Pancreas Center [1247], whose published mean is 8.3 m (Chemotherapy: Gemzar+5-FU+Leukovorin). The parametric fit of the control shows a unified cohort (see Fig. 4.198).

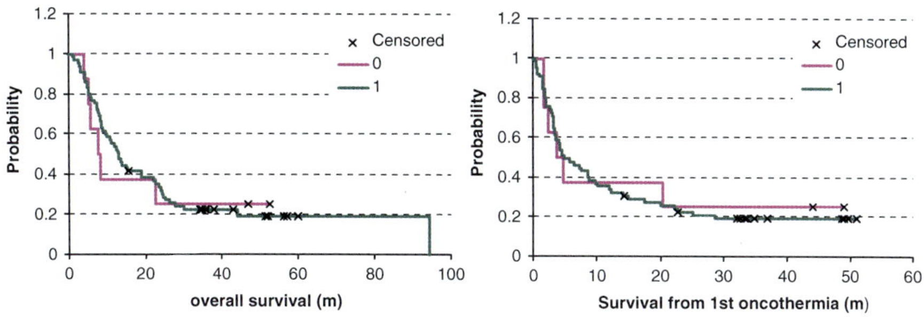

Fig. 4.196 The survivals' dependence on the surgery [(0) – no; (1) – yes]

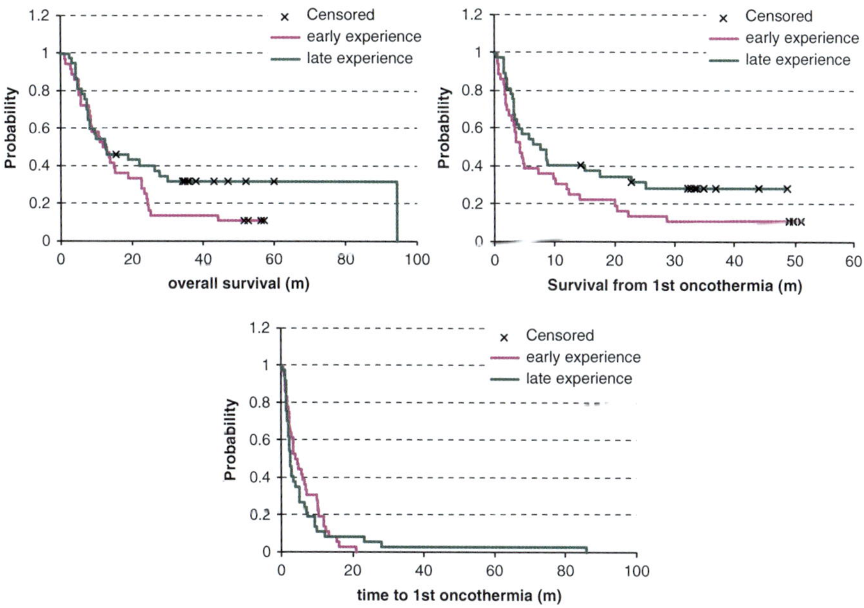

Fig. 4.197 The experience of the treating personnel did not modify the results significantly ("early experience" = the treatment was started earlier than the median time of the study; "late experience" = the treatment was started later than the median time of the study)

The comparison of the KM survival curves of the control and the active arms demonstrates the cogently significant difference ($p < 10^{-4}$) (see Fig. 4.199).

4.8.4 Comparison of Pancreas Efficacy Studies I and II

Through the comparison of the two Hungarian studies [1248], the parallels enforce the validity of the data.

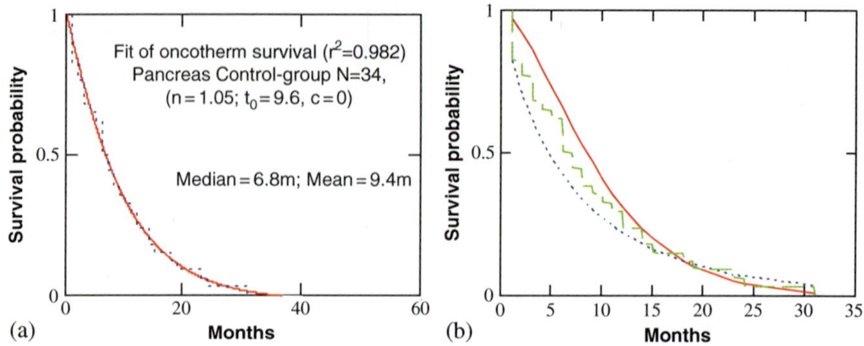

Fig. 4.198 The fit is simple Weibull (**a**), its decomposition [equal additions of the subcurves (**b**) is inside the confidence interval of the Kaplan–Meier fit

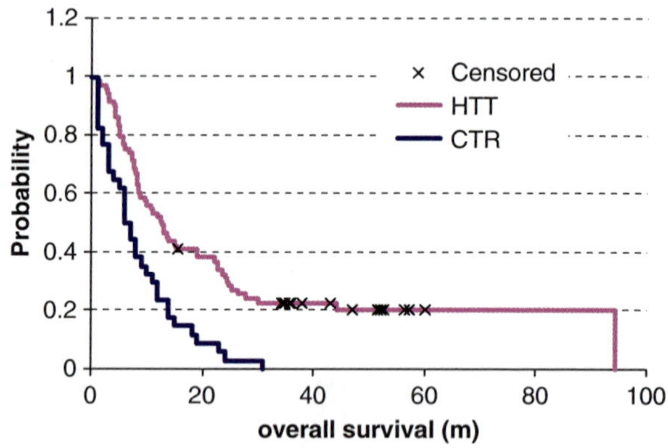

Fig. 4.199 The comparison of the Kaplan–Meier survival curves of the results from HTT and the historical control, collected for HTT by the same treating physician who works at HTT and in St. Borbala Hospital

The age distribution of in total $n=99$ patients was near to normal (see Figs. 4.200); and no outlier were present. The median age was 60 years (24–79), the mean age was 60 years (Std.err= 1.1). In the spectrum of the PFY a shift to the elderly patients was present (see Fig. 4.201). The gender distribution was 47/52 female/male (47/53%), and no significant difference could be measured between the locations (see Fig. 4.202). The PFY/HTT patients' ratio is 61/197 (24/76%).

74 and 88% of the patients had distant metastases in the HTT and PFY groups, respectively (see Fig. 4.203). Patients were heavily pretreated (see Fig. 4.204), in PFY chemotherapy, in HTT surgery was the most frequent modality.

The elapsed time to first oncothermia from the first diagnosis was identical ($p=0.69$) in the two locations, see Fig. 4.205.

The oncothermia treatment was provided twice a week, the treatment number on average was more in PFY than in HTT procedures (see Fig. 4.206).

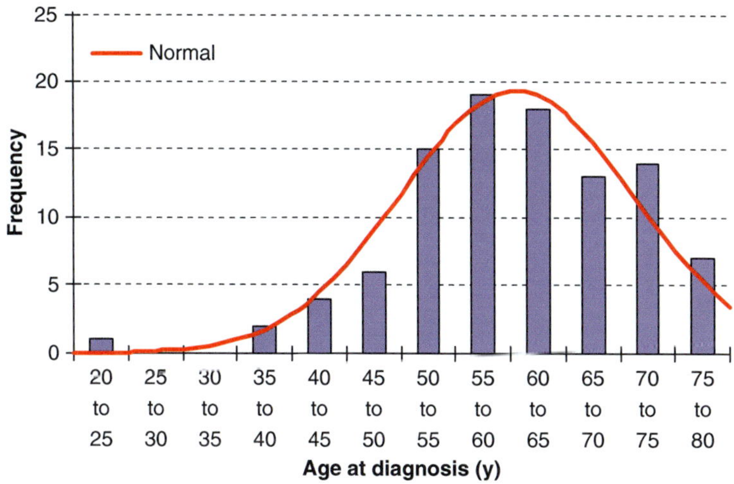

Fig. 4.200 Age distribution of lung tumor patients ($n = 99$)

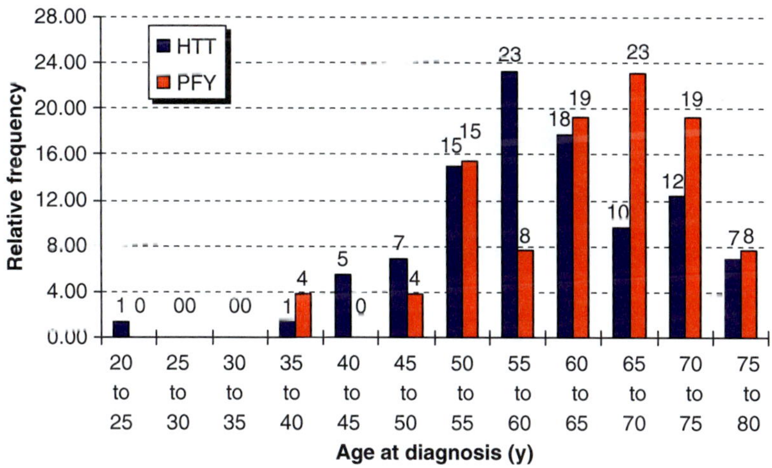

Fig. 4.201 The age distribution differences in the given clinics

The overall survival (OS) and the survival from the first oncothermia treatment (OSO) are shown in Figs. 4.207, 4.208. Neither of the measured parameters differed from each other ($p = 0.38$ and 0.39, respectively).

Survival after the treatment was not different in the two location ($p = 0.34$, see Fig. 4.209).

Results show the identical survival parameters in the two independent places. The yearly survival rate is also not significantly different (see Fig. 4.210)

The results could be well compared to the available SEER [47] and Eurocare-3 [53] data. The comparison of the yearly survival rate is shown in Fig. 4.211.

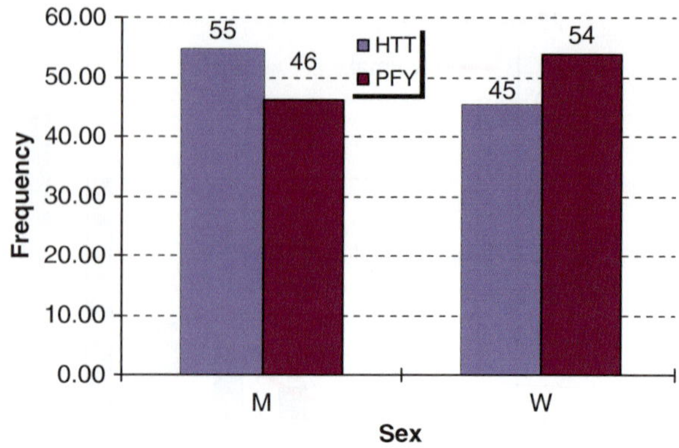

Fig. 4.202 The gender distribution in the given clinics

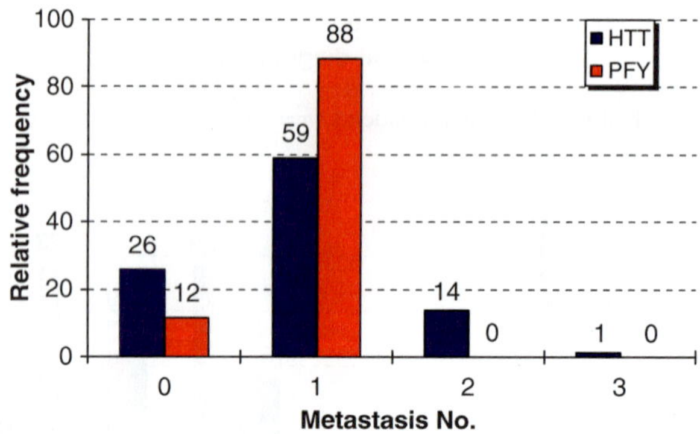

Fig. 4.203 Number of metastases of the patients involved in the study

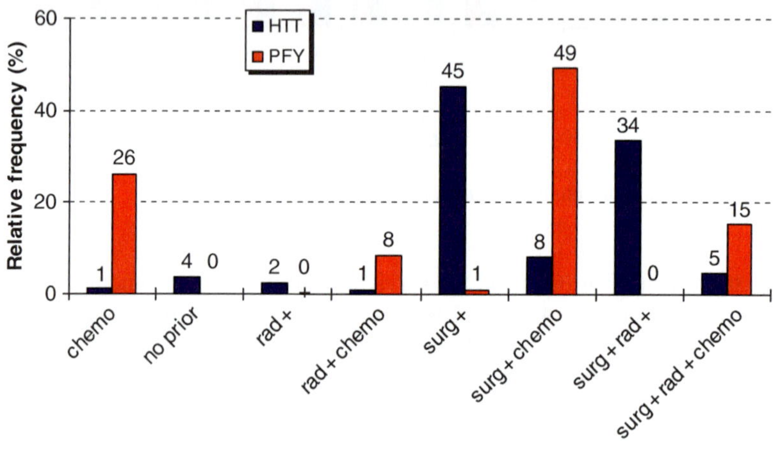

Fig. 4.204 Pre-treatment distribution

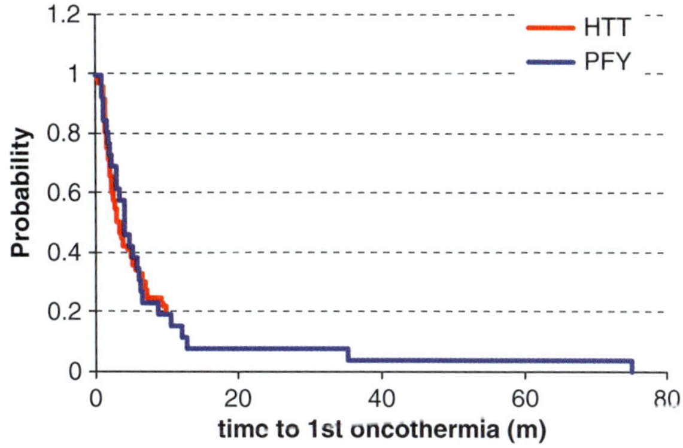

Fig. 4.205 The distribution of the elapsed time to the first oncothermia treatment

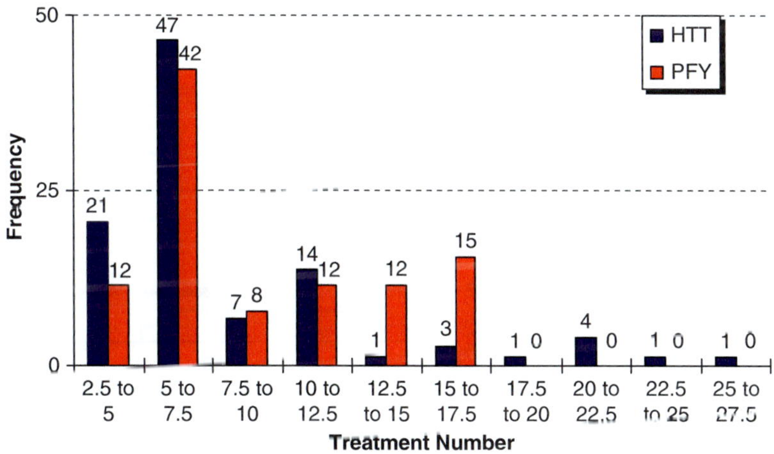

Fig. 4.206 The number of treatments for the patients in the study

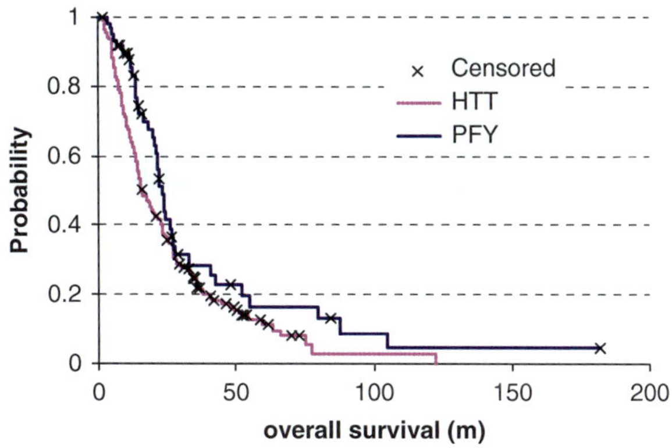

Fig. 4.207 The overall survival comparison of the studies

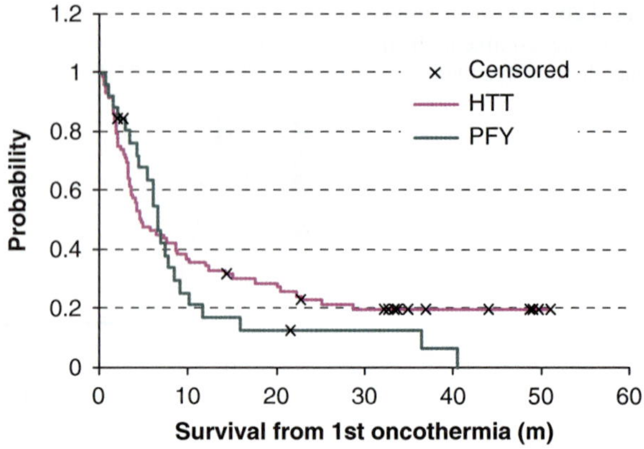

Fig. 4.208 The oncothermia survival comparison of the studies. No significant difference could be observed

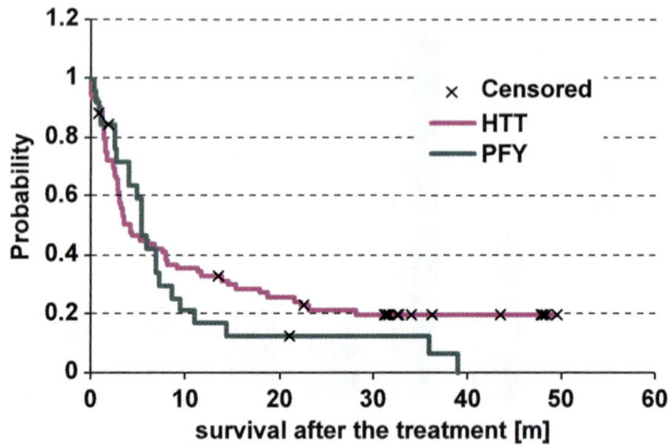

Fig. 4.209 The follow-up time does not differ in the two studies either

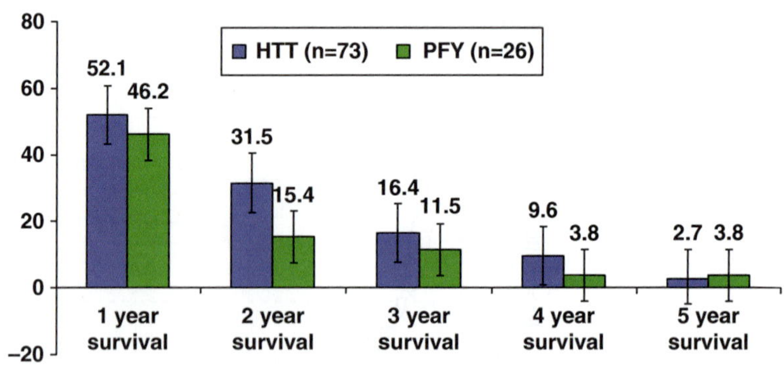

Fig. 4.210 The yearly survivals correspond well in the two studies

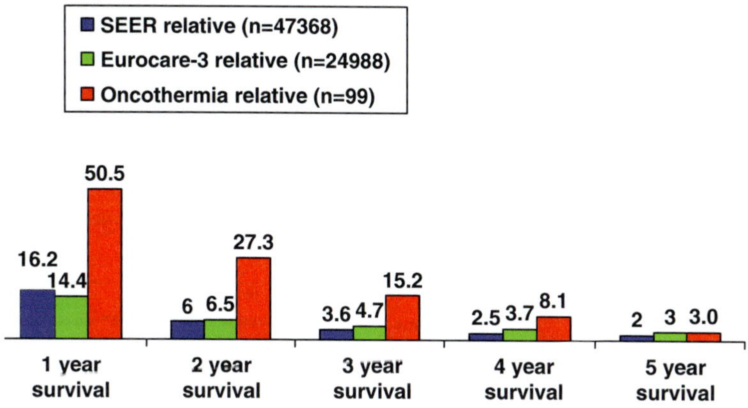

Fig. 4.211 The comparison of the results with SEER and Eurocare-3 data in first 5 years survival rate (%)

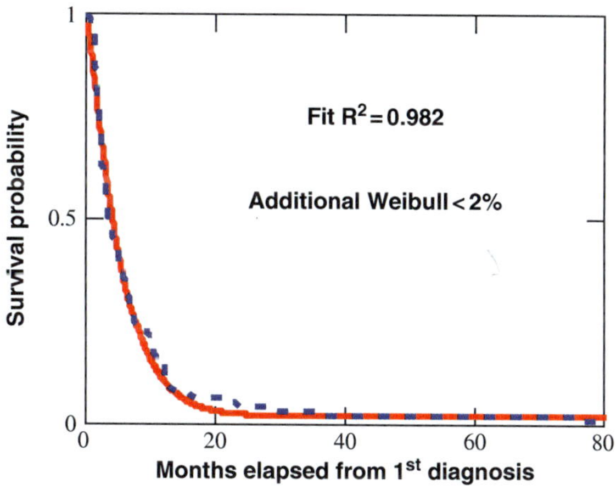

Fig. 4.212 The proper cohort construction is supported by the proper fit of a single Weibull on the elapsed time from first diagnosis to first oncothermia. (The additional Weibull correction is only 2%)

The gain in the first few years is obvious, while the difference gradually vanishes approaching the 5th year. The reason is the difference in the treated patients. When the patient has a long survival, his/her oncothermia treatment starts only at the end of the available conventional treatments; the patient receives oncothermia only in a small fraction of his/her survival time, therefore the survival time mostly does not depend on the oncothermia applied at the end. While in the case of the short survivals a considerable lifetime depends on the oncothermia application.

The parametric evaluation of the united studies (meta-analysis) shows an accurate fit of the Weibull function for the elapsed time from the first diagnosis to the first oncothermia (see Fig. 4.212); which indicates the homogeneity of the inclusion of patients in the study.

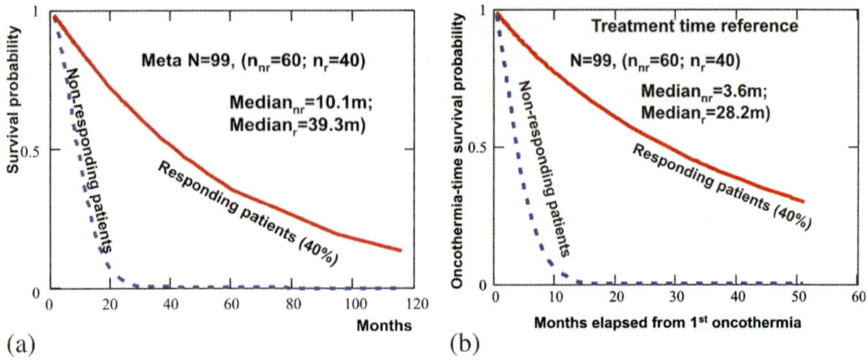

Fig. 4.213 The parametric division of the survival curves for overall survival (**a**) and for oncothermia survival (**b**)

The parametric division to find the group of responders shows only 40%, and significantly differs from the nonresponders (see Fig. 4.213).

4.8.5 Pancreas Efficacy Study III

Strong evidence for the oncothermia application was given by Clinic St. Georg (Bad Aibling, Germany) and published in two subsequent papers [1249, 1250]. Distribution of the $n = 30$ patients by their age is shown in Fig. 4.214. The aggressive, mostly even palliative inoperable patients were treated in the second line. The first-line treatment was chemotherapy (see Fig. 4.215). The oncothermia was concomitantly applied with Mitomycin-C (8 mg/m^2; 1–5 days) + 5-fluorouracil (5-FU, 500 mg/m^2) + Leukovorine (calcium folinate, 200 mg/m^2). Oncothermia was applied on treatment days 1, 3, 5, and 10 for 60 min each. The equivalent temperature of the targeted tumor tissue was 42–44°C over 90% of the treatment time. Oncothermia treatment was repeated every 4 weeks until next progression.

The local clinical response was high in the cases of stage IV patients (see Fig. 4.216).

The PD local response was observed in the cases, when only 1–2 oncothermia treatments were made (see Fig. 4.217).

The survival time from the first oncothermia treatment (median 7.5 m, mean 13.05 m) and its parametric separation for responders and nonresponders is shown in Fig. 4.218. (The overall survival time has no complete data set for its exact evaluation.)

Calculating the direct response (CR+PR) shows good correspondence with the parametric separation (see Fig. 4.219).

It is interesting to see the prognostic value of the CA-19-9 tumor marker. The division of patients into responders and nonresponders gives a little more positive information than the real clinical response, but its difference is by far not significant (see Fig. 4.220). The median of the two curves is only slightly different, but the longer survivals (counted to the mean values) have definite differences.

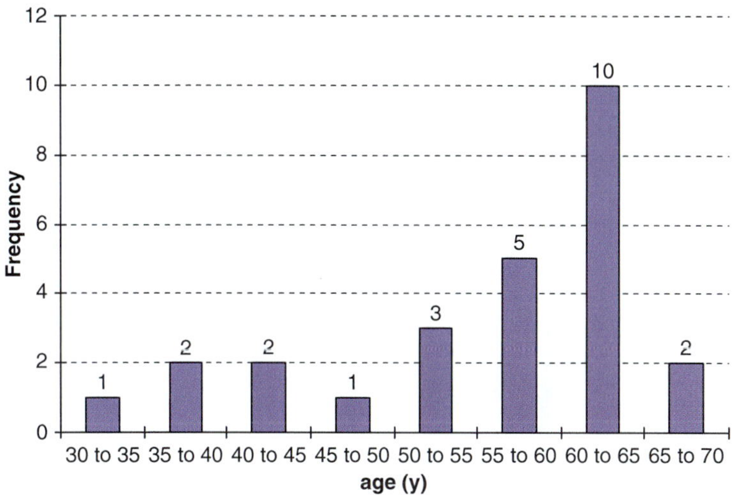

Fig. 4.214 The age distribution of the patients involved in the study

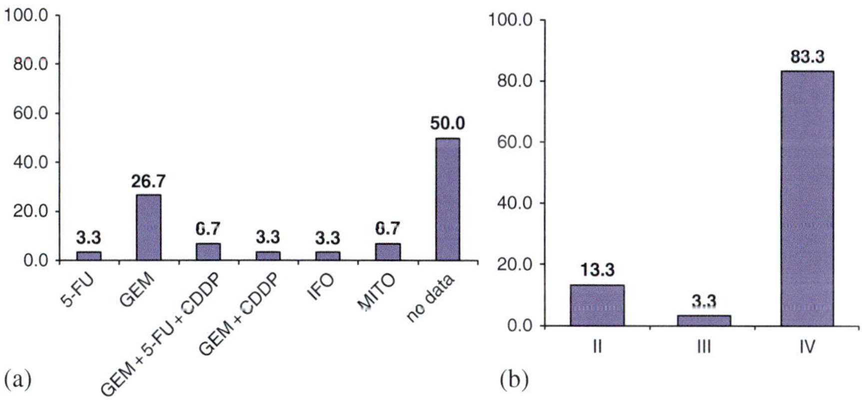

Fig. 4.215 The pre-treatment chemotherapies (**a**) and the stages at the first oncothermia treatment (**b**) of patients involved in the study

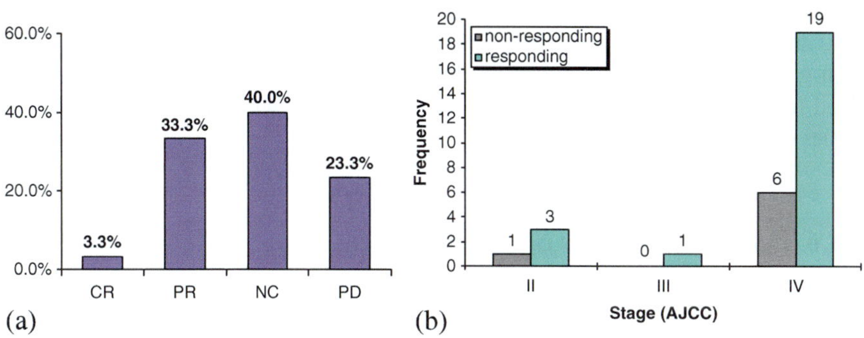

Fig. 4.216 Local clinical responses (**a**) and their distribution by stages ("responding"= CR+PR+NC; "nonresponding"=PD) of the patients involved in the study (**b**)

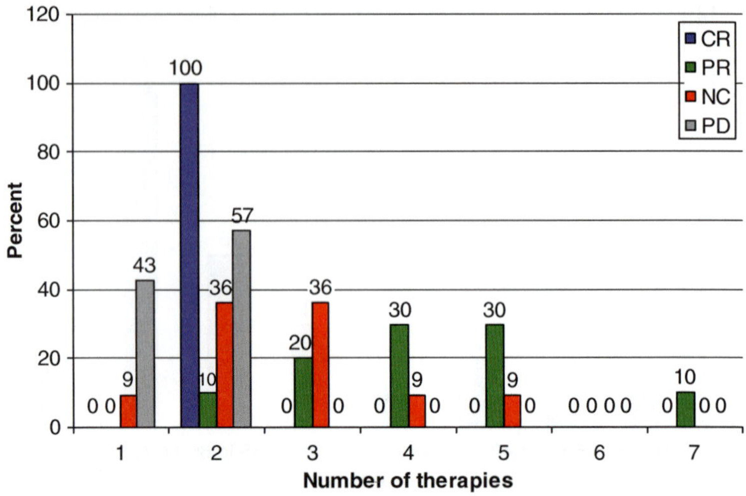

Fig. 4.217 Local response by the number of oncothermia treatments (Percentages of categories. PD occurred only at low treatment numbers)

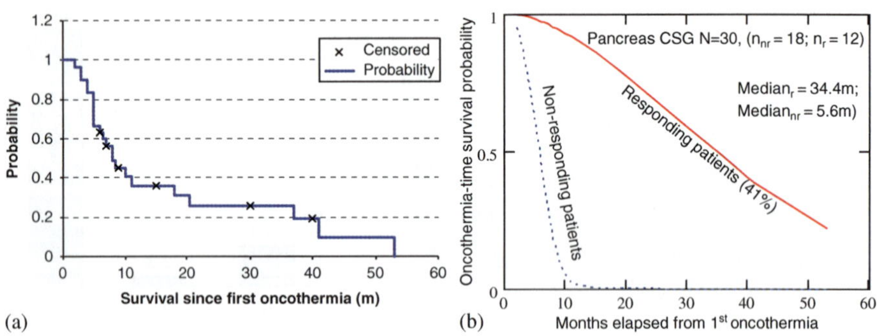

Fig. 4.218 Survival plot of oncothermia treatment

The usual toxicity counted by laboratory values is shown for hemoglobin (anemia) and for leukocytes by the response reactions, see Fig. 4.221. Interestingly the high-grade toxicity has a good response rate. This tendency is even more emphasized for patients with no-response to oncothermia, showing no change in normal values of thrombocytes, see Fig. 4.222.

4.8.6 Pancreas Efficacy Study IV

Another additional retrospective oncothermia trial ($n = 42$) was performed in the VeraMed Clinic (Investigator: Dr. M. Kalden) [1251]. The trial included heavily pre-treated, advanced-stage pancreas carcinoma cases, see Fig. 4.223.

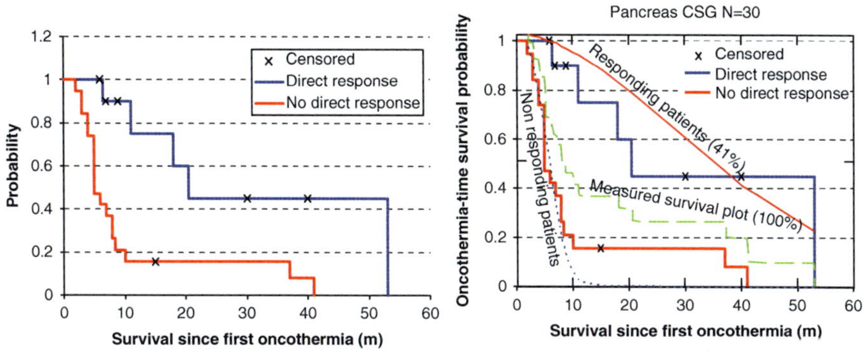

Fig. 4.219 Significant correspondence of the measured and calculated separation of the patient's survivals by their local response

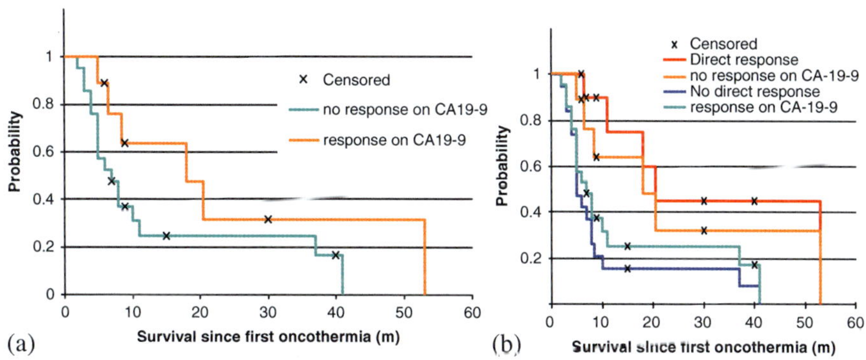

Fig. 4.220 The prognostic value of CA-19-9 tumor marker in the present pancreas study. The survival is significantly separated into responders and nonresponders on CA-19-9 (**a**). In comparison with the local clinical response separation, (**b**), the tumor marker is a little more optimistic, but not significantly

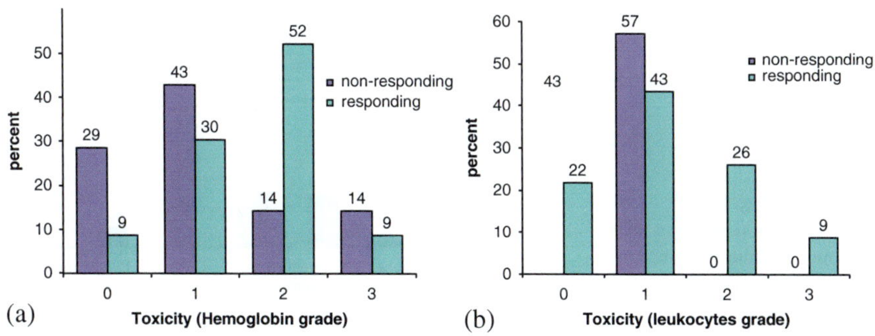

Fig. 4.221 The hemotoxicity of the patients by their responses; Hemoglobin (**a**) and Leukocytes (**b**)

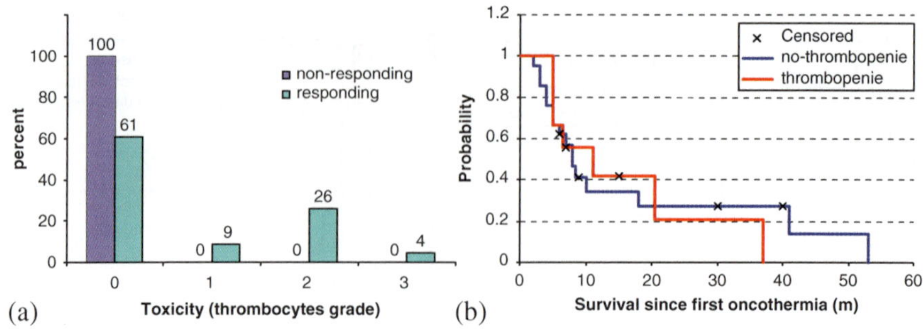

Fig. 4.222 The thrombocytes correlating well with responses. Interestingly, the "no thrombocyte toxicity" is only seen in patients with no response (**a**) but the survival has no significant difference (**b**)

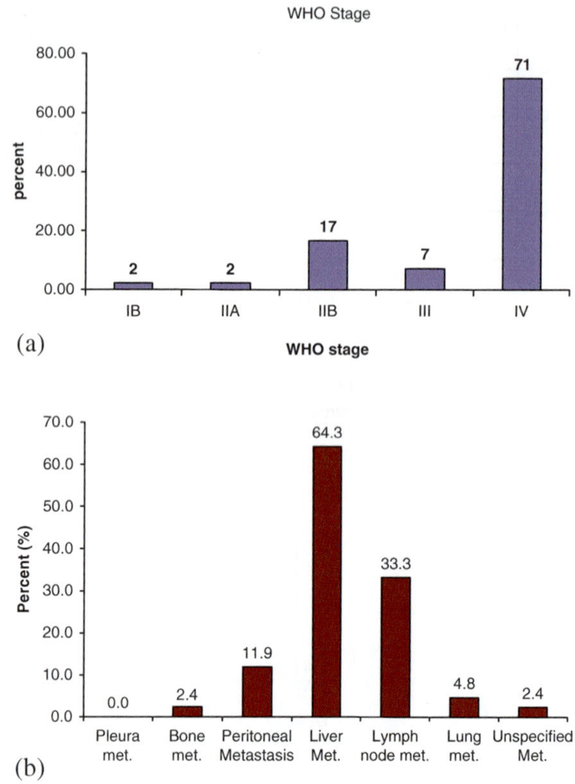

Fig. 4.223 The stage distribution of patients at the first oncothermia treatment (**a**) and the status of metastases (**b**)

The local clinical response had no complete remission, but all together it was over half of the patients (54.8%), see Fig. 4.224. The importance of the multiple oncothermia treatments in these cases is proven by the remission distribution.

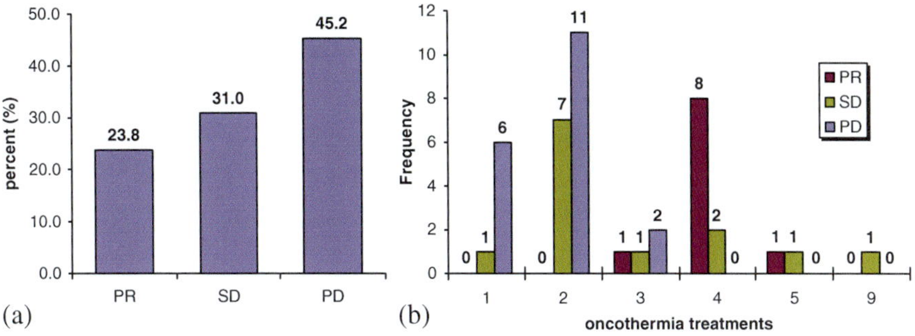

Fig. 4.224 The local clinical response of the patients (**a**), and their response by the number of oncothermia treatments (**b**)

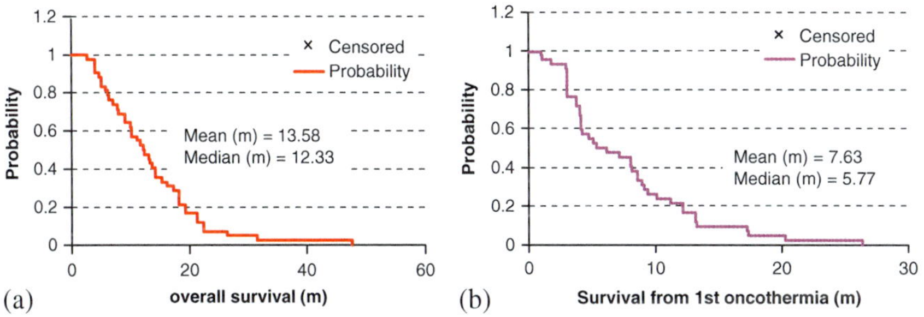

Fig. 4.225 Overall (**a**) and oncothermia (**b**) survivals of the additional retrospective study ($n = 42$)

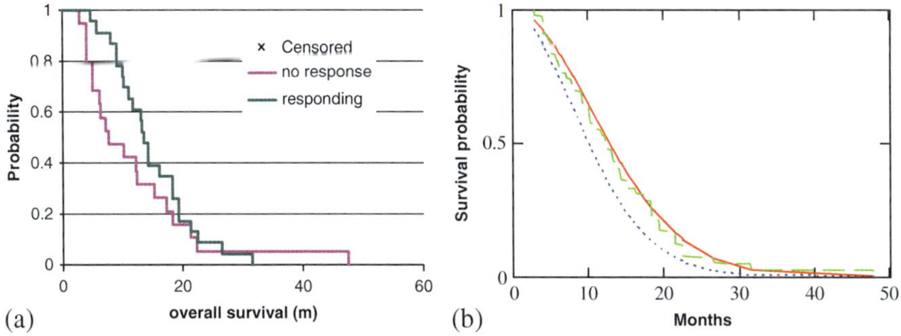

Fig. 4.226 The overall survival by local clinical response (**a**) and its parametric evaluation from the original survival curve (see Fig. 4.225) (the *dashed line* is the original, the *thin solid* and *dot lines* are the decomposed plots) (**b**)

Both the overall and oncothermia survivals support the previous studies (see Fig. 4.225). The overall survival is well distinguishable by the local remissions, but not significant ($p > 0.3$), see Fig. 4.226. The parametric evaluation has again good and significant correspondence with the measured data, see Fig. 4.227.

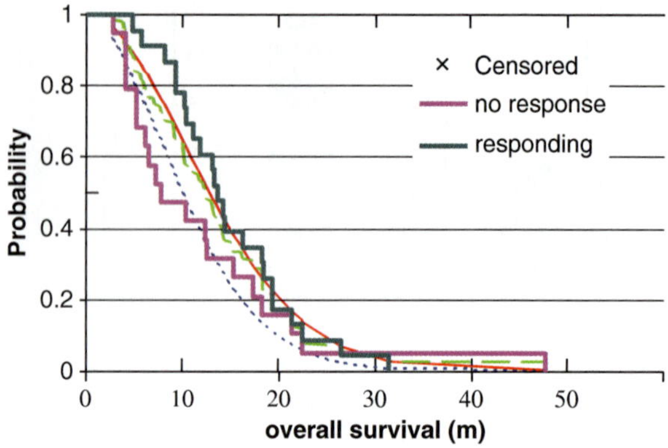

Fig. 4.227 Comparison of the plots of Fig. 4.225. The acceptable fit of the real and theoretical decomposition is demonstrated

4.8.7 Other Oncothermia Pancreas Studies and Their Comparison

Additional clinical results for oncothermia therapy for pancreas carcinoma could be cited. One of these studies was performed in the Praxis at Klinikum Nord, Nürnberg, Germany (referred to as PKN) ($n = 13$) [1252]. The investigators were Prof. Dr. H. Renner and I. Albrecht. They used trimodal therapy {RT+CHT+OTH}, [CHT=Gemzar]. The trial showed median survival 11.9 m (4.4–211.4), mean survival 29.2 m (std.Err.15.35). Its overall survival plot is shown in Fig. 4.228.

For clear evidence of the results let us compare the first-year survivals and the median survivals obtained in various clinics in comparison to the historical control and the large databases. The result is shown in Figs. 4.229, 4.230. The result convincingly demonstrates the significant difference between the oncothermia and the general retrospective data.

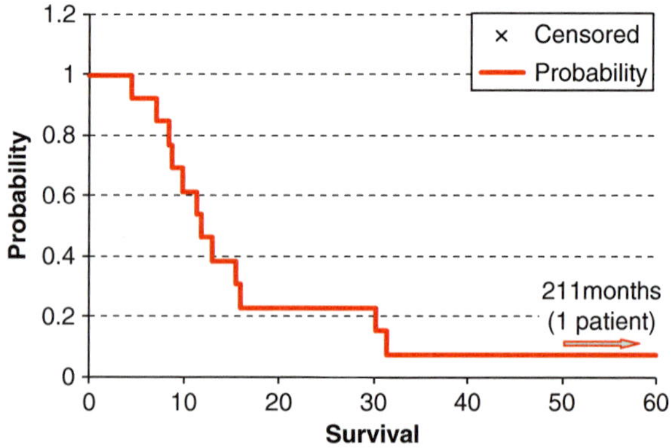

Fig. 4.228 Kaplan–Meier survival plot of trimodal therapy of pancreatic cancer

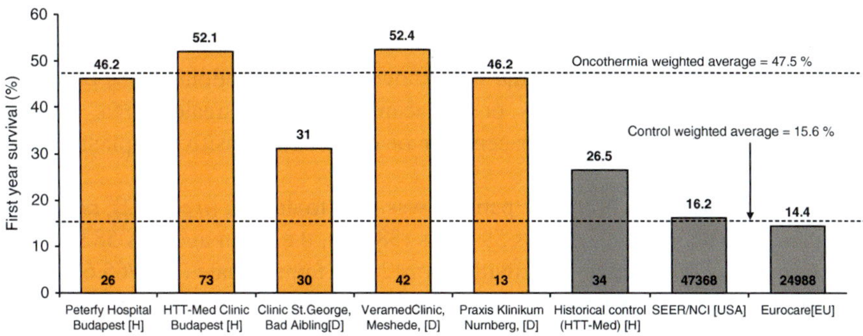

Fig. 4.229 Comparison of the first-year survival rates (%) in various clinics (the number of the patients involved in the study is shown at the bottom of the columns)

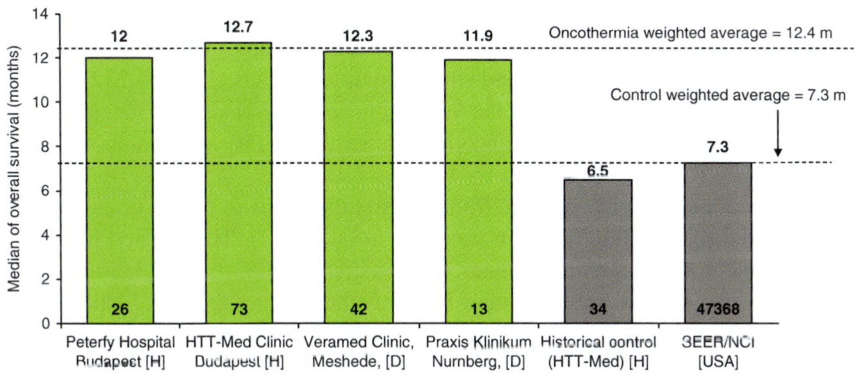

Fig. 4.230 Comparison of the median of overall survival (months) in various clinics (The number of the patients involved in the study is shown at the bottom of the columns)

4.9 Lung Studies

4.9.1 Oncothermia Lung Study I

This study represents an open-label, single-arm, monocentric study, ($n = 61$) carried out in Peterfy Hospital, Budapest, Hungary, [referred to as PFY]. Investigator: Dr. T. Magyar. The involved patients were analyzed according to an ITT schedule. Recruiting time was from April 1997 to August 2002, altogether 64 months. The primary endpoints of the study were the overall survival time (OS) and the survival time from the first oncothermia treatment. The dates of exitus were checked by the National Death Register, so actual and accurate data were collected. The final check of the deaths was December, 2003. Inclusion criteria were: (1) Inoperable or sub-totally resected, or recurrent primary pancreas tumor, (2) progression after radio- and/or chemotherapy, (3) KPS $\geq$ 30 and the inclusion was irrespective of the localization of the lesion in the lung. Patients started the oncothermia process in their late/advanced stages, where most of them had failed to respond to any of the applied conventional therapies.

Exclusion criteria were only the well-known contraindications of the oncothermia method (metallic implants or replacements in the treated area, missing heat-sense in the treated area, pacemaker or other field-sensitive implants in the patient).

The age distribution of $n = 61$ patients was near to normal ($p = 0.82$); no outlier was present. The median age was 58 years (38–77), the mean age was 58.97 years (Std.err= 1.17). The gender distribution was 21/40 female/male (34.4/65.6%). The ratio of the elderly (>68 years) patients was 21.3%. Oncothermia was performed in two/three sessions per week. Treatment time per session was 60 min. The power was gradually and linearly raised depending on the patient's tolerance from 40–80 W to 100–150 W. The applied average energy was 300 kJ/treatment (250–450). The applied applicators were 3.1 dm^2 and 7.1 dm^2, depending on the tumor volume.

Most of the patients (49, 80.3%) had distant metastases. They were heavily pretreated; everybody received at least one chemotherapy and underwent other treatments (surgery, radio- or second chemotherapy, see Figs. 4.231, 4.232).

The actual staging was made at the first diagnosis (44% were in advanced [WHO IIIb or IV] stages and at the first oncothermia treatment 75% were in an advanced stage).

The median of the elapsed time from the first diagnosis to the first oncothermia was 8 m (0.4–172), while its mean was 16.3 m (st.err.3.1). The elapsed time ratio to the overall survival was more than 50% [median 59.9%, (6.5–99.1), mean 59.4 (st.err.3.5)]; the patients received their first oncothermia in the second half of their survival time.

The oncothermia treatment was provided twice a week, the treatment number was on average 8.1 (st.err.0.55) and its median 8 (2–23).

The KM plots of the overall survival [median 16.4 m, (1.7–181.9); mean 25.6 m, (st.err.3.8)] and the survival from the first oncothermia treatment [median 5.7 m,

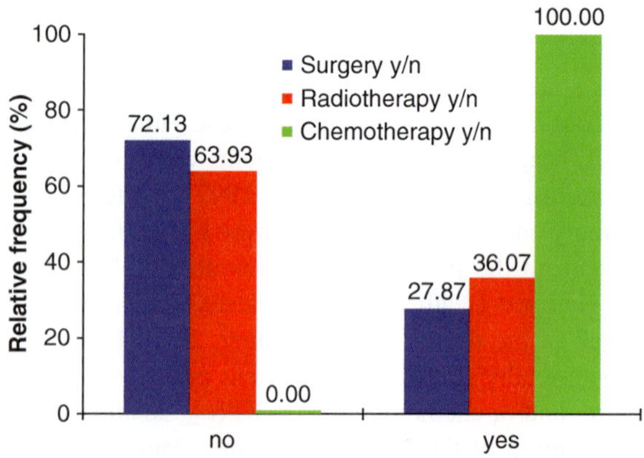

Fig. 4.231 Pretreatments of the patients

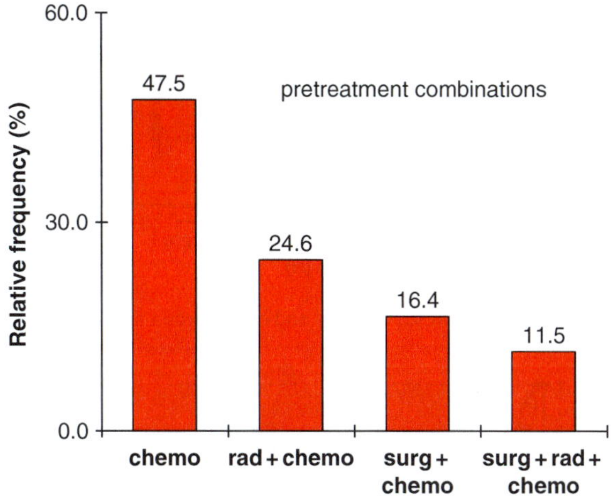

Fig. 4.232 Pre-treatment combinations

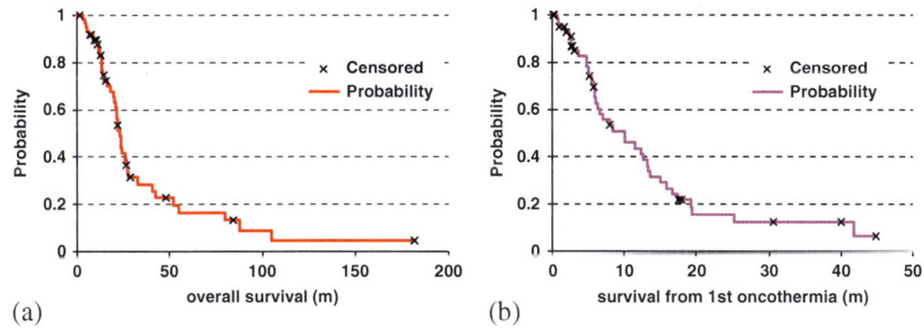

Fig. 4.233 Overall (**a**) and oncothermia treatment time (**b**) survivals by Kaplan–Meier plot of the patients in the PFY study

(0.1–44.9); mean 9.2 m, (st.err.1.3)] are shown in Fig. 4.233. For elderly patients neither the overall survival nor the oncothermia survival were different ($p \sim 0.68$).

Naturally, the survival was significantly different for patients without or with metastases, ($p = 0.0003$ and 0.031 for overall survival and oncothermia survival, respectively), see Fig. 4.234.

The elapsed time to the first oncothermia shows an important parameter. Namely, this is of course smaller ($p = 0.0019$) for the patients with advanced disease in their first diagnosis ($n = 34$, median, 13.0 m [1.5–142]; mean 24.0 m, [st.err.5.2]; and $n = 27$, median, 6.5 m [0.4–19.9]; mean 6.67 m, [st.err.0.83] for nonadvanced and advanced, respectively). Although, the opposite was registered ($p = 0.14$) when the staging at the first oncothermia was studied ($n = 15$, median, 4.10 m [1.5–29.3]; mean 8.9 m [st.err.2.3]; and $n = 46$, median, 8.3 m [0.4–142]; mean 18.78 m, [st.err.4.0]; for nonadvanced and advanced, respectively).

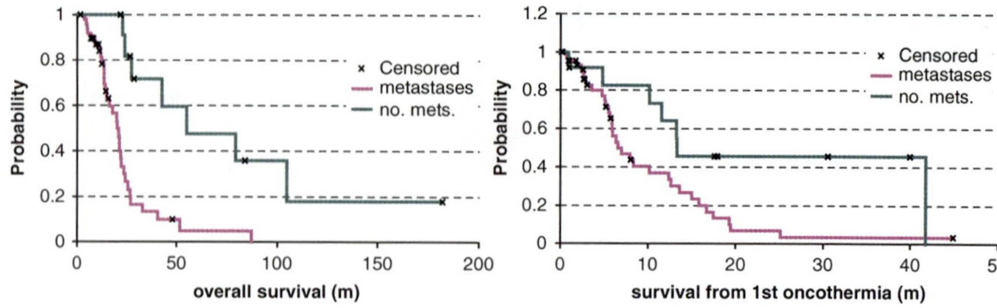

Fig. 4.234 Overall survival (**a**) and oncothermia survival (**b**) of the patients with metastases

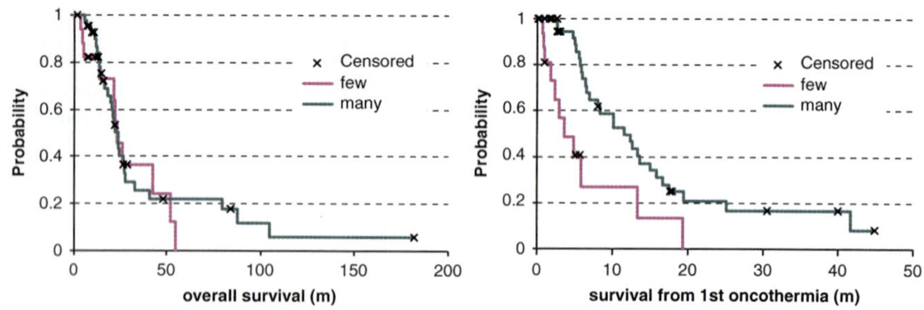

Fig. 4.235 The various survival times for patients depending on the treatment session time ("few" lower than the median number, "many" higher than the median number of the treatments)

This tendency is more obvious to register the ratio of overall survival and of oncothermia survival to the elapsed time till first oncothermia. This divides the patients into "early oncothermia" and "late oncothermia" categories. The overall survival shows the expected result: the low survivals ($p = 0.0065$) begin the oncothermia quicker [$n = 31$, median, 16.4 m (4.7–79.7); mean 19.62 m, (st.err.2.61)] than the long survivals, [$n = 30$, median, 17.4 m (1.7–182); mean 31.7 m, (st.err.7.07)]. While the oncothermia survival was the opposite ($p = 0.073$): the early start [$n = 31$, median, 8.4 m (2.4–44.9) mean 12.7 m, (st.err.1.9)] showed longer survival, than the late, [$n = 30$, median, 2.7 m (0.1–40.0); mean 5.6 m, (st.err.1.6)].

The number of treatments does not influence the overall survival significantly ($p = 0.61$), but the oncothermia survival ($p = 0.0023$) and the follow-up time after the last oncothermia ($p = 0.01$) depend well on the number of oncothermia treatments, see Fig. 4.235.

Again the surgical pretreatment was especially ($p = 0.0005$) important for the longer survival (see Fig. 4.236), but the other pretreatments did not affect significantly either the overall survival or the oncothermia survival rates.

The effect of the experience of the treating medical personnel on the data before and after the median time of the study was also measured. Measurements based on early experience ($n_{ee} = 33$) showed an overall survival median of 22.3 m, (1.7–181) mean 33.7 (st.err.6.4); oncothermia survival median 8.0 m, (0.1–45) mean

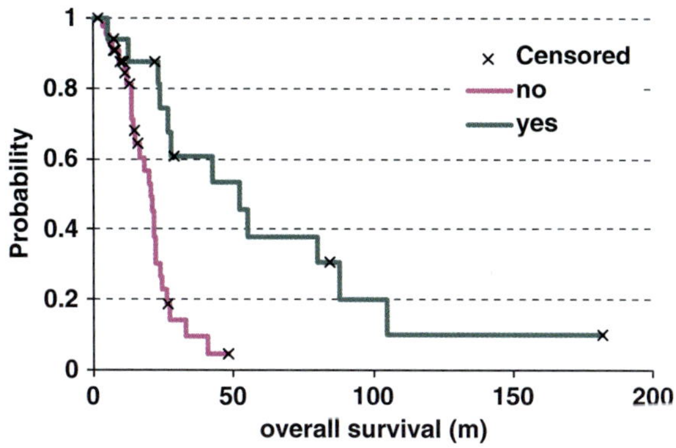

Fig. 4.236 Effect of the pre-treatment operation is significant considering the overall survival

11.6 (st.err.2.07); and elapsed time to first oncothermia median 10.3 m, (1.5–142) mean 22.1 (st.err.5.3). For late experience ($n_{le} = 28$) the data were: overall survival median 12.3 m, (3.6–51.9) mean 15.9 (st.err.2.2); oncothermia survival median 5.0 m, (0.1–25.1) mean 6.37 (st.err.1.24); elapsed time to first oncothermia median 5.9 m, (43–77) mean 61.1 (st.err.1.8). The differences between the early and late experiences are significant in the case of overall survival ($p = 0.028$) and elapsed time to first oncothermia ($p = 0.012$), but not significant in oncothermia survival ($p = 0.19$). The significantly better survivals in the first half of the study time compared to the second half probably originate from the fact, that the patient spectrum had been shifted to the more advanced side. In the early experience the ratio of the advanced cases was 33%, while in the late experience it advanced to 57%, but both of them increased (76 and 75%, respectively) when measured at the first oncothermia treatment. (The nearly equal percentage of advanced cases in both categories (increasing from very different starts) indicates the assumption that the patients start the oncothermia treatment at nearly the same stage irrespective of their elapsed time from the first diagnosis to the first oncothermia.

4.9.2 Oncothermia Lung Study II

The age distribution of $n = 197$ patients was acceptably normal ($p = 0.59$); no outlier was present. The median age was 57 years (16–84), the mean age was 56.71 years (Std.err= 0.77). The gender distribution was 62/135 female/male (31.5/68.5%). The ratio of the elderly (>68 y) patients was 20.3%.

Most of the patients (157, 79.7%) had distant metastases (one, two, and three metastases were observed for 101, 43, and 13 patients, respectively). They were heavily pretreated; most of them (93.4%) underwent surgery and subsequent radiation therapy (see Figs. 4.237, 4.238).

The actual staging was made at the first diagnosis (46.2% were in advanced [WHO IIIb or IV] stages) and at the first oncothermia treatment they were at a more advanced status.

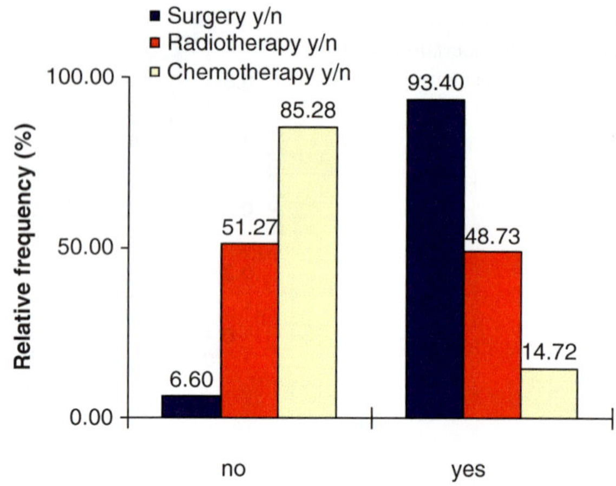

Fig. 4.237 Pre-treatment distribution of HTT patients

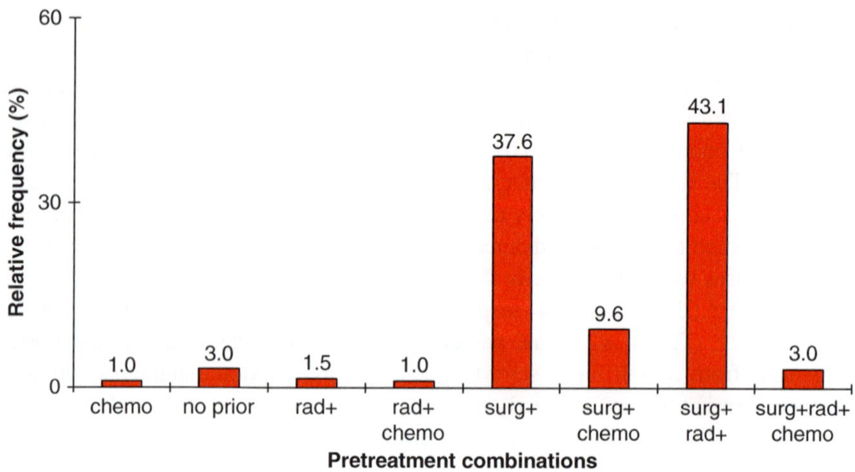

Fig. 4.238 Pre-treatment combinations

The median of the elapsed time from the first diagnosis to the first oncothermia was 5.5 m (0.2–111.3), while its mean was 10.6 m (st.err.1.0). The elapsed time ratio to the overall survival was close to 50% [median 45.4%, (1.6–96.7), mean 45.7 (st.err.3.9)].

The oncothermia treatment was provided twice a week, the treatment number was on average 7.9 (st.err.0.4) and its median 6 (3–40). The median treatment time was 60 min, (45–135) and the mean was 69.6 min (st.err.1.3), while the median equivalent temperature was 52 (43–59) and its mean was 51.4 (st.err.0.3). Note that the equivalent temperature is not the real temperature. It is the calculated value from the actual energy absorption and the impedance, meaning of the actual destruction rate, which is as high, as it would have been in the purely temperature-oriented case.

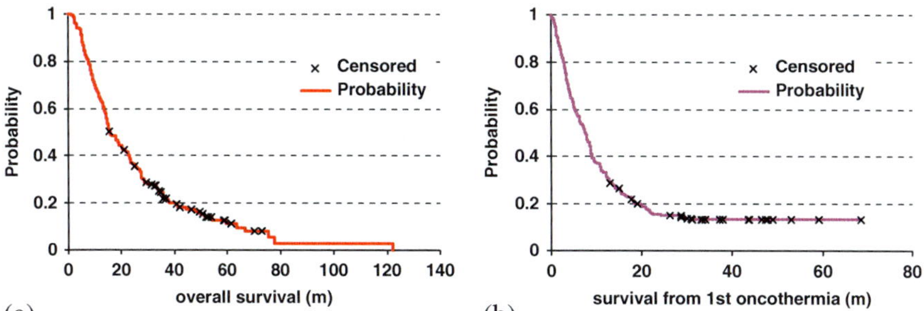

(a) (b)

Fig. 4.239 Overall survival (**a**), and survival from the first oncothermia (**b**) for the patients entered in the HTT study

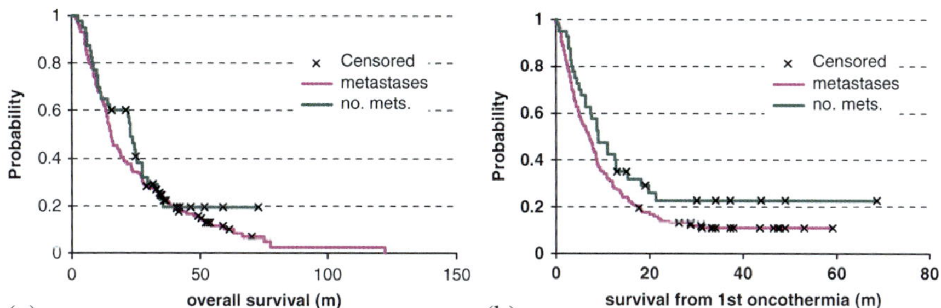

(a) (b)

Fig. 4.240 The effect of metastases on the overall survival (**a**) and oncothermia survival (**b**) for HTT patients

The KM plots of the overall survival [median 15.6 m, (1.1–122.1); mean 22.4 m, (err.1.31)] and the survival from the first oncothermia treatment [median 7.57 m, (0.1–68.6); mean 11.8 m, (st.err.0.91)] are shown in Fig. 4.239. For elderly patients neither the overall survival nor the oncothermia survival were different ($p \sim 0.37$ and 0.49, respectively).

The differences between patients without or with metastases in their overall survival and oncothermia survival were not significant ($p = 0.33$ and 0.07 for overall survival and oncothermia survival, respectively) Fig. 4.240.

The number of treatments significantly influences the overall survival ($p = 0.048$) and the oncothermia survival ($p = 0.00046$) and the follow-up time after the last oncothermia ($p = 0.0017$) very much depends on the number of oncothermia treatments.

Again the surgical pretreatment was especially ($p = 0.0005$) important for longer survival in terms of overall survival ($p = 0.005$) and oncothermia survival ($p = 0.016$) (see Fig. 4.241), but the other pretreatments did not affect significantly either the overall survival or the oncothermia survival rates.

We studied the effect of the experience of the treating medical personnel on the data before and after the median time of the study. For the early experience

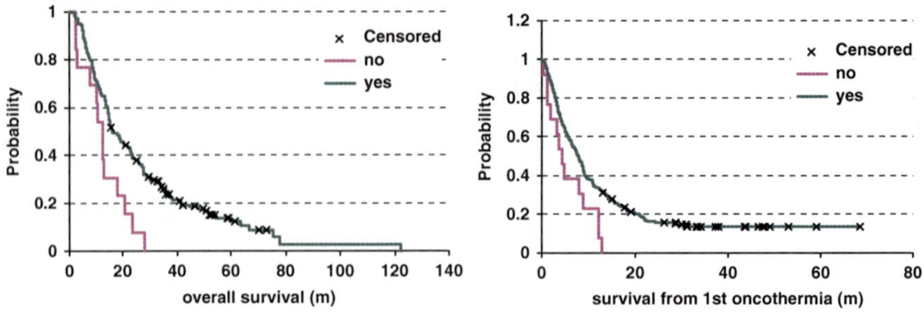

Fig. 4.241 Effect of surgical pretreatments on the overall survival (**a**) and oncothermia survival (**b**)

($n_{ee} = 94$) the data were: overall survival median 15.3 m, (2.4–122.1) mean 24.0 (st.err.2.17); oncothermia survival median 7.2 m, (0.3–68.6) mean 11.8 (st.err.1.5); median of elapsed time to first oncothermia 5.37 m, (0.4–111.3) mean 12.2 (st.err.1.8). For the late experience ($n_{le} = 103$) the data were: overall survival median 15.83 m, (1.1–77.7) mean 21.0 (st.err.1.5); oncothermia survival median 8.13 m, (0.1–43.9) mean 11.8 (st.err.1.1); elapsed time to first oncothermia median 5.6 m, (0.2–64.8) mean 9.1 (st.err.1.1). The differences between the early and late experiences are not significant in the case of overall survival ($p = 0.85$), oncothermia survival ($p = 0.17$), and elapsed time to first oncothermia ($p = 0.21$).

4.9.3 Meta-Analysis of Oncothermia Lung Studies

The age distribution of the altogether $n = 258$ patients was near to normal ($p = 0.71$); and no outlier was present. The median age was 57 years (16–84), the mean age was 57.2 years (Std.err= 0.7). In the spectrum of the PTF a shift to the elderly patients was present (see Fig. 4.242). The overall gender distribution was 83/175 female/male (32/68%), and no significant difference could be measured between the locations. The ratio of the elderly (>68 years) patients was 20.5% (20.3 and 21.3% in PFY and HTT, respectively). The PFY/HTT patient ratio was 61/197 (24/76%).

Eighty-percent of the patients had distant metastases in both study locations (see Fig. 4.243) and half of them were in advanced stages at the first diagnosis of the disease (see Fig. 4.244). The patients were heavily pretreated (see Fig. 4.245), and in PFY chemotherapy, in HTT surgery was the most frequent modality.

The median elapsed time to first oncothermia from the first diagnosis (ETO) was significantly ($p = 0.028$) shorter in HTT than in PFY (Fig. 4.246).

The oncothermia treatment was provided twice a week, the number of treatments on average was more in PFY than in HTT procedures (see Fig. 4.247).

The overall survival (OS) and the survival from the first oncothermia treatment (OSO) are shown in Fig. 4.248a,b. The overall survivals are significantly lower in the HTT case ($p = 0.044$) but in terms of the oncothermia survival there are no significant differences ($p = 0.53$). Survival after the treatment was not different in the

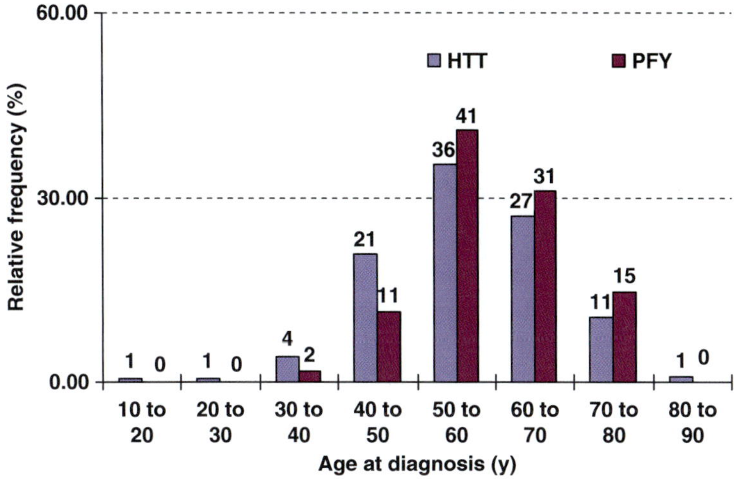

Fig. 4.242 Comparison of the age distribution in the given studies

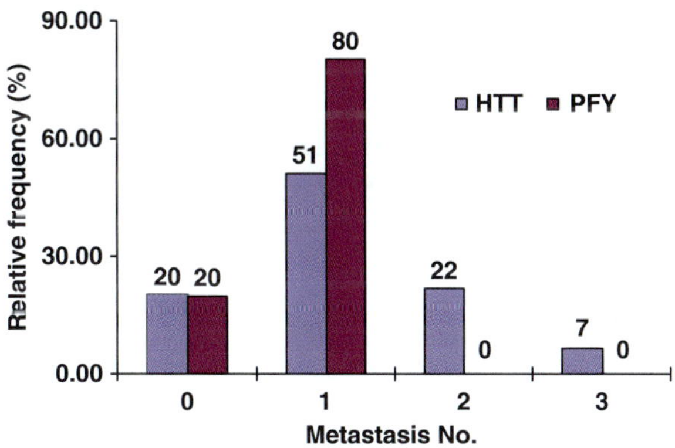

Fig. 4.243 Comparison of metastatic cases

two locations ($p = 0.55$). However, for elderly patients neither the overall survival nor the oncothermia survival were different ($p \sim 0.38$ and 0.86, respectively), see Fig. 4.248c.

In both locations most of the patients reported subjective improvement of their QoL. No extra toxicity or safety problems were observed during the treatments.

The above two studies were performed using the same guidelines but in entirely independent hospitals, with no overlap in medical personnel. The two retrospective data sets can be regarded as independent. The study of the expertise of the personnel in the two locations was the same, and their training was enough to allow for an optimal performance from the very start of the treatment.

The patients' pretreatments were mostly dominated by surgery and chemotherapy in HTT and PFY, respectively. In addition the elapsed time to first oncothermia

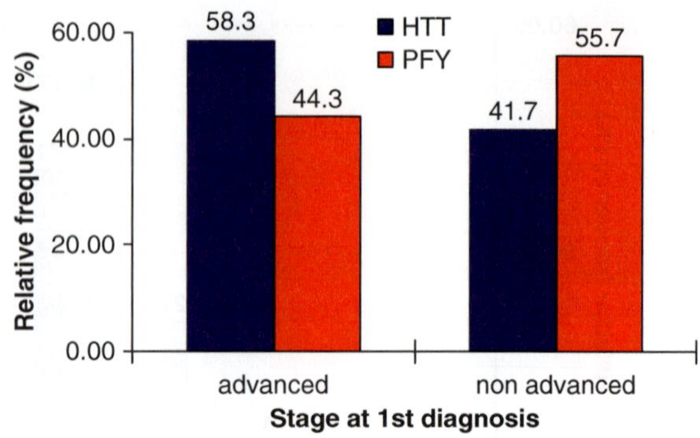

Fig. 4.244 Staging differences

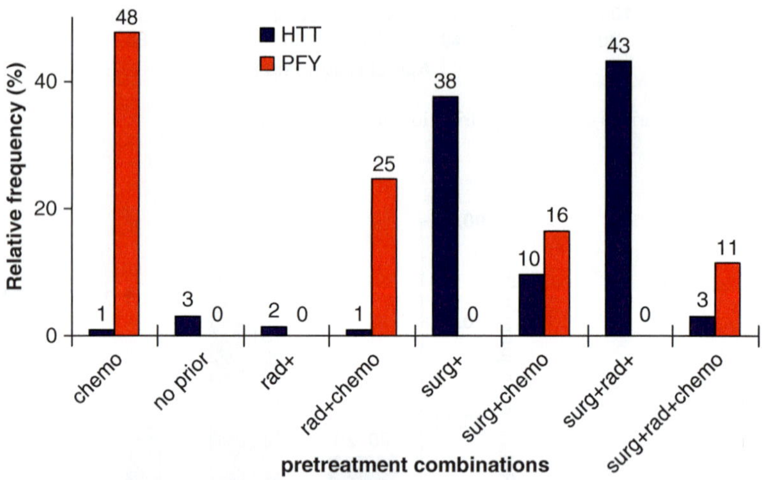

Fig. 4.245 Pre-treatment combinations show the different emphases in the treatment strategies

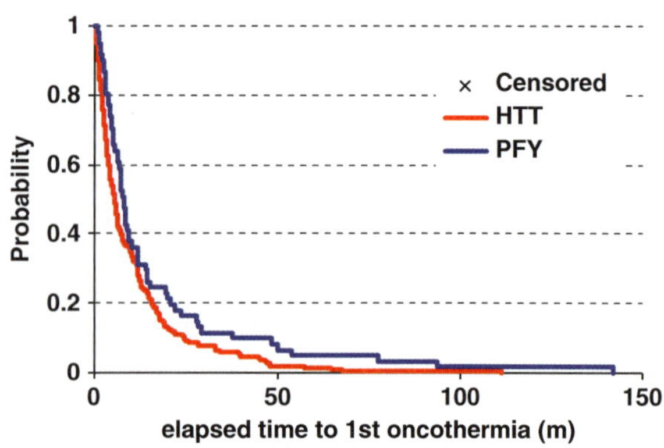

Fig. 4.246 Elapsed time to first oncothermia is significantly shorter for HTT patients

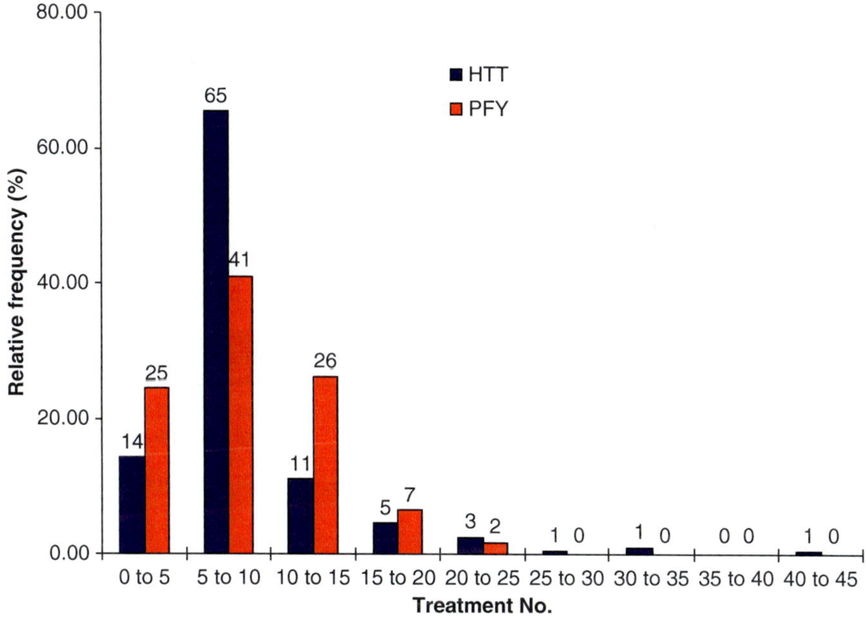

Fig. 4.247 The treatment numbers in the two institutions

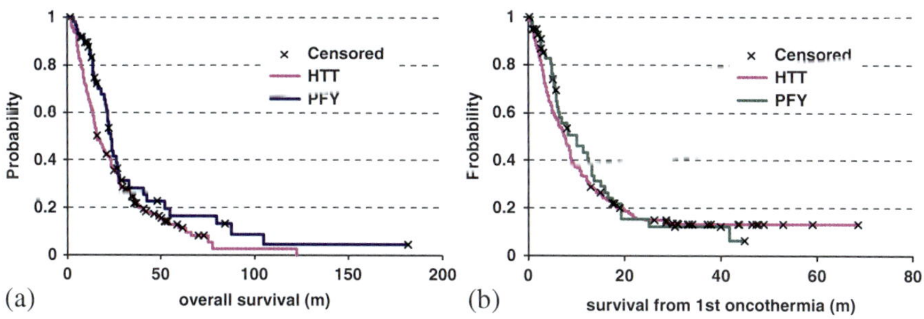

Fig. 4.248 The difference between the overall (**a**) and oncothermia survival (**b**), for patients involved in the studies

was significantly different with an earlier start of oncothermia in HTT, and surprisingly the overall survival was also significantly lower. It looks as if the patients treated by HTT were more advanced at their first diagnosis (more metastases were detected) than the PFY counterparts, which explains the difference. Despite the difference in overall survival, the oncothermia survival does not differ significantly between the two locations. The yearly survival rates could be regarded as identical ($p > 0.99$) within the measurement error (see Fig. 4.249). This could be an indication of an oncothermia leveling action as well.

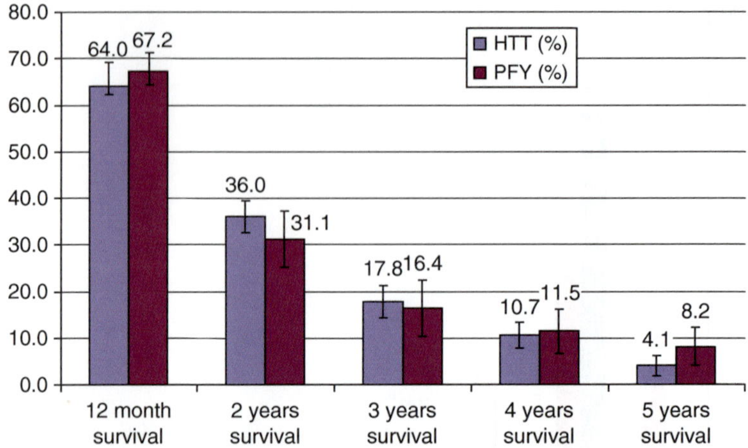

Fig. 4.249 Yearly survivals of the patients in the two institutions (no significant differences exist)

4.9.4 Comparison to Historical Control

A historical (retrospective) control ($n=53$) is for comparison. The overall survival KM plot shows significant benefit from the oncothermia ($p=0.033$) (Fig. 4.250). Median 15.8 m (1–182) and mean 23.1 m (St.err.1.3) for oncothermia and 14.0 m (1–84), 18.5 m (St.err.2.3) for the historical control.

The common pool of the PFY and HTT data ($n=258$) is decomposed (see Appendix 35) for responders and nonresponders, see Fig. 4.251.

Patients were divided into subgroups of advanced (III, IIIa, IIIb, IV) ($n=140$) and not-advanced (I, Ia, Ib, II, IIa, IIb) ($n=77$) stages. (Data were not available for a smaller group ($n=41$), their data were not used in this evaluation.) It is not surprising that the two subgroups are significantly different ($p=0.038$) by their survival, see Fig. 4.252.

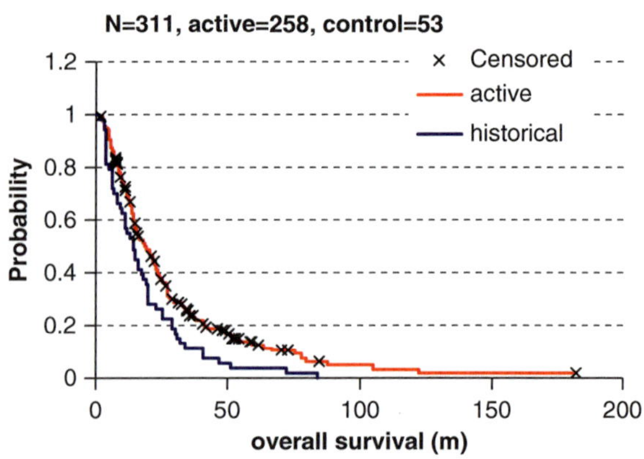

Fig. 4.250 Kaplan–Meier plot for the historical and active arms in the study. The difference is significant ($p=0.033$)

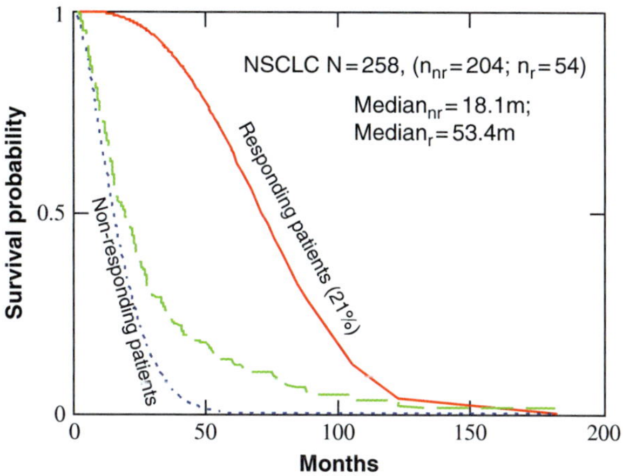

Fig. 4.251 Decomposition of the common database

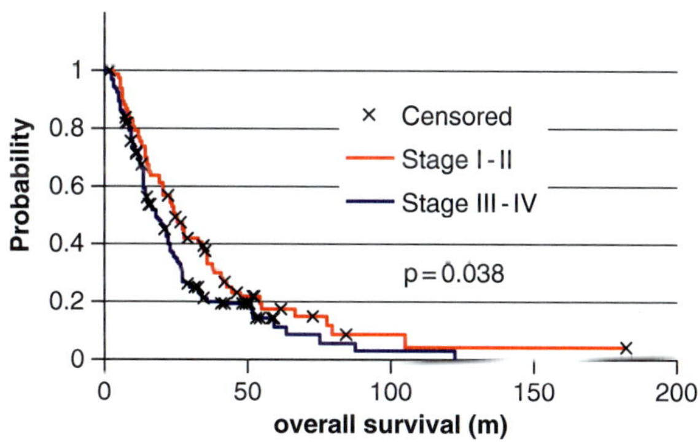

Fig. 4.252 Survival difference between ($p = 0.038$) advanced and non-advanced NSCLC subgroups

The non-advanced subgroup (Stages I, Ia, Ib, II, IIa, IIb) has all together 87 patients (77 active, 10 control) and the advanced subgroup (Stages III, IIIa, IIIb, IV) 183 patients (140 active, 43 control). On making a comparison of both the subgroups with the relevant subgroup of the historical control, the effect of oncothermia becomes more pronounced in advanced cases (III+IV) (active arm: median 14.7 months, [$n = 132$], control arm: median 11.0 months [$n = 43$]) The result is significant, $p = 0.023$, compared to the non-advanced case subgroup, where the differences were not large, and were not significant (see Fig. 4.253).

Analyzing both the subgroups parametrically, the ratio of responders in the non-advanced group is only 17%, while in the advanced group it is 88% (Figs. 4.254 and 4.255).

The results could be compared well to the available SEER [47] and Eurocare [1253] data, see Fig. 4.256. The yearly survival rate is definitely much higher

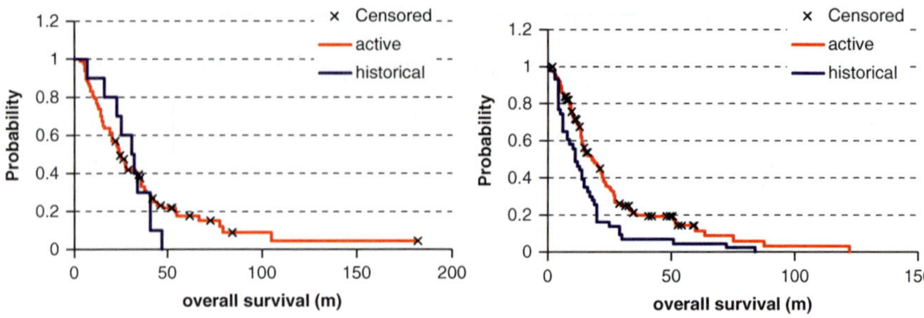

Fig. 4.253 Comparison of subgroups with the relevant part of the historical data (Non-advanced subgroup: $n = 87$, (77, 10), $p = 0.63$; advanced subgroup $n = 183$, (140, 43), $p = 0.0017$)

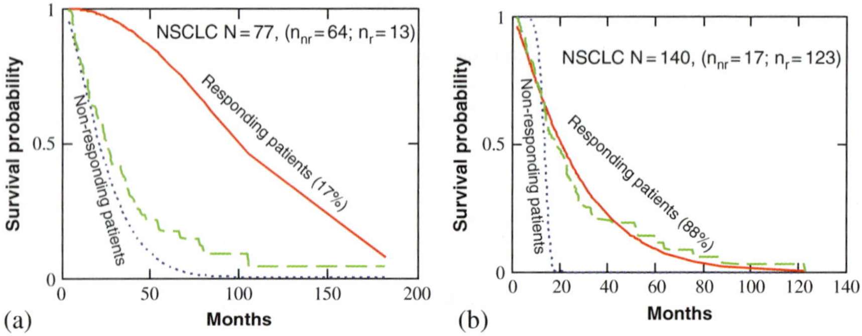

Fig. 4.254 The non-advanced cases (**a**) $n = 77$ are less responding than the advanced ones (**b**) $n = 140$

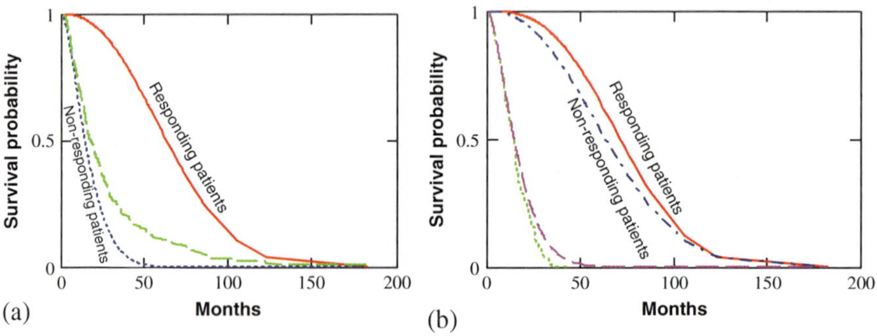

Fig. 4.255 The distribution together with the historical control ($n = 311$) only slightly differs from the distribution without the control group ($n = 258$)

(significant) in the first 3 years than the database average. This result is remarkable taking into consideration the advanced patient spectrum of oncothermia-treated patients. The decrease of the difference by years is probably due to the very small influence on the longer survivals of the oncothermia applied at a late stage for a short time. Oncothermia starts in the most rapid cases in earlier stage. Consequently the all survival is strongly influenced by the oncothermia treatment in these cases.

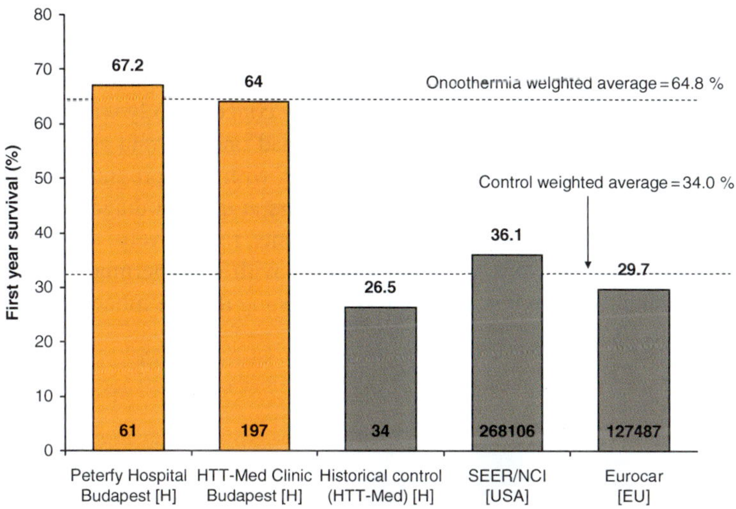

(a)

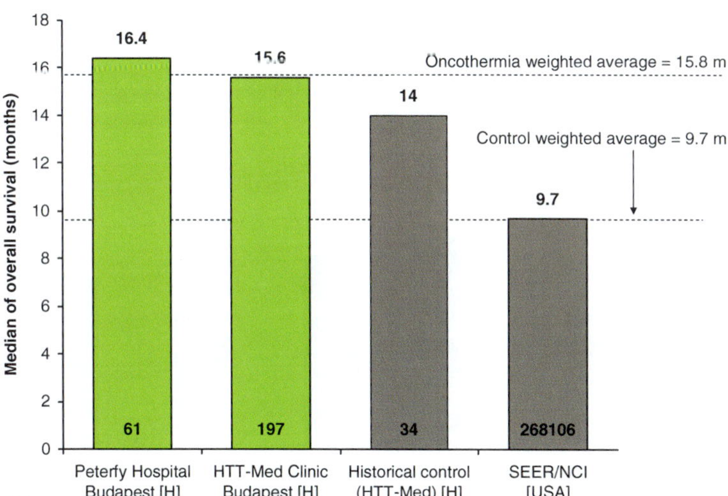

(b)

Fig. 4.256 The comparison of the first year (**a**) and the median (**b**) survival results with SEER and Eurocare data

This is supported by the fact that despite the significantly lower elapsed time to first oncothermia the survivals are not notably different in the two institutions.

4.10 Liver Studies

4.10.1 Study of Liver Metastases of Colo-Rectal Origin

One of the earliest studies of oncothermia on colo-rectal metastases to liver ($n = 80$) was published in 1999 [1254] (Biomed Clinic, Bad Bergzabern, Germany; Investigator: Dr. D. Hager). Histology of the tumor is adeno-carcinoma. Prior liver resection was carried out in 16% of patients. Prior chemotherapies were unsuccessful. 37.5% of patients had palliative chemotherapy concomitantly with oncothermia (dominantly [65%] 5-FU+Leukovorine), others had oncothermia as monotherapy. Many patients also had metastases other than liver, with poor prognosis factors (see Fig. 4.257). The CA 19-9 and CEA tumor markers at the first diagnosis were over 24 [U/l] and 2.5 [ng/l], in 80 and 70% of the patients, respectively.

The median survival was significantly higher with oncothermia than expected without this treatment (see Fig. 4.258). It is interesting that the monotherapy results

(a)

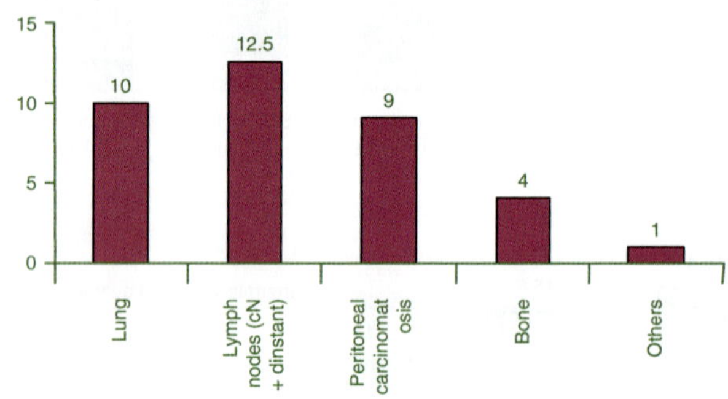

(b)

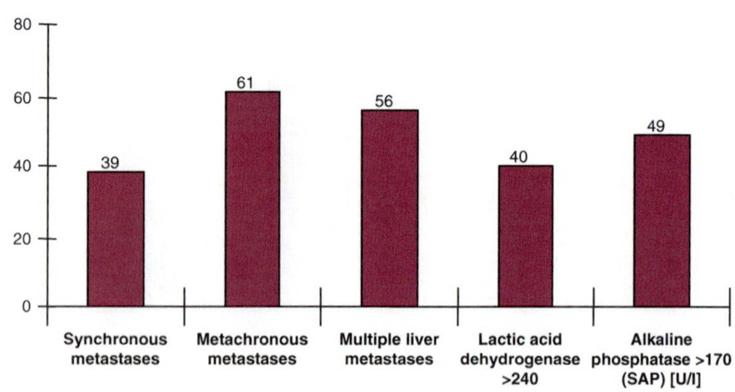

Fig. 4.257 Additional metastases (%) (**a**) and the prognosis factors (%) (**b**) of patients involved in the study

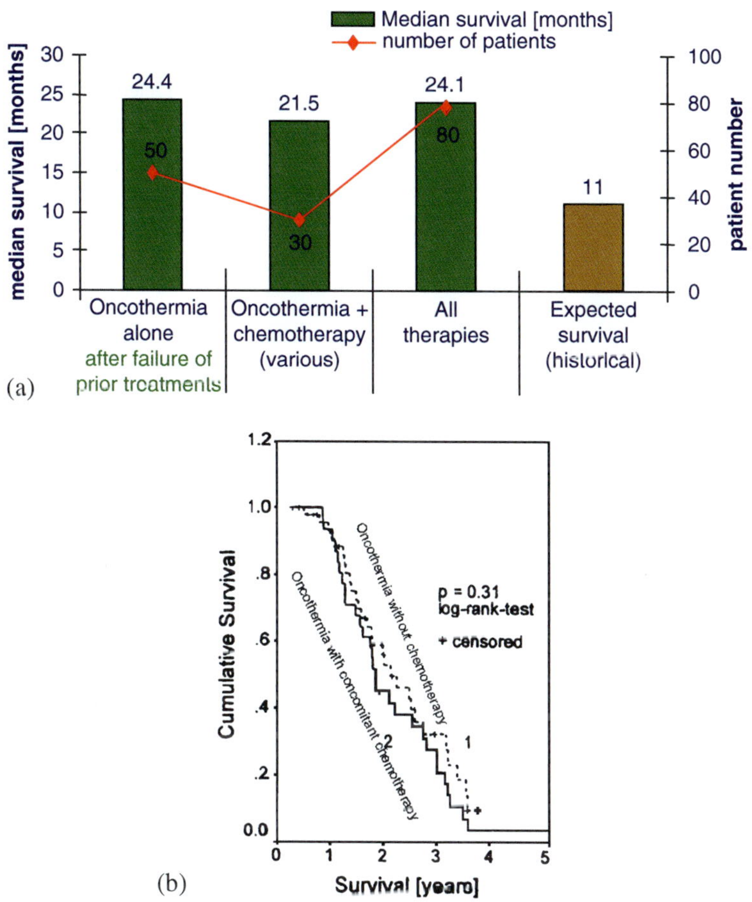

Fig. 4.258 The median survival (**a**) and the Kaplan–Meier survival plot (**b**) for patients having complementary or monotherapy with oncothermia

were higher, but of far not significantly ($p = 0.31$). The yearly survivals are also better than expected, see Fig. 4.259.

4.10.2 Study of Advanced Liver Metastases of Colo-Rectal Origin II

A study was carried out on advanced liver metastases of colo-rectal carcinoma origin ($n = 22$) at the Department of Oncology, Spedali Civili, Brescia, Italy [1255] (Investigator: Prof. V.D. Ferrari). The stage of the patients was C (BCLC classification). Patient's characteristics: nonoperable: 15/22, (68%), portal vein thrombosis: 16/22, (72%), distant metastasis (other than liver) was observed in 9% of the patients. The concomitant chemotherapy was Oxalyplatine (50 mg/m2) for

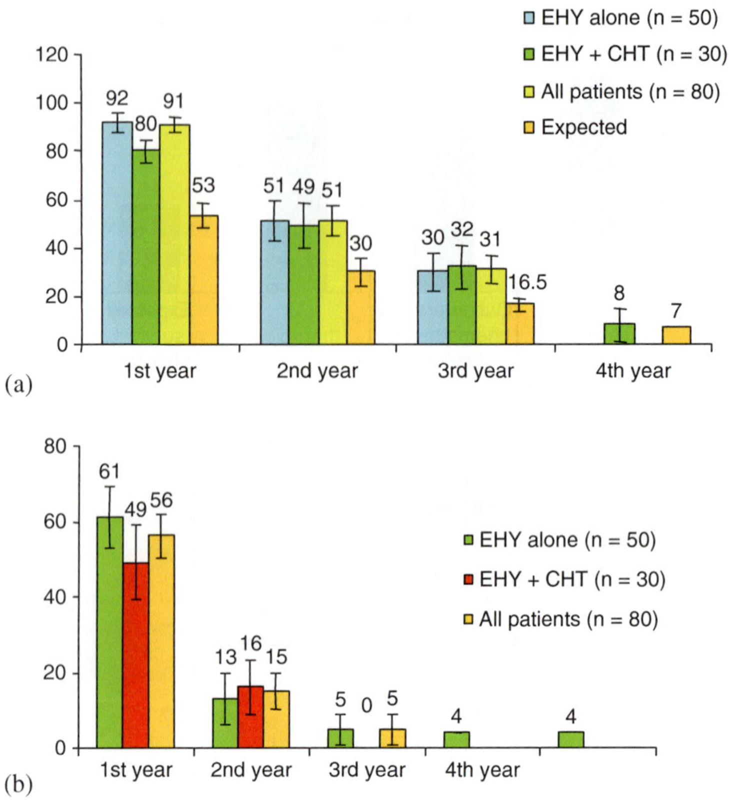

Fig. 4.259 The yearly survivals from the first diagnosis (**a**) and from the first oncothermia treatment (**b**) (EHY=oncothermia, carried out with the EHY2000 device)

64% of the patients. The oncothermia treatment protocol was: 60 min/session, 2 sessions/week, 10 sessions/cycle, [median 1.5 cycle (1–4)], 80–140 W, (41–47°C). The local clinical response of liver metastases was 28%, see Fig. 4.260; the QoL was reported as better for 50% of the patients. The local toxicity of oncothermia was relatively high (32%) (see Fig. 4.261).

4.10.3 Comparison Study of Treatment Lines of Colo-Rectal Liver Metastases

This study is devoted to a comparison of first-line (without oncothermia) and second-line (with oncothermia) therapies for colo-rectal cancer liver metastasis ($n = 15$) [1256]. (Investigators: Prof. H. Kirchner and Dr. P. Panagiotou. Department of Hematology and Oncology, Siloah Hospital, Hannover, Germany.) Oncothermia

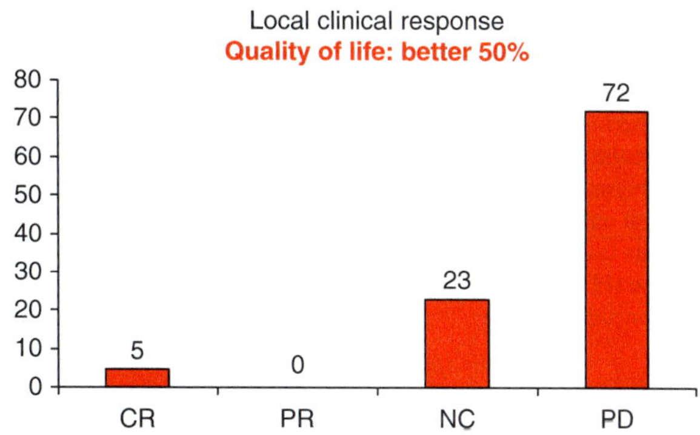

Fig. 4.260 The local clinical response of liver metastases for the patients involved in the study

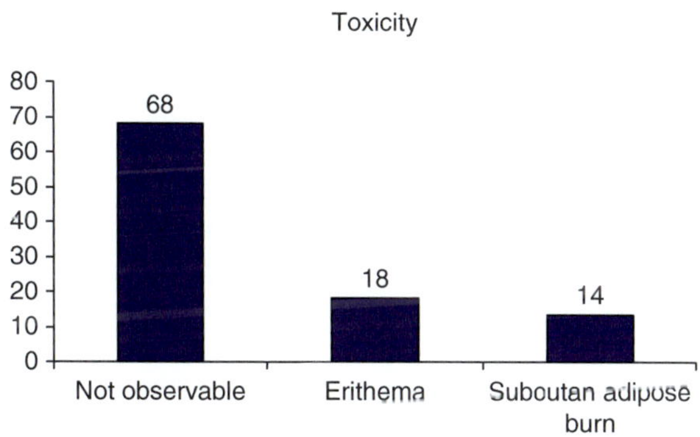

Fig. 4.261 Local toxicity of oncothermia measured in the trial ($n = 22$)

Table 4.17 Protocol of the second-line treatment in the study

Week				1							2							3				4
Days	1	2	3	4	5	6	7	8	9	10	11	12	13	14	15	16	17	18	19	20	21	22
Irinotecan	I							I														Repetition
Capecitabine	C	C	C	C	C	C	C	C	C	C	C	C	C	C								
Oncothermia	O		O		O			O		O		O			O		O					

was applied after the first-line treatment had failed, in the second line only. The treatment protocol is shown in Table 4.17. The local response after the second line was better than after the first one, see Fig. 4.262, without extra toxicity for the patients, see Fig. 4.263.

The median survival was 23 months, while the historical expectation was 10–20 months. An important observation was noted in the study: if a progression occurred initially or recidive after a stable phase, it was observed outside the oncothermia applicator in neraly 80%.

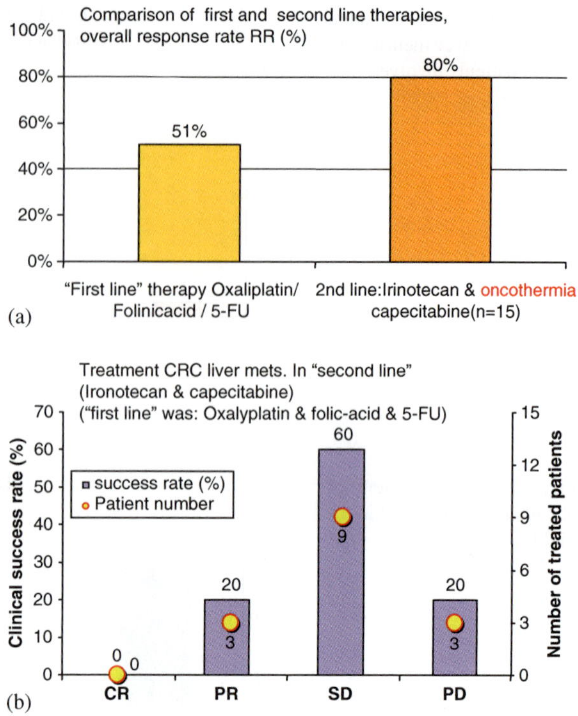

Fig. 4.262 The local response was higher in the second line (complementary oncothermia application), than in the first where oncothermia was not applied (**a**), and only 20% of the patients had progressive disease (PD) in second line with oncothermia, (**b**)

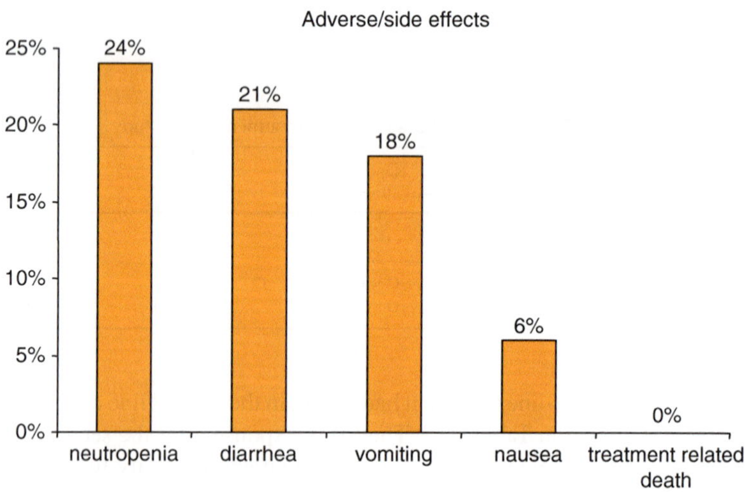

Fig. 4.263 Adverse/side effects were still the expected, while chemotherapy (XELIRI) is administered without oncothermia

4.10.4 Study of Platinum Derivatives with Oncothermia for Liver Metastases of Colo-Rectal Origin

This first-line, phase II study ($n = 30$) was devoted to comparing the effect of platinum derivatives on liver metastases from colo-rectal cancer origin [1257]. (Investigator: Prof. G. Fiorentini, Department of Oncology, St. Giuseppe Hospital, Empoli (Florence), Italy.) The median survival time was 22 m (10–34), while the median relapse time was 9 m (6–18). All the platinum derivatives show 20% response rate and 50% improving the QoL (KPS). The main side effect was anxiety reduction (83% of the patients), nausea and vomiting were 13.3% while the other side effects were under 10%. A definite oncothermia side effect (erythematic + mild adipose burn) was observed in 6.7%. The independent study of Oxalyplatine + oncothermia ($n = 12$) and of Cisplatin + oncothermia ($n = 18$) shows definite differences, see Fig. 4.264. The local response rate was definitely higher for Cisplatin, while the other benefits show significantly lower results, the side effects also differ significantly.

4.10.5 Study of Liver Metastases of Rectal Origin

This study was carried out on advanced, non-operable rectal carcinoma ($n = 65$) and its liver metastases ($n = 29$) [1258]. (Investigators: Prof. E. Mako and Prof. Z. Vigvary, Department of Radiology, Semmelweis University, Budapest.) Oncothermia was applied 2–3×/week with concomitant chemotherapy. Adriamycin and/or Mitomycin C + Gelaspon (selective artery chemoembolization was applied for liver metastases). Concomitant radiotherapy was LDR Cs-137. Additionally intra-luminar cryotherapy was used for rectal treatment. Regular control was made by abdominal ultrasound and CT-scan. Local clinical responses for the primary and metastatic lesions are shown in Fig. 4.265.

4.10.6 Study of Liver Metastases of Various Origins

An investigation of advanced, refractory liver metastases ($n = 25$) of various origins was carried out by HTT-Med Clinic Budapest [1259] (Investigators: Dr. A. Varkonyi and Dr. A. Dani). The treatments were applied as monotherapy. Medians of KM survivals (see Fig. 4.266) are 20.5 and 7.0 m for overall and oncothermia survivals, respectively.

4.10.7 Study of Very Advanced Liver Metastases of Various Origins: Comparison of Complementary Therapies

This study ($n = 28$) was devoted to the comparison of complementary oncothermia to radiotherapy and chemotherapy, as well as to monotherapy applications for liver

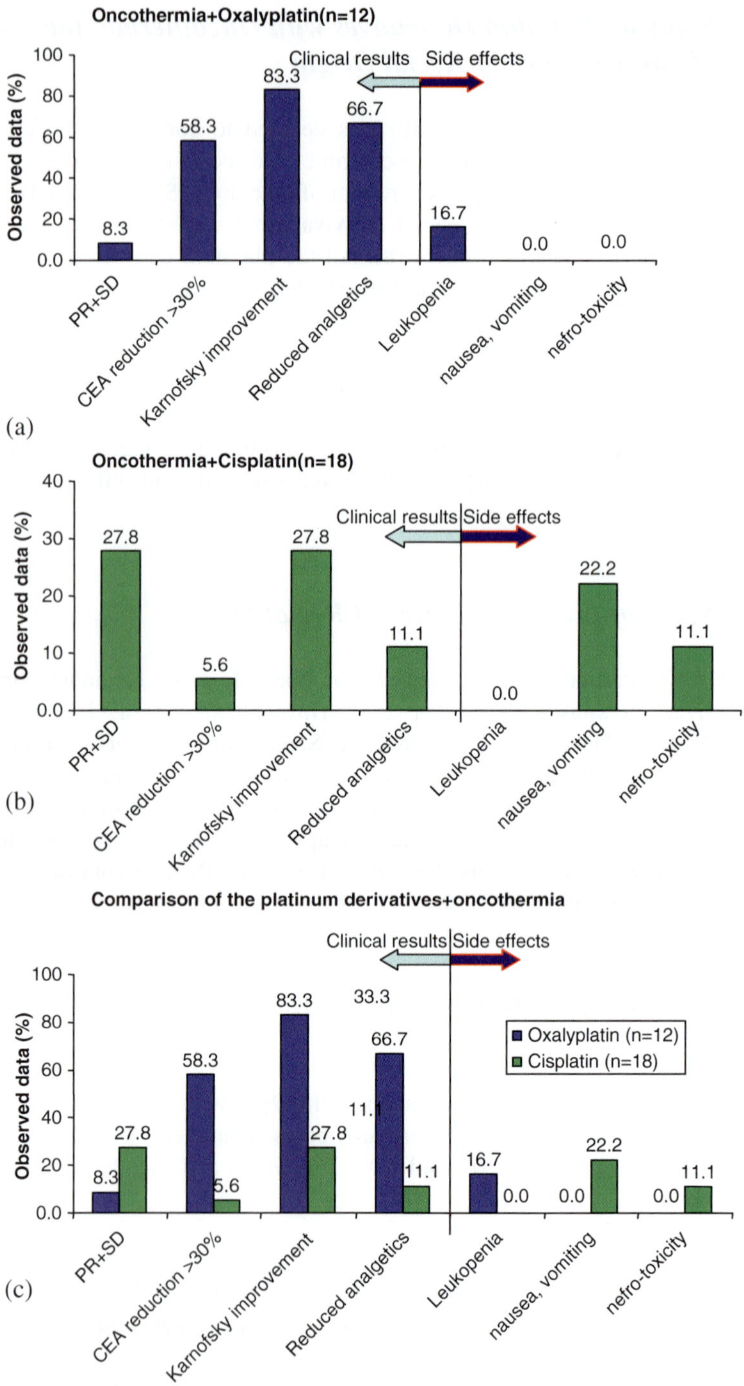

Fig. 4.264 The results for oncothermia concomitantly applied with Oxalyplatine (**a**) and Cisplatin (**b**)

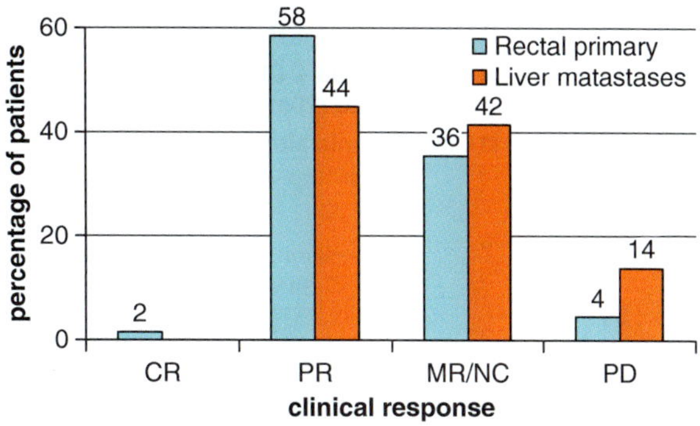

Fig. 4.265 Local clinical response for advanced, inoperable rectum tumors ($n = 65$) and its liver metastases ($n = 29$)

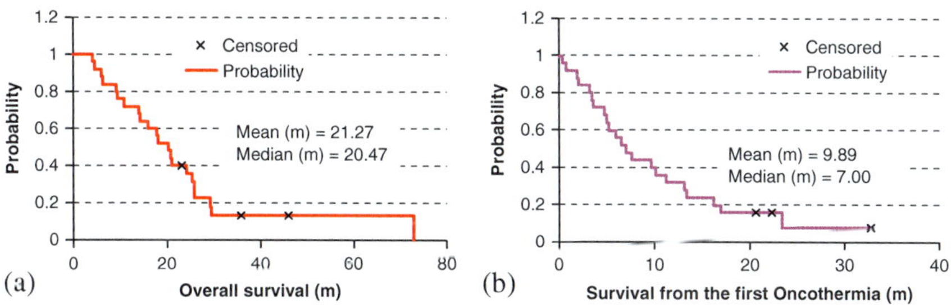

Fig. 4.266 Kaplan–Meier plots for overall (**a**) and for oncothermia (**b**) survivals

Table 4.18 The applied protocol

Therapy/week-days	Mon.	Tue.	Wed.	Thu.	Fri.	Number of patients
Radiotherapy +	RT	RT	RT	RT	RT	16
Oncothermia		OT		OT		
Chemotherapy +			CHT			8
Oncothermia		OT		OT		
Oncothermia monotherapy		OT		OT		4

metastases from various kinds of primary tumors [1260]. (Investigator: Prof. H. Aydin; Clinic and Institute of Radio-Oncology, Zentralkrankenhaus Reinkenheide, Bremerhaven, Germany.) Table 4.18 shows the treatment protocol. Oncothermia was applied two times a week. Concomitant radiotherapy: 10 MV, 1.5–1.8 Gy fractional radiation 5×/week, overall dose: 21–24 Gy; concomitant chemotherapy: Vinorelbine (20 mg/m^2/week). Local control (overall response) was 81%, 38%, and 25% for oncothermia + radiotherapy, for oncothermia + chemotherapy, and for oncothermia as monotherapy, respectively (see Fig. 4.267).

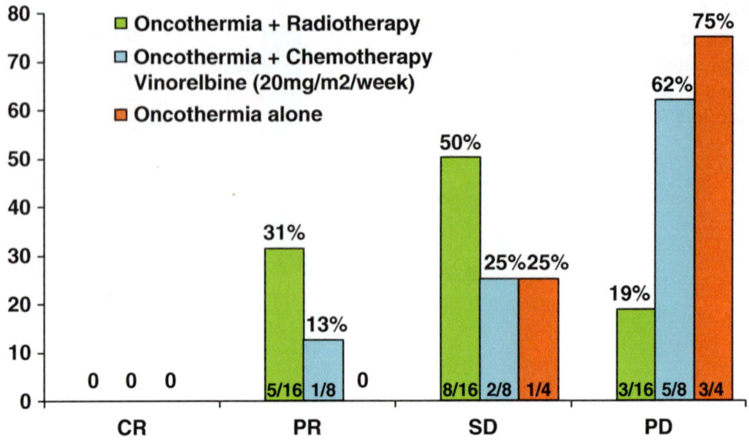

Fig. 4.267 Local clinical response for different treatment protocols (number of patients involved in the study is shown at the *bottom of the columns*)

4.11 Comparison of Studies of Liver Metastases

Comparing the above-described studies (see Fig. 4.268), similarities and differences appear which are probably due to the problems of the patient's cohorts and different pre- and concomitant treatments. The oncothermia protocol was unified (2–3-times a week, 60 min/session).

4.12 Gynecological (Pelvic) Cancer Study

4.12.1 Ovary Study

This ovary study ($n = 27$) was performed in Peterfy Hospital, Budapest (Investigator: Dr. T. Magyar). The inclusion criteria are as usual: the high-line treatment for refractory ovarian cancer. The medians are 37.8 and 20.4 m (mean 55 and 23 m) for overall and oncothermia survivals, see Fig. 4.269. Oncothermia was applied weekly 2–3 times with 6–10 treatments using a large 30-cm diameter electrode, to cover the disseminated malignancy.

The parametric decomposition shows medians 132.7 and 19.4 m for responders (67%) and for nonresponders (33%), respectively, see Fig. 4.270.

The first-year survival rate was 100%; in comparison the SEER and Eurocare data are 66.9% and 65.4%, respectively. The second-year survival rate (70.4%) also transcends the large databases (SEER 49.3%, Eurocare 50.8%).

4.12.2 Uterine Corpus Cancer

Only a low number of patients were collected in the study of uterine cancer ($n = 9$) (Peterfy Hospital, Budapest. Investigator: Dr. T. Magyar). The inclusion criteria are

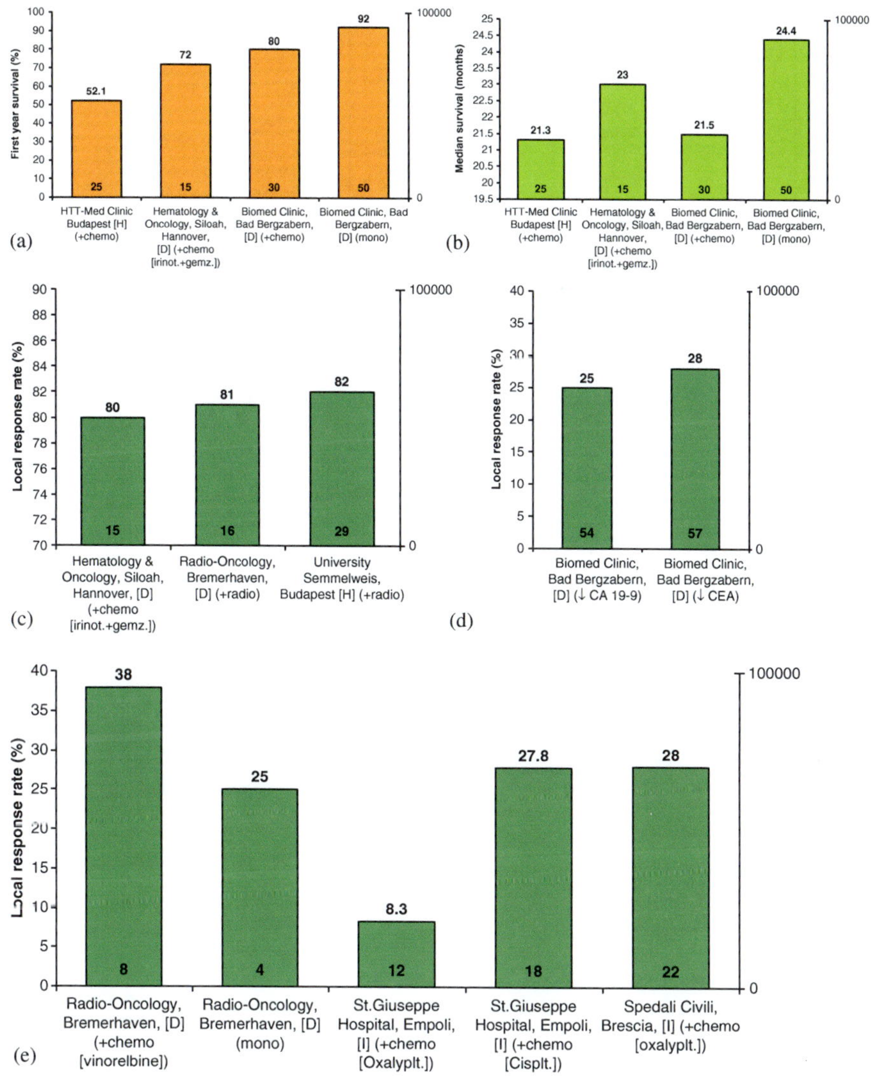

Fig. 4.268 Comparison of the parameters of various oncothermia studies of metastatic liver. The first-year survival (**a**) has large variations but the median of overall survival (**b**) shows homogeny for obtained results. The local response rate (panels **c–e**) shows a non-unified data set. (The response rate is measured by the tumor size (**c**, **e**), while the data on panel (**d**) is measured by the decrease of the relevant tumor markers) (Number of patients involved in the study is shown at the *bottom of the columns*)

the high-line treatment for refractory uterus (corpus) cancer. The medians are 61.5 and 6.13 m (mean 52.3 and 9.2 m) for overall and oncothermia survivals, Fig. 4.271. Oncothermia was applied weekly 2–3 times with 4–8 treatments using a standard 20-cm diameter electrode.

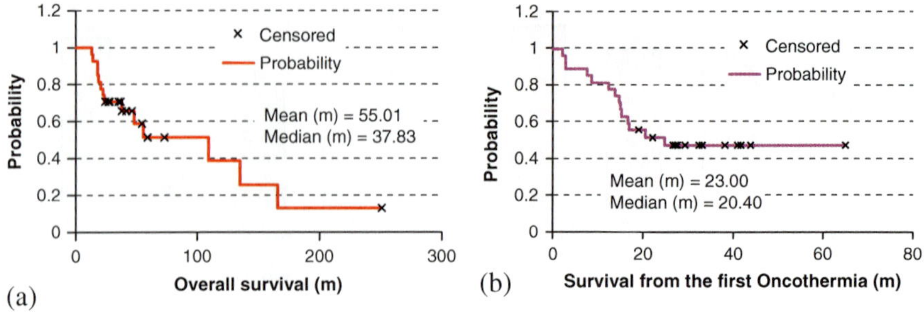

Fig. 4.269 The Kaplan–Meier plot of overall (**a**) and oncothermia (**b**) survivals of ovarian cancer

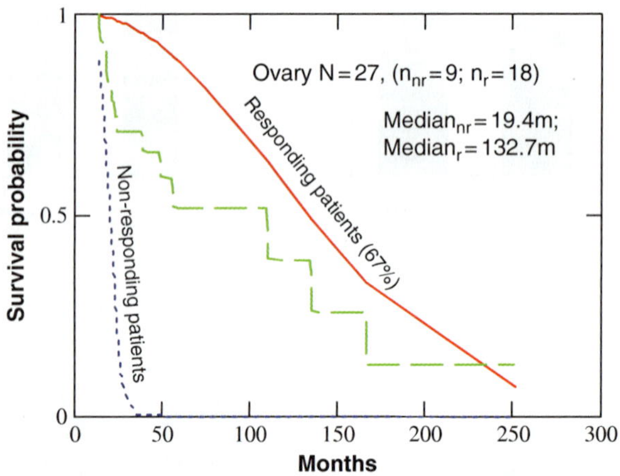

Fig. 4.270 The parametric decomposition of the overall survival of ovarian cancer (the measured plot is the *dashed-line*)

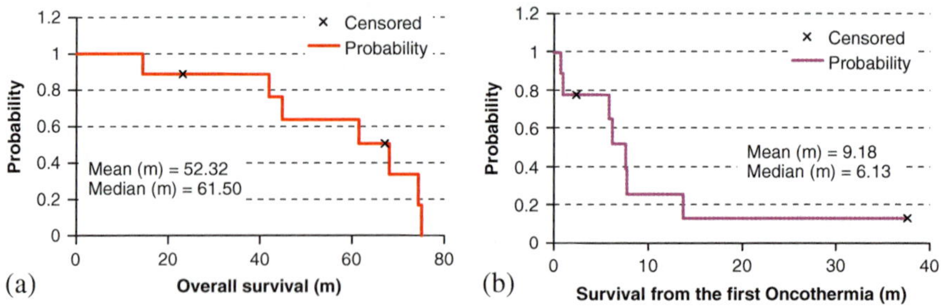

Fig. 4.271 Kaplan–Meier plot of overall (**a**) and oncothermia (**b**) survivals of cancer of corpus uterus

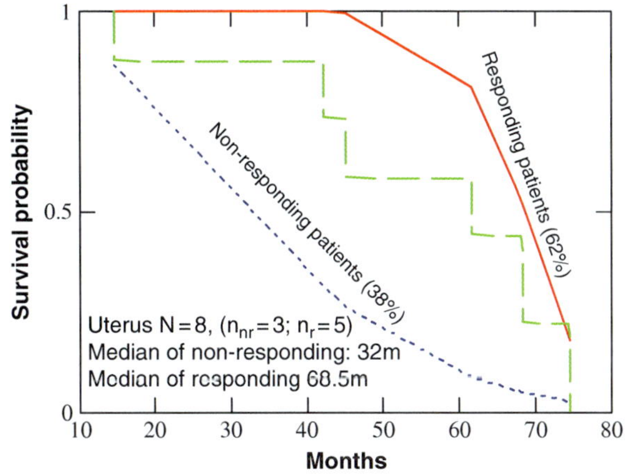

Fig. 4.272 The parametric decomposition of the overall survival of uterine cancer (corpus) (the measured plot is the *dashed-line*)

The parametric decomposition shows medians 68.5 and 32.0 m for responders (62%) and for nonresponders (38%), respectively, see Fig. 4.272.

The first-year survival rate was 100%; in comparison the SEER and Eurocare data are 86.3% and 87.6%, respectively. The second-year survival rate (77.8%) also transcends the large databases (SEER 77.0%, Eurocare 80.0%).

4.12.3 Uterine Cervix

Advanced cervical cancer (stage FIGO IIb–Iva) was studied ($n = 38$). (Peterfy Hospital, Budapest. Investigator: Dr. T. Magyar.) The medians are 27.6 and 3.8 m (mean 31.2 and 8.56 m) for overall and oncothermia survivals, see Fig. 4.273. Oncothermia was applied weekly 2–3 times with 4–8 treatments using a standard 20-cm diameter electrode.

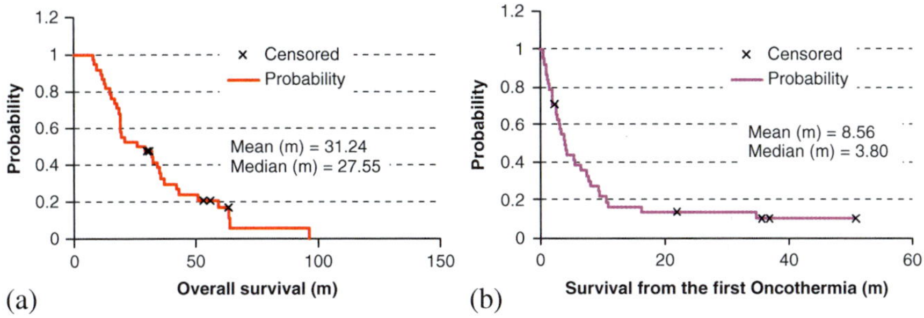

Fig. 4.273 Kaplan–Meier plot of overall (**a**) and oncothermia (**b**) survivals of cancer of corpus uterus

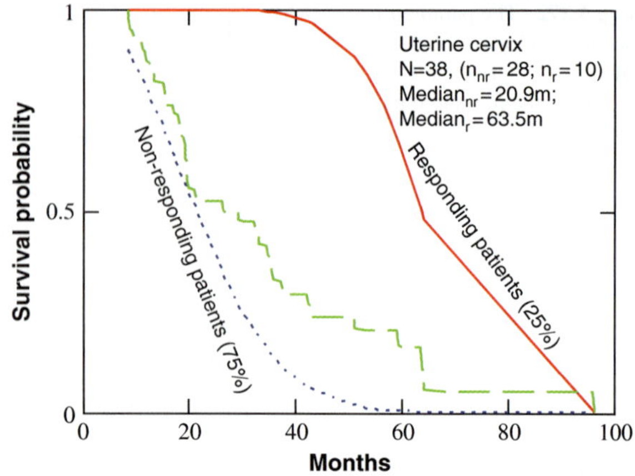

Fig. 4.274 The parametric decomposition of the overall survival of uterine cancer (corpus) (the measured plot is the *dashed-line*)

The parametric decomposition shows medians 63.5 and 20.9 m for responders and for nonresponders, respectively, see Fig. 4.274. The responders by the parametric decomposition were only 25% ($n_r = 10$) of the total treated patients ($n = 38$).

First-year survival rate was 86.8%, (In comparison the SEER and Eurocare data are 82.0% and 83.0%, respectively.)

4.12.4 Comparison of Oncothermia in Pelvic Gynecology

Comparing the gynecological data, the first-year survival is better at all the treated sites than the SEER data (see Fig. 4.275). The median survival shows high success for ovarian and uterus oncothermia treatments, but the cervix is less successful (Fig. 4.276). This discrepancy is perhaps related to the well-known fact that the human papilloma viruses can be well treated in most cases, but the oncothermia patients are out of the conventionally controllable regime of the disease. The difficult situation for cervix is indicated by the parametric evaluation also, where the responding patients are in the minority (25%), see Fig. 4.276. The high response rate for ovarian cases indicated by the parametric decomposition (see Fig. 4.277) is probably due to the high selectivity of oncothermia at the cellular level, which has central importance in the case of the ovarian malignancies, which are promptly disseminated in the pelvic volume.

4.13 Breast Study

Hyperthermia is feasible for refractory, far advanced breast tumors, however it is limited by inflammatory lesions [1261]. Oncothermia (with its lower temperature increase as usual) is able to treat the inflammatory mamma-carcinoma.

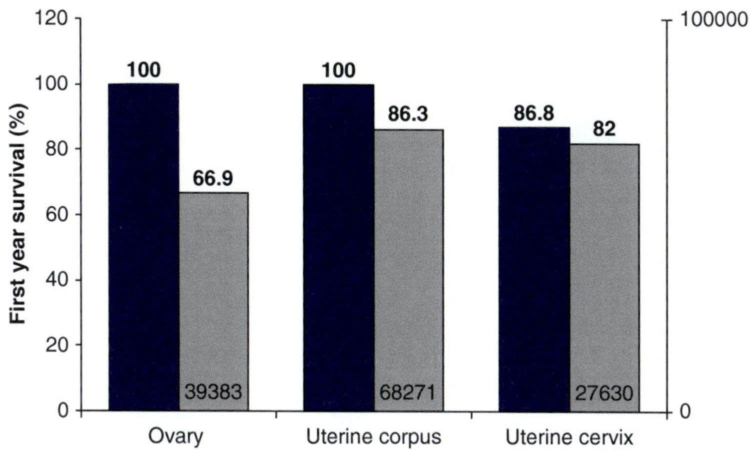

Fig. 4.275 First-year survival (%) of various pelvic gynecological malignancies treated by oncothermia. SEER data are shown for comparison (*right-hand side columns*) (number of patients involved in the study is shown at the *bottom of the columns*)

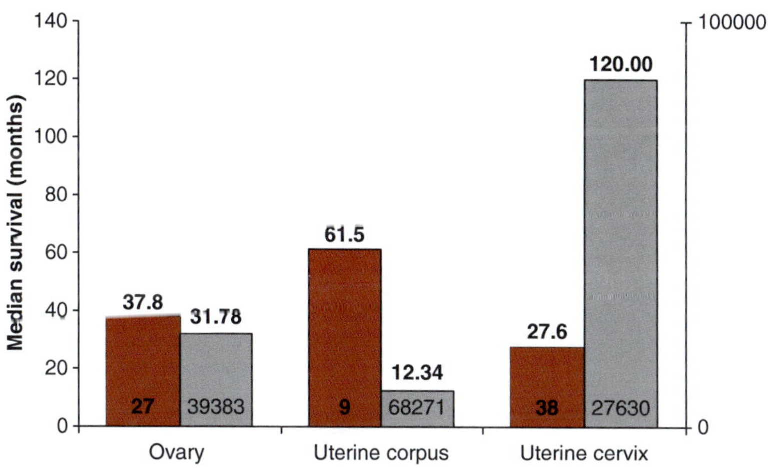

Fig. 4.276 Median survival times (months) for various pelvic gynecological sites. SEER data are shown for comparison (*right-hand side columns*) (number of patients involved in the study is shown at the *bottom of the columns*)

Advanced breast cancer was studied ($n = 103$). (HTT-Med Clinic, Budapest. Investigator: Dr. A. Dani.) The medians are 52.1 and 16.5 m (mean 67.4 and 23.6 m) for overall and oncothermia survivals, see Fig. 4.278. Oncothermia was applied weekly 2–3 times with 8–10 treatments using a standard 20-cm diameter electrode.

The parametric decomposition shows medians 274.8 and 10.9 m for responders and for nonresponders, respectively, see Fig. 4.279. The responders by the parametric decomposition were less than half, 35% ($n_r = 4$) of the total treated patients ($n = 103$). This large difference focuses our attention well on the in any case

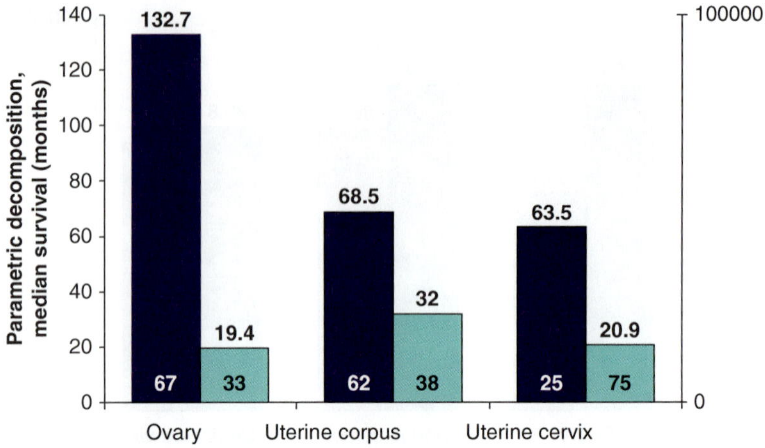

Fig. 4.277 Results of the parametric fit of the pelvic gynecological malignancies (percentages of responders/nonresponders are shown at the *bottom of the columns*)

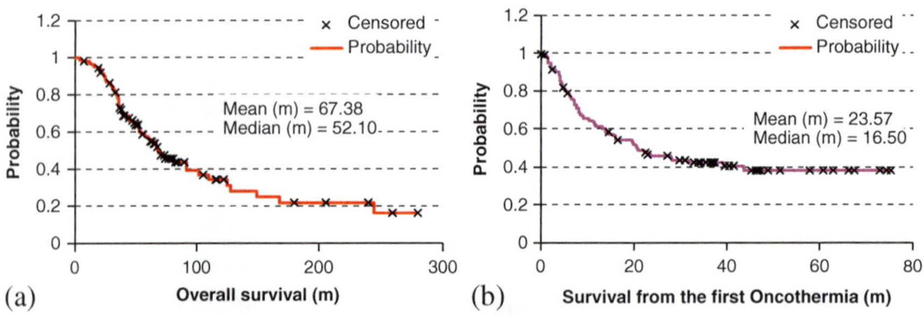

Fig. 4.278 The measured survival plot for overall (**a**) and for oncothermia (**b**) times

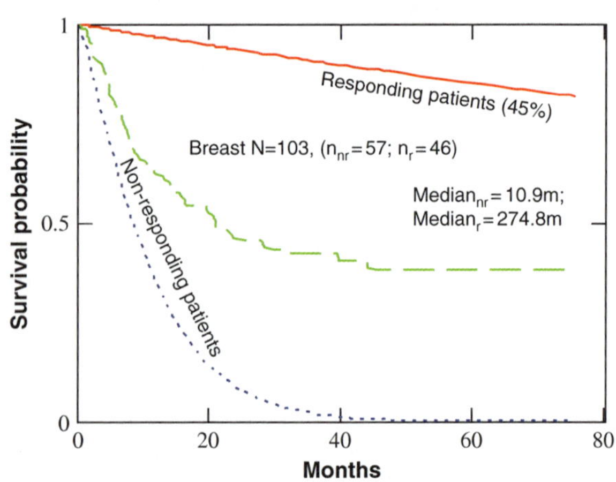

Fig. 4.279 Parametric decomposition of the oncothermia breast study (the measured plot is the *dashed-line*)

high survival probability of breast cancers [median registered in SEER is 120 m, ($n=278{,}784$)] These patients are in fact completely cured (over 10-years survival), which is shown also in the decomposition of the oncothermia study results. The large cure rate is indicated by the 25-year follow-up possibility in overall survival, and by the high value of the survival probability (~0.4) on the KM plot of oncothermia after 5 years follow-up (see Fig. 4.278b).

The first-year survival rate for oncothermia was 97.1%; in comparison the SEER and Eurocare data are 89.0% and 91.7%, respectively.

4.14 Esophagus Study

4.14.1 Esophagus Study I

Advanced esophagus cancer was studied ($n=12$) [1262]. (HTT-Med Clinic and Peterfy Hospital, Budapest. Investigators: Dr. T. Magyar, Dr. A. Varkonyi and Dr. A. Dani.) The medians are 28.5 and 8.6 m (mean 34.4 and 14.8 m) for overall and oncothermia survivals, Fig. 4.280. Oncothermia was applied weekly 2–3 times with 10–12 treatments using 10- and 20-cm diameter electrodes.

The parametric decomposition shows medians 29.4 and 8.5 m for responders and for nonresponders, respectively (Fig. 4.281). The responders by the parametric decomposition were less than half, 35% ($n_r=4$) of the total treated patients ($n=12$).

First-year survival rate for oncothermia was 41.7%; in comparison the SEER and Eurocare data are 31.9 and 30.3%, respectively. In the second-year survival the difference between the database-registered data and that measured by oncothermia became larger (see Fig. 4.282).

4.14.2 Esophagus Study II

This small controlled study ($n=7$) was performed on definitely inoperable ($R2$) patients [1263]. Praxis at Klinikum Nord, Nürnberg, Germany. Investigator:

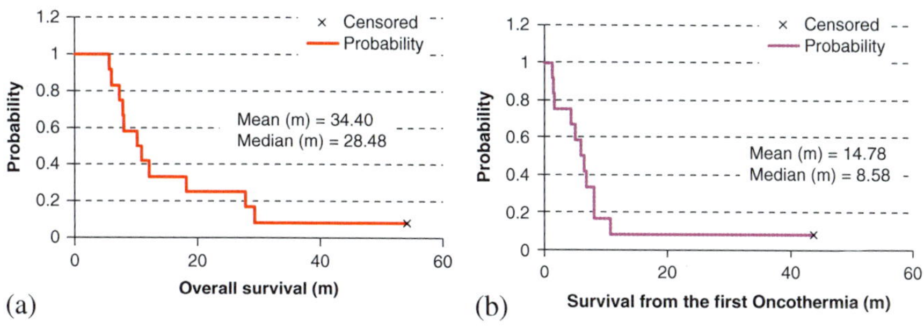

Fig. 4.280 Plots of overall (**a**) and oncothermia (**b**) survivals

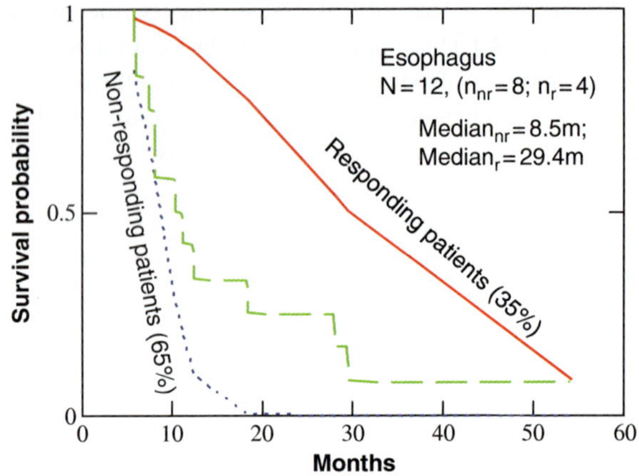

Fig. 4.281 Parametric decomposition of the overall survival in the esophagus study. The original measured curve is *dashed*

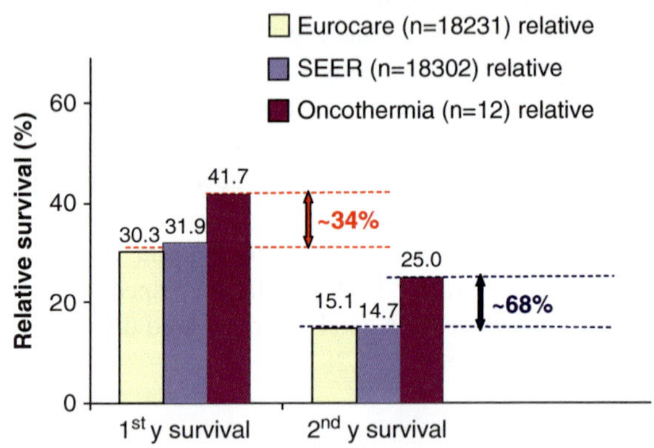

Fig. 4.282 The relative survival (%) in first and second years, in comparison with the large databases

Prof. H. Renner.) Trimodal therapy was administered: radiotherapy: 45+5 Gy, (fractional); Chemotherapy: 5-FU+Mitomicine-C (2×); Oncothermia: 60 min, diam. 30 cm (8–10×). The local clinical response rate of the treatment was excellent (100%), no progress of the disease was observed (Fig. 4.283). The good PR by imaging (CT) was histology complete necrotic tissue only.

Median time of overall survival was 6.8 m (mean was 7.5 m), see Fig. 4.284. The separation of the survival to no-change (NC) and active response (CR+PR) shows a certain (but not significant ($p = 0.14$) difference between the subgroups, see Fig. 4.285.

4.15 Stomach Study

Advanced stomach cancer ($n = 68$) was studied in a bicentral study (HTT-Med Clinic and Peterfy Hospital, Budapest. Investigators: Dr. T. Magyar, Dr. A. Varkonyi

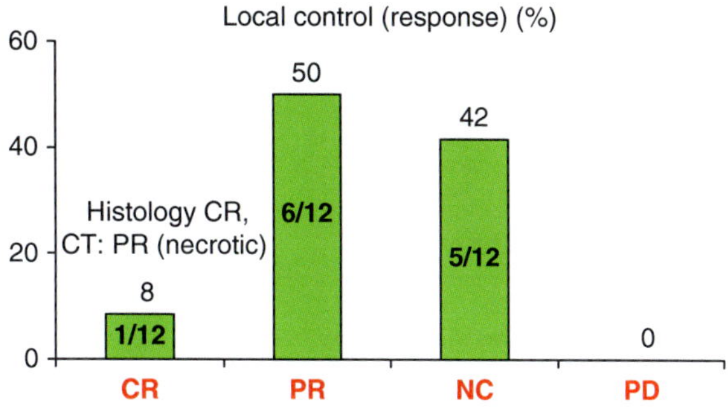

Fig. 4.283 Local clinical response in the esophagus study by oncothermia

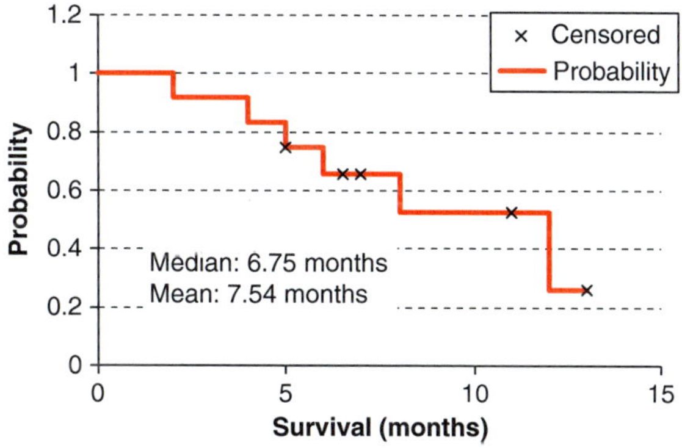

Fig. 4.284 Kaplan–Meier plot of overall survival

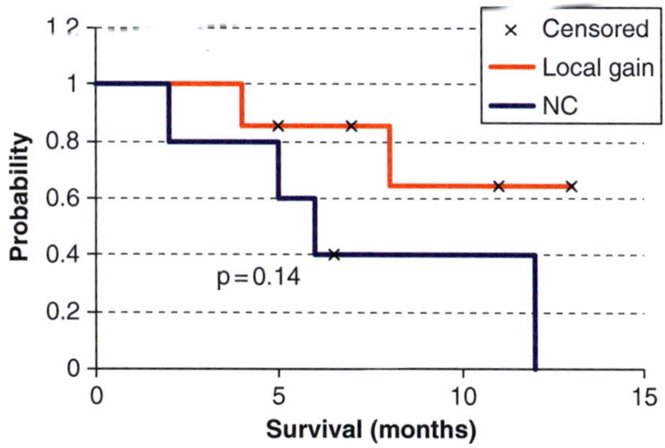

Fig. 4.285 The NC local clinical response has shorter survival

and Dr. A. Dani.) [1262]. The median survival time is 14.4 m (mean 20.5 m), while the median time from the start of oncothermia therapy was 4.8 m (mean: 9.6 m), see Fig. 4.286. Oncothermia was applied weekly 2–3 times with 8–10 treatments using 20-cm diameter electrodes.

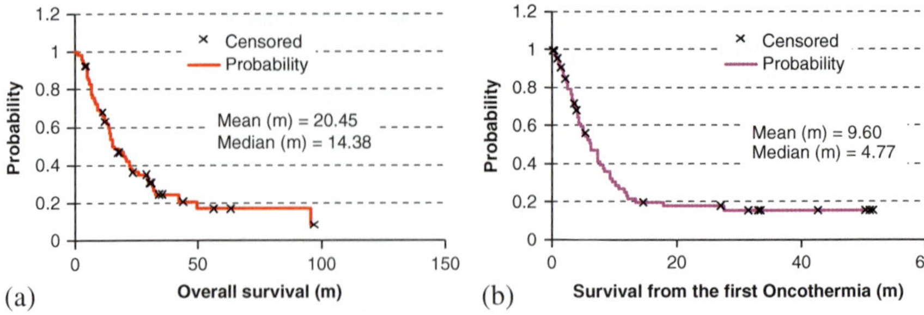

Fig. 4.286 Survival plots for overall (**a**) and for oncothermia (**b**) results for stomach

Fig. 4.287 Comparison of the first-year survivals from the two locations and the large databases

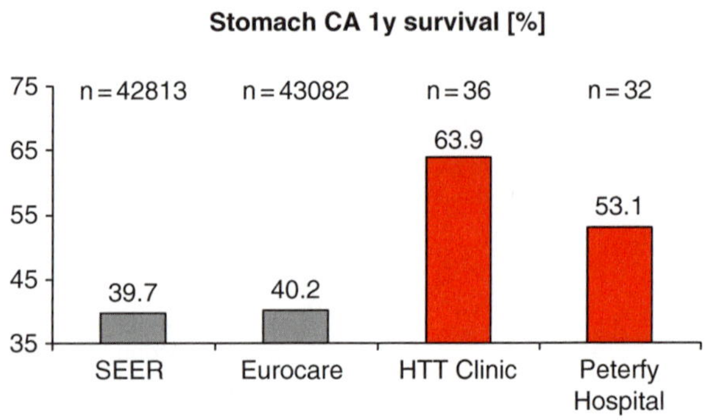

First-year survival rate for oncothermia was 58.8%; in comparison the SEER and Eurocare data are 39.7 and 40.2%, respectively. For second-year survival the difference between the database-registered data and that measured by oncothermia became shallow (oncothermia: 29.4%, SEER 23.8%, Eurocare: 27.8%). In the comparison of the two locations the difference is significant, but these also differ significantly from the large databases, see Fig. 4.287.

4.16 Colo-Rectal Studies

4.16.1 Pre-Operative Oncothermia for Rectum Carcinoma

This study was devoted to pre-operative application of oncothermia for liver metastases from rectum carcinoma [1263]. (Praxis at Klinikum Nord, Nürnberg, Germany. Investigator: Prof. H. Renner.). The studied primary tumors were inoperable ($R2$) rectum carcinoma, ($n = 7$). A trimodal therapy was applied: radiotherapy: 45+5 Gy, (fractional), chemotherapy: 5-FU/Mitomicine-C (2×), Oncothermia: 60 min, diam. 30 cm (8–10×). Result: after oncothermia all patients became eligible for operation. The results of operation (Fig. 4.288) were excellent: 71% of

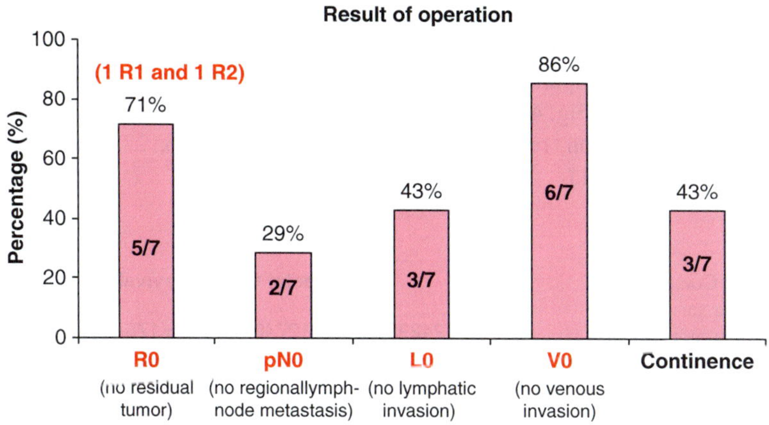

Fig. 4.288 The result of the operations made post oncothermia on the patients that were previously inoperable

patients were in a condition for complete resection ($R0$) while one was partially resected ($R1$) and one was not successfully operated, (remained $R2$).

4.16.2 Colo-Rectal Carcinoma Study

Advanced, heavily pretreated (failed pretreatments), colo-rectal cancer ($n = 218$) was studied in a bicentral study (HTT-Med Clinic and Peterfy Hospital, Budapest. Investigators: Dr. T. Magyar, Dr. A. Varkonyi and Dr. A. Dani.) [1264]. Patients were categorized with rectum ($n = 92$), with colon ($n = 114$), and with rectosigmoid junction (sigma, $n = 12$) carcinomas.

The median survival time is 28.5 m (mean 34.4 m), while the median time from the start of oncothermia therapy was 8.6 m (mean: 14.8 m), see Fig. 4.289. Oncothermia was applied weekly 2–3 times with 6–12 treatments using 20-cm diameter electrodes.

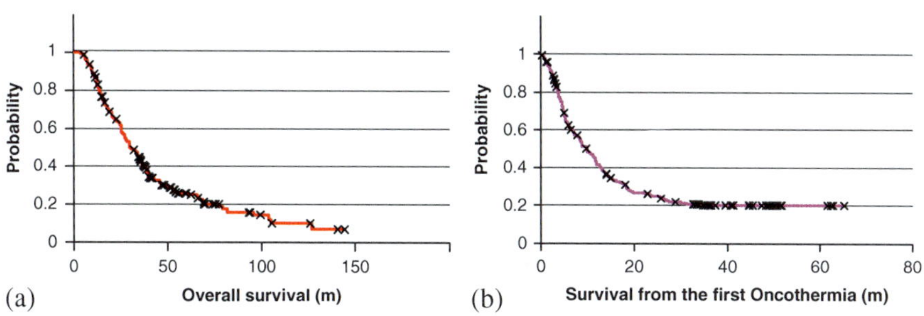

Fig. 4.289 Survival plots of all the colo-rectal patients. Overall survival (**a**) and oncothermia survival (**b**) are shown

First-year survival rate for oncothermia was 84.9%; in comparison the SEER and Eurocare data are 72.0 and 68.9%, respectively. The first-year survival in the two institutions are comparable (87.5 m ($n=178$) and 81.4 m ($n=40$) for HTT and PFY, respectively, see Fig. 4.290).

The median of colon, rectum, and sigma cohorts are 25.6, 27.4, and 28.0 m, respectively (Fig. 4.291).

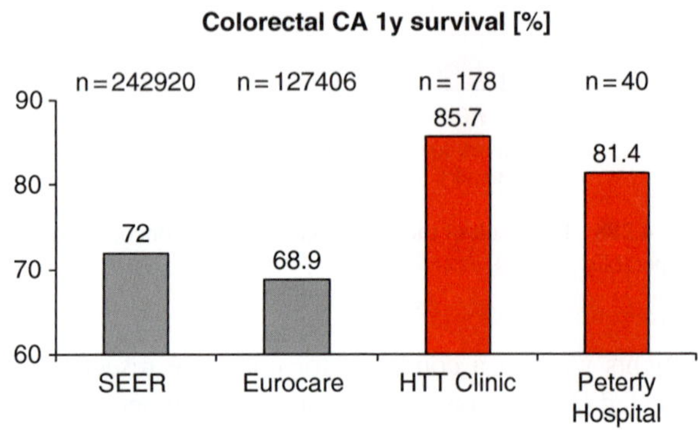

Fig. 4.290 Comparison of the first-year survival of colo-rectal tumor patients. Oncothermia was applied for advanced cases only

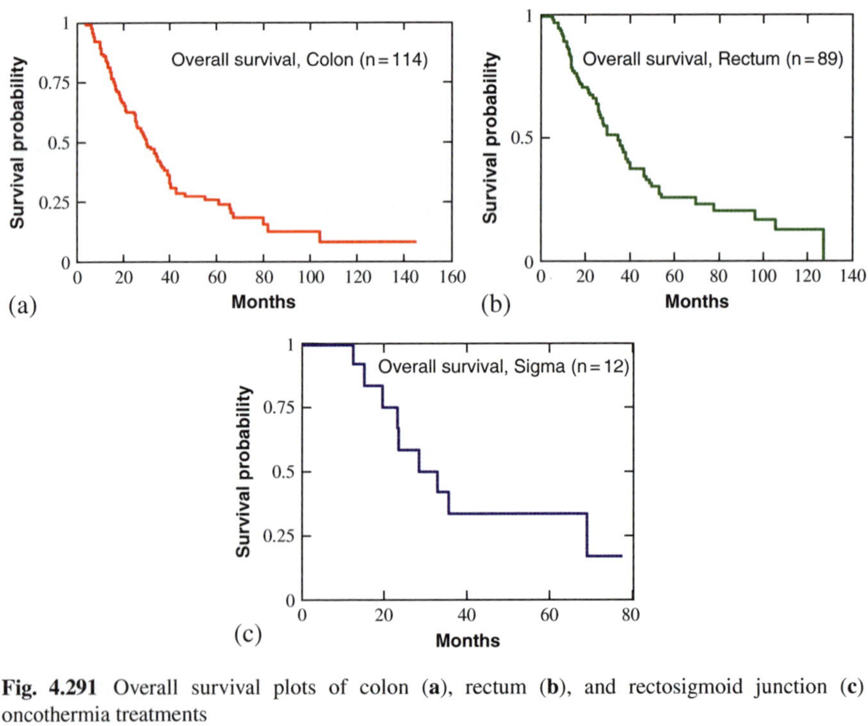

Fig. 4.291 Overall survival plots of colon (**a**), rectum (**b**), and rectosigmoid junction (**c**) oncothermia treatments

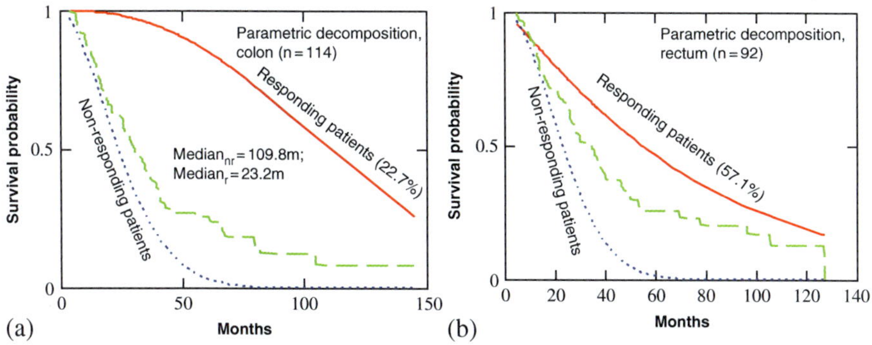

Fig. 4.292 Parametric decomposition of colon (**a**) and rectal (**b**) sites of oncothermia treatment

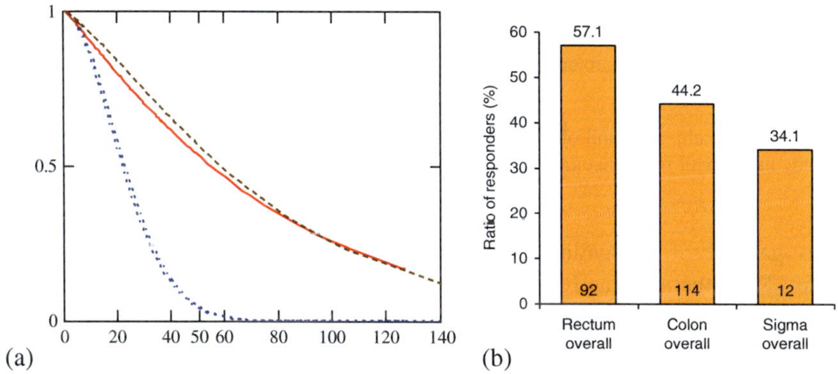

Fig. 4.293 Comparison of the decomposed colon and rectum survival plots (**a**) (the *solid* and *dashed lines* are the responders of colon and rectum sites, respectively. The *dots* are the non-responders from both groups). The main difference between them is the responders ratio (**b**)

The parametric decomposition shows medians 59.5 and 21.4 m for responders and for nonresponders in the case of colon, and 54.3 and 22.6 m for responders and for nonresponders in the case of rectum, respectively. Ratio of responders by the parametric decomposition were 44.2 and 57.1% for colon and rectum, respectively (see Fig. 4.292). Their comparison shows (see Fig. 4.293) an almost identical non-responders survival curve and a slightly different responders survival curve. This indicates the very similar effects of oncothermia at both sites.

4.17 Bone Studies

4.17.1 Refractory Bone Metastases Complementary to Radiotherapy

This small study ($n = 11$) was carried out in the Clinic and Institute of Radio-Oncology, Zentralkrankenhaus Reinkenheide, Bremerhaven, Germany

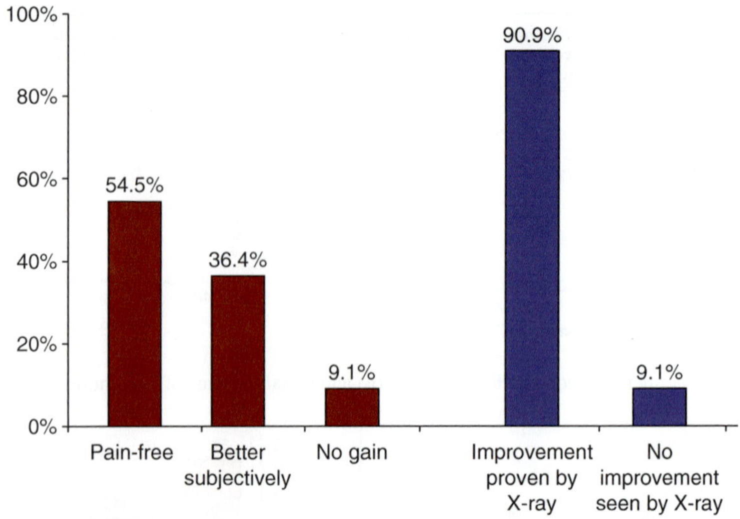

Fig. 4.294 The subjective and objective evaluation of the patients. 10 of 11 patients involved in the study had benefit from oncothermia

(Investigator: Prof. H. Aydin). Oncothermia was applied complementary to radiotherapy. Radiotherapy was 10 MV, 1.5–1.8 Gy fractional radiation 5×/week, overall dose: 21–24 Gy and oncothermia was given 2×/week [1260]. The subjective opinions of patients and the objective measurements of their status were identical: no improvement was seen only at 9% (1/11) of the patients (see Fig. 4.294).

4.17.2 Monotherapy for Advanced Bone Metastases

Advanced bone metastatic tumors were studied ($n=6$) [1264] (HTT-Med Clinic, Budapest, Hungary; Dr. A. Varkonyi and Dr. A. Dani). The medians are 40.1 and 15.4 m (mean 41.3 and 19.1 m) for overall and oncothermia survivals, see Fig. 4.295. Oncothermia was applied weekly 2–3 times with 10–12 treatments using 10 and 20-cm diameter electrodes.

First-year survival rate for oncothermia was 100%; in comparison the SEER and Eurocare data are 72.0 and 68.9%, respectively.

Nyiro ($n=6$)

4.17.3 Osteosarcoma Study

The bone is a good isolator so the effect of oncothermia on this site is not a simple process. This relatively large study ($n=62$) was devoted to clarifying the effect of oncothermia on bone osteosarcoma cases [1265] (Investigator: Dr. F. Douwes,

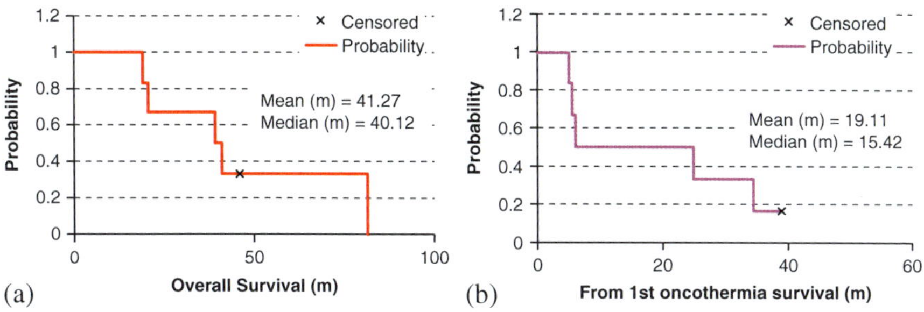

Fig. 4.295 Survival plots for overall (**a**) and oncothermia (**b**) times

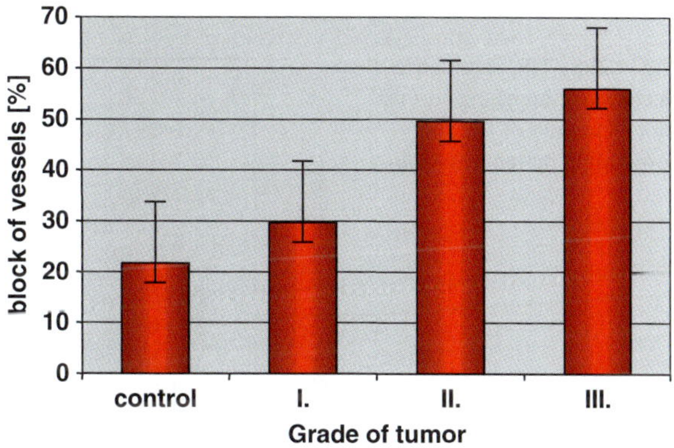

Fig. 4.296 Vasocontraction of tumor-supplying vessels by oncothermia treatment

Clinic St. Georg, Bad Aibling, Germany). The location of the sarcoma in 72.6% of the patients was bones in knees. The treatment was firstly surgery, followed by post-operative radiotherapy 20–36 Gy. This was followed by chemotherapy: (Doxorubicin 30–60 mg/m^2, Adriamycin or Cisplatin 3 × 30 mg/m2), and then oncothermia (2×/week all together 6 ×60 min) was applied. The result: 39.3% subtotal 35.7 total devitalization of the tumor. The mechanism was clarified by measurement of the blood perfusion, vasocontraction of blood vessels (sec Table 4.1.) supplying the tumor. It was shown clearly that oncothermia attacks the supplying blood vessels, (Section 4.1.9) and it is more effective when the tumor is more advanced (see Fig. 4.296). This observation correlates well with other studies where oncothermia has increased efficacy in more advanced cases.

4.18 Kidney Study

Advanced kidney cancer was studied ($n = 39$) [1264]. (HTT-Med Clinic and Peterfy Hospital, Budapest. Investigators: Dr. T. Magyar, Dr. A. Varkonyi and Dr. A. Dani.) The medians are 35.9 m and 10.1 m (mean 49.2 and 14.7 m) for overall and

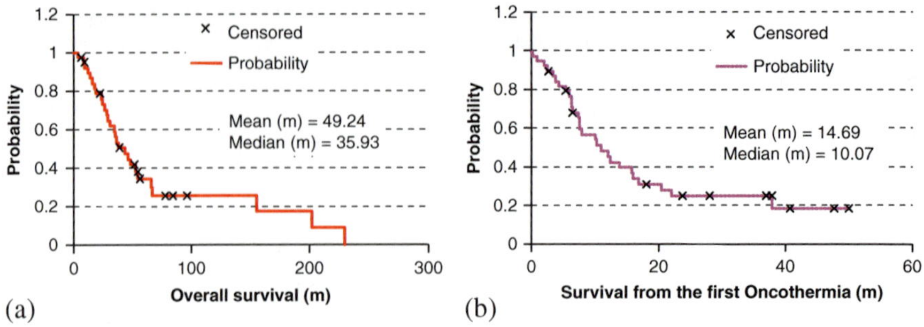

Fig. 4.297 Kaplan–Meier plots for overall (**a**) and oncothermia (**b**) survivals

Fig. 4.298 The parametric decomposition of the survival plot. The ratio of the responding patients is 48%

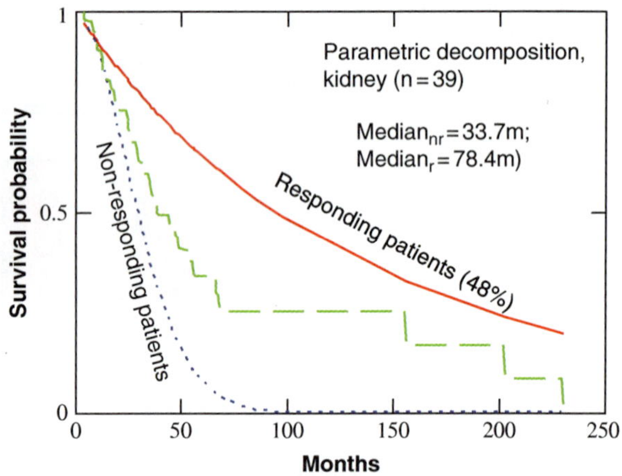

oncothermia survivals (Fig. 4.297). Oncothermia was applied weekly 2–3 times with 10–12 treatments using 20-cm diameter electrodes.

The parametric decomposition shows medians 78.4 and 33.7 m for responders and for nonresponders, respectively, see Fig. 4.298. The responders by the parametric decomposition were less than half, 48% ($n_r = 19$) of the total treated patients ($n = 39$).

First-year survival rate for oncothermia was 84.6%; in comparison the SEER and Eurocare data are 67.5 and 70.9%, respectively.

4.19 Head and Neck Study

Various advanced head and neck cases ($n = 64$, see Table 4.19) were collected [1264]. (HTT-Med Clinic and Peterfy Hospital, Budapest. Investigators: Dr. T. Magyar, Dr. A. Varkonyi and Dr. A. Dani.) The medians are 26.1 and 7.4 m (mean 41.9 and 15.7 m) for overall and oncothermia survivals, see Fig. 4.299

Table 4.19 Identification of head and neck localizations included in the study

Localization	ICD code	Number of patients
Lip	C00	1
Tongue	C01–C02	17
Oral cavity	C03, C05–C06,9	7
Floor of mouth	C04	3
Salivary glands	C08	1
Tonsil	C09	6
Oropharynx	C10	8
Nasopharynx	C11	4
Hypopharynx	C12–C13,9	8
Pharynx/other buccal	C14	9

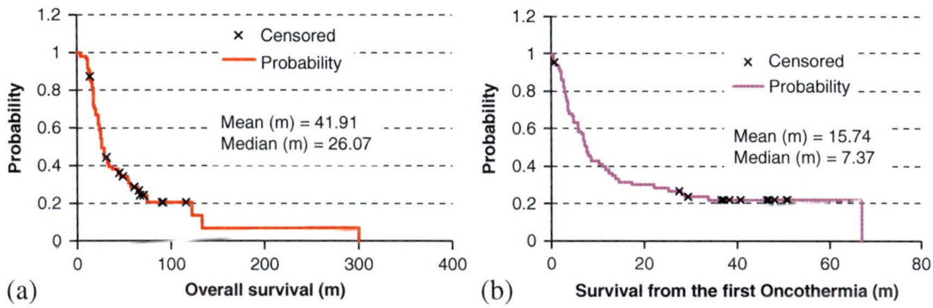

Fig. 4.299 Overall (**a**) and oncothermia (**b**) survival plots

Oncothermia was applied weekly 2–3 times with 8–12 treatments using 10-cm diameter electrodes.

First-year survival rate for oncothermia was 92.2%; in comparison the SEER and Eurocare data are 74.9 and 67.4%, respectively.

4.20 Urinary Bladder Malignancies

Advanced urinary bladder cancer was studied ($n = 18$) [1262]. (HTT-Med Clinic, Budapest, Hungary. Investigators: Dr. A. Varkonyi and Dr. A. Dani.) The medians are 36.5 and 9.9 m (mean 47.2 and 20.2 m) for overall and oncothermia survivals, see Fig. 4.300. Oncothermia was applied weekly 2–3 times with 10–12 treatments using 20-cm diameter electrodes.

The parametric decomposition shows medians 42.0 and 22.6 m for responders and for nonresponders, respectively, see Fig. 4.301. The responders by the parametric decomposition had high percentages, 73% ($n_r = 13$) of the total treated patients ($n = 18$).

First-year survival rate for oncothermia was 85%; in comparison the SEER data is 81.5%.

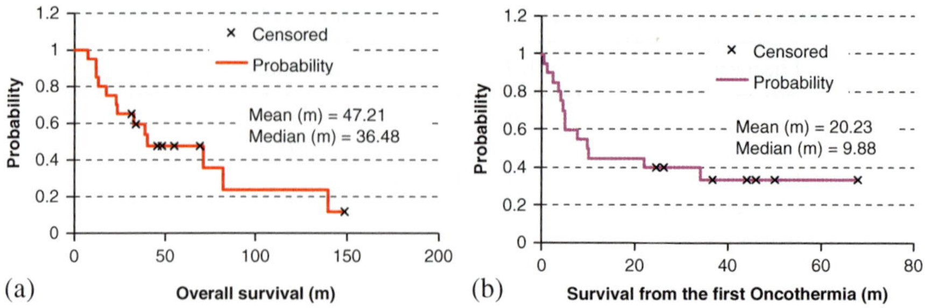

Fig. 4.300 Kaplan–Meier plots for overall (**a**) and oncothermia (**b**) survivals

Fig. 4.301 Parametric decomposition of urinary bladder overall survival curve

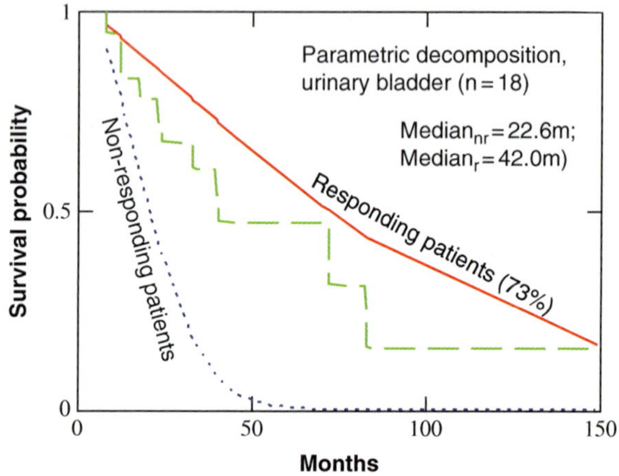

4.21 Soft-Tissue Malignancies

Advanced soft-tissue cancer was studied ($n = 16$) [1262]. (HTT-Med Clinic, Budapest, Hungary. Investigators: Dr. A. Varkonyi and Dr. A. Dani.) The medians are 35.9 and 13.3 m (mean 46.8 and 16.3 m) for overall and oncothermia survivals, Fig. 4.302. Oncothermia was applied weekly 2–3 times with 10–12 treatments using 20-cm diameter electrodes.

The parametric decomposition shows medians 115.3 and 31.3 m for responders and for nonresponders, respectively, see Fig. 4.303. The responders by the parametric decomposition are 31% ($n_r = 5$) of the total treated patients ($n = 16$).

First-year survival rate for oncothermia was 100%; in comparison the SEER and Eurocare data are 76.7 and 77.7%, respectively.

4.22 Prostate Study

Advanced prostate cancer was studied ($n = 18$) [1262]. (HTT-Med Clinic, Budapest, Hungary. Investigators: Dr. I. Philip.) The medians are 38.8 and 10.9 m (mean 46.5

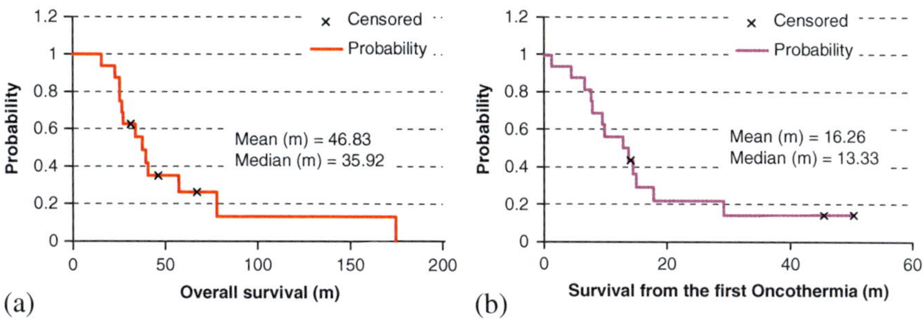

Fig. 4.302 Overall (**a**) and oncothermia (**b**) survival plots

Fig. 4.303 Parametric decomposition of overall survival for responding and non-responding patients

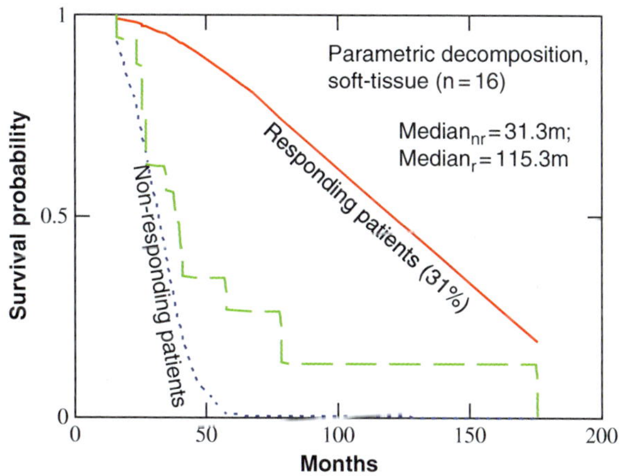

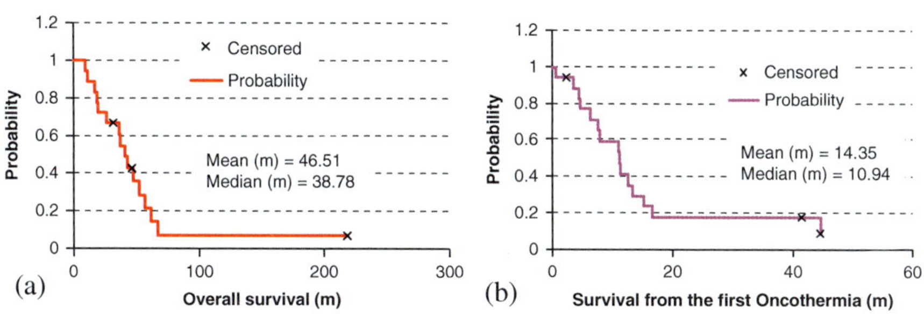

Fig. 4.304 Overall (**a**) and oncothermia (**b**) survivals

and 14.4 m) for overall and oncothermia survivals, see Fig. 4.304. Oncothermia was applied weekly 2–3 times with 10–12 treatments using 20-cm diameter electrodes.

First-year survival rate for oncothermia was 88.9%; in comparison the SEER and Eurocare data are 85.9 and 82.7%, respectively.

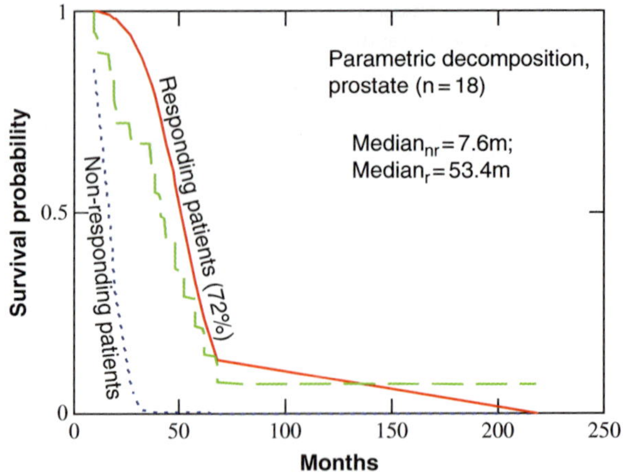

Fig. 4.305 Parametric decomposition of the prostate survivals for responders and nonresponders for oncothermia treatment

The parametric decomposition shows medians 42.0 and 22.6 m for responders and for nonresponders, respectively, see Fig. 4.305. The responders by the parametric decomposition had high percentages, 72% ($n_r = 13$) of the total treated patients ($n = 18$).

4.23 Oncothermia Perspectives

It is feasible to apply oncothermia as a routine treatment modality in advanced cases, when the "gold standards" have failed or are no longer applicable, or when boosting/resensitization is necessary. The results for first-year survival ratio in oncothermia above that of the SEER database (see Fig. 4.306) represent a real promise for future applications. The survival benefits shown by median survival in oncothermia for gastrointestinal diseases (see Fig. 4.307). It works perfectly in the case when the "gold standards" are less effective, but is less perfect, when the conventional methods are strong (like in the case of colo-rectal localizations). The enhancement is satisfactorily high in the localizations, where the median survival time is relatively low, causing earlier start of oncothermia.

Grouping the first-year survivals results by the relative benefit over SEER, we can distinguish three groups of localizations for oncothermia applications (see Fig. 4.308): a group where oncothermia offers especially good results (like liver, pancreas, lung, esophagus, brain, etc.), a group showing moderate improvement (like head and neck, kidney, ovary), and a group where the additional benefit is around 20% (like rectum, colon, skin, breast). However, in all the cases we have better results when oncothermia is applied, and even in the group showing mild improvement over the standards the advantage of oncothermia begins when the standard treatments are no longer applicable.

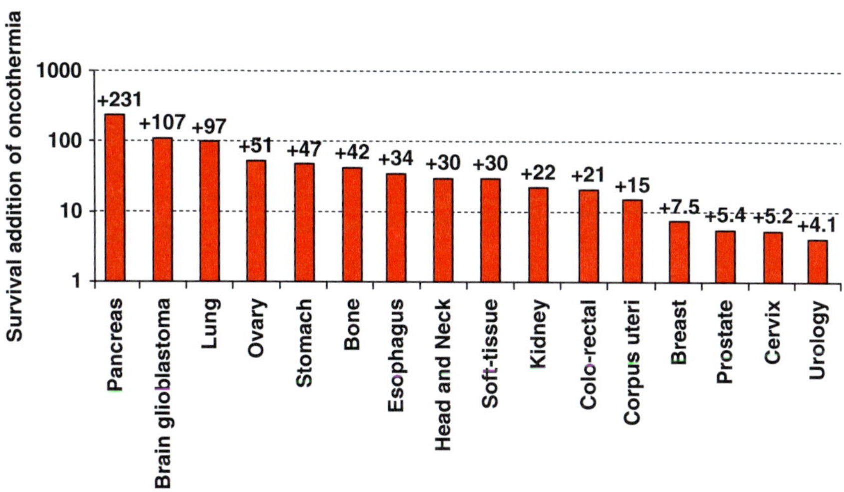

Fig. 4.306 Additional percentages to SEER data by oncothermia in the first-year survivals

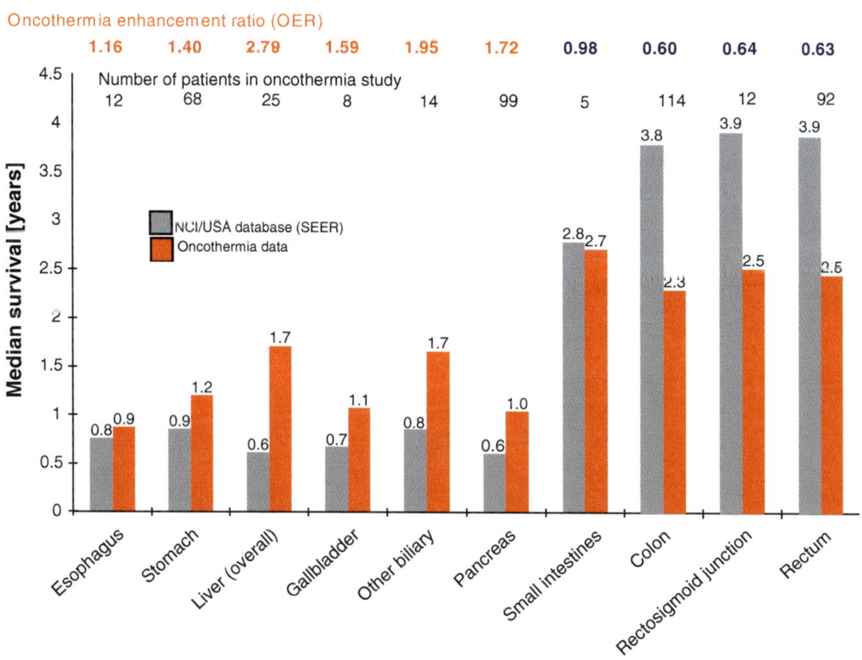

Fig. 4.307 Comparison of oncothermia results (median survival) to SEER database for gastrointestinal tumors

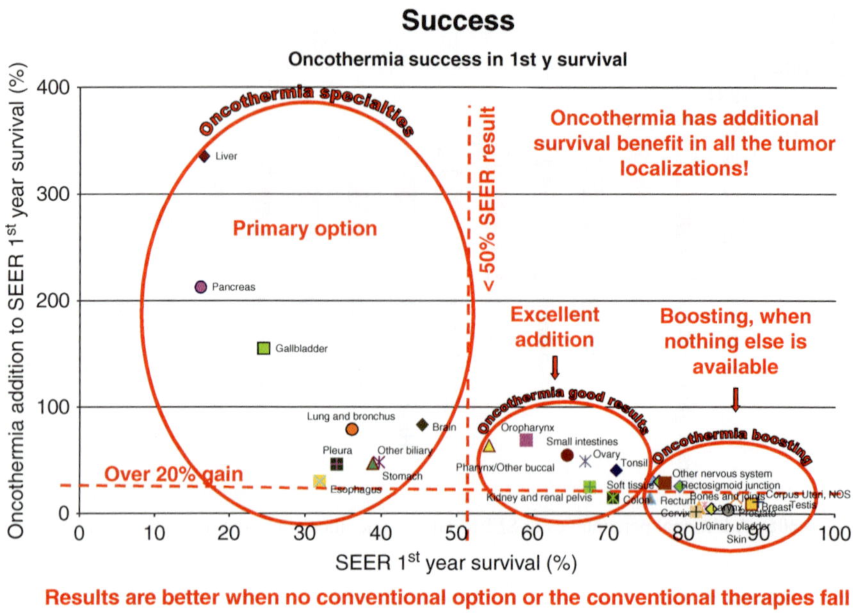

Fig. 4.308 Oncothermia benefit on first-year survival, relative to SEER data

Oncothermia definitely has a future in curative and palliative care in oncology. It is a complementary method, applicable only in cases where improvement of the gold standards is necessary for variable reasons.

There is a long path before us to make oncothermia a well-established routine among the standard modalities. However, the journey has begun and with the support of smart researchers and clinicians oncothermia will reach its goal.

Appendixes

Appendix 1: Entropy and Temperature

The temperature-proportional part of the energy (average energy of the particles) is the internal energy (U) of an object. This is the part that is represented by the kinetic energy of the particles involved in the structure. The other part of the energy can be represented by the volumetric mechanical energy of the system pV. The total energy of such a system is called enthalpy (H):

$$H = U + pV \tag{A.1.1}$$

Enthalpy is a very useful type of energy for living objects afterall life is studied under constant ambient pressure; consequently the change in enthalpy may be calculated from the change in the temperature-proportional internal energy and the volume-proportional change of mechanical energy:

$$\Delta H = mc\Delta T + p\Delta V \tag{A.1.2}$$

where m is the mass of the system, c the specific heat, and Δ denotes the change of the actual parameter. Naturally, if there were other variables in the system, than Eqs. (A.1.1) and (A.1.2) would have additional terms.

Entropy by definition is the part of a system's heat/energy that is not available to perform work [1266]. The general definition of entropy is not simple. The present definition is the "thermodynamic entropy" introduced by Clausius, Kelvin, and Carnot in the middle of the nineteenth century [1267].

This is the energy blocked for heating, making the actual temperature. It shows how many heat (energy) is absorbed in the system at the actual temperature:

$$\text{change of entropy} = \frac{\text{requested heat (energy) } [J]}{\text{temperature } [K]} \quad \left\{ \Delta S = \frac{\delta Q}{T} \left[\frac{J}{K} \right] \right\} \tag{A.1.3}$$

where ΔS is the change of entropy by δQ heat-energy intake at temperature T. (Simply put, the $T\Delta S$ product is the heat-energy content of the system.) Entropy

A. Szasz et al., *Oncothermia: Principles And Practices*,
DOI 10.1007/978-90-481-9498-8, © Springer Science+Business Media B.V. 2011

is measured in [J/K]. In this negative definition the entropy measures the dispersed energy, i.e. measures how much energy was spread out in a particular process. Dispersal of energy is generally present in all spontaneous processes, so the entropy definitely increases in every particular unaided event. The value of the introduced entropy could therefore solely measure the direction of progress in a system, measure how the energy is spontaneously redistributed among a growing number of energy levels, and measure the growing randomization of movements in the system. A fallen ball gradually loses its potential energy and stops somewhere on the ground (i.e. it reaches its equilibrium at some point). Its original energy was distributed into the heating of the ball and the ground, and other factors (e.g. sound) also removed energy. Originally every particle in the ball was falling with the same velocity and in the same direction. After the ball has stopped the particles average velocity will be greater than before the ball fell, but the particles are moving in disordered directions. Order–disorder phenomena could lead to another definition of entropy. This definition characterizes the disorder, and it is calculated by the logarithm of the number, which shows all possible realizations of the same arrangement at the actual temperature [1268]. This "logical entropy" was used to characterize the order–disorder state, and it has no physical unit. This entropy was used by Shannon for his information theory [1269]. The two definitions could be differentiated as "thermal" and "configurational" entropy. In any case entropy is an "anthropomorphic concept" [1270], which could help us to understand the relation between individual interactions and multiple interactions (order of magnitude of Avogadro's number) in large, many-body systems.

Entropy describes simple observations. A spontaneous ordering of the disoriented movements in the ball leading to the ball lifting itself up again is not possible; however, this unexpected process has enough energy to be performed, but it is spontaneously never ever able to occur. We can arrange a constrained order of the randomly moving particles again, but the system has to be cooled down, i.e. we need to take away the individual random energies of the involved particles. The other possibility is to ignore the obtained random, individual movements of the involved particles (without respect for its actual temperature), and add an extra collective movement to all of them by moving the whole system in a definite direction, giving new (kinetic) energy to the whole. This kinetic energy could work or could be "lost" for further work by the random internal movements. The other trivial example is two bodies with different temperatures touching each other. It is a rule of nature that the cooler body will warm and the warmer body will cool until an equal temperature (i.e. equilibrium) is reached; spontaneously the opposite is never possible despite their energetic condition allows it. The exact rule is: the entropy of a closed system (isolated from outside interactions, i.e. from new energy-sources) could never decrease spontaneously.

In closed (energetically isolated from the environment) systems any spontaneous processes divide the available energy before the process begins into two parts: one part that remains for possible further work and another part of the energy that is trapped in the system, which is unusable later, i.e. not available for further work (see Fig. A.1.1).

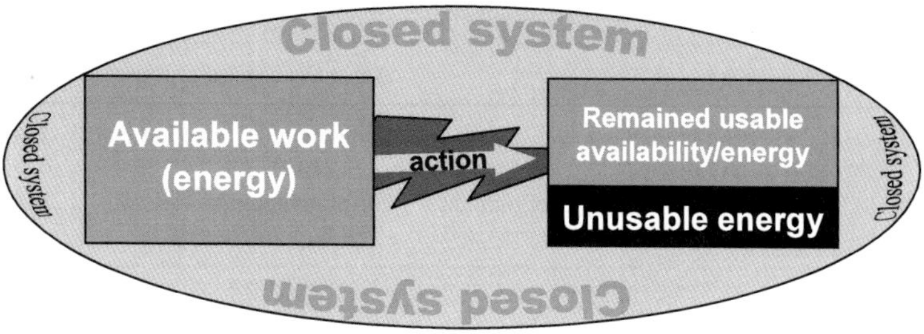

Fig. A.1.1 Energy categories after a spontaneous process in a closed system

The two processes are not always connected. Spontaneity cannot be deduced from disorder or entropy alone. There are processes which the energy in a system prefers (*exothermic*, energy is liberated) and these may be a stronger driving force than the entropy increase, therefore these processes could lead to an entropy decrease. For example, water forming from hydrogen and oxygen is an exothermic process:

$$2H_2 + O_2 \Leftrightarrow 2H_2O + \text{energy} \tag{A.1.4}$$

The liberated enthalpy change is $\Delta H = 571.6$ kJ/mol and the entropy change is the opposite: $T\Delta S = -97.4$ kJ/mol. Ultimately the full free energy is the difference between these: $\Delta G = 474.2$ kJ/mol. [The whole process was activated (ignited) by $E_{a\text{–}H+O} = 7.2$ kJ/mol (~600°C)]. The ΔG value is called Gibbs free energy which is:

$$G = H - TS \tag{A.1.5}$$

G changes under isothermal conditions as:

$$\Delta G = \Delta H - T\Delta S \tag{A.1.6}$$

Under isothermal (T=const.) and isobar (p = const.) conditions the spontaneity occurs when the Gibbs free energy is not positive: $\Delta G \leq 0$. The general driving force of spontaneous biological (chemical/physical) processes (under isothermal and isobar conditions) is the minimizing of Gibbs free energy. In the process of seeking energy minima, two independent parts of ΔG must be considered: the ΔH enthalpy-change and the $T\Delta S$ heat-energy content (average kinetic energy in a given internal order of the system) are both important and independently modify the minimizing process (see Table A.1.1) (Note an endothermic process could also be spontaneous above a definite temperature, when the driving force towards disorder is overwhelmed by the energy-reduction forces.)

Seeking energy minima is a plausible request, supported by everyday experience. The entropy-containing second part is not so trivial. We have to understand the other kind of energy content: i.e. order, the internal arrangement of a given system. We

Table A.1.1 Condition of spontaneity for most of the biological (isotherm/isobar) processes

$\Delta G = \Delta H - T\Delta S$	$\Delta S > 0$	$\Delta S < 0$	Name
$\Delta H < 0$	Spontaneously exergonic (every T)	Spontaneously exergonic at $T < \Delta H/\Delta S$	Exotherm
$\Delta H > 0$	Spontaneously exergonic at $T > \Delta H/\Delta S$	Unspontaneously endergonic (every T)	Endotherm

know that to obtain order in a particle set we have to put in work/energy to order the generally disordered arrangement. Some examples could be given: The individual pages of a written essay could easily become jumbled if we drop them. To order the pages again needs definite work. Also significant energy has to be expended to order liquid water molecules in a very definite order to obtain solid ice. A well-ordered system could become spontaneously disordered (sugar dissolves in water, a liberated gas spontaneously mixes in air, etc.), and work would be necessary to re-establish the original order. Other examples of the spontaneous direction of internal order–disorder phenomena could be derived from the general one-way behavior of our life, the "arrow of time."

However, in some processes the liberated energy from the ΔH enthalpy change could be enough to produce order, and can be used for the work of ordering. For example, the mixture of H_2 and O_2 gases has significant disorder, but after ignition such a large amount of enthalpy liberates that it can order the molecules and make liquid (which is better ordered than gas) H_2O as a final product. The $T\Delta S$ term of ΔG is definitely the same energy component as ΔH. However, the enthalpy change is the "usual" available energy, while $T\Delta S$ is the energy that is stored in the actual structure (ΔS is the pure structural "skeleton" and T is the average energy of the units in the structure) and/or in the actual interactions. Naturally it is a part, which spontaneously grows, which is never again available for use in spontaneous processes. Changes in chemical bonds in a reaction could have a Janus-phase, i.e. seeking stability (spontaneity) in the system: if the chemical bond is more ordered in an isothermal process, than it has to have energy investment, however, if the reaction is exothermic, it could be a spontaneous process without any extra work being put in; the liberated exothermic energy meets the requested energy for the ordering. Also, if an exothermic chemical reaction could take place, but the work ($T\Delta S$) which would be necessary for the actual ordering of the product is greater, then the reaction can not be performed without outside energy investment. The balances of ΔG and its terms (with some characteristic examples), are shown in Fig. A.1.2.

A complex system could have many different chemical species and their reactions have various driving forces. The change in Gibb's free energy from the change in concentration (molecular atomic/molecular number, n) of a given species (lets say the i-th) when no other reactions occur under isothermic and isobaric conditions is the chemical potential,

$$\mu_i = \left(\frac{\Delta G}{\Delta n_i}\right)_{p,T,n_j(j\neq i)} \qquad \text{(A.1.7)}$$

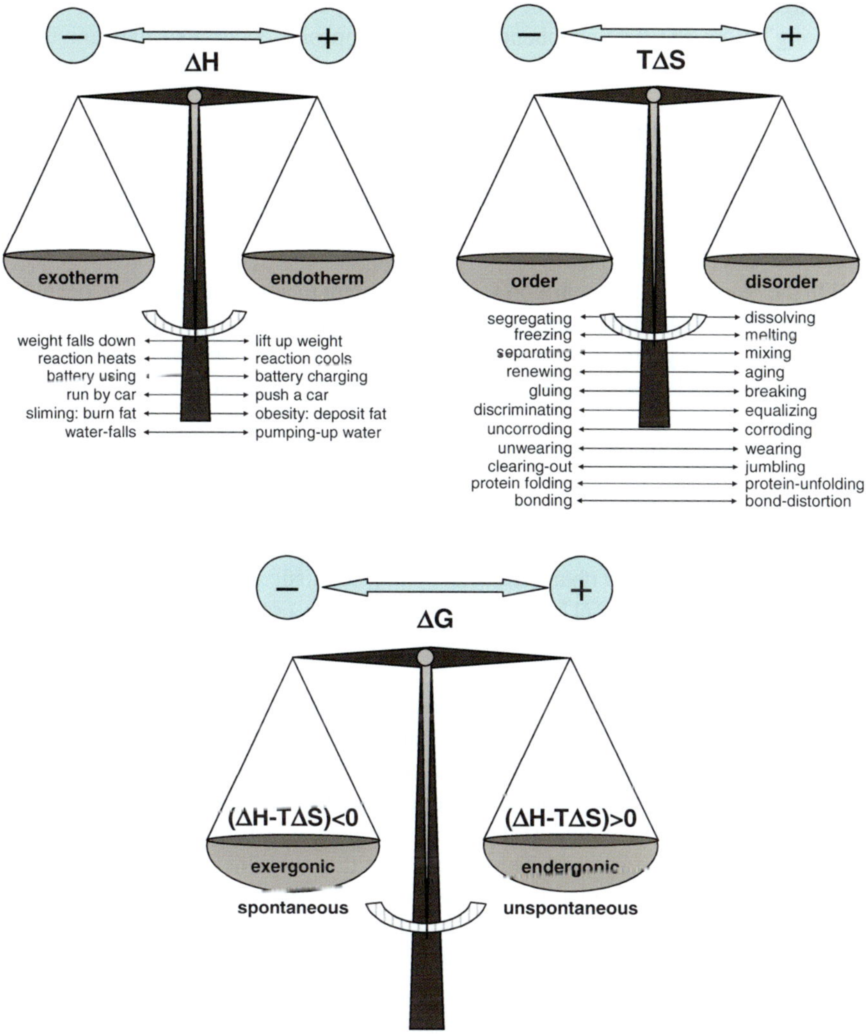

Fig. A.1.2 The balances of Gibbs free energy and its terms with some characteristic examples

where the necessary conditions, (the values kept constant during the process), are noted in the subscripts.

The phase transitions, (e.g. melting/freezing, evaporation/condensation, ferro-/para-magnetism), depending on their kind and order, could have a remarkable effect on the energy balance. These energies (usually called "latent heat") have to be taken into account when the actual phase changes.

We learned above, that the energy invested in a system could be used for simple heating (temperature rise) or it could change chemical bonds and/or change the ordering/structure of the actual arrangement. Naturally the energy part, which rearranges the structure, will not increase the temperature, so it will be apparently "missing" from the heating phenomena. The opposite could also happen: if the invested energy ignites a process, which itself liberates significant energy through disordering (or via a chemical reaction) than the temperature of the system will be higher than expected from the invested energy alone.

The biological energy "combustion" is the general driving force for the development of more and more sophisticated and accommodated biosystems. This evolution enables the living system to be more and more independent of its environment. Life organizes and stabilizes itself.

On the one hand, such evolution forces the system to adapt and improve, developing a perfect "combustion" ability, but on the other hand, self-reproduction is rather conservative: the system sustains itself at the same level. Slight changes in this conservative process lead to evolution, controlled only by changing the conditions and environmental demands.

A possible uncontrolled "combustion" of a chosen unit, will consume the energy of others in its near vicinity, blocking continuation of this process. This situation can be solved by a positive feedback to limit the extra demand. This is a special self-organized collectivity. From the chemical point of view, the living process is a highly (self-) organized charge transfer with a sophisticated energy accumulation and self-reproduction [1271]. To maintain the living process, nutrition has to be transported to the living unit (e.g. cell) or the living system moves itself to the source of nutrition.

If we put energy into a system, forcing it to carry out a process at odds with its spontaneous direction (e.g. lift up a ball, cool a room, produce order in our workplace, etc.) we increase the Gibbs free energy by force (an endergonic process). Among isochoric (unchanged volume) conditions it is a definite decrease of the entropy, which works against the averaging, puts an extra and definite property (information) into the system. This process (with double negation) increases the negative entropy, and is called in short negentropy [639].

While the entropy as a property of a system/substance qualifies the energy of the system/substance that is no longer available to perform useful work; the negentropy is a property of the system that shows how far it is from equilibrium (from the spontaneous final stage) or in the order–disorder meaning of entropy the negentropy characterizes the order, the information carried by the given system, and it increases by the complexity of a given unit, see Fig. A.1.3.

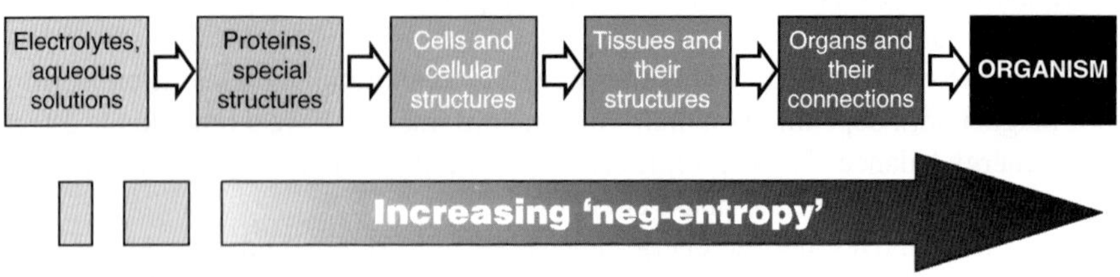

Fig. A.1.3 The building blocks of living systems and the increased negentropy

Besides biology, negentropy is widely used in various modern sciences, such as in information theory, computer science, communications theory, organization/risk theory, and in many statistical considerations, etc.

Appendix 2: Thermodynamics – Onsager's Relation

In every case we may introduce the densities of extensive and intensive parameters (x_i and y_i) as well. (In thermal equilibrium the densities are constant over the whole system).

$$X_i = \int_V x_i\left(\underline{r}, t\right)dV \quad \text{and} \quad Y_i = \int_V y_i\left(\underline{r}, t\right)dV \tag{A.2.1}$$

All the extensive densities must satisfy the balance of the sources and flows:

$$\frac{\partial x_i\left(\underline{r}, t\right)}{\partial t} + \operatorname{div}\left[\underline{j}_i\left(\underline{r}, t\right)\right] = y_i\left(\underline{r}, t\right) \tag{A.2.2}$$

where

$$\underline{j}_i = \frac{dX_i}{dt} = \sum_{k=1}^{n} A_{i,k}\operatorname{grad}(Y_k) \tag{A.2.3}$$

This equation is known as Onsager's law, stating the various interacting driving forces, and so the $A_{l,\,k}$ matrix characterizes how the k-th intensive affects the flow of the i-th extensive. The flow of any extensive is driven by the set of gradients of the intensives. This non-equilibrium thermodynamic behavior is an essential character of bioenergetics [1272].

The intensives could have their gradient only from the nonhomogeneities of extensive parameters. Consequently:

$$\operatorname{grad}(Y_k) = \sum_{l=1}^{n} \frac{\partial Y_k}{\partial x_k}\operatorname{grad}(x_l) \tag{A.2.4}$$

Hence

$$\underline{j}_i = \sum_{k=1}^{n} B_{i,k}\operatorname{grad}(x_k) \tag{A.2.5}$$

where

$$B_{i,l} = \sum_{k=1}^{n} A_{i,k}\frac{\partial Y_k}{\partial \rho_k} \tag{A.2.6}$$

The results clearly show all flows originate from the gradient (nonhomogeneity) of the extensive parameters, which is in good correspondence with the entropy law. The flows (currents) exist till the gradient exists, and vanish when the homogeneity is re-established.

Appendix 3: Self-Similarity and Bioscaling

Below we show that the self-similarity of biological systems produces the "fourth" dimension, so only by a simple self-similar assumption can we rigorously obtain the ¾ scaling of metabolic power for systems with circulation networks, under the condition when the main driving force available is the maximum of metabolic power. Simply speaking this scaling is the direct consequence of the optimizing of available metabolic energy consumption in every volumes and every scales by "fuel" transport. The fourth dimension is the effective transport surface (the blood-vessel network is calculated) relative to its one dimensional (like a pipe system) behavior. We may derive it from the self-similarity of the organism having L_0 characteristic length. As we have seen the energizing of processes is a crucial factor for living phenomena. Organization and all its peculiarities follow the general physical rules. A so-called symmorphosis was proposed [1273], hypothesizing that animals evolved their structures and functional capacities economically to satisfy the maximum physiological requirements. According to this we suppose the characteristic parameters are self-similar functions of L_0.

$$l \propto {L_0}^{a_l}, \quad a \propto {L_0}^{a_a}, \quad v \propto {L_0}^{a_v} \tag{A.3.1}$$

where l is the average length of the circulation system, a is the effective transport surface of the full transport system (it is proportional to the available metabolic activity), and v is the volume of the transporting fluid (blood). a_l, a_a, and a_v are the fractal dimensions of the given parameters. Furthermore, we suppose these dimensions are larger than 1, 2, and 3, respectively. (The character is fractal-like.) Hence:

$$l \propto {L_0}^{1+\varepsilon_l}, \quad a \propto {L_0}^{2+\varepsilon_a}, \quad v \propto {L_0}^{3+\varepsilon_v} \tag{A.3.2}$$

and the volume is proportional to the surface by length:

$$v \propto a \times l \tag{A.3.3}$$

In consequence the exponents are not independent:

$$\varepsilon_v = \varepsilon_l + \varepsilon_a \tag{A.3.4}$$

The trivial assumption is the volume of blood (transporting fluid) is proportional to the full mass of the organism (M):

$$v \propto M \tag{A.3.5}$$

So, we obtain from (A.3.2):

$$L_0 \propto v^{\frac{1}{3+\varepsilon_v}} \rightarrow a \propto v^{\frac{2+\varepsilon_a}{3+\varepsilon_v}} \tag{A.3.6}$$

Considering Eqs. (A.3.4), (A.3.5), and (A.3.6) the effective surface is a definite power-function of the mass of the organism:

$$a \propto M^{\frac{2+\varepsilon_a}{3+\varepsilon_l+\varepsilon_a}} \tag{A.3.7}$$

The result is clear: when the transport system has no fractal structure ($\varepsilon_l = 0$, $\varepsilon_a = 0$, and $\varepsilon_v = 0$), the effective surface of the transport (which is proportional to the available metabolism) has a scaling exponent of $2/3$. Evolution, however, chooses the best available solution for the given conditions. In consequence, the surface for the active metabolism must be as large as possible to give the highest possibility of energizing the system. Thus requesting the maximum of a:

$$a\left(\varepsilon_l, \varepsilon_a\right) \propto M^{\frac{2+\varepsilon_a}{3+\varepsilon_l+\varepsilon_a}} = \max \tag{A.3.8}$$

The solution is simple:

$$\varepsilon_l = 0, \quad \varepsilon_a = 1 \tag{A.3.9}$$

so

$$a \propto P_{\mathrm{met}} \propto M^{\frac{3}{4}} \tag{A.3.10}$$

The metabolic power (P_{met}) has exponent $p = 3/4$ scaling when the request of evolution (the maximal metabolic activity) is counted. The metabolic rate (R_{met}) is the metabolic power in unit mass:

$$R_{\mathrm{met}} \propto \frac{P_{\mathrm{met}}}{M} \propto M^{-\frac{1}{4}} \tag{A.3.11}$$

Of course the original dimensionalities with this result are:

$$l \propto L_0{}^1, \quad a \propto L_0{}^3, \quad v \propto L_0{}^4 \tag{A.3.12}$$

So the mass of the blood (transport electrolyte for metabolism) has four dimensions. Life in this meaning is "four dimensional" (in addition to the space dimensions the self-organized highest energy consumption is the fourth dimension), where metabolic exchange processes proceed on fractal surfaces. The metabolic power and its fluctuation have universal scaling as well [1274]. The scaling is shown for a large number of living structures and processes [1275]. The optimization was rigorously formulated by the scaling idea, and can be discussed in a universal frame

[1276], even in relation to sub-cellular energy consumption and mitochondria and respiratory complexes [713].

The above calculation shows a structural, geometrical constrain for the living organisms. However, life is more complex than its geometrical structure determines alone. The metabolic power depends on not only the active transport surface, but also on the flow rate, which could modify transport on active surfaces of the same size. In the above discussion we supposed a unified flow for every organism.

The transported mass of nutrition (N_t) has to be a product of the heartbeat (f_h) and the nutrition mass in a single heartbeat (N_{st}) and has to be proportional to the actual metabolic power ($N_{st} \propto M$). Hence

$$N_t = N_{st} f_h \propto P_{met} \propto M^{3/4} \tag{A.3.13}$$

Suppose the nutrition delivery in a single heartbeat is proportional to the mass of the organism, we obtain:

$$f_h \propto \frac{P_{met}}{N_{st}} \propto \frac{M^{3/4}}{N_{st}} \propto \frac{M^{3/4}}{M} \propto M^{-1/4} \tag{A.3.14}$$

So, the heart rates are well scaled by the power of $P = -\frac{1}{4}$ power of the body-mass in mammals [1277]. The heartbeat and the metabolic rate have the same mass-scaling dependence. Surprisingly the statistical value of the heartbeat in a life-time does not change in relation to life expectancy or the mass of an organism and is pretty stable for mammals at around $n_{hb/lt} \approx (7.3 \pm 5.6) \cdot 10^8$ heartbeat/lifetime [1278], which supports the unifying delivery of the nutrition, but many other factors could modify this picture. In consequence the full lifetime (T_{lt}) is also scaled by the $P = \frac{1}{4}$ power law [1278]:

$$T_{lt} = n_{hb/lt} / f_h \propto M^{1/4} \tag{A.3.15}$$

Appendix 4: Energy Supply by Demand

All living reactions are surface controlled. The life-reaction rate on these surfaces depends on the surface size and the transport flow, but also depends on the actual status of the subunit, which needs to be supplied. The actual surface has to allow various reaction rates depending on the actual tissue demand, which is of course different in resting state than under active load. The idea of metabolic level boundaries (MLB) [1279] was developed for these conditions. In resting state the energy demand is minimal and the supply could follow the $p = \frac{2}{3}$ scaling exponent, which is geometrically the lowest available rate. In full demand however, the actual body part (organ or whole body) needs maximal available energy supply, which is proportional to its mass, so the scaling exponent is $p = 1$ in this case [1280]. In this case the actual demand decides the metabolic power and not the geometry. Of course both extremes are not ideal for the living object, and would not follow well evolutionary requests. What is optimal?

Under normal conditions the actual surface of the metabolic reactions must allow both demands properly. So in healthy objects the exponent of metabolic scaling (p) by the mass ranges $2/3 \leq p \leq 1$. A huge set of data on the scaling exponents of various living species shows a wide range of p ($-1.25 < p < 2$ [!], $0.61 < p_{mean} < 1$; confidence mean: $0.06 < CI_{mean} < 0.22$) [711].

The challenge is the situation, when the developed active surface is not able to supply the actual demands. Such a situation can originate from two different sources:

1. The actual active surface is normal, but the new demand increased abnormally. In this case the supply surface follows the evolution-requested exponent of 3 ($\varepsilon_a = 1$) in the self-similar conditions in (A.3.1), but the requested length and volume are not adequate, so $\varepsilon_l > 1$ and $\varepsilon_v > 1$. Because of the self-similar conditions (A.3.4), denote the non-zero ε_l by ε_{lm}:

$$l \propto L_0^{1+\varepsilon_{lm}}, \quad a \propto L_0{}^3, \quad v \propto L_0{}^{4+\varepsilon_{lm}} \tag{A.4.1}$$

   Consequently, the surface for reactions behaves by the power law:

$$a \propto P_{met} \propto M^{\frac{2+\varepsilon_a}{3+\varepsilon_{lm}+\varepsilon_a}} = M^{\frac{3}{4+\varepsilon_{lm}}} \tag{A.4.2}$$

   So the "dimension" of the reaction request is $(4+\varepsilon_{lm}) > 4$. The ε_{lm} is measurable by the fractal dimension of the vessel structures [1281]. Suppose $\varepsilon_{lm} = 0.3$, the scaling exponent is $p = 0.69$. The micro-vessel fractal dimension (MFD) [which is equivalent to $(1+\varepsilon_{lm})$] for renal cell carcinoma ranges between 1.30 and 1.66 [1282], and correlates well with the tumor micro vessel density (MVD) [1282]. At $\varepsilon_{lm} = 0.478$, the scaling exponent is the well-known $p = 2/3$. (Note, there are numerous controversies surrounding the fractal evaluation of vascular networks [1283]).
2. The demand is normal, only the active surface had been suppressed abnormally. In this case the supply volume follows the evolution-requested exponent of 4 ($\varepsilon_v = 1$) under the self-similar conditions in (A.3.1), but the requested length and surface are not adequate for the required work, so $\varepsilon_l > 1$ and $\varepsilon_a < 1$. Because of the self-similar conditions (A.3.4), denote the non-zero ε_l by ε_{lm}:

$$l \propto L_0{}^{1+\varepsilon_{lm}}, \quad a \propto L_0{}^{3-\varepsilon_{lm}}, \quad v \propto L_0{}^4 \tag{A.4.3}$$

   Consequently, the surface for reactions behaves by the power law:

$$a \propto P_{met} \propto M^{\frac{2+\varepsilon_a}{3+\varepsilon_l+\varepsilon_a}} = M^{\frac{3-\varepsilon_{lm}}{4}} \tag{A.4.4}$$

   Here the "dimension" of the reaction request is 4, but the actual conditions are less than optimal. ε_{lm} here again is also measurable by the fractal dimension of the structures [1284, 1285]. Using $\varepsilon_{lm} = 0.41$ [1284], the scaling exponent is $p = 0.65$; and $\varepsilon_{lm} = 0.28$ [1285], the scaling exponent is $p = 0.68$. At $\varepsilon_{lm} = 0.33$, the scaling exponent is the well-known $p = \frac{2}{3}$.

Both situations lead to under-energized tissue, and down-regulation of the exponent of the power-law scaling. Measuring the scaling exponent of metabolism together with the fractal dimension of the supplying microvessels, we could make a decision in regard to the two possible paths of metabolic supply above. This has diagnostic value concerning actual deviations from normal.

This thinking supposes the same reactions in every demand condition, i.e. the available reactants always carry out the same chemical reactions. However, in metabolic processes, as we have seen, there are at least two main categories of metabolism: the anaerobic and the aerobic one.

The chemical reactions and so the metabolic rates are of course temperature-dependent. Their temperature dependence close to equilibrium could be followed by the Arrhenius function. Temperature modification of the scaling properties was first proposed by Gillooly et al. in 2001 [566]. They simply considered the Arrhenius factor in the factor of the mass-scaling:

$$P_{\text{met}} \propto M^{\frac{3}{4}} \cdot e^{-\frac{E_a}{kT}} \tag{A.4.5}$$

Which fits well with the experimental data they presented, considering the E_a activation energy in the interval of 0.6–0.7 eV (average 0.65 eV). Although this approach was heavily criticized [1286–1288], they defended the idea [624], but the debate continues [1289]. The body-temperature data are uncorrelated with the body mass [1290], the challenge is to find an appropriate grouping to categorize the organisms.

Appendix 5: Oncogenic Growth

On the basis of scaling theory a general model for oncogenic growth was introduced [1291], and discussed [1292–1295]. The oncogenic growth in these publications relates to ideal nutrition supply. However cancer growth, at least at larger sizes, never occurs under conditions of optimal nutrition supply, and the cells are intensively competing for the available energy sources. The number of cancerous cells (N_c) is the difference of newly produced cells (P) and the perished drop-off cells (N) in unit time:

$$\frac{dN_c}{dt} = P - N \tag{A.5.1}$$

It is a realistic assumption that the perished cells are proportional to the full cell number in a unit time:

$$N = \lambda N_c \tag{A.5.2}$$

where λ is the cell-death rate and $\lambda^{-1} = T$ is the average life-span of a cell in the tumor.

In consequence, the balance of the cell number with time is

$$\frac{dN_c}{dt} = P - \lambda N_c \tag{A.5.3}$$

The energy balance is determined by the transported energy-flux delivered by the blood stream. The energy-transport current intensity (I) is divided into two parts, a part that will be used to produce new cells and another to keep the living cells alive. Hence:

$$\begin{aligned} I &= N_c B_c + E_c P = N_c B_c + E_c \left(\tfrac{dN_c}{dt} + N \right) \\ &= N_c B_c + E_c \left(\tfrac{dN_c}{dt} + \lambda N_c \right) \end{aligned} \tag{A.5.4}$$

Where B_c is the metabolic rate of a cell, and E_c is the necessary metabolic energy to create a new cell. Consequently:

$$E_c \frac{dN_c}{dt} = I - N_c \left(B_c + \lambda E_c \right) \tag{A.5.5}$$

As we have seen above the metabolic energy can be scaled by exponent p, which is in most cases $2/3 \le p \le 1$ (in ideal conditions: $p = 3/4$):

$$I = B_0 \cdot N_c^p, \tag{A.5.6}$$

So we obtain:

$$E_c \frac{dN_c}{dt} = B_0 * N_c^p - N_c \left(B_c + \lambda E_c \right) \tag{A.5.7}$$

Multiply this by the average mass of a single cell (m_c) and we obtain the energy balance for the full tumor mass (m):

$$\begin{aligned} \frac{dm}{dt} &= am^p - bm \\ b &= \tfrac{B_c}{E_c} + \lambda \quad a = \frac{B_0 * {m_c}^{1-p}}{E_c} = \frac{B_0 {m_c}^{1-p}}{B_c} \end{aligned} \tag{A.5.8}$$

This balance was similarly formulated before [1296]. The mass has a maximal limit M, therefore:

$$aM^p - bM = 0 \rightarrow M = \left(\frac{a}{b} \right)^{\frac{1}{1-p}} = \left(\frac{B_0 {m_c}^{1-p}}{B_c + \lambda E_c} \right)^{\frac{1}{1-p}} \tag{A.5.9}$$

In the ideal case [1291] $p = 3/4$, so the ideally produced mass $M_{id} = (a/b)^4$, consequently:

$$M = \left(\frac{a}{b}\right)^{\frac{1}{1-p}} = (M_{\mathrm{id}})^{\frac{4}{(1-p)}} \tag{A.5.10}$$

Substituting (A.5.10) to (A.5.8) we obtain:

$$\frac{dm}{dt} = am^p\left(1 - \left(\frac{m}{M}\right)^{1-p}\right) \to \frac{d\left(\frac{m}{M}\right)^{1-p}}{dt} = \frac{\frac{1}{1-p}a}{M^{1-p}}\left(1 - \left(\frac{m}{M}\right)^{1-p}\right) \tag{A.5.11}$$

this has a sigmoid solution:

$$\left(\frac{m}{M}\right)^{1-p} = 1 - \left(1 - \left(\frac{m_0}{M}\right)^{1-p}\right)e^{-at/\frac{1}{1-p}M^{1-p}}$$

$$= 1 - e^{-at/\frac{1}{1-p}M^{1-p}+\ln\left(1-\left(\frac{m_0}{M}\right)^{1-p}\right)} = 1 - e^{-\tau}, \tag{A.5.12}$$

$$\tau = at/\frac{1}{1-p}M^{1-p} - \ln\left(1 - \left(\frac{m_0}{M}\right)^{1-p}\right)$$

where m_0 is the mass at the start of the tumor (probable a few times m_c). The ratio (r) of the energy expended on keeping cells alive is:

$$r = \frac{N_c B_c}{I} = \frac{B_c m}{m_c B_0 m^p} = \frac{b}{a}m^{1-p} = \left(\frac{m}{M}\right)^{1-p} = 1 - e^{-\tau} \tag{A.5.13}$$

While, of course, the ratio of the energy maintaining the new cells is $R=(1 - r)= e^{-\tau}$. Assuming the average density of the colony in the experiments of Bru et al. [725], the scaling law could be determined by the characteristic lengths, which are in the ideal case (at $p=3/4$, [1291]), $m \propto L^4$, consequently:

$$\left(\frac{m}{M}\right)^{1-p} = \left(\frac{L^4}{L_0^4}\right)^{1/4} = \frac{L}{L_0} = 1 - e^{-\tau} \tag{A.5.14}$$

where L_0 is the asymptotic size of the colony. If it is logically assumed that L_0 »L, then the Taylor expansion of τ [from Eq. (A.5.12)] could be stopped at its second term, so (A.5.14) will have linear dependence as it was measured indeed [725]:

$$L(t) = \frac{a}{4K}t - L_0 \ln\left(1 - \frac{L(\tau = 0)}{L_0}\right) \tag{A.5.15}$$

However, if the energy supply is not ideal (which is the case in almost all the developed tumors in vivo), then the results of Bru et al. [725] do not support the general ontogenetic growth [1291]. It is shown in a generalized model [1297], the fractal surface and the covered volume are scaled pretty rigorously.

An interesting energetic consideration was formulated on the basis of the limited nutrition in the tissue [1298]. A demanding cell culture artificially placed on the area of the tumor successfully competed for the energy resources in competition with cancer cells supplied from the same source. The experimental in silico results have not been checked yet.

Appendix 6: Basic Bioelectromagnetics

The interacting electromagnetic fields (magnetic induction **B** and electric field strength **E**) are described by the Maxwell equations adapted to the actual medium. The medium is characterized by material parameters such as the electric permittivity (ε) and magnetic permeability (μ). The four Maxwell equations describe the basic physical laws (energy and material/charge conservations) by their changes (derivatives) in space: notations div and rot describe the volume/source and surface/whirl conditions, respectively). The equations are:

1. The effective charge (ρ_{eff}) is the source of the electric field:

$$\mathrm{div}\mathbf{E} = \rho_{\mathrm{eff}} \tag{A.6.1}$$

2. The whirls of the electric field originate from the change of magnetic field in time:

$$\mathrm{rot}\mathbf{E} = -\frac{\partial \mathbf{B}}{\partial t} \tag{A.6.2}$$

3. The magnetic field originates from dipoles (no magnetic monopoles exist in the Maxwell description):

$$\mathrm{div}\mathbf{B} = 0 \tag{A.6.3}$$

4. The whirls of magnetic fields have two sources, the effective electric current (j_{eff}) and the change of electric field in time:

$$\mathrm{rot}\mathbf{B} = \mu_0\varepsilon_0\frac{\partial \mathbf{E}}{\partial t} + \mathbf{j}_{\mathrm{eff}} \tag{A.6.4}$$

where, ε_o and μ_o are universal constants (values describing the vacuum), the absolute permittivity and magnetic permeability, and are respectively:

$$\varepsilon_0 \cong \left[(36\pi)^{-1} \cdot 10^{-9}\right], \quad \mu_0 = \left[(4\pi) \cdot 10^{-7}\right] \tag{A.6.5}$$

The effective charge and current values depend on the free charge density (ρ) and their current densities (j), as well as the polarization (**P**) (order of electric dipoles) and magnetization (**M**) (order of magnetic dipoles) vectors of the material:

$$\rho_{\mathrm{eff}} = \frac{\rho}{\varepsilon_0} - \mathrm{div}\mathbf{P} \quad \mathbf{j}_{\mathit{eff}} = \mathbf{j} + \frac{\partial \mathbf{P}}{\partial t} + \mathrm{rot}\mathbf{M} \tag{A.6.6}$$

Assuming that the polarization depends only on the external electric field and the magnetization on the external magnetic field:

$$\mathbf{P} = \varepsilon_0 \alpha_e \mathbf{E} \quad \mathbf{M} = \frac{\alpha_m}{1+\alpha_m}\mathbf{B} \tag{A.6.7}$$

where α_e and α_m are the electric and magnetic susceptibilities, respectively. The susceptibilities show how the outside fields affect the order of the dipoles in the material. In this way we have the relative electric permittivity and relative magnetic permeability $\varepsilon_r = 1 + \alpha_e$ and $\mu_r = 1 + \alpha_m$, respectively. The energy density (ρ_W) of the electromagnetic radiation [1149] is:

$$\rho_W = \frac{\varepsilon_0 \varepsilon_r}{2}\mathbf{E}^2 + \frac{1}{2\mu_0 \mu_r}\mathbf{B}^2 \tag{A.6.8}$$

In a real situation the bioelectromagnetic interactions are more complex, and the measurement of the absorbed energy is necessary. The energy absorption is characterized by the specific absorption rate (SAR, [W/kg]), identifying the absorbed energy in unit mass:

$$\mathrm{SAR} = \frac{d}{dt}\left(\frac{dW}{dm}\right) = \frac{d}{dt}\left(\frac{dW}{\rho(dV)}\right) \tag{A.6.9}$$

The simplest case is the current flow through the given tissue, when the well-known Joule heat is developed. This value (taking into account the well-known Ohm's law) is:

$$q_e = \frac{1}{2}\sigma E_{\mathrm{local}}^2 = \rho \mathrm{SAR} \tag{A.6.10}$$

where E_{local} is the locally presented electric field in the tissue and σ is the electric conductivity.

In the case of radiofrequency currents the energy source (A.6.10) of the bioheat equation (A.22.3) mainly derives from the dielectric loss:

$$q_e = 2\pi f \varepsilon_0 \mathrm{Im}(\varepsilon_r) E_{\mathrm{local}}^2 = \rho \mathrm{SAR}_{RF} \tag{A.6.11}$$

where f is the actual frequency. (In many descriptions we use for simplicity the notation of $\omega = 2\pi f$ [circular frequency.]) The temperature gain by the absorbed energy (if the physiologic modifications are neglected) is shown in Fig. A.6.1.

The complex amplitude of the energy delivery (energy radiation, e.g. Poynting vector ($\mathbf{S}$) expressed in $\mathrm{W/m^2}$) can be calculated as

$$\mathbf{S} = \frac{1}{\mu_0}\mathbf{E} \times \mathbf{B} \tag{A.6.12}$$

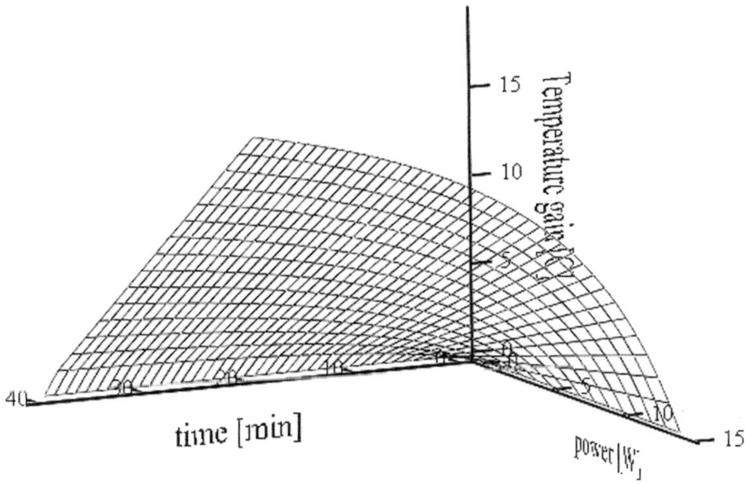

Fig. A.6.1 Result of a model calculation of the temperature gain in homogeneous media with stabilized surface temperature (constant environment)

The electromagnetic energy absorbed by the treated part and converted into heat equals the surface (**A**) integral of the vector's average value calculated for the boundary surface of the treated part. Therefore,

$$P = \frac{1}{2}\mathrm{Re}\left[\oint \mathbf{S} \cdot d\mathbf{A}\right] \qquad \text{(A.6.13)}$$

Appendix 7: Bioimpedance of Cells and Tissues

Generally we may write the dielectric constant in a complex number (denoted ε^*). The real part of it is the permittivity (ε), while the imaginary one is the conductivity (σ). Both are functions of the frequency (ω), and also (due to mathematical transformations) the imaginary part is inversely proportional to ω:

$$\varepsilon^*(\omega) = \varepsilon(\omega) - i\frac{\sigma_s(\omega)}{\omega} \quad \left\{i = \sqrt{-1}\right\} \qquad \text{(A.7.1)}$$

The permittivity has two extremes: the value at zero frequency (static value ε_s), and the value at extremely high ("infinite") frequency (ε_∞). Between the two values the function of the permittivity is supposed to be a monotonic sigmoid-like curve [1299] (see Fig. A.7.1), which has an inflexion point at the "characteristic frequency" ω_0, consequently:

$$\varepsilon(\omega) = \varepsilon_\infty + \frac{\varepsilon_s - \varepsilon_\infty}{1 + i\omega\tau}, \quad \tau = 1/\omega_0 \qquad \text{(A.7.2)}$$

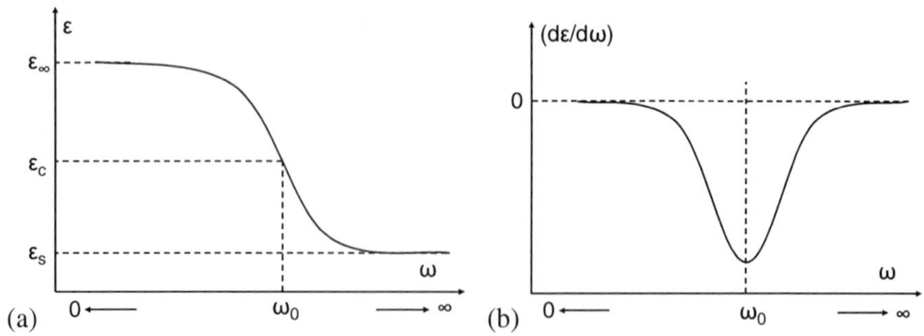

Fig. A.7.1 The supposed Sigmoid curve of dispersion of permittivity for the Debay relaxation (**a**) and its derivative (**b**)

when τ is usually called "relaxation time." Therefore, the relative complex dielectric constant is:

$$\varepsilon_{\mathrm{r}}^{*}(\omega) = \varepsilon_{\mathrm{r}\infty} + \frac{\varepsilon_{\mathrm{rs}} - \varepsilon_{\mathrm{r}\infty}}{1 + i\omega\tau} - i\frac{\sigma_{\mathrm{s}}}{\omega\varepsilon_0} \qquad \text{(A.7.3)}$$

This corresponds to the schematics shown in Fig. A.7.2 [1300]. In this model the real and imaginary part of the permittivity is:

$$\varepsilon' = \varepsilon_\infty + \frac{\varepsilon_{\mathrm{s}} - \varepsilon_\infty}{1 + (\omega\tau)^2}, \quad \varepsilon'' = \frac{\sigma_{\mathrm{s}}}{\omega\varepsilon_0} + \frac{\omega\tau(\varepsilon_{\mathrm{s}} - \varepsilon_\infty)}{1 + (\omega\tau)^2} \qquad \text{(A.7.4)}$$
$$\left\{\varepsilon' = \mathrm{Re}\left(\varepsilon^*\right), \quad \varepsilon'' = -\mathrm{Im}(\varepsilon^*)\right\}$$

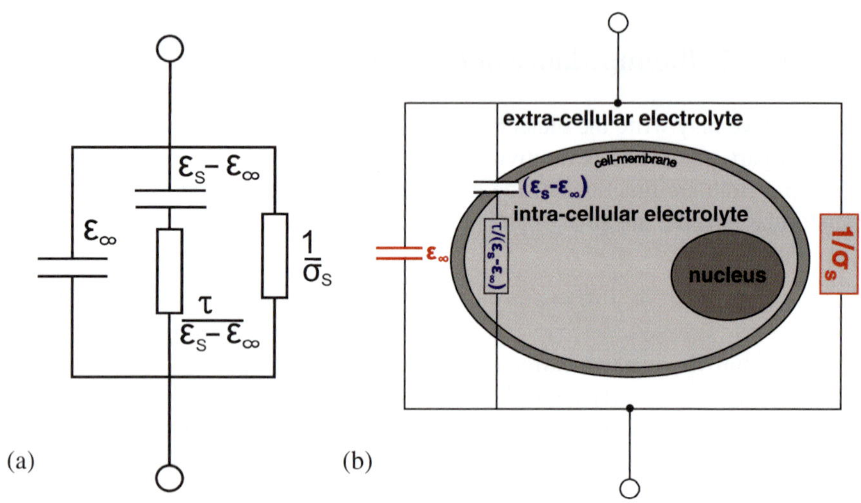

Fig. A.7.2 The relevant circuit of the Debay relaxation model [1300]. The values of the resistances and capacities are shown (**a**). The possible image of the cellular parameters in the Debay relaxation frame (**b**)

Therefore, the absorbed energy in a unit volume in the Debay relaxation approach:

$$s = \tfrac{1}{2}Ej^*, \\ w = \mathrm{Im}(s) = -\tfrac{1}{2}\omega\varepsilon' E^2, \; p = \mathrm{Re}(s) = \tfrac{1}{2}\omega\varepsilon'' E^2 \tag{A.7.5}$$

where j^* is the current (complex number), s is the complete (apparent) energy absorption, w is the reactive power (oscillating in-and-out by double frequency), [has no real heating], and p is the real heating power (Joule heat).

The relaxation theory of dense cellular solutions has not yet been solved. An approximation for our case could be derived based on [1300], considering the equivalent electric circuit on Fig. A.7.2:

$$\begin{aligned} \varepsilon_s - \varepsilon_\infty &= \frac{9c_{\text{cell}} r_{\text{cell}} C_{\text{memb}}}{4\varepsilon_0 \left[1 + r_{\text{cell}} \frac{1}{R_{\text{memb}}} \left(\frac{1}{\sigma_{\text{int}}} + \frac{1}{2\sigma_{\text{ext}}}\right)\right]} \\ \sigma_s &= \sigma_{\text{ext}} \left[1 - \frac{3c_{\text{cell}}}{2} \frac{1 + r_{\text{cell}} \frac{1}{R_{\text{memb}}} \left(\frac{1}{\sigma_{\text{int}}} + \frac{1}{\sigma_{\text{ext}}}\right)}{r_{\text{cell}} \frac{1}{R_{\text{memb}}} \left(\frac{1}{\sigma_{\text{int}}} + \frac{1}{2\sigma_{\text{ext}}}\right)}\right] \\ \tau &= r_{\text{cell}} C_{\text{memb}} \frac{\sigma_{\text{int}} + 2\sigma_{\text{ext}}}{2\sigma_{\text{int}}\sigma_{\text{ext}} + r_{\text{cell}} \frac{1}{R_{\text{memb}}} (\sigma_{\text{int}} + 2\sigma_{\text{ext}})} \\ \varepsilon_\infty &= \varepsilon_{\text{ext}} \left[1 - 3c_{\text{cell}} \frac{\varepsilon_{\text{ext}} - \varepsilon_{\text{int}}}{2\varepsilon_{\text{ext}} + \varepsilon_{\text{int}}}\right] \\ C_{\text{memb}} &= \frac{\varepsilon_{\text{memb}}\varepsilon_0}{d_{\text{memb}}} \quad [F/m^2], \quad R_{\text{memb}} = \frac{d_{\text{memb}}}{\sigma_{\text{memb}}} \quad [\Omega m^2] \end{aligned} \tag{A.7.6}$$

where the indices refer to the intra-cellular, extra-cellular or membrane parameters, r_{cell} is the average radius of the cell, c_{cell} is the average concentration of the cells in the studied tissue (suspension), d_{memb}, C_{memb} and R_{memb} are the membrane thickness, the membrane capacity and its resistance, respectively. Using average data [1302]: the conductivities $\sigma_{\text{int}} = 0.3$ S/m, $\sigma_{\text{ext}} = 1.0$ S/m, $\sigma_{\text{memb}} = 0.3 \cdot 10^{-7}$ S/m; the permittivity: $\varepsilon_{\text{int}} = 60$, $\varepsilon_{\text{ext}} = 80$, $\varepsilon_{\text{memb}} = 3$, the membrane thickness: $d_{\text{memb}} = 7.5 \cdot 10^{-9}$ m, average radius of the cell: $r_{\text{cell}} = 10^{-5}$ m, average concentration of the cell: $c_{\text{cell}} = 0.5$. The results: the membrane capacity and resistance: $C_{\text{memb}} = 3.54 \cdot 10^{-3}$ F/m^2, $R_{\text{memb}} = 0.25$ Ωm^2, $\sigma_s = 0.25$ S/m, $\varepsilon_\infty = 69.1$, $(\varepsilon_s - \varepsilon_\infty) = \Delta\varepsilon = 4.5 \cdot 10^3$, $\tau = 1.36 \cdot 10^{-7}$ s. The dispersion curves for permittivity and conductivity are shown in Fig. A.7.3.:

Simplifying (A.7.6) taking into account $r_{\text{cell}}/R_{\text{memb}} \ll \sigma_{\text{int}}$:

$$\begin{aligned} \varepsilon_s - \varepsilon_\infty &= \frac{9c_{\text{cell}} r_{\text{cell}} C_{\text{memb}}}{4\varepsilon_0} \\ \sigma_s &= \sigma_{\text{ext}} \left[1 - \frac{3c_{\text{cell}}}{2}\right] \\ \tau &= r_{\text{cell}} C_{\text{memb}} \left(\frac{1}{\sigma_{\text{int}}} + \frac{1}{2\sigma_{\text{ext}}}\right) \\ \varepsilon_\infty &= \varepsilon_{\text{ext}} \left[1 - 3c_{\text{cell}} \frac{\varepsilon_{\text{ext}} - \varepsilon_{\text{int}}}{2\varepsilon_{\text{ext}} + \varepsilon_{\text{int}}}\right] \end{aligned} \tag{A.7.7}$$

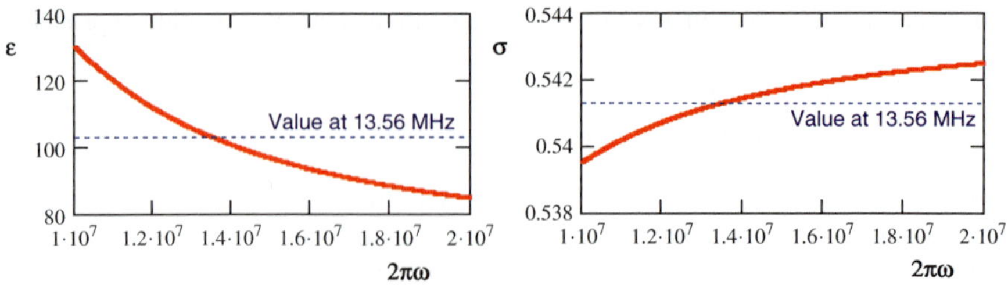

Fig. A.7.3 The dispersion curves of the permittivity and conductivity, calculated by Debay relaxation. The values at 13.56 MHz are: $\varepsilon 13.56 = 102.5$; $\sigma 13.56 = 0.54$

These equations were used in the experimental study of mitochondria [1301].

The completed model shows well the dominance of the extra-cellular conductivity in the current flow at RF like 13.56 MHz. Applying the previous picture to the cellular electrolyte conduction, a simpler circuit had been introduced [741], (see Fig. A.7.4) to identify the membrane capacity by its dielectric constant ($\varepsilon_{\text{memb}}$) and the resistance of the intra- and extra-cellular electrolytes by their inverse conductivity ($1/\sigma_{\text{int}}$ and $1/\sigma_{\text{ext}}$):

This approximation is fitted to the dispersion experiments. This simple picture ignores the capacitive part of the extra-cellular electrolyte, which could be pretty high ($\varepsilon_{\text{ext}} \approx 80$) [1302]. However, at low frequency there is no difference; both circuits behave in a simple resistance fashion ($1/\sigma_{\text{s}}$ and $1/\sigma_{\text{ext}}$ for four- {see Fig. A.7.2} and for three-*element* {see Fig. A.7.4} circuits, respectively). The difference exists at high frequencies, where in the four-element picture ε_{∞} will be dominant, while the three-element picture remains resistive ($1/\sigma_{\text{int}}$ and $1/\sigma_{\text{ext}}$ are joined in parallel).

We may add a resistance to the three-component model, taking into account the resistance of the membrane as well (see Fig. A.7.5).

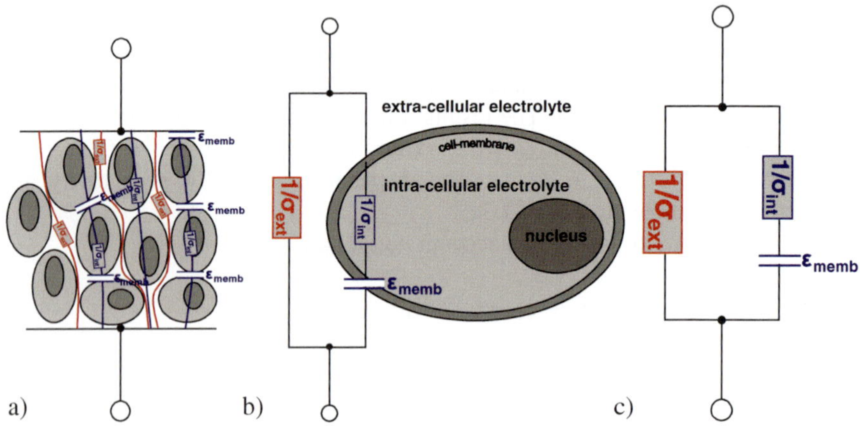

Fig. A.7.4 Electric modeling of cellular impedance. Tissue (**a**) single cell (**b**) circuit equivalent of the cell (**c**)

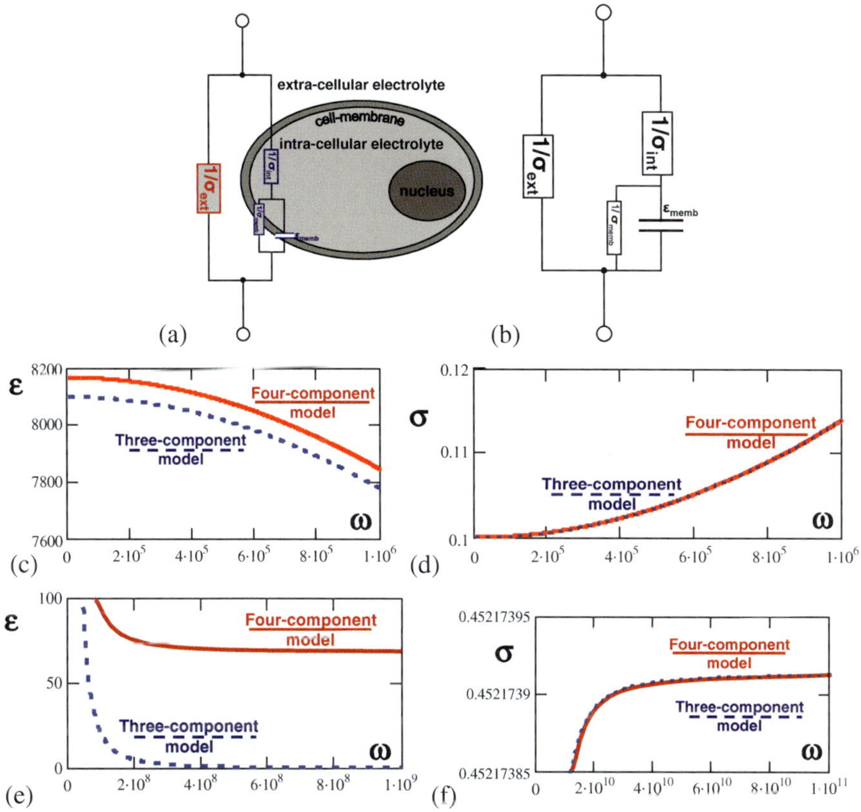

Fig. A.7.5 The completed "three-component model" in cellular components (**a**) and its electric circuit equivalent. (**b**) Comparison of three- and four-component models at low frequency (**c–d**) and high frequency (**e–f**) permittivity (**c, e**) and conductivity (**d, f**), respectively

The experimental data could be well described by the three-component model, supposing the special membrane capacity dependence of the frequency [790]:

$$C_{\text{memb}} = C_0 \left(\frac{\omega_c}{\omega} \right)^{\alpha} \tag{A.7.8}$$

where C_0 is the membrane capacity on the circular-frequency ω_c and the exponent $\alpha \approx 0.5$.

Therefore:

$$\omega_c = \frac{1}{C_0 \left(R_{\text{ext}} + R_{\text{int}} \right)} \tag{A.7.9}$$

where $R_{\text{ext}} = (1/\sigma_{\text{ext}})$ and $R_{\text{int}} = (1/\sigma_{\text{int}})$. Hence the impedance:

$$Z(\omega) = \frac{\frac{R_{ext}R_{int}}{R_{ext}+R_{int}} i \left(\frac{\omega}{\omega_c}\right)^{1-\alpha} + R_{ext}}{1 + i\left(\frac{\omega}{\omega_c}\right)^{1-\alpha}} \quad \text{and so} \quad \mathrm{Im}\, Z(\omega) = -\frac{\frac{R_{ext}^2}{R_{ext}+R_{int}}\left(\frac{\omega}{\omega_c}\right)^{1-\alpha}}{1 + \left(\frac{\omega}{\omega_c}\right)^{2(1-\alpha)}} \tag{A.7.10}$$

The imaginary part on the characteristic frequency:

$$\mathrm{Im}Z(\omega_c) = -\frac{{R_{ext}}^2}{2(R_{ext}+R_{int})} \tag{A.7.11}$$

This modification makes a more sensitive evaluation of the measured data (see Fig. A.7.6).

Studying the differences between the malignant and healthy tissues, there the exponents deviate: $\alpha_{(malignant)} = 0.56$ $\alpha_{(healthy)} = 0.47$ [766]; and the characteristic frequency of the function of ImZ(ω_c) was measured as significantly different [766]; $\omega_{c(malignant)} > \omega_{c(healthy)}$ [1305, 766]. An index (I_R) was introduced [766] to distinguish between malignant tissue and its healthy counterpart.

$$I_R(\omega_c) = \frac{\arctan\left[\left(\frac{\omega_c}{2\pi\cdot 10^6}\right)^{\alpha-1} - \arctan\left(\frac{\omega_c}{2\pi\cdot 10^5}\right)^{\alpha-1}\right] + p}{\arctan\left[\left(\frac{\omega_c}{2\pi\cdot 10^5}\right)^{\alpha-1} - \arctan\left(\frac{\omega_c}{2\pi\cdot 10^4}\right)^{\alpha-1}\right] + p}, \quad \text{where } p = \frac{k(R_{ext}+R_{int})(1-\alpha)\ln(10)}{R_{ext}^2} \tag{A.7.12}$$

The index I_R could be measured, and this makes it possible to distinguish the malignant tissue. It is important to see that the tissue parameters are the resistivity of the intra- and extra-cellular electrolytes. (The index from average data of various investigated tissues are: $I_{R(malignant)} = 1.8$, $I_{R(healthy)} = 0.68$, which shows significant difference.)

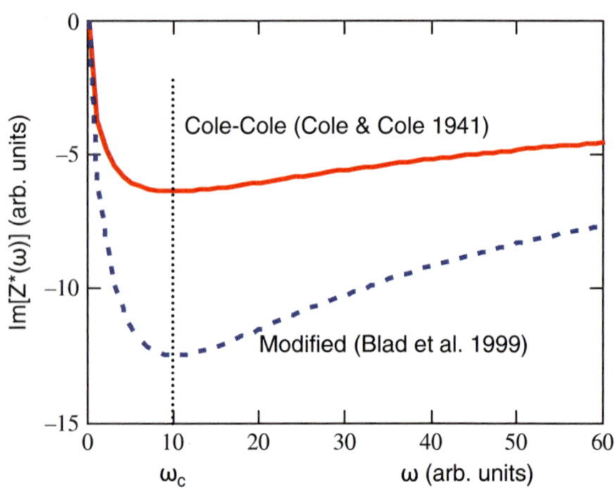

Fig. A.7.6 Comparison of the imaginary part of the impedance in the original Cole–Cole arc-model [1303, 1304] and the modified one [766], using the same parameters for both (α=0.4). The modified model has more sensitivity around the characteristic frequency

The bio-objects are radically non-homogenous systems. Because of the variation of the dielectric properties a space-charge could be accumulated from the outside current sources.

The first principles for the conduction:

$$\operatorname{div}\left(\varepsilon\overline{E}\right)=\rho,\quad \bar{j}=\sigma\overline{E},\quad \frac{\partial\rho}{\partial t}+\operatorname{div}\bar{j}=0\Rightarrow \operatorname{div}\left(i\omega\varepsilon+\sigma\right)\overline{E}=0 \qquad \text{(A.7.13)}$$

Therefore:

$$\overline{E}\cdot\operatorname{grad}\varepsilon+\varepsilon\operatorname{div}\overline{E}=\rho,\quad \overline{E}\cdot\operatorname{grad}\sigma^{*}+\sigma^{*}\operatorname{div}\overline{E}=0,\quad \sigma^{*}=\left(i\omega\varepsilon+\sigma\right) \qquad \text{(A.7.14)}$$

In consequence:

$$\left[\operatorname{grad}\varepsilon+\operatorname{grad}(\ln\sigma^{*})\right]\cdot\overline{E}=\rho \qquad \text{(A.7.15)}$$

So, the value of the space-charge is proportional to the field strength and to the gradient of the permittivity and logarithm of conductivity. The gradient of the permittivity does not give a space-charge when it is perpendicular to the electric field. This phenomenon introduces a directional dependent behavior of the tissue. Perpendicular gradients have no effect, while the parallel gradient (the change in the direction of the current) has the maximal available interfacial polarization. The simplest arrangement for maximal polarization is shown in Fig. A.7.7 for a unit-surface part of the layers.

In this case the equation describing the action:

$$\begin{aligned}&J=\tfrac{\partial D}{\partial t}+J_{\text{cond}}=\left(i\omega\varepsilon+\sigma\right)E=\sigma^{*}\tfrac{U}{d},\\&\sigma^{*}=i\omega\varepsilon+\sigma=i\omega\left(\varepsilon-i\tfrac{\sigma}{\omega}\right)=i\omega\left(\varepsilon'-i\varepsilon''\right)\\&\left\{\varepsilon'=\operatorname{Re}\left(\varepsilon^{*}\right)=\varepsilon,\quad \varepsilon''=\operatorname{Im}\left(\varepsilon^{*}\right)=\tfrac{\sigma}{\omega}\right\}\end{aligned} \qquad \text{(A.7.16)}$$

The impedance ($Z(\omega)$) and the conductance ($Y(\omega)$) (by Ohm's law):

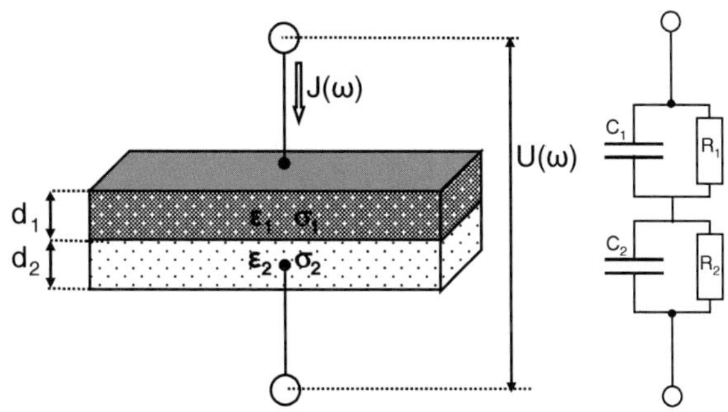

Fig. A.7.7 Unit surface part of layered dielectric materials

$$Z(\omega) = \frac{U(\omega)}{J(\omega)} = \frac{d}{\sigma^*(\omega)}, \quad Y(\omega) = \frac{J(\omega)}{U(\omega)} = \frac{\sigma^*(\omega)}{d} \tag{A.7.17}$$

The impedances are additional:

$$Z(\omega) = \frac{d_1}{\sigma_1^*(\omega)} + \frac{d_2}{\sigma_2^*(\omega)} \tag{A.7.18}$$

The complex conductivity of the system ($\sigma^*(\omega)$) could be like:

$$Z = \frac{d_1}{\sigma_1^*} + \frac{d_2}{\sigma_2^*} = \frac{d_1 + d_2}{\sigma^*} \tag{A.7.19}$$

therefore the complex material constants

$$\begin{aligned} \sigma^* &= \frac{\sigma_1^* \sigma_2^*}{a\left(\sigma_2^* - \sigma_1^*\right) + \sigma_1^*}, \\ \varepsilon^* &= \frac{\sigma^*}{j\omega} = \frac{\sigma_1^* \sigma_2^*}{j\omega\left[a\left(\sigma_2^* - \sigma_1^*\right) + \sigma_1^*\right]} \\ a &= \frac{d_1}{d_1 + d_2} \end{aligned} \tag{A.7.20}$$

From these the resulting parameters are:

$$\begin{aligned} \varepsilon^* &= \varepsilon_\infty + \frac{\varepsilon_s - \varepsilon_\infty}{1 + j\omega\tau} + \frac{\sigma_s}{j\omega} \\ \varepsilon_\infty &= \frac{\varepsilon_1 \varepsilon_2}{a\left(\varepsilon_2 - \varepsilon_1\right) + \varepsilon_1} \\ \varepsilon_s &= \frac{\left(\sigma_1 \varepsilon_2 - \sigma_2 \varepsilon_1\right)^2 a\left(1 - a\right)}{\left[a\left(\varepsilon_2 - \varepsilon_1\right) + \varepsilon_1\right]\left[a\left(\sigma_2 - \sigma_1\right) + \sigma_1\right]^2} + \varepsilon_\infty \\ \tau &= \frac{a\left(\varepsilon_2 - \varepsilon_1\right) + \varepsilon_1}{a\left(\sigma_2 - \sigma_1\right) + \sigma_1} \\ \sigma_s &= \frac{\sigma_1 \sigma_2}{a\left(\sigma_2 - \sigma_1\right) + \sigma_1} \end{aligned} \tag{A.7.21}$$

This shows a very important consequence: the inhomogeneous dielectric material could show higher dielectric parameters than any of its components has. So-called Maxwell–Wagner dispersion occurs on the interfaces of two different dielectrics, like that arranged in Fig. A.7.7. In this arrangement we have the classical Debay dispersion, even without any dipole relaxation in the dielectrics. The dispersion is created by the conductance of the imperfect dielectrics, and so the interface is charged by this conductivity as we saw above. When by usual capacitor/resistor notation (see Fig. A.7.7b.) the actual relaxation time is $\tau = RC$, the common values can be calculated:

$$\tau = \left(\frac{1}{R_1+R_2}\right) R_1 R_2 (C_1 + C_2), \quad \{\tau_1 < \tau < \tau_2, \text{ when } \tau_1 < \tau_2\}$$

$$C = \left(\frac{1}{R_1+R_2}\right) \frac{\tau_1 + \tau_2 - \tau + \omega^2 \tau \tau_1 \tau_2}{1 + \omega^2 \tau^2} \tag{A.7.22}$$

$$R = \left(\frac{1}{R_1+R_2}\right) \frac{1 + \omega^2 [\tau (\tau_1 + \tau_2) - \tau_1 \tau_2]}{1 + \omega^2 \tau^2}$$

The low and high frequency limits differ, and

$$R(\omega \sim 0) > R(\omega \to \infty) \quad C(\omega \sim 0) > C(\omega \to \infty) \tag{A.7.23}$$

The full complex impedance in this picture :

$$Z^*(\omega) = \frac{R_{\text{ext}} + R_{\text{int}} R_{\text{ext}} C_{\text{memb}} i\omega}{1 + (R_{\text{int}} + R_{\text{ext}}) C_{\text{memb}} i\omega} \tag{A.7.24}$$

which is a half-circle on the complex Z-plane (see Fig. A.7.8a), which has a center $\overline{K}$ and the radius R:

$$\overline{K} = \frac{2R_{\text{int}} R_{\text{ext}} + R_{\text{ext}}^2}{2(R_{\text{int}} + R_{\text{ext}})}, \; R = \frac{{R_{\text{ext}}}^2}{2(R_{\text{int}} + R_{\text{ext}})} \tag{A.7.25}$$

The points ω=0 and $\omega \to \infty$ fix $Z^*(0)=R_{\text{ext}}$ and $Z^*(\infty)=R_{\text{ext}}R_{\text{int}}/(R_{\text{ext}}+R_{\text{int}})=R_\infty$ respectively. The standard form of $Z(\omega)$ by this notation is:

$$Z^*(\omega) = R_\infty + \frac{R_0 - R_\infty}{1 + (R_{\text{int}} + R_{\text{ext}}) C_{\text{memb}} i\omega} = R_\infty + \frac{R_0 - R_\infty}{1 + i\frac{\omega}{\omega_c}}$$
$$\omega_c = \frac{1}{(R_{\text{int}} + R_{\text{ext}}) C_{\text{memb}}} \tag{A.7.26}$$

Hence:

$$Z^*(\omega = \omega_c) = \frac{1}{2}(R_0 + R_\infty) - j\frac{1}{2}(R_0 - R_\infty) \tag{A.7.27}$$

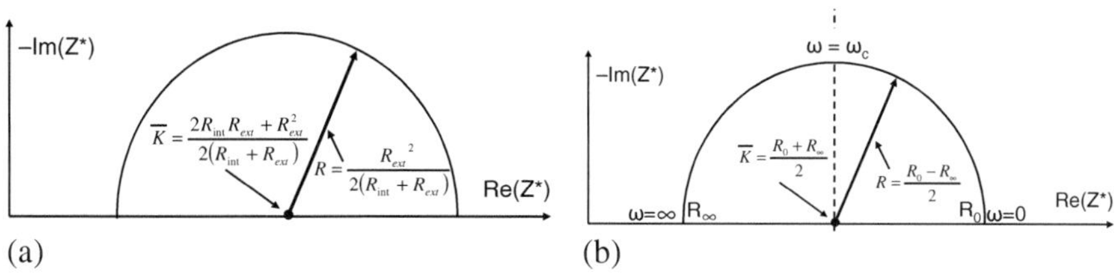

Fig. A.7.8 The semi-circular pattern of Z* by the frequency

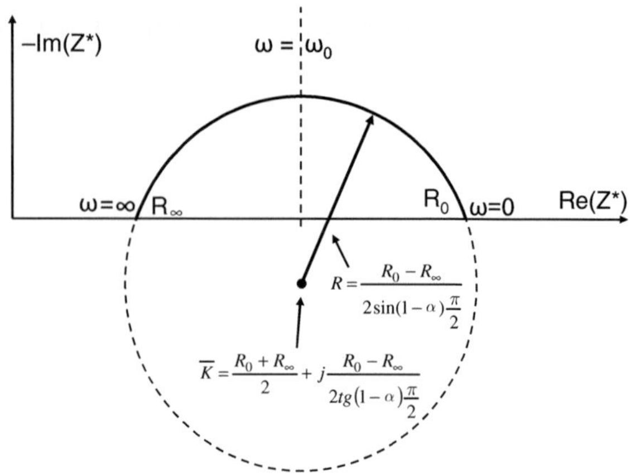

Fig. A.7.9 The Cole–Cole arc-pattern (At α=0 it returns the well-known Debay relaxation)

Which has a special peculiarity, namely its real part is the center of the circle, while the absolute value of the imaginary part is the radius of the circle (see Fig. A.7.8b).

Measuring numerous dielectrics Cole and Cole recognized: the reality is not a half circle [1303, 1304]. The measurements were given only an arc, the center of the circle was not on the real axis. The modified Cole–Cole impedance introduces a fourth parameter (α to R_0, R_∞ and ω_c), which characterizes the actual material (tissue) in [0,1) interval:

$$Z^*(\omega) = R_\infty + \frac{R_0 - R_\infty}{1 + \left(j\frac{\omega}{\omega_c}\right)^{1-\alpha}} = \frac{R_0 + i^{1-\alpha}\frac{R_\infty}{\omega_c^{1-\alpha}}\omega^{1-\alpha}}{1 + i^{1-\alpha}\frac{1}{\omega_c^{1-\alpha}}\omega^{1-\alpha}} \tag{A.7.28}$$

and so the pattern remains a circle (Fig. A.7.9), only the center and the radius of the circle became:

$$\overline{K} = \frac{R_0 + R_\infty}{2} + j\frac{R_0 - R_\infty}{2tg\,(1-\alpha)\,\frac{\pi}{2}},\ R = \frac{R_0 - R_\infty}{2\sin(1-\alpha)\frac{\pi}{2}} \tag{A.7.29}$$

Appendix 8: Boundary Conditions

The initial data can be easily specified by the actual initial body temperature (T_o), assuming its homogeneity in the malignant area and surroundings. (Note: sometimes this is not the case because of the higher tumor temperature due to its increased metabolic rate.) However, the boundary conditions are essential to solve (A.22.3) at least numerically, but they are complicated to determine. Three types of boundary conditions may be considered:

The boundary surface of the targeted area has a constant temperature. This assumption is supported by the observation that the functioning of tissue is normal outside the area, thus the increase in temperature augments the perfusion rate and – as a result – keeps the healthy tissue section at a constant temperature.

A boundary layer is formed on the surface of the targeted area. This could be physiologically constructed by a transient tissue between the malignant and healthy areas. The considerable increase of perfusion in the healthy tissue section close to the boundary surface presents another problem, as the extent of this should be known as well.

The transition of conductive heat-flux density and temperature is continuous on the boundary surface of the target area. The consideration of this boundary condition means that the bioheat equation (A.22.3) shall be solved both for the target area and for the area outside the target, and the two solutions shall be matched at the boundary of the two areas under consideration. This is a very complicated task, mainly in the case where the surroundings can not be considered as infinitely large. Therefore, a newer boundary condition is needed.

These conditions – with the additional problem of knowing the SAR distribution in situ– clearly show that the solution of the bioheat equation [1114] is not realistic for practical use. However, note the actual driving forces in the bioheat equation are always the temperature gradients in space and time; and so no assumption of the homogenous temperature could be valid.

The schematic structure of the complex conditions of the treated biomatter is shown in Fig. A.8.1. The multi-layered structure is the most common from the point of view of the chosen loco-regional treatment. The layers represent the various tissues, for example skin epidermis, skin capillary bed, fatty tissue, muscle tissue, tumor tissue, etc. It may include non-layered structures like blood vessels, bone structures, lymph nodes, etc. The tissue (and the transmitter material, which is in most of the cases air or water) is characterized from the electromagnetic point of view by the complex impedance function, which depends on the basic material constants, such as, on the dielectric permittivity (ε), the magnetic permeability (μ), and the conduction (σ) of the tissue. All other parameters – like the attenuation constant (α), the phase constant (β), the propagation constant (γ), the penetration depth (δ), and the actual wavelength in the tissue (λ), as well as the reflection ratio (Γ) – can be calculated [1151] from these basic material constants ε, μ, and σ (f is the frequency of the applied treatment):

$$\alpha_i = \omega\sqrt{\frac{\mu_i\varepsilon_i}{2}}\left[\sqrt{1+\left(\frac{\sigma_i}{\omega\varepsilon_i}\right)^2}-1\right]^{1/2}, \quad \beta = \omega\sqrt{\frac{\mu_i\varepsilon_i}{2}}\left[\sqrt{1+\left(\frac{\sigma_i}{\omega\varepsilon_i}\right)^2}+1\right]^{1/2}$$

$$\gamma_i = \alpha_i + i\beta_i \quad \delta_i = \frac{1}{\alpha_i} \quad \lambda_i = \frac{2\pi}{\beta_i} = \frac{1}{f}\sqrt{\frac{2}{\mu_i\varepsilon_i}}\left[\sqrt{1+\left(\frac{\sigma_i}{\omega\varepsilon_i}\right)^2}+1\right]^{-1/2} \tag{A.8.1}$$

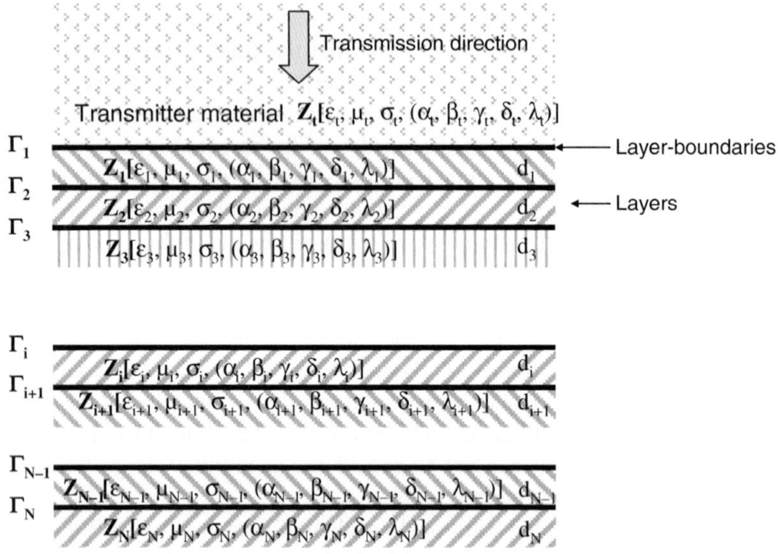

Fig. A.8.1 Layered structure of the treated body part. The layers represent different tissues (e.g. skin and its structures, adipose tissues, muscle tissue, tumor tissue, etc., noted by numeration), characterized by their "*d*" thicknesses and "*Z*" impedances, which depends on the "Γ" reflection coefficient, "ε" dielectric constant, "μ" magnetic permeability, "σ" conduction, "α" attenuation constant, "β" phase constant, "γ" propagation constant, "δ" penetration depth, "λ" wavelength in the medium

and the reflection and the impedances may be calculated [221, 1151] in the *i*th layer as:

$$\begin{aligned} \Gamma_i &= \left(\frac{j\gamma_{i-1}}{\omega\mu_{i-1}} - \frac{j\gamma_i}{\omega\mu_i}\right)\left(\frac{j\gamma_{i-1}}{\omega\mu_{i-1}} + \frac{j\gamma_i}{\omega\mu_i}\right)^{-1} \\ Z_i &= Z_{i-1}\zeta_i\left(\frac{1+\varsigma_i\zeta_i}{1+\varsigma_i/\zeta_i}\right) \quad \left[\zeta_i = \left(\frac{1+\Gamma_i}{1-\Gamma_i}\right)\right] \quad \left[\varsigma_i = th\gamma_{i+1}d_i\right] \end{aligned} \tag{A.8.2}$$

The interaction description is complicated enough, but it is definitely much more complex in reality, where the perpendicularity, homogeneity, radiation condition, source geometry etc. might modify the results. This is the reason why for the calculation of real cases only numerical approximations can be considered. A minor possible simplification is that in the natural situation (without artificially added materials) the biomaterial has no magnetic properties, as its relative magnetic permeability is equal to one with high accuracy.

We have further complications when trying to keep in hand both the rather complex bioelectric interactions and the highly sophisticated mathematical apparatus: if our aim is to accomplish a deep heating condition we face many problems physically and physiologically as well. The main problematic points are:

The incident energy has a frequency-dependent exponential decay in the depth, the distance from the boundary in the case of plane waves [1152]:

Fig. A.8.2 The penetration depth of the radiative electromagnetic field and absorbed energy in homogeneous matter. The definition of the penetration is the distance where the intensity weakens to 36%. The next distance will decrease the intensity again by 36%, which is 13% of the original incident beam. (The next depth has only 4.7% etc.)

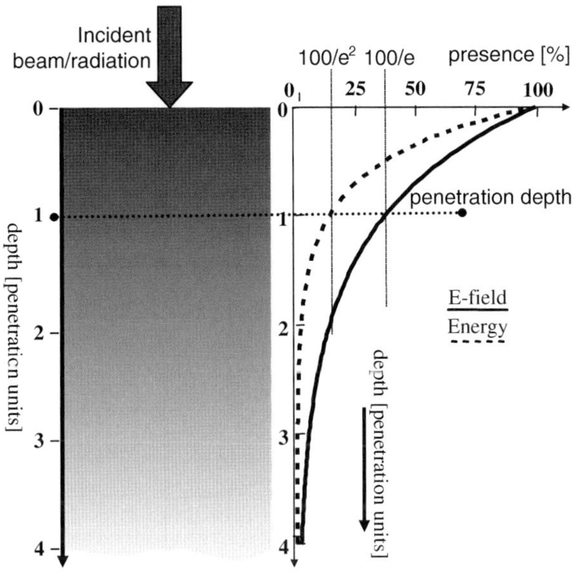

$$d(z) = d_0 e^{-\frac{z}{\delta}} \quad \left\{ \delta = \frac{1}{\sqrt{\pi f \mu \sigma}} \right\} \tag{A.8.3}$$

The penetration is defined by the depth where the field intensity decreases by 1/e-part (about 36%) of the incident beam [δ in (A.8.3)] (see Fig. A.8.2).

The penetration of the field into the targeted material depends significantly on the applied antenna arrangement and the material properties as well. The planar waves at the applied – relatively low (1–15 MHz) – frequency generated by capacitive-coupled antennas penetrate into biosystems by 14–20 cm [1151, 1152] depending on the actual electric conductivity of the biotissue. The penetration depth for the planar waves is shown in Fig. A.8.3.

The incident energy beam reflects on the surfaces and internal boundaries as well. Note that the reflection at the air–skin boundary is very high (it is more than 0.8 at 100 MHz and about 0.6 at 10 GHz [1306]).

In the case of imperfect dielectrics like living tissues, the penetration depth in homogeneous matter depends on the permittivity as well [1152]. Because of the frequency dispersion of ε, the penetration depth depends more complexly on the frequency than in the simple case in (A.8.3), (see Fig. A.8.4). All of these changes represent the radiative incidence on the surface of the material.

$$\delta = \frac{1}{2\pi f \left[\frac{\varepsilon\mu}{2} \left(\sqrt{1 + \left(\frac{\sigma}{2\pi f \varepsilon} \right)^2} - 1 \right) \right]^{1/2}} \tag{A.8.4}$$

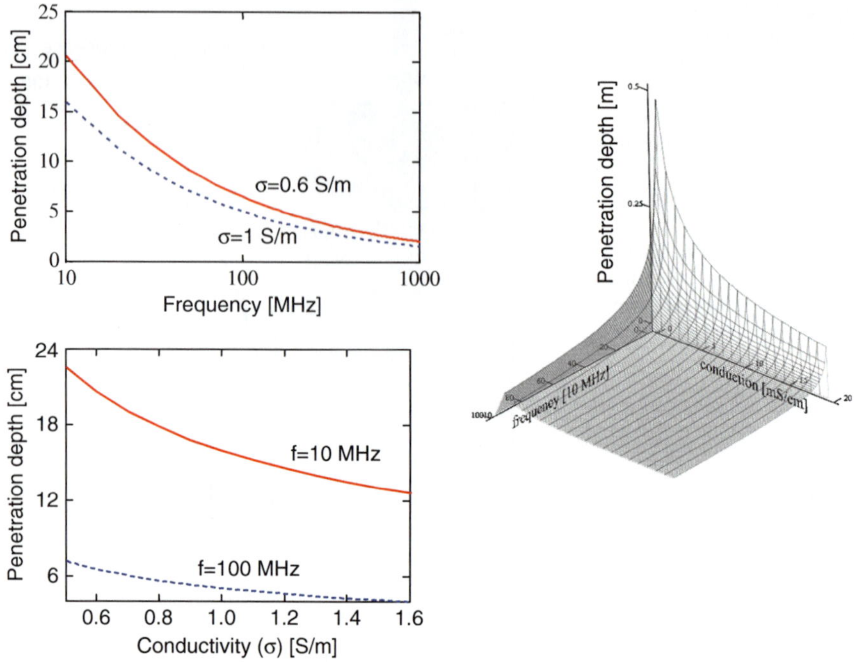

Fig. A.8.3 Penetration depth of planar waves into the tissue versus the frequency and the conductivity. The depth is approximated from simplified conditions of the imperfect conductor (The shaded area represents the most common values for muscle tissue)

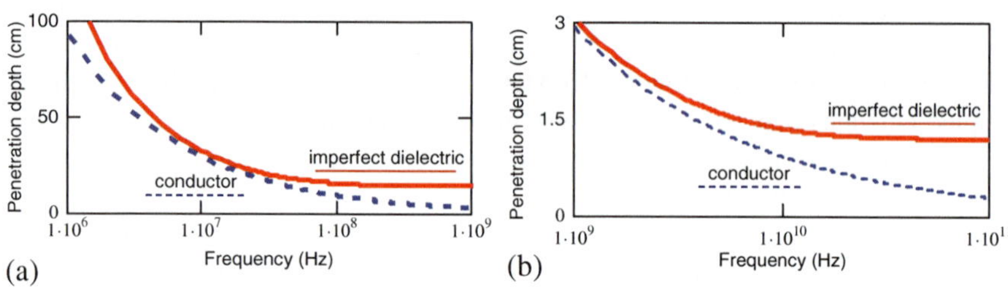

Fig. A.8.4 Comparison of the estimations of penetration depth for conductor and imperfect dielectrics in the MHz range (**a**) and in the GHz range (**b**)

Appendix 9: Direct Current Applications

The applied DC current is in fact an "ion separation" effect in the non-ionizing energy region: the current in the aqueous electrolytes separates the solvated ions by their electric charges. The quality parameter is the applied charge load (in coulombs, C), which in this meaning corresponds well with the old unit measuring the ion-pair production, 1 R (roentgen) = $2.58*10^{-4}$ C/kg. The charge which loads the target tissue polarizes the electrolytes and the large molecules (proteins) in it.

This polarization could change the pH conditions in the tissue. The destroyed tissue is proportional with the logarithm of the applied charge [1307]. This non-trivial charge dependence could be explained on the non-equilibrium thermodynamic basis, again in the frame of the non-temperature dependent (NTD) effect, (Section 3.2.6). The intra- and extra-cellular electrolytes are separated by the cellular membrane, through which various extensive quantities could exchange, like the entropy (*S*), particles (molecules, ions, their number is *N*), and charge (*q*). In our case we concentrate on the change of pH, so *N* is the number of protons (or hydronium [H_3O^+] molecules), and the *i* index denotes the different electrolytes. Then the energy balance [according to Eq. (3.31)] is:

$$\frac{dU_i}{dt} = T\frac{dS_i}{dt} + \mu_i\frac{dN_i}{dt} + \varphi_i\frac{dq_i}{dt} \quad (i = 1,2) \tag{A.9.1}$$

and the charge balance:

$$\frac{dq_i}{dt} = ZF\frac{dN_i}{dt} \quad (i = 1,2) \tag{A.9.2}$$

where *F* is the Faraday charge (*F=96,500 C*), and *Z* is the ionization number. Therefore, the united balance is:

$$\frac{dS_i}{dt} = \frac{1}{T}\frac{dU_i}{dt} - \frac{(\mu_i + ZF\varphi_i)}{T}\frac{dN_i}{dt} \quad (i = 1,2) \tag{A.9.3}$$

and the exchange

$$dN_1 = -dN_2 := dN \quad dU_1 = -dU_2 := dU \tag{A.9.4}$$

In consequence:

$$\frac{dS}{dt} := \frac{d(S_1 + S_2)}{dt} = -\frac{\mu_1 - \mu_2}{T}\frac{dN}{dt} - \frac{\varphi_1 - \varphi_2}{T}\left(ZF\frac{dN}{dt}\right) = -\frac{\Delta\mu}{T}J_N - \frac{\Delta\varphi}{T}J_q,$$
$$\Delta\mu = \mu_1 - \mu_2, \quad \Delta\varphi = \varphi_1 - \varphi_2, \quad J_N = \frac{dN}{dt}, \quad J_q = ZF\frac{dN}{dt} \tag{A.9.5}$$

Onsager's cross-effects [according to Eq. (A.2.3)]:

$$J_N = -L_{11}\frac{\Delta\mu}{T} - L_{12}\frac{\Delta\varphi}{T}, \quad J_q = -L_{21}\frac{\Delta\mu}{T} - L_{22}\frac{\Delta\varphi}{T} \tag{A.9.6}$$

Considering that

$$L_{12} = L_{21} <<< L_{11}, \quad L_{12} = L_{21} <<< L_{22} \tag{A.9.7}$$

We get:

$$J_N = -L_{11}\frac{\Delta\mu}{T} \tag{A.9.8}$$

The chemical potential μ in aqueous solutions [335]:

$$\mu_i = \mu_{i0} + RT \ln c_i, \quad (i = 1, 2) \tag{A.9.9}$$

So the intra- and extra-cellular hydronium balance:

$$\frac{\Delta\mu}{T} = R \ln c_{\mathrm{int\,H_3O^+}} - R \ln c_{\mathrm{ext\,H_3O^+}} \tag{A.9.10}$$

Hence:

$$J_N = -L_{11} \frac{\Delta\mu}{T} = L_{11} R \ln c_{\mathrm{ext\,H_3O^+}} - L_{11} R \ln c_{\mathrm{int\,H_3O^+}} \tag{A.9.11}$$

Because of the assumption that the pH change defines the denaturation of proteins and the charge is proportional with the generated hydronium ions:

$$V_{\mathrm{destroyed}} \propto \ln Q \tag{A.9.12}$$

where $V_{\mathrm{destroyed}}$ is the destroyed volume of the tissue, and Q is the applied complete charge for the treatment. Therefore, the distortion is proportional to the logarithm of the charge load, as was experimentally observed [1307].

$$V_{\mathrm{destroyed}} = k_{\mathrm{d}} \ln Q + c_{\mathrm{d}} \Rightarrow V_{\mathrm{destroyed}} = k_{\mathrm{d}} \ln \frac{Q}{Q_0} \tag{A.9.13}$$

where k_{d} and c_{d} are constants, and Q_0 is the charge deposited at the start of the measurement.

This result has a serious consequence:

$$d\,V_{\mathrm{destroyed}} \propto \frac{dQ}{Q} \Rightarrow \frac{dV_{\mathrm{destroyed}}}{dt} \propto \frac{1}{Q}\frac{dQ}{dt} = \frac{I_{\mathrm{treat}}}{Q} \tag{A.9.14}$$

the number of actually destroyed cells by a unit charge decreases by the absolute value of the applied charge. The lethal change with time is directly proportional to the applied current (I_{treat}) but inversely with the applied charge.

The effect of course has temperature dependence by the Arrhenius law. In the reaction-kinetic model [579] balance between the concentration of the vivid (N_{V}) and damaged (N_{D}) cells is:

$$N_{\mathrm{V}} \underset{\beta}{\overset{\alpha}{\rightleftarrows}} N_{\mathrm{D}} \quad \{\alpha\rangle\rangle\beta\} \tag{A.9.15}$$

So the concentrations:

$$\frac{d\,[N_{\mathrm{V}}]}{dt} = -\alpha\,[N_{\mathrm{V}}] + \beta\,[N_{\mathrm{D}}], \quad \frac{d\,[N_{\mathrm{D}}]}{dt} = \alpha\,[N_{\mathrm{V}}] - \beta\,[N_{\mathrm{D}}] \tag{A.9.16}$$

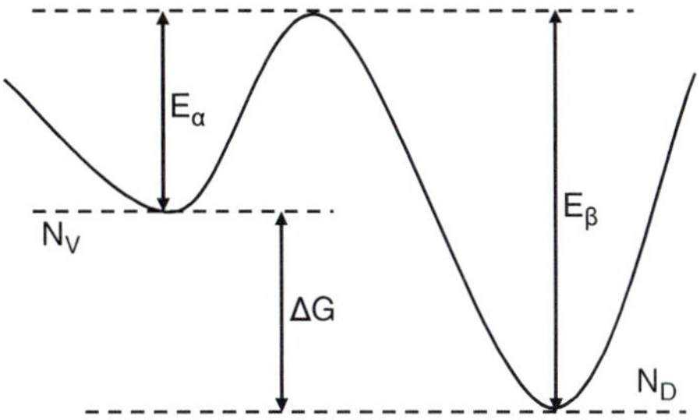

Fig. A.9.1 Active- (excited-) state

and by the Arrhenius law:

$$\alpha = Ce^{-\frac{E_\alpha}{RT}}, \quad \beta = Ce^{-\frac{E_\beta}{RT}} \tag{A.9.17}$$

This exciting process is shown in Fig. A.9.1.

$$\frac{[N_D]}{[N_V]} = \frac{\alpha}{\beta} = e^{-\frac{E^\alpha}{RT}+\frac{E^\beta}{RT}} = e^{-\frac{\Delta G}{RT}}, \quad \frac{[N_D]}{[N_V]+[N_D]} = \frac{\alpha}{\alpha+\beta} = \frac{e^{-\frac{\Delta G}{RT}}}{1+e^{-\frac{\Delta G}{RT}}} \tag{A.9.18}$$

Because of (A.9.9), and knowing the pH changed via hydronium ions (protons):

$$\frac{\Delta G}{RT} = \frac{\Delta\mu}{RT} = \ln c_{1H_3O^+} - \ln c_{2H_3O^+} = -\Delta pH \tag{A.9.19}$$

and so the distortion efficacy η_D due to the pH change is (see Fig. A.9.2):

$$\eta_D(\Delta pH) := \frac{[N_D]}{[N_V]+[N_D]} = \frac{e^{\Delta pH}}{1+e^{\Delta pH}} \tag{A.9.20}$$

Note, contrary to the Arrhenius law, which was used, this effect is NTD. However, in the full picture (when the pH is only one factor of many) the ionic exchange has temperature dependence, but also NTD is involved [1308].

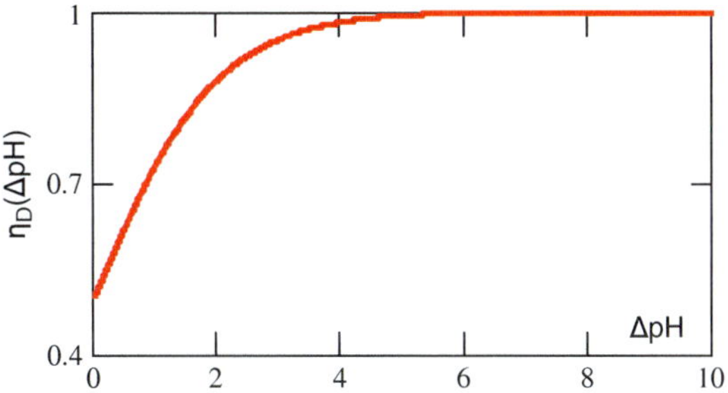

Fig. A.9.2 Efficacy of the cellular distortion versus change of pH. Practically the distortion is complete at ΔpH=4

Appendix 10: Development of Edema

In aqueous electrolytes the ionic transports (like the injury current) create a mass-transport of water molecules by Onsager's cross-effects [Eq. (A.2.3)], the change of electric potential ($\Delta\varphi$) accompanied by the hydrodynamical pressure (Δp) and initiate electric (J_e) and mass (J_m) currents as:

$$\begin{aligned} J_m &= L_{11}\Delta p + L_{12}\Delta\varphi \\ J_e &= L_{21}\Delta p + L_{22}\Delta\varphi \end{aligned} \tag{A.10.1}$$

The L_{ik} Onsager's phenomenological constants are symmetric and due to the entropy law the constants of the pure effects are positive. The mixed indexed coefficients could be negative also, but according to the experimental data they are always considerable lower than the main effects. Therefore:

$$L_{12} = L_{21}, \quad L_{11} > 0, \quad L_{22} > 0, \quad L_{11}L_{22} \geq {L_{12}}^2 \tag{A.10.2}$$

It is realistic to assume the extra-cellular space has constant pressure, all changes are quickly equalized, so:

$$J_m = \frac{L_{12}}{L_{22}} J_e \tag{A.10.3}$$

Hence the J_m and J_e could not necessarily flow in the same direction. A realistic quantitative assumption is $J_e = 10$ nA and $L_{12}/L_{22} = 10^{-4}$. In this case:

$$J_m = \frac{L_{12}}{L_{22}} J_e = 10^{-11}\frac{m^3}{s} = 3600\ 10^{-2} = 36\frac{mm^3}{h} \tag{A.10.4}$$

This is a considerable value to deliver water parallel by the flow of the injury current. Consequently, the electric field affects edema development via the NTD process.

The heat flow (heat current) can also be considered in the frame of Onsager's cross-effects; then a new current (heat-current J_q) has to be additional and driven by the ΔT temperature gradient:

$$\begin{aligned} J_m &= L_{11}\Delta p + L_{12}\Delta\varphi + L_{13}\Delta T \\ J_e &= L_{21}\Delta p + L_{22}\Delta\varphi + L_{23}\Delta T \\ J_q &= L_{31}\Delta p + L_{32}\Delta\varphi + L_{33}\Delta T \end{aligned} \tag{A.10.5}$$

The symmetry and entropy conditions:

$$\begin{aligned} &L_{12} = L_{21}\ L_{13} = L_{31}, \quad L_{23} = L_{32} \\ &L_{11} > 0, \quad L_{22} > 0, \quad L_{33} > 0 \\ &L_{11}L_{22} \geq {L_{12}}^2, \quad \begin{vmatrix} L_{11} & L_{12} & L_{13} \\ L_{21} & L_{22} & L_{23} \\ L_{31} & L_{32} & L_{33} \end{vmatrix} > 0 \end{aligned} \tag{A.10.6}$$

In consequence both the electric field and the temperature affects the transport of the water, and influences the development of edema.

Appendix 11: Warburg Impedance

There are high concentrations of Cl^- and H^+ (in H_3O^+ form) in extra-cellular electrolyte. In the case of non-charged electrodes both ions are hydrate-jacketed, meaning the polar water molecules arrange a shell around the naked ions; resulting in reduced ionic charge and decreased ion mobility. Without the external electric field, the ions of the electrolyte have zero drift velocity. Drift velocity and conduction in the electrolyte will be introduced via application of voltage to the electrodes. Between the current density (j) and the polarity-dependent drift velocities ($v^{\pm}$) of the ions

$$j = n^+ z^+ e v^+ + n^- z^- e v^- \tag{A.11.1}$$

where $n^{\pm}$ are the number of ions, $z^{\pm}$ the ionizing degree, and e is the electron charge. The current does not appear immediately, some time is necessary to obtain the non-zero velocity of ions because of their inertia and the viscosity of the solution. This means that just at the moment of application of voltage no current appears, so it looks as if the resistance of the electrolyte is infinite. After this moment, the resistance decreases to a finite value as the ions attain their stationary drift velocity.

The above-described process is significantly modified by diffusion and by the electrode process. When the hydrated ion arrives at the cathode it loses its hydrate-jacket for the charge exchange between the ion and electrode, and therefore the generation of current could start. Hence the first step of the process is the dehydration, which also takes some time. Until the hydration takes place the electrode has a hydrogen ion coating, namely, we might observe the generation of hydrogen gas at the electrode. This has a concentration potential that – in accordance with the Chatelier–Braun principle – is opposite to the external potential that generates the process. If the hydrogen exchanges electrons with the electrode then we get hydrogen molecules. This process increases the potential of the electrodes and does not allow an increase of the current. Then the diffusion of hydrogen gas begins into the electrode material, which helps to build up current again. The mathematical evaluation of the above-described electrochemical process is very complicated, and only a few cases have exact solutions [1309]. For a large plain electrode we get [1309]:

$$Z(t) = \frac{K}{\sqrt{\pi\, t}} \tag{A.11.2}$$

where K is a constant value which depends on the diffusion coefficient and the directly influenceable electrode surface. The above impedance is called the Warburg impedance. In the case of sinusoidal supply and in a stationary case the Warburg impedance can be expressed as:

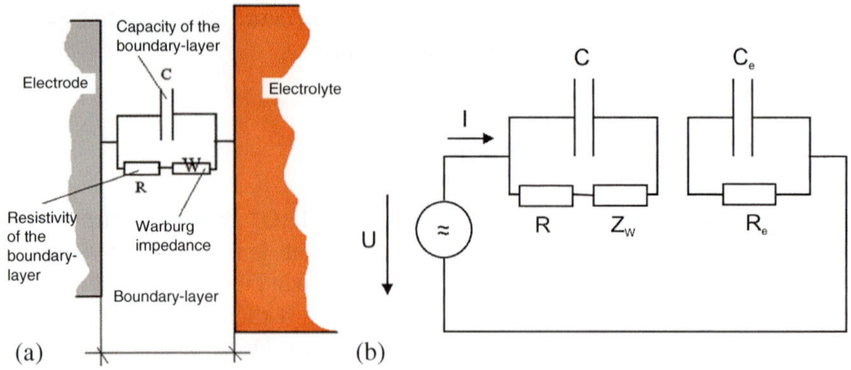

Fig. A.11.1 The impedance changes at the surface of the electrode immersed in an electrolyte (**a**); the full equivalent circuit of the electrolyte effect (Warburg effect) (**b**)

$$Z(\omega) = \frac{K}{\sqrt{s = j\omega}} = K\frac{1-j}{\sqrt{\omega}} \tag{A.11.3}$$

Therefore, indeed, the Warburg impedance is infinite in the case of constant current and we get zero resistance if the frequency is very high:

$$\begin{aligned} Z(\omega = 0) &= \infty \\ Z(\omega = \infty) &= 0 \end{aligned} \tag{A.11.4}$$

The electric circuit of this effect is shown in Fig. A.11.1.

In the case of small C (when it does not short-cut the $(R+Z_W)$ resistivity:

$$\begin{aligned} &\tau \tfrac{d}{dt}\left(R_w(t)i(t)\right) + \left(R_w + R_e\right) i(t) = \tau \tfrac{du}{dt} + u \\ &u = U_0 \sin(\omega t) \\ &\tau = R_e C_e \end{aligned} \tag{A.11.5}$$

The current density and voltage changes with time are shown in Fig. A.11.2. The time to reach equilibrium at 13.56 MHz is 5.8 min.

The effect modifies the Cole–Cole diagram in the low frequency region (see Fig. A.11.3). This requests an appropriate time-dependent current, which has to be fitted to the time constants of the electrode–electrolyte interface and the processes there.

For sinusoidal supply the complex Warburg impedance (Z_W) [1309] has inverse-square-root dependence on the frequency (f) in the case of a planar electrode [1310]:

$$Z_W(f) = A\frac{(1-i)}{\sqrt{f}} \tag{A.11.6}$$

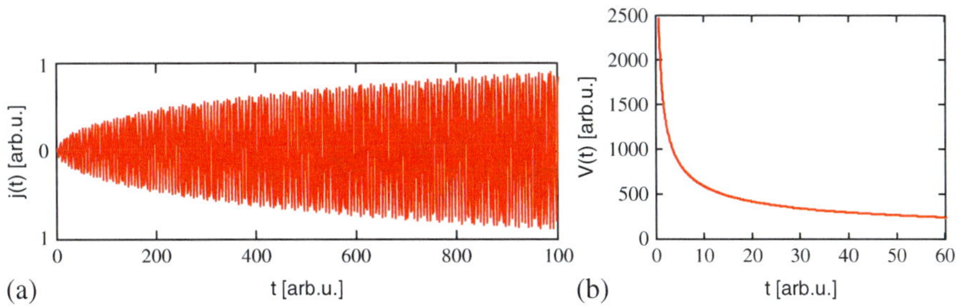

Fig. A.11.2 The time to be in charge-equilibrium is about 5.8 min on the surface of an indifferent electrode (The units are shown in arbitrary values, only as a demonstration of the effect)

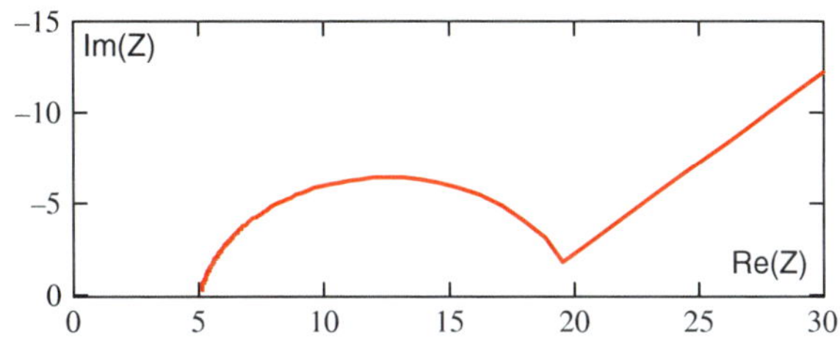

Fig. A.11.3 Modification of the Cole–Cole diagram by the Warburg effect

where A is a constant. Usually the problems of the Warburg impedance are avoided by multi-electrode applications and as high a frequency as possible. However, the error-current caused by the scattered capacity in multi-electrode solutions (which grows linearly with the frequency) could be high enough to modify the experimental results. A compromise is always necessary to choose an accurate solution. The noise also could be a limiting parameter of accurate measurements. At low frequencies the large Warburg resistance produces significant electric noise, (the electric noise on a resistance is proportional to the value of the actual resistivity) [1311], while the high frequencies stimulate other noise sources to be active.

Appendix 12: Cell-Membrane Permeability

Denote the density of the i-th component at x by $C_i(x)$. Then the transport current by Fick's diffusion law [1312]:

$$j_{id} = -D_i \frac{dC_i(x)}{dx} \tag{A.12.1}$$

where D_i is the diffusion constant of the i-th species. The drift velocity is:

$$v_i\text{drift} = -\beta_i \frac{\psi_{\text{memb}}}{\xi} = -\frac{kT}{D_i}\frac{\psi_{\text{memb}}}{\xi} \tag{A.12.2}$$

where Z_i and β_i are the ionizing degree and motility of the i-th component, ψ_{memb} is the membrane potential and ξ is the thickness of the membrane. So the drift current:

$$j_{i\text{drift}} = -C_{io}(x)\frac{kT}{D_i}\frac{\psi_{\text{memb}}}{\xi}Z_i e \tag{A.12.3}$$

where e is the electron charge. So the full current density of the i-th component is:

$$j_i = -D_i\frac{dC_i(x)}{dx} - C_{io}(x)\frac{kT}{D_i}\frac{\psi_{\text{memb}}}{\xi}Z_i e \qquad (i = 1, 2, ...) \tag{A.12.4}$$

From where the electric current through the membrane is:

$$\begin{aligned} j_{ei} &= \frac{Z_i^2 e^2 D_i}{kT\xi}\frac{C_i^{\text{ext}} e^{-\frac{Z_i e\psi_{\text{memb}i}}{kT}} - C_i^{\text{int}}}{1 - e^{-\frac{Z_i e\psi_{\text{memb}i}}{kT}}}\psi_{\text{memb}} \\ &= e^2 \varsigma_i N_{\text{L}}\frac{C_i^{\text{ext}} e^{-\frac{Z_i e\psi_{\text{memb}i}}{kT}} - C_i^{\text{int}}}{1 - e^{-\frac{Z_i e\psi_{\text{memb}i}}{kT}}}\psi_{\text{memb}} \qquad \left\{\varsigma_i = \frac{Z_i^2 e^2 D_i}{kT\xi N_{\text{L}}}\right\} \end{aligned} \tag{A.12.5}$$

where C_i^{ext} and C_i^{int} are the concentrations of the chemicals on the extra- and intra-cellular side of the membrane. The introduced ς_i is the membrane permeability on the i-th component of the electrolytes. ($N_{\text{L}} = 6{\cdot}10^{23}$, Loschmidt number.) Furthermore, the Goldman–Hodgkin–Katz equation [200] is as follows:

$$\psi_{\text{memb}} = \frac{kT}{e}\ln\left(\frac{\sum_{i=1}^{N}\varsigma_i\, C_i^{\text{ext}}}{\sum_{i=1}^{N}\varsigma_i\, C_i^{\text{int}}}\right) = \frac{kT}{e}\ln\left(\frac{\sum_{i=1}^{N}\frac{\varsigma_i}{\varsigma_K}\, C_i^{\text{ext}}}{\sum_{i=1}^{N}\frac{\varsigma_i}{\varsigma_K}\, C_i^{\text{int}}}\right) \tag{A.12.6}$$

When we have any phase transition (first kind) in the cell (e.g. order–disorder transition), the concentration changes as:

$$\Delta C_i^{\text{int}} = -C_i^{\text{int}}\frac{\Delta V}{V} \tag{A.12.7}$$

where V is the volume of the phase before the transition and ΔV is the change of the volume due to the transition. Then from (A.12.6)

$$\Delta\psi_{\text{memb}} = -\frac{kT}{e}\ln\left(1 - \frac{\Delta V}{V}\right) \tag{A.12.8}$$

and if $(\Delta V/V) \ll 1$, then:

$$\Delta\psi_{\text{memb}} = \frac{kT}{e}\frac{\Delta V}{V} \tag{A.12.9}$$

Assuming T=300 K and using the same thinking for the membrane permeability:

$$\Delta\psi_{\text{memb}} = 25.88\frac{\Delta V}{V}\ [mV] \text{ and } \Delta\psi_{\text{memb}} = 56\Delta\left(\frac{\varsigma_i}{\varsigma_k}\right)\ [mV] \tag{A.12.10}$$

Appendix 13: Stochastic Processes – Pink Noise

In a simple example the growth of a tumor can be described deterministically in a way that its change over time (ΔM_{tumor}) is proportional with its actual mass, (M_{tumor}) and the time interval of the observation (Δt):

$$\Delta M_{\text{tumor}}(t) = kM_{\text{tumor}}(t)\,\Delta t \tag{A.13.1}$$

where k is a constant. This leads to a simple differential equation and the well-known exponential solution with the mass (M_0) measured at the start of observation:

$$\frac{dM_{\text{tumor}}(t)}{dt} = kM_{\text{tumor}}(t) \Rightarrow M_{\text{tumor}}(t) = M_0e^{kt} \tag{A.13.2}$$

If life would be so easily deterministic, then the initial task of oncology to make prognoses would be very simple. However the process is stochastic, so we have to calculate the development of the tumor step-by-step. To follow the development of the tumor we choose the observation time interval (Δt) as the time to produce an additional cell or eliminate a single cell from the mass. So the probability to add a cell to the tumor at time t during the interval Δt will be proportional with $kM(t)\,\Delta t$, as we assumed originally in (A.13.1). (Note we consider only a single cell development and not an arbitrary mass). Then the probability equation with the added and eliminated cells in Δt will be:

$$P_{\text{M}}(t+\Delta t) = P_{\text{M}}(t) + k(M-1)\,\Delta t P_{\text{M}-1}(t) - kM\Delta t P_{\text{M}}(t) \tag{A.13.3}$$

so the probability depends on the cell added to the tumor from the previous time interval and the one eliminated at the actual time. In a differential equation formulation:

$$\frac{dP_{\text{M}}(t)}{dt} = k(M-1)\,P_{\text{M}-1}(t) - kMP_{\text{M}}(t) \tag{A.13.4}$$

When we start from a single cell ($P_{\text{M}}(0)$=1 if M_0=1, and $P_{\text{M}}(0)$=0 in every other case), the solution of (A.13.4) for $M \geq M_0$ cases is:

$$P_{\text{M}}(t) = \binom{M-1}{M-M_0} e^{-kM_0t}\left(1-e^{-kt}\right)^{M-M_0} \tag{A.13.5}$$

Compare (A.13.2) and (A.13.5), how they are different! The deterministic approach is continuous in time, running in real values, while the stochastic approach jumps in integers, building up the tumor mass step-by-step. The deterministic equation gives a concrete result, while the stochastic one shows "only" probability. It is interesting to see, that the deterministic result is a special case of the stochastic one, when the probabilities are always deterministic: $P_M(t)=M_{tumor}(t)$ irrespective of the actual number of steps. Anyway, the mean (or expected size of the tumor) is by its definition:

$$\langle M_{tumor}(t)\rangle = \sum_{M=M_0}^{\infty} MP_M(t) = M_0 e^{kt} \tag{A.13.6}$$

so the deterministic solution!

In order to simulate the stochastic processes, we have to base our investigation on the Fourier transform approach. Let us denote the time-dependent function (the process) by $x(t)$. Its Fourier transform is defined by [1313]:

$$X(f) = \frac{1}{\sqrt{2\pi}} \int_{-\infty}^{\infty} x(t)\, e^{-j2\pi f t} dt = F\{x(t)\} \tag{A.13.7}$$

It is easy to prove by means of the above definition that the Fourier transform of the function $x(a\,t)$, where 'a' is an arbitrary complex number, is:

$$F\{x(at)\} = \frac{1}{a} X\left(\frac{f}{a}\right) \tag{A.13.8}$$

Let us define the work of the $x(t)$ process by

$$W = \int_{-\infty}^{\infty} x^2(t) dt \tag{A.13.9}$$

It follows from Parseval's formula [1313] that this work may be evaluated by

$$W = \int_{-\infty}^{\infty} x^2(t) dt = \int_{-\infty}^{\infty} S(f) df \tag{A.13.10}$$

where the so-called spectral density function $S(f)$ is

$$S(f) = \frac{|X(f)|^2}{2\pi} \tag{A.13.11}$$

as a function of frequency f. It can be proved that the spectral density is the even function of the frequency [1313], i.e. $S(f)=S(-f)$.

By definition, a stationary random process has indefinite duration. To introduce a modified density spectrum, consider a finite segment of the random process $x(t)$ of duration $2T$, defined by:

$$x_T = \begin{cases} x(t), & -T \le t \le T \\ 0, & \text{otherwise} \end{cases} \quad \text{and} \quad \lim_{T\to\infty} x_T(t) = x(t) \tag{A.13.12}$$

The Fourier transform of $x_T(t)$ has the form of

$$X(f,T) = \frac{1}{\sqrt{2\pi}} \int_{-T}^{T} x(t) e^{-j2\pi ft} dt \tag{A.13.13}$$

and

$$F\{x_T(at)\} = \frac{1}{a} X\left(\frac{f}{a}, T\right) \tag{A.13.14}$$

In this case Parseval's formula can be expressed as:

$$\lim_{T\to\infty} \frac{1}{2T} \int_{-T}^{T} x^2(t)dt = \int_{-\infty}^{\infty} S(f)df \quad \left\{ S(f) = \frac{1}{2\pi} \lim_{T\to\infty} \frac{|X(f,T)|^2}{2T} \right\} \tag{A.13.15}$$

$S(f)$ is the so-called power density spectrum in any randomly stationary case.

The living processes are basically self-similar [1314, 1315], so it is convenient to define the self-similarity of a stochastic process. A stochastic process is said to be self-similar if the effective power of the stochastic process representation $x(t)$ equals the effective power of the representation $x(at)$ defined over time scale $[at]$, for every a positive scalar, i.e.:

$$\lim_{T\to\infty} \frac{1}{2T} \int_{-T}^{T} x^2(t)dt = \lim_{T\to\infty} \frac{1}{2T} \int_{-T}^{T} x^2(at)d(at) \tag{A.13.16}$$

And so from (A.13.8) and (A.13.12), we get

$$a \int_{-\infty}^{\infty} \frac{1}{a^2} S\left(\frac{f}{a}\right) df = \int_{-\infty}^{\infty} S(f)df \tag{A.13.17}$$

Also, for the power spectral density function, the functional equation may be expressed as follows

$$S\left(\frac{f}{a}\right) = a\,S(f) \tag{A.13.18}$$

for every positive scalar a and every scalar f. To solve this equation, we assume that $f>0$ and set for a the value $a=f$. Hence:

$$S(f) = \frac{S(1)}{f} \quad \text{(A.13.19)}$$

On the other hand, if $f < 0$ then $f=-|f|$, and

$$\frac{1}{a}S\left(\frac{f}{a}\right) = \frac{1}{a}S\left(-\frac{|f|}{a}\right) = S(f) \quad \text{(A.13.20)}$$

Let us set for 'a' the value $a=|f|$ and take into account that the power density function is even, so we obtain the $1/f$ spectrum, or "pink noise":

$$S(f) = \frac{S(1)}{|f|} \quad \text{(A.13.21)}$$

Appendix 14: Autocorrelation

In accordance with the ergodic hypothesis [1316], the autocorrelation function of a stationary random process $x(t)$ can be defined as

$$R_{xx}(\tau) = \lim_{T\to\infty} \frac{1}{2T}\int_{-T}^{T} x(t)x(t+\tau)dt \quad \{R_{xx}(\tau) = R_{xx}(-\tau)\} \quad \text{(A.14.1)}$$

The relation between the autocorrelation function and the power density spectrum can be expressed by the Fourier transform of the autocorrelation function (Wiener–Khinchine theorem [1313]), namely:

$$R_{xx}(f) = \frac{1}{\sqrt{2\pi}}\int_{-\infty}^{\infty} R_{xx}(\tau)\, e^{-j2\pi f\tau} d\tau \text{ and } R_{xx}(\tau) = \frac{1}{\sqrt{2\pi}}\int_{-\infty}^{\infty} R_{xx}(f)\, e^{j2\pi f\tau} df \quad \text{(A.14.2)}$$

From these (considering [1313, 1317]) we may conclude

$$R_{xx}(\tau) = \int_{-\infty}^{\infty} S(f)\, e^{j2\pi f\tau} df = \int_{-\infty}^{\infty} \frac{S(1)}{|f|} e^{j2\pi f\tau} df = \frac{\sqrt{2\pi}\, S(1)}{|\tau|} \quad \text{(A.14.3)}$$

Appendix 15: Dissipative Systems

Considering that the quantum theory of dissipative systems is not adequately worked out, we are going to stay within the range of the classical theory. We suppose that the pieces of information necessary for the communication are carried by the analogue signals describing the physico-chemical state of the individual cells. Furthermore, we are going to suppose that the state of coaching biological subsystems can be represented by the self-similar Markov processes. Gillespie could show that from this assumption the equation describing the dynamics of processes can be concluded. This is the generalized Langevin equation [1318]:

$$\frac{dX_i}{dt} = A_i(X_j, t) + D_i^{\frac{1}{2}}(X_j, t)\Gamma(t) \quad (i = 0, 1, 2, ..., N-1) \tag{A.15.1}$$

where

$$\Gamma(t) = \lim_{dt \to 0} \mathbf{N}(0, dt^{-1}) \tag{A.15.2}$$

is the so-called white noise with zero mean value, infinite dispersion, and normal distribution. Let us decompose the A_i (X_j, t) function into three parts:

$$A_i(X_j, t) = f_i(t) + A_i(X_i) + \sum_{k=0}^{N-1} c_{ik}X_k \tag{A.15.3}$$

where the c_{ik} elements form a cyclic matrix.

$$\overline{C} = \begin{bmatrix} c_0 & c_1 & \cdots\cdots\cdots c_{N-2} & c_{N-1} \\ c_{N-1} & c_0 & . & c_{N-2} \\ . & & & \\ & . & . & . \\ . & & & \\ c_1 & c_2 & . & c_0 \end{bmatrix} \tag{A.15.4}$$

$A_i(x_i)$ can be nonlinear and $f_i(t)$ is the time function generated by the internal active processes of the cell. It is reasonable to assume that $A_i(x_i)$ is identical for each cell, and in the same way, we may suppose that D_i is constant for each cell. The latter can be justified by the fact that each cell is to be found in the same heat container. We did not assume any confinement for the $f_i(t)$ function. The proposed equation is a generalization of the model of the coupled damped oscillators, which showed [1319] that the stochastic resonance is included in the forms of motion. We are going to examine a case where the social signal has low amplitude; therefore, the non-linear members can be neglected. Then (A.15.1):

$$\frac{dX_i}{dt} = f_i(t) + \sum_{k=0}^{N-1} c_{ik}X_k + D^{\frac{1}{2}}\Psi(t) \quad (i = 0, 2, ..., N-1) \tag{A.15.5}$$

Now, we are going to prove that among the modes belonging to the eigenvectors of matrix (A.15.4) of Eq. (A.15.1) there are modes of zero noise spectrum. It is well known that any cyclic matrix can be diagonalized by the transformation matrix [1320], that is

$$T = \frac{1}{\sqrt{N}} \begin{bmatrix} 1 & 1 & \cdots & 1 & \cdots & 1 \\ 1 & a & \cdots & a^i & \cdots & a^{N-1} \\ \vdots & & & & & \\ 1 & a^j & \cdots & a^{ji} & \cdots & a^{j(N-1)} \\ & & \vdots & & & \\ 1 & a^{N-1} & \cdots & a^{(N-1)i} & \cdots & a^{(N-1)^2} \end{bmatrix} \quad \text{(A.15.6)}$$

where $a=e^{i2\pi/N}$. Applying this transformation to the Eq. (A.15.5) we can obtain:

$$\frac{dx_{si}}{dt} = \lambda_i x_{si} + f_{si}(t) + \Gamma_{si}(t) \quad (i = 0, \ldots, N-1) \quad \text{(A.15.7)}$$

Here the new coordinates and the eigenvalues of the cyclic matrix are [1319]

$$\begin{aligned} x_{si} &= \frac{1}{\sqrt{N}} \sum_{k=0}^{N-1} a^{-ik} x'_k, \ \Gamma_{si}(t) = D^{\frac{1}{2}} \frac{1}{\sqrt{N}} \sum_{k=0}^{N-1} a^{-ik} \Gamma(t), \\ \lambda_j &= \sum_{k=0}^{N-1} a^{jk} c_k, \quad f_{si}(t) = \frac{1}{\sqrt{N}} \sum_{k=0}^{N-1} a^{-ik} f_i(t) \ \ (j = 0, \ldots, N-1) \end{aligned} \quad \text{(A.15.8)}$$

Let us consider any one of the new

$$\Gamma_{si}(t) = D^{\frac{1}{2}} \frac{1}{\sqrt{N}} \sum_{k=0}^{N-1} a^{-ik} \Gamma(t) \quad \text{(A.15.9)}$$

noise components for which $k \neq 0$ (non-zero order component). Let us take the Fourier transform thereof and consider that the amplitudes are unitary in the white-noise spectrum. Then we obtain

$$\Gamma_{si}(t) = D^{\frac{1}{2}} \frac{1}{\sqrt{N}} \sum_{k=0}^{N-1} a^{-ik}, \quad k \neq 0 \quad \text{(A.15.10)}$$

On the other hand we know that

$$\sum_{k=0}^{N-1} a^{-ik} = 0 \quad \text{(A.15.11)}$$

In consequence, every non-zero order mode is noiseless, because:

$$\Gamma_{si}(t) = 0, \quad k \neq 0 \quad \text{(A.15.12)}$$

Consequently the zero-order signals are noiseless and also the thermal noises do not limit these informations.

Appendix 16: Stochastic Resonance

To understand the underlying physics of stochastic resonance let us consider a bifurcative potential well, according to Fig. 3.41. The noise constrains the particle to randomly oscillate in this double well, and the time of the one or the other well is a probability variable of time ($P(t)$) with an exponential distribution function:

$$P(t) - \frac{1}{T_K} e^{-\frac{t}{T_K}} \tag{A.16.1}$$

The value of T_K occupation time (the so-called Kramer's time-scale), is determined by a generalized Arrhenius factor weighted by an average f frequency:

$$T_K(D) = \frac{1}{f} e^{-\frac{\Delta U}{D}} \tag{A.16.2}$$

Where ΔU is the depth of the potential well and D is the average energy of the noise (if it is thermal, than $D=kT$). If the particle in the double well is under periodic force $A\cos(\Omega t)$ (where the amplitude A is small compared with ΔU; $A << \Delta U/D$), then the coupled wells periodically fluctuate up and down in opposite phase (see Fig. A.16.1).

In the time $t=0$ the jump from right to left is more probable then in the time $t=\pi/2\Omega$, and the opposite is true in the time of $t=\pi/\Omega$. This means that the weak periodic signal (much weaker than the activation energy) synchronizes the jumps, but of course in a stochastic – and not deterministic – way. The stochastic resonance occurs at D^*, when Kramer's time is half of the period-time of the weak deterministic signal:

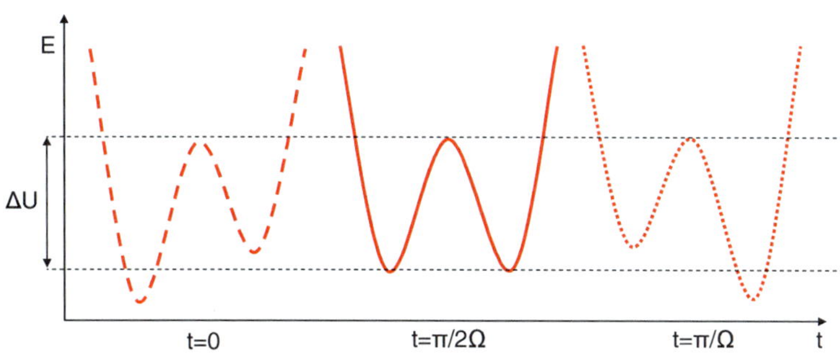

Fig. A.16.1 Changes in the bistable potential-well over elapsed time (one time period of the exciting signal is $T=2\pi/\Omega$)

$$D^* = \frac{\Delta U}{\ln\left(\frac{\pi f}{\Omega}\right)} \tag{A.16.3}$$

Consequently, the distribution function of the jumping time in the case of the noise modulated with a weak periodic signal will not be a rigorously monotonic decreasing function as in (A.16.2), but will have definite periods of maxima:

$$t_{\max} = (2n+1)\frac{\pi}{\Omega} \tag{A.16.4}$$

and a considerable amplification of the weak periodic signal could be observed depending on the D^* noise. The amplification increases also with decreasing frequencies (at constant A amplitude), and the amplification increases by the decreasing A amplitude at the same signal frequency. These have thresholds, where the resonance disappears (window phenomenon). Probably this is the reason of the Adey-window [1025].

With the more rigorous calculations using Focker–Plank equations where the amplification of the system (η) is definitely determined at the f_r=(Ω/2 π) frequencies, the spectral power density function has needle-like (similar to the Dirac delta) peaks at frequencies (n f_r) where n is an integer; and the power of these peaks depends on the A amplitude and D noise-power.

The stochastic resonance could easily be shown on a simple dynamic system, an over-smoothed non-linear oscillator, which can be described by the motion equation:

$$\frac{dx}{dt} = -\frac{dU(x)}{dx} + F(t) + aA(x)\cos(\Omega t + \varphi_0) \tag{A.16.5}$$

where the $U(x)$ potential has two stable and one instable (the middle) minima at X^-, X^+, and X_0, (see Fig. A.16.2); $F(t)$ is a random force (assumed for simplicity as a white noise with normal distribution and with D power), $A(x)\cos(\Omega t+\varphi_0)$ is the exciting deterministic (periodic) signal, a is a constant.

The random force constrains the oscillation between the two stable states, so it jumps through the potential barrier:

$$\Delta U^{\pm} = U(X_0) - U\left(X^{\pm}\right) \tag{A.16.6}$$

The kinetics of the transition is determined by the height of the barrier ($\Delta U^{\pm}$), and the D power of the noise. The $\tau^{\pm}$ time is the stay in the actual minima (and the $1/\tau^{\pm}=r^{\pm}$ transition rate). It is determined from Kramer's equation (A.16.2):

$$\frac{1}{\tau^{\pm}} = r^{\pm} = \frac{1}{2\pi}\sqrt{-\left.\frac{\partial^2 U(x)}{\partial x^2}\right|_{x=X_0} \left.\frac{\partial^2 U(x)}{\partial x^2}\right|_{x=X^{\pm}}}\, e^{-\frac{\Delta U^{\pm}}{D}} \tag{A.16.7}$$

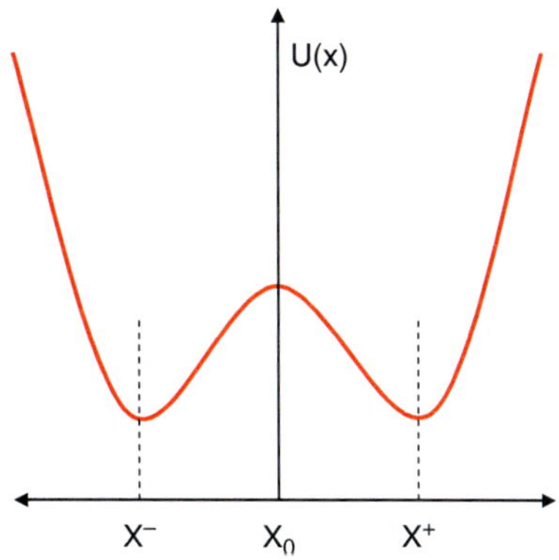

Fig. A.16.2 The characterization parameters of the bistable potential-well

The periodic potential will be modulated with time and its value oscillates at the place $X^{\pm}$ in opposite phase due to the periodic force:

$$W(x,t) = U(x) - aG(x)\cos(\Omega t + \varphi_0) \left\{ A(x) = \frac{dG(x)}{dx} \right\} \tag{A.16.8}$$

When D is small, the Flocker–Plank equation in adiabatic approach in X^{+} and X^{-} points is reduced on the master equation of the probabilities:

$$\frac{dp_+(t)}{dt} = r_-(t)\,p_-(t) - r_+(t)\,p_+(t) \quad \{p_+(t) + p_-(t) = 1\} \tag{A.16.9}$$

The values of $r^{\pm}$ could be determined from Eq. (A.16.7), substituting W instead of U. When $A(x)$=1, the potential became (as shown in Fig. 3.41) very simple:

$$U(x) = -\lambda\frac{x^2}{2} + \frac{x^4}{4} \quad \left\{\lambda > 0, \quad \left[X_0 = 0, \quad X^{\pm} = \pm\lambda^{\frac{1}{2}}\right]\right\} \tag{A.16.10}$$

Hence from (A.16.9) we get:

$$\delta p(t) = A(D)\cos(\Omega t + \varphi + \psi) \tag{A.16.11}$$

where the $A(D)$ amplitude and the ψ phase-shift are:

$$\begin{aligned} A(D) &= a\frac{\lambda}{D}\frac{r(D)}{\sqrt{r^2(D)+\frac{\Omega^2}{4}}}, \quad \psi = -artg\left(\frac{\Omega}{2r(D)}\right) \\ r(D) &= r_+ = r_- = \frac{\lambda}{\sqrt{2\pi}}e^{-\frac{\lambda^2}{2D}} \end{aligned} \tag{A.16.12}$$

The amplitude has a resonance-like behavior (see Fig. A.16.3), in the case where D=kT we are speaking about thermal noise. The maximum depends on D noise

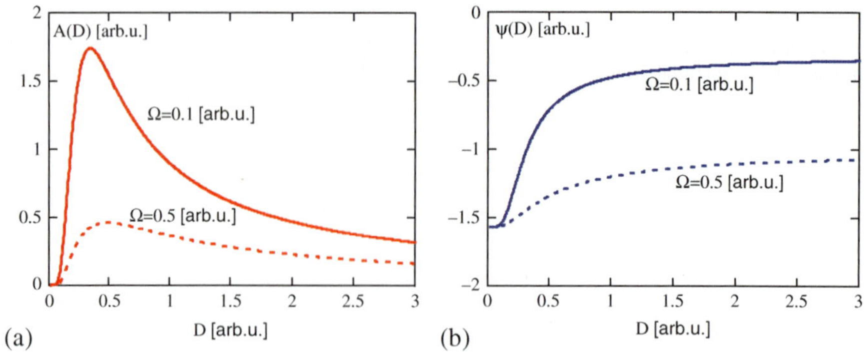

Fig. A.16.3 The amplitude (**a**) and phase (**b**) resonance in stochastic processes

amplitude or at D=const. the maximum is determined by the frequency. This is the typical frequency-amplitude window formulated previously from experiments [1025].

Appendix 17: Resonance of enzymatic reactions

Let us assume two certainly stable confirmation states of an actual enzymatic reaction: $E_1 = A$ and $E_2 = B$, with concentrations $[A]$ and $[B]$, respectively. These are the results of given chemical reactions, so

$$A \underset{\beta}{\overset{\alpha}{\rightleftarrows}} B \qquad \text{(A.17.1)}$$

Then the reaction (A.17.1) could be written quantitatively:

$$\frac{d[A]}{dt} = -\alpha\,[A] + \beta[B] \quad \frac{d[B]}{dt} = \alpha\,[A] - \beta[B] \qquad \text{(A.17.2)}$$

where α and β are the rates of forwarded and back-warded reactions, and both are governed by the Arrhenius law with E^{α} and E^{β} activation energies:

$$\alpha = Ce^{-\frac{E^{\alpha}}{RT}}, \quad \beta = Ce^{-\frac{E^{\beta}}{RT}} \qquad \text{(A.17.3)}$$

Under the effect of the periodic outside electric field:

$$E^{\alpha} = E^{\alpha}{}_0 + a\cos\Omega t, \quad E^{\beta} = E^{\beta}{}_0 - b\cos\Omega t \qquad \text{(A.17.4)}$$

With probability notations $[T]$=$[A]$+$[B]$, $p_1 = [A]/[T]$, and $p_2 = [B]/[T]$ (anyway, p_1 and p_2 are the probabilities of the state A and B, respectively) (A.17.2) looks like:

$$\frac{dp_1}{dt} = -\alpha p_1 + \beta p_2 \quad \frac{dp_2}{dt} = \alpha p_1 - \beta p_2 \quad \{p_1 + p_2 = 1\} \tag{A.17.5}$$

And consequently:

$$\frac{dp_1}{dt} = -(\alpha + \beta)p_1 + \beta \tag{A.17.6}$$

Therefore, the time variation of the probability of the A-state:

$$\frac{dp_1}{dt} = -C\left(e^{-\frac{E_0^\alpha}{RT}} + e^{-\frac{E_0^\beta}{RT}}\right)p_1 + \frac{Cb}{RT}e^{-\frac{E_0^\alpha}{RT}} \cos \Omega t \tag{A.17.7}$$

Supposing

$$\frac{Cb}{RT}e^{-\frac{E_0^\beta}{RT}} = \frac{Ca}{RT}e^{-\frac{E_0^\alpha}{RT}} \tag{A.17.8}$$

Note this approach was introduced by McNamara [1321]. From (A.17.7) the amplitude of p_1 is:

$$A(RT) = \frac{\frac{b}{RT}e^{-\frac{E_0^\alpha}{RT}}}{\sqrt{\left(e^{-\frac{E_0^\alpha}{RT}} + e^{-\frac{E_0^\beta}{RT}}\right)^2 + \Omega^2}} \tag{A.17.9}$$

Result (A.17.9) is identical with (A.16.12). Every catalytic reactions has its resonant frequency. Consequently large number of resonances exists.

Appendix 18: Demodulation by Stochastic Resonance

Let us start with (A.17.5), which anyway describes a two-state Markovian process, supposing the coefficients could be modified by an outside electric field. Let us study the ionic channel via the (A.17.5) equation, when A corresponds to the open state and B to the closed one. In the stationery case

$$\frac{dp_i}{dt} = 0 \text{ so } \frac{p_1}{p_2} = \frac{\beta}{\alpha} \tag{A.18.1}$$

The activation energy (barrier) E_0 fluctuates in membrane transport [1322]. Counting the deviation ($\pm\Delta E$) of the two states from the reference (equality) value and supposing Boltzmann distribution:

$$\frac{p_1}{p_2} = \frac{\beta}{\alpha} = \frac{e^{-\frac{E_0+\Delta E}{kT}}}{e^{-\frac{E_0-\Delta E}{kT}}} \tag{A.18.2}$$

In consequence the reaction kinetics follows the Arrhenius law:

$$\alpha = Ce^{-\frac{E_0-\Delta E}{kT}}, \quad \beta = Ce^{-\frac{E_0+\Delta E}{kT}} \tag{A.18.3}$$

Therefore, substituting it into (A.17.5):

$$2\tau\frac{dp_1}{dt} = -(e^{\frac{\Delta E}{kT}} + e^{-\frac{\Delta E}{kT}})p_1 + e^{-\frac{\Delta E}{kT}} \quad \left\{2\tau = \left(Ce^{-\frac{E_0}{kT}}\right)^{-1}\right\} \tag{A.18.4}$$

Supposing the ($\pm\Delta E$) is small, apply the first term of the Taylor expansion:

$$\frac{dp_1}{dt} = -\frac{1}{\tau}p_1 + \frac{1}{2\tau} - \frac{1}{2\tau}\frac{\Delta E}{kT} \tag{A.18.5}$$

In the case of harmonic excitation of the energy [e.g. sinus-type like $\Delta E = a\sin(\omega t)$], we obtain:

$$\frac{dp_1}{dt} = -\frac{1}{\tau}p_1 + \frac{1}{2\tau} - \frac{1}{2\tau}\frac{a}{kT}\sin\omega t \tag{A.18.6}$$

For the resonance concentrate on the harmonic term of the sum:

$$\frac{dp}{dt} + \frac{1}{\tau}p = -\frac{1}{2\tau}\frac{a}{kT}\sin\omega t \tag{A.18.7}$$

Its solution depends on the thermal noise kT as follows:

$$\hat{p}(kT) = \frac{a}{2kT}\frac{1}{\sqrt{\tau^2\omega^2+1}} = \frac{a}{2kT}\frac{1}{\sqrt{1+\left(\frac{e^{\frac{E_0}{kT}}}{2C}\omega\right)^2}} \tag{A.18.8}$$

This has a sharp maximum depending on the noise energy (see Fig. A.18.1).

This means there is a noise energy, where the probability of the reaction is resonantly high, and also at the fixed noise energy there is a frequency when the probability is resonantly high.

When we amplitude-modulated the signal with low frequency, we get by the Taylor expansion from (A.18.4):

$$2\tau\frac{dp_1}{dt} = -\left(2+\left(\frac{\Delta E}{kT}\right)^2\right)p_1 + 1 - \frac{\Delta E}{kT} + \frac{1}{2}\left(\frac{\Delta E}{kT}\right)^2 \tag{A.18.9}$$

When we modulate the amplitude by the $m<1$ modulation-depth with $\Omega \ll \omega$ modulation frequency:

$$\Delta E = a(1 + m\cos\Omega t)\sin\omega t \tag{A.18.10}$$

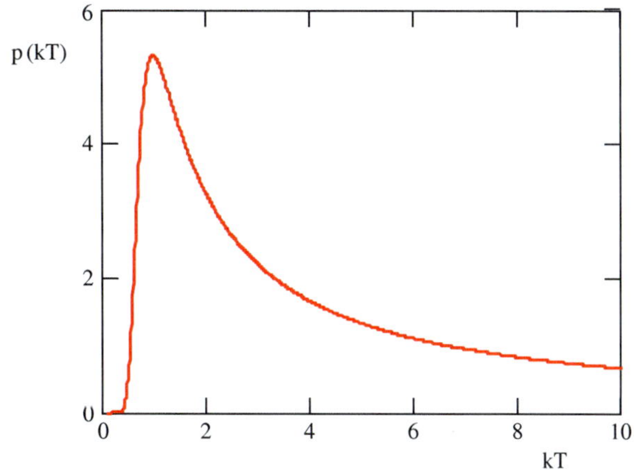

Fig. A.18.1 The stochastic resonance by thermal noise

Then:

$$2\tau \frac{dp_1}{dt} = -\left(2 + \left(\frac{a}{kT}\right)^2 (1 + m\cos \Omega t)^2 \sin^2 \omega t\right) p_1 \\ +1 - \frac{a}{kT}(1 + m\cos \Omega t)\sin \omega t + \frac{1}{2}\left(\frac{a}{kT}\right)^2 (1 + m\cos \Omega t)^2 \sin^2 \omega t \tag{A.18.11}$$

We see a slow change (from the modulation) and a quick fluctuation (from the carrier frequency) in the probability function of time.

To filter the effect of the modulation make an averaging on the high frequency carrier, and substitute it with its time-average, while the low frequency slow change could be regarded as constant, then the modulation is described by:

$$2\tau \frac{dp_1}{dt} = -\left(2 + \left(\frac{a}{\sqrt{2}kT}\right)^2 (1 + m\cos \Omega t)^2\right) p_1 + 1 + \frac{1}{2}\left(\frac{a}{\sqrt{2}kT}\right)^2 (1 + m\cos \Omega t)^2 \tag{A.18.12}$$

Using a small modulation depth (m<<1), Eq. (A.18.12) became linear:

$$2\tau \frac{dp_1}{dt} = -\left(2 + \left(\frac{a}{\sqrt{2}kT}\right)^2 (1 + 2m\cos \Omega t)\right) p_1 + 1 + \frac{1}{2}\left(\frac{a}{\sqrt{2}kT}\right)^2 (1 + 2m\cos \Omega t) \tag{A.18.13}$$

We are looking for resonance, so study the effect of the modulated exciting function:

$$2\tau \frac{dp}{dt} + \left[2 + \left(\frac{a}{\sqrt{2}kT}\right)^2 + \left(\frac{a}{kT}\right)^2 m\cos \Omega t\right] p = \left(\frac{a}{\sqrt{2}kT}\right)^2 m\cos \Omega t \tag{A.18.14}$$

Introducing the formulae:

$$F(p) = 2\tau \frac{dp}{dt} + \left[2 + \left(\frac{a}{\sqrt{2}kT}\right)^2 + \left(\frac{a}{kT}\right)^2 m\cos \Omega t\right] p \tag{A.18.15}$$

We apply successive approximation to solve (A.18.14):

$$\begin{aligned} F(p_1) &= \left(\frac{a}{\sqrt{2}kT}\right)^2 m\cos\Omega t \\ F(p_2) &= f(p_1) \\ F(p_i) &= f(p_{i-1}) \end{aligned} \tag{A.18.16}$$

Hence:

$$\begin{aligned} &F(p_1) + F(p_2) + \cdots + F(p_i) = F(p_1 + p_2 + \cdots + p_i) \\ &= \left(\frac{a}{\sqrt{2}kT}\right)^2 m\cos\Omega t + f(p_1) + f(p_2) + \cdots + f(p_{i-1}) \\ &= \left(\frac{a}{\sqrt{2}kT}\right)^2 m\cos\Omega t + f(p_1 + p_2 + \cdots + p_{i-1}) \end{aligned} \tag{A.18.17}$$

If the present successive approximation is convergent than the solution is obtained when $i \rightarrow \infty$, and Eq. (A.18.17) approaches the original:

$$F(p) = \left(\frac{a}{\sqrt{2}kT}\right)^2 m\cos\Omega t + f(p) \tag{A.18.18}$$

The first approximation is the solution of the simple differential equation:

$$2\tau\frac{dp_1}{dt} + \left[2 + \left(\frac{a}{\sqrt{2}kT}\right)^2 + \right]p_1 = \left(\frac{a\sqrt{m}}{\sqrt{2}kT}\right)^2 \cos\Omega t \tag{A.18.19}$$

This has a solution:

$$\hat{p}_1(RT) = \left(\frac{a\sqrt{m}}{\sqrt{2}kT}\right)^2 \frac{1}{\sqrt{\left[2 + \left(\frac{a}{\sqrt{2}kT}\right)^2\right]^2 + \left(\frac{\Omega}{C}e^{\frac{E_0}{RT}}\right)^2}} \tag{A.18.20}$$

This is again a resonant solution (like in Fig A.18.1).

Appendix 19: Non-Linear Effects of Energy Flow

Focusing of energy flow by non-linear optics examined with the assumption that the material is isotropic. In the isotropic case, the constitutive equations are covariant against the orthogonal transformation. It follows from this assumption that the constitutive equations can be generated by the isotropic $\boldsymbol{f_i}$ vector base of

$$\overline{E},\ \overline{\overline{H}}\,\overline{E},\quad \overline{\overline{H}}^2\overline{E}, \tag{A.19.1}$$

and

$$\overline{H},\ \overline{E}\otimes\ \overline{\overline{H}}\,\overline{E}-\overline{\overline{H}}\,\overline{E}\otimes\overline{E},\quad \overline{E}\otimes\overline{\overline{H}}^2\,\overline{E}-\overline{\overline{H}}^2\,\overline{E}\otimes\overline{E} \tag{A.19.2}$$

for the polar and axial vector, respectively, in the form of

$$\bar{\mathbf{J}}=\sum\alpha_i\bar{\mathbf{f}}_i \tag{A.19.3}$$

where the α_i scalar quantities are the function of the isotropic scalar invariants [1323]:

$$\overline{E}^2,\ \overline{E}\cdot\overline{\overline{H}}^2\,\overline{E},\quad tr(\overline{\overline{H}}^2) \tag{A.19.4}$$

In our case, $\overline{\overline{H}}$ is the asymmetrical tensor of the magnetic field strength vector, $\otimes$ is the symbol of the dyadic multiplication. In the case of a second-order approximation this means that the polarization vector is a function of the

$$\overline{P}=\varepsilon_0\kappa\overline{E}+a\overline{E}^2\overline{E} \tag{A.19.5}$$

field strength. Numerous non-linear effects evolve from this dependence: the Kerr effect, the soliton solution, and the self-focusing process. Consequently, if we take into account the approximation of

$$\bar{j}=\sigma\ \overline{E}+\gamma\ \overline{\overline{H}}\,\overline{E} \tag{A.19.6}$$

in the constitutive equation of conduction then we are able to explain the effect of magnetic field on the conduction that has been observed also experimentally.

Let the magnetic field be negligible, and let us expand in series the coefficient of (A.19.1) by the isotropic scalar invariant of (A.19.3) up to the second-order. Then the vector of polarization depends on the field strength of

$$\overline{P}=\varepsilon_0\kappa\overline{E}+a\overline{E}^2\overline{E} \tag{A.19.7}$$

From this we get the displacement vector of

$$\overline{D}=\varepsilon\,\overline{E}=\varepsilon_0\overline{E}+\overline{P}=[\varepsilon_0(1+\kappa)\ +a\,\overline{E}^2]\overline{E} \tag{A.19.8}$$

Hence, the permittivity can be expressed as

$$\varepsilon\ =\varepsilon_0(1+\kappa)\ +a\,\overline{E}^2 \tag{A.19.9}$$

and the refractive coefficient equals

$$n=\sqrt{\frac{\varepsilon}{\varepsilon_0}}=\sqrt{1+\kappa+\frac{a\,\overline{E}^2}{\varepsilon_0}}\approx\sqrt{1+\kappa}\left(1+\frac{a\,\overline{E}^2}{2\,\varepsilon_0\,(1+\kappa)}\right) \tag{A.19.10}$$

where we supposed that the non-linear correction of the refractive coefficient is inconsiderable. Therefore, the refractive coefficient equals

$$n = n_0 + n_2\,\overline{E}^2 \tag{A.19.11}$$

Let the electric wave propagate in the form of a beam, and let us suppose that the current density distribution of the energy is nonuniform in the beam. Let us take that the value of this is lower at the border of the beam and constant positive according to (A.19.11). As a result of this, the wave propagates more rapidly at the border than inside, and self-focusing might occur. For this, we need the electric output propagating in the beam to be higher than the

$$P_{\text{thr}} = \frac{\pi\,\varepsilon_0 c\,\lambda^2}{8n_2} \tag{A.19.12}$$

threshold output. Otherwise, the beam diverges.

Then we might suppose according to the above rough illustration that the self-focusing is possible, and it has an output limit.

Appendix 20: Principle of Minimal Energy-Dispersion

We briefly show this in a simple parallel resistor circuit (see Fig. A.20.1), where the currents are distributed on the lowest "wasted" energy arrangement, they minimize the energy dissipation.

Indeed, the power on the resistors:

$$P = I_1{}^2 R_1 + I_2{}^2 R_2 \tag{A.20.1}$$

the currents are conserved:

$$I = I_1 + I_2 \tag{A.20.2}$$

then the disturbance of the energy-dissipation is:

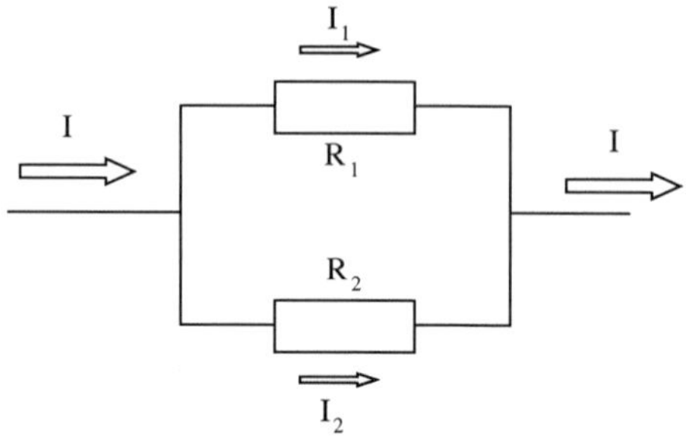

Fig. A.20.1 Current division by parallel resistors

$$\begin{aligned}\delta P &= (I_1{}^* + \delta I_1)^2 R_1 + (I_2{}^* + \delta I_2)^2 R_2 - (I_1{}^*)^2 R_1 + (I_2{}^*)^2 R_2 \\ &= (2I_1{}^* R_1 \delta I_1 + 2I_2{}^* R_2 \delta I_2) + \left(\delta I_1^2 R_1 + \delta I_2^2 R_2\right) = \delta I_1^2 R_1 + \delta I_2^2 R_2\end{aligned} \quad \text{(A.20.3)}$$

where the possible variation of the quantities is denoted by δ. And because of the energy conservation $\delta I_1 = -\delta I_2$. It is trivial that $\delta P \geq 0$. Hence the used energy is always larger in the perturbed case, and minimal, when no perturbation ($\delta I_1 = \delta I_2 = 0$) is applied (shortest path).

Appendix 21: Charge Inhomogeneities

The energy conservation could be expressed by the energy (U) of the electromagnetic field

$$U = \frac{1}{2}\left(\varepsilon_0 \overline{E}^2 + \frac{1}{\mu_0}\overline{B}^2\right) \quad \text{(A.21.1)}$$

and so the change of the energy and the divergence of **S** are in balance of the current flow:

$$\frac{\partial U}{\partial t} + \operatorname{div}\left(\overline{S}\right) = -\overline{j}\overline{E} \quad \text{(A.21.2)}$$

where j is the total current density. Because of the divergence of S this energy balance is invariant to adding rotation of any $\boldsymbol{F}$ fields to S, because $\operatorname{div}(\operatorname{rot}(\boldsymbol{F})) = 0$. This invariance of adding a curl of any fields could cause some inconsistence in further descriptions.

Despite of the similarities in the electric field, this solution however, differs from the invasive electrode application. In the invasive case the electrode reaction was directly involved in the process providing effective charge exchange between the electrolyte and the electrode material. In the present non-invasive case no such electrode reaction could happen, so this kind of protein denaturalization is excluded. The other factor (change the pH by the outside electric field), could be possible if it affects the local ionic concentrations in the immediate neighborhood of the proteins. To guess this effect, approximate the timing of the steady-state situation near the large molecules. From the basic equations (charge conservation and Maxwell equations) we get:

$$\frac{\partial \rho}{\partial t} + \nabla \overline{j} = 0, \quad \nabla \overline{D} = 0, \quad \overline{D} = \varepsilon \overline{E}, \quad \overline{j} = \sigma \overline{E} \quad \text{(A.21.3)}$$

Hence:

$$\frac{\partial \rho}{\partial t} + \frac{\sigma}{\varepsilon}\rho = 0 \quad \text{(A.21.4)}$$

The solution of (A.21.4):

$$\rho = \rho_0 e^{-\frac{t}{\tau}}, \quad \tau = \sigma^{-1}\varepsilon \qquad \text{(A.21.5)}$$

Consequently the generated charge is stable during τ time, which is approximated numerically as:

$$\tau = \sigma^{-1}\varepsilon \approx 200\varepsilon_0 \approx 1\,ns \qquad \text{(A.21.6)}$$

Appendix 22: Pennes' Equation

The thermodynamic energy balance defines the actual heat exchange, including the electromagnetic energy as well. The energy change in time ($\partial U/\partial t$), its flow in space is: grad(I_U)=$\sum(\partial I_U/\partial x_i)$, and its sources $q_U = q_e + q_m + q_b$, where I_U is the energy current density, q_e, q_m, and q_b are the energy source/sink from electric, metabolic, and blood perfusion origins, respectively. The well-known balance equation [576] is:

$$\frac{\partial U}{\partial t} + \sum_{i=1}^{3} \frac{\partial I_U}{\partial x_i} = q_e + q_m + q_b \qquad \text{(A.22.1)}$$

As is well known from [335], the current density of the heat flow is proportional to the temperature gradient (heat diffusion):

$$I_U = -\lambda \text{grad}\,(T) = -\lambda \sum_{i=1}^{3} \frac{\partial T}{\partial x_i} \qquad \text{(A.22.2)}$$

By substituting (A.22.2) into (A.22.1) and by using (A.22.3) we obtain the so-called bioheat (Pennes') equation [1114] which serves for the description of tissue heating and internal energy transport. It has the form

$$\rho c \frac{\partial T}{\partial t} = \lambda \sum_{i=1}^{3} \frac{\partial^2 T}{\partial x_i^2} + \rho_b c_b v_b (T_b - T) + q_e + q_m \qquad \text{(A.22.3)}$$

where T is the temperature of tissue; ρ and c are the mass density and the specific heat of tissue, respectively; λ is the coefficient of thermal conduction; $\rho_b c_b v_b$ is the blood perfusion; T_b is the blood temperature (it is in fact a negative value at the loco-regional treatment, cooling term, due to $T_b < T$; while in the whole body treatment it will be positive). The q_e and q_m are the metabolic energy and the external energy-exchange, respectively. q_e could be heat-sink or source, (in our hyperthermia case it is absorbed from external sources).

Appendix 23: Pennes' Equation is Revised

We are going to refer to these parameters as internal variables of a given cellular composition. The state of the examined cancerous tissue can be expressly characterized by the u internal energy (per unit volume) and by the n_1, n^*, n_2, molar numbers of the intact, the thermally exited and the dead (necrotic or apoptotic) cell clusters, respectively. For the sake of simplicity, the ionic composition of tissue is considered as constant. Let us introduce the internal energy of tissue as the state-function of the above state-parameters:

$$u = u(T, n_1, n^*, n_2) \tag{A.23.1}$$

The u internal energy of the tissue fulfils the first law of thermodynamics [576]:

$$\frac{\partial u}{\partial t} = -div\boldsymbol{J}_q + p = -div\boldsymbol{J}_q + p_e + \rho_b c_b w\,(T - T_b) \tag{A.23.2}$$

where $\boldsymbol{J}_q$ is the density of heat flux of thermal conduction, while p_{e} is the heat input of unit volume generated by external actions (in our case, the electric energy heat-input) and the last term is the blood perfusion. This energy balance is the theoretical basis of the Pennes equation (A.22.3) and its various modifications [1324].

We might introduce the s=s(T, n_1, n^*, n_2) state function of entropy, the change of which consists of two parts:

$$\frac{\partial s}{\partial t} = \frac{\partial s_e}{\partial t} + \frac{\partial s_i}{\partial t} \tag{A.23.3}$$

where $\frac{\partial s_e}{\partial t}$ and $\frac{\partial s_i}{\partial t}$ denote the entropy change – the entropy production – of the irreversible external and internal processes (e.g. cell destruction, heat conduction, etc.), respectively. Of course we have to study the entropy changes due to the overall modifications; including the structural-, chemical-, and phase-parameters. Through the change of the entropy the entire complex process could be controlled.

The external entropy change satisfies the

$$\frac{\partial s_e}{\partial t} = -div\left(\frac{\mathbf{J_q}}{T}\right) + \frac{p}{T} \tag{A.23.4}$$

Carnot–Clausius condition, while the internal term of (A.23.3) fulfils the Clausius–Durhem inequality [576]:

$$\frac{\partial s_i}{\partial t} \geq 0 \tag{A.23.5}$$

We apply also the Gibbs- and Maxwell-relations well known from the stationary thermodynamics [576]:

$$\frac{\partial s}{\partial t} = \frac{\partial s}{\partial u}\frac{\partial u}{\partial t} + \frac{\partial s}{\partial n_1}\frac{\partial n_1}{\partial t} + \frac{\partial s}{\partial n^*}\frac{\partial n^*}{\partial t} + \frac{\partial s}{\partial n_2}\frac{\partial n_2}{\partial t}$$

$$= \frac{1}{T}\frac{\partial u}{\partial t} + \frac{\mu_1}{T}\frac{\partial n_1}{\partial t} + \frac{\mu^*}{T}\frac{\partial n^*}{\partial t} + \frac{\mu_2}{T}\frac{\partial n_2}{\partial t} \tag{A.23.6}$$

$$\frac{1}{T} = \frac{\partial s}{\partial u}, \quad \frac{\mu_1}{T} = \frac{\partial s}{\partial n_1}, \quad \frac{\mu^*}{T} = \frac{\partial s}{\partial n^*}, \quad \frac{\mu_2}{T} = \frac{\partial s}{\partial n_2}$$

where μ_1, μ^*, μ_2 are the chemical potentials of the insentient, the excited (by hyperthermia), and the lethally modified (by apoptosis or necrosis) cell clusters, respectively.

For the cell transformation occurring during the hyperthermia process the intact cells (A_1) have been transformed to their excited (A^*) and their lethal (necrotic or apoptotic) (A_2) form. The following formal "reaction equations" could be constructed:

$$A_1 \underset{k_1}{\overset{k}{\rightleftarrows}} A^* \tag{A.23.7}$$

$$A^* \xrightarrow{k_2} A_2 \tag{A.23.8}$$

The first equation means that the insentient cancerous cell moves into the A^* excited transient state via thermal excitation, and from this state it might return to its original state as well. The second reaction is the non-reversible cell death. The k, k_1, k_2 quantities above the arrows denotes the specific reaction rates of processes. The first has a chemical equilibrium, the second hasn't; it is permanently active till the energy is sufficient for the action. Suppose, the producing A^* is a stationer, which might happen, if k was much larger than the other two rate-constants (which is likely in the hyperthermia process). So the production of the excited cells is constantly reproducing the deaths by A_2; the decomposition reaction advances with constant velocity. Next, for simplicity, we are going to examine a case like this.

Let us define the ζ_1 and ζ_2 generalized reaction coordinates as independent variables. Then, the $\partial\zeta_i/\partial t$ reaction velocities relating to each of the above reactions can be introduced by using the following equations:

$$\begin{aligned} -\frac{\partial n}{\partial t} &= \frac{\partial_1 n}{\partial t} = \frac{\partial_1 n^*}{\partial t} = \frac{\partial \zeta_1}{\partial t} \\ -\frac{\partial_2 n^*}{\partial t} &= \frac{\partial_2 n_2}{\partial t} = \frac{\partial n_2}{\partial t} = \frac{\partial \zeta_2}{\partial t} \end{aligned} \tag{A.23.9}$$

The subscripts 1 and 2 at derivatives denote the process by (A.23.7) and (A.23.8), respectively.

The actual form of entropy production [1325] – which is of primary importance in the Onsager non-equilibrium thermodynamics – can be determined from the (A.23.2), (A.23.3), (A.23.4), (A.23.6), and (A.23.9) equations:

$$\begin{aligned}\frac{\partial s_{\mathrm{i}}}{\partial t} &= \frac{\partial s}{\partial t} - \frac{\partial s_{\mathrm{e}}}{\partial t} = \mathbf{J_q}\cdot \mathrm{grad}\frac{1}{T} + \left(\frac{\mu^*}{T} - \frac{\mu_1}{T}\right)\frac{\partial \zeta_1}{\partial t} + \left(\frac{\mu_2}{T} - \frac{\mu^*}{T}\right)\frac{\partial \zeta_2}{\partial t}\\ &= \mathbf{J_q}\cdot \mathrm{grad}\frac{1}{T} + A_1\frac{\partial \zeta_1}{\partial t} + A_2\frac{\partial \zeta_2}{\partial t} \geq 0\\ A_1 &= \left(\frac{\mu^*}{T} - \frac{\mu_1}{T}\right), \quad A_2 = \left(\frac{\mu_2}{T} - \frac{\mu^*}{T}\right)\end{aligned} \tag{A.23.10}$$

The corresponding affinities are denoted by A_i. Hereby, we have got the entropy production as the bilinear expressions of the

$$\begin{aligned}\boldsymbol{J} &= \left[J_{\mathrm{q}1}, J_{\mathrm{q}2}, J_{\mathrm{q}3}, \frac{\partial \zeta_1}{\partial t}, \frac{\partial \zeta_2}{\partial t}\right]^T\\ \boldsymbol{X} &= \left[\left(\mathrm{grad}\frac{1}{T}\right)_1, \left(\mathrm{grad}\frac{1}{T}\right)_2, \left(\mathrm{grad}\frac{1}{T}\right)_3, A_1, A_2\right]^T\end{aligned} \tag{A.23.11}$$

thermodynamic currents ($\boldsymbol{J}$) and dissipative forces ($\boldsymbol{X}$), where T is the symbol of transposition.

The expressions of the reaction equations (A.23.7) and (A.23.8) entropy production (A.23.10) allow a process, where there is no heat conduction and the reaction of (A.23.7) results in equilibrium, while (A.23.8) does not, namely, the

$$\mathrm{grad}\left(\frac{1}{T}\right) = 0, \quad A_1 = 0, \quad A_2 \neq 0 \tag{A.23.12}$$

conditions are valid. (The reaction is performed under isotherm conditions producing constant excited cells in the system.) Instead of the n_1, n^*, n_2 variables, which are not independent in accordance with (A.23.7) and (A.23.8), the internal energy and entropy with the independent reaction coordinates have the

$$u = u(T, \zeta_1, \zeta_2), \quad s = s(T, \zeta_1, \zeta_2) \tag{A.23.13}$$

caloric state equations. Now, the Gibbs- and Maxwell-relations of (A.23.6) are replaced by the following expression:

$$\begin{aligned}\frac{\partial s}{\partial t} &= \frac{\partial s}{\partial u}\frac{\partial u}{\partial t} + \frac{\partial s}{\partial \zeta_1}\frac{\partial \zeta_1}{\partial t} + \frac{\partial s}{\partial \zeta_2}\frac{\partial \zeta_2}{\partial t} = \frac{1}{T}\frac{\partial u}{\partial t} + A_1\frac{\partial \zeta_1}{\partial t} + A_2\frac{\partial \zeta_2}{\partial t}\\ \frac{1}{T} &= \frac{\partial s}{\partial u}, \; A_1 = \frac{\partial s}{\partial \zeta_1}, \; A_2 = \frac{\partial s}{\partial \zeta_2}\end{aligned} \tag{A.23.14}$$

In order to get the definite form of (A.23.13) let us expand the entropy in series and stop at the second term. Then:

$$s = s_0(T) + \sum_{i=1}^{2} c_i\zeta_i - \frac{1}{2}\sum_{i,j=1}^{2} c_{ij}\zeta_i\zeta_j \tag{A.23.15}$$

Whereas, pursuant to the reaction equation (A.23.7) in equilibrium if $A_1 = 0$, thus, from this and from the above relationship we get that

$$\frac{\partial s}{\partial \zeta_1} = c_1 - \sum_{j=1}^{n} c_{1j}\zeta_j = 0, \quad \frac{\partial s}{\partial \zeta_2} = c_2 - \sum_{j=1}^{n} c_{2j}\zeta_j \neq 0 \tag{A.23.16}$$

From these we have a unique solution for ζ_1 at an arbitrary ζ_2 if

$$c_{12} = c_{21} = c_{22} = 0 \tag{A.23.17}$$

With this, the expression of entropy from (A.23.15) will be:

$$s = s_0(T) - c_1\zeta_1 - c_2\zeta_2 + \frac{1}{2}c_{11}\zeta_1^2 \tag{A.23.18}$$

Considering that according to the first Maxwell-relation of (A.23.14) $\partial u = T\partial s$, we get from the above expression that the partial time-derivative of the internal energy will be as follows:

$$\begin{aligned}
&\tfrac{\partial u}{\partial t} = T\tfrac{\partial s_0(T)}{\partial T}\tfrac{\partial T}{\partial t} + T\left(c_1 - c_{11}\zeta_1\right)\tfrac{\partial \zeta_1}{\partial t} + Tc_2\tfrac{\partial \zeta_2}{\partial t} = \rho c\tfrac{\partial T}{\partial t} + r_1\tfrac{\partial \zeta_1}{\partial t} + r_2\tfrac{\partial \zeta_2}{\partial t} \\
&\rho c = T\tfrac{\partial s_0(T)}{\partial T}, \quad r_1 = T\left(c_1 - c_{11}\zeta_1\right), \quad r_2 = Tc_2
\end{aligned} \tag{A.23.19}$$

where ρ is the mass-density, c is the coefficient of specific heat, and r_1, r_2 are the reaction heats of the definite cell transformations of (A.23.7) and (A.23.8).

The material equations of transport processes can be formed by using the Onsager theory. Consequently, the relationship between the thermodynamic forces and currents (A.23.11) is linear [576]. In accordance with the Curie theory only the currents and forces of the same tensor order can interfere [576], therefore, the Onsager constitutive equations will have the following form:

$$\begin{aligned}
&\mathbf{J_q} = \lambda \mathrm{grad}\tfrac{1}{T} = -\tfrac{\lambda}{T^2}\mathrm{grad}T = -\kappa\,\mathrm{grad}T \\
&\tfrac{d\zeta_1}{dt} = L_{11}A_1 + L_{12}A_2, \quad \tfrac{d\zeta_2}{dt} = L_{21}A_1 + L_{22}A_2
\end{aligned} \tag{A.23.20}$$

Here the L_{mn}, $(m, n = 1, 2)$ are the Onsager coefficients, which might also depend on temperature and reaction coordinates. The Onsager theory of non-equilibrium thermodynamics is not able to bring forward more. Reaction kinetics has to be used for the definition of the form of L_{mn}, $(m, n = 1, 2)$.

In the interest of a simple discussion we are going to ignore the conduction coefficients of mixed indices, which – in general – can be done as the cross-effects are usually weaker than the primary processes. With this, also the coupling of reaction equations will be abolished. We use the

$$\frac{d\zeta_1}{dt} = \Lambda_1\left(e^{\frac{\mu^*}{RT}} - e^{\frac{\mu}{RT}}\right), \quad \frac{d\zeta_2}{dt} = \Lambda_2\left(e^{\frac{\mu_1}{RT}} - e^{\frac{\mu^*}{RT}}\right) \tag{A.23.21}$$

chemical reaction kinetics equation [311]; where R is the universal gas constant. We are going to examine how the above equation can be linearized and, hereby, how it can be harmonized with the Onsager theory. Let us transform the equation as follows:

$$\frac{d\zeta_1}{dt} = \Lambda_1 e^{\frac{\mu}{RT}} \left(e^{\frac{\mu^* - \mu}{RT}} - 1 \right) \tag{A.23.22}$$

Then, supposing that the power index of the exponential member is small, let us expand it in series and stop at the second-order:

$$\begin{aligned} \tfrac{d\zeta_1}{dt} &= \Lambda_1 e^{\frac{\mu}{RT}} \left(e^{\frac{\mu^* - \mu}{RT}} - 1 \right) \cong \Lambda_1 e^{\frac{\mu}{RT}} \left(1 + \tfrac{\mu - \mu^*}{RT} - 1 \right) \\ &= \tfrac{\Lambda_1 e^{\frac{\mu}{RT}}}{R} \left(\tfrac{\mu - \mu^*}{T} \right) = L_{11} \left(\tfrac{\mu - \mu^*}{T} \right) = L_{11} A_1 \end{aligned} \tag{A.23.23}$$

Consequently, the Onsager conduction coefficient has Arrhenius-type temperature dependence, which is well observed in many experimental works [1326].

$$L_{11} = \frac{\Lambda_1 e^{\frac{\mu}{RT}}}{R} \tag{A.23.24}$$

In accordance with the above Onsager material equations (A.23.20), let us consider the

$$\begin{aligned} &\mathbf{J_q} = \lambda \operatorname{grad} \tfrac{1}{T} = -\tfrac{\lambda}{T^2} \operatorname{grad} T = \kappa \operatorname{grad} T \\ &\tfrac{d\zeta_1}{dt} = L_{11} A_1, \quad \tfrac{d\zeta_2}{dt} = L_{22} A_2 \\ &L_{11} - \tfrac{\Lambda_1 e^{\frac{\mu}{RT}}}{R}, \quad L_{22} = \tfrac{\Lambda_2 e^{\frac{\mu^*}{RT}}}{R} \end{aligned} \tag{A.23.25}$$

form. Here the conduction coefficients show a dependence again similar to the Arrhenius law, which has been observed also with hyperthermia [239, 240].

The material equations (A.23.25) can be further concretized if we take the form of entropy (2.23.18) and the definition of activities (A.23.14) into consideration. Then, we get after some transformations that

$$\begin{aligned} &\mathbf{J_q} = \lambda \operatorname{grad} \tfrac{1}{T} = -\tfrac{\lambda}{T^2} \operatorname{grad} T = \kappa \operatorname{grad} T \\ &\tfrac{d\zeta_1}{dt} = -L_{11} \left(c_{11} \zeta_1 - c_1 \right) = -\tfrac{1}{\tau} \left(\zeta_1 - \zeta_{1e} \right), \quad \tfrac{d\zeta_2}{dt} = L_{22} c_2 \\ &\tau = \tfrac{1}{L_{11} c_{11}}, \quad \zeta_{1e} = \tfrac{c_1}{c_{11}} \end{aligned} \tag{A.23.26}$$

Here the τ time constant and ζ_{1e} equilibrium value of reaction depend on the temperature. The excitation activated cancerous cells are approaching exponentially their steady state condition. The second equation describes the cell death, which practically means that the cell destruction takes place linearly by decomposition, and its temperature dependence is governed by the Arrhenius law.

When we substitute the caloric equation (A.23.19) and – from (A.23.26) – the Fourier law into the internal energy balances (A.23.2) then we have the heat equation valid during the process of hyperthermia:

$$\rho c\frac{\partial T}{\partial t}+T\left(c_1-c_{11}\zeta_1\right)\frac{\partial \zeta_1}{\partial t}+Tc_2\frac{\partial \zeta_2}{\partial t}-\text{div}(\lambda \text{grad}T)+\rho_b c_b w\left(T-T_b\right)=p_e \tag{A.23.27}$$

We have also from (A23.26) the two

$$\frac{d\zeta_1}{dt}=-\frac{1}{\tau}\left(\zeta_1-\zeta_{1e}\right),\quad \frac{d\zeta_2}{dt}=L_{22}c_2 \tag{A.23.28}$$

reaction equations. Consequently, if we want to determine the temperature of tissue and the level of cell destruction then we have to solve the system of equations comprising the above three equations, which are coupled because the coefficients of (A.23.28) are temperature-dependent. By using the second equation of (A.23.28) the Eq. (A.23.27) can be simplified, and we get the following equations to be solved:

$$\begin{aligned}&\rho c\tfrac{\partial T}{\partial t}+T\left(c_1-c_{11}\zeta_1\right)\tfrac{\partial \zeta_1}{\partial t}+TL_{22}c_2^2-\text{div}(\lambda \text{grad}T)\\&+\rho_b c_b w\left(T-T_b\right)=p_e\\&\tfrac{d\zeta_1}{dt}=-\tfrac{1}{\tau}\left(\zeta_1-\zeta_{1e}\right)\end{aligned} \tag{A.23.29}$$

The above-derived modified system of the Pennes equation (A.23.29) contains already the cell destruction energy, so it provides a full picture of the energy usage. The ζ_1 and ζ_2 generalized reaction parameters provide additional terms and an independent equation to the classical, equilibrium Pennes equation.

The right way to choose the dose is if we suppose that the relevant quantities are the energy necessary for the cell destruction. This is derived from the expression of entropy production (A.23.10). From this, the entropy production resulting from cell destruction is:

$$\left(\frac{\partial s_i}{\partial t}\right)_{tr}=A_1\frac{\partial \zeta_1}{\partial t}+A_2\frac{\partial \zeta_2}{\partial t}\geq 0 \tag{A.23.30}$$

And the power dissipation is T-times higher:

$$p_{tr}=T\left(\frac{\partial s_i}{\partial t}\right)_{tr}=T\left(A_1\frac{\partial \zeta_1}{\partial t}+A_2\frac{\partial \zeta_2}{\partial t}\right)\geq 0 \tag{A.23.31}$$

The proposed dose can be calculated as the time integral of the power:

$$D\left(\bar{r},t\right)=\int_0^t T\left(\frac{\partial s_i}{\partial t}\right)_{tr}dt=\int_0^t T\left(A_1\frac{\partial \zeta_1}{\partial t}+A_2\frac{\partial \zeta_2}{\partial t}\right)dt \tag{A.23.32}$$

If we substitute here the expression of activities from (A.23.14) and the kinetic equation (A.23.26) we get that the first member is in connection with the activation and the second one with the cell death.

$$D(\bar{r},t) = \int_0^t L_{11}(c_1 - c_{11}\zeta_1)^2 dt + \int_0^t L_{22}c_2^2 dt \tag{A.23.33}$$

To make a comparison of the dose defined by the entropy with the experimentally used empirical dose at reference CEM43°C, let us examine the form of this dose at the reference temperature of 43°C. In this case the first term of (A.23.33) can be calculated in a simple way, so we get:

$$D|_{T=T_{43}} = \frac{c_1^2}{2c_{11}}\left(1 - e^{-\frac{2t}{\tau}}\right) + L_{22}c_2^2\Big|_{T=T_{43}} t \tag{A.23.34}$$

If we neglect the first term of (A.23.34) then – similarly to the CEM43°C empirical dose – we get a dose proportional to time.

$$D|_{T=T_{43}} = L_{22}c_2^2\Big|_{T=T_{43}} t \tag{A.23.35}$$

For another constant temperature this dose has the

$$D|_T = L_{22}c_2^2\Big|_T t \tag{A.23.36}$$

form. These two doses are equivalent if

$$L_{22}c_2^2\Big|_{T=T_{43}} t = L_{22}c_2^2\Big|_T t_{\text{eq}} \tag{A.23.37}$$

From this, the equivalent treatment time is:

$$t_{\text{eq}} = \frac{L_{22}c_2^2|_{T=T_{43}}}{L_{22}c_2^2|_T} t = \frac{\Lambda_2(T_{43})\,[c_2(T_{43})]^2 e^{\frac{\mu^*}{RT_{43}}}}{\Lambda_2(T)\,[c_2(T)]^2 e^{\frac{\mu^*}{RT}}} t = f(T, T_{43})\, t \tag{A.23.38}$$

If the temperature wasn't constant during the actual therapy (denote its duration by t), the equivalent treatment time is an integral:

$$t_{\text{eq}} = \int_0^t f(T, T_{43})\, dt \tag{A.23.39}$$

The presently derived thermodynamic dose is connected to the empirical CEM43°C one.

Accordingly, if we carry out the treatment under the above conditions at the given temperature then the treatment efficacy can be characterized by the time alone as a dose. In other words: if the temperature is constant and no memory-effect exists (namely, the system does not change because of the treatment) then the dose can be deduced from the time. The proposed reference temperature is 43°C [1327].

What is the situation in the case of other temperatures [1328]? On the basis of experiments carried out on cell cultures – for comparison with the constant temperature treatment performed at the reference temperature – the following dose function was proposed [1329]:

$$t_{\mathrm{eq}} = \int_0^t f(T_{\mathrm{c}}, T(\bar{r}, \tau), \tau) d\tau = \int_0^t R^{[T_{\mathrm{c}} - T(\bar{r},\tau)]} d\tau = \int_0^t R^{[43 - T(\bar{r},\tau)]} d\tau \qquad \text{(A.23.40)}$$

where

$$R = \begin{cases} 0, & \text{if } T < 39\,°C \\ 0.25, & \text{if } 39\,°C \le T < 43\,°C \\ 0.5, & \text{if } T \ge 43\,°C \end{cases} \qquad \text{(A.23.41)}$$

This introduction is given from the experimental definitions. (In vitro experiments: in the cell culture the same quantity of cell death shall belong to the given temperature, the dose definition is equivalent [1328]. As the basis of the measurement is the time, therefore, this basically means the definition of the equivalent time.) In this case, practically, the discontinuity kink of the Arrhenius plot has been considered, which results from the different levels of thermotolerance.

However, we have again three problems:

- the kink of the Arrhenius graph depends on the applied chemotherapy [1330, 336];
- the kink of the Arrhenius graph depends on the prehistory and dynamics of treatment [1145–1148];
- the Arrhenius graph gives different time doses for the different points of the target (because of its non-homogeneous structure).

These conditions could be considered as a "memory effect", because the stimulating force and the real change caused are not simultaneous but of course, are correlating. The solution of the problem is made difficult by the fact that we are not able to find two identical individuals, two identical tumors of the same structure and composition (the tumor is not a homogenous structure), two identical thermotolerances etc.

The clinical practice can not do anything with these problems. To solve at least the last point, a distribution characteristic had to be introduced instead of the spatial distribution of temperature. This is the temperature for which 90% of temperature values measured in the target is higher, namely, it is the so-called 10th percentile of the measured temperature. Its symbol is T_{90} [1331]. This is a statistical characteristic which can be specified by using a lot of measuring points under the efficacy of 75% and empirical treatment dose. This is not an adequate matching. By this, we get the definition of CEM43°C T_{90}:

$$t_{\text{eq}} = \int_0^t R^{[43 - T_{90}(\tau)]} d\tau \qquad \text{(A.23.42)}$$

Together with its unsubstantial theory, unfortunately, this result correlates slightly with the clinical practice [509]. For the performed experiments the PCR (pathological complete response) has shown 75% correlation with the CEM43°C T_{90} in the case of high-grade soft-tissue sarcomas treated with thermo-radiotherapy [509]. Large discrepancies could be expected between the empirical dose and PCR in these cases, when the time of the memory effect is not negligible in comparison with the treatment time. Due to the systemic and other physiological effects, the inaccuracy of the empirical dose in living system is unfortunately always considerable. The empirical dose-dependence was proven by a canine randomized trial only in low and high dose significance [1332]. The redefinition of the thermal dose concept looks to be a further step [1333].

We have theoretically deduced by very general conditions that the cell destruction reaction rate fulfils the Arrhenius law. In accordance with this, we are going to show that under certain conditions the Separeto–Dewey empirical formula can be deduced from the Arrhenius law.

Let us start from the isothermal cell destruction at 43°C. Then, the number of destructed cells during time t will be as follows.

$$A\left(T_{43}\right) e^{-\frac{H_{43}}{kT_{43}}} t \qquad \text{(A.23.43)}$$

Let us take another isothermal treatment of temperature T and examine the section of the same slope of the Arrhenius graph. (Of course, it is quite complicated in practice as the slope depends on pre-treatment conditions [1144].) Let us suppose that for the same degree of cell destruction we need t_{eq} time. As the process is isothermal, we might define the same death rate for the other temperature [509]:

$$A\left(T_{43}\right) e^{-\frac{H_{43}}{kT_{43}}} t = A\left(T\right) e^{-\frac{H_{43}}{kT}} t_{\text{eq}} \qquad \text{(A.23.44)}$$

From this we get that

$$t_{\text{eq}} = \frac{A(T_{43})}{A(T)} e^{-\frac{H_{43}}{kT_{43}} + \frac{H_{43}}{kT}} t = \frac{A(T_{43})}{A(T)} e^{\frac{H_{43}}{kT_{43}T}(T_{43}-T)} t \cong \left\langle \left\{ \left[\frac{A(T_{43})}{A(T)}\right]^{\frac{k(T_{43})^2}{H_{43}}} e \right\}^{\frac{H_{43}}{k(T_{43})^2}} \right\rangle^{(T_{43}-T)}$$
$$= R^{(T_{43}-T)} \qquad \text{(A.23.45)}$$

The error of substitution of temperature T with the T_{43} absolute temperature corresponding to 43°C is equal to $\left(T_{43} - T\right) / \left(T_{43}\right)^2$. As this is a very small value, the applied approximation is good. The approximation is not so good if we compare two treatments for which the slope of the Arrhenius graph is different (we should not forget that these in vitro examinations do not have any in vivo dynamics).

Let us derive the Separeto–Dewey empirical formula from the thermodynamic dose! Start with (A.23.38) and considering the Onsager constants from (A.23.24) and (A.23.25) we get:

$$t_{eq} = \frac{L_{22}c_2^2\big|_{T=T_{43}}}{L_{22}c_2^2\big|_T}t = \frac{\Lambda_2\left(T_{43}\right)\left[c_2\left(T_{43}\right)\right]^2 e^{\frac{\mu^*}{RT_{43}}}}{\Lambda_2\left(T\right)\left[c_2\left(T\right)\right]^2 e^{\frac{\mu^*}{RT}}}t, \quad \left\{t_{eq} := R^{\left(T_{43}-T\right)}t\right\} \quad \text{(A.23.46)}$$

This expression formally corresponds with the equivalent treatment time derived from the empirical dose. So if the energy taken for the activation and distortion is neglected, the energy- and empirical-doses are identical, and the parameters have to fulfill the following condition:

$$\frac{\Lambda_2\left(T_{43}\right)\left[c_2\left(T_{43}\right)\right]^2}{A\left(T_{43}\right)} = \text{const.} \quad \text{(A.23.47)}$$

This means that the present non-equilibrium thermodynamics describes well the given processes. Also it seems that the present energy dose is more general, it can consider the memory effects of the actual processes, and with this fits better to the reality.

It is interesting to observe the changes caused by the different factors (pH, low glucose level, thermotolerance) [509]:

- The slope of the lower section of the Arrhenius graph (the section belonging to the higher temperature) will be approximately identical to the slope of the section belonging to the lower temperature.
- The A proportionality factor changes in Eq. (A.23.43).

In accordance with the above deduction, both effects influence the accuracy of the empirical formula (A.23.40). For example, in the case of thermotolerance only one constant will be lower in (A.23.41)! Regarding the qualitative effect, the jump in the survival graph – which can be observed as a function of treatment time and shows significant deviations for the treatments of different temperatures [551] – will lessen.

An additional problem is that in the case of hyperthermia treatment the target temperature is obtained from the solution of Pennes' equation. In this case the temperature distribution is calculated numerically, and from this we get the T_{90} temperature and the CEM43°C T_{90} dose. To verify the temperature the non-invasive MRI measurement is in use [1334, 1335], for which presenting the physiological effects (e.g. blood perfusion, heat-tolerance, cell destruction, etc.) on the applied reference phantoms is problematic.

In consequence of the above, the form of the heat-conduction equation (Pennes or Pennes-like equation) affects the definition and the definability of the adequate dose. The original Pennes equation does not describe the reality well, does not consider the energy (heat) consumed by the cell-distortion processes. (This cell disruption however is the main goal of the entire process.) Because of the missing energy, the dose calculated from the Pennes equation is always higher than the reality. The difference between the temperatures calculated from the Pennes equation (T_P) and the temperature derived from our present work could be calculated by perturbation theory as a simple approach. This could be accurate in the case where the energy intake by the cell disruption is much less than the overall energy absorption. (Of course this is not an optimal treatment, when most of the energy is not expended to carry out the desired job, but this could be a good approximation of the reality in most of the treatment cases.) Stopping at the first term of the perturbation approximation, the deviation of the real and the Pennes-calculated temperature is:

$$\Delta T = (T - T_P) = -\frac{T_P L_{22} (T_P) [c_2 (T_P)]^2}{\rho_b c_b w} \qquad \text{(A.23.48)}$$

The relative change could be estimated from the experimental dose-function. The correction at 43°C for arbitrary temperature T, based on Eq. (A.23.48), is:

$$\frac{(\Delta T)_T}{(\Delta T)_{43}} = \frac{T_{PT}}{T_{P43}} \frac{L_{22} (T_{PT}) [c_2 (T_{PT})]^2}{L_{22} (T_{P43}) [c_2 (T_{P43})]^2} \frac{\rho_b c_b w_{43}}{\rho_b c_b w_T} \qquad \text{(A.23.49)}$$

where the T and 43 subscripts denote the values calculated at arbitrary T and at 43°C. (The extra P subscript denotes the value obtained from the original Pennes equation.)

Using the results from the dose-idea

$$t_{eq} = \frac{L_{22} (T_{PT}) [c_2 (T_{PT})]^2}{L_{22} (T_{P43}) [c_2 (T_{P43})]^2} t .= R^{(T_{43}-T)} t \qquad \text{(A.23.50)}$$

where R is shown in (A.23.41), therefore:

$$\frac{(\Delta T)_T}{(\Delta T)_{43}} = \frac{T_{PT}}{T_{P43}} \frac{\rho_b c_b w_{43}}{\rho_b c_b w_T} \frac{1}{R^{(T_{43}-T)}} \qquad \text{(A.23.51)}$$

Example: Let us compare two temperature points:, 41°C (314 K) and 45°C (318 K). In the first interval (41–43°C) $R_{41} = 0.25$ and in (43–45°C) $R_{45} = 0.5$. The blood-perfusion rate at T_{41} $w_{b41} = 20$ ml/min/100 g, and at T_{45} is $w_{b45} = 5$ ml/min/100 g [13]. Hence, according to Eq. (A.23.51), when at temperature 45°C we have the inaccuracy $\Delta T_{45C} = 0.2$°C due to the chemical reactions; the correction for T=41°C became $\Delta T_{41C} = 3.2$°C. This is rather large in the hyperthermia case.

Appendix 24: Self-Focusing

Considering a disk-shaped electrode and heating arrangement with cylindrical symmetry with axis Z (see Fig. A.24.1), we describe the electromagnetic properties

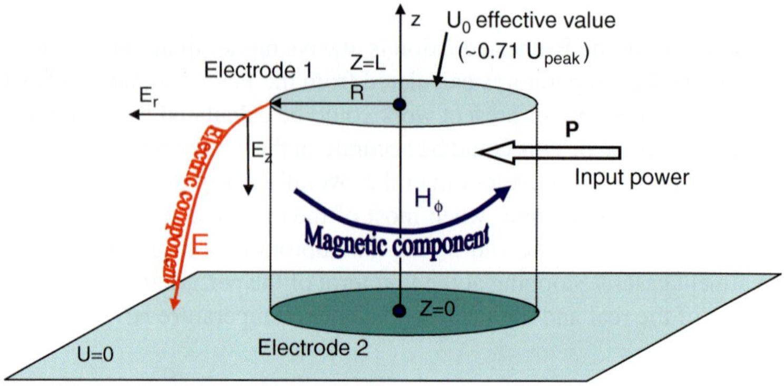

Fig. A.24.1 Electromagnetic field distribution in asymmetric electrode arrangement

of oncothermia heating. In this case the electric field could be described by the E_z, E_r, H_φ parameters, where $\boldsymbol{E}$ and $\boldsymbol{H}$ are the electric and magnetic fields and the indexes denote their cylindrical components. The Maxwell equations using the periodic (sinus) signal conditions:

$$H_\varphi = \frac{\chi}{\mu} C J_1(\chi r) e^{-iaz}, \quad E_r = \frac{a\chi}{\omega\mu\varepsilon'} C J_1(\chi r) e^{-iaz}, \quad E_z = \frac{\chi^2}{j\omega\mu\varepsilon'} C J_0(\chi r) e^{-iaz}$$
$$\chi^2 = \omega^2\mu\varepsilon' - a^2, \quad \varepsilon' = \frac{\sigma}{i\omega} + \varepsilon \tag{A.24.1}$$

where C and a constants will be determined from the actual physical conditions, and J_i are the i-th Bessel functions, with the following definitions:

$$J_0(\chi r) = 1 - \frac{(\chi r)^2}{2^2} + \frac{(\chi r)^4}{(2 \;\; 4)^2} - \frac{(\chi r)^6}{(2 \;\; 4 \;\; 6)^2} + \ldots$$
$$J_1(\chi r) = \frac{\chi r}{2}\left[1 - \frac{(\chi r)^2}{2 \cdot 4} + \frac{(\chi r)^4}{2 \cdot 4^2 \cdot 6} - \frac{(\chi r)^6}{2 \cdot (4 \cdot 6)^2 \cdot 8} + \ldots\right] \tag{A.24.2}$$

By integration on the electrode r radius we get the effective current of the electrode:

$$I = \oint H_\varphi R d\varphi = 2\pi R H_\varphi(R) = 2\pi R \frac{\chi}{\mu} C J_1(\chi R) \tag{A.24.3}$$

and the current determines the U_0 potential between the electrodes and it determines the C:

$$U_0 = \int_0^L E_z dz = \frac{\chi^2 C}{\mu\sigma \; \varepsilon'} \frac{e^{-iaL} - 1}{a} \rightarrow C = U_0 \frac{\mu \; \omega \; \sigma}{\chi^2} \frac{a}{e^{-iaL} - 1} \tag{A.24.4}$$

and the apparent power absorbed by the target is:

$$S = U_0 I^* = U_0{}^2 \frac{\sigma 2\pi R}{L}\left(\left(1+\frac{i\omega\varepsilon}{\sigma}\right)\frac{iaL}{1-e^{-iaL}}\right)^*\left(\frac{J_1(\chi R)}{\chi}\right)^* \tag{A.24.5}$$

where * denotes the complex conjugation. The parameters in (A.23.26) are measurable, consequently a (which is the reciprocal value of the penetration depth L_0 [$a=1/L_0$]), could be determined, and the power in the fifth-order approximation by Bessel functions is:

$$S \cong U_0{}^2 \frac{\sigma\pi R^2}{L}\left(\left(1+\frac{i\omega\varepsilon}{\sigma}\right)\frac{iaL}{1-e^{-iaL}}\right)^*\left[1-\frac{R^2\chi^2}{8}+\frac{R^4\chi^4}{256}\right]^* \tag{A.24.6}$$

Introducing the DC resistivity as $R_0=L/(\sigma\pi R^2)$ we get the picturesque form:

$$S \cong \frac{U_0{}^2}{R_0}\left(\left(1+\frac{i\omega\varepsilon}{\sigma}\right)\frac{iaL}{1-e^{-iaL}}\right)^*\left[1-\frac{R^2\chi^2}{8}+\frac{R^4\chi^4}{256}\right]^* \tag{A.24.7}$$

From this the electrode impedance:

$$Z = \frac{L}{\sigma\pi R^2}\frac{1}{\left(\left(1+\frac{i\omega\varepsilon}{\sigma}\right)\frac{iaL}{1-e^{-iaL}}\right)^*\left[1-\frac{R^2\chi^2}{8}+\frac{R^4\chi^4}{256}\right]^*} \tag{A.24.8}$$

and the real – imaginary powers are:

$$P = \mathrm{Re}\left(\frac{U_0{}^2}{Z}\right), \quad Q = \mathrm{Im}\left(\frac{U_0{}^2}{Z}\right) \tag{A.24.9}$$

The specific absorption rate (SAR) could be easily determined from these equations. The dissipation is originated from both fields, consequently:

$$\begin{aligned} p_r &= \sigma|E_r|^2 = \sigma U_0{}^2\left|\frac{a}{e^{-jaL}-1}\right|^2\left|\frac{J_1(\chi r)}{\chi}\right|^2 e^{-2jaz} \\ p_z &= \sigma|E_z|^2 = \sigma U_0{}^2\left|\frac{ja}{e^{-jaL}-1}\right|^2 |J_0(\chi r)|^2 e^{-2jaz} \\ \mathrm{SAR}(r,z) &= p_r + p_z \end{aligned} \tag{A.24.10}$$

The equation of heat-conduction also could be calculated. The blood-perfusion coefficient at 13.56 MHz is one order of magnitude larger than the coefficient of the heat-conduction [1336]. So the heat equation has a simple form:

$$\frac{\partial T^*(r,z,t)}{\partial t} = \frac{1}{\rho c}\mathrm{SAR}(r,z) - \frac{\rho_b c_b m_b}{\rho c}T^*(r,z,t) - \frac{1}{\rho c}\mathrm{SHR}(r,z,t) \tag{A.24.11}$$

where SHR is the Specific Heating Rate, which is controlled by the bolus on the electrode:

$$\mathrm{SHR}(r,z,t) = A(r)e^{-\frac{z}{L_H}}(1 - be^{-\frac{t}{\tau_H}}) \tag{A.24.12}$$

where L_H is the penetration depth of the cooling, $A(r)$ is the distribution along the radius, b is a constant, compensating the starting heat-loss, τ_H is cooling relaxation time. Hence the equation to solve is:

$$\frac{\partial T^*(r,z,t)}{\partial t} = \frac{1}{\rho c}\mathrm{SAR}(r,z) - \frac{\rho_b c_b m_b}{\rho c}T^*(r,z,t) - \frac{1}{\rho c}\mathrm{SHR}(r,z,t) \tag{A.24.13}$$

Using a Laplace transformation:

$$(s+\tau)\,T^*(r,z,s) - T^*(r,z,0) = \frac{1}{s}\frac{1}{\rho c}\mathrm{SAR}(r,z) - \frac{1}{\rho c}\mathrm{SHR}(r,z,s)$$
$$\tau^{-1} = \frac{\rho_b c_b m_b}{\rho c} \tag{A.24.14}$$

where $T^*(r,z,0)$ is the starting temperature distribution. From this finally we get:

$$T^*(r,z,t) = \frac{\tau}{\rho c}\mathrm{SAR}(r,z) - \left[\frac{\tau}{\rho c}\mathrm{SAR}(r,z) - T^*(r,z,0)\right]e^{-\frac{t}{\tau}}$$
$$-\frac{\tau A(r)}{\rho c}e^{-\frac{z}{L_H}}\left(1 - e^{-\frac{t}{\tau}}\right) - \frac{A(r)}{\rho c}e^{-\frac{z}{L_H}}\frac{b}{\tau^{-1}-\tau_H^{-1}}\left(e^{-\frac{t}{\tau_H}} - e^{-\frac{t}{\tau}}\right), \tag{A.24.15}$$

The $A(r)exp(-z/L_H)$ shows the absorbed heat-power under the electrode from the cooling procedure. This is the energy, which is wasted from the forwarded electric heating when the cooling water takes it away from the system. The full power is proportional with P_c:

$$P_c = \int_0^L\int_0^R A(r)e^{-\frac{z}{L_H}}2\pi r\,dr\,dz = \left(1 - e^{-\frac{L}{L_H}}\right)L_H A_{\mathrm{average}}R^2\pi;$$
$$A_{\mathrm{average}} = \frac{I_q}{R^2\pi}\frac{1}{\left(1 - e^{-\frac{L}{L_H}}\right)L_H} \tag{A.24.16}$$

For simplicity, supposing $b=0$, then:

$$T^*(r,z,t) = \frac{\tau}{\rho c}\mathrm{SAR}(r,z) - \left[\frac{\tau}{\rho c}\mathrm{SAR}(r,z) - T^*(r,z,0)\right]e^{-\frac{t}{\tau}}$$
$$-\frac{\tau A(r)}{\rho c}e^{-\frac{z}{L_H}}\left(1 - e^{-\frac{t}{\tau}}\right) \tag{A.24.17}$$

Appendix 25: Dynamism of Temperature on the Membrane

The actual energy balance is:

$$\rho_i c_i V \frac{dT}{dt} = I_q + P_m \quad \text{(A.25.1)}$$

where $\rho_i c_i$ is the specific heat capacity of the cell, V is the cell volume, P_m is the metabolic heat-power. The I_q heat current contains a metabolic and an electric term:

$$I_q = I_{qm} + I_{qe} \quad \text{(A.25.2)}$$

where I_{qm} =–P_m≈const., so the current (I_{qe}) and its density (j_{qe}) are:

$$I_{qe} = \rho_i c_i V \frac{dT}{dt}, \quad j_{qe} = \rho_\imath c_\imath \frac{V}{A} \frac{dT}{dt} \quad \text{(A.25.3)}$$

where A is the surface area of the cell. Its numerical value at 1 K/s heating rate is equal to j_{qe}≈140 W/m^2, which is significant [334]. The temperature difference between the two sides of the cell membrane can be calculated by the expression $\Delta T = j_q/\alpha \approx 0.007$ K. Consequently, the actual gradient on the 7-nm thick membrane amounts to 10^6 K/m, which is a remarkably large value! It is well above the natural heat flow induced by metabolism. The metabolism generates only $j_{qcell} \approx 1.510^{-3}$ W/m^2≈–$j_{qe}10^{-5}$, five orders of magnitude smaller than the constrained opposite heating, so the metabolic temperature gradient through the membrane is negligible.

Let us represent the cellular membrane with a plane-parallel wall with thickness d having the temperature T_e and T_i on one and the other side, respectively, where $T_e > T_i$. In this case the density of heat-flux j_q is [1337]

$$j_q = -\lambda \cdot \text{grad}\,(T) \cong -\lambda \frac{(T_e - T_i)}{d} \quad \text{(A.25.4)}$$

where λ and d are the thermal conductivity and the thickness of the membrane, respectively. The temperature change in time and space is governed by the

$$\frac{\partial T}{\partial t} = \frac{\lambda}{\rho c} \left(\frac{\partial^2 T}{\partial x^2} + \frac{\partial^2 T}{\partial y^2} + \frac{\partial^2 T}{\partial z^2} \right) \quad \text{(A.25.5)}$$

expression where ρ and c are the density and the specific heat of the membrane, respectively. With a simple calculation we obtain for the time necessary for the penetration of heat into the cell:

$$t_p = \frac{d^2 \rho c}{2\lambda} \quad \text{(A.25.6)}$$

The diffusion coefficients of the chemical constituents throughout the cell membrane have an order of magnitude of 10^{-9} m^2/s [1338]. Because of the massive ionic exchange through the membrane, the main energy transfer is connected to this process. Therefore, we assume that the typical thermal diffusivity has the same order of magnitude as the diffusion coefficient, i.e. $a = \lambda/(\rho c) \approx 10^{-9}$ m^2/s. If the thickness of the cell membrane (d) equals approximately 10^{-8} m then for the heat penetration time we get

$$\Delta t = \frac{d^2}{2a} \cong 10^{-7} s \tag{A.25.7}$$

this indicates rather rapid temperature equalization throughout the membrane.

Appendix 26: Changes of the Membrane Potential

The diffusion-constrained mass transport of the i-th component at x place with $C_i(x)$ density can be followed by the Fick law:

$$j_{id} = -D_i \frac{dC_i(x)}{d\,x} \tag{A.26.1}$$

where D_i is the diffusion constant. The drift velocity of the i-th component with Z_i ionization grade, and β_i mobility on a membrane having ξ_{memb} thickness is:

$$v_i^{\text{drift}} = -\beta_i \frac{\psi_{\text{memb}}}{\xi_{\text{memb}}} = -\frac{kT_{\text{memb}}}{D_i} \frac{\psi_{\text{memb}}}{\xi_{\text{memb}}} \tag{A.26.2}$$

where T_{memb} is the temperature of the membrane. Considering the ionic concentration as $C^{\text{ion}}{}_i$ and denoting the electron charge with e, the entire current from the diffusion and drift components is:

$$j_i = -D_i \frac{dC_i(x)}{dx} - C_i^{\text{ion}}(x) \frac{kT_{\text{memb}}}{D_i} Z_i e \frac{\psi_{\text{memb}}}{\xi_{\text{memb}}} \qquad (i = 1, 2, ...) \tag{A.26.3}$$

Introducing the ω_i membrane permeability and the concentrations on the intra- ($C^{\text{i}}{}_i$) and extra-cellular ($C^{\text{e}}{}_i$) sides of the membrane, the solutions of (3.5) are:

$$\begin{aligned} j_{\text{ei}} &= \frac{Z_i^2 e^2 D_i}{kT_{\text{memb}} \xi_{\text{memb}}} \frac{C_i^{(e)} e^{-\frac{Z_i e \psi_{\text{memb}}}{kT}} - C_i^{(i)}}{1 - e^{-\frac{Z_i e \psi_{\text{memb}}}{kT}}} \psi_{\text{memb}} \\ &= e^2 \omega_i N_{\text{L}} \frac{C_i^{(e)} e^{-\frac{Z_i e \psi_{\text{memb}}}{kT}} - C_i^{(i)}}{1 - e^{-\frac{Z_i e U_{\text{memb}}}{kT}}} \psi_{\text{memb}} \quad \omega_i = \frac{Z_i^2 e^2 D_i}{kT_{\text{memb}} \xi_{\text{memb}} N_{\text{L}}} \end{aligned} \tag{A.26.4}$$

Knowing [200] the ratio of the permeabilities of Na^+ and Cl^- ions to K^+:

$$\omega_{\text{K}} : \omega_{\text{Na}} : \omega_{\text{Cl}} = 1 \ : 0.04 \ : 0.45 \tag{A.26.5}$$

we reduce the formula to relative current density:

$$j_i^{\text{rel}} := \frac{j_{ei}}{e^2 \omega_K N_L} = \psi_{\text{memb}} \frac{\omega_i}{\omega_K} \frac{C_i^{(e)} e^{-\frac{Z_i e \psi_{\text{memb}}}{kT}} - C_i^{(i)}}{1 - e^{-\frac{Z_i e \psi_{\text{memb}}}{kT}}} \tag{A.26.6}$$

The cell-membrane potential as a function of temperature may be calculated from the Goldman–Hodgkin–Katz (GHK) equation [1339].

$$\psi_{\text{memb}} = \frac{kT}{e} \ln \frac{\omega_{\text{Na}} C_{\text{Na}}^{(e)} + \omega_K C_K^{(e)} + \omega_{\text{Cl}} C_{\text{Cl}}^{(e)}}{\omega_{\text{Na}} C_{\text{Na}}^{(i)} + \omega_K C_K^{(i)} + \omega_{\text{Cl}} C_{\text{Cl}}^{(i)}} \tag{A.26.7}$$

(The GHK equation was first proven on nerve cells, but this describes the basic dynamic phenomena of membrane stabilization for every eukaryote system. The static part of membrane stability is described by the Donan approach.) The temperature dependence of the membrane potential is trivially linear. Its slope (S_T):

$$S_T := \frac{d\psi_{\text{memb}}}{dT_{\text{memb}}} = \frac{R}{F} \ln \frac{\frac{\omega_{\text{Na}}}{\omega_K} C_{\text{Na}}^{(e)} + C_K^{(e)} + \frac{\omega_{\text{Cl}}}{\omega_K} C_{\text{Cl}}^{(i)}}{\frac{\omega_{\text{Na}}}{\omega_K} C_{\text{Na}}^{(i)} + C_K^{(i)} + \frac{\omega_{\text{Cl}}}{\omega_K} C_{\text{Cl}}^{(e)}} \tag{A.26.8}$$

Appendix 27: Membrane Damage by Increasing Pressure

According to Onsager's theorem, the heat flow is also coupled to the volume (mass) transport. (The entire complex phenomenon is simplified on the intensive pairs only.) The relevant Onsager relations are:

$$J_V = L_{VV} \Delta p + L_{Vq} \frac{\Delta T}{T}, \quad J_{\text{q}} = L_{qV} \Delta p + L_{qq} \frac{\Delta T}{T} \tag{A.27.1}$$

The membrane permeability is much higher for water than for ions, so the main transported component in this coupling is water. The dynamic equilibrium hydrostatic pressure can be determined by the pressure equality on both sides of the membrane [576], so:

$$\frac{\Delta p}{\Delta T} = -\frac{Q^*}{\overline{V} T_i} \tag{A.27.2}$$

where Q^* is the transport heat and V is the molecular volume of the water. The transport-heat value for 1 liter water under normal conditions (room temperature, normal pressure) is equal [1340] to $Q^*_{\text{H2O}} = 71.2$ kJ, consequently:

$$\frac{\Delta p}{\Delta T} = -1.32 \cdot 10^7 \frac{Pa}{\text{K}} = -132 \frac{\text{bar}}{\text{K}} \tag{A.27.3}$$

This means that the $\Delta T=0.01$ K temperature difference generates $\Delta p=1.32$ bar$=1.32\cdot10^5$Pa pressure, which is a large value. Consequently, the lateral tensile stress rises due to the lateral dilatation generated by the radial mechanical pressure from the electric field, and the hydrostatic pressure difference tries to balance it. The lateral tensile stress calculated from the above non-equilibrium transports using $D=10$ μm cell diameter and $\xi=10$ nm membrane thickness results in

$$\sigma_{ny} = \frac{D\Delta p}{4\xi} = 1.32\frac{10\,\mu\text{m}}{410\,\text{nm}} \approx 3\cdot10^7\text{Pa} = 30\text{MPa} \tag{A.27.4}$$

which is a huge value? Comparing this value to the regular conditions the induced scalar pressure (from the Maxwell stress-tensor [1149]) is $\sigma_r=\varepsilon E^2\approx 310^3$Pa, which is too small to be relevant and negligible in relation to the transport pressure. The maximal tolerable lateral tensile stress amounts to $\sigma_{\max}=(2\text{–}20)\cdot10^5$ Pa [1341], which is smaller by two orders of magnitude than the temperature-induced stress. The increase of the osmotic pressure also contributes as an additional factor to the damage of cellular membrane. The increase of the extra-cellular matrix (ECM) temperature boosts the electrochemical potentials in the ECM electrolyte. The chemical potentials are: $\mu_i=\mu_{i0}+V_ip+RT\cdot\ln(\gamma c_i)+Z_iF\psi_{\text{memb}}$, where V_i is the molar volume, μ_{i0} is the initial chemical potential, p is the pressure, γ is the activity coefficient, c_i is the concentration and Z_i is the ionizing state of the i-th component. F is the Faraday number (96,500 C) and ψ_{memb} is the membrane potential. With the increasing j_{Na}, both j_{K} and j_{Cl} decrease, therefore, in stationary membrane state the intra-cellular concentrations $C_k^{(i)}$ ($k=\text{Na}^+$, K^+, Cl^-) also change. This process increases the osmotic pressure in the intra-cellular liquid, while decreases it in extra-cellular electrolyte.

Appendix 28: Dynamics of Adherent Bonds

Study for simplicity only two interacting adherins A and B belonging to the cells 1 and 2, respectively. Then the formal reaction kinetics:

$$A + B \underset{k_\text{r}}{\overset{k_\text{f}}{\rightleftarrows}} AB \tag{A.28.1}$$

where k_f and k_r are the association and dissociation reaction rates, respectively. When the surface concentrations of the A, B and bonded AB species are C_A, C_B, and C_{AB}, then:

$$\frac{dC_{AB}}{dt} = k_\text{f}C_AC_B - k_\text{r}C_{AB} \tag{A.28.2}$$

When the complete concentration are C_{AT} and C_{BT}:

$$C_A = C_{AT} - C_{AB}, \quad C_B = C_{BT} - C_{AB} \tag{A.28.3}$$

Therefore

$$\frac{dC_{AB}}{dt} = k_{\mathrm{f}}\,(C_{AT} - C_{AB})(C_{BT} - C_{AB}) - k_{\mathrm{r}} C_{AB} \tag{A.28.4}$$

The bonded states in the stationer case:

$$k_{\mathrm{f}} C_{AB}^2 - [k_{\mathrm{f}}(C_{AT} + C_{BT}) + k_{\mathrm{r}}]\, C_{AB} + k_{\mathrm{f}} C_{AT} C_{BT} = 0 \tag{A.28.5}$$

Therefore, if C_{AB} is small, then the number of adherent bonds in the stationer case:

$$C_{AB\mathrm{stac}} \cong \frac{k_{\mathrm{f}} C_{AT} C_{BT}}{k_{\mathrm{f}}(C_{AT} + C_{BT}) + k_{\mathrm{r}}} \tag{A.28.6}$$

Supposing the cells have the same capacity of the transmembrane proteins: $C_{AT} \approx C_{BT} = C$:

$$C_{AB\mathrm{stac}} \cong \frac{k_{\mathrm{f}} C^2}{2k_{\mathrm{f}} C + k_{\mathrm{r}}} \tag{A.28.7}$$

When the association or dissociation dominates we get:

$$C_{AB\mathrm{stac}} = \frac{C}{2} \quad (2k_{\mathrm{f}} C >> k_{\mathrm{r}})$$

$$C_{AB\mathrm{stac}} = \frac{k_{\mathrm{f}} C^2}{k_{\mathrm{r}}} = \frac{C^2}{\frac{k_{\mathrm{r}}}{k_{\mathrm{f}}}} = \frac{C^2}{k_{\mathrm{D}}}, \quad (2k_{\mathrm{f}} C << k_{\mathrm{r}}) \tag{A.28.8}$$

where k_{D} is the reaction constant of association: $k_{\mathrm{D}} = k_{\mathrm{r}}/k_{\mathrm{f}}$. The association reaction constant determines well the number of bonds (see Fig. A.28.1).

The dissociation is approached by the Arrhenius-type function ([1165]):

$$k_{\mathrm{r}} = e^{-\frac{E_0 - \gamma f}{kT}} \tag{A.28.9}$$

where E_0 is the activation energy of the adherent connection, λ is the width of the bonding barrier, while f is the force on one single bond. The expression shows the dissociation with increasing temperature.

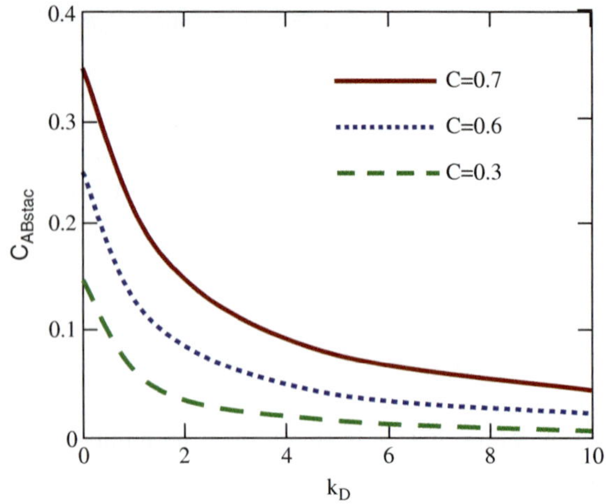

Fig. A.28.1 Arrhenius plot by various reaction constants

Appendix 29: Modulation Effect

For the description let us denote the deterministic $\Omega = 2\pi\nu$; $\nu = 13.56$ MHz carrier signal by $s(t)$ with U_0 amplitude, while the modulating pink noise is $z(t)$ which makes m modulation depth. In this case the full potential $u(t)$ of the signal on the applicator is:

$$\begin{aligned} u(t) &= s(t) + n(t) \\ s(t) &:= U_0 \sin \Omega t \\ n(t) &:= mz(t)U_0 \sin \Omega t \end{aligned} \tag{A.29.1}$$

The autocorrelation function of $u(t)$ with long averaging time:

$$u(t)\otimes u(t+\tau) := \lim_{T\to\infty} \frac{1}{T}\int_{-T}^{T} u(t)u(t+\tau)dt \cong s(t)\otimes s(t+\tau) + n(t)\otimes n(t+\tau) \tag{A.29.2}$$

Obtaining the power-spectrum, the Fourier transformation of the noise:

$$F\{n(t)\} := \frac{j}{2} mU_0 \left[Z(|\omega - \Omega|)\right] = \frac{j}{2} mU_0 \left[\frac{1}{j\sqrt{|\omega - \Omega|}}\right] \tag{A.29.3}$$

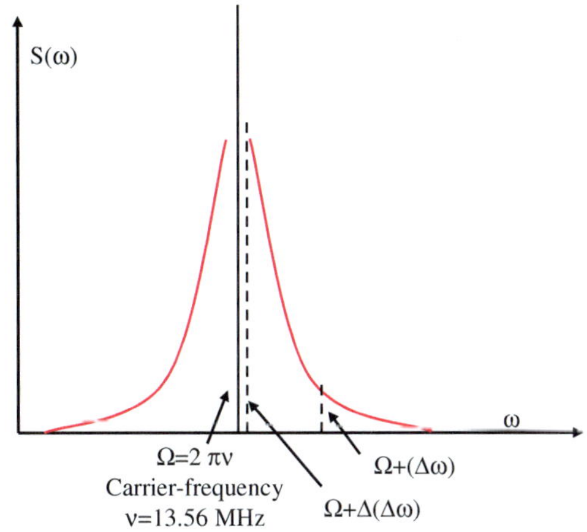

Fig. A.29.1 Modulation widens the spectrum of frequency at the carrier

Hence the power-spectrum of the potential on the applicator (see Fig. A.29.1):

$$S(\omega) := F\{u(t) \otimes u(t+\tau)\} := U_0^2 \delta(\omega - \Omega) + \frac{m^2 U_0^2}{4} \frac{1}{|\omega - \Omega|} \tag{A.29.4}$$

So the effective potential in the interval [{Ω+$\Delta(\Delta\omega)$} – {Ω+($\Delta\omega$)}] could be calculated by the Wiener–Khintchine theorem [1342]:

$$\Delta U_{\mathrm{eff}}^2 = \int\limits_{\Omega}^{\Omega+\Delta\omega} \Phi_{uu}(\omega) d\omega = \int\limits_{\Omega+\Delta(\Delta\omega)}^{\Omega+\Delta\omega} \left(U_0^2 \delta(\omega - \Omega) + \frac{m^2 U_0^2}{4} \frac{1}{|\omega - \Omega|} \right) d\omega$$

$$= U_0^2 + \frac{m^2 U_0^2}{4} \ln \frac{\Delta\omega}{\Delta(\Delta\omega)} \tag{A.29.5}$$

and the average effective field at the electrode distance d:

$$\Delta E_{\mathrm{eff}} = \frac{\Delta U_{\mathrm{eff}}}{d} = \frac{1}{d} \sqrt{U_0^2 + \frac{m^2 U_0^2}{4} \ln \frac{\Delta\omega}{\Delta(\Delta\omega)}} \tag{A.29.6}$$

The term multiplied by ln[$\Delta\omega/\Delta(\Delta\omega)$] could be extremely large, at the limited power availability. The pink-noise spectrum alone has a spectrum centered on zero frequency, while the modulated signal shifts the pink-noise spectrum to the ν=13.56 MHz center, see Fig. A.29.2.

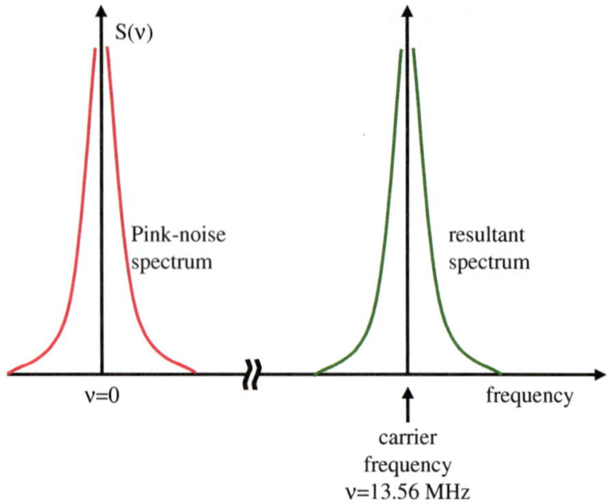

Fig. A.29.2 Frequency shift by pint-noise modulation

Appendix 30: Components of Cell Destruction

The HSP molecules have two translocations of the HSP molecules [1343]: they have a function in the nuclei and also they have a function in cytoplasm and membrane induced by the membrane sensory rafts [1344]. The stress effect is the resultant of the forces of the electric field (see Fig. A.30.1):

$$[E(x+l) - E(x)]\,\mathrm{grad}\mu \cong l\,\mathrm{grad}E(x)\,\mathrm{grad}\mu \cong \frac{l}{\Delta l}\Delta\mu\,\mathrm{grad}E(x) \qquad \text{(A.30.1)}$$

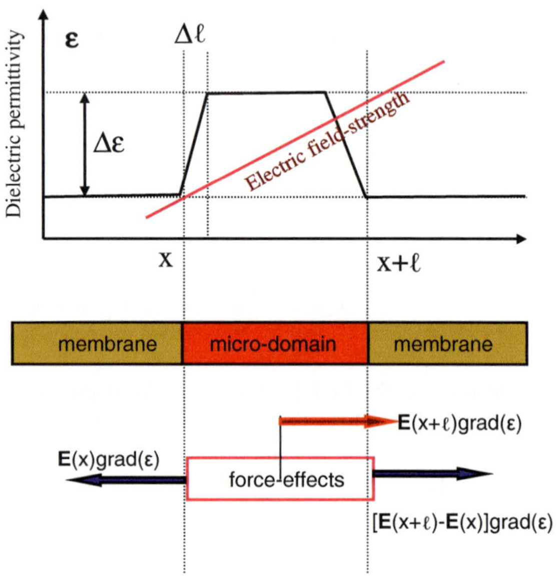

Fig. A.30.1 Electric field changes at the microdomain

This force will move the microdomains to the areas of more stress. The other force mechanically deforms the rafts:

$$E(x)\text{grad}\mu \cong \frac{\Delta\mu}{\Delta l}E(x) \tag{A.30.2}$$

Together with these the locally dense SAR_t heats up the touching point:

$$\text{SAR}_t = \sigma E^2 \tag{A.30.3}$$

Hence there are two stress-components: the field [Eqs. (A.30.1) and (A.30.2)] and the heat [Eq. (A.30.3)], proportional linearly and squared to the field itself.

The current densities (j) and currents (I) from electrodes have two components; capacitive ($\boldsymbol{j}_{\text{cap}}$ and I_{cap}) and ohmic ($\boldsymbol{j}_{\text{ohm}}$ and I_{ohm}) in the local electric field ($\boldsymbol{E}$) (see Fig. A.30.2):

$$\begin{aligned} &\underline{j}_{\text{cap}} = i\omega\underline{E}, \qquad \underline{j}_{\text{ohm}} = \sigma\underline{E} \\ &I_{\text{cap}} = \int_A \underline{j}_{\text{cap}} d\underline{A} = i\omega\int_A \varepsilon\underline{E}d\underline{A}, \quad I_{\text{ohm}} = \int_A \underline{j}_{\text{ohm}} d\underline{A} = \int_A \sigma\underline{E}d\underline{A} \end{aligned} \tag{A.30.4}$$

Hence:

$$I = I_{\text{ohm}} + I_{\text{cap}} = \int_A \left(\underline{j}_{\text{ohm}} + \underline{j}_{\text{cap}}\right) d\underline{A} = \int_A (\sigma + i\omega\varepsilon)\,\underline{E}d\underline{A} \tag{A.30.5}$$

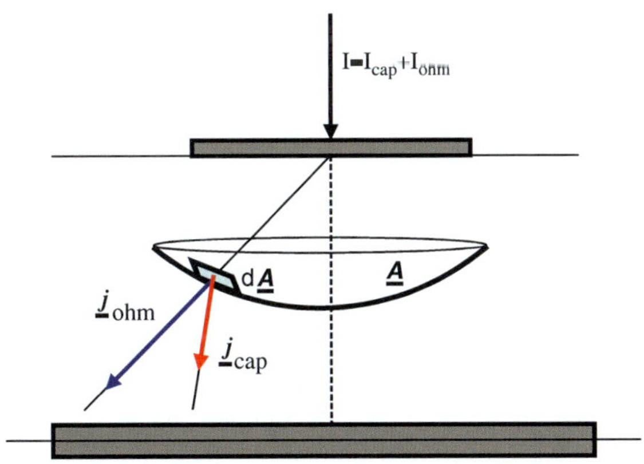

Fig. A.30.2 The ohmic current propagates straight while the capacitive deflects on the boundary

Appendix 31: Experimental Conditions In Vitro

The in vitro studies were performed in a special cell-culture chamber. It has relatively large plane-parallel surfaces (see Fig. A.31.1a), immersing the adherent cell-cultures on the coverslip (see Fig. A.31.1b) into the chamber filled with medium.

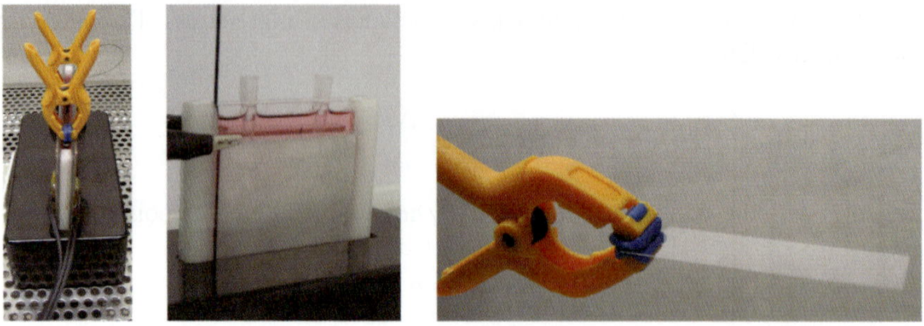

(a) (b) (c)

Fig. A.31.1 The in vitro chamber (**a**) (the overall view of the chamber and its holder; (**b**) the chamber side-view, (**c**) the coverslip with the cell culture with a holder to immerse it into the medium

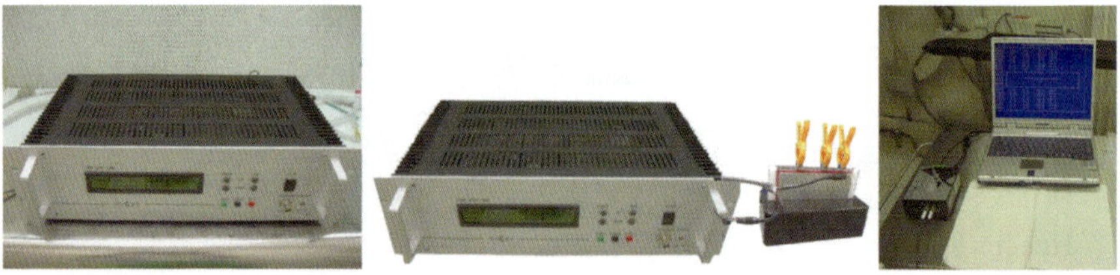

Fig. A.31.2 The treatment setup

The chamber was directly connected to the laboratory device (see Fig. A.31.2), where the temperature was measured in situ real time in the chambers.

As was described before, oncothermia is selective for the higher conductivity and higher permittivity of the extra-cellular matrix of malignant tissue. To measure this selectivity a coculture experiment was performed. A431 squamous carcinoma cell lines were developed with connective healthy fibroblasts in a common cell culture (human malignant melanoma model). The result of the treatment is spectacular: the healthy cells remain intact, while the aggressively malignant A431 melanoma cells (tumorgenetic creatocytes) were destroyed (see Fig. 4.37) [767]. An important control, if the cells are less malignant (HaCaT cell line, immortal creatocytes, see Fig. 4.38) the selection is less effective, as the selection is based on the products of the malignant activity (ionic concentration).

A comparative study of conventional hyperthermia and oncothermia was performed in vitro. The cell samples were submersed into a medium between glass-sides covered by hotplates or by electrodes (see Fig. A.31.3).

The basis of the comparison was the temperature, which was carefully measured by Luxtron fluoroptical sensors with no metallic components near the samples. Because of the importance of temperature as a control of the identical treatments we have to verify the identity of the macroscopic temperature with the microscopic one. The microscopic (subcellular) temperature was verified by transfected Luciferase (as a molecular thermometer). The model was a HEK293 (human embryonic kidney) cell line, and the Luciferase was cotransfected with GFP non-temperature

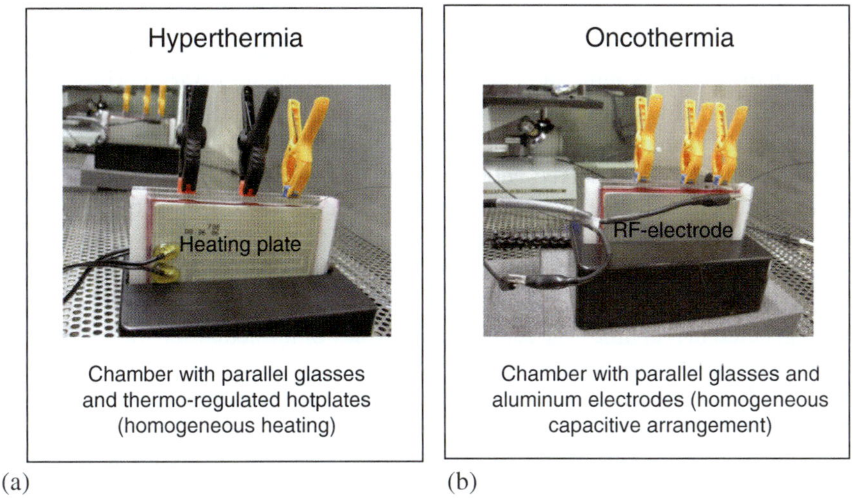

Fig. A.31.3 The in vitro chambers for comparison studies: (**a**) the hyperthermia solution; (**b**) the oncothermia solution

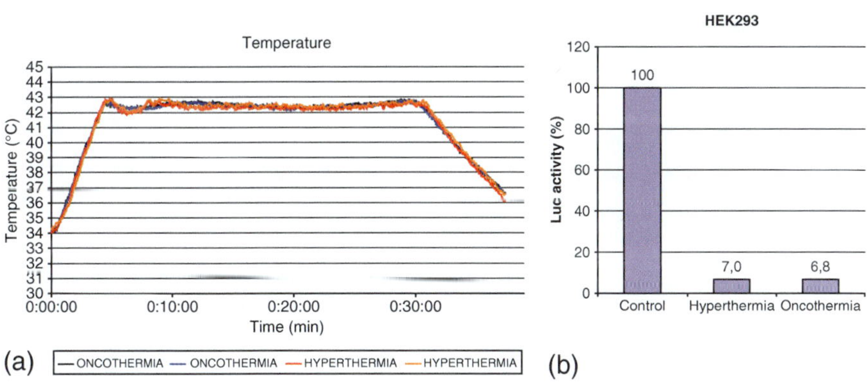

Fig. A.31.4 The macroscopic temperature measurement (Luxtron sensor, **a**) and the microscopic (Luciferase activity, **b**) temperature comparison of the hyperthermia and oncothermia measurements

sensitive protein for reference [1345]. The macroscopic and microscopic temperature measurements are shown on Fig. A.31.4. The identity is obvious.

All the comparison studies were carefully managed in terms of their dynamics as well. We kept not only the temperature identical in the samples, but the dynamics of heating and cooling were also controlled and kept equivalent (see Fig. A.31.4a).

Appendix 32: Experimental Conditions In Vivo

Oncothermia was performed in the classical capacitive arrangement. A plane-parallel condenser provided the energy input into the mice placed in between the electrodes. The arrangement was asymmetric [1346]. The large, rectangular-shaped grounded electrode was 6 × 12 cm in size; the upper electrode was round in shape with a diameter of 18 mm, and can be adjusted into the necessary position above the treated animal with a flexible arm. The appropriate electromagnetic coupling is assured by the water bolus (a latex bolus), filled with distilled water. The cooling of the upper electrode is solved with a cooling unit (fully separated from the water bolus) operating with continuously circulating water, cooled with Peltier units. The output of the cooling unit is controllable. An ultra-fast tuning system makes possible a sudden automatic adjustment in any reactions. Parallel to the four temperature data the tuning status (standing wave ratio), the forwarded and reflected powers are also registered in real time. The sampling frequency of the actual software is 60/min. The experimental setup is shown in Fig. A.32.1.

To study the effect of the temperature and the field-effects we compared the heating methods applying classical infrared (hyperthermia) and electric field (oncothermia) heating [580]. The classical hyperthermia treatment was provided by a specially built infrared radiation (IR) emitter which fitted the size of the tumors.

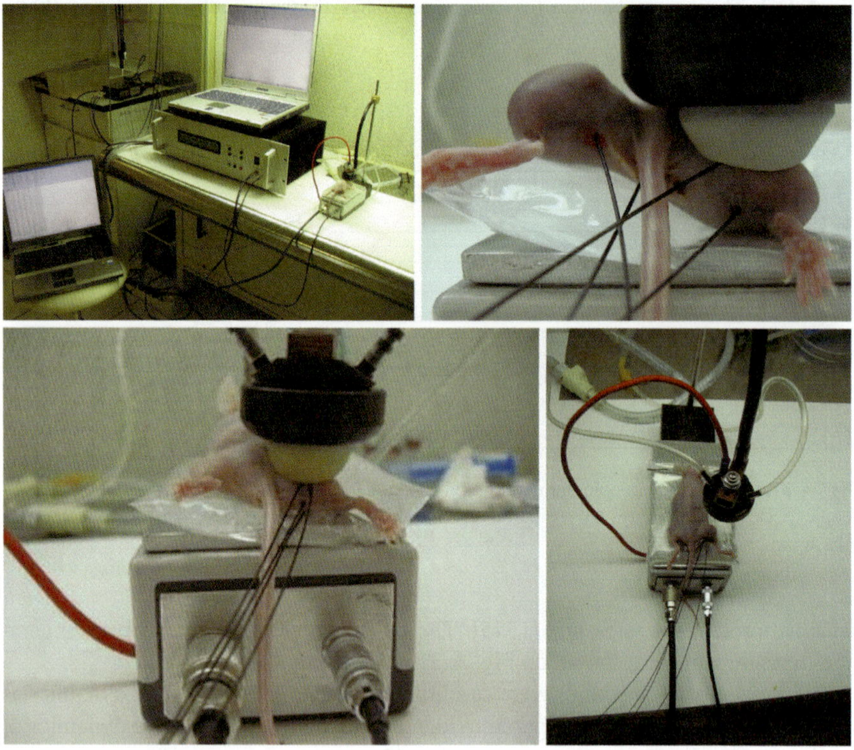

Fig. A.32.1 Experimental setup of the in vivo experiments. The optical fiber temperature sensors are placed intratumoral, (*thin black cables* inserted into the tumor). Every animal had their own untreated control tumor in a symmetrical position {Institute: Department of Pharmacology and Toxicology, Faculty of Veterinary Science, St. Istvan University, Budapest, Hungary, Investigators: Dr. G. Andocs}

The IR emitter was composed of a 5 cm in diameter round-shaped metal reflector and eight tiny infrared light-emitting bulbs. Both treatments were controlled by accurately measured intratumoral temperature using the fluoroptical method. The IR heater was positioned above the treated tumor and its radiated energy was controlled by the measured temperature, adjusted to maintain a steady-state temperature plateau during the treatment. To avoid cooling of the mice, the temperature of the water bed (polyethylene sachet) was controlled; maintained at 37°C during the treatment. Photos of the treatment setups are shown in Fig. A.32.2.

The desired temperature was very accurately measured taking care of the precious equivalence of the actual temperature in all phases of the process. The actual temperature of the oncothermia treatment was adjusted by the absorbed radiofrequency power, while the classical hyperthermic heat therapy was adjusted by the IR-radiation intensity. The heating-up and cooling-down periods were also kept identical to control the dynamic physiological parameters (like the synthesis of heat-shock proteins [HSPs]). Each tumor-temperature was measured intratumorally to control the core temperature, furthermore one sensor controlled the systemic temperature (rectal) and one was placed on the outer surface over the treated tumor to avoid skin burn.

The method used to calculate the killing rate of the treatments is a morphological comparison based on the observed pathological differences. As the living part is undergoing intensive proliferation microscopically it could easily be distinguished from the necrotic part containing the dead tumor cells.

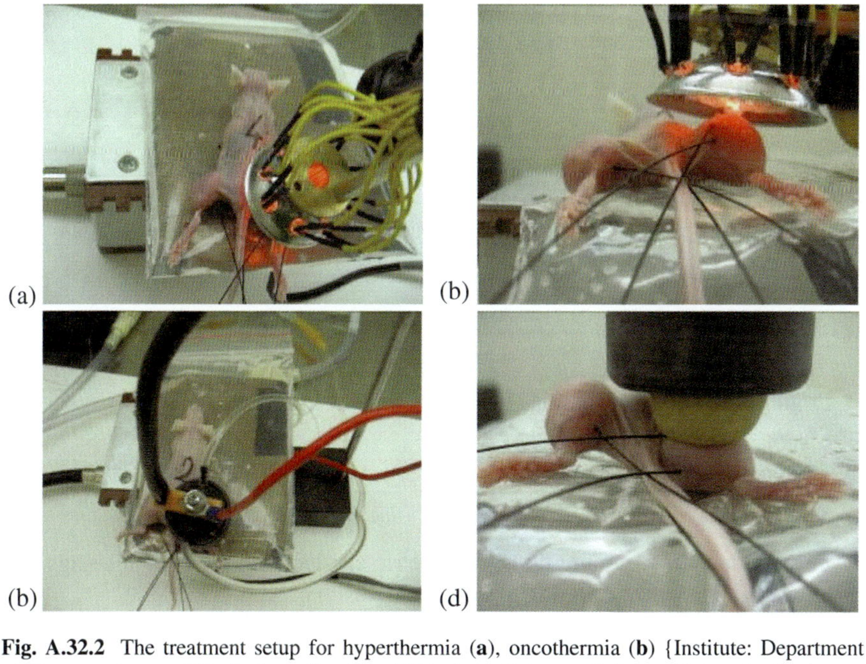

Fig. A.32.2 The treatment setup for hyperthermia (**a**), oncothermia (**b**) {Institute: Department of Pharmacology and Toxicology, Faculty of Veterinary Science, St. Istvan University, Budapest, Hungary, Investigators: Dr. G. Andocs}

Appendix 33: Considerations for Experimental Setup In Vivo

The applied power is carefully designed to make the laboratory conditions most similar to the clinical applications. However both the geometries and the physiological conditions are very different in the case of humans and small animals. The studies used the absorbed RF-power to reach the desired temperature, while the applied voltage was necessary to study the electric-field effect. However the power and the electric field are not independent, they are definitely correlated through the applied load, which is the treated body. To make the proper comparable projection a theoretical consideration was necessary.

Let us start again with the widely applied thermal description of the heating in vivo is the Pennes' bioheat equation. The electrical heating power (q_e) [1347] is

$$q_e = \sigma |\boldsymbol{E}|^2 \tag{A.33.1}$$

where $\boldsymbol{E}$ is the actual value of field strength and σ is the conductivity of the tissue. It is reasonable to assume that the degree of blood perfusion is proportional to the ΔV blood volume transported by one beat multiplied by the number of heartbeats per unit time and divided by the mass:

$$w \propto \frac{\Delta V f_{\text{heartbeat}}}{M} \tag{A.33.2}$$

The number of heartbeats is proportional to the $-1/4^{\text{th}}$ power of body mass while the value of ΔV to the body mass itself [1348, 1349]. Consequently:

$$w \propto M^{\frac{1}{4}} \tag{A.33.3}$$

The real treatment uses the effective field strength values (E_{eff}), which is always lower than the maximum at the peak. Supposing the body and blood temperature as well as the density and specific heat data are the same in the case of humans and mice then we get

$$\frac{w_{\text{mice}}}{w_{\text{human}}} = \frac{q_{\text{mice - eff}}}{q_{\text{human - eff}}} \tag{A.33.4}$$

From the above two equations we get for the electric powers that

$$\left(\frac{M_{\text{mice}}}{M_{\text{human}}}\right)^{\frac{1}{4}} = \frac{q_{\text{mice - eff}}}{q_{\text{human - eff}}} \tag{A.33.5}$$

In consequence, assuming the same electric conductivity for human beings and mice, we have for the proportion of field strengths that

$$\frac{E_{\text{mice - eff}}}{E_{\text{human - eff}}} = \left(\frac{M_{\text{mice}}}{M_{\text{human}}}\right)^{\frac{1}{8}} \cong 2.7 \tag{A.33.6}$$

Therefore, to achieve the same thermal effect in the case of mice we have to approx. triplicate the effective field strength. The above calculation relates to the thermal effect. Now, let us deal with the electric field effect! The field-governed processes intervene in the membrane transport of cells. In the case of oncothermia the actuating parameter is the electric field strength. However, this can change the distribution of membrane potential. If we suppose the same cell-sizes then for the induction of the same field effect we need the same field strength. The living organism is not able to react immediately to all changes. We might reasonably suppose that the time constant can be compared with the single mixing time of blood in the organism. From this assumption we get that

$$\tau \approx \frac{V_{blood}}{\Delta V}\frac{1}{f_{\text{heartbeat}}} \qquad \text{(A.33.7)}$$

We might suppose that V_{blood} and ΔV are proportional to the body mass. From this we have the interesting result that

$$\frac{\tau_{\text{mice}}}{\tau_{\text{human}}} = \left(\frac{M_{\text{mice}}}{M_{\text{human}}}\right)^{\frac{1}{4}} \qquad \text{(A.33.8)}$$

according to which the number of heartbeats is proportional to the $-1/4$ th power of body mass. From this and by using the previous masses we get that

$$\frac{\tau_{\text{mice}}}{\tau_{\text{human}}} = \left(\frac{M_{\text{mice}}}{M_{\text{human}}}\right)^{\frac{1}{4}} \approx 1/7 = 0{,}147 \qquad \text{(A.33.9)}$$

This approx. factor can be checked: the heartbeat of mouse is ~500/min, while in the case of human beings this is ~72/min. Namely, the physiological reaction of mouse is seven-times faster! In the case of an adult person the period of blood exchange is about 1 min, therefore, the physiological time constant is: $T_{\text{physiol}} \approx$ *1–2* min. For the mouse we get: 10–20 s. This means that for a human being the integrated duty time shall exceed 1–2 min, while for a mouse this period is 10–20s. Comparing all of the above parameters, we get projected parameters from human to mouse treatment as collected in Table A.33.1 (The data could vary slightly of course depending on the actual treatment.)

Similar calculations could be made on the other animals, using the above scaling law. Further in vivo experiments were guided by the above calculation so as to be similar to the situation in humans.

Table A.33.1 Comparison of the treatment parameters of humans and mice

	Weight (kg)	Electrode diameter (cm)	Treatment time (min)	Treatment power (W)	Treatment energy (kJ)
Human	65	30	60	130–150	468–540
Mice	0.02–0.03	1.5–2.0	20–30	6–8	7.2–14.4

Appendix 34: Validation of Morphologic Evaluation

We first validated the control group and the control tumor method on the same animal. Using untreated animals, the two tumors were compared in every animal without any treatments (see example in Fig. A.34.1). By accurate comparison we established the accuracy of the symmetrical controlling method. The difference between the tumors in two femoral sides was less than 6%, the Pearson correlation was 0.88. So the distribution of the tumors in was identical. The measured ratio of the dead-cell volume to the full tumor size by the applied morphological method was 6.1% on average, ($p>0.93$ [$W_+ = 15$, $W_- = 13$] using Wilcoxon matched-pairs signed-rank). In consequence the method correctly and with high significance measures the dead-cell ratio in the tumor volume.

Together with this step, we compared the control tumors in all of the included animals from all four groups. We saw pretty good correspondence in all the control tumors of the animals. The average ratio of the death volume in the full tumor mass was 6.3% ($p>0.72$, Student t-test). We concluded that the control group could be used for overall reference. Furthermore, to decrease the fluctuation error caused by the individual animals, we used their own control for comparison, and the statistics of their ratio were evaluated.

The tumor sizes are summarized in Table A.34.1

The treatment evaluation process was simply to compare the death-cell volume to the full tumor in every sample and the ratio was compared to the control tumor, deducting the dead-volume ratio of the non-treated (normothermia) reference tumor from the treated one. This value was used to characterize the efficacy of the actual treatment.

The method to calculate the killing rate of the treatments is a morphological comparison based on the observed pathological differences. Histological tumor samples were imaged using the MiraxScan (3DHISTECH, Hungary) dedicated slide scanner. According to the microscopic analysis the living part being in intensive proliferation could be easily distinguished from the necrotic part containing the dead tumor cells. It could also easily be contoured with the help of the MiraxViewer (3DHISTECH, Hungary) image managing and analyzing software (example in Fig. A.34.2). The

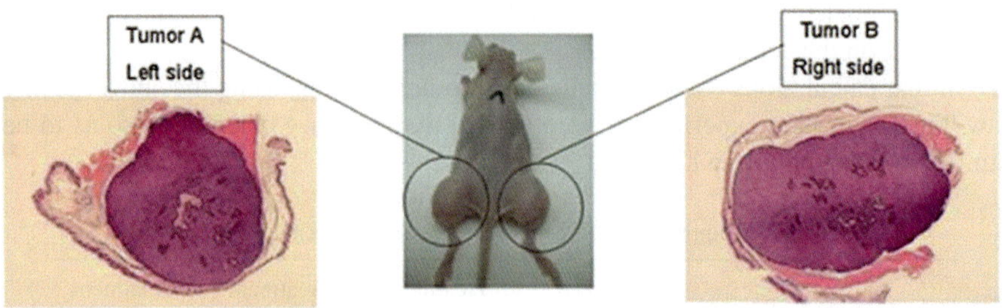

Fig. A.34.1 An example of the control validation {Institute: Department of Pharmacology and Toxicology, Faculty of Veterinary Science, St. Istvan University, Budapest, Hungary, Investigators: Dr. G. Andocs}

Table A.34.1 The distribution of tumor sizes in the study (seven tumors in each category)

Tumor size []	Group 1 (control)		Group 2 (hyperthermia 42°C)		Group 3 (oncothermia 42°C)		Group 4 (oncothermia 38°C)	
	Left	Right	Left (control)	Right (active)	Left (control)	Right (active)	Left (control)	Right (active)
Mean	115,378	121,606	78,521	84,728	7,056	87,213	42,107	43,743
Minimum	63,197	40,915	52,833	58,621	69,805	60,071	23,507	31,141
Maximum	159,226	155,246	99,792	96,861	101,086	115,280	57,940	68,159
Confidence [±] (95%)	85,472	90,085	58,168	62,766	64,491	64,607	31,193	32,405
Wilcoxon matched-pairs signed-rank	*P*>0.57		*P*>0.57		*P*>0.99		*P*>0.81	

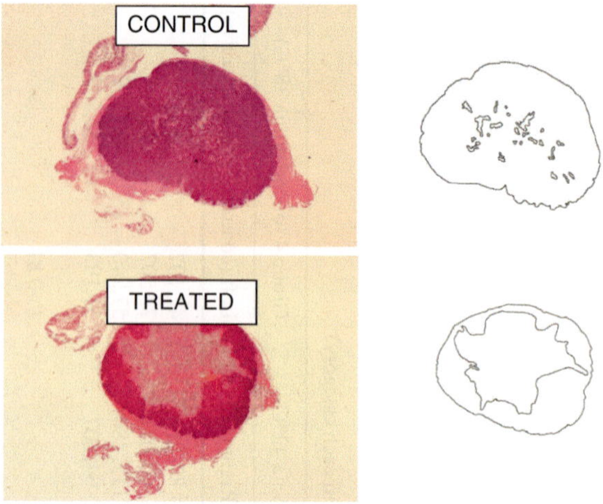

Fig. A.34.2 Typical morphological evaluation by area measurement

areas of the living and dead tumor parts were defined by area calculations, assuming the equal average density of the dead and living cells in the calculated area. We calculated their percentage correlated to the whole area of the tumor cross-section. With those results we compared the change in the dead part of the control and treated tumor originating from the same animal.

The method of course measures the cross-section, which also makes it possible to guess the volume by a spherical assumption. In this case the power of $\frac{2}{3}$ of the ratio of the dead-cell areas measured in the cross-sections corresponds to the volume ratio of the morphologically distinguished cells. The area ratio in all of the cases was larger than unity, so the volume ratio has to be higher than what is obtained from the areas. We could state that the volume ratio of the killing efficacy definitely is not less than the calculated ratio from the cross-sectional areas. On this basis, for simplicity, the obtained cross-sectional areas are considered to characterize the quantitative situation in the tumor, so we compared these values only. Twenty-four hours later the single-treatment animals were sacrificed and both the control and treated tumors were removed and studied in pairs. All the removed tumors were cut accurately at their centerline, and fixed in 4% buffer formalin. After that standard histological samples were made stained with hematoxylin-eosin.

Appendix 35: Evaluation of Survival Study with Single Arm

Oncothermia survival studies are problematic due to the missing control arm. This is a problem in general, when the treatment targets are advanced, mostly refractory, relapsed malignancies in high treatment lines, when the only way forward is sequential treatment. The sequential trial [1350, 1209], is well known, and applied frequently in the case of small trials [1351].

Evaluating single-arm treatments has numerous challenges. Data losing their references can lead to misinterpretations. The living variability of the personal and tumor cases is the main possible stumbling block. This also may occur when the reference arm is not a cohort with the active one. To be sure the two arms form identical cohorts a randomization is necessary.

However, information on the effect of sequential treatment is also present in the single arm, only to mine it is a rather complex task. Some definite points have to be fixed to solve this problem at least approximately.

The sequential trial (like oncothermia) is applied to the same patient in sequences. In this approach the development of the patient is measured and documented. Of course we have no idea of development should the actual sequence not be applied, so we are not able to measure the changes quantitatively.

However, we have some qualitative assumptions:

- The patient begins a new sequence should the previous sequence not have achieved a result (or not a satisfactory result). This condition is generally valid, as there is no reason to start new therapy when the previous has worked satisfactorily. (In some cases due to psychology or other factors a successful therapy could be abandoned, but we assume this is less than 5% of all treatments.)
- We suppose that no worsening of the patient's stage due to the applied therapy in the actual sequence. Excluding the direct negative effects of the actual treatment sequence has to be checked in other seperate study (safety, dose escalation, adverse/side effects, etc.). When in an independent study the complete effect shows no side effects, or shows easily distinguishable ones from the adverse effects of the previous treatments, we may handle this sequence as positive.
- It could be a negative addition, when we apply wrong sequence. This means the sequence that should have been applied could be better than the one actually applied. In this case there no actual worsening caused by the sequence, but the overall therapy results could be worse then would be possible using the actual state-of-art in the given case.
- The new sequence does not block the possibility of subsequent sequences, the actual therapy does not exclude the patients from other possible therapies.
- The effect of the new sequence affects the survival curve, so the studied Kaplan–Meier plot includes the information. (Example: should the effect improve quality-of-life but not survival, the sequence can not be studied by survival curves.)
- The sequence is medically controlled in the same way as in previous therapies. No uncontrolled "side therapies" are in use.
- Oncothermia satisfies the above criteria.
- Oncothermia is applied in the stage after "gold standards" fail. Its application is clearly intended in advanced stages, after failure of previous treatments.
- The no-harm status of oncothermia is proven by its long-term application (over 20 years). The rare (3–8%) surface erythematic redness of (<3%) adipose burn is easily distinguishable from the symptoms of the malignant disease, and are easily controlled. Oncothermia is a complementary treatment, presently no exclusion is known for concomitant application. Oncothermia is applied in most cases as one of the last "remaining opportunities" of treatment

- Oncothermia has addition to survival curves. The survivals when oncothermia is involved are different from any historical controls, so its affect is present in the survival plot.
- Application of supplementary treatments is hard to control. This is not a problem peculiar to oncothermia, but a definite complication for all other clinical studies, especially when the treatment is on an outpatient basis.

The sequenced steps in the case of oncothermia are shown in Fig. A.35.1. First diagnosis and treatment start also marks the start of the study. Treatments made during the elapsed time are not considered as steps; this time interval is regarded as one unit. Oncothermia survival is the time from the beginning of oncothermia treatment, while overall survival covers all the time the patient has been followed. Of course the last info (when the patient had been controlled lost and censored from this time or died) and the evaluation time could be identical in some patients. In most cases the overall survival is much greater than the oncothermia survival. The last info could be a check-up, an examination for other reasons, or death.

The best mining of the data would be when the non-parametric Kaplan–Meier survival plot could be parameterized. A proper parameterization could be based on one of the inherent properties of survival, the self-similarity.

Self-similarity means that all the measurable quantities of the studied survival pattern depend on the scale. Assuming r is the linear size of the scale and L is a scale-dependent behavior: $L(r)$. Elongate the scale by ar times, $L(r)$ transforms

$$L(ar) = kL(r) \tag{A.35.1}$$

where the factor k depends only on a. Generally, we assume the power-type functional relationship, like

$$L(r) = Ar^{\alpha} \tag{A.35.2}$$

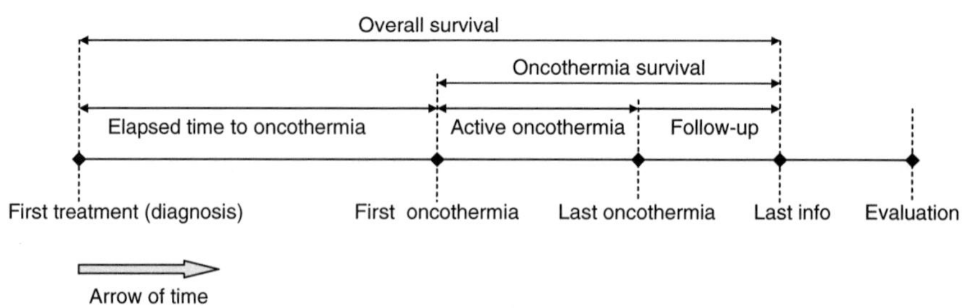

Fig. A.35.1 The time sequences of oncothermia studies. The time between the first-ever treatment and the first oncothermia is complex, involving numerous pretreatments. It is regarded here as one step this changes in the form of

$$L\,(ar) = Ar^{\alpha}a^{\alpha} = a^{\alpha}L\,(r) \tag{A.35.3}$$

Hence

$$\lg L\,(r) = \lg A + \alpha \lg r \tag{A.35.4}$$

the lg-lg scale is a straight line with a gradient equal to the exponent. The power-type equation is the consequence of some simple and very general assumptions. The Eq. (A.35.4) defines the self-similarity, where any a and k independent from r. Hence k is a general function of a, therefore,

$$L\,(ar) = k\,(a)\,L\,(r) \tag{A.35.5}$$

From this we get a new and more general definition of self-similarity, namely, we call the

$$L : r \in I_r \mapsto L\,(r) \in R^1 \tag{A.35.6}$$

mapping self-similar, if $L(ar)$ may be expressed in a more general form:

$$L\,(ar) = k\,(a)\,G\,(r) \tag{A.35.7}$$

this is, the scale-transformed function can be separated in the form of a product.

Let us suppose $a = 1$. Then, from (3.8):

$$L\,(r) = k\,(1)\,G\,(r) = cG\,(r) \tag{A.35.8}$$

where $c = k\,(1)$ is a constant. Consequently, $L\,(r)$ and $G\,(r)$ differ only in the constant multiplicator. On the basis of the latter property (3.8) can be rewritten:

$$L\,(ar) = \frac{k\,(a)}{k\,(1)}L\,(r) = \eta\,(a)\,L\,(r) \tag{A.35.9}$$

where $\eta(a)=k(a)/k(1)$. We may repeat the same procedure for the case of $r=1$. As a result of this, the following relationship is valid for any u,v:

$$L\,(uvr) = \eta\,(uv)\,L\,(r) = \eta\,(u)\,\eta\,(v)\,L\,(r) \tag{A.35.10}$$

Therefore, the first function in (A.35.10) satisfies the equation

$$\eta\,(uv) = \eta\,(u)\,\eta\,(v) \tag{A.35.11}$$

namely it is a special self-similar function. The general solution of this function equation is difficult. We may choose $a = r$, $r = 1$ for the common part of variables

a and r, if $r = 1$ occurs in it. Herewith, the solution can be derived from the relationship (A.35.11), where $C = L(1)$ is a constant:

$$L(r) = \frac{k(a)}{k(1)} L(r) = \eta(r) L(1) = C\eta(r) \qquad \text{(A.35.12)}$$

Consequently, if $L(r)$ is a continuous function on this common set, then it is true also for $\eta(r)$. In this case the solution of function equation (3.13) is known for positive r values [1352]:

$$\eta(r) = Kr^{\alpha} \qquad \text{(A.35.13)}$$

where K and α are constants. Hence, we get from (A.35.13) that

$$L(r) = Ar^{\alpha} \qquad \text{(A.35.14)}$$

where A is another constant. In consequence of these the so-called power-law principle can be applied in the general case for the description of self-similarity. The concept of self-similarity may be extended to the processes as well. Let us take a process described by the function:

$$f : t \in I = [0, \infty) \mapsto f(t) \in R \qquad \text{(A.35.15)}$$

Let us choose the time t expressed with the help of an earlier t' time:

$$t = \tau t' \qquad \text{(A.35.16)}$$

Our statement is that the process described by the function f is self-similar, if

$$f(t) = f(\tau\, t') = g(\tau)\, h(t') \quad \forall t, t' \in I = [0, \infty) \qquad \text{(A.35.17)}$$

Since we proved earlier that the solution of the above function equation has the form of

$$f(t) = abt^{n}, g(\tau) = a\tau^{n}, \quad h(t') = bt'^{n} \qquad \text{(A.35.18)}$$

thus we may see that the self-similarity can be written in a simpler form of

$$f(\tau t') = f(\tau) f(t') \quad \forall t, t' \in I = [0, \infty) \qquad \text{(A.35.19)}$$

This means that the value of the physical quantity of the process can be established in a later point of time by the extension or shrinking of an earlier value, where the amount of extension or shrinking depends only on the ratio of the two points of time. Therefore, the ratio of function values is the function of the ratio of points of time. Of course, not every process is self-similar in nature. Those processes deserve attention which can be transformed to a self-similar process, when f is not self-similar, but the composition of $(u \circ f)$ is self-similar by applying a suitably chosen u

transformation, namely:

$$u\left[f\left(\tau t'\right)\right] = g\left(\tau\right)h\left(t'\right) \quad \forall t, t' \in I = [0, \infty) \tag{A.35.20}$$

The u function is called the comparison function. A wide group of processes differ in terms of mathematics; however, it can be transformed into a self-similar one with the help of a comparison function. Introducing two distribution functions: the lifetime distribution function, and the distribution function of the survival probability. Let T be the stochastic variable defined on the set of individuals (lifetime). The lifetime distribution function is the probability that the lifetime is less than or equal to t, namely

$$F(t) = P\{T \leq t\} \tag{A.35.21}$$

Thus, the survival probability distribution can be defined by

$$\overline{F}(t) = 1 - F(t) \tag{A.35.22}$$

Namely, the probability that the T lifetime is higher than t can be expressed in the form of

$$\overline{F}(t) = 1 - F(t) = P\{T > t\} \tag{A.35.23}$$

The derivative of the lifetime distribution function is the

$$f(t) = \frac{dF(t)}{dt} \tag{A.35.24}$$

probability density, therefore, the average lifetime:

$$\overline{T} = \int_0^\infty tf(t)dt = \int_0^\infty [1 - F(t)]\, dt = \int_0^\infty \overline{F}(t)dt \tag{A.35.25}$$

The $q(t)dt$ death rate is the probability that in the case of survival of t length of time, the death occurs at $(t{+}\Delta t)$. The probability that in the case of survival of t length of time the death occurs at $(t{+}\Delta t)$ has conditional probability:

$$\frac{\overline{F}(t + dt)}{\overline{F}(t)} \tag{A.35.26}$$

Therefore, the probability that in the case of survival of t length of time the death occurs at $(t{+}\Delta t)$ is

$$1 - \frac{\overline{F}(t + dt)}{\overline{F}(t)} = -\frac{d\frac{\overline{F}(t)}{dt}}{\overline{F}(t)}dt = \frac{\frac{d[1-F(t)]}{dt}}{\overline{F}(t)}dt = \frac{f(t)}{F(t)}dt \tag{A.35.27}$$

where we used the relationships (A.35.27) and (A.35.27). On the other hand (A.35.27) defines the death rate, therefore

$$q(t)\,dt = 1 - \frac{\overline{F}(1+dt)}{\overline{F}(t)} = -\frac{d\frac{\overline{F}(t)}{dt}}{\overline{F}(t)}dt = -\frac{\frac{d[1-F(t)]}{dt}}{\overline{F}(t)}dt = \frac{f(t)}{F(t)}dt \tag{A.35.28}$$

From this:

$$q(t) = -\frac{d\frac{\overline{F}(t)}{dt}}{\overline{F}(t)} = \frac{f(t)}{F(t)} \tag{A.35.29}$$

Now, let us take that the time function of death rate is self-similar, then, as we showed earlier it takes the form of

$$q(t) = at^{u_1-1} = (u_0)^{u_1}u_1t^{u_1-1} \tag{A.35.30}$$

From the relationship (A.35.30) we may determine the survival probability distribution function:

$$\int_0^t q(t')dt' = -\ln\overline{F}(t) \rightarrow \overline{F}(t) = e^{-\int_0^t q(t')dt'} \tag{A.35.31}$$

Substituting the self-similar death rate we get the well-known Weibull distribution [1353]:

$$\overline{F}(t) = e^{-\int_0^t (u_0)^{u_1}u_1t'^{u_1-1}dt'} = e^{-(u_0)^{u_1}t^{u_1}} = e^{-(u_0t)^{u_1}} \tag{A.35.32}$$

this describes the distribution function, where the death rate is self-similar. The Weibull distribution function has been used for a long time for survival description in gerontology [1354, 1355] and in oncology [1356] as well.

Let us reshape Eq. (A.35.30) defining the death rate with the help of (A.35.27) into the form:

$$q(t) = -\frac{d\frac{\overline{F}(t)}{dt}}{\overline{F}(t)} = -\frac{\frac{d[1-F(t)]}{dt}}{1-F(t)} \tag{A.35.33}$$

From this we get the differential equation:

$$\frac{dF(t)}{dt} = q(t)[1-F(t)] \tag{A.35.34}$$

Substituting the death rate in the form of (A.35.34) we get the Avrami differential equation [1357–1359]:

$$\frac{dF}{dt} = (u_0)^{u_1} u_1 t^{u_1-1} (1-F) = u_0 u_1 (u_0 t)^{u_1-1} (1-F) \tag{A.35.35}$$

The solution is the Avrami function:

$$F(t) = 1 - e^{-(u_0 t)^{u1}} \tag{A.35.36}$$

this in our case is equal to the lifetime distribution function. The universal applicability of the Avrami function was recognized much earlier [1360–1362].

A concept that is analogous with the time constant is the death rate in the form of

$$q(t) = at^{u_1-1} = \left(\frac{t}{\left(\frac{1}{a}\right)^{\frac{1}{u_1-1}}} \right)^{u_1-1} = \left(\frac{t}{\left(\frac{1}{(u_0)^{u_1} u_1}\right)^{\frac{1}{u_1-1}}} \right)^{u_1-1} \tag{A.35.37}$$

on the basis of which we may see that the natural scale of the function variation is

$$\Theta = \left(\frac{1}{(u_0)^{u_1} u_1} \right)^{\frac{1}{u_1-1}} \tag{A.35.38}$$

In summary, we proved that the similarity concept verified by the power-type functional relationship introduced to the self-similarity is more general than we have used before. The surprising generality shows the self-similar properties of the survival curve. This guess is confirmed by the fact that the processes can be transformed to a self-similar process by use of an appropriate mapping method, and the individual transforming functions, although their mathematical forms are different, differ hardly from each other. The distribution function – as shown above in (A.35.32), – is approximated by the Weibull function, [$W(t)$] which is used frequently for survival approximations [1363, 1211, 1215, 1216]. For a clear description the u_0 and u_1 parameters are denoted by ($1/t_0$) and n:

$$W(t) = e^{-(u_0 t)^{u_1}} = e^{-\left(\frac{t}{t_0}\right)^n} \tag{A.35.39}$$

The Weibull function has two parameters for one curve, t_0 is the scale parameter, and n is the shape parameter. We proved above that the death rate is self-similar also and can be described by the Weibull function. With this approach the non-parametric Kaplan–Meier plot could be described with appropriate accuracy by a parametric function (the Weibull function has two parameters for one curve, t_0 and n). The widely accepted hypothesis check evaluation of the double-arm survival curves

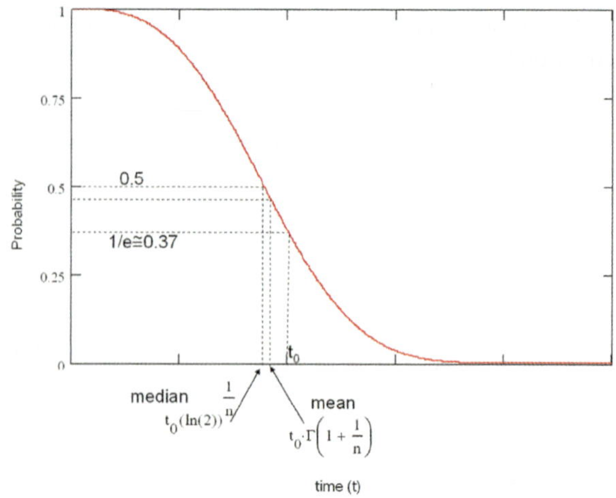

Fig. A.35.2 The special points of the Weibull function: the median, the mean, and the value ($1/e \approx 0.37$), where $t = t_0$.

[1364, 115, 1212, 1211] could be approximated by the single-arm fitting by self-similar assumptions. The survival curves (when it is a distribution of a single cohort) fit well to the Weibull function, and could be described by one single Weibull distribution (two parameters). The median and the mean are calculable from the parametric formula, (see Fig. A.35.2):

$$\begin{aligned} \text{median}\,[W(t)] &= t_0[\ln(2)]^n \\ \text{mean}\,[W(t)] &= t_0\Gamma\left(1+\tfrac{1}{n}\right) \end{aligned} \tag{A.35.40}$$

(This makes it possible to generate routinely the Weibull function for the Kaplan–Meier plot by knowing its median and mean.) The ratio of the median and mean depends on only the n form-factor, in a rigorously monotonic way in our interval of interest (Fig. A.35.3). Some functions with various parameters are shown in Fig. A.35.4. The function has its inflexion point (where the tendency of decreasing changes) in $t = t_0$ at the $1/e$ (≈ 0.37) value (see Fig. A.35.5). The derivative in this point is proportional to $-n$. (The derivative there is exactly $-n/e$ [$\approx -0.37n$].) Therefore, the parametric evaluation could be well checked at the $t = t_0$ point.

When some of the patients are cured, the function approaches the number of the cured patients at the end of the study. For this case, when the ratio of the cured patients is c, then the function (see Fig. A.35.6):

$$W^{(\text{cure})}(t) = c + (1-c)e^{-\left(\frac{t}{t_0^{(\text{cure})}}\right)^{n^{(\text{cure})}}} \tag{A.35.41}$$

In this case the median:

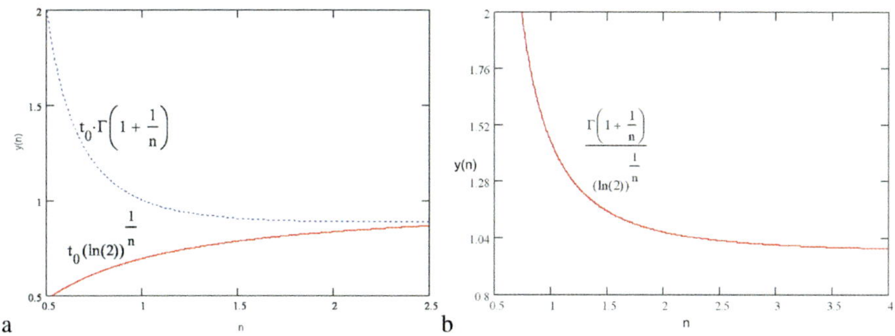

Fig. A.35.3 Dependence of mean and median (**a**) and their ratio (**b**) on the shape-factor n

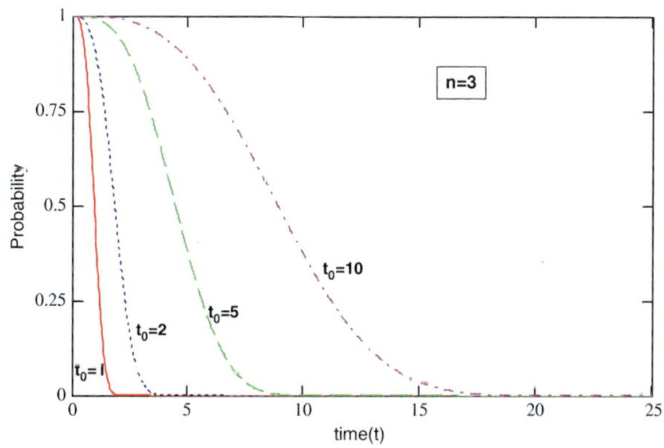

Fig. A.35.4 Weibull distribution with various parameters

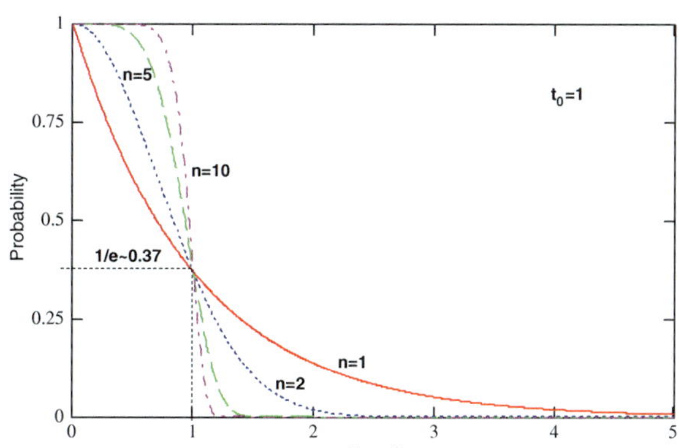

Fig. A.35.5 The inflexion point of the Weibull function at $t=1$

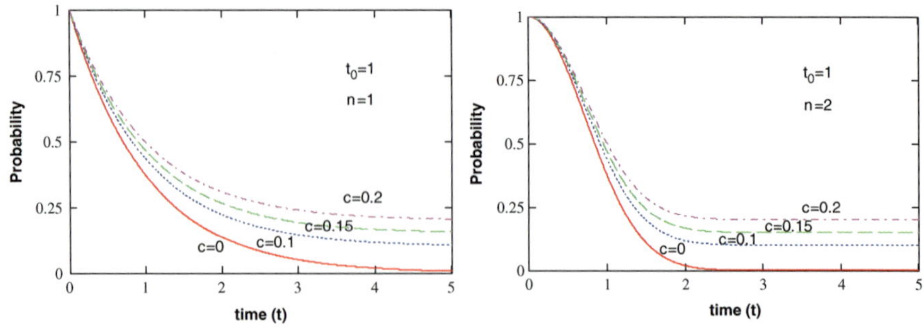

Fig. A.35.6 The Weibull function when a definite ratio of the patients is cured

$$\text{median}\left[W^{(\text{cure})}(t)\right] = t_0^{(\text{cure})}\left[\ln\left(1 + \frac{1}{1-2c}\right)\right]^{1/n^{(\text{cure})}} \tag{A.35.42}$$

However, in real cases, the survival curve with the cure rate could be poorly approximated by the Weibull function. The patients were sorted into two subcohorts: the responders and nonresponders. The responder's subgroup creates the c cured ratio at the end of the study time, while the nonresponders were probably lost before.

With this we have a couple of extra assumptions.

We assume the time of the first oncothermia is determined by the actual stage of the patient, so the point, when the previous treatments fail. This point is an inclusion criterion, and unifies the oncothermia cohort.

We assume the patients who died soon after becoming involved in the oncothermia process, are nonresponders. (A case which was worsened by oncothermia has not been observed in 20 years). This selection of nonresponders is supported by:

1. those patients were not reacting to oncothermia,
2. those patients had only a short time period in which to receive the appropriate oncothermia treatment dose,
3. those patients were dropped too early to allow a follow-up on their state.

The approximation at point 2. includes small error possibility by the patients who respond to the treatment, only the time of measuring their reaction was short identifying it.

This concept could be realized by fitting the measured Kaplan–Meier survival curve ($KM(t)$) with a function $S(t)$ composed of two Weibull functions [with parameters denoted by superscripts (r) and (nr)], describing the responders and nonresponders by a composite ratio C, respectively:

$$KM(t) \approx S(t) = (1-C)e^{-\left(\frac{t}{t_0^{(r)}}\right)^{n^{(r)}}} + Ce^{-\left(\frac{t}{t_0^{(nr)}}\right)^{n^{(nr)}}} \tag{A.35.43}$$

In this case the ratio of the cured patients is:

$$c = S(T) = (1 - C)e^{-\left(\frac{T}{t_0^{(r)}}\right)^{n^{(r)}}} + Ce^{-\left(\frac{T}{t_0^{(nr)}}\right)^{n^{(nr)}}} \qquad \text{(A.35.44)}$$

where T is the running-time of the study, (see Fig. A.35.7).

With these assumptions we study a split of the original cohort distribution, splitting it into two groups: responding and non-responding patients. The Weibull approach [1365] is divided into two different distributions [1366, 1367], composed linearly, one in which the treatment had no or minor influence and one where the treatment was effective. The weighted addition of the curves reconstructs the original. The "inclusion criteria" for patients to oncothermia treatment is when the "gold standards" are no longer eligible. These criteria could be checked by studying the elapsed time to the first oncothermia from the first diagnosis. The time from the first diagnosis to the first oncothermia has to be a cohort (when the inclusion of the patients to oncothermia had identical criteria) consequently it has to be characterized by a single-Weibull parametric formulation.

The process is performed for oncothermia survival first (five parameters are considered for the best fit: the two Weibull curves t_0 and n for each, and their composite ratio C (ratio of the nonresponders), which fixes the patients by their numbers into two groups (Eq. A.35.45). The residual will be automatically obtained from this fit, which is the value of the survival-fit at the maximal survival time $[S^{(OT)}(t_{\max})]$.

$$S^{(OT)}(t) = (1 - C)\exp\left[-\left(\frac{t}{t_0^{(r)}}\right)^{n^{(r)}}\right] + C\exp\left[-\left(\frac{t}{t_0^{(nr)}}\right)^{n^{(nr)}}\right] \qquad \text{(A.35.45)}$$

The same composite ratio is applied for overall survival also (Eq. A.35.46):

$$S^{(\Sigma)}(t) = (1 - C)\exp\left[-\left(\frac{t}{\tilde{t}_0^{(r)}}\right)^{\tilde{n}^{(r)}}\right] + C\exp\left[-\left(\frac{t}{\tilde{t}_0^{(nr)}}\right)^{\tilde{n}^{(nr)}}\right] \qquad \text{(A.35.46)}$$

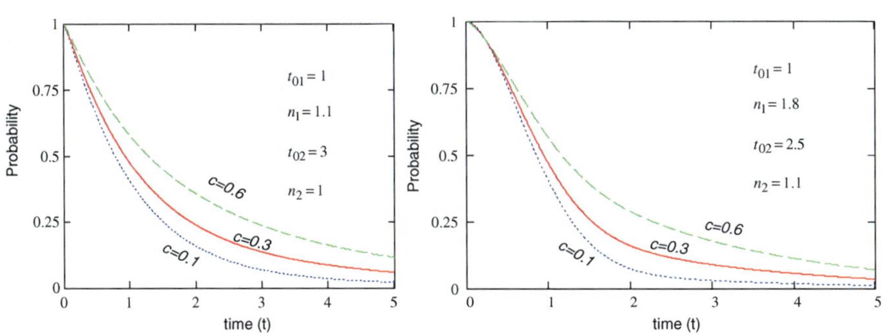

Fig. A.35.7 The fitting curves at various c-values

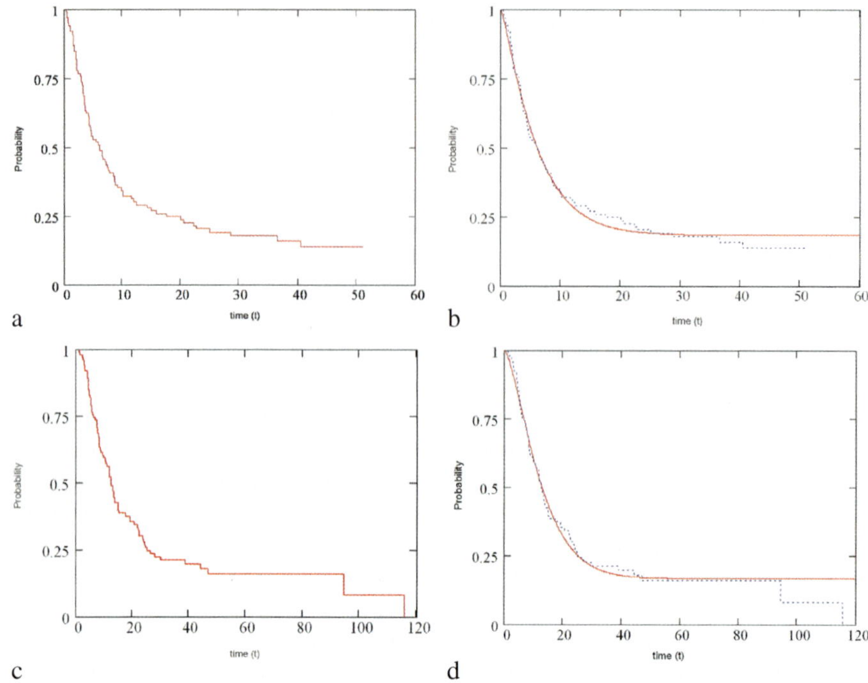

Fig. A.35.8 The original Kaplan–Meier curves for oncothermia (**a**) and overall (**c**) survivals and their fits according to (A.35.41). The parameters are $n = 1.14$, $t_0 = 6.34$, $c = 0.18$ and $n = 1.38$, $t_0 = 13.74$, $c = 0.17$ for oncothermia and overall survivals, respectively

Let us study an actual example of the pancreas trial ($n = 99$) [1368]. The original survival curves could be fitted by (A.35.41) (see Fig. A.35.8).

Better fits could be achieved by parametric decomposition of the survivals. The decomposition significantly divides the cohort of advanced, inoperable pancreas-cancer patients into two subgroups (responders and nonresponders) in oncothermia survival (see Fig. A.35.9). Keeping the C composite parameter, the fit and decomposition of the overall survival is available (see Fig. A.35.10).

The "inclusion criteria" for the patients to oncothermia treatment is when the "gold standards" are no longer eligible. These criteria could be checked by studying the elapsed time to the first oncothermia from the first diagnosis. The time from the first diagnosis to the first oncothermia has to be a cohort (when the inclusion of the patients to oncothermia had identical criteria) consequently it has to be characterized by a Weibull parametric formulation, where the two distributions are close, or $C^{(\text{start})}$ is small. Indeed, if you fit and decompose [by (A.35.47)] the curve of elapsed time from the first diagnosis to the start of oncothermia, the definite dominance of one single curve is shown (see Fig. A.35.11). This shows our "inclusion criteria" are really a valid cohort-forming condition.

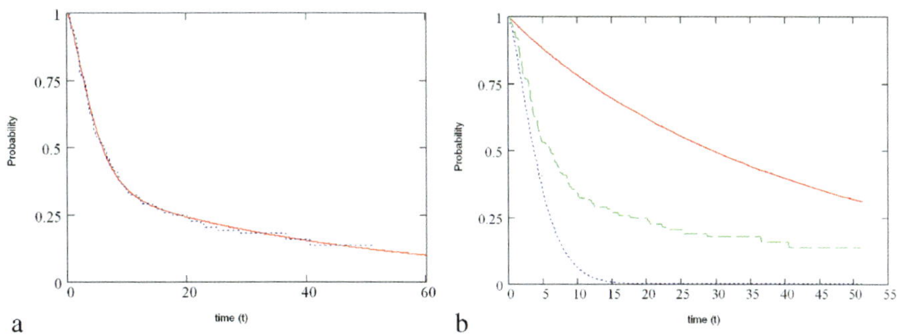

Fig. A.35.9 Full fit (**a**) of the curve by parameters using (A.35.45): $n^{(r)}$=0.95, $t_0{}^{(r)}$=43.13, $n^{(nr)}$=1.34, $t_0{}^{(nr)}$=4.75, C=0.61. The decomposition curves (**b**) show the significant difference of responders (39%) and nonresponders (61%)

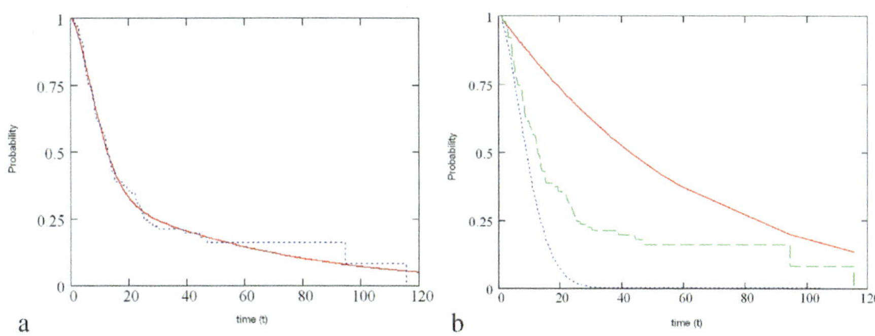

Fig. A.35.10 Full fit (**a**) of the curve by parameters using (A.35.46): $n^{(r)}$=1.06, $t_0{}^{(r)}$=59.94, $n^{(nr)}$=1.67, $t_0{}^{(nr)}$=11.00, C=0.61. The decomposition curves (**b**) show the significant difference of responders (39%) and nonresponders (61%) [Applying the same ratio as in oncothermia survival]

$$S^{(\text{start})}(t) = \left(1 - C^{(\text{start})}\right) \exp\left[-\left(\frac{t}{\hat{t}_0^{(1)}}\right)^{\hat{n}^{(1)}}\right] + C^{(\text{start})} \exp\left[-\left(\frac{t}{\hat{t}_0^{(2)}}\right)^{\hat{n}^{(2)}}\right] \tag{A.35.47}$$

Further control could be given by studying the historical control of the pancreas treatment from the same investigator ($n = 34$), who did the oncothermia treatments. The Weibull decomposition fit (see Fig. A.35.12) produces statistically identical curves, no possibility to detect any significant differences in decomposition, it is a cohort. Comparison of the nonresponders in overall survival and the control group shows remarkable correspondence (see Fig. A.35.13), this supports again the validity of the decomposition.

The same could be observed in another study (non-small-cell lung cancer [NSCLC]) [1369], where the distribution together with the historical control ($n = 311$) only slightly differs from the distribution of nonresponders without the control group ($n = 258$), see Fig. A.35.14.

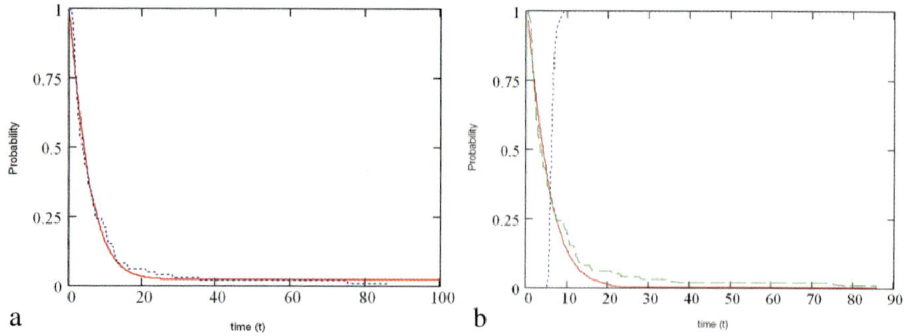

Fig. A.35.11 Full fit (**a**) of the curve by parameters using (A.35.47): $n^{(r)}=1.17$, $t_0^{(r)}=5.51$, $n^{(nr)}=-14.02$, $t_0^{(nr)}=6.00$, $C=0.021$. The decomposition curves (**b**) show the absolute dominance of a single Weibull distribution (its modification is 2.1%)

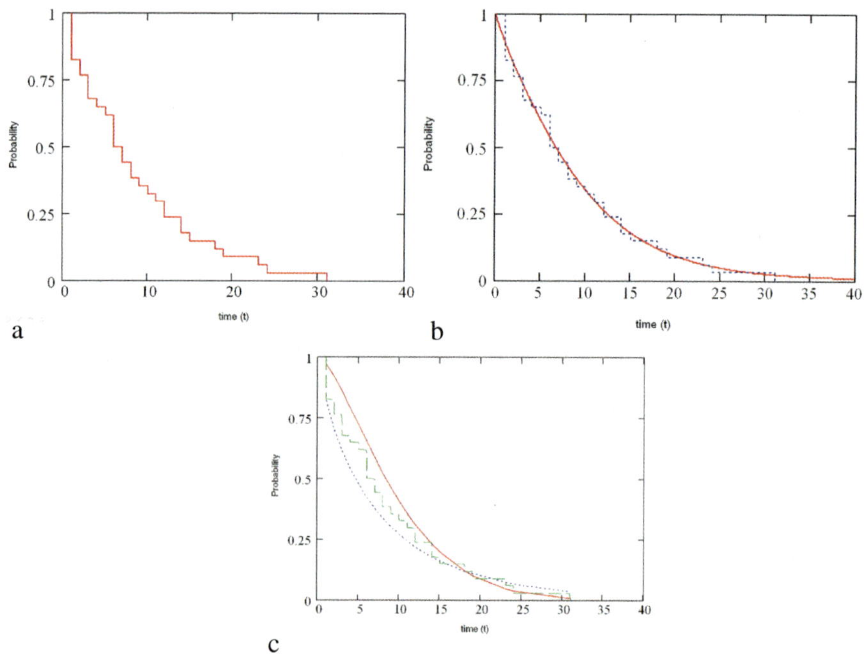

Fig. A.35.12 The Kaplan–Meier survival curve of the historical control (**a**) and its fit by two Weibull functions (**b**). Parameters: $n^{(1)}=1.47$, $t_0^{(1)}=10.76$, $n^{(2)}=0.83$, $t_0^{(2)}=7.19$, $C=0.50$. The decomposition curves are statistically identical (**c**)

Another prospective study [1370], had measured the local clinical response and the survival time in the same trial. The direct response (CR+PR) shows good, significant correspondence with the parametric separation (see Fig. A.35.15).

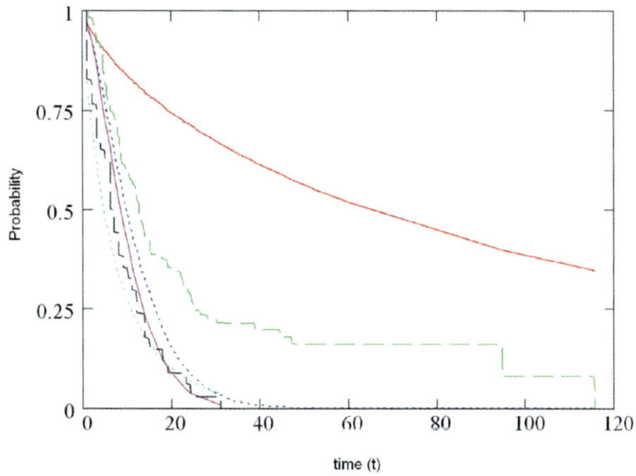

Fig. A.35.13 Comparison of Fig. A.35.10b. and Fig. A.35.12c shows remarkable correlation of the historical control with the nonresponders in overall survival

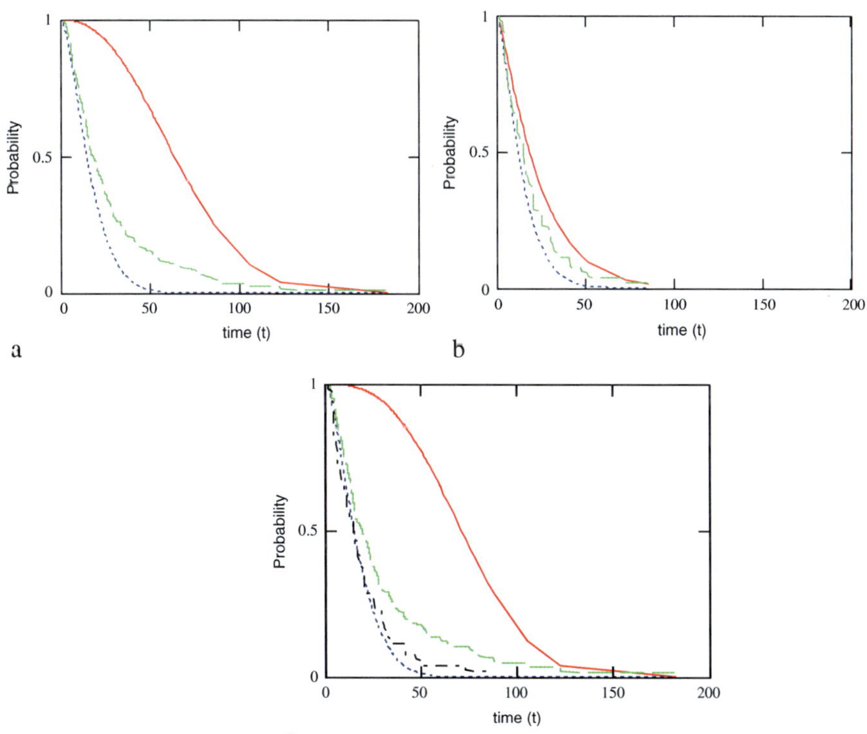

Fig. A.35.14 The Kaplan–Meier plot of a single-arm study (n=258) of non-small-cell lung cancer and its decomposition, [C=0.79.4, (**a**)] and decomposition of the historical control (n=53), from the same investigator [C=0.55, (**b**)] and their comparison (**c**)

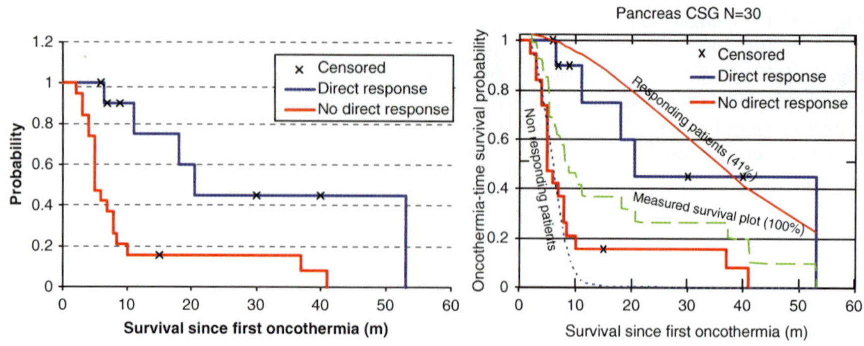

Fig. A.35.15 Significant correspondence of the measured and calculated separation of the patient's survivals by their local response

Acknowledgment

The authors are grateful to Dr. Z. Szaszne-Csih for her outstanding support and help in all the parts of this complex work. For help in arranging the typed materials and pictures is highly appreciated to Ms. Erdelyi A.

Special thanks go to Dr. Vincze Gy. for help in formulation of numerous theoretical challenges.

Authors express their gratitude to the smart and talented physicians and researchers who helped in the collection of these materials: Prof. Dr. Aydin H., Prof. Dr. Bogdahn U., Prof. Dr. Ferrari; Prof. Dr. Fiorentini; Prof. Dr. Hau P., Prof. Dr. Herzog A., Prof. Dr. Galfi P., Prof. Dr. Groenemeyer DHW., Prof. Dr. Kampinga HH., Prof. Dr. Kim SJ., Prof. Dr. Kirchner H., Prof. Dr. Lang I., Prof. Dr. Lee DY., Prof. Dr. Mako E., Prof. Dr. Renner H., Prof. Dr. Sommer H., Prof. Dr. Wehner H., Prof. Dr. Yoon SH., Dr. Andocs G., Dr. Baier J., Dr. Balogh L., Dr. Brenner Y., Dr. Brockmann W-P., Dr. Brunner G., Dr. Buettner C., Dr. Fonyad L., Dr. Dani A., Dr. Dank M., Dr. Douwes F., Dr. Csejtei A., Dr. Hager D., Dr. Holzhauer P., Dr. Jakab Cs., Dr. Juestock J., Dr. Kalden M., Dr. Magyar T., Dr. Migeod F., Dr. Patonay L., Dr. Piko B., Dr. Rubovszky G., Dr. Sahinbas H., Dr. Saupe H., Dr. Szucs M., Dr. Varkonyi A., Dr. Wismeth C., Mr. Gnadig B., Mr. Lorencz P., and Mr. Rajeczky Z.

The authors are grateful to the following Institutes:

1st Department of Pathology and Experimental Cancer Research, Semmelweis University, Budapest, Hungary

BioMed Clinic, Bad Bergzabern, Germany

Cell Stress Laboratory, Gronningen University, The Netherlands

Clinical "New Hope", Tel Aviv, Israel

Department of Biotechnics, Faculty of Engineering, St. István University, Budapest, Hungary

Department of Oncology, Pandi K. Hospital, Gyula, Hungary

Department of Pathology, Faculty of Veterinary Science, St. István University, Budapest, Hungary

Department of Pharmacology and Toxicology, Faculty of Veterinary Science, Budapest, Hungary

Division of Hematology-Oncology, Department of Internal Medicine, Samsung Changwon Hospital, Sungkyunkwan University, Seoul, Korea
Fachklinik Dr. Herzog, Nidda, Germany
Fachklinik Hornheide, Muenster, Germany
Gynecologic Cancer Center, Bundang CHA Hospital, CHA University, Seoul, Korea
Gynecology Clinic, Ludwig Maximillian University, Munich, Germany
HTT-Med Polyclinics, Budapest, Hungary
Institute of Microtherapy, University Witten Herdecke, Bochum, Germany
KangNam Severance Hospital, Yonsei University College of Medicine, Seoul, Korea
National Institute of Oncology, Budapest, Hungary
National Research Institute for Radiobiology and Radiohygiene, Budapest, Hungary
Neurology Clinic, Regensburg University, Germany
Department Oncology, Peterfy Hospital, Budapest, Hungary
Oncotherm GmbH, Troisdorf, Germany
Oncotherm Innovation and Trade Ltd., Paty, Hungary
Ospitale Civili, Brescia, Italy
Praxis Clinic of Radiooncology, Klinikum Nuernberg, Germany
Radio Oncology Centrum, Kecskemet, Hungary
St. Georg Klinik, Bad Aibling, Germany
St. Giuseppe Hospital, Empoli, Italy
St. Istvan University, Budapest, Hungary
Veramed Clinic, Brannenburg, Germany
Veramed Clinic, Meschede, Germany

This work was partly supported by Hungarian Economic Development Center Ltd., Budapest, Hungary.

References

1. Streffer C, Van Beuningen D, Dietzerl F et al (1978) Cancer therapy by hyperthermia and radiation. Urban and Schwarzenberg, Baltimore, MD
2. Hornback NB (1984) Hyperthermia and cancer: human clinical trial experience. CRC Press, Boca Raton, FL
3. Gautherie M, Albert E (eds) (1982) Biomedical thermology. Alan R. Liss, New York, NY
4. Anghileri LJ, Robert J (1986) Hyperthermia in cancer treatment. Vol. 1–3. CRC Press, Boca Raton, FL
5. Field SB, Franconi C (eds) (1987) Physics and technology of hyperthermia. NATO ASI series, Martinus Nijhoff Publ. Dordrecht, Boston, MA
6. Urano M, Douple E (eds) (1988) Hyperthermia and oncology, Vol. 1. Thermal effects on cells and tissues. VSP BV, Utrecht, The Netherlands
7. Urano M, Douple E (eds) (1989) Hyperthermia and oncology, Vol. 2. Biology of thermal potentiation of radiotherapy. VSP BV Utrecht, The Netherlands
8. Gautherie M (ed) (1990) Methods of hyperthermia control. Springer, Berlin
9. Gautherie M (ed) (1990) Biological Basis of oncological thermotherapy. Springer, Berlin
10. Gautherie M (ed) (1990) Interstitial endocavitary and perfusional hyperthermia. Springer, Berlin
11. Urano M, Douple E (eds) (1992) Hyperthermia and oncology, Vol. 3. Interstitial hyperthermia: physics, biology and clinical aspects. VSP BV, Utrecht, The Netherlands
12. Seegenschmiedt MH, Sauer R (1993) Interstitial and intracavitary thermoradiotherapy. Springer, Berlin
13. Matsuda T (ed) (1993) Cancer treatment by hyperthermia, radiation and drugs. Taylor & Francis, Washington, DC
14. Urano M, Douple E (eds) (1994) Hyperthermia and oncology, vol. 4. Chemopotentiation by hyperthermia. VSP BV, Utrecht, The Netherlands
15. Seegenschmiedt MH, Fessenden P, Vernon CC (eds) (1996) Thermoradiotherapy and Thermochemotherapy, Vol. 1. Biology, physiology and physics. Springer, Berlin
16. Seegenschmiedt MH, Fessenden P, Vernon CC (eds) (1996) Thermo-radiotherapy and thermo-chemotherapy, Vol. 2. Clinical applications. Springer, Berlin
17. Kosaka M, Sugahara T, Schmidt KL et al (eds) (2001) Thermotherapy for neoplasia, inflammation, and pain. Springer, Tokyo
18. Ellis LM, Curley SA, Tanabe KK (2004) Radiofrequency ablation of cancer. Springer, Berlin
19. Baronzio GF, Hager ED (eds) (2006) Hyperthermia in cancer treatment: a primer. Springer, Berlin; Landes Bioscience, Austin, TX
20. Kufe D et al (2003) Cancer medicine. BC Decker, Hamilton, ON
21. Perez CA, Brady LW, Halperin EC et al (2004) Principles and practice of radiation oncology. 4th ed. Lippincott Williams and Wilkins, Philadelphia, PA

22. DeVita VT, Hellman S Jr, Rosenberg SA (2004) Cancer: principles and practice of oncology. 7th edition. Lippincott Williams and Wilkins, Philadelphia, PA
23. Diamandopoulos GT (1996) Cancer: an historical perspective. Anticancer Res 16(4A):1595–1602
24. Gallucci BB (1985) Selected concepts of cancer as a disease: from the Greeks to 1900. Oncol Nurs Forum 12(4):67–71
25. Harvey AM (1974) Early contributions to the surgery of cancer: William S. Halsted, Hugh H. Young and John G. Clark. Johns Hopkins Med J 135(6):399–417
26. Kardinal C, Yarbro J (1979) A conceptual history of cancer. Semin Oncol 6(4):396–408
27. Contran R, Kumar V, Collins T et al (1989) Robbins pathologic basis of disease, 4th ed. Philadelphia, WB Saunders
28. Weinberg RA (1999) One renegade cell. How cancer begins. Basic Books, Parseus Book Group, New York, NY
29. Cairns J (1975) The Cancer Problem. Scientific American 233(5):64–72, 77–78
30. Reya T, Morrison SJ, Clarke MF et al (2001) Stem cells, cancer, and cancer stem cells. Nature 414(6859):105–111
31. Singh SK, Clarke ID, Terasaki M et al (2003) Identification of a cancer stem cell in human brain tumors. Cancer Res 63:5821–5828
32. Balkwill F, Charles K, Mantovani A (2005) Smoldering and polarized inflammation in the initiation and promotion of malignant disease. Cancer Cell 7(3):211–217
33. De Visser KE, Eichten A, Coussens LM (2006) Paradoxical roles of the immune system during cancer development. Nat Rev Cancer 6(1):24–37
34. Inagaki J, Rodroguez V, Bodey GP (1974) Causes of death in cancer patients. Cancer Philadelphia 33(2):568–573
35. Welch HG, Schwartz LM, Woloshin S (2000) Are increasing 5-year survival rates evidence of success against cancer? JAMA 283(22):2975–2978
36. Morgan G, Ward R, Barton M (2004) The contribution of cytotoxic chemotherapy to 5-year survival in adult malignancies. Clin Oncol (R Coll Radiol) 16(8):549–560
37. Bailair JC, Smith EM (1986) Progress against cancer? N Engl J Med 314(19):1226–1232
38. Holleb AI (1986) Progress against cancer? A broader view. CA Cancer J Clin 36:243–244
39. Schoenbach VJ (1999) Descriptive studies and surveillance-solutions. www.epidemiolog.net/evolving/DescriptiveStudiesSolns.pdf. Cited 12 September 2007
40. Bailar JC, Gornik HL (1997) Cancer undefeated. N Engl J Med 336(22):1569–1574
41. Pauling L (1989) Quoted by Moss RW: the cancer syndrome. Equinox Press, New York, NY
42. The failed War on Cancer. People Against Cancer. http://www.peopleagainstcancer.com/pdfs/news/20080916n2.pdf. Editorial 12 July 2008
43. Arnold K (2003) Statistics offer insights into progress against cancer. Journal of the National Cancer Institute 95(17):1266–1267
44. Honore BE, Lleras-Muney A (2006) Bounds in completing risks models and the war on cancer. Econometrica 74(6):1675–1698
45. National Cancer Institute (2003) SEER Cancer Statistics Review. http://seer.cancer.gov/csr/1975_2003/results_merged/topic_year_lost.pdf. 1975–2003
46. National Cancer Institute (2003) Plans and Priorities for Cancer Research. http://plan2003.cancer.gov/inbrief/execsum.htm
47. Surveillance, Epidemiology, and End Results (SEER) (2000) National Cancer Institute. www.seer.cancer.gov
48. Colotta F, Mantovani A (eds) (2007) Targeted therapies in cancer: myth or reality? American Institute of Physics (AIP)
49. Frost & Sullivan Market Intelligence (2003) 1996, 2002–2008. Reuters Business Insight, Health Care

50. Goldberg R, Pitts P (2006) A working paper of the 21st century FDA task force. Prescription for progress: the critical path to drug development. http://www.manhattan-institute.org/html/fda_task_1.htm. Cited 11 Sept 2007
51. National Center for Health Statistics (2007) Hyattsville. http://www.cdc.gov/nchs/fastats/lcod.htm. Cited 12 Sept 2007
52. BBC News. Thursday, 9 November 2000, 18:08 GMT
53. EUROCARE-3 (1983–1994) European Cancer Database. www.eurocare.org/profiles/index.html
54. Research America Health Poll, Charlton Research Company. www.researchamerica.org/polldata/2006/longevity_fullresults. Cited 2006
55. Private Healthcare UK. http://www.privatehealth.co.uk/hospitaltreatment/treatment-news/april-2007/britons-fear-cancer. Cited 12 September 2007
56. Fear from AIDS rivals worry over cancer (1987) The New York Times. www.query.nytimes.com/gst/fullpage.html?sec=health&res9BODE1DD1E3DF93AA15752C1A961948260. Cited 12 Sept 2007
57. Harima Y, Nagata K, Harima K et al (2001) A randomized clinical trial of radiation therapy versus thermoradiotherapy in stage IIIB cervical carcinoma. Int J Hyperthermia 17(2):97–105
58. Feig BW, Berger DH, Fuhrman GM (2006) The M.D. anderson surgical oncology handbook. Lippincott Williams & Wilkins, New York, NY
59. Warburg O (1956) On the origin of cancer cells. Science 123(3191):309–314
60. Anderson MW, Reynolds SH, Maronpot RM (1992) Role of proto-oncogene activation in carcinogenesis. Environ Health Perspect 98:13–24
61. Meden H, Marx D, Fattahi A et al (1994) Elevated serum levels of a c-erbB-2 oncogene product in ovarian cancer patients and in pregnancy. J Cancer Res Clin Oncol 120(6):378–381
62. Attie-Bitach T, Abitbol M, Gérard M (1998) Expression of the RET proto-oncogene in human embryos. Am J Med Genet 80(5):481–486
63. Quenby SM, Gazvani MR, Brazeau C et al (1999) Oncogenes and tumor suppressor genes in first trimester human fetal gonadal development. Mol Hum Reprod 5(8):737–741
64. Okada Y et al (1996) Expression of fos family and jun family proto-oncogenes during corneal epithelial wound healing. Curr Eye Res 15(8):824–832
65. Stiles CD (1985) The biological role of oncogenes-insights from platelet-derived growth factor: Rhoads Memorial Award lecture. Cancer Res 45:5215–5218
66. Fiala ES, Sohn OS, Wang CX et al (2005) Induction of preneoplastic lung lesions in guinea pigs by cigarette smoke inhalation and their exacerbation by high dietary levels of vitamins C and E. Carcinogenesis 26(3):605–612
67. Platz EA, De Marzo AM (2004) Epidemiology of inflammation and prostate cancer. J Urol 171(2):S36–S40
68. Murthy NS, Mathew A (2000) Risk factors for pre-cancerous lesions of the cervix. Eur J Cancer Prev 9(1):5–14
69. Molloy RM, Sonnenberg A (1997) Relation between gastric cancer and previous peptic ulcer disease. Gut 40(2):247–252
70. Friess H, Lu Z, Graber HU et al (1998) Bax, but not bcl-2, influences the prognosis of human pancreatic cancer. Gut 43(3):414–421
71. Warburg O (1996) Oxygen, the creator of differentiation, biochemical energetics. In: Warburg O (1996) The prime cause and prevention of cancer. Revised lecture at the meeting of the Nobel-Laureates on June 30 1966, Lindau, Lake Constance, Germany. Academic, New York, NY
72. Schulz TJ, Thierbach R, Voigt A et al (2006) Induction of oxidative metabolism by mitochondrial frataxin inhibits cancer growth. J Biol Chem 281(2):977–981
73. Miles KA, Williams RE (2008) Warburg revisited: imaging tumor blood flow and metabolism. Cancer Imaging 8:81–86

74. Heiden MGV, Cantley LC, Thompson CB (2009) Understanding the Warburg Effect: The Metabolic Requirements of Cell Proliferation. Science 324(5930): 1029–1033
75. Garber K (2004) Energy boost: The Warburg effect returns in a new theory of cancer. J Nat Canc Inst 96(24):1805–1806
76. Seyfried TN, Mukherjee P (2005) Targeting energy metabolism in brain cancer: review and hypothesis. Nutr Metab 2:30–38
77. Xiaolong M, Riordan NH (2006) Cancer is a functional repair tissue. Med Hypotheses 66(3):486–490
78. Tan BT, Park CY, Ailles Le et al (2006) The cancer stem cell hypothesis: a work in progress. Lab Inves 86(12):1203–1207
79. Paguirigan A, Beebe DJ, Liu B et al (2006) Mammary stem and progenitor cells: Tumor precursors? Eur J Cancer 42(9):1225–1236
80. Houghton JM et al (2007) Stem cells and cancer. Sem Cancer Biol 17(3):191–203
81. Weinberg RA (1998) One Renegade Cell: How Cancer Begins. Basic Books, Member of the Perseus Books Group, New York, NY
82. Sell S (2004) Stem cell origin of cancer and differentiation therapy. Crit Rev Oncol Hematol 51(1):1–28
83. Krause M, Prager J, Zhou X et al (2007) EGFR-TK inhibition before radiotherapy reduces tumor volume but does not improve local control: differential response of cancer stem cells and nontumorigenic cells? Radiother Oncol 83(3):316–325
84. Jordan CT (2004) Cancer stem cell biology: from leukemia to solid tumors. Curr Opin Cell Biol 16(6):708–712
85. Tang C, Ang BT, Pervaiz S (2007) Cancer stem cell: target for anti-cancer therapy. FASEB J 21:3777–3785
86. Marx J (2007) Molecular Biology: Cancer's Perpetual Source? Science 317(5841): 1029–1031
87. Neuzil J, Stantic M, Zobalova R et al (2007) Tumor-initiating cells vs. cancer 'stem' cells and CD133: What's in the name? Biochem Biophys Res Commun 355(4):855–859
88. Kartner N, Ling V (1989) Multidrug resistance in Cancer. Scientific Am 260(3):44–51
89. West BJ (2006) Where medicine went wrong: Rediscovering the Path to Complexity. World Scientific, London
90. Skrabanek P (1994) The death of human medicine. The Social Affairs Unit, Crowley Esmonde Ltd, Suffolk
91. Fox R (1994) Foreword in book: Skrabanek P. The Death of Humane Medicine. Social Affairs Unit, Suffolk
92. Horgan J (1997) The end of Science. Broadway books, New York, NY
93. Milton R (1994) Forbidden Science. Fourth Estate, London
94. Greider K (2003) The big fix. Public Affairs, New York, NY
95. Kassirer JP (2005) On the take. Oxford University Press, New York, NY
96. Moss RW (1999) The cancer industry. Equinox Press, New York, NY
97. Moss RW (1995) Questioning chemotherapy. Equinox Press, New York, NY
98. Baron J (1996) Do no harm. In: Messick DM, Tenbrunsel AE (eds) Codes of conduct: behavioural research into business ethics, Russell Sage Foundation, New York, NY, pp.197–213
99. US Food and Drug Administration (FDA). Critical Path Initiative. http://www.fda.gov/ScienceResearch/SpecialTopics/CriticalPathInitiative/default.htm. Cited 02 Sept 2010
100. Green S, Benedetti J, Crowley J (2003) Clinical trials in oncology. Chapman & Hall/CRC, Boca Raton, FL
101. McGovern DPB, Valori RM, Summerskill WSM et al (2001) Evidence-based medicine. Bios Scientific, Oxford
102. Sackett DL, Rosenberg WM, Gary JA et al (1996) Evidence based medicine: what it is and what it isn't. BMJ 312(7023):71–72

103. Centre for Evidence-Based Medicine in Oxford (2009) http://www.cebm.net/index.aspx?o=1025
104. Glantz SA (1997) Primer of bio-statistics. McGraw Hill, Health Professions Division, New York, NY
105. The Office of Cancer Complementary and Alternative Medicine (OCCAM). National Cancer Institute (NCI). http://www.cancer.gov/CAM. Cited 02 Sept 2010
106. Eisenberg DM, Kessler RC, Foster C et al (1993) Unconventional Medicine in the United States – Prevalence, Costs, and Patterns of Use. N Engl J Med 328(4):246–252
107. American Public (2005) Complementary and Alternative Medicine in the United States. National Academic Press. http://www.nap.edu/catalog.php?record_id=11182
108. Edwards SJ, Braunholtz DA, Lilford RJ et al (1998) Ethical issues in the design and conduct of randomized controlled trials. Health Tech Assess NHS R&D HTA Program 15(2):1–146
109. Kolata GB (1977) Clinical trials: Methods and Ethics are debated. Science 198(4322):1127–1131
110. Gifford F (1986) The conflict between randomized clinical trials and the therapeutic obligation. J Med Philos 11(4):347–366
111. Benson K, Hartz AJ (2000) A comparison of observational studies and randomized, controlled trials. NEJM 342(25):1878–1886
112. Concato J, Shah N, Horwitz RI (2000) Randomized, controlled trial, observational studies, and the hierarchy of research design. NEJM 342(25):1887–1892
113. Kennedy D (2004) Clinical trials and public trust. Science 306(5702):1649
114. DiTrocchio F (1994) Der grosse schwindel – betrug und falschung in der wissenschaft. Campus Verlag, Frankfurt
115. Gardner MJ, Altman DG (1989) Statistics with Confidence. British Medical Journals Book, British Medical Journal, London
116. Jaynes ET (2003) Probability theory: the logic of science. Cambridge University Press, Cambridge
117. Gauler TC, Weihrauch TR (1997) Placebo-ein wirksames und ungefährliches Medikament? Urban & Schwarzenberg, Baltimore, MD
118. Kaptchuk TJ (1998) Intentional ignorance: A history of blind assessment and placebo controls in medicine. Bull Hist Med 72(3):389–433
119. Stron FC (1999) The history of the double-blind test. J Pharm Pharmacol 51:237–239
120. Pappercorn JM, Weeks JC, Cook EF et al (2004) Comparison of outcomes in cancer patients treated within and outside clinical trials: conceptual framework and structured review. The Lancet 363:263–270
121. Giacchetti S, Perpoint B, Zidani R et al (2000) Phase III multicenter randomized trial of oxaliplatin added to chronomodulated fluorouracil-leucovorin as first-line treatment of metastatic colorectal cancer. J Clin Oncol 18(1):136–147
122. Possinger K, Kaufmann M, Coleman R (1999) Phase II study of gemcitabine as first-line chemotherapy in patients with advanced or metastatic breast cancer. Anticancer Drugs 10(2):155–162
123. Carmichael J, Walling J (1996) Phase II activity of gemcitabine in advanced breast cancer. Semin Oncol 23:77–81
124. Krcyzanowska MK, Pintilie M, Tannock IF (2003) Factors associated with failure to publish large randomized trials presented at an oncology meeting. JAMA 290(4):495–501
125. Lexchin J, Bero LA, Djulbegovic B et al (2003) Pharmaceutical industry sponsorship and research outcome and quality: systematic review. BMJ 326(7400):1167–1170
126. Melander H, Ahlqvist-Rastad J, Meijer G et al (2003) Evidence b(i)ased medicine—selective reporting from studies sponsored by pharmaceutical industry: review of studies in new drug applications. BMJ 326:1171–1173
127. Stelfox HT, Chua G, O'Rourke K et al (1998) Conflict of interests in the debate over calcium-channel antagonists. NEJM 338(2):101–106

128. Choudhry NK, Stelfox HT, Detsky AS (2002) Relationships between authors of clinical practice guidelines and the pharmaceutical industry. JAMA 287(5):612–617
129. Baird P, Downie J, Thompson J (2002) Clinical trials and industry. Science 297:2211
130. Brecht B (1935) Leben des Galilei. In: Losey J (1975) Galileo, Grove Press, New York, NY
131. Djulbegovic B, Guyatt GH, Aschroft RE (2009) Epistemologic inquiries in evidence-based medicine. Cancer Cont 16(2):158–168
132. Vlachos PK, Gelfand AE (1996) Issues in the Bayesian Design of clinical trials with multiple categorical endpoints. Proceedings of the ASA Section on Bayesian Statistical Science. http://citeseer.ist.psu.edu/60418.html. Cited 0 October 2007
133. Qian J (1994) A Bayesian Weibull Survival Model. PhD Thesis, Duke University, http://citeseer.ist.psu.edu/cache/papers/cs/922/ftp:zSzzSzftp.isds.duke.eduzSzpubzSzThe seszSzjiang.pdf/qian94bayesian.pdf. Cited 02 October 2007
134. Stangl D, Berry D (1998) Bayesian Statistics in Medicine: Where Are We and Where Should We Be Going? Invited Paper, Sankhya, Series B 60:176–195. http://citeseer.ist.psu.edu/401035.html. Cited 02 October 2007
135. Thall P, Simon RM, Estey EH (1996) New statistical strategy for monitoring safety and efficacy in single-arm clinical trials. J. Clin. Oncology 14:296–303
136. Evans CH, Ildstad ST (eds) (2001) Small clinical Trials: Issues and challenges. Institute of Medicine, National Academy Press, Washington, DC
137. Medical dictionary (2008) http://www.medterms.com/script/main/art.asp?articlekey=3848
138. The free dictionary (2008) http://ww.thefreedictionary.com/hyperthermia
139. National Cancer Institute (2008) http://www.cancer.gov/Templates/db_alpha.aspx?CdrID=468790
140. Oncology Encyclopedia (2008) http://www.answers.com/topic/hyperthermia
141. Free Medical Dictionary (2008) http://www.free-medical-dictionary.com
142. Coley WB (1893) The treatment of malignant tumors by repeated inoculations of erysipelas with a report of ten original cases. Am J Med Sci 105:487–511
143. Medline Plus. http://www.nlm.nih.gov/medlineplus/ency/article/001315.htm. Cited 02 October 2007
144. Dewhirst MW, Prosnitz L, Thrall D et al (1997) Hyperthermic Treatment of Malignant Diseases: Current Status and a View Toward the Future. Sem Oncol 24(6): 616–625
145. Urano M, Douple E (eds) Hyperthermia and Oncology: Vol. 1. (1988) Thermal effects on cells and tissues. VSP BV Utrecht, The Netherlands. Vol. 2. (1989) Biology of thermal potentiation of radiotherapy, VSP BV Utrecht The Netherlands Vol. 3. (1992) Interstitial Hyperthermia: Physics, biology and clinical aspects, VSP BV Utrecht The Netherlands. (1994) Vol. 4. Chemopotentiation by hyperthermia VSP BV Utrecht The Netherlands
146. Seegenschmiedt MH, Fessenden P, Vernon CC (eds) (1996) Thermo-radiotherapy and Thermo-chemotherapy, Vol. 1. Biology, physiology and physics. Vol. 2. Clinical application. Springer, Berlin
147. International Journal of Hyperthermia. (The official Journal of the North American Hyperthermia Society, European Society for Hyperthermic Oncology, Asian Society of Hyperthermic Oncology) Taylor & Francis, ISSN 0265–6736
148. van der Zee J, Gonzalez Gonzalez D, van Rhoon GC et al (2000) Comparison of radiotherapy alone with radiotherapy plus hyperthermia in locally advanced pelvic tumors: a prospective, randomised, multicentre trial. Dutch Deep Hyperthermia Group. Lancet 355(9210):1119–1125
149. Wust P, Hildebrandt B, Sreenivasa G et al (2002) Hyperthermia in combined treatment of cancer. Lancet Oncol 3(8):487–497
150. Overgaard J, Gonzalez Gonzalez D, Hulshof MC et al (1995) Randomised trial of hyperthermia as adjuvant to radiotherapy for recurrent or metastatic malignant melanoma. European Society for Hyperthermic Oncology. Lancet 345(8949):540–543

151. PubMed (2007) http://www.ncbi.nlm.nih.gov/entrez/query.fcgi?DB=pubmed. search profile: (cancer OR tumor OR oncology OR neoplasm OR malignant) AND (hyperthermia OR heat-therapy OR thermotherapy) NOT (malignant-hyperthermia OR fever); limits: (clinical trial OR randomized controlled trial) Sept 2007
152. Kufe DW, Bast RC, Hait W et al (2006) (eds) Cancer Medicine Holland-Frei – Cancer Medicine 7. American Association for Cancer Research, BC Decker, Philadelphia, PA
153. Ara G, Anderson RR, Mandel KG (1989) Irradiation of pigmented melanoma cells with high-intensity pulsed radiation generates acoustic waves and kills the cells. Laser Surg Med 10:52–59
154. Findlay RP, Dimbylow PJ (2005) Effects of posture on FDTD calculations of specific absorption rate in a voxel model of the human body. Phys Med Biol 50:3825–3835
155. Joo E, Szasz A, Szendro P (2005) Metal-framed spectacles and implants and specific absorption rate among adults and children using mobile phones at 900/1800/2100 MHz. Electromagn Biol Med 25(2):103–112
156. Jianging W, Mukaide N, Fujiwara O (2003) FTDT calculation of organ resonance characteristics in an anatomically based human model for plane-wave exposure. Proc Conf Environ Electromagn 4–7 November 2003, pp 126–129
157. Radiofrequency radiation dosimetry handbook (1997) Armstrong Laboratory, USAF School of Aerospace Medicine, AFSC. http://niremf.ifac.cnr.it/docs/HANDBOOK/chp1.htm. Cited 02 October 2007
158. Nielsen OS, Horsman M, Overgard J (2001) A future for hyperthermia in cancer treatment? Eur J Cancer 37(13):1587–1589
159. van der Zee J (2002) Heating the patient: a promising approach? Annals of Oncology 13:1173–1184
160. Devrient W (1950) Überwärmungsbäder. A. Marcus & E. Weber's Verlag, Berlin
161. Hoff F (1957) Fieber, Unspezifische Abwehrvorgänge, Unspezifische Therapie. Georg Thieme, Stuttgart
162. Lampert H (1948) Überwärmung als Heilmittel. Hippokrates, Stuttgart
163. Schmidt KL (1987) Hyperthermie und Fieber. Hippokrates, Stuttgart
164. Heckel M (1990) Ganzkörperhyperthermie und Fiebertherapie – Grundlagen und Praxis. Hippokrates, Stuttgart
165. Heckel M (1992) Fiebertherapie und Ganzkörper-HT, Bessere Verträglichkeit und Effizienz durch thermoegulatorisch ausgewogene, kombinierte Anwendung beider Verfahren. ThermoMed 14–19
166. Hildebrandt B, Drager J, Kerner T et al (2004) Whole-body hyperthermia in the scope of von Ardenne's systemic cancer multistep therapy (sCMT) combined with chemotherapy in patients with metastatic colorectal cancer: a phase I/II study. Int J Hyperthermia 20: 317–333
167. Ardenne A. Von, Wehner H: (2005) Extreme Whole-Body Hyperthermia with Water-Filtered Infrared-A Radiation. Eurekah Bioscience Collection, Oncology, Landes Bioscience, Austin, TX
168. Wust P, Riess H, Hildebrandt B (2000) Feasibility and analysis of thermal parameters for the whole-body hyperthermia system IRATHERM. Int J Hyperthermia 4:325–339
169. Dahl O, Dalene R, Schem BC (1999) Status of clinical hyperthermia. Acta Oncol 38(7):863–873
170. Senior K (2001) Hyperthermia and hypoxia for cancer-cell destruction. Lancet Oncology 2:524–525
171. Szasz A, Szasz O, Szasz N (2001) Electro-hyperthermia: a new paradigm in cancer therapy. Deutsche Zeitschrift fur Onkologie 33:91–99
172. Szasz A, Szasz N, Szasz O (2003) Hyperthermie in der Onkologie mit einem historischen Uberblick. Wissenschaft & Forschung, Deutsche Zeitschrift für Onkologie 35: 140–154

173. Abe M, Hiraoka M, Takahashi M et al (1986) Multi-institutional studies on hyperthermia using an 8-MHz radiofrequency capacitive heating device (thermotron RF-8) in combination with radiation for cancer therapy. Cancer 58:1589–1595
174. Szasz A, Szasz O, Szasz N (2006) Physical background and technical realization of hyperthermia. In: Baronzio GF, Hager ED (2006) Hyperthermia in cancer treatment: a primer. Springer, Landes Bioscience, Austin, TX
175. Seegenschmiedt MH, Vernon CC (1995) A Historical Perspective on Hyperthermia in Oncology. In: Seegenschmiedt MH, Fessenden P, Vernon CC (eds) (1995) Thermoradiotherapy and Thermochemotherapy Vol. 2: Clinical Applications, Springer, Berlin, pp 3–46
176. Holt JAG (1988) Microwaves are not hyperthermia. Radiograp 35(4):151–162
177. Shinc J (Chairman) (2005) Review of the use of microwave therapy for the treatment of patients with cancer. Final report to the minister for health and ageing, http://www.nhmrc.gov.au/_files_nhmrc/file/publications/synopses/nh67exec.pdf
178. Liboff AR (2003) Ion Cyclotron Resonance in Biological Systems: Experimental Evidence. In: Stavroulakis P (ed) Biological effects of electromagnetic fields. Springer, Berlin
179. Szendro P, Vincze G, Szasz A (2001) Pink noise behaviour of the bio-systems. Eur Biophys J 30(3):227–231
180. Rosch PJ, Markov MS (2004) Bioelectromagnetic medicine. Marcel Dekker, New York, NY
181. Torchilin VP (2006) Multifunctional nanocarriers. Adv Drug Deliv Rev 58:1532–1555
182. Brannon-Peppas L, Blanchette JO (2004) Nanoparticle and targeted systems for cancer therapy. Adv Drug Deliv Rev 56:1649–1659
183. Henderson MA, Pettigrew RT (1971) Induction of controlled hyperthermia in treatment of cancer. The Lancet 1275–1277
184. Gebbers N, Hirt-Burri N, Scaletta C, Hoffmann G, Applegate LA (2007) Water-filtered infrared-A radiation (wIRA) is not implicated in cellular degeneration of human skin. GMS German Med Sci 5:1–14
185. Heckel M. (1990) Ganzkörperhyperthermie und Fiebertherapie – Grundlagen und Praxis. Hippokrates Verlag, Stuttgart
186. Ardenne von A, Wehner H, (2005) Extreme Whole-Body Hyperthermia with Water-Filtered Infrared-A Radiation. NCBI Bookshelf, Landes Bioscience, Madame Curie Bioscience database, http://www.ncbi.nlm.nih.gov/bookshelf/br.fcgi?book=eurekah&part=A46701
187. Srobl B. (2006) 6 Jahre Ganzkörperhyperthermie kombiniert mit Chemotherapie beim Ovarialkarzinom, Dolphin-0- und -1-Studie. Hyperthermia Symposium, 22–23, September
188. Kelleher DK, Thews O, Rzeznik J, Scherz A, Salomon Y, Vaupel P (1999) Water-filtered infrared-A radiation: a novel technique for localized hyperthermia in combination with bacteriochlorophyll-based photodynamic therapy. In J Hyperthermia 15:467–474
189. Granmt D (1904) The Galvano-Cautery in the Treatment of Intra-Laryngeal Growths. J Laryngol Rhinol Ontol 19:294–297
190. Short History of Bioelectrics (2009), http://www.pulsedpower.eu/bioelectrics/bio_02_main.html
191. Kratzer GL, Onsanit T (2007) Fulguration of selected cancers of the rectum. Report of 27 cases. Diseases of Colon and Rectum, 15:431–435
192. LeVeen HH, Wapnick S, Piccone V et al (1976) Tumor eradication by radiofrequency therapy. JAMA 235(20):2198–2200
193. Short JG, Turner PF (1980) Physical Hyperthermia and cancer therapy. Proc IEEE 68: 133–142
194. Storm FK, Morton DL, Kaiser LR (1982) Clinical radiofrequency hyperthermia: a review. Natl Cancer Inst Monogr 61:343–350
195. Jo S, Sugahara T, Yamamoto I (1994) Clinical response of hyperthermia using heating equipment Thermotron-RF8 in Japan. Biomed Eng Appl Basis Commun 6:340–362

196. Hiraki Y, Nakajo M (1998) Effectiveness of RF capacitive hyperthermia combined with radiotherapy for stages III and IV oro-hypopharyngeal cancers: a non-randomized comparison between thermoradiotherapy and radiotherapy. Int J Hyperthermia 14(6):593–594
197. Harima Y, Nagata K, Harima K et al (2000) Bax and Bcl-2 Protein Expression following Radiation Therapy versus Radiation plus Thermoradiotherapy in Stage IIIB Cervical Carcinoma. Cancer 88(1):132–138
198. Lee CK, Song CW, Rhee JG et al (1995) Clinical experience using 8 MHz radiofrequency capacitive hyperthermia in combination with radiotherapy: results of a phase I/II study. Int J Radiat Oncol Biol Phys 32(3):733–745
199. Masunaga SI, Hiraoka M, Akuta K et al (1994) Phase I/II trial of preoperative thermoradiotherapy in the treatment of urinary bladder cancer. Int J Hyperthermia 10(1):31–40
200. Weiss TF (1996) Cellular Biophysics. Bradford Book, MIT Press, Cambridge, MA
201. Rand RW, Snow HD, Brown WJ (1982) Thermomagnetic Surgery for Cancer. J Surg Res 33:177–183
202. Matsuki H, Satoh T, Murakami K (1990) Local hyperthermia based on soft heating method utilizing temperature sensitive ferrite-rod. IEEE Trans Magn 26:1551–1553
203. Gilchrist RK, Medal R, Shorey WD et al (1957) Selective inductive heating of lymph nodes. Ann Surg 146(4):596–606
204. Hoshino T, Sato T, Masai A (1994) Conduction System Ablation using Ferrite rod for cardiac arrhythmia. IEEE Trans Magn 30:4689–4691
205. Gordon RT, Hines JR, Gordon D (1979) Intracellular hyperthermia: a biophysical approach to cancer treatment via intracellular temperature and biophysical alteration. Med Hypotheses 5(1):83–102
206. Rabin Y (2002) Is intracellular hyperthermia superior to extracellular in the thermal sense? Int. J. Hyperthermia 18(3):194–202
207. Jordan A, Scholz R, Wust P, Faehling H, Felix R (1999) Magnetic fluid hyperthermia (MFH): Caner treatment with AC magnetic field induced excitation of biocompatible supermagnetic nanoparticles. J Magn Magnetic Materials 201:413–419
208. Nishide Oleson JR (1985) The role of magnetic induction techniques for producing hyperthermia. In: Anghileri LJ, Robert J (eds) Hyperthermia in Cancer Treatment, Vol. II. CRC Press, Boca Raton, Fl, pp 141–154
209. Nishide SM, Ueno S (1993) A method of localized hyperthermia by using a figure-of-eight coil and short-circuit rings. Eng Med Biol Soc Proc 15th Ann Int Conf IEEE, pp 1447–1448
210. Iojo M, Murakami A, Sato F et al (2001) Consideration of handy excitation apparatus for the inductive hyperthermia. IEEE Trans Magn 37(1):2944–2946
211. Taylor LS (1978) Devices for microwave hyperthermia. In: Streffer C, vanBeuningen D, Dietzel F (eds) Cancer Therapy by Hyperthermia and Radiation, Urban & Schwarzenberg, Baltimore, MD, pp 115–121
212. Turner PF (1984) Regional hyperthermia with an annular phase array. IEEE Trans Biomed Eng BME-31:106–111
213. Gonzalez-Gonzalez D, van Dijk JDP, Oldenburger F (1992) Results of combined treatment with radiation and hyperthermia in 111 patients with large of deep-seated tumors. In: Germeg EW (ed) Hyperthermia oncology, Vol. 1, Vol. 2. Tucson, AZ, p 415
214. Myerson RJ, Leybovich L, Emami B et al (1991) Phantom studies and preliminary clinical experience with the BSD2000. Int J Hyperthermia 7(6):937–951
215. Wust P, Fahling H, Wlodarczyk W (2001) Antenna arrays in the sigma-eye applicator: Interactions and transforming networks. Med. Phys 28, 1793–1805
216. Wust P, Felix R, Deuflhard P (1999) Kunstliches Fieber gegen Krebs. Spektrum der Wissenschaft Dezember 78–84
217. Issels R (1999) Hyperthermia combined with chemotherapy – biological rationale, clinical application, and treatment results. Onkologie 22(5):374–381
218. Issels RD, Adbel-Rahman S, Wendtner C-M et al (2001) Neoadjuvant chemotherapy combined with regional hyperthermia (RHT) for locally advanced primary or recurrent

high-risk adult soft-tissue sarcomas (STS) of adults: long-term results of a phase II study. Eur J Cancer 37:1599–1608

219. Head JF, Wang F, Lipari CA et al (2000) The important role of Infrared Imaging in Breast cancer. IEEE Engineering in Medicine and Biology Magazine 19(3):52–57
220. Weiss TF (1996) Cellular Biophysics. Transport, Vol. 1. MIT Press, Cambridge
221. Matay G, Zombory L (2000) Physiological effects of radiofrequency radiation and their application for medical biology. Muegyetemi Kiado, Budapest, p 80
222. Gautherie M (1982) Temperature and blood-flow patterns in breast cancer during natural evolution and following radiotherapy. Biomedical Thermology, Alan R. Liss, New York, NY, pp 21–24
223. Vaupel P, Kallinowski F, Okunieff P (1989) Blood flow, oxygen and nutrient supply, and microenvironment of human tumors: a review. Cancer Res 49(23):6449–6465
224. Dudar TE, Jain RK (1984) Differential response of normal and tumor microcirculation to hyperthermia. Cancer Res 44(2):605–612
225. Song CW, Lokshina A, Rhee JG et al (1984) Implication of blood-flow in hyperthermic treatment of tumors. IEEE Trans Biomed Eng 31(1):9–16
226. Song CW, Choi IB, Nah BS et al (1995) Microvasculature and Perfusion in Normal Tissues and Tumors. In: Seegenschmiedt MH, Fessenden P, Vernon CC (eds) Thermoradiomet Thermochemother 1:139–156
227. Song CW, Park H, Griffin RJ (2001) Theoretical and experimental basis of hyperthermia. In: Kosaka M, Sugahara T, Schmidt KL et al (eds) Thermotherapy for neoplasia, inflammation, and pain, Springer, Tokyo, pp 394–407
228. Takana Y (2001) Thermal responses of microcirculation and modification of tumor blood flow in treating the tumors. In: Kosaka M, Sugahara T, Schmidt KL et al (eds) Theoretical and experimental basis of Hyperthermia. Thermotherapy for neoplasia, inflammation, and pain. Springer, Tokyo, pp 408–419
229. Guy AW, Chou CK (1983) Physical aspects of localized heating by radio-waves and microwaves. In: Storm, K.F. (ed.) Hyperthermia in cancer therapy, GK Hall Medical Publishers, Boston, MA
230. Vaupel P (1990) Pathophysiological mechanism of hyperthermia in cancer therapy. In: Gautherie M (ed) Methods of hyperthermia control, clinical thermology, Springer, Berlin, pp 73–134
231. Gottstein U (1969) Störungen des Hirnkreislaufes und zerebralen Stoffwechsels durch Hypoglykämie. In: Quandt J (ed) Die zerebralen Durchblutungsstörungen des Erwachsenenalters, Volk und Gesundheit, Berlin, pp 857–867
232. Vaupel P, Hammersen F (Hrsg) (1982) Mikrozirkulation in malignen Tumoren. Karger, Basel
233. Song CW (1984) Effect of Local hyperthermia on blood-flow and microenvironment: a review. Cancer Res 44(10 Suppl):4721s–4730s
234. Hahn GM (1987) Blood-flow. In: Field, S.B., Franconi, C. (eds) Physics and Technology of hyperthermia, No. 127, pp 441–446. NATO ASI Series, Series E: Applied Sciences, Martinus Nijhoff Publishers, Dordrecht, Boston, MA
235. Song CW, Choi IB, Nah BS et al (1996) Microvasculature and perfusion in Normal tissues and tumors. In: Seegenschmiedt MH, Fessenden P, Vernon CC (eds) Thermo-radiotherapy and thermo-chemotherapy, vol. 1, pp. 139–156. Biology, physiology and physics. Springer, Berlin
236. Ardenne M von, Reitnauer PG (1980) Vergleichende photoelektronische Registrierung des Einstromes von Evanslbau in das Blutgefäßsystem von Normalgewebe und von Tumoren mit selektiv ausgelöster Hämostase. Arch Geschwulstforsch. 50:443–462
237. Vaupel PW, Kelleher DK (1996) Metabolic status and reaction to heat of normal and tumor tissue. In: Seegenschmiedt, M-H-, Fessenden, P. Vernon, C.C. (eds) Thermo-radiotherapy and Thermo-chemotherapy. Biology, physiology and physics, Vol. 1. Springer, Berlin, pp 157–176

238. Dewey WC, Hopwood LE, Sapareto SA et al (1977) Cellular response to combination of hyperthermia and radiation. Radiology 123(2):463–474
239. Lindholm C-E (1992) Hyperthermia and radiotherapy. Ph.D. Thesis, Lund University, Malmo, Sweden
240. Hafstrom L, Rudenstam CM, Blomquist E et al (1991) Regional hyperthermic perfusion with melphalan after surgery for recurrent malignant melanoma of the extremities. J Clin Oncol 9:2091–2094
241. Pence DM, Song CW (1986) Effect of heat on blood-flow. In: Anghileri LJ, Robert J (eds) Hyperthermia in cancer treatment, Vol. II. CRC Press, Boca Raton, FL, pp 1–17
242. Reinhold HS (1987) Effects of hyperthermia on tumor microcirculation. In: Field SB, Franconi C (eds) Physics and technology of hyperthermia, No. 1270, NATO ASI Series, Series E: Applied Sciences, Martinus Nijhoff Publishers, Dordrecht, pp 458–469
243. Reinhold HS, Blachiewicz B, Berg-Blok A (1978) Decrease in tumor microcirculation during hyperthermia. In: Streffer C, vanBeuningen D, Dietzel F et al (eds) Cancer therapy by hyperthermia and radiation, Urban & Schwarzenberg, Munich, pp 231–232
244. Endrich B, Hammersen F (1986) Morphologic and hemodynamic alterations in capillaries during hyperthermia. In: Anghileri LJ, Robert J. (eds) Hyperthermia in cancer treatment, Vol II, Ch. 2. CRC Press, Boca Raton, FL, pp 17–47
245. Zant G van (1986) Effects of hyperthermia on hematopoietic tissues. In: Anghileri LJ, Robert J (eds) Hyperthermia in Cancer Treatment, Vol II, Ch 4. CRC Press, Boca Raton, FL, pp 59–73
246. Hales JRS, Hirata K (1986) Aspects of circulatory responses in animals pertinent to the use of hyperthermia in cancer treatment. In: Anghileri LJ, Robert J (eds) Hyperthermia in Cancer Treatment, Vol II, Ch 3. CRC Press, Boca Raton, FL, pp 49–58
247. Ingram DL, Mount LE (1975) Man and animals in hot environments, topics in environmental physiology and medicine. Springer, Berlin
248. Doi O, Kodama K, Higashiyama M et al (1993) Postoperative chemothermotherapy for locally advanced lung cancer with carcinomatous pleuritis. In: Matsuda T (ed) Cancer treatment by hyperthermia, radiation and drugs, Taylor & Francis, London; Washington, Chapter 31, pp.338–352
249. Ardenne M von, Reitnauer PG (1977) Krebs-Mehrschritt-Therapie und Mikrozirkulation. Krebsgeschehen 9:134–149
250. Streffer C (1990) Biological basis of thermotherapy (with special reference to Oncology). In: Gautherie M (ed) Biological basis of oncologic thermotherapy. Springer, Berlin, pp 1–72
251. Law MP (1988) The response of normal tissues to hyperthermia. In: Urano M, Douple E (eds) Hyperthermia and oncology, Vol. 1. Thermal effects on cells and tissues, VSP BV Utrecht The Netherlands, pp 121–159
252. Prescott DM (1996) Manipulation of physiological parameters during hyperthermia. In: Seegenschmiedt MH, Fessenden P, Vernon CC (eds) Thermo-radiotherapy and Thermo-chemotherapy, Vol. 1. Biology, physiology and physics, Springer, Berlin, pp 177–189
253. Baronzio GF, Gramaglia A, Baronzio A et al (2006) Influence of tumor microenvironment on thermoresponse: biologic and clinical implications. In: Baronzio GF, Hager ED (eds) Hyperthermia in cancer treatment: a primer, Landes Bioscience, Springer, New York, NY, pp 62–86
254. Roca C, Primo L (2006) Hyperthermia and angiogenesis, results and perspectives. In: Baronzio GF, Hager ED (eds) Hyperthermia in Cancer treatment: a primer, Landes Bioscience, Springer, New York, NY pp 87–93
255. Kelleher DK, Vaupel P (2006) Vascular effects of localized hyperthermia. In: Baronzio GF, Hager ED (eds) Hyperthermia in Cancer treatment: a primer, Landes Bioscience, Springer, New York, NY, pp 94–104
256. Ardenne M von (1975) Gesetzmäßigkeiten der Substratversorgung, der Zellkinetik und der Therapiemechanismen im Interkapillarraum der Krebsgewebe. Z Naturforsch 30c:91–106

257. Overgaard J, Nielsen OS, Lindegaard JC (1987) Biological basis for rational design of clinical treatment with combined hyperthermia and radiation. In: Field, S.B., Franconi, C., (eds) Physics and Technology of Hyperthermia, No. 127. NATO ASI Series, E: Applied Sciences, Martinus Nijhoff Publ. Dordrecht, Boston, pp 54–79
258. Ardenne M von (1986) The present developmental state of cancer multistep therapy (CMT): selective occlusion of cancer tissue capillaries by combining hyperglycemia with two-stage regional or local hyperthermia using the CMT Selectrotherm technique. In: Anghileri LJ, Robert J (eds) Hyperthermia in Cancer Treatment, Vol. III., CRC Press Inc., Boca Raton, FL, pp 1–24
259. Ardenne M von (1990) Oxygen Multistep Therapy, Physiological and Technical Foundations. Georg Thieme Verlag, Thieme Medical Publishers, Stuttgart, NY
260. Ardenne M von (1997) Systemische Krebs-Mehrschritt-Therapie, Hyperthermie und Hypergykamie als Therapiebasis, Grundlagen, Konzeption, Technik, Klinik. HippoctaresVerlag, Stuttgart
261. McGuire BJ, Secomb TW (2003) Estimation of capillary density in human skeletal muscle based on maximal oxygen consumption rates. Am J Physiol Heart Circ Physiol 285:H2382–H2391
262. Heilbrunn LV (1923) The colloid chemistry of protoplasm. Am J Physiol 64:481–498
263. Yatvin MB, Dennis WH (1978) Membrane lipid composition and sensitivity to killing by hyperthermia, Procaine and Radiation. In: Streffer C, vanBeuningen D, Dietzel F et al (eds) Cancer Therapy by Hyperthermia and Radiation, Urban & Schwarzenberg, Baltimore, Munich, pp 157–159
264. Bowler K, Duncan CJ, Gladwell RT et al (1973) Cellular heat injury. Comp Biochem Physiol 45A:441–450
265. Belehradek J (1957) Physiological aspects of heat and cold. Annu Rev Physiol 19:59–82
266. Wallach DFH (1978) Action of Hyperthermia and ionizing radiation on plasma membranes In: Streffer C, vanBeuningen D, Dietzel F et al (eds) Cancer therapy by hyperthermia and radiation. Urban & Schwarzenberg, Baltimore, MD, pp 19–28
267. Nishida T, Akagi K, Tanaka Y (1997) Correlation between cell killing effect and cell-membrane potential after heat treatment: analysis using fluorescent dye and flow cytometry. Int J Hyperthermia 13(2):227–234
268. Weiss TF (1996) Cellular Biophysics. Electrical properties, Vol. 2. MIT Press, Cambridge
269. Ricardo R, Gonzalez-Mendez R, Hahn GM (1989) Effects of hyperthermia on the intercellular pH and membrane potential of Chinese hamster ovary cells. Int J Hyperthermia 5:69–84
270. Mikkelsen RB, Verma SP, Wallach DFH (1978) Hyperthermia and the membrane potential of erythrocyte membranes as studied by Raman Spectroscopy. In: Streffer C, vanBeuningen D, Dietzel F et al (eds) Cancer Therapy by Hyperthermia and Radiation. Urban & Schwarzenberg, Baltimore, MD, pp 160–162
271. Hahn GM (1990) The heat-shock response: effects before, during and after gene activation. In: Gautherie M (ed) Biological basis of oncologic thermotherapy. Springer, Berlin, pp 135–159
272. Hodgkin AL, Katz B (1949) The effect of temperature on the electrical activity of the giant axon of the squid. J Physiol 108:37–77
273. Kabakov AE, Gabai VL (1997) Heat shock proteins and cytoprotection: ATP-deprived mammalian cells. (Series: Molecular Biology Intelligence Unit). Springer, Heidelberg
274. Keszler G, Csapo Z, Spasokoutskaja T et al (2000) Hyperthermy increase the phosporylation of deoxycytidine in the membrane phospholipid precursors and decrease its incorporation into DNA. Adv Exper Med Biol 486:333–337
275. Dikomey E, Franzke J (1992) Effect of heat on induction and repair of DNA strand breaks in X-irradiated CHO cells. Int J Radiat Biol 61(2):221–233
276. Hayashi S, Kano E, Hatashita M et al (2001) Fundamental aspects of hyperthermia on cellular and molecular levels. In: Kosaka M, Sugahara T, Schmidt KL et al (eds) Springer, Tokyo, pp 335–345

277. Okumura Y, Ihara M, Shimasaki T et al (2001) Heat inactivation of DNA-dependent protein kinase: possible mechanism of hyperthermic radio-sensitization. In: Kosaka M, Sugahara T, Schmidt KL et al (eds) Thermotherapy for Neoplasia, Inflammation, and Pain. Springer, Tokyo, pp 420–423
278. Shen RN, Lu L, Young P et al (1994) Influence of elevated temperature on natural killer cell activity, lymphokine-activated killer cell activity and lecitin-dependent cytotoxicity of human umbilical cord blood and adult blood cells. Int J Radiat Oncol Biol Phys 29: 821–826
279. Srivastava PK, DeLeo AB, Old LJ (1986) Tumor Rejection Antigens of Chemically Induced Tumors of Inbred Mice. Proc Natl Acad Sci USA 38(10):3407–3411
280. Csermely P, Schnaider T, Soti C et al (1998) The 90 kDa Molecular Chaperone Family: Structure, Function and Clinical Applications. A comprehensive review. Pharmacol Ther 79(2):129–168
281. Pothmann R (1991) TENS Transkutane elektrische Nervenstimulation in der Schmerztherapie. Hippokrates Verlag GmbH, Stuttgart
282. Gonzalez-Gonzalez, D (1996) Thermo-radiotherapy for tumors of the lower gastro-intestinal tract. In: Seegenschmiedt MH, Fessenden P, Vernon CC (eds) Thermo-radiotherapy and thermo-chemotherapy. Biology, physiology and physics, Vol. 1. Springer, Heidelberg, pp 105–119
283. Latchman DS (1999) Stress proteins. Springer, Berlin
284. Soti C, Csermely P (1998) Molecular chaperones in the etiology and therapy of cancer. Pathol Oncol Res 4(4):316–321
285. Ferrarini M, Heltai S, Zocchi MR et al (1992) Unusual expression and localization of heat shock proteins in human tumor cells. Int J Cancer 51(4):613–619
286. Protti MP, Heltai S, Bellone M et al (1994) Constitutive expression of the heat shock protein 72 kDa in human melanoma cells. Cancer Lett 85(2):211–216
287. Gress TM, Muller-Pillasch F, Weber C et al (1994) Differential expression of heat shock proteins in pancreatic carcinoma. Cancer Res 54(2):547–551
288. Xu M, Wright WD, Higashikubo R et al (1996) Chronic thermotolerance with continued cell proliferation. Int J Hyperthermia 12(5):645–660
289. Pirity M, Hever-Szabo A, Venetianer A (1996) Overexpression of P-glycoprotein in Heta and/or Drug Resistant Hepatoma Variants. Cytotechnology 19(3):207–214
290. Santin AD, Hermonat PL, Ravaggi A et al (1998) The Effects of Irradiation on the Expression of a Tumor Rejection Antigen (Heat Shock Protein GP96) in Human Cervical Cancer. Int Radiat Biol 76(6):699–704
291. Morgan J, Whitaker JE, Oseroff AR (1998) GRP78 Induction by Calcium Ionophore Potentiates Photodynamic Therapy Using the Mitochondrial Targeting Dye Victoria Blue BO. Photocem Photobiol 67(1):155–164
292. Punyiczki M, Fesus L (1998) Heat shock and apoptosis: the two defense systems of the organisms may have overlapping molecular elements. Ann NY Acad Sci 951: 67–74
293. Sapozhnikov AM, Ponomarev ED, Tarasenko TN et al (1999) Spontaneous apoptosis and expression of cell-surface heat-shock proteins in cultured EL-4 lymphoma cells. Cell Prolif 32(6):363–378
294. Huot J, Roy G, Lambert H, Landry J (1992) Co-induction of HSP27 Phosphorylation and Drug Resistance in Chinese Hamster Cells. Int J Oncol 1:31–36
295. Goodman R, Blank M (1999) The induction of stress proteins for cytoprotection in clinical applications. First International Symposium on Nonthermal Medical/Biological Treatments Using Electromagnetic Fields and Ionized Gases, ElectroMed'99, Norfolk VA, USA, 12–14 April 1999
296. Liu FF, Bezjak A, Levin W et al (1996) Assessment of palliation in women with recurrent breast cancer. Int J Hyperthermia 12(6):825–826

297. Hupp TR, Meek DW, Midgley CA et al (1992) Regulation of the specific DNA binding function of p53. Cell 71(5):875–886
298. Chen CF, Chen Y, Dai K et al (1996) A New Member of the HSP90 Family of molecular chaperones interacts with the retinoblastoma protein during mitosis and after heat shock. Mol Cell Biol 16(9):4691–4699
299. de Pomarai D, Daniels C, David H et al (2000) Non-thermal heat-shock response to microwaves. Nature 405(6785):417–418
300. Bukau B, Horwich AL (1998) The HSP70 and HSP60 chaperone machines. Cell 92(3):351–366
301. Watanabe M, Suzuki K, Kodama S et al (1995) Normal human cells at confluence get heat resistance by efficient accumulation of hsp72 in nucleus. Carcinogenesis 16(10): 2373–2380
302. Rosner GL, Clegg ST, Prescott DM (1996) Estimation of cell survival in tumors heated to nonuniform temperature distributions. Int J Hyperthermia 12(82):223–239
303. Kraybill W, Olenki T (2002) A phase I study of fever-range whole body hyperthermia (FR-WBH) in patients with advanced solid tumors: correlation with mouse models. Int J Hyperthermia 18(3):253–266; Burd R, Dziedzic ST (1998) Tumor Cell Apoptosis, Lymphocyte Recruitment and Tumor Vascular Changes Are Induced by Low Temperature, Long Duration (Fever-Like) Whole Body Hyperthermia. J of Cell physiol 177:137–147
304. Toyota N, Strebel FR, Stephens LC et al (1997) Long duration – mild whole body hyperthermia with cisplatin: tumor response and kinetics of apoptosis and necrosis in a metastatic mammary adenocarcinoma. Int J Hyperthermia 13(5):497–506
305. Sakaguchi Y, Makino M, Kaneko T et al (1994) Therapeutic efficacy of long duration – low temperature whole body hyperthermia when combined with tumor necrosis factor and carboplatin in rats. Cancer Res 54:2223–2227
306. Ostberg JR, Repasky EA (2000) Use of mild, whole body hyperthermia in cancer therapy. Immunol Invest 29(2):139–142
307. Appenheimer MM, Qing C, Gipard RA et al (2005) Impact of fever-range thermal stress on lymphocyte-endothelial adhesion and lymphocyte trafficking. Immunol Invest 34(3): 295–323
308. Shah A, Unger E, Bain MD et al (2002) Cytokine and adhesion molecule expression in primary human endothelial cells stimulated with fever-range hyperthermia. Int J Hyperthermia 18(6):534–551
309. Ostberg JR, Gellin C, Patel R, Repasky EA (2001) Regulatory potential of fever-range whole body hyperthermia on langerhans cells and lymphocytes in an antigen-dependant cellular immune response. J Immunol 167:2666–2670
310. Burd R, Dziedzic TS, Xu Y et al (1998) Tumor cell apoptosis, lymphocyte recruitment and tumor vascular changes are induced by low temperature, long duration (Fever-Like) whole body hyperthermia. J Cell Physiol 177(1):137–147
311. Pilling MJ, Seakins PW (1995) Reaction kinetics. Oxford Science, Oxford University Press, Oxford
312. Hobbie RK (2001) Intermediate physics for medicine and biology. Biological Physics Series, AIP Press, Woodbury, NY; Springer, Berlin
313. Law MP (1978) The relationship between heating time and temperature for hyperthermia alone or combined with radiation. In: Streffer C, vanBeuningen D, Dietzel F et al (eds) Cancer Therapy by hyperthermia and radiation, Urban & Schwarzenberg, Baltimore, MD, pp 222–224
314. Oleson JR (1995) Review Eugene Robertson Special Lecture, Hyperthermia from the clinic to the laboratory: a hypothesis. Int J Hyperthermia 11(3):315–322
315. Overgaard J (1989) The current and potential role of hyperthermia in radiotherapy. Int J Rad Oncol Biol Phys 16(3):535–549
316. van der Zee J, Truemiet-Donker AD, The SK et al (1988) Low-dose reirradiation in combination with hyperthermia: a palliative treatment for patients with breast cancer recurring in previously irradiated areas. Int J Rad Oncol Biol Phys 15(6):1407–1413

317. Vernon CC, Harrison M (1990) Hyperthermia with low-dose radiotherapy for recurrent breast carcinoma. The Lancet 336(8727):1383
318. icher JI, Al-Bussam N, Wolfstein RS (2006) Thermotherapy with curative intent – breast, head, and neck, and prostate tumors. Deutsche Zeitschrift fur Oncologie 38:116–122
319. Urano M, Kuroda M, Nishimura Y (1999) For the clinical application of thermochemotherapy given at mild temperatures. Int J Hyperthermia 15(2):79–107
320. Wiedermann GJ, Feyerabend T, Mentzel M et al (1994) Thermochemotherapie: grunde fur die kombinationsbehandlung mit hyperthermia und chemotherapie. Focus Mul 11:44–50
321. LeVeen HH, Rajagopalan PR, Vujic I et al (1984) Radiofrequency thermotherapy, local chemotherapy, and arterial occlusion in the treatment of non-resectable cancer. Am Surg 50(2):61–65
322. Okamura K, Nakashima K, Fukushima Y et al Hyperthermia with low dose chemotherapy for advanced non-small-cell lung cancer. http://www.isshin.or.jp/okamura/awaji2004/awaji1.html. Cited 02 Sept 2010
323. Franchi F, Grassi P, Ferro D et al (2007) Antiangiogenic metronomic chemotherapy and hyperthermia in the palliation of advanced cancer. Eur J Cancer Care 16(3):258–262
324. Molls M (1992) Hyperthermia – the actual role in radiation oncology and future prospects. Strahlentherapie und Oncologie 168:183–190
325. Seegenschmiedt MH, Feldmann HJ, Wust P (1995) Hyperthermia – its actual role in radiation oncology. Strahlentherapie und Oncologie 171:560–572
326. Emami B, Scott C, Perez CA et al (1996) Phase III study of interstitial thermoradiotherapy compared with interstitial radiotherapy alone in the treatment of recurrent or persistent human tumors: a prospectively controlled randomized study by radiation therapy oncology group. Int J Rad Oncol Biol Phys 34(5):1097–1104
327. Wust P, Rau B, Gemmler M et al (1995) Radio-thermotherapy in multimodal surgical treatment concepts. Onkologie 18(2):110–121
328. Urano M, Douple E. (eds) (1992) Hyperthermia and Oncology: Volume .2. Biology of thermal potentiation of radiotherapy. VSP BV, Utrecht, The Netherlands
329. Hehr T, Wust P, Bamberg M et al (2003) Current and potential role of thermoradiotherapy for solid tumors. Onkologie 26(3):295–302
330. Roti JL, Laszlo A (1988) The effects of hyperthermia on cellular macromolecules. In: Urano M, Douple E (eds) Hyperthermia and Oncology Vol 1, Thermal effects on cells and tissues, VSP Utrecht, The Netherlands, pp 13–56
331. Ohno T, Sakagami T, Shiomi M et al (1993) Hyperthermia therapy for deep regional cancer: thermochemotherapy, a combination of hyperthermia with chemotherapy. In: Matsuda T (ed) Cancer treatment by hyperthermia, radiation and drugs, Taylor&Francis, London, pp 303–316
332. Ohtsubo T, Kano E, Hayashi S et al (2001) Enhancement of cytotoxic effects of chemotherapeutic agents with hyperthermia in vitro. In: Kosaka M, Sugahara T, Schmidt KL et al (eds) Thermotherapy for neoplasia, inflammation, and pain. Springer, Tokyo, pp 451–455
333. Kawasaki S, Asaumi J-I, Shibuya K et al (2001) Recent aspects of elucidating the cellular basis of thermochemotherapy. In: Kosaka M, Sugahara T, Schmidt KL et al (eds) Thermotherapy for neoplasia, inflammation, and pain. Springer, Tokyo, pp 424–432
334. Szasz A, Vincze Gy, Szasz O et al (2003) An energy analysis of extracellular hyperthermia. Magneto- and electro-biology 22(2):103–115
335. Lupis CHP (1983) Chemical thermodynamics of materials. Wiley, North Holland
336. Urano M (1994) Thermochemotherapy: from in vitro and in vivo experiments to potential clinical application. In: Urano M, Douple E (eds) Hyperthermia and oncology, VSP Utrecht, Tokyo, 4:169–204
337. Wiederman GJ, Siemens HJ, Mentzel M et al (1993) Effects of Temperature on the Therapeutic Efficacy and Pharmacokinetics of Ifosamide. Cancer Res 53(18): 4268–4272

338. Hager D, Dziambor H, Hoehmann D et al (2002) Survival and quality of life of patients with advanced pancreatic cancer. ASCO Annual Meeting, Abstract:2359
339. Feyerabend T, Wioedemann GJ, Jager B et al (2001) Local hyperthermia, radiation and chemotherapy in recurrent breast cancer is feasible and effective except for inflammatory disease. Int J Radia Oncol Biol Phys 49(5):1317–1325
340. Masunaga S, Hiraoka M, Akuta K et al (1990) Non-Randomized Trials of Thermoradiotherapy versus Radiotherapy for Preoperative Treatment of Invasive Urinary Bladder Cancer. J Jpn Soc Ther Radiol Oncol 2:313–320
341. Rau B, Wust P, Hohenberger P et al (1998) Preoperative Hyperthermia Combined with Radiochemotherapy in Locally Advanced Rectal Cancer – A Phase II Clinical Trial. Ann Surg 227(3):380–389
342. Kodama K, Doi O, Higashyama M et al (1993) Long-term results of postoperative intrathoracic chemo-thermotherapy for lung cancer with pleural dissemination. Cancer 72(2):426–431
343. Pearson AS, Izzo F, Fleming RYD et al (1999) Intraoperative radiofrequency ablation of cryoablation for hepatic malignancies. Am J Surg 178(6):592–598
344. Kouloulias VE, Kouvaris JR, Nikita KS et al (2002) Intraoperative hyperthermia in conjunction with multi-schedule chemotherapy (pre- intra- and post operative), by-pass surgery, and post-operative radiotherapy for the management of unresectable pancreatic adenocarcinoma. Int. J Hyperthermia 18:233–252
345. Ohtsuru A, Braiden V, Cao Y (2001) Cancer Gene Therapy in Conjunction with Hyperthermia Under the Control of Heat-Inducible Promoter. In: Kosaka M, Sugahara T, Schmidt KL (eds) Thermotherapy for Neoplasia, Inflammation, and Pain. Springer .Tokyo, pp 464–470
346. Gaber MH, Wu NZ, Hong K et al (1996) Thermosensitive liposomes: extravasation and release of contents in tumor microvascular networks. Int J Radiat Oncol Biol Phys 36(5):1177–1187
347. Blackburn LV, Galoforo SS, Corry PM et al (1998) Adenoviral-mediated transfer of heat-inducible double suicide gene into prostate carcinoma cells. Cancer Res 58(7):1358–1362
348. Huang Q, Hu JK, Zhang L et al (2000) Heat-induced gene expression as a novel targeted cancer gene therapy strategy. Cancer Res 60(13):3435–3439
349. Yerushalmi A, Shani A, Fishelovitz Y et al (1986) Local microwave hyperthermia in the treatment of carcinoma of the prostate. Oncology 43(5):299–305
350. Piantelli M, Tatone D, Castrilli G et al (2001) Quercetin and tamoxifen sensitize human melanoma cells to hyperthermia. Melanoma Res 11:469–476
351. Oleson JR, Calderwood SK, Coughlin CT, Dewhirst MW, Gerweck LE, Gibbs FA, Kapp DS (1988), Biological and Clinical Aspects of Hyperthermia in Cancer Therapy. Am J Clin Oncol 11:368–380
352. Henderson BW, Waldow SM, Potter WR, Dougherty TJ (1985) Interaction of photodynamic therapy and hyperthermia: tumor response and cell survival studies after treatment of mice in vivo. Cancer Res 45:6071–6077
353. Lohr F Hu K, Huang Q, Zhang L, Samulski T, Dewhirst M, Li C (2000) Enhancement of radiotherapy by hyperthermia-regulated gene therapy. Int J Radiat Oncol Biol Phys 48:1513–1518
354. Skitzki JJ, Repasky EA, Evans SS (2009) Hyperthermia as an immunotherapy strategy for cancer. Curr Opini Invest Drugs 10:550–558
355. Vertrees RA, Jordan JM, Zwischenberger JB (2007) Hyperthermia and chemotherapy: the science. In: Hlem CW, Edwards RP (eds) current clinical oncology, intraperitoneal cancer therapy. Humana Press, Totowa NJ
356. Takahashi M, Hiraoka M, Nishimura Y et al (1993) Clinical results of thermoradiotherapy for deep-seated tumors. In: Matsuda T (ed) Cancer treatment by hyperthermia, radiation and drugs. Taylor & Francis, Oxford, pp 227–239

357. Hiraoka M, Jo S, Akuta K et al (1987) Radiofrequency capacitive hyperthermia for deep-seated tumors – I. Studies on thermometry. Cancer 60:121–127
358. Myerson RJ, Scott CB, Emami B et al (1996) A phase I/II study to evaluate radiation therapy and hyperthermia for deep-seated tumors: a report of RTOG 89-08. Int J Hyperthermia 4:449–459
359. Tsukiyama I, Kajiura Y, Ogino T et al (1993) Clinical results of thermoradiotherapy for superficial and shallow-seated tumors. In: Matsuda T (ed) Cancer treatment by hyperthermia, radiation and drugs, Taylor & Francis, London, pp 216–228
360. Simina P, van der Yee J, Wondergem J et al (1994) Effect of hyperthermia on the central nervous system: review. Int J Hyperthermia 10:1–30
361. Matsumi N, Matsumoto K, Mishima N et al (1994) Thermal damage threshold of brain tissue: Histological study of heated normal monkey brains. Neurol Med Chir (Tokyo) 34(4):209–215
362. Haveman J, Siminia P, Wondergem J et al (2005) Effects of hyperthermia on the central nervous system: what was learnt from animal studies? Int J Hyperthermia 21: 473–487
363. Seed PK, Stea B (1996) Thermoradiotherapy for Brain Tumors. In: Seegenschmiedt MH, Fessenden P, Vernon CC (eds) Thermoradiotherapy and thermochemotherapy, Vol. 2. Clinical applications. Springer, Berlin
364. Selker RG, Eddy MS, Deutsch M et al (1995) On the development of an interstitial radiation protocol for a multicenter consortium. Experience with permanent low-dose rate and temporary high-dose rate 125I implants in 'failed' and 'newly diagnosed' glioblastoma patients: quality assurance methodology and a possible future adjuvant for therapeutic enhancement. J Neurooncol 26(2):141–155
365. Stea B, Rossman K, Kittelson J et al (1994) Interstitial irradiation versus interstitial thermoradiotherapy for supratentorial malignant gliomas: a comparative survival analysis. Int J Radiat Oncol Biol Phy 30(3):591–600
366. Nakajima T, Roberts DW, Ryan TP et al (1993) Pattern of response to interstitial hyperthermia and brachytherapy for malignant intracranial tumor: a CT analysis. Int J Hyperthermia 9(4):491–502
367. Iacono RP, Stea B, Lulu BA et al (1992) Template-guided stereotactic implantation of malignant brain tumors for interstitial thermoradiotherapy. Stereotact Funct Neurosurg. 59(1–4):199–204
368. Sneed PK, Stauffer PR, Gutin PH et al (1991) Interstitial irradiation and hyperthermia for the treatment of recurrent malignant brain tumors. Neurosurgery 28(2):206–215
369. Stea B, Cetas TC, Cassady JR et al (1990) Interstitial thermoradiotherapy of brain tumors: preliminary results of a phase I clinical trial. Int J Radiat Oncol Biol Phys 19(6):1463–1471
370. Sneed PK, Gutin PH, Stauffer PR et al (1992) Thermoradiotherapy of recurrent malignant brain tumors. Int J Radiat Oncol Biol Phys 23(4):853–861
371. Edwards DK, Stupperich TK, Baumann CK et al (1991) Volumetric interstitial hyperthermia: role of the critical care nurse. Focus Crit Care 18(1):35–39, 42, 45–50
372. Salcman M, Samaras MJ (1983) Interstitial microwave hyperthermia for brain tumors. Results of a phase-1 clinical trial. J Neurooncol 1(3):225–236
373. Fan M, Ascher PW, Schrottner O et al (1992) Interstitial 1.06 Nd:YAG laser thermotherapy for brain tumors under real-time monitoring of MRI: experimental study and phase I clinical trial. J Clin Laser Med Surg 10(5):355–361
374. Kahn T, Harth T, Bettag M et al (1997) Preliminary experience with the application of gadolinium-DTPA before MR imaging-guided laser-induced interstitial thermotherapy of brain tumors. J Magn Reson Imaging 7(1):226–229
375. Borok TL, Winter A, Laing J et al (1988) Microwave hyperthermia radiosensitized iridium-192 for recurrent brain malignancy. Med Dosim 13(1):29–36
376. Moran CJ, Marchosky JA, Wippold FJ et al (1995) Conductive interstitial hyperthermia in the treatment of intracranial metastatic disease. J Neurooncol 26(1):53–63

377. Jordan A, Scholz R, Maier-Hauff K et al (2001) Presentation of a new magnetic field therapy system for the treatment of human solid tumors with magnetic fluid hyperthermia. J Magn Magn Materials 225(1–2):118–126
378. Stahl H, Wust P, Maier-Hauff K et al (1995) The use of an early postoperative interstitial-hyperthermia combination therapy in malignant gliomas. Strahlenther Onkol 171(9): 510–524
379. Hulshof MCCM, Raaymakers BW, Lagendijk JJW et al (2004) A feasibility study of interstitial hyperthermia plus external beam radiotherapy in glioblastoma multiforme using the Multi Electrode Current Source (MECS) system. Int J Hyperthermia 20(5):451–463
380. Sneed PK, Stauffer PR, McDermott MW et al (1998) Survival benefit of hyperthermia in a prospective randomized trial of brachytherapy boost +/– hyperthermia for glioblastoma multiforme. Int. J. Radiat. Oncol. Biol. Phys. 40(2):287–295
381. Pontiggia P, Duppone Curto F, Rotella G (1995) Hyperthermia in the treatment of brain metastases from lung cancer. Experience on 17 cases. Anticancer Res 15(2):597–601
382. Tanaka R, Gan To Kagaku Ryoho (1988) Radiofrequency hyperthermia in malignant brain tumors: clinical trials. Gan To Kagaku Ryoho 15(4 Pt 2–2):1370–1375
383. Tanaka R, Kim CH, Yamada N et al (1987) Radiofrequency hyperthermia for malignant brain tumors: preliminary results of clinical trials. Neurosurgery 21(4):478–483
384. Guthkelch AN, Carter LP, Cassady JR et al (1991) Treatment of malignant brain tumors with focused ultrasound hyperthermia and radiation: results of a phase I trial. J Neurooncol 10(3):271–284
385. Ley-Valle A (2003) Non invasive intracranial hyperthermia with Electric Capacitive Transference -ECT- Intratumoral and cerebral thermometry results. Neurocirugia (Astur) 14(1):41–45
386. Falk RE, Moffat FL, Lawler M et al (1986) Combination therapy for respectable and unresectable adenocarcinoma of the pancreas. Cancer 57(3):685–688
387. Gonzalez-Cao M, Salgado E, Rodriguez J (2001) Docetaxel (D) with gemcitabine (GEM) in metastatic pancreatic cancer. Proc Am Soc Clin Oncol 20:A2274
388. Kakolyris S, Stathopoulos G, Tsavaris N et al (1999) First-line treatment with docetaxel (D) and gemcitabine (G) in patients with advanced pancreatic cancer: a multicenter Phase II study. Proc Am Soc Clin Oncol 18:A960
389. Kindler HL (2002) The Pemetrexed/Gemcitabine Combination in Pancreatic Cancer. CANCER Supplement 95(4):928–932
390. Yamada S, Takai Y, Nemoto K et al (1992) Intraoperative Radiation Therapy Combined with Hyperthermia against Pancreatic Carcinoma. Tohoku J Exp Med 166(3):395–401
391. Shibamoto Y, Nishimura U, Abe M (1996) Intraoperative radiotherapy and hyperthermia for unresectable pancreatic cancer. Hepatogastroenterology 43(8):326–332
392. Kouloulias VE, Nikita KS, Kouvaris JR et al (2002) Intraoperative hyperthermia and chemo-radiotherapy for inoperable pancreatic carcinoma. Eur J Cancer Care 11(2): 100–107
393. Matsui Y, Nakagawa A, Kamiyama Y et al (2000) Selective thermocoagulation of unresectable pancreatic cancers by using radiofrequency capacitive heating. Pancreas 20:14–20
394. Hyperthermia with chemotherapy for locally advanced or metastatic pancreas cancer (2007) University of Texas, Health Science Center, Houston, NCT001178763, US National Institute of Health, http://clinicaltrials.gov/ct/show/NCT00178763;jsessionid=3B4AB11CCF3D35F792D6D7CF18AC8EE2?order=5 Cited 02 October 2007
395. Peters SO, Stoltz AS, Bakshandeh A (2006) Analysis of preclinical and clinical data on the role of hyperthermia with gemcitabine and carboplatin on pancreatic adenocarcinoma. Journal of Clinical Oncology, ASCO Annual Meeting Proceedings Part I. No. 18S, 24:14123
396. Hiraoka M, Masunaga S, Nishimura Y et al (1992) Regional hyperthermia combined with radiotherapy in the treatment of lung cancer. Int. J. Radiat. Oncol. Biol. Phys. 22(5): 1009–1014

397. Imada H, Nomoto S, Tomimatsu A et al (1999) Local control of nonsmall cell lung cancer by radiotherapy combined with high power hyperthermia using an 8 MHz RF capacitive heating device. Jpn J Hyperthermia Oncol 15(2):19–24
398. Karasawa K, Muta N, Nakagawa K et al (1994) Thermoradiotherapy in the treatment of locally advanced non-small cell lung cancer. Int J Radiat Oncol Biol Phys 30(5): 1171–1177
399. Sakurai H, Hayakawa K, Mitsuhashi N et al (2002) Effect of hyperthermia combined with external radiation therapy in primary non-small cell lung cancer with direct bony invasion. Int J Hyperthermia 18(5):472–483
400. Sakao S, Takiguchi Y, Nemoto K et al (2002) Thermoradiotherapy for local control of chest wall invasion in patients with advanced non-small cell lung cancer. Int J Clin Oncol 7:343–348
401. Shinn KS, Choi IB, Kim IA et al (1996) Thermoradiotherapy in the Treatment of Locally Advanced Nonsmall Cell Lung Cancer. J Korean Soc Ther Radiol Oncol 14(2): 115–122
402. Higashiyama M, Doi O, Kodama K et al (1994) Intrathoracic chemothermotherapy following panpleuropneumonectomy for pleural dissemination of invasive thymoma. Chest 105(6):1884–1885
403. Mitsumori M, Zeng ZF, Oliynychenko P (2007) Regional hyperthermia combined with radiotherapy for locally advanced non-small cell lung cancers: a multi-institutional prospective randomized trial of the International Atomic Energy Agency. Int J Clin Oncol 12:192–198
404. Hettinga JVE, Lemstra W, Meijer C et al (1994) Hyperthermic potentiation of cisplatin toxicity in a human small cell lung carcinoma cell line and a cisplatin resistant subline. Int J Hyperthermia 10:795–805
405. Vertrees RA, Das GC, Popov VL et al (2005) Synergestic interaction of hyperthermia and Gemcitabine in lung cancer. Cancer Biol Ther 4:1144–1155
406. Yang H, Jiang G, Fu X et al (2005) Radiotherapy and hyperthermia for NSCLC. ASCO Annual Meeting, No. 7289
407. Yokoyama M, Nitta S (1993) Systemic hyperthermia to patients with advanced pulmonary cancer. In: Matsuda T (ed) Cancer Treatment by Hyperthermia, Radiation and Drugs, Taylor & Francis, London, pp 376–381
408. Chhajed PN, Tamm M (2003) Radiofrequency heat ablation for lung tumors: potential applications. Med Sci. Monit 9:ED5–7
409. Okuma T, Matsuoka T, Yamamoto A et al (2007) Factors contributing to cavitation after CT-guided percutaneous radiofrequency ablation for lung tumors. J Vasc Interv Radiol 18(3):399–404
410. Weigel C, Kirsch M, Mensel B et al (2004) Precutaneous laser-induced thermotherapy of lung metastases: experience gained during 4 years. Radiologe 44:700–707
411. Matsuzaki Y, Edagawa M, Shimizu T et al (2004) Intrapleural hyperthermic perfusion with chemotherapy increases apoptosis in malignant pleuritis. Ann Thorac Surg 78: 1769–1772
412. Sekins KM, Leeper DB, Hoffman JK et al (2004) Feasibility of lung cancer hyperthermia using breathable perfluorochemical PFC liquids, Part I: Convective hyperthermia. Int. J. Hyperthermia 20:252–277, Part II: Ultrasound hyperthermia. Int J Hyperthermia 20: 278–299
413. Kondo M, Oyamada H, Yoshikawa T (2000) Therapeutic effects of chemoembolization using degradable starch microspheres and regional hyperthermia on unresectable hepatocellular carcinoma. In Selected papers on hyperthermia using Thermotron-RF8, April 2000
414. Kakehi M (1990) Multiinstitutional clinical studies on hyperthermia combined with radiotherapy or chemotherapy in advanced cancer of deep-seated organs. Int J Hyperthermia 6:719–740

415. Yoshikawa T, Oyamada H, Ichikawa H, Naito Y, Ueda S, Tainaka K, Itani K, Seto O, Sugino S, Kondo M. (1989) Antitumor effect and indication of chemoembolization using degradable starch microspheres and regional hyperthermia. J Jpn Soc Cancer Ther 24: 786–792
416. Nagata Y, Hiraoka M, Nishimura Y et al (1997) Clinical results of radiofrequency hyperthermia for malignant liver tumors. Int J Radiat Oncol Biol Phys 38:359–365
417. Alexander HR, Libutti SK, Pingpank JF et al (2003) Hyperthermic isolated hepatic perfusion using melphalan for patients with ocular melanoma metastatic to liver. Clin Cancer Res 9:6343–6349
418. Nagata Y, Hiraoka M, Akuta K et al (1990) Radiofrequency thermotherapy for malignant liver tumors. Cancer 65(8):1730–1736
419. Kasianenko IV, Osinsky SP, Pivnyuk VM et al (2000) Thermochemotherapy for liver metastases in patients with mammary carcinoma and gastrointestinal tumors. Oncol Ukr 2:34–36
420. Yamamoto K, Tanaka Y (1997) Radiofrequency capacitive hyperthermia for unresectable hepatic cancers. J Gastroenterol 32:361–366
421. Seong J, Lee HS, Han KH et al (1994) Combined Treatment of Radiotherapy and Hyperthermia for Unresectable Hepatocellular Carcinoma. Yonsei Med J 35(3): 252–259
422. Pacella CM, Valle D, Bizzarri G et al (2006) Percutaneous laser ablation in patients with isolated unresectable live metastases from colorectal cancer: Results of a phase II study. Acta Oncologica 45:77–83
423. Vogl TJ, Mack MG, Blazer JO et al (2003) Liver metastases: Neoadjuvant Downsizing with transarterial chemoembolization before laser-induced thermotherapy. Radiology 229(2):457–464
424. Goldberg SN, Gazelle GS, Solbiati L, Livraghi T, Tanabe KK, Hahn PF, Mueller PR (1998) Ablation of Liver Tumors Using Percutaneous RF Therapy. AJR 170:1023–1028
425. Ohguri T, Imada H, Yahara K et al (2004) Effect of 8-MHz radiofrequency-capacitive regional hyperthermia with strong superficial cooling for unresectable or recurrent colorectal cancer. Int J Hyperthermia 20(5):465–475
426. Nishimura Y, Hiraoka M, Abe M (1993) Thermoradiotherapy of locally advanced colorectal cancer. In: Matsuda T (ed) Cancer Treatment by Hyperthermia, Radiation and Drugs. Taylor & Francis, London, pp 278–289
427. Nishimura Y, Hiraoka M, Akuta K et al (1992) Hyperthermia combined with radiation therapy for primarily unresectable and recurrent colorectal cancer. Int J Rad Onc Biol 23(4):759–768
428. Berdov BA, Menteshashvili GZ (1990) Thermoradiotherapy of patients with locally advanced carcinoma of the rectum. Int J Hyperthermia 6(5):881–890
429. Hildebrandt B, Wust P, Drager J et al (2004) Regional pelvic hyperthermia as an adjunct to chemotherapy (oxaliplatin, folinic acid, 5-fluorouracil) in pre-irradiated patients with locally recurrent rectal cancer: a pilot study. Int J Hyperthermia 20(4):359–369
430. Rau B, Wust P, Tilly W et al (2000) Preoperative radiochemotherapy in locally advanced or recurrent rectal cancer: regional radiofrequency hyperthermia correlates with clinical parameters. Int J Radiat Oncol Biol Phys 48(2): 381–391
431. Ohno S, Tomoda M, Tomisaki S et al (1997) Improved Surgical Results After Combining Preoperative Hyperthermia with Chemotherapy and Radiotherapy for Patients with Carcinoma of the Rectum. Dis Colon Rectum 40(4):401–406
432. Rau B, Wust P, Gellermann J et al (1998) Phase-II-Studie zur praoperativen Radio-Chemo-Thermo-Therapie beim lokal fortgeschrittenen Rektum-Karzinom. Strahlenther Onkol 174(11):556–565
433. Takahashi M, Fujimoto S, Kobyashi K et al (1994) Clinical outcome of intraoperative pelvic hyperthermochemotherapy for patients with Dukes' C rectal cancer. Int J Hyperthermia 10(6):749–754

434. Kitamura K, Kuwano H, Watanabe M et al (1995) Prospective randomized study of hyperthermia combined with chemoradiotherapy for esophageal carcinoma. J Surg Oncol 60(1):55–58
435. Sucimachi K, Inokuchi K (1986) Hyperthermochemo-radiotherapy and esophageal carcinoma. Semin Surg Oncol 2(1):38–44
436. Kitamura K, Sugimachi K (1996) Thermoradiotherapy and chemotherapy for esophageal tumors. In: Seegenschmiedt MH, Fessenden P, Vernon CC (eds) Thermoradiotherapy and thermochemotherapy. Clinical applications of Hyperthermia, Vol 2. Springer, Berlin, pp 85–94
437. Li D-J, Hou B-S (1987) Preliminary Report on the Treatment of Esophageal Cancer by Intraluminal Microwave Hyperthermia and Chemotherapy. Cancer Treat Rep 71(11): 1013–1019
438. Sakamoto T, Katoh H, Shimizu T et al (1997) Clinical results of treatment of advanced esophageal carcinoma with hyperthermia in combination with chemoradiotherapy. Chest 112(6):1487–1493
439. Albregts M, Hulshof M, Zum Vörde Sive Vörding PJ et al (2004) A feasibility study in oesophageal carcinoma using deep loco-regional hyperthermia combined with concurrent chemotherapy followed by surgery. Int J Hyperthermia 20(6):647–659
440. Yahara K, Imada H, Nomoto S et al (2004) Thermoradiotherapy for recurrent esophageal carcinoma. Jpn J Hyp Oncol 20(1):1–8
441. Matsuda H, Tsutsui S, Morita M et al (1992) Hyperthermo-chemo-radiotherapy as a definitive treatment for patients with early esophageal carcinoma. Am J Clin Oncol (CCT) 15(6):509–514
442. Sugimachi K, Kitamura K, Baba K et al (1992) Hyperthermia combined with chemotherapy and irradiation for patients with carcinoma of the oesophagus – a prospective randomized trial. Int J Hyperthermia 8(3):289–295
443. Kuwano H, Matsuura H, Masaki M et al (1993) Hyperthermia combined with chemotherapy and irradiation for the treatment of patients with carcinoma of the oesophagus and the rectum. In: Matsuda T (ed) Cancer Treatment by Hyperthermia, Radiation and Drugs. Taylor & Francis, London, pp 353–364
444. Matsuda H, Kuwano H, Tsutsui S et al (1992) A complete response of an advanced oesophageal carcinoma treated with hyperthermic chemotherapy: a case report. Int J Hyperthermia 8(4):423–429
445. Hou B-S, Xiong Q-B, Li DJ (1989) Thermo-Chemo-Radiotherapy of Esophageal Cancer – A Preliminary Report of 34 Cases. Cancer 64(9):1777–1782
446. Onoyama Y et al (1993) Clinical experience in hyperthermic treatment for head and neck tumor. In: Matsuda T (ed) Cancer treatment by hyperthermia, radiation and drugs, chapter 22. Taylor & Francis, London, pp 241–249
447. Kazumoto T, Kato S, Sakura M et al (2002) A case of locally advanced hypopharyngeal cancer treated with curative resection after thermoradiotherapy. Jpn J Hyp Oncol 18(2): 99–107
448. Datta NR, Bose AK, Kapoor HK et al (1990) Head and neck cancers: result of thermoradiotherapy versus radiotherapy. Int J Hyperthermia 6(3):479–486
449. Valdagni R, Amichetti M, Pani G (1998) Radical radiation alone versus radical radiation plus microwave hyperthermia for N3 (TNM-UICC) neck nodes: a prospective randomized clinical trial. Int J Radiat Oncol Biol Phys 15(1):13–24
450. Valdagni R, Amichetti M (1995) Thermoradiotherapy for head and neck tumors. In: Seegenschmiedt MH, Fessenden P, Vernon CC (eds) Thermoradiotherapy and thermochemotherapy, Vol. 2. Clinical applications. Springer, Berlin, pp 49–67
451. Amichetti M, Graiff C, Fellin G et al (1993) Cisplatin, hyperthermia, and radiation (trimodal therapy) in patients with locally advanced head and neck tumors: a phase I–II study. Int J Radiat Oncol Biol Phys 26:801–807

452. Valdagni R, Amichetti M (1993) Report of long-term follow-up in a randomized trial comparing radiation therapy and radiation therapy plus hyperthermia to metastatic lymph nodes in stage IV head and neck patients. Int J Radiat Oncol Biol Phys 28:163–169
453. Nakabayashi T, Mochiki E, Kamiyama Y et al (2003) Efficacy of Intraperitoneal Chemohyperthermia for Gastric Cancer Patients with Peritoneal Carcinomatosis. Jpn J Hyperthermic Oncol 19(4):195–200
454. Nagata Y, Hiraoka M, Nishimura Y et al (1995) Clinical experiences in the thermoradiotherapy for advanced gastric cancer. Int J Hyperthermia 11(4):501–510
455. Shchepotin IB, Evans SR, Chorny V et al (1994) Intensive preoperative radiotherapy with local hyperthermia for the treatment of gastric carcinoma. Surg Oncol 3(1):37–44
456. Fu QG, Meng FD, Shen XD et al (2002) Efficacy of intraperitoneal thermochemotherapy and immunotherapy in intraperitoneal recurrence after gastrointestinal cancer resection. World J Gastroenterol 8(6):1019–1022
457. Hamazoe R, Maeta M, Koga S (1993) Efficacy of intraperitoneal hyperthermic chemotherapy in preventing peritoneal recurrence of gastric cancer. In: Matsuda T (ed) Cancer Treatment by Hyperthermia, Radiation and Drugs. Taylor & Francis, New York, NY, pp 328–338
458. Mizuguchi N, Mukojima T, Itoh M et al (2002) Two long-term survival cases of gastric carcinoma treated with hyperthermo-chemo-radiotherapy. Jpn J Hyperthermic Oncol 18(2):93–98
459. Bornstein BA, Zouranjian PS, Hansen JL et al (1992) Irradiation alone or combined with hyperthermia in the treatment of recurrent carcinoma of the breast in the chest wall: a nonrandomized comparison. Int J Hyperthermia 2(2):179–187
460. van der Zee J, Vernon CC (1996) Thermoradiotherapy for Advanced and Recurrent Breast Tumors. In: Seegenschmiedt MH, Fessenden P, Vernon CC (eds) Thermoradiotherapy and Thermochemotherapy, Vol. 2. Clinical Applications. Springer, Berlin, pp 35–48
461. Sherar M, Liu FF, Levin W et al (1997) Relationship between thermal dose and outcome in thermoradiotherapy treatments for superficial recurrences of breast cancer: data from a phase III trial. Int J Radiat Oncol Biol Phys 39(2):371–380
462. Vernon CC, Hand JW, Field SB et al (1996) Radiotherapy with or without hyperthermia in the treatment of superficial localized breast cancer: results from five randomized controlled trials. (International Collaborative Hyperthermia Group) Int J Radiat Oncol Biol Phys 35(4):731–744
463. Fuwa N, Morita K (1993) Combined treatment of radiotherapy and local hyperthermia using 8 MHz RF capacitive heating for advanced carcinoma of the breast. In: Matsuda T (ed) Cancer Treatment by Hyperthermia, Radiation and Drugs, Taylor & Francis, London., pp 252–260
464. Raymond U (2002) Clinical results using 8 MHz radiofrequency capacitive hyperthermia and radiotherapy for recurrent breast carcinoma. Rex Healthcare Cancer Center, Raleigh, N-Carolina
465. Masunaga S, Hiraoka M, Takahashi M et al (1990) Clinical results of thermoradiotherapy for locally advanced and/or recurrent breast cancer – comparison of results with radiotherapy alone. Int J Hyp 6(3):487–497
466. Kouloulias VE, Dardoufas CE, Kouvaris JR et al (2002) Liposomal doxoruicin in conjunction with reirradiation and local hyperthermia treatment in recurrent breast cancer: a Phase I/II Trial. Clin Cancer Res 8:374–382
467. Bornstein BA, Zouranjian PS, Hansen JL et al (1992) Local Hyperthermia, Radiation Therapy, and Chemotherapy in Patients with Local-Regional Recurrence of Breast Carcinoma. Int J Radiat Oncol Biol Phys 25:79–85
468. Vargas HI, Dooley WC, Gardner RA et al (2004) Focused Microwave Phased Array Thermotherapy for Ablation of Early-Stage Breast Cancer: Results of Thermal Dose Escalation. Ann Surg Oncol 11(2):139–146

469. Pontiggia P, Curto FC, Sabato A et al (1995) Is metastatic breast cancer refractory to usual therapy curable? Biomed Pharmacother 49:79–82
470. Nishimura Y, Hiraoka M, Jo S et al (1989) Radiofrequency (RF) capacitive hyperthermia combined with radiotherapy in the treatment of abdominal and pelvic deepseated tumors. Radiother Oncol 16(2):139–149
471. Hiraoka M. et al (1996) Thermoradiotherapy for upper abdominal tumors. In: M.H. Seegenschmiedt et al (eds) Thermoradiotherapy and thermochemotherapy, Vol. 2. Springer, Berlin
472. Elias D, Sideris L, Pocard M et al (2004) Efficacy of intraperitoneal chemohyperthermia with oxaliplatin in colorectal peritoneal carcinomatosis. Preliminary results in 24 patients. Ann Oncol 15:781–785
473. Elias D, Bonnay M, Puizillou JM et al (2002) Heated intra-operative intraperitoneal oxaliplatin after complete resection of peritoneal carcinomatosis: pharmacokinetics and tissue distribution. Ann Oncol 13:267–272
474. Elias D, Raynard B, Farkhondeh F et al (2006) Peritoneal carcinomatosis of colorectal origin – Long-term results of intraperitoneal chemohyperthermia with oxaliplatin following complete cytoreductive surgery. Gastroenterol Clin Biol 30:1200–1204
475. Elias D, Matsuhisa T, Sideris L et al (2004) Heated intra-operative intraperitoneal oxaliplatin plus irinotecan after complete resection of peritoneal carcinomatosis: pharmacokinetics, tissue distribution and tolerance. Ann Oncol 15:1558–1565
476. Overgaard J, Gonzalez DG, Hulshof MCCM et al (1996) Hyperthermia as an adjuvant to radiation therapy of recurrent or metastatic malignant melanoma. A multicentre randomized trial by the European Society for Hyperthermic Oncology. Int J Hyperthermia 12:3–20
477. Overgaard J (1995) Thermoradiotherapy of Malignant Melanoma. In: Seegenschmiedt MH, Fessenden P, Vernon CC (eds) Thermoradiotherapy and Thermochemotherapy, Vol. 2. Clinical Applications, Springer, Berlin; Telos, New York, NY, pp 69–83
478. Kim JH, Hahn EW, Tokita N (1978) Combination hyperthermia and radiation therapy for cutaneous malignant melanoma. Cancer 41(6):2143–2148
479. Kim JH, Hahn EW, Ahmed SA (1982) Combination hyperthermia and radiation therapy for malignant melanoma. Cancer 50:478–487
480. Jones EL, Oleson JR, Prosnitz LR et al (2005) Randomized trial of hyperthermia and radiation for superficial tumors. J Clin Oncol 23:3079–3085
481. Lindholm CE, Kjellen E, Nilsson P et al (1987) Microwave-induced hyperthermia and radiotherapy in human superficial tumors: clinical results with a comparative study of combined treatment versus radiotherapy alone. Int J Hyperthermia 3:393–411
482. Kim JH, Hahn EW, Tokita N et al (1977) Local Tumor Hyperthermia in Combination with Radiation Therapy – 1. Malignant Cutaneous Lesions. Cancer 40:161–169
483. Marmor JB, Hahn GM (1980) Combined radiation and hyperthermia in superficial human tumors. Cancer 46:1986–1991
484. Seegenschmiedt MH, Sauer R (1993) External thermoradiotherapy in superficial tumors: basic considerations and results. J Jpn Soc Ther Radiol Oncol 5:303–326
485. Egawa S et al (1989) A randomized clinical trial of hyperthermia and radiation versus radiation alone for superficially located cancers. J Jpn Soc Ther Radiol Oncol 1:135–140
486. Gibbs FA (1995) Thermoradiotherapy for Genitourinary and Gynecological tumors. In: Seegenschmiedt MH, Fessenden P, Vernon CC (eds) Thermoradiotherapy and thermochemotherapy, Vol. 2. Clinical applications, Springer, Berlin; Telos, New York, NY
487. Sekiba K, Hasegawa T, Kobashi Y (1993) Hyperthermic treatment for gynaecological malignancies. In: Matsuda T (ed) Cancer Treatment by Hyperthermia, Radiation and Drugs. Taylor & Francis, London, pp 261–270
488. Rietbroek RC, Schilthuis MS, Bakker PJM et al (1997) Phase II Trial of Weekly Locoregional Hyperthermia and Cisplatin in Patients with a Previously Irradiated Recurrent Carcinoma of the Uterine Cervix. Cancer 79(5):935–943

489. Jones EL, Samulski TV, Dewhirst, MV et al (2003) A pilot phase II trial of concurrent radiotherapy, chemotherapy, and hyperthermia for locally advanced cervical carcinoma. Cancer 98(2):277–282
490. Tsuda H, Tanaka M, Manabe T et al (2003) Phase I study of combined radiation, hyperthermia and intra-arterial carboplatin for local recurrence of cervical cancer. Ann Oncol 14:298–303
491. Westermann AE, Jones EL, Schem BC et al (2005) First results of triple-modality treatment combining radiotherapy, chemotherapy, and hyperthermia for the treatment of patients with stage IIB, III, and IVA cervical carcinoma. Cancer 104:763–770
492. van der Zee J, Gonzalez DG (2002) The Dutch Deep Hyperthermia Trial: results in cervical cancer. Int J Hyperthermia 18:1–12
493. Prosnitz L, Jones E et al (2002) Counterpoint: Test the value of hyperthermia in patients with carcinoma of the cervix being treated with concurrent chemotherapy and radiation. Int J Hyperthermia 18:13–18
494. van der Zee J, Koper PCM, Lutgens LCHW et al (2002) Point-counterpoint: What is the optimal trial design to test hyperthermia for carcinoma of the cervix? Point: Addition of hyperthermia or cisplatin to radiotherapy for patients with cervical cancer, two promising combinations – no definite conclusions. Int J Hyperthermia 18(1):19–24
495. Tilly W, Gellermann J, Graf R et al (2005) Regional hyperthermia in conjunction with definitive radiotherapy against recurrent or locally advanced prostate cancer T3pN0M0. Strahlenther Onkol 181(1):35–41
496. Algan O, Dosmire H, Hynynen K et al (2000) External Beam Radiotherapy and Hyperthermia in the Treatment of Patients with Locally Advanced Prostate Carcinoma. Cancer 89:399–403
497. Kalapurakal JA, Mittal BB, Sathiaseelan V (2001) Re-irradiation and external hyperthermia in locally advanced, radiation recurrent, hormone refractory prostate cancer: a preliminary report. Br J Radiol 74:745–751
498. Bdesha AS, Bunce CJ, Kelleher JP et al (1993) Transurethral microwave treatment for benign prostatic hypertrophy: a randomized controlled clinical trial. BMJ 306: 1293–1296
499. Petrovich Z, Pike MC, Boyd SD et al (1996) Transurethral hyperthermia for benign prostatic hyperplasia: long term results. Int J Hyperthermia 12:595–606
500. Rosette DeLa, Floratos DL, Severens JL et al (2003) Transurethral resection vs microwave thermotherapy of the prostate: a cost-consequences analysis. BJU International 92:713
501. Hisazumi H, Nakajima K, Koshida K et al (1993) 8 MHz RF-Hyperthermia for deep-seated urological malignancies. In: Matsuda T (ed) Cancer Treatment by Hyperthermia, Radiation and Drugs, Taylor & Francis, London, pp 270–277
502. Masunaga SI, Hiraoka M, Akuta K et al (1990) Non-randomized trials of thermoradiotherapy versus radiotherapy for preoperative treatment of invasive urinary bladder cancer. J Jpn Soc Ther Radiol Oncol 2:313–320
503. Moskovitz B, Meyer G, Kravtzov A et al (2005) Thermo-chemotherapy for intermediate or high-risk recurrent superficial bladder cancer patients. Ann Oncol 16:585–589
504. Hiraoka M, Nishimura Y, Nagata Y et al (1995) Clinical results of thermoradiotherapy for soft tissue tumors. Int J Hyperthermia 11:365–377
505. Egawa S, Tsukiyama I, Kajiura Y et al (1993) Clinical results of hyperthermia combined with radiation or chemotherapy for soft-tissue sarcomas. In: Matsuda T (ed) Cancer Treatment by Hyperthermia, Radiation and Drugs. Taylor & Francis, New York, NY, pp 290–299
506. Rossi CR, Foletto M, Filippo FD et al (1999) Soft Tissue Limb Sarcomas. Cancer 86: 1742–1749
507. Otsukaka T et al (2001) Clinical results of thermoradiotherapy for soft tissue tumors. Int J Clin Oncol 6:253–258

508. Prosnitz LR, Maguire P, Anderson JM et al (1999) The Treatment of high-grade soft tissue sarcomas with preoperative thermoradiotherapy. Int J Radiat Oncol Biol Phys 45: 941–949
509. Maguire PD, Samulski TV, Prosnitz LR et al (2001) A phase II trial testing the thermal dose parameter CEM43°T90 as a predictor of response in soft tissue sarcomas treated with pre-operative thermoradiotherapy. Int J Hyperthermia 17:283–290
510. Rossi CR, Deraco M, Simone MD et al (2004) Hyperthermic intraperitoneal intraoperative chemotherapy after cytoreductive surgery for the treatment of abdominal sarcomatosis. Cancer 100(9):1943–1950
511. Wioedemann GJ, Robins HI, Gutsche S et al (1996) Ifosfamide, carboplatin and etoposide (ICE) combined with 41.8 degrees C whole body hyperthermia in patients with refractory sarcoma. Eur J Cancer 32A:888–892
512. Wioedemann GJ, d'Oleire F, Knop E et al (1994) Ifosfamide and carboplatin combined with 41.8 degrees C whole-body hyperthermia in patients with refractory sarcoma and malignant teratoma. Cancer Res 54:5346–5350
513. Wioedemann GJ, Robins HI, Katschinski DM et al (1997) Systemic hyperthermia and ICE chemotherapy for sarcoma patients: rationale and clinical status. Anticancer Res 17: 2899–2902
514. Issels RD, Prenninger SW, Nagele A et al (1990) Ifosfamide plus etoposide combined with regional hyperthermia in patients with locally advanced sarcomas: a phase II study. J Clin Oncol. 8:1818–1829
515. Leopold K, Issels RD (1995) Thermoradiotherapy and thermochemotherapy for sarcomas. In: Seegenschmiedt MH, Fessenden P, Vernon CC (eds) Thermoradiotherapy and Thermochemotherapy, Vol. 2. Clinical Applications, Springer, Berlin; Telos, New York, NY, pp 147–158
516. Wendtner CM, Abdel-Rahman S, Krych M et al (2002) Response to neoadjuvant chemotherapy combined with regional hyperthermia predicts long-term survival for adult patients with retroperitoneal and visceral high-risk soft tissue sarcomas. J Clin Oncol 20:3156–3164
517. Issels RD, Abdel-Rahman S, Salat C et al (1998) Neoadjuvant Chemotherapy Combined with Regional Hyperthermia (RHT) Followed by Surgery and Radiation in Primary and Recurrent High-Risk Soft Tissue Sarcomas (HR-STS) of Adults (Updated Report). 23rd National Cancer Congress of the German Cancer Society, Berlin
518. Issels RD, Mittermuller J, Gerl A et al (1991) Improvement of local control by regional hyperthermia combined with systemic chemotherapy (ifosfamide plus etoposide) in advanced sarcomas: updated report on 65 patients. J Cancer Res Clin Oncol 4:S141–S147
519. Issels RD, Bosse D, Abdel-Rahman S et al (1993) Preoperative systemic etoposide/ifosfamide/doxorubicin chemotherapy combined with regional hyperthermia in high-risk sarcoma: a pilot study. Cancer Chemother Pharmacol 31:S233–S237
520. Wendtner CM, Adbel-Rahman S, Baumert J et al (2001) Treatment of primary, recurrent or inadequately resected high-risk soft-tissue sarcomas (STS) of adults: results of a phase II pilot study (RHT-95) of neoadjuvant chemotherapy combined with regional hyperthermia. Eur J Cancer 37(13):1609–1616
521. Issels RD, Abdel-Rahman S, Falk MH et al (1999) Initial response to neoadjuvant chemotherapy combined with regional hyperthermia (RHT) in high-risk soft tissue sarcomas (HR-STS) of adults and its correlation to survival parameters. Eur J Cancer 35:266
522. Issels RD, Bosse D, Strack M et al (1993) Weichteiltumoren: Indikation und Ergebnisse der Hyperthermie. Chirurg 64:461–467
523. Issels RD, Lindner LH, Wust P et al (2008) A phase III randomized prospective trial of neoadjuvant chemotherapy with or without regional hyperthermia (RHT). EORTC 62961, ESHO RHT95 Intergroup Trial; Eur Jf Cancer

524. Hornbach NB (1987) Is the community radiation oncologist ready for clinical hyperthermia? RadioGraphics 7:139–141
525. Smythe WR, Mansfield PF (2003) Hyperthermia: has its time come? Ann Surg Oncol 10:210–212
526. Szasz A (2006) What is against the acceptance of hyperthermia? Die Naturheilkunde Forum-Medizine 83:3–7
527. Oleson J.R (1991) Progress in hyperthermia? Int J Radiat Oncol Biol Phys 20:1147–1164
528. Oleson, J.R (1993) Prostate cancer: hot, but hot enough? Int J Radiat Oncol Biol Phys 26:369–370
529. Storm FK (1993) What happened to hyperthermia and what is its current status in cancer treatment? J Surg Oncol 53:141–143
530. Brizel DM (1998) Where there's smoke, is there fire? Int J Hyperthermia 14:593–594
531. Sneed PK, Dewhirst MW, Samulski T et al (1998) Should interstitial thermometry be used for deep hyperthermia? Int J Radiat Oncol Biol Phys 40:1205–1212
532. Oleson JR (1989) If we can't define the quality, can we assure it? Int J Radiat Oncol Biol Phys 16:879
533. Vasanthan A, Mitsumori M, Part JH et al (2005) Regional hyperthermia combined with radiotherapy for uterine cervical cancers: a multiinstitutional prospective randomized trial of the international atomic energy agency. Int J Rad Oncol Biol Phys 61, 145–153
534. Fatehi D, van der Zee J, van der Wal E et al (2006) Temperature data analysis for 22 patients with advanced cervical carcinoma treated in Rotterdam using radiotherapy, hyperthermia and chemotherapy: a reference point is needed. Int J Hyperthermia 22:353–363
535. Walker A, McCallum HM, Wheldon TE et al (1978) Promotion of metastasis of C3H mouse mammary carcinoma by local hyperthermia. Br J Cancer 38(4):561–563
536. Dickson JA, Ellis HA (1976) The influence of tumor volume and the degree of heating on the response of the solid Yoshida sarcoma to hyperthermia. Cancer Res 36(3):1188–1195
537. Hahn EW, Alfiery AA, Kim JH (1979) The significance of local tumor hyperthermia/radiation on the production of disseminated disease. Int J Radiat Oncol Biol Phys 5:819–823
538. Ando K, Urano M, Kenton L et al (1987) Effect of thermo-chemotherapy on the development of spontaneous lung metastases. Int J Hyperthermia 3(5):453–458
539. McChesney Gillette S, Dewhirst MW et al (1992) Response of canine soft tissue sarcomas to radiation or radiation plus hyperthermia: randomized phase II study. Int J Hyperthermia 8(3):309–320
540. Lord PF, Kapp DS, Morrow D (1981) Increased skeletal metastases of spontaneous canine osteosarcoma after fractionated systemic hyperthermia and local x-irradiation. Cancer Res 41:4331–4334
541. Sminia P, Jansen W, Haveman J et al (1990) Incidence of tumors in the cervical region of the rat after treatment with radiation and hyperthermia. Int. J Radiat Biol 57(2):425–436
542. Hei TK, Hall EJ, Kushner S et al (1986) Hyperthermia chemotherapeutic agents and oncogenic transformation. Int J Hyperthermia 2(3):311–320
543. Komatsu K, Miller RC, Hall EJ (1988) The oncogenetic potential of a combination of hyperthermia and chemotherapy. Br J Cancer 57:59–63
544. Miller RC, Roizin-Towle L, Komatsu K (1989) Interaction of heat with X-rays and cis-platinum; cell lethality and oncogenetic transformation. Int J Hyperthermia 5:697–705
545. Valdagni R, Knapp DS, Valdagni C (1986) N3 (TNM-UICC) metastatic neck nodes managed by combined radiation-therapy and hyperthermia: clinical results and analysis of treatment parameters. Int J Hyperthermia 2(2):189–200
546. Howard GCW, Sathiaseelan V, Freedman L et al (1987) Hyperthermia and radiation in the treatment of superficial malignancies: an analysis of treatment parameters response and toxicity. Int J Hyperthermia 3(1):1–8
547. Ben-Yosef R, Kapp DS (1992) Persistent and/or late complications of combined radiation therapy and hyperthermia. Int J Hyperthermia 8(6):733–745

548. Osinsky S, Ganul V, Protsyk V et al (2004) Local and regional hyperthermia in combined treatment of malignant tumors: 20 years experience in Ukraine. The Kadota Fund International Forum, Awaji, Japan, 15–18 June 2004
549. Seegenschmiedt MH, Klautke G, Seidel R et al (1995) Clinical Practice of interstitial thermoradiotherapy. In: Seegenschmiedt MH, Fessenden P, Vernon CC (eds) Thermoradiotherapy and thermochemotherapy Vol. 2. Springer, Heidelberg, pp 207–262
550. Waterman FM (1995) Invasive thermometry techniques. In: Seegenschmiedt MH, Fessenden P, Vernon CC (eds) Thermoradiotherapy and thermochemotherapy Vol. 1. Springer, Heidelberg, pp 331–360
551. Field SB (1987) Biological Aspects of Hyperthermia, Physics and Technology of Hyperthermia. In: Field SB, Franconi C (eds) NATO ASI Series, E. Applied Sciences, No.127. Martinus Nijhoff Publ. Dordrecht, pp 19–53
552. Jones E, Dewhirst M, Vujaskovic Z (2003) Hyperthermia improves the complete response rate for superficial tumors treated with radiation: results of a prospective randomized trial testing the thermal dose parameter CEM 43°T90. Int J Rad Oncol Biol Phys 57: S253–S254
553. Holt JAG (1977) Increase in X-ray sensitivity of cancer after exposure to 434 MHz electromagnetic radiation. J Bioeng 1:479–485
554. Blank M (1999) Coupling of AC Electric Fields to Cellular Processes. In: Abstracts of First International Symposium on Nonthermal Medical/Biological Treatments Using Electromagnetic Fields and Ionized Gases, ElectroMed'99, Norfolk VA, USA, 12–14 April 1999
555. Young RA (1990) Stress proteins and immunology. Ann Rev Immunol 8:401–420
556. Fajardo LF, Prionas SD, Kowalsky J et al (1988) Hyperthermia Inhibits angiogenesis. Radiation Research 114(2):297–306
557. Hetts SW (1998) To die or not to die. JAMA 279:300–307
558. Kopper L, Fesus L (eds) (2002) Apoptosis. Medicina, Budapest
559. Walser EM (2005) Percutaneous laser ablation in the treatment of hepatocellular carcinoma with a tumor size of 4 cm or smaller. J Vasc Interv Radiol 16(11):1427–1429
560. Mulcahy RT, Gould MN, Hidvegi E et al (1981) Hyperthermia and surface morphology of P388 ascites tumor cells: Effects on membrane modifications. Int J Radiat Biol 39(1): 95–106
561. Lepock JR, Cheng KW, Al-Qysi H et al (1983) Thermotropic lipid and protein transitions in Chinese hamster lung cell membranes: relationship to hyperthermia killing. Can J Biochem Cell Biol 61(6):421–427
562. Richier S, Sabourault C, Courtiade J et al (2006) Oxidative stress and apoptotic events during thermal stress in the symbiotic sea anemone, Anemonia visidis. FEBS J 273(18):4186–4198
563. Multhoff G (2002) Activation of natural killer cells by heat shock protein 70. Int J Hyperthermia 18(6):576–585
564. Guyton AC, Hall JE (2000) Textbook of Medical Physiology. W.B. Saunders Co., Philadelphia, London
565. Calder WA (1974) Consequences of body-size for avian energetics. In: Paynter RA (ed) Avian energetics, Nuttal Ornithological Club, Cambridge, MA, pp 41–51
566. Gillooly JF, Brown JH, West GB et al (2001) Effects of size and temperature on metabolic rate. Science 293(5538):2248–2251
567. Gautherie M (1982) Temperature and blood-flow patterns in breast cancer during natural evolution and following radiotherapy. Prog Clin Biol Res 107:21–64
568. Benzinger TH (1959) On physical heat regulation and the sence of temperature in man. Proc Natl Acad Sci USA 45(4):645–659
569. Erdmann B, Lang J, Seebass M (1998) Optimization of temperature distributions for regional hyperthermia based on a nonlinear heat transfer model. Ann NY Acad Sci 858(11):36–46

570. Morowitz HJ (1992) Beginnings of cellular life: metabolism recapitulates biogenesis. Yale University Press, New Haven, CT, p 69
571. Glaser R (1999) Biophysics. Springer, Berlin
572. Hobbie RK (1997) Intermediate physics for medicine and biology, Biological physics series. Springer, Berlin; AIP Press, New York, NY
573. Markovitch O, Agmon N (2007) Structure and energetics of the hydronium hydration shells. J Phys Chem A. 111(12), 2253–2256
574. Jeffrey GA (1997) An Introduction to Hydrogen Bonding (Topics in Physical Chemistry). Oxford University Press, Oxford
575. Szent-Gyorgyi A (1978) The living state and cancer. Marcel Dekker, New York, NY
576. Katchalsky A, Curran PF (1967) Non-equilibrium thermodynamics in biophysics. Harvard University Press, Cambridge, MA
577. Boltzmann L (1872) Weitere studien uber das warmegleihtgewicht unter gas moleculen. Wiener Berichte 66:275–370
578. Arrhenius S (1915) Quantitative laws in biological chemistry. G. Bell, London
579. Jackson MB (2006) Molecular and cellular biophysics. Cambridge University Press, Cambridge
580. Szasz O, Andocs G, Szasz A (2008) Thermally induced effects in oncothermia treatment. Symposium on Biophysical Aspects of Cancer, Electromagnetic mechanisms (in memoriam H. Froelich), Prague, 1–3 July 2008
581. Truhlar DG, Kohen A (2001) Convex Arrhenius plots and their interpretation. Proc Natl Acad Sci 98(3):848–851
582. Tolman RC (1920) Statistical mechanics applied to chemical kinetics. J Amer Chem Soc 42:2506–2528
583. Vattulainen I, Merikoski J, Ala-Nissila T et al (1997) Non-Arrhenius behaviour of surface diffusion near a phase-transition boundary. Phys Rev Lett 79(2):257–260
584. Nelson P (2004) Biological Physics. WH Freeman and Company, New York, NY
585. Wright NT (2003) On a relationship between the Arrhenius parameters from thermal damage studies. J Biol Eng 125(2):300–304
586. Huang K (2005) Lectures on statistical physics and protein folding. World Scientific, Singapore
587. Pelzer H, Wigner E (1932) Uber die Geschwindigkeitskonstante von Austauschreaktionen. Z Phys Chem B 15:445–471
588. Eyring H (1935) The Activated Complex in Chemical Reactions. J Chem Phys 3:107–115
589. Laidler KJ, King MC (1983) The development of transition-state theory. J Phys Chem 87:2657–2664
590. Pollak E, Talkner P (2005) Reaction rate theory: what it was, where is it today, and where is it going? Chaos 15(2):26116–26117
591. Miller WH (1993) Beyond transition-state theory: a rigorous quantum theory of chemical reaction rates. Acc Chem Res 26:174–181
592. Berne BJ, Borkovec M (1998) Classical theories of reaction dynamics: transition-state theory, spatial diffusion controlled reaction, and the energy diffusion limit. J Chem Soc Faraday Trans 94:2717–2723
593. Ward CA (1977) Effect of concentration on the rate of chemical reactions. J Chem Phys 67:229–235
594. Moore WJ, Pearson RG (1981) Kinetics and Mechanisms. Wiley, New York, NY
595. Hou KC, Palmer HB (1965) The kinetics of thermal decomposition of diacetylene in a flow system. J Phys Chem 69(3):858–862
596. Kolmogorov NN (1937) On the statistical theory of the crystallization of metals. Bull Acad Sci UssR Math Ser 1:355–359
597. Joshnson WA, Mehl PA (1939) Reaction kinetics in processes of nucleation and growth. Trans Amer Inst Mining (Metal) Engrs 135:416

598. Avrami MA (1941) Kinetics of phase change Parts: I. J. Chem. Phys. 7, 1103 (1939); Avrami. M.A.: Kinetics of phase change Parts: II. J. Chem. Phys. 8, 212 (1940); Avrami, M.A.: Kinetics of phase change Parts: III. J Chem Phys 9:1103
599. Augis JA, Bennett JE (1978) Calculation of the Avrami parameters fer heterogenous solid state reactions using a modification of the Kissinger method. J Therm Anal 13:285–291
600. Levine LE, Narayan KL, Kelton KF (1997) Finite size corrections for the Johnson-Mehl-Avrami-Kolmogorov equation. J Mater Res 12(1):124–131
601. Cope FW (1977) Detection of phase transitions and cooperative interactions by Avrami analysis of sigmoid biological time curves for muscle, nerve, growth, firefly, and infrared phosphorescence, of green leaves, melanin and cytochrome C. Physiol Chem Phys 9(4–5): 443–459
602. Cope FW (1977) Solid State physical replacement of Hodgkin-Huxley theory. Phase transformation kinetics of axonal potassium conductance. Physiol Chem Phys 9(2): 155–160
603. Cope FW (1977) The kinetics of biological phase transitions manifested by sigmoid time curves: a review of approaches. Physiol. Chem. Phys. 8(6):519–527
604. Cope FW (1980) Avrami analysis of electrical switching in hydrated melanin suggest dependence on a phase transition. Physiol Chem Phys 12:537–538
605. Suckjoon J, Bechhoefer J (2005) Nucleation and growth in one dimension. II. Application to DNA replication kinetics. Phys Rev E 71(1):011909
606. Haiyang Z (2005) Reconstructing DNA replication kinetics from small DNA fragments. Simon Fraser University, China
607. Binney JJ, Dowrick NJ, Fisher AJ et al (1992) The theory of critical phenomena. An introduction to the renormalization group. Oxford Science Publications, Oxford
608. Wolfram S (2002) A new kind of Science. Wolfram Media, Champaign, IL
609. Camazine S, Deneubourg JL, Franks NR et al (2003) Self-organization in biological systems. Princeton Studies in Complexity. Princeton University Press, Princeton, NJ
610. Streater RF (1999) Onsager relations in statistical dynamics. Open Syst Info Dyn 6(1): 87–100
611. Data from Swedish National Food Administration. http://www.slv.se. Cited 02 Sept 2010
612. Burton RF (1994) Physiology by numbers. Cambridge University Press, Cambridge
613. Alberts B et al (1994) Molecular Biology of the Cell, (3rd edition). Garland, New York, NY
614. Voet D, Voet JG (1990) Biochemistry. Wiley, New York, NY
615. Cotterill R (2005) Biophysics: an introduction. Wiley, San Francisco, CA
616. Darnell J, Lodish H, Baltimore D (1986) Molecular Cell Biology. Scientific American Books, New York, NY
617. Montgomery C, Reilly JJ, Jackson DM et al (2005) Validation of energy intake by 24-hour multiple pass recall: comparison with total energy expenditure in children aged 5–7 years. Br J Nutr 93:671–676
618. Nicholls DG, Lindberg O (1973) Brown-adipose-tissue mitochondria. The influence of albumin and nucleotides on passive ion permeabilities. Eur J Biochem 37(3):523–530
619. Andrianantoandro E, Basu S, Karig DK et al (2006) Synthetic biology: new engineering rules for am emerging discipline. Mol Syst Biol 2:1–14
620. Editorial (2007) Meanings of 'life'. Nature 447:1031–1032
621. Mathews CK, van Holde KE (1990) Biochemistry. The Benjamin/Cummings, Redwood City, CA
622. Gorshkov VG (1995) Physical and biological bases of life-stability: Man, Biota, Environment. Springer, Berlin
623. Winberg GG (1976) Dependence of energetic metabolism on body mass in water ectothermic animals. J General Biol 37:56–71
624. Gillooly JF, Allen AP, Savage VM et al (2006) Response to Clarke and Fraser: effects of temperature on metabolic rate. Funct Ecol 20:400–404

625. Blier PU, Guderley HE (1993) Mitochondrial activity in rainbow trout red muscle: the effect of temperature on the ADP-dependence of ATP synthesis. J Exp Biol 176(1): 145–158
626. Starritt EC, Angus D, Hargreaves M (1999) Effect of short-term training on mitochondrial ATP production rate in human skeletal muscle. J Appl Physiol 86:450–454
627. Kauffman SA (1993) The origins of order: self-organization and selection in evolution. Oxford University Press, Oxford
628. Levinthal C (1968) Are there pathways for protein folding? Extrait du Journal de Chimie Physique 65(1):44–45
629. Schneider SH (1989) The greenhouse effect: science and policy. Science 243(4892): 771–781
630. Whittaker RH, Likens GE (1975) The biosphere and man. In: Leith H, Whittaker RH (eds) Primary productivity of the biosphere. Springer, Berlin
631. Clarke A (2006) Temperature and the metabolic theory of ecology. Funct Ecol 20(2): 405–412
632. Odum EP (1983) Basic ecology. Saunders College Pub, New York, NY
633. Feynman PR, Leighton RB, Sands M (1963) The Feynman lectures on physics. Addison-Wesley Publ Co Reading, MA, & Caltech, CA
634. Szent-Gyorgyi A. (1941) Towards a new biochemistry? Science 93(2426):609–611
635. Szent-Gyorgyi A (1946) Internal photo-electric effect and band spectra in proteins. Nature 157:875–875
636. Cope FW (1971) Evidence from activation energies for superconductive tunneling in biological systems at physiological temperatures. Physiol Chem Phys 3:403–410
637. Liboff AR (1985) Geomagnetic cyclotron resonance in living cells. J Biol Phys 13(4): 99–102
638. Liboff AR (2003) Ion cyclotron resonance in biological systems: experimental evidence. In: Stavroulakis P (ed) Biological effects of electromagnetic fields. Springer, Berlin, pp 6–113
639. Schrodinger E (1967) What is life? Cambridge University Press, Cambridge, United Kingdom
640. Haken H (1987) Self-Organization and Information. Phys Script 35(3):247–254
641. Sornette D (2000) Chaos, Fractals, Self-Organization and Disorder: Concepts and Tools. Springer, Berlin
642. Walleczek J (ed) (2000) Self-organized biological dynamics & nonlinear control. Cambridge University Press, Cambridge
643. Deering W, West BJ (1992) Fractal physiology. IEEE Engg Med Biol 11(2):40–46
644. West BJ (1990) Fractal Physiology and Chaos in Medicine. World Scientific, Singapore
645. Bassingthwaighte JB, Leibovitch LS, West BJ (1994) Fractal Physiology. Oxford University Press, Oxford
646. Musha T, Sawada Y (eds) (1994) Physics of the living state. IOS Press, Amsterdam
647. Brown JH, West GB (eds) (2000) Scaling in biology. Oxford University Press, Oxford
648. Brown JH, West GB, Enquist BJ (2005) Yes, West, Brown and Enquist's model of allometric scaling is both mathematically correct and biologically relevant. Funct Ecol 19(4):735–738
649. West GB, Brown JH (2005) The origin of allometric scaling laws in biology from genomes to ecosystems: towards a quantitative unifying theory of biological structure and organization. J Exp Biol 208:1575–1592
650. Albert R (2005) Scale-free networks in cell biology. J Cell Sci 118(Pt 21):4947–4957
651. Kleiber M (1947) Body size and metabolic rate. Physiol Rev 27:511–541
652. West GB, Brown JH, Enquist BJ (1999) The fourth dimension of life: fractal geometry and allometric scaling of organisms. Science 284(5420):1677–1679
653. Semenza GL (2008) Tumor metabolism: cancer cells give and take lactate. J Clin Invest 118(12):3835–3837

654. Stine KE: On-line review: Energy metabolism and cancer. http://personal.ashland.edu/kstine/Research/Energy%20and%20disease.pdf
655. Rathmell JC, Newgard CB (2009) Biochemistry. A glucose-to-gene link. Science 324(5930):1021–1022
656. Rosch PJ, Markov MS (2004) Bioelectromagnetic medicine. Marcell Decker, New York, NY
657. Reid B, McCaig CD, Zhao M et al (2005) Wound healing in rat cornea: the role of electric currents. FASEB J 19:379–386
658. Carrasco D, Weih F, Bravo R (1994) Developmental expression of the mouse c-rel proto-oncogene in hematopoietic organs. Development 120(10):2991–3004
659. Caubet JF, Bernaudin JF (1988) Expression of the c-fos proto-oncogene in bone, cartilage and tooth forming tissues during mouse development. Biol Cell 64(1):101–104
660. Workalemahu G et al (2003) Human 1 × 3-T lymphocytes express and synthesize connective tissue growth Factor: effect of IL-15 and TGF-131 and comparison with NO-T lymphocytes. J Immunol 170:153–157
661. Blotnick S, Peoples GE, Freeman MR et al (1994) T lymphocytes synthesize and export heparin-binding epidermal growth factor-like growth factor and basic fibroblast growth factor, mitogens for vascular cells and fibroblasts: differential production and release by CD4′ and CD8′ T cells. PNAS 91(8):2890–2894
662. Son Y, Hong H, Kim J (2004) Identification of substance-p as an early inductive cytokine of corneal wound and its possible role in the mobilization of mesenchymal stem cell and corneal wound healing. Invest Ophthalmol Vis Sci 45:1423
663. Lavon N, Yanuka O, Benvenisty N (2004) Differentiation and isolation of hepatic-like cells from human embryonic stem cells. Differentiation 72(5):230–238
664. Noszczyk BH, Majewski ST (201) p63 expression during normal cutaneous wound healing in humans. Plast Reconstr Surg 108(5):1242–1250
665. Shin BJ, Ching SS (2003) A case of limbal stem cell deficiency in a patient with chronic mucocutaneous candiddiasis. Invest Ophthalmol Vis Sci 44:1359
666. Houghton J, Stoicov C, Nomura S et al (2004) Gastric cancer originating from bone marrow-derived cells. Science 306(5701):1568–1571
667. Ouahes N, Phillips TJ, Park HY (1998) Expression of c-fos and c-Ha-ras protooncogenes is induced in human chronic wounds. Derrnatol Surg 24(12):1354–1357, Discussion 1358
668. Huang S, Trujillo JM, Chakrabarty S (1992) Proliferation of human colon cancer cells: role of epidermal growth factor and transforming growth factor. Int J Cancer 52(6):978–986
669. Dahiya R, Lee C, Haughney PC et al (1996) Differential gene expression of transforming growth factors alpha and beta, epidermal growth factor, keratinocyte growth factor, and their receptors in fetal and adult human prostatic tissues and cancer cell lines. Urology 48(6):963–970
670. Mizuno K, Sone S, Orino E et al (1994) Autonomous expressions of cytokine genes by human lung cancer cells and their paracrine regulation. Jpn J Cancer Res 85(2):179–186
671. Zhang H, Vutskits L, Pepper MS et al (2003) VEGF is a chemoattractant for FGF-2-stimulated neural progenitors. J Cell Biol 163(6):1375–1384
672. Cicuttini FM, Begley CG, Boyd AW (1992) The effect of recombinant stem cell factor (SCF) on purified CD34-positive human umbilical cord blood progenitor cells. Growth Factors 6(1):31–39
673. Meyer NA, Muller MJ, Herndon DN (1994) Nutrient support of the healing wound. New Horiz 2(2):202–214
674. Printz C (2001) Spontaneous regression of melanoma may offer insight into cancer immunology. J Natl Cancer Inst 93(14):1047–1048
675. Turner AM, Zsebo KM, Martin F et al (1992) Nonhematopoietic tumor cell lines express stem cell factor and display c-kit receptors. Blood 80(2):374–381
676. Oehr P, Biersack HJ, Coleman RE (eds) (2004) PET and PET-CT in Oncology. Springer, Berlin

677. Larson SM (2004) Positron Emission Tomography-Based Molecular Imaging in Human Cancer: Exploring the Link between Hypoxia and Accelerated Glucose Metabolism. Clin Cancer Res 10(7):2203–2204
678. Nelson DL, Cox MM (2005) Lehninger principles of biochemistry fourth edition. W.H. Freeman and Company, New York, NY, p 543
679. Hsu PP, Sabatini DM (2008) Cancer metabolism: Warburg and beyond. Cell 134(5): 703–707
680. Szent-Gyorgyi A (1978) How new understandings about the biological function of ascorbic acid may profoundly affect our lives. Executive Health 14(8):1–4
681. Szasz A, van Noort D, Scheller A et al (1994) Water states in living systems. I. Structural aspects. Physiol. Chem. Phys. 26(4), 299–322
682. Eisenberg D, Kauzmann W (1969) Structure and properties of water. Clarendon Press, Oxford
683. Nelson GM, Enderby JE (1985) Water and aqueous solutions. Adam Hilger, Bristol
684. Maryan M, Szasz A, Szendro P et al (2005) Synergetic model of the formation of non-crystalline structures. Journal of Non-Crystalline Solids 351(2):189–193
685. Nemethy G, Scheraga HA (1962) Structure of water and hydrophobic bonding in proteins. I. A model for the thermodynamic properties of liquid water. J Chem Phys 36:3382–3400
686. Marjan M, Kurik M, Kikineshy A et al (1999) Two structure model of liquid water. Modelling Simul Mater Sci Eng 7(3):321–331
687. Urquidi J (2001) Theoretical studies on liquid water. PhD Thesis Texas Tech University. http://etd.lib.ttu.edu/theses/available/etd-07312008-31295017999581/unrestricted/31295017999581.pdf
688. Pauling L (1959) The structure of water. In: Hadzi D, Thompson H (eds) Hydrogen bonding, Pergamon Press Ltd, London, pp 1–6
689. Szasz A (1991) An electronically driven instability: the living state. Physiol. Chem. Phys. Med. NMR 23(1):43–50
690. Agmon N (1995) The Grotthuss mechanism. Chem Phys Lett 244(5): 456–462
691. Markovitch O, Agmon N (2007) Structure and energetics of the hydronium hydration shells. J Phys Chem A. 111(23):2253–2256
692. Maryan MI, Kikineshy A, Szendrő P et al (2001) Modeling of the dissipative structure of water. Acta Technologica Agriculturae Slovaca Universitas Agriculturae Nitriae 3:77–80
693. Szendro P, Koltay J, Szasz A et al (1999) Is the structure of water convertible in a physical way? Hungarian Agric Engg 12:43–45
694. Fu-Ming Tao (2003) Solvent Effects of Individual Water Molecules. In: Buch V, Devilin JP (eds) Water in confining geometries, cluster physics. Springer, Berlin, pp 79–99
695. Mott NF, Jones H (1958) The theory of the properties of metals and alloys. Dover Publ Inc, New York
696. Cope FW (1969) Nuclear magnetic resonance evidence using D2O for structured water in muscle and brain. Biophys J 9(3):303–319
697. Damadian R (1971) Tumor detection by nuclear magnetic resonance. Science 171(3976):1151–1153
698. Cope FW (1975) A review of the applications of solid state physics concepts to biological systems. J Biol Phys 3(1):1–41
699. Hazlewood CF, Nichols BL, Chamberlain NF (1969) Evidence for the existence of a minimum of two phases of ordered water in skeletal muscle. Nature 222(195):747–750
700. Hazlewood CF, Chang DC, Medina D et al (1972) Distinction between the Preneoplastic and Neoplastic State of Murine Mammary Glands. Proc Natl Acad Sci USA 69(6): 1478–1480
701. Szent-Gyorgyi, A (1980) The living state and cancer. Physiol Chem Phys 12, 99–110
702. Gniadecka M, Nielsen OF, Wulf HC (2003) Water content and structure in malignant and benign skin tumors. J Mol Struc 661–662:405–410

703. Beall PT et al (1979) Water-relaxation times of normal, preneoplastic, and malignant primary cell cultures of mouse mammary gland. In: 23rd Annual Meeting of the Biophysical Society, Atlanta, Georgia, USA, 26–28 February 1979
704. Chung SH, Cerussi AE, Klifa C et al (2008) In vivo water state measurements in breast cancer using broadband diffuse optical spectroscopy. Phys Med Biol 53:6713–6727
705. Fiskum G (2000) Mitochondrial participation in ischemic and traumatic neural cell death. J Neurotrauma 17(10):843–855
706. Ichas F, Mazat JP (1998) From calcium signaling to cell death: two conformations for the mitochondrial permeability transition pore. Switching from low- to high- conductance state. Biochimica et Biophysica Acta 1366:33–50
707. Bonnet S, Archer SL, Allalunis-Turner J et al (2007) A Mitochondria-K+ Channel Axis Is Suppressed in Cancer and Its Normalization Promotes Apoptosis and Inhibits Cancer Growth. Cancer Cell 11(1):37–51
708. Chidanbaram, R., Ramanadham, M (1991) Hydrogen bonding in biological molecules-an update. Physica B 174(1–4):300–305
709. Szent-Gyorgyi A (1960) Introduction to submolecular biology. Academic, London
710. Akerlof GA, Shiller RJ (2009) Animal spirits. Princeton University Press, Oxford
711. Moses ME, Hou C, Woodruff WH et al (2008) Revisiting a model of oncogenic growth: estimating model parameters from theory and data. Am Nat 171(5):632–645
712. Pamatmat MM (2005) Measuring aerobic and anaerobic metabolism of benthic infauna under natural conditions. J Exp Zool 228(3):405–413
713. West GB, Woodruf WH, Born JH (2002) Allometric scaling of metabolic rate from molecules and mitochondria to cells and mammals. Proc Natl Acad Sci USA 99(1): 2473–2478
714. Lane N (2006) Mitochondria: Key to Complexity. In: Martin W (ed) Origins of mitochondria and hydrogenosomes, Chapter 2. Springer, Heidelberg
715. Szent-Gyorgyi A (1998) Electronic biology and cancer. Marcel Dekker, New York, NY
716. Weinberg RA (1998) One renegade cell. Basic Books, A Member of the Perseus Books Group, New York, NY
717. Kurakin A (2009) Scale-free flow of life: on the biology, economics, and physics of the cell. Theor. Biol Med Model 6:6
718. Getling AV, Rayleigh-Benard C (1998) Structures and dynamics. World Scientific, Singapore
719. Hochachka PW (1999) The metabolic implications of intracellular circulation. Proc Natl Acad Sci USA 96(22):12233–12239
720. Coulson RA (1986) Metabolic rate and the flow theory: a study in chemical engineering. Comp Biochem Physiol A Comp Physiol 84(2):217–229
721. Szent-Gyorgyi A (1968) Bioelectronics, A Study on Cellular Regulations, Defense and Cancer. Acad. Press, New York, London
722. Brown MF, Gratton TP, Stuart JA (2007) Metabolic rate does not scale with body mass in cultured mammalian cells. Am J Physiol Regul Integr Comp Physiol 292: R2115–R2121
723. Jun J Pepper JW, SAVAGE VM et al (2003) Allometric scaling of ant foraging trail networks. Evol Ecol Res 5(2):297–303
724. Drasdo D, Höhme S (2005) A single-cell-based model of tumor growth in vitro: monolayers and spheroids. Phys Biol 2:133–147
725. Bru A, Albertos S, Luis Subiza et al (2003) The Universal dynamics of tumor growth. Biophysical J 85(5):2948–2961
726. Buceta J, Galeano J (2005) Comments on the article "the universal dynamics of tumor growth" by A. Bru et al. Biophysical J 88(5):3734–3736
727. Bru A, Pastor JM, Fernaud I et al (1998) Super-rough dynamics on tumor growth. Phys Rev Lett 81(18):4008–4011
728. Williams JM (2002) Thermal and Nonthermal Mechanisms of the Biological Interaction of Microwaves. http://arxiv.org/ftp/physics/papers/0102/0102007.pdf

729. Lv Y-G, Liu J (2003) A theoretical way of distinguishing the thermal and non-thermal effects in biological tissues subject to EM radiation. Forschung im Ingenieurwesen 67(6):242–253
730. Lovely RH, Mizumori SJY, Johnson RB et al (1983) Subtle consequences of exposure to weak microwave fields: are there nonthermal effects? In: Adair ER (ed) Microwaves and thermoregulations. Academic, New York, NY, pp 401–429
731. European Cooperative Action on Science and Technology COST281, Germany's Forschungsgemeinschaft Funk e.V. (FGF), State Ministry of Environment and Transport (2005) "Subtle temperature elevations by weak RF-fields and their possible influences on processes of thermoregulation in cells and organisms." Research Association for Radio Applications, Stuttgart, Germany, 21–23 November 2005
732. Schwan HP (1957) Electrical properties of tissue and cell suspensions. Adv Biol Med Phys 5:147–209
733. Cole KS (1968) Membranes, ions and impulses. University of California Press, Berkeley, LA
734. Pethig R, Kell DB (1987) The passive electrical properties of biological systems: their significance in physiology, biophysics and biotechnology. Phys Med Biol 32:933–977
735. Pennock BE, Schwan HP (1969) Further observations on the electrical properties of haemoglobin bound water. J Phys Chem 73:2600–2610
736. Schwan HP (1963) Determination of biological impedances. In: Physical Techniques in Biological Research, vol. 6, pp 323–406, Academic Press, New York, NY
737. Pethig R (1984) Dielectric properties of biological materials: biophysical and medical application. IEEE Trans Electr Insulation E1–19(5):453–474
738. Pliquett F, Pliquett U (1992) Tissue impedance, measured by pulse deformation. In: Lahtinen T (ed) The 8th International Conference on Electrical Bio-impedance, University of Kuopio, Finland, 28–31 July 1992, pp 179–181
739. Loft SM et al (1992) Bioimpedance and cancer therapy. In: Lahtinen T (ed) The 8th International Conference on Electrical Bio-impedance, University of Kuopio, Finland, 28–31 July 1992, pp 119–121
740. Baker LE (1989) Principle of the Impedance Technique. IEEE Engg Med Biol Mag 8(1):11–15
741. Min M, Parve T, Ronk A et al (2007) Synchronous sampling and demodulation in an instrument for multifrequency bioimpedance measurement. Instrum Meas IEEE Trans 56(4):1365–1372
742. Scholz B, Anderson R (2000) On electrical impedance scanning – principles and simulations. Erectromedica 68:35–44
743. Barnett A (1940) Electrical method for studying water metabolism and translocation in body segments. Proc Soc Exp Biol Med 44:142–147
744. Nyboer J, Bango S, Barnett A et al (1940) Radiocardiograms – the electrical impedance changes of the heart in relation to electrocardiograms and heart sounds. J Clin Invest 19:963–966
745. Giu PJ et al (eds) (1999) Electrical bioimpedance methods and biotechnology: annals of the new york academy of sciences. N Y Acad Sci
746. Grimnes S, Martinsen OG (2000) Bioimpedance and Bioelectricity Basis. Academic, New York, NY
747. Salinari S, Bertuzzi A, Mingrone G et al (2002) New bioimpedance model accurately predicts lower limb muscle volume: validation by magnetic resonance imaging. Am J Physiol Endocrinol Metab 282(4):E960–E966
748. Smit HJ, Vonk Noordegraaf A, Roeleveld RJ et al (2002) Epoprostenol-induced pulmonary vasodilatation in patients with pulmonary hypertension measured by electrical impedance tomography. Physiol Meas 23(1):237–243
749. Molinari M, Cox SJ, Blott BH et al (2002) Comparison of algorithms for non-linear inverse 3D electrical tomography reconstruction. Physiol Meas 23(1):95–104

750. Cherepenin V, Karpov A, Korjenevsky A et al (2002) Preliminary static EIT images of the thorax in health and disease. Physiol Meas 23(1):33–41
751. Holder D (ed) (2002) Biomedical applications of electrical impedance tomography. Physiol Meas 23(1):3
752. Malich A, Boehm T, Facius M et al (2001) Differentiation of mammographically suspicious lesions: evaluation of breast ultrasound, MRI mammography and electrical impedance scanning as adjunctive technologies in breast cancer detection. Clin Radiol 56(4):278–283
753. Cherepenin V, Karpov A, Korjenevsky A et al (2001) A 3D electrical impedance tomography (EIT) system for breast cancer detection. Physiol Meas 22(1):9–18
754. Har-Shai Y, Glickman YA, Siller G el al (2005) Electrical impedance scanning for melanoma diagnosis: a validation study. Plast Reconstr Surg 116(3):782–790
755. Kyle UG, Bosaeus I, De Lorenzo AD et al (2004) Bioelectrical impedance analysis – part I: review of principles and methods. Clin Nutr 23(5):1226–1243
756. Kyle UG, Bosaeus I, De Lorenzo AD et al (2004) Bioelectrical impedance analysis – part II: utilization in clinical practice. Clin Nutr 23(5):1430–1453
757. FDA approval date, 4/16/1999, http://www.cancer.org/docroot/NWS/content/NWS_1_1×_FDA_Approves_T_Scan.asp, http://www.newsrx.com/newsletters/Medical-Letter-on-the-CDC-and-FDA/2005-06-05/0605200533328DC.html
758. Peirce A (tutor) IAM-CSC-PIMS senior undergraduate math modeling workshop. The Institute of Applied Mathematics (IAM), The University of British Columbia
759. Metherall P (1998) Three-dimensional electrical impedance tomography of the human thorax. PhD Thesis, University of Sheffield
760. Babaeizadeh S, Brooks DH, Isaacson D (2007) 3-D Electrical impedance tomography for piecewise constant domains with known internal boundaries. IEEE Trans Biomed Engg 54(1):2–10
761. Mikac U, Demsar F, Beravs K et al (2001) Magnetic Resonance Imaging of alternating electric currents. Magn Reson Imaging 19(6):845–856
762. Joy M, Scott G, Henkelman M (1989) In-Vivo Detection of Applied Electric Currents by Magnetic Resonance Imaging. Magn Reson Imaging 7(1):89–94
763. Seersa I, Beravs K, Dodd NJF et al (1997) Electric current Imaging of Mice tumors. MRM 37:404–409
764. Scott GC, Joy MLG, Armstrong RL et al (1995) Electromagnetic considerations for RF current density imaging. IEEE Trans Med Imaging 14(3):515–524
765. Sha L, Ward ER, Story B (2002) A review of dielectric properties of normal and malignant breast tissue. IEEE Southeast Con 457–462
766. Blad B, Wendel P, Jönsson M et al (1999) An electrical impedance index to distinguish between normal and cancerous tissues. J Med Engg Tech 23(2):57–62
767. Brunner G (2007) Elektrohyperthermie von Hautkrebbszellen: Neue Ergebnisse zu potentiellen molekularen Wirkungsmechanismen. Hyperthermie Symposium, Cologne, Germany, 19–20 October 2007
768. McRae DA, Esrick MA, Mueller SC (1997) Non-invasive, in-vivo electrical impedance of EMT-6 tumors during hyperthermia: correlation with morphology and tumor-growth delay. Int J Hyperthermia 13(1):1–20
769. Esrick MA, McRae DA (1994) The effect of hyperthermia induced tissue conductivity changes on electrical impedance temperature mapping. Phys Med Biol 39: 133–144
770. McRae DA, Esrick MA (1992) The dielectric parameters of excised EMT-6 tumors and their change during hyperthermia. Phys Med Biol 37:2045–2058
771. McRae DA, Esrick MA, Mueller SC (1999) Changes in the non-invasive, in vivo electrical impedance of the xenografts during the necrotic cell-response sequence. Int J Radiat Oncol Biol Phys 43:849–857
772. Dissado LA, Alison JM, Hill RM et al (1995) Dynamic scaling in the dielectric response of excised EMT-6 tumors undergoing hyperthermia. Phys Med Biol 40:1067–1084

773. Szendro P, Vincze G, Szasz A (2001) Bio-response on white-noise excitation. Electromagn Biol Med 20(2):215–229
774. Häfner H-M, Brauer K, Eichner M et al (2005) Wavelet analysis of cutaneous blood flow in melanocytic skin lesions. J Vasc Res 42(1):38–46
775. Blad B, Baldetorp B (1996) Impedance spectra of tumor tissue in comparison with normal tissue; a possible clinical application for electric impedance tomography. Physiol Meas 4A(17):A105–A115
776. Dubois JM, Rouzaire-Dubois B (2004) The influence of cell-volume changes on tumor cell proliferation. Eur Biophys.J 33(3):227–232
777. Dissado LA (1990) A fractal interpretation of the dielectric response of animal tissues. Phys Med Biol 35:1487–1503
778. El-Lakkani A (2001) Dielectric response of some biological tissues. Bioelectromagnetics 22(4):272–279
779. Loewenstein WR, Kanno Y (1967) Intercellular communication and tissue growth. J Cell Biol 33(2):225–234
780. Loewenstein WR (1999) The touchstone of life, Molecular information, cell communication and the foundations of life. Oxford University Press, Oxford, pp 298–304
781. Soler AP, Miller RD, Laughlin KV et al (1999) Increased tight junctional permeability is associated with the development of colon cancer. Carcinogenesis 20(8):1425–1431
782. Haemmerich D, Staelin ST, Tsai JZ et al (2003) In vivo electrical conductivity of hepatic tumors. Physiol Meas 24:251–260
783. Smith SR, Foster KR, Wolf GL (1986) Dielectric properties of VX-2 carcinoma versus normal liver tissue. IEEE Trans Biomed Eng 33(5):522–524
784. Jossinet J (1998) The impedivity of freshly excised human breast tissue. Physiol Meas 19:61–75
785. Jossinet J, Schmitt M (1999) A review of parameters for the bioelectrical characterization of breast tissue. Electrical Bioimpedance Methods: Appl Med Biotechnol 873:30–41
786. Chauveau N, Hamzaoui L, Rochaix P et al (1999) Ex vivo discrimination between normal and pathological tissues in human breast surgical biopsies using bioimpedance spectroscopy. Ann N Y Acad Sci 873:42–50
787. Tosso S, Piccoli A, Gusella M et al (2000) Altered tissue electric properties in lung cancer patients as detected by bioelectric impedance vector analysis. Nutrition 16(2):120–124
788. Tosso S, Piccoli A, Gusella M et al (2003) Bioimpedance vector pattern in cancer patients without disease versus locally advanced or disseminated disease. Nutrition 19(6): 510–514
789. Glickman YA, Filo O, David M et al (2003) Electrical Impedance Scanning: a new approach to skin cancer diagnosis. Skin Res Technol 9(3):262–268
790. Siemens Impedance Tomograph TRANS-SCAN, http://findarticles.com/p/articles/mi_m3498/is_6_62/ai_54753349
791. Skourou C, Hoopes PJ, Strawbridge RR et al (2004) Feasibility studies of electrical impedance spectroscopy for early tumor detection in rats. Physiol Meas 25(1):335–346
792. Chillcott TC, Coster HG (1999) Electrical impedance tomography study of biological processes in a single cell. Ann N Y Acad Sci 873:269–286
793. McRae DA, Esrick MA (1996) Deconvolved electrical impedance spectra track distinct cell morphology changes. IEEE Trans Biomed Eng 43(6):607–618
794. Gershing E (1999) Monitoring temperature-induced changes in tissue during hyperthermia by impedance methods. In: Riu PJ, Rosell J, Bragos R et al (eds) Electrical bioimpedance methods. Ann New York Acad Sci 873:13–20
795. McRae DA, Esrick MA (1993) Changes in electrical impedance of skeletal muscle measured during hyperthermia. Int J Hyperthermia 9(2):247–261
796. Shchepotin IB, McRae DA, Shabahang M et al (1997) Hyperthermia and verapamil inhibit the growth of human colon cancer xenografts in vivo through apoptosis. Anticancer Res 17(3C):2213–2216

797. Casas O, Bragós R, Riu PJ et al (1999) In vivo and in situ ischemic tissue characterization using electrical impedance spectroscopy. Ann N Y Acad Sci 873:51–58
798. Schafer M, Kirlum H-J, Schleger C et al (1999) Dielectric properties of skeletal muscle during ischemia in the frequency-range from 50 Hz to 200 MHz. Ann NY Acad Sci 873: 59–64
799. Gheorghiu M, Gersing E, Gheorghiu E (1999) Quantitative analysis of impedance spectra of organs during ischemia. Ann N Y Acad Sci 873:65–71
800. Sabah NH (200) Rectification in Biological Membranes. IEEE Engg Med Biol 19(1): 106–113
801. Osterman KS, Paulsen KD, Hoopes PJ (1999) Application of linear circuit models to impedance spectra in irradiated muscle. Ann N Acad Sci 873:21–29
802. Santini MT, Cametti C, Zimatore G et al (1995) A dielectric relaxation study on the effects of the antitumor drugs Lomidamineand Rhein on the membrane electrical properties of Erlich ascites tumor cells. Anticancer Res 15:29–36
803. Keese CR, Wegener J, Walker S et al (2004) Electrical wound-healing assay for cells in vitro. Proc Nat Acad Sci USA 101(6):1554–1559
804. Kyle AH, Chan CTO, Minchinton AI (1999) Characterization of three-dimensional tissue culture using electrical impedance spectroscopy. Biophys Soc 76(5):2640–2648
805. De Blasio BF, Laane M, Walkmann T et al (2004) Combining optical and electrical impedance techniques for quantitative measurement of confluence in MDCK-I cell cultures. BioTechniques 36(4):650–662
806. Overview of ECIS Method (ECIS^{TM} or Electric Cell-substrate Impedance Sensing). In: Giaever I Nobel-laureate (leaded). www.biophys.com. Cited 02 Sept 2010
807. Avitall B, Mughal K, Hare J et al (1997) The effects of electrode-tissue contact on radiofrequency lesion generation. Pacing Clin Electrophysiol 20(12 Pt 1):2899–2910
808. Schmidt D, Trubenbach J, Konig CW et al (2003) Radiofrequency ablation ex vivo: comparison of the efficacy impedance control mode versus manual control mode by using internally cooled clustered electrode. Fortschr Geb Rontgenstr Neuen Bildgeb Verfahr 175(7):967–972
809. National Institute of Health Technology Assessment Conference Statement (1994) Bioelectric Impedance Analysis in Body Composition Measurement. USA, 12–14 December 1994
810. Bowen WD, Beck CA, Iverson SJ (1999) Bioelectrical Impedance analysis as a means of estimating total body water in grey seals. Can J Zool 77(3):418–422
811. Talluri T, Lietdke RJ, Evangelisti A et al (1999) Fat-free mass qualitative assessment with bioelectric impedance analysis. Ann N Y Acad Sci 873:94–98
812. Goovaerts HG, Faes Th-JC, de Valk-de Roo GW et al (1999) Estimation of extracellular volume by two frequency measurement. Ann N Y Acad Sci 873(1):99–104
813. Muftuler TL, Hamamura MJ, Birgul O et al (2006) In vivo MRI electrical impedance tomography (MREIT) of tumors. Tech Cancer Res Treat 5(4):381–387
814. Joy ML (2004) MR current density and conductivity imaging: the state of the Art. Conf Proc IEE Eng Med Biol Soc 7:5315–5319
815. Oh SH, Lee BI, Woo EJ et al (2003) Conductivity and current density image reconstruction using harmonic Bz algorithm in magnetic resonance electrical impedance tomography. Phys Med Biol 48(19):3101–3116
816. Sersa I, Beravs K, Dodd NJ et al (1997) Electric current Imaging of Mice tumors. Magn Reson Med 37(3):404–409
817. Joines WT, Zhang Y, Li C et al (1994) The measured electrical properties of normal and malignant human tissues from 50 to 900 MHz. Med Phys 21(4):547–550
818. Inagaki T, Bhayani SB, Allaf M et al (2004) Tumor capacitance: electrical measurement of renal neoplasia. J Urol 172(2):454–457

819. Nalcioglu O (2006) Magnetic Resonance Electric Impedance Tomography – A novel technique for cancer imaging. Reference section, Business briefing: future directions in imaging, http://citeseerx.ist.psu.edu/viewdoc/summary?doi=10.1.1.86.1820
820. Aberg P, Geladi P, Nicancer I et al (2004) Skin bioimpedance – electronic views of malignancies. Proc XII ICEBI, Gdansk, pp 79–82
821. Fukumoto T, Eom GM, Ohba S et al (2007) Temporal resolution of the skin impedance measurement in Frequency-Domain method. IEEE Trans Biomed Engg 54(1):170–173
822. Sapper A, Wegener J, Janshoff A (2006) Cell motility probed by noise analysis of thickness shear mode resonators. Anal Chem 15;78(14):5184–5191
823. Sapper A, Reiss B, Janshoff A et al (2006) Adsorption and fluctuations of giant liposomes studied by electrochemical impedance measurements. Langmuir 22(2):676–680
824. Mentzel HJ, Malich A, Kentouche K et al (2003) Electrical impedance scanning – application of this new technique for lymph node evaluation in children. Rediatr Radiol 33(7):461–466
825. Gupta D, Lammersfeld CA, Dahlk S et al (2004) Phase angle, determined by bioelectrical impedance, as a prognostic indicator in advanced colorectal cancer. J Clin Oncology 2004 ASCO Annual Meeting Proc 22:(14S):3689
826. Emtestam L, Nicander I, Stenström M et al (1998) Electrical impedance of nodular basal cell carcinoma: a pilot study. Dermatology 197(4):313–316
827. Gonzalez-Correra CA, Brown BH, Smallwood RH et al (1999) Virtual biopsies in Barrett's Esophagus using an impedance probe. Ann N Y Acad Sci 873:313–321
828. Freeman T (2007) Non-invasive test reduces need for biopsies. SpectRx Company – LightTouch, Medicalphysicsweb, 24 January 2007
829. Goff LL, Lecuit T (2009) Phase transition in a cell. Science 324(5935):1654–1655
830. Milazzo G (1988) Applications of bioelectrochemistry in medicine. Bioelectrochem Bioenerget 19:191–205
831. Galvas-Medici R, Day-Magdaleno SR (1976) Extremely low frequency, weak electric fields affect schedule-controlled behaviour of monkeys. Nature 261(5557):256–259
832. Adey WR (1993) Biological effects of electromagnetic fields. J Cell Biochem 51(4): 410–416
833. Randerson J (2007) Electrosmog in the clear with scientists. The Guardian http://www.guardian.co.uk/technology/2007/jan/18/guardianweeklytechnologysection4. Cited 18 Jan 2007
834. Oschman LJ (2000) Energy Medicine The Scientific Basis. Churchill Livingstone, Edinburgh London New York, NY
835. Rampl I (2008) New biophysical field attacks cancer cells. http://www.rampl.eu
836. Rein G, Tiller WA (1996) Anomalous information storage in water: spectroscopic evidence for non-quantum informational transfer. Proc. 3rd Int Symp New Energy, Denver, Colorado, 24–28 April 1996, p 365
837. Meyl K (2001) Scalar waves: theory and experiments. J Sci Explor 15(2):199–205
838. Tiller WA (1999) Alternative medicine. Subtle energies. Sci Med 6(3), http://tillerfoundation.com/subtle-energies.html
839. Oschman LJ (2005) A review of energy medicine: the scientific basis. Skeptic Magazine 11(3):89–93
840. Takebe H, Shiga T, Kato M et al (2001) Biological and Health Effects from Exposure to Power-line Frequency Electromagnetic Fields. IOS Press, Amsterdam
841. Bruhn GW (2000) Commentary on the Chapter "Scalar Waves" in "Energy Medicine – The Scientific Basis". http://www.mathematik.tu-darmstadt.de/~bruhn/Commentary-Oschman.htm
842. Bruhn GW (2001) On the existence of K. Meyl's Scalar Waves. J Sci Explor 15(2):206–210
843. Blank M, Goodman R (2009) Electromagnetic fields stress living cells. Pathophysiology 16(2–3):71–78

844. Schwan HP (1989) Frequency selective propagation of extracellular electrical stimuli to intracellular compartments. In: Adey WR, Lawrence AF (eds) Nonlinear electrodynamics in biological systems, Plenum Press, London, pp 327–338
845. Velizarov S, Raskmark P, Kwee S (1999) The effects of radiofrequency fields on cell proliferation are non-thermal. Bioelectrochem Bioenerget 48(1):177–180
846. Gimsa J, Marszalek P, Loewe U et al (1991) Dielectrophoresis and electrorotation of neurospora slime and murine myeloma cells. Biophys J 60(4):749–760
847. Zhao M, Forrester JV, McCaig AD (1999) A small, physiological electric field orients cell division. Proc Natl Acad Sci USA 96(9):4942–4946
848. Kamp F, Astumian RD, Wesrethoff HV (1988) Coupling of vectorial proton flow to a biochemical reaction by local electric interactions. Proc Natl Acad Sci USA 85(11): 3792–3796
849. Tsong TY, Chang C-H (2003) Ion pump as Brownian motor: theory of electroconformational coupling and proof of ratchet mechanism for Na,K ATPase action. Physica A. 321(1), 124–138
850. Clejan S, Ide C, Walker C et al (1996) Electromagnetic field induced changes in lipid second messengers. J Lipid M
851. Chenguo Y, Dengbin Mo, Chengxiang L et al (2007) Study of transmembrane potentials of inner and outer membranes induced by pulsed-elecric-field model and simulation. IEEE Trans Plasma Sci 35(5):1541–1549
852. Marszalek P, Liu D-S, Tsong TY (1990) Schwan equation and transmembrane potential induced by alternating electric field. Biophys J 58(4):1053–1058
853. Broude N et al (1994) Correlation between the amplitude of plasma membrane fluctuations and the response of cells to electric and magnetic fields. Bioelectrochem Bioenerget 33: 19–23
854. Johns LD (2002) Nonthermal effects of therapeutic ultrasound: the frequency resonance hypothesis. J Athletic Training 37 (3):293–299
855. Feril Jr LB, Kondo T, Zhao Q-L et al (2002) Enhancement of hyperthermia-induced apoptosis by non thermal effects of ultrasound. Cancer Lett 178(1):63–70
856. Review of the use of microwave therapy for the treatment of patients with cancer (2005) Review Committee on Microwave Cancer Therapy, National Health and Medical Research Council, Australia
857. dePomerai D, Danniells C, David H et al (2000) Non-thermal heat-shock response to microwaves. Nature 405(6785):417–418
858. de la Hoz A, Díaz-Ortiz A, Moreno A (2007) Review on non thermal effects of microwave irradiation in organic synthesis. J Microw Power Electromagn Energy 41(1):44–64
859. Daniells C, Duce I, Thomas D et al (1998) Transgenic nematodes as biomonitors of microwave-induced stress. Mutation Res 399(1):55–64
860. dePomerai DI, Dawe A, Djerbib L et al (2002) Growth and maturation of the nematode Caenorhabditis elegans following exposure to weak microwave fields. Enzyme Microbial Tech 30(1):73–79
861. dePomerai DI, Smith B, Dawe A et al (2003) Microwave radiation can alter protein conformation without bulk heating. FEBS Lett 543(1–3):93–97
862. Xie T-D, Sun L, Tsong TY (1990) Study of mechanisms of electric field-induced DNA transfection. I. DNA entry by surface binding and diffusion through membrane pores. Biophys J 58(1):13–19
863. Xie T-D, Tsong TY (1990) Study of mechanisms of electric field-induced DNA transfection II. Transfection by low-amplitude, low-frequency alternating electric fields. Biophys J 58(4):897–903
864. Xie T-D, Tsong TY (1992) Study of mechanisms of electric field-induced DNA transfection III. Electric parameters and other conditions for effective transfection. Biophys J 63(1): 28–34

865. Xie T-D, Sun L, Zhao HG et al (1992) Study of mechanisms of electric field-induced DNA transfection. IV. Effects of DNA topology on cell uptake and transfection efficiency. Biophys J 63(4):1026–1031
866. Xie T-D, Tsong T-Y (1993) Study of mechanisms of electric field-induced DNA transfection. V. Effects of DNA Topology on Surface Binding, Cell Uptake, Expression, and Integration into Host Chromosomes of DNA in the Mammalian Cell. Biophys J 65(4):1684–1689
867. Blank M (2008) Protein and DNA reactions Stimulated by Electromagnetic Fields. Electromagn Biol Med 27(1):3–23
868. Blank M (1993) Biological effects of electromagnetic fields. Bioelectrochem Bioenerget 32:203–210
869. Blank M (1995) Biological effects of environmental electromagnetic fields: molecular mechanisms. BioSystems 35(2–3):175–178
870. Blank M (2005) Do Electromagnetic Fields Interact with Electrons in the Na,K-ATPase? Bioelectromagnetics 26(8):677–683
871. Foster K (2006) Comment Concerning ''Do Electromagnetic Fields Interact With Electrons in Na,K-ATPase?'' (Bioelectromagnetics 26:677–683 [2005]). Bioelectromagnetics 27(5):335; discussion 336
872. Levin M (2003) Bioelectromagnetics in Morphogenesis. Bioelectromagnetics 24(5): 295–315
873. Sowers AE (1984) Characterization of Electric Field-induced Fusion in Erythrocyte Ghost Membranes. J Cell Biol 99(6):1989–1996
874. Marszalek P, Tsong TY (1995) Cell Fission and Formation of Mini Cell Bodies by High Frequency Alternating Electric Field. Biophys J 68(4):1218–1221
875. Cleary SF, Liu L-M, Garber F (1985) Erythrocyte Hemolysis by Radiofrequency Fields. Bioelectromagnetics 6(3):313–322
876. Liu D-S, Astumian RD, Tsong TY (1990) Activation of Na+ and K+ Pumping Modes of (Na,K)-ATPase by an Oscillating Electric Field. J Biol Chem 265(13):7260–7267
877. Walleczek J (1992) Electromagnetic field effects on cells of the immune system: the role of calcium signaling. FASEB J 6:3177–3185
878. Cho MR, Thatte HS, Silvia MT et al (1999) Transmembrane calcium influx induced by ac electric fields. FASEB J 13:677–683
879. Ho M-W, Popp F-A, Warnke U (eds) (1994) Bioelectrodynamics and biocommunication. World Scientific, Singapore
880. Bernardi P, D'Inzeo G (1989) Physical mechanisms for electromagnetic interaction with biological systems. In: Lin JC (ed) Electromagnetic interaction with biological systems, Plenum Press, New York, NY, pp 179–214
881. Markov MS (1994) Biological effects of extremely low frequency magnetic fields. In: Ueno S (ed) Biomagnetic stimulation, Plenum Press, London, pp 91–104
882. Bauerus Koch CL, Sommarin M, Persson BR et al (2003) Interaction between weak low frequency magnetic fields and cell membranes. Bioelectromagnetics 24(6):395–402
883. Benett, W.R. Jr (1994) Cancer and power lines. Phys Today 47:23–29
884. Portier CJ, Wolfe MS (1998) Assessment of Health Effects from Exposure to Power-Line Frequency Electric and Magnetic Fields. National Institute of Environmental Health Sciences, Research Triangle Park, NIH Publication No. 98–3981
885. Harland JD, Liburdy RP (1997) Environmental magnetic fields inhibit the antiproliferation action of tamoxifen and melatonin in a human breast cancer cell line. Bioelectromagnetics 18:555–562
886. Ahlbom A, Day N, Feychting M et al (2000) A pooled analysis of magnetic fields and childhood leukemia. Br J Cancer 83(5):692–698
887. Greenland S, Sheppard AR, Kaune WT et al (2000) A pooled analysis of magnetic fields, wire codes, and childhood leukemia. Epidemiology 11(6):624–634

888. Blackman CF, Benane SG, House DE (2001) The influence of 1.2μT, 60 Hz magnetic fields on melatonin- and tamoxifen induced inhibition of MCF-7 cell growth. Bioelectromagnetics 22(2):122–128
889. Plotnikov A, Fishman D, Tichler T et al (2004) Low electric field enhanced chemotherapy can cure mice with CT-26 colon carcinoma and induce anti-tumor immunity. Clin Exp Immunol 138(3):410–416
890. Caubet R, Pedarros-Caubet F, Chu M et al (2004) A Radio Frequency Electric Current Enhances Antibiotic Efficacy against Bacterial Biofilms. Antimicrobal Agents Chemother 48(12):4662–4664
891. Tkalec M, Malaric K, Pavlica M et al (2009) Effects of radiofrequency electromagnetic fields on seed germination and root meristematic cells of Allium cepa. Mutation Res 672(2):76–81
892. Andocs G, Vincze Gy, Szasz O et al (2009) Effect of Curl-Free Potentials on Water. Electromagn Biol Med 28(2):166–181
893. Liu LM, Cleary SF (1995) Absorbed Energy Distribution From Radiofrequency Electromagnetic Radiation in a Mammalian Cell Model: Effect of Membrane-Bound Water. Bioelectromagnetics 16(3):160–171
894. Grant EH (1982) The Dielectric Method of Investigating Bound Water in Biological Material: an Appraisal of the Technique. Bioelectromagnetics 3(1):17–24
895. Essex, C.G (1977) Five-component dielectric dispersion in bovine serum albumin solution. Phys Med Biol 22(6):1160–1167
896. Oleinikova A, Sasisanker P, Weingartner H (2004) What Can Really Be Learned from Dielectric Spectroscopy of Protein Solutions? A Case Study of Ribonuclease A J Phys Chem B 108(24):8467–8474
897. Gimsa J, Wachner D (1998) A unified resistor-capacitor model for impedance, dielectrophoresis, electrorotation and induced transmembrane potential. Biophys J 75(2): 1107–1116
898. Blank M, Soo L (1987) Surface free energy as the potential in oligomeric equilibria: prediction of hemoglobin disaggregation constant. Bioelectrochem Bioenerg 231(3): 349–360
899. Rodríguez-De la Fuente AO, Alcocer-González JM, Antonio Heredia-Rojas J et al (2009) Effect of 60 HZ electromagnetic fields on the activity of hsp70 promoter: an in vitro study. Cell Biol Int 33(3):419–423
900. Rossi RM, Somji S, Garrett SH et al (2002) Expression of hsp 27, hsp 60, hsc 70, and hsp 70 stress response genes in cultured human urothelial cells (UROtsa) exposed to lethal and sublethal concentrations of sodium arsenite. Environ Health Perspect 110(12):1225–1232
901. Albers R, van der Pijl A, Bol M et al (1996) Stress Proteins (HSP) and Chemical-Induced Autoimmunity. Toxicol Appl Pharmacol 140(1):70–76
902. Yasui S, Maeda T, Kameda T (1998) Differential responsiveness against mechanical stress between osteoblastic cells and periodontal ligament cells. Odontology 86(1):64–71
903. Goodman R, Blank M, Lin H et al (1994) Increased levels of Hsp70 transcripts induced when cells are exposed to low-frequency electromagnetic-fields. Bioelectrochem Bioenerget 33(2):115–120
904. Repasky E, Issels R (2002) Physiological consequences of hyperthermia: heat, heat shock proteins and the immune response. Int J Hyperthermia 18(6):486–489
905. Gross CC (2004) Membrane-Bound Hsp70 an Activating Ligand for NK Cells. PhD Theses, Ludwig-Maximilians-University, Munich
906. Horvath I, Multhoff G, Sonnleitner A et al (2008) Membrane-associated stress proteins: More than simply chaperones. Biochimica et Biophysica Acta 1778(7–8):1653–1664
907. Multhoff G, Botzler C, Wiesnet M et al (1995) A stress-inducible 72-kDa heat-shock protein (HSP72) is expressed on the surface of human tumor cells, but not on normal cells. Int J Cancer 61(2):272–279

908. Gehrmann M, Liebisch G, Schmitz G et al (2008) Tumor-Specific Hsp70 Plasma Membrane Localization Is Enabled by the Glycosphingolipid Gb3. PLoS One 2;3(4):e1925–e1933
909. Multhoff G, Botzler C, Jennen L et al (1997) Heat shock protein 72 on tumor cells: a recognition structure for natural killer cells. J Immunol 158(9):4341–4350
910. Ronchi R, Marano L, Braidotti P et al (2004) Effects of broad band electromagnetic fields on HSP70 expression and ischemia-reperfusion in rat hearts. Life Sci 75(16):1925–1936
911. Han L, Lin H, Head M et al (1998) Application of magnetic field-induced heat shock protein 70 for presurgical cytoprotection. J Cell Biochem 71(4):577–583
912. Fleishner M, Johnson JD (2005) Endogenous extra-cellular heat shock protein 72: releasing signal(s) and functions. Int J Hyperthermia 21(5):457–471
913. Tytell M (2005) Release of heat shock proteins (HSPs) and the effects of extracellular HSPs on neural cells and tissues. Int J Hyperthermia 21(5):445–455
914. Mahrour N, Pologea-Moraru R, Moisescu MG et al (2005) In vitro increase of the fluid-phase endocytosis induced by pulsed radiofrequency electromagnetic fields: importance of the electric field component. Biochimica et Biophysica Acta 1668(1):126–137
915. Szasz N, Sen H, Grodzinsky S et al (2003) Electric field regulation of chondrocyte biosynthesis in agarose gel constructs. 49th Annual Meeting of the Orthopaedic Research Society, Poster #0672, http://www.ors.org/web/Transactions/49/0672.PDF
916. Garcia AM, Szasz N, Trippel SB et al (2003) Transport and binding of insulin-like growth factor I through articular cartilage. Arch Biochem Biophys 415(1):69–79
917. Fanning PJ, Emkey G, Smith RJ et al (2003) Mechanical Regulation of Mitogen Activated Protein Kinase Signaling in Articular Cartilage. J Biol Chem 278(51):50940–50948
918. Szasz N (2003) Electric field regulation of chondrocyte proliferation, biosynthesis and cellular signaling. PhD Theses, MIT, Cambridge
919. Johnson MT, Waite LR, Nindl G (2004) Noninvasive treatment of inflammation using electromagnetic fields: current and emerging therapeutic potential. Biomed Sci Instrum 40:469–474
920. Zecca L et al (1985) The effect of alternating magnetic fields on experimental inflammation in the rat. Bioelectrochemistry and Bioenergetics 14:39–43
921. Fernandez MI, Watson PJ, Rowbotham DJ et al (2007) Effect of pulsed magnetic field therapy on pain reported by human volunteers in a laboratory model of acute pain. Br J Anaesthesia 99(2):266–269
922. Becker RO, Selden G (1985) The body electric. Morrow., New York
923. Becker RO (1990) Cross Currents. Jeremy P Tarcher, Los Angeles, CA
924. Nordenstrom BWE (1983) Biologically Closed Electric Circuits: Clinical, experimental and theoretical evidence for an additional circulatory system. Nordic Medical Publications, Stockholm, Sweden
925. Nordenstrom BWE (1998) Exploring BCEC-systems, (Biologically Closed Electric Circuits). Nordic Medical Publications, Stockholm, Sweden
926. Nordenström BEW (1978) Preliminary clinical trials of electrophoretic ionization in the treatment of malignant tumors. IRCS Med Sci 6:537
927. Nordenström BEW (1985) Electrochemical treatment of cancer. Ann Radiol 28:128-129
928. Watson BW (1991) Reappraisal: The treatment of tumors with direct electric current. Med Sci Res 19:103–105
929. Samuelsson L, Jonsson L, Stahl E (1983) Percutaneous treatment of pulmonary tumors by electrolysis. Radiologie 23(6):284–287
930. Miklavcic D, Sersa G, Kryzanowski M et al (1993) Tumor treatment by direct electric current, tumor temperature and pH, electrode materials and configuration. Bioelectr Bioeng 30(26):209–220
931. McCaig, C.D., Rajnicek, A.M., Song, B. et al (2005) Controlling cell behaviour electrically: current views and future potential. Physiol Rev 85(3):943–978
932. Astumian D (1991) Direct electric field effects and sequential process in biosystems. Bioelectrochem Bioenerget 25:455–462

933. Katzberg AA (1974) The induction of cellular orientation by low-level electrical currents. Ann New York Acad Sci 238:445–450
934. Mycielska ME, Djamgoz MBA (2004) Cellular mechanisms of direct-current electric field effects: galvanotaxis and metastatic disease. J Cell Sci 117:1631–1639
935. Sersa G, Miklavcic D, Cemazar M et al (2008) Electrochemotherapy in treatment of tumors. Eur J Surg Oncol 34(2):232–240
936. The first international conference on the topic was in Beijing, China, 20–22 October 1992 (200 Chinese and 30 foreign participants, one-hundred-thirty-six papers were presented), from that time in every second year regularly held, special international organization (IABC) organized with the center in USA
937. Xin Y-L (1994) Organization and Spread electrochemical therapy (ECT) in China. Eur J Surg S-574:25–30; Xin Y-L (1994) Advances in the treatment of malignant tumors by electrochemical therapy (ECT) Eur J Surg S–574:31–36
938. Matsushima Y, Takahashi E, Hagiwara K et al (1994) Clinical and experimental studies of anti-tumoral effects of electrochemical therapy (ECT) alone or in combination with chemotherapy. Eur J Surg S-574:59–67
939. Chou CK et al (1999) Development of Electrochemical treatment at the City of Hope (USA). In: Bersani (ed) Electricity and magnetism in biology and medicine. Kluwer Academic/Plenum Pub., New York, pp 927–930
940. Xin Y, Xue F, Ge B et al (1997) Electrochemical treatment of lung cancer. Bioelectromagnetics 18(1):8–13
941. Robertson GS, Wemyss-Holden SA, Dennison AR et al (1998) Experimental study of electrolysis-induced hepatic necrosis. Br J Surg 85(9):1212–1216
942. Jaroszeski MJ, Coppola D, Pottinger C et al (2001) Treatment of hepatocellular carcinoma in a rat model, using electrochemotherapy. Eur J Cancer 37(3):422–430
943. Holandino C, Veiga VF, Rodriques ML et al (2001) Direct current decreases cell viability but not P-glucoprotein expression and function in human multidrug resistant leukemic cells. Bioelectromagnetics 22(7):470–478
944. Susil R, Semrov D, Miklavcic D (1998) Electric field-induced transmembrane potential depends on cell density and organization. Electro Magnetobiol 17(3):391–399
945. Barker AT, Jaffe LF, Vanable JW Jr (1982) The glabrous epidermis of cavies contains a powerful battery. Am J Physiol 242(3):R358–R366
946. Song B, Zhao M, Forrester J et al (2004) Nerve regeneration and wound healing are stimulated and directed by an endogenous electrical field in vivo. Journal of Cell Science 117 (Pt 20):4681–4690
947. Carbon M, Wübbeler G, Mackert B-M et al (2004) Non-invasive magnetic detection of human injury currents. Clin Neurophysiol 115(5):1027–1032
948. Reid B, Nuccitelli R, Zhao M (2007) Non-invasive measurement of bioelectric currents with a vibrating probe. Nature Protocols 2:661–669
949. Mackert B-M, Mackert J, Wübbeler G et al (1999) Magnetometry of injury currents from human nerve and muscle specimens using superconducting quantum interferences devices. Neurosci Lett 262(3):163–166
950. Song B, Zhao M, Forrester JV et al (2002) Electrical cues regulate the orientation and frequency of cell division and the rate of wound healing in vivo. PNAS 99(21): 13577–13582
951. Zhao M (2009) Electrical fields in wound healing—An overriding signal that directs cell migration. Semin Cell Dev Biol 20(6):674–682
952. James AM, Ambrose EJ, Lowick JHB (1956) Differences between the electrical charge carried by normal and homologous tumor cells. Nature 177(4508):576–577
953. Binggeli R, Weinstein RC (1986) Membrane potentials and sodium channels: hypotheses for growth regulation and cancer formation based on changes in sodium channels and gap junctions. J Theor Biol 123(4):377–401
954. Pu J, McCaig CD, Cao L et al (2007) EGF receptor signalling is essential for electric-field-directed migration of breast cancer cells. J Cell Sci 120:3395–403

955. Meng X, Riordan NH (2006) Cancer is a functional repair tissue. Med Hypotheses 66(3):486–490
956. Bard AJ, Faulkner LR (2000) Electrochemical Methods, Fundamentals and Applications. Wiley, New York, NY
957. Ahmed NAG, Calderwood JH, Fröhlich H et al (1975) Evidence for collective magnetic effects in an enzyme likelihood of room temperature superconductive regions. Phys Lett 53(2):129–130
958. Clark AD, Dunne LJ (1979) Search for superconducting regions in lysozyme. Physiol Chem Phys 11(6):535–536
959. Cope FW (1978) Discontinuous magnetic field effects (Barkhausen noise) in nucleic acids as evidence for room temperature organic superconduction. Physiol Chem Phys 10: 233–245
960. Goldfein S (1974) Some evidence for high-temperature superconduction in cholates. Physiol Chem Phys 6:261–268
961. Cope FW (1974) Enhancement by high electric fields of superconduction in organic and biological solids at room temperature and a role in nerve conduction? Physiol Chem Phys 6:405–410
962. Halpern EH, Wolf AA (1970) Speculations of superconductivity in biological and organic systems. Adv cryogenic engg 17:109–115
963. Milani M et al (1991) Superconductive and Josephson-like behaviour of cells. La Radiologia Medica 81:51–55
964. Wolf AA, Halpern EH (1976) On a class of organic superconductors: a summary of findings. The Institute of Electrical and Electronics Engineers Inc, USA
965. Szasz A (1991) A short-range electronic instability in high Tc superconductors. In: Zipper E, Manka R, Maska M (eds) Strongly Correlated Electron systems & High-Tc superconductivity, World Scientific, Singapore
966. Szasz A (1991) Electronically driven short-range lattice instability: possible role of superconductive pairing. J Superconduct 4(1):3–15
967. Szasz A (1991) Electronically driven short-range lattice instability: cluster effects in superconductors. J Superconduct 4(1):17–27
968. Szasz A (1993) Fullerene superconductivity by short-range order instability. J Superconduct 6(2):99–106
969. Szasz A, Kopayev YA, DasGupta A (1991) On the electronic structure driven pairing mechanism in high-Tc superconductors. Phys Lett 152:361–366
970. Szasz A, DasGupta A (1991) Some general remarks on metastability and high-Tc superconductivity. J Superconduct 4:189–191
971. Horn LW (1993) A novel method for the observation of membrane transporter dynamics. Biophys J 64(1):281–289
972. Astumian RD (1993) Effects of time-dependent electric fields on membrane transport. Biophys J 64(1):7–8
973. Barnes F, Kwon Y (2005) A theoretical Study of the Effects of RF Fields in the vicinity of membranes. Bioelectromagnetics 26(2):118–124
974. Groves JT, Boxer SG, McConnell MH (1997) Electric field-induced reorganization of two-component supported bilayer membranes. Proc. Natl Acad Sci USA 94(25):13390–13395
975. Groves JT, Boxer SG, McConnell HM (1998) Electric field-induced critical demixing in lipid bilayer membranes. Proc Natl Acad Sci USA 95(3):935–938
976. Esser AT, Smith KC, Gowrishankar TR et al (2007) Towards Solid Tumor Treatment by Irreversible Electroporation: Intrinsic Redistribution of Fields and Currents in Tissue. Tech Cancer Res Treat 6(4):261–273
977. Wentian Li (1989) Spatial 1/f spectra in open dynamical systems. Europhys Lett 10: 395–400
978. Schlesinger MS (1987) Fractal time and 1/f noise in complex systems. Ann N Y Acad Sci 504:214–228

979. Bak P, Tang C, Wieserfeld K (1988) Self-organized criticality. Phys Rev A 38:364–373
980. Frohlich H (1983) Coherence in Biology. In: Frohlich H, Kremer F (eds) Coherent excitations in biological systems. Springer, Berlin, pp 1–5.
981. Frohlich H (ed) (1988) Biological coherence and response to external stimuli. Springer, Berlin
982. Tuckerman ME, Laasonen K, Sprik M et al (1995) Ab initio molecular dynamics simulation of the solvation and transport of hydronium and hydroxyl ions in water. J Chem Phys 103:150
983. Tuckerman ME, Laasonen K, Sprik M et al (1995) Laasonen K, Sprik M, Parrinello M: Ab initio molecular dynamics simulation of the solvation and transport of H3O+ and OH- ions in water. J Phys Chem 99(16):5749
984. Marx D, Tuckerman ME, Hutter J et al (1999) The nature of the hydrated excess proton in water. Nature 397:601–604
985. Pnevmatikos SN, Tsironis GP (1989) Protonic conductivity: a new application of soliton theory. J Physique Colloque 50:C3-3-C3-10
986. Davydov AS (1979) Solitons in molecular systems. Physica Scripta 20:1387–1395
987. Davydov AS (1982) Biology and quantum mechanics. Pergamon Press, Oxford
988. Davydov AS (1978) Solitons bioenergetics and mechanism of muscle contraction. Int J Quantum Chem 16(1):5–17
989. Peierls RE (1955) Quantum theory of solids. Clarendon, Oxford
990. Chothia C (1989) Polyhedra for helical proteins. Nature 337:204–205
991. Murzin AG, Finkelstein AV (1988) General architecture of the alpha-helical globule. J Mol Biol 204(3):749–769
992. Bistolfi F (1991) Biostructures and Radiation Order Disorder. Edizioni Minerva Medica, Torino
993. Szasz A, Fabian DJ (1988) On electronic structure and metastability. Solid State Commun 65(10):1085–1088
994. Smirnov, B.M (1995) Icosahedral clusters with pair interaction of atoms. Chem Phys Lett 232(4), 395–400
995. Widom M (1988) Short- and long-range icosahedral order in crystals, glass and quasicrystals. In: Jaric MV (ed) Aperiodicity and Order, Introduction to Quasi Crystals, Acad. Press, New York, pp 59–110
996. Porkert M (1974) The theoretical foundations of Chinese medicine. MIT Press, Cambridge, MA
997. Jones NS (2007) Brown Noise can make Pink Noise. http://www.physics.ox.ac.uk/cm/cmt/jones/recent/brown2pink.pdf
998. Lamsweerde-Gallez van D (1981) Phase transitions induced by coloured voltage noise in nerve membranes. Bioelectrochem Bioenerget 8:179–187
999. Bersani F (1999) Electricity and magnetism in biology and medicine. Kluwer, Plenum Publishers, New York, NY
1000. Gilden DL, Thornton T, Mallon MW (1995) 1/f noise in human cognition. Science 267(5205):1837–1839
1001. Voss RF (1993) 1/f Noise and fractals in DNA-base Sequences. In: Crilly AJ, Earnshaw RA, Jones H (eds) Applications of Fractals and Chaos. Springer, Berlin, pp 7–20
1002. Holden AV, Hyde J, Zhang H (1993) Computing with the Unpredictable: Chaotic Dynamics and Fractal Structures in the Brain. In: Crilly AJ, Earnshaw RA, Jones H (eds) Applications of fractals and Chaos. Springer, Berlin
1003. Adair RK (1991) Constraints on biological effects of weak extremely-low frequency electromagnetic fields. Phys Rev A 43(2):1039–1048
1004. Weaver JC, Astumian RD (1990) The response of living cells to very week electric fields: The thermal noise limit. Science 247(4941):459–462
1005. Kaune WT (2002) Thermal Noises Limit on the sensitivity of cellular membranes to power frequency electric and magnetic fields. Bioelectromagnetics 23(8):622–628

1006. Vincze Gy, Szász A, Szasz N (2005) On the thermal noise limit of cellular membranes. Bioelectromagnetics 26(1):28–35
1007. Raff MC (1992) Social controls on cell survival and death. Nature 356(6368):397–400
1008. Schrodinger E (1951) Science and humanism. Cambridge University Press, Cambridge
1009. von Neumann J (1959) The computer and the brain. Yale University Press, New Haven
1010. Loewenstein WR, Kanno Y (1967) Intercellular communications and tissue growth, I. Cancerous Growth. J Cell Biol 33:225–234
1011. Galeotti T, Borrello S, Minotti G et al (1986) Membrane alterations in cancer cells. Role of oxi radicals. Ann N Y Acad Sci 488:468–480
1012. Hameroff SR (1988) Coherence in the cytoskeleton: implications for biological information processing. In: Froelich H (ed) Biological coherence and response to external stimuli. Springer, Berlin, pp 242–266
1013. Janmey P (1995) Cell membranes and the cytoskeleton. In: Lipowsky R, Sackinan E (eds) Handbook of biological physics Vol I, Elsevier, New York
1014. Del Giudice E et al (1988) Structures, correlations and electromagnetic interactions in living matter. In: Froelich H (ed) Biological coherence and response to external stimuli. Springer, Berlin, pp 49–64
1015. Cho MR, Thatte HS, Lee RC et al (1996) Reorganization of microfilament structure induced by ac electric fields. FASEB J 10(13):1552–1558
1016. Zsoldos I, Szendro P, Watson L et al (2001) Topological Correlation in amorphous structures. Comp Mater Sci 20(1):28–36
1017. Vincze, G., Zsoldos, I., Szasz, A (2004) On the Aboav-Weaire law. J Geomet Phys 51(1), 1–12
1018. Zsoldos I, Szasz A (1999) Appearance of collectivity in two-dimensional cellular structures. Comp Material Science 15(4):441–448
1019. Maryan, M.I., Kikineshi, A.A., Szasz, A (2001) Self-organizing processes and dissipative structure formation in the non-crystalline materials. Phys Chem Stat Sol 2(4): 585–593
1020. Puck TT, Marcus PI, Cieciura SJ (1956) Clonal growth of mammalian cells in vitro: growth characteristics of colonies from single HeLa cells with and without a "feeder" layer. J Exp Med 103(2):273–283
1021. Taylor AF, Tinsley MR, Wang F et al (2009) Dynamical quorum sensing and synchronization in large populations of chemical oscillators. Science 323(5914):614–617
1022. Caer, G. Le (1991) Topological models of cellular structures. J Phys A Math Gen 24: 1307–1317
1023. Puck TT, Marcus PI (1955) A rapid method for viable cell titration and clone production with HeLa cells in tissue culture: the use of X-irradiated cells to supply conditioning factors. Proc Natl Acad Sci USA 41(7):432–437
1024. Marko M (2005) "Biological windows": a tribute to WR Adey. environmen 25(2–4):67–74
1025. Adey WR (1984) Nonlinear, nonequilibrium aspects of electromagnetic field interactions at cell membranes. In: Adey WR, Lawrence AF (eds) Nonlinear Electrodynamics in Biological Systems, Plenum Press, New York, NY, pp 3–22
1026. Adey WR (1990) Joint actions of environmental nonionizing electromagnetic fields and chemical Pollution in Cancer Promotion. Environ Health Perspect 86:297–305
1027. Blackman CF, Kinney LS, House DE et al (1989) Multiple power-density windows and their possible origin. Bioelectromagnetics 10(2):115–128
1028. Markin VS, Tsong TY (1991) Frequency and concentration windows for the electric activation of a membrane active transport system. Biophysics J 59(6):1308–1316
1029. Lin JC (1989) Electromagnetic interaction with biological systems. Pergamon Press, London
1030. D'Inzeo G, Galli A, Palombo A (1993) Further investigations on non-thermal effects referring to the interaction between ELF fields and transmembrane ionic fluxes. Bioelectrochem Bioenerget 30(8):93–102

1031. Obo M, Konishi S, Otaka Y et al (2002) Effect of magnetic field exposure on calcium channel currents using patch clamp technique, Bioelectromagnetics 23(4):306–314
1032. Lednev VV (1991) Possible mechanism for the influence of weak magnetic fields on biological systems. Bioelectromagnetics 12(2):71–75
1033. Blanchard JP, Blackman CF (1994) Clarification and application of an ion parametric resonance model for magnetic field interactions with biological systems. Bioelectromagnetics 15(3):217–238
1034. Del Giudice E, Fleischmann M, Preparata G et al (2002) On the "Unreasonable" Effects of ELF Magnetic Fields Upon a System of Ions. Bioelectromagnetics 23(7):522–530
1035. Jacobson JI (1991) A look at the possible mechanism and potential of magneto therapy. J Theor Biol 149(1):97–119
1036. McLoad BR, Liboff AR (1986) Dynamic characteristic of membrane ions in multifield configurations of low-frequency electromagnetic radiation. Bioelectromagnetics 7(2): 177–189
1037. Vincze G, Szasz A, Liboff AR (2008) New theoretical treatment of ion-resonance phenomena. Bioelectromagnetics 29(5):380–386
1038. Bawin SM, Kazmarek LK, Adey WR (1975) Effects of modulated VHF fields on central nervous system. Ann N Y Acad Sci 247:74–81
1039. Morariu VV, Ciorba D, Neamtu S (2000) Life in zero magnetic field. I. In vitro human blood aging. Electro Magnetobiol 19:289–302
1040. Ciortea LI, Morariu VV, Todoran A et al (2001) Life in zero magnetic field. II. Effect on zinc and copper in human blood serum during in vitro aging. Electro Magnetobiol 20(2):127–139
1041. Blackman CF, Benane SG, Rabinowitz JR et al (1985) Role for the magnetic field in the radiation induced efflux of calcium ions from brain tissue in vitro. Bioelectromagnetics 6(4):327–337
1042. Zhadin MN, Novikov VV, Barnes FS et al (1998) Combined action of static and alternating magnetic fields on ionic current in aqueous glutamic acid solution. Bioelectromagnetics 19(1):41–45
1043. Betancur C, Dell'Omo G, Alleva E (1994) Magnetic field effects on stress-induced analgesia in mice: modulation by light. Neurosci Lett 182(2):147–150
1044. Sartucci F, Bonfiglio L, Del Seppia C et al (1997) Changes in pain reception and pain-related somatosensory evoked potentials in humans produced by exposure to oscillating magnetic field. Brain Res 769(2):362–366
1045. Choleris E, Del Seppia C, Thomas AW et al (2002) Shielding, but not zeroing of the magnetic field reduces stress-induced analgesia in mice. Proc Biol Sci 269(1487): 193–201
1046. Fitzsimmons RJ, Ryaby JT, Magee FP et al (1994) Combined magnetic fields increased net calcium flux in bone cells. Calcif Tissue Int 55(5):376–380
1047. Fitzsimmons RJ, Ryaby JT, Mohan S et al (1995) Combined magnetic fields increased insulin-like growth factor-II in Te-85 human osteosarcoma bone cell cultures. Endocrinology 136(7):3100–3106
1048. Gammaitoni L, Hanggi P, Jung P et al (1998) Stochastic resonance. Rev Mod Phys 70: 223–287
1049. Tsong TY, Chang C-H (2007) A Markovian engine for a biological energy transducer: the catalytic wheel. Bio Systems 88(3):323–333
1050. Savageau MA (1998) Development of fractal kinetic theory for enzyme-catalysed reactions and implications for the design of biochemical pathways. Biosystems 47(1–2):9–36
1051. Astumian RD, Chock PB, Tsong TY et al (1987) Can free energy be transduced from electric noise? Proc Natl Acad Sci USA 84(2):434–438
1052. Tsong TY, Astumian RD (1988) Electroconformational Coupling: How Membrane-Bound ATPase Transduces Energy from Dynamic Electric Fields. Annu Rev Physiol 50: 273–290

1053. Astumian RD (1994) Electroconformational coupling of membrane proteins. Ann N Y Acad Sci 720(1):136–140
1054. Markin VS, Liu D, Rosenberg MD et al (1992) Resonance transduction of low level periodic signals by an enzyme: an oscillatory activation barrier model. Biophys J 61(4):1045–1049
1055. Tinoco I, Sauer K, Wang JC et al (2002) Physical chemistry. Principles and applications in biological sciences, 4th ed. Prentice, NJ
1056. Xie TD, Chen Y, Marszalek P et al (1997) Fluctuation-driven directional flow in biochemical cycle: further study of electric activation of Na,K pumps. Biophys J 72(6): 2496–2502
1057. Astumian RD (1997) Thermodynamics and kinetics of a Brownian motor. Science 276(5314):917–922
1058. Astumian RD, Bier M (1994) Fluctuation driven ratchets: molecular motors. Phys Rev Lett 72(11):1766–1769
1059. Xie TD, Marszalek P, Chen YD et al (1994) Recognition and processing of randomly fluctuating electric signals by Na,K – ATPase. Biophys J 67(3):1247–1251
1060. Tsong TY, Xie TD (2002) Ion pump as molecular ratchet and effects of noise: electric activation of cation pumping by Na,K-ATPase. Appl Phys A 75(2):345–352
1061. FGF, University of Rostock, COST 281 (org.) Workshop on "Proposed mechanisms for the interaction of RF-signals with living matter, – Demodulation in biological systems", Rostock, Germany, 11–13 September 2006
1062. Astumian RD, Weaver JC, Adair RK (1995) Rectification and signal averaging of weak electric fields by biological cells. Proc Natl Acad Sci USA 92(9):3740–3743
1063. Silny J (2007) Demodulation in tissue, the relevant parameters and the implications for limiting exposure. Health Physics 92(6):604–608
1064. Sultan MF, Cain CA, Tompkins WAF (1983) Immunological Effects of Amplitude-Modulated Radio Frequency Radiation: B Lymphocyte Capping. Bioelectromagnetics 4(2):157–165
1065. Lynes B (ed) (2005) Cancer Solutions. Rife, Energy medicine and Medical Politics, Elsmere Pr
1066. Westerhoff HV, Tsong TY, Chock PB et al (1986) How enzymes can capture and transmit free energy from an oscillating electric field. Proc Natl Acad Sci USA 83(13):4734–4738
1067. Seegers JC, Engelbrecht CA, Papendorp van DH (2001) Activation of signal-transduction mechanisms may underlie the therapeutic effects of an applied electric field. Med Hypotheses 57(2):224–230
1068. Barbault1 A, Costa FP, Bottger B et al (2009) Amplitude-modulated electromagnetic fields for the treatment of cancer: Discovery of tumor-specific frequencies and assessment of a novel therapeutic approach. J Exp Clin Cancer Res 28(1):51–61
1069. Kirson ED, Dbal V, Rochlitz C et al (2006) Treatment of locally advanced solid tumors using alternating electric fields (TTFields) – a translational study. Clinical Research 17: Phase II and III Adult Clinical Trials 1. In: Proceedings of American Association Cancer Research, Volume 47, Abstract #5259
1070. National Institutes of Health US, Low Levels of Electromagnetic Fields to Treat Advanced Cancer (ADLG3). http://clinicaltrials.gov/ct2/show/NCT00805337
1071. Kirson ED, Giladi M, Gurvicz Z et al (2009) Alternating electric fields (TTFields) inhibit metastatic spread of solid tumors to the lungs. Clin Exp Metastasis 26(7):633–640
1072. Schwan HP (1982) Nonthermal cellular effects of electromagnetic fields: ac-field induced ponderomotoric forces. Br J Cancer 45(5):220–224
1073. Cho MR, Thatte HS, Lee RC et al (1994) Induced redistribution of cell surface receptors by alternating current electric fields. FASEB J 8(10):771–776
1074. Kirson ED, Gurvich Z, Schneiderman R et al (2004) Disruption of Cancer Cell Replication by Alternating Electric Fields. Cancer Res 64(9):3288–3295

1075. Kirson ED, Dbaly V, Tovarys F et al (2007) Alternating electric fields arrest cell proliferation, in animal tumor models and human brain tumors. Proc Nat Acad Sci USA 104(24):10152–10157
1076. Czyza J, Nokoklova T, Schuderer J et al (2004) Non-thermal effects of power-line magnetic fields (50 Hz) on gene expression levels of pluripotent embryonic stem cells—the role of tumour suppressor p53. Mutation Res 557(1):63–74
1077. Prashanth KS, Chouhan TRS, Nadiger S (2009) Effect of 50 Hz electromagnetic fields on acid phosphatase activity. Af J Biochem Res 3(3):060–065
1078. Kotnik T, Miklavcic D (2000) Second-order model of membrane electric field induced by alternating external electric fields. IEEE Trans Biomed Engg 47(8):1074–1082
1079. Balzano Q, Sheppard A (2003) RF Nonlinear interactions in living cells – I: Nonequilibrium thermodynamic theory. Bioelectromagnetics 24(7):473–482
1080. Beuter A, Glass L, Mackey MC et al (eds) (2003) Nonlinear dynamics in physiology and medicine. Interdisciplinary applied mathematics, vol. 25. Springer, Berlin
1081. Tesla N (1905) Art of transmitting electrical energy through the natural medium. United States Patent 787,412 April 1905
1082. Aharonov Y, Bohm D (1959) Significance of electromagnetic potentials in quantum theory. Phys Rev 115(3):485–491
1083. Marjan MI, Szasz A (2000) Self-organizing processes in non-crystalline materials: from lifeless to living objects. OncoTherm Kft, Budapest
1084. Pontiggia F, Colombo G, Micheletti C et al (2007) Anharmonicity and self-similarity of the free energy landscape of protein G. Phys Rev Lett 98(4):048102
1085. Ochiai T, Nacher JC, Akutsu T (2005) Symmetry and dynamics in living organisms: The self-similarity principle governs gene expression dynamics. Biomolecules 1–27, http://arxiv.org/abs/q-bio/0503003
1086. Hegyi G, Vincze Gy, Szasz A (2007) Axial-vector interaction with bio-systems. Electr Biol Med 26(2):107–118
1087. Sharipov F (2006) Onsager-Casimir reciprocal relations based on the Boltzmann equation and gas-surface interaction: Single gas. Phys. Rev. E 73(2):026110.1–026110.8
1088. Konopinsky EJ (1978) What the electromagnetic vector potential describes. Am J Phys 46(5):499–502
1089. Reno VR, Nutini LG (1963) Effect of Magnetic fields on tissue respiration. Nature 198:204–205
1090. Wolf AA (1981) On a unified theory of cancer etiology and treatment based on the superconduction double-dipole model. Physiol Chem Phys 13(6):493–510
1091. Easterly CE (1981) Cancer link to magnetic field exposure: a hypothesis. Am J Epidem 114(2):169–175
1092. Bak P, Tang C, Wiesenfeld K (1987) Self-organized criticality: An explanation of 1/f noise. Phys Rev Lett 59(4):381–384
1093. Goldberger AL, Amaral LA, Hausdorff JM et al (2002) Fractal dynamics in physiology: Alterations with disease and aging. PNAS Colloquium 99(1):2466–2472
1094. Haken H (1977) Synergetics. Springer, Berlin
1095. Gubarev FV, Sodolsky L, Zakharov VI (2001) On the significance of the vector potential squared. Phys Rev Lett 86(11):2220–2222
1096. Szász A, Vincze G, Andocs G et al (2009) Do Field-Free Electromagnetic Potentials Play a Role in Biology? Electromagn Biol Med 28(2):135–147
1097. Szász A, Vincze Gy, Szasz O et al (2009) Effect of Curl-Free Potentials on Water. Electromagn Biol Med 28(2):166–181
1098. Gelinas RC (1984) Apparatus and method for demodulation of a modulated curl-free magnetic vector potential field. United States Patent 4,429,280 31 Jan 1984
1099. Pekar R (1996) Die Perkutane Galvano-Therapie bei Tumoren- Schwachstrombehandlung von zugänglichen Neoplasmen und ihre vitale Hybridisation in Theorie und Praxis, Verlag W. Maudrich, Berlin

1100. Pekar R (2002) Die perkutane Bio-Elektrotherapie bei Tumoren (The percutaneous bio electrical therapy for tumors). Verlag W. Maudrich, Berlin
1101. Pekar R, Korpan NN (2001) Krebs, Die biologische und medizinische Tragödie. Bio-Onkologie, Verlag Wilhelm Maudrich, Wien-Munchen-Bern, p. 266
1102. Ling X-Y (1994) Advances in the treatment of malignant tumors by electrochemical therapy (ECT). Eur J Surg Suppl 574:S31–S36
1103. Quan K (1994) Analysis of the clinical effectiveness of 144 cases of soft tissue and superficial malignant tumors treated with electrochemical therapy. Eur J Surg Suppl 574:S45–49, Scandinavian University Press
1104. Song Y (1994) Electrochemical therapy in the treatment of malignant tumors on the body surface. Eur J Surg Suppl 574:S41–43, Scandinavian University Press
1105. Senn E (1990) Elektrotherapie. Thieme Verlag, Stuttgart
1106. Song CW, Rhee JG, Lee CK et al (1986) Capacitive heating of phantom and human tumors with an 8 MHz radiofrequency applicator (Thermotron RF-8). Int J Radiation Oncol Biol Phys 12(3):365–372
1107. Sugahara T et al (2001) Physiology and psychology in hyperthermic oncology: contribution of clinical oncologists in development of heating of deep-seated tumors. In: Kosaka M, Sugahara T, Schmidt KL, Simon E (eds) Thermotherapy for neoplasia, inflammation, and pain, Springer-Verlag, Tokyo, pp 480–483
1108. Hyperthermia – Preface (ed) (2000) Selected papers on hyperthermia using Thermotron-RF8. pp 193–196
1109. Hiraoka M, Mitsumori M, Nagata Y et al (2001) Current status of clinical hyperthermic oncology in Japan. In: Kosaka M, Sugahara T, Schmidt KL, Simon E (eds) Thermotherapy for neoplasia, inflammation, and pain, Springer-Verlag, Tokyo, pp 471–479
1110. Masunaga SI, Ono K, Mitsumori M et al (1996) Clinical usefulness of determining the rate of thermal clearance within heated tumors. Jpn J Clin Oncol 26(6):428–437
1111. Szasz A, Szasz N, Szasz O (2004) Hyperthermia in der oncology: eine aktuell beforschte behandlungsmethode. Integrative Onkologie 1:19–27
1112. Szasz A (2003) Elektromagnetische Hyperthermieverfahren: die kapazitive Kopplung. Forum Komplementare Onkologie Hyperthermie, 4:III–IX
1113. Dewhirst MW (1995) Thermal dosimetry. In: Seegenschmiedt MH, Fessenden P, Vernon CC (eds), Thermo-radiotherapy and thermo-chemotherapy, vol. 1. Biology, Physiology and Physics. Springer, Berlin, pp 123–136
1114. Pennes HH (1948) Analysis of tissue and arterial blood temperatures in the resting human forearm. J. Appl. Phys. 1, 93–122
1115. Liu J (2000) Temperature monitoring and heating optimization in cancer hyperthermia. Prog Nat Sci 10(10):787–793
1116. Deng ZS, Liu J (2002) Analytical study on bioheat transfer problems with spatial or transient heating on skin surface or inside biological bodies. J Biomech Eng 124(6):638–649
1117. Wren J, Karlsson M, Loyd D (2001) A hybrid equation for simulation of perfused tissue during thermal treatment. Int J Hyp 17(6):483–498
1118. Davalos RD, Rubinsky B, Mir LM (2003) Theoretical analysis of the thermal effects during in vivo tissue electroporation. Bioelectrochemistry 61:99–107
1119. Bagaria HG, Johnson DT (2005) Transient solution to the bioheat equation and optimization for magnetic fluid hyperthermia treatment. Int J Hyperthermia 21(1):57–75
1120. Gao B, Langer S, Corry PM (1995) Application of the time-dependent Green's function and Fourier transforms to the solution of the bioheat equation. Int J Hyp 11(2):267–285
1121. Wissler EH (1998) Pennes' 1948 paper revisited. J Appl Physiol 85:35–41
1122. Nelson DA (1998) Invited editorial on "Pennes' 1948 paper revisited". J Appl Physiol 85:2–3
1123. Charny CK, Weinbaum S, Levin RL (1990) An evaluation of the Weinbaum-Jiji equation for normal and hyperthermic conditions. J Biomech Eng 112(1):80–87
1124. Najarian S, Pashaee A (2001) Improvement of the Pennes Equation in the analysis of heat transfer phenomenon in blood perfused tissues. Biomed Sci Instrum 37:185–190

1125. Szasz A, Vincze Gy (2006) Dose concept of oncological hyperthermia: Heat-equation considering the cell destruction. J Cancer Res Ther 2(4):171–181
1126. Szasz A, Vincze Gy (2007) Hyperthermia, a modality in the wings. J Cancer Res Ther 3(1):56–66
1127. Vujaskovic Z et al (2004) A Randomized trial of Hyperthermia and Radiation for Superficial Tumors. Presentation and Abstracts for The Kadota Fund International Forum, Awaji Yumebutai, Japan, 15–18 June 2004
1128. Wissler EH (1998) Pennes' 1948 paper revisited. J. Appl Physiol 85:35–41
1129. Nelson DA (1998) Invited editorial on „Pennes' 1948 paper revisited". J Appl Physiol 85(1):2–3
1130. Adair RK (2003) Biophysical limits on athermal effects of RF and microwave radiation. Bioelectromagnetics 24(1):39–48
1131. McLeod BR, Liboff AR, Smith SD (1992) Electromagnetic gating in ion channels. J Theor Biol 158:15–31
1132. Durney CM (1987) Electromagnetic regional heating. In: Field SB, Franconi C (eds) Physics and Technology of Hyperthermia, NATO ASI Series, Martinus Nijhoff Publ, Dordrecht, pp 241–249
1133. Robert J, Escanye JM, Brunotte F et al (1986) Physical basis of hyperthermia. In: Anghileri LJ, Robert J (eds) Hyperthermia in cancer treatment, CRC Press Inc., Boca Raton, FL, 2:2–16
1134. Gabriel C, Gabriel S (1996) Complation of the dielectric properties of body tissues atrf and microwave frequencies. Final Report for the Period 15 Dec 1994–14 Dec 1995, Prepared for AFOSR/NL Bolling AFB DC 20332-0001. http://niremf.ifac.cnr.it/docs/DIELECTRIC/Report.html
1135. Gabriel S, Lau RW, Gabriel C (1996) The dielectric properties of biological tissue: II. Measurements in the frequency range 10 Hz to 20 GHz. Phys Med Biol 41:2251–2269
1136. Hober R (1912) Eine Methode die elektrische Leitfahigkeit im innern von Zellen zu messen, Pflugers Arch. Gesamte Physiol 133, 237–238 (1910):148, 189–191
1137. Smith SR, Foster KR (1985) Dielectric properties of low-water-content tissues. Phys Med Biol 30:965–970
1138. Hasted JB (1973) Aqueous Dielectrics. Chapman and Hall, London, p 222
1139. Dickson JA, Calderwood SK (1980) Temperature range and selective sensitivity of tumors by hyperthermia: a critical review. Ann NY Acad Sci 335:180–205
1140. Murata T, Akagi K, Ostapenko VV et al (1998) Relevance of a new impedance matching, or Substrap, method for the reduction of pain during hyperthermia. Acta Oncologica 37: 485–488
1141. Chatterjee I, Hagmann M, Gandhi O (1981) An empirical relationship for electromagnetic energy absorption in man for near-field exposure condition. IEEE Trans Microwave Theory Tech 29(11):1235–1238
1142. Iskander M, Olson S, MacCalmont J (1987) Near-field absorption characteristics of models in the resonance frequency range. IEEE Trans Microwave Theory Tech 35(8): 776–779
1143. Bhowmick P, Coad JE, Bhowmick S et al (2004) Bischof JC: In vitro assessment of the efficacy of thermal therapy in human benign prostatic hyperplasia. Int J Hyperthermia 20:421–439
1144. Lindegaard JC (1992) Thermosensitization induced by step-down heating. Int J Hyperthermia 8(5):561–582
1145. Hasegawa T, Gu Y-H, Takahashi T et al (2001) Enhancement of hyperthermic effects using rapid heating. In: Kosaka M, Sugahara T, Schmidt KL et al (eds) Thermotherapy for Neoplasia, Inflammation, and Pain, Springer, Tokyo, pp 439–444
1146. van Rijn J, van den Berg J, Wiegant FA et al (1994) Time-temperature relationships for step-down heating in normal and thermotolerant cells. Int J Hyperthermia 10: 643–652

1147. Henle KJ, RotiRoti JL (1988) Response of cultured mammalian cells to hyperthermia. In: Urano M, Douple E (eds) Hyperthermia and Oncology, Vol. 1, VSP, Utrecht, Tokyo, pp 57–82
1148. Konings AWT (1995) Interaction of heat and radiation in vitro and in vivo. In: Seegenschmiedt MH, Fessenden P, Vernon CC (eds) Thermo-radiotherapy and thermo-chemotherapy, vol.1. Biology, Physiology and Physics, Springer, Berlin, 89–102
1149. Jackson JD (1999) Classical Electrodynamics. John Wiley & Sons Inc, New York
1150. Cloude S (1995) An introduction to electromagnetic wave propagation & antennas. UCL Press Ltd, London
1151. Rao NN (1994) Elements of Engineering Electromagnetics. Prentice Hall International, London
1152. Polk C, Postow E (1996) Handbook of Biological Effects of Electromagnetic Fields. CRC Press, Tokyo, p 15
1153. Kotnik T, Miklavcic D (2000) Theoretical evaluation of the distributed power dissipation in biological cells exposed to electric field. Bioelectromagnetics 21:385–394
1154. Weinbaum S, Jiji LM (1985) A new simplified bioheat equation for the effect of blood-flow on local average tissue temperature. J Biomech Eng 107:131–139
1155. Fujita S, Tamazawa M, Kuroda K (1998) Effects of blood perfusion rate on the optimization of RF-capacitive hyperthermia. IEEE Trans Biomed Engg 45:1182–1186
1156. Parolini I, Federici C, Raggi C, Lugini L, Palleschi S, De Milito A, Coscia C, Iessi E, Logozzi M, Molinari A, Colone M, Tatti M, Sargiacomo M, Fais S (2009) Microenvironmental pH is a key factor for exosome traffic in tumor cells. J Biol Chem 284:34211–34222
1157. Stoy RD, Foster KR, Schwan HP (1982) Dielectric properties of mammalian tissues from 0.1 to 100 MHz: a summary of recent data. Phys Med Biol 27:501–513
1158. Foster KR, Schepps JL (1981) Dielectric properties of tumor and normal tissues at radio through microwave frequencies. J Microwave Power 16:107–119
1159. Andocs G (2005) Electro-hyperthermia in the veterinary practice. Hyperthermie Symposium, Cologne, Germany, 14–16 October 2005
1160. Brunner G (2006) Zellulare Hyperthermie-effecte in vitro in einem progressionsmodell des lattenepithelkarzinoms. Hyperhermia Seminar, Cologne, Germany, 22–23 September 2006
1161. Brunner G, Erkell L (2008) Cellular and molecular effects of electrohyperthermia in a cell model of skin cancer progression. 10th International Congress on Hyperthermic Oncology, ICHO'2008, Munich 9–12 April 2008
1162. Andocs G (2008) Front page illustration of Forum Medizine. Forum Hyperthermia, November 2008
1163. Mosmann T (1983) Rapid colorimetric assay for cellular growth and survival: Application to proliferation and cytotoxicity assays. J Immunol Methods 65:55–63
1164. Leckband D (2008) Beyond structure: mechanism and dynamics of intercellular adhesion. Biochem Soc Trans 36(Pt 2):213–220
1165. Bell GI (1978) Models for the specific adhesion of cells to cells. Science 200:618–627
1166. Yap AS, Brieher WM, Gumbiner BM (1997) Molecular and functional analysis of cadherin-based adherens junctions. Ann Rev Cell Dev Biol 13:119–146
1167. Andocs G, Szasz A, Somossy Z (2006) Change of adherent connections after heat-treatment on HepG2 hepato-cellular carcinoma in vitro. 36th Membrane Transport Conference, Sumeg, Hungary, 24 June 2006
1168. West GB, Brown JH, Enquist BJ (1999) The Four Dimensions of Life: Fractal Geometry and Allometric Scaling of Organisms. Science 284(5420):1677–1679
1169. Andocs G, Szasz O, Szasz A (2009) Oncothermia treatment of cancer: from the laboratory to clinic. Electromagn Biol Med 28(2):148–165
1170. van Gijn ME, Snel F, Cleutjens JP et al (2001), Overexpression of Components of the Frizzled-Dishevelled Cascade Results in Apoptotic Cell Death, Mediated by b-Catenin. Exp Cell Res 265(1):46–53

1171. Andocs G (2007) Recent results on oncothermia. Hyperthermie Symposium, Cologne, Germany, 19–20 October 2007
1172. Andocs G, Szasz O. Szasz A. (2009) In vitro and in vivo evidences of effects of modulated rf-conducting heating. 25th Annual Meeting of the European Society for Hyperthermic Oncology, ESHO, Verona, June 4–6
1173. Bremnes RM, Veve R, Hirsch FR et al (2002) The E-cadherin cell–cell adhesion complex and lung cancer invasion, metastasis, and prognosis. Lung Cancer 36(2): 115–124
1174. Shinitzky M (1984) Membrane fluidity in malignancy: adversative and recuperative. Biochem Biophys Acta 738:251–261
1175. Park HG, Han SI, Oh SY et al (2005) Cellular responses to mild heat stress. Cell Mol Life Sci 62:10–23
1176. Experiments were performed by Lorus Inc, Toronto, Canada, (Courtesy Dr. Yoon Lee and Mr. RC Perralta (DVM) unpublished results
1177. Andocs G, Renner H, Balogh L et al (2009) Strong synergy of heat and modulated electromagnetic field in tumor cell killing, Study of HT29 xenograft tumors in a nude mice model. Strahlentherapie und Onkologie 185:120–126
1178. Urano M, Douple E (eds) (1988–1994) Hyperthermia and Oncology – Vol.1. Thermal effects on cells and tissues. VSP BV Utrecht, The Netherlands (1988), Hyperthermia and Oncology – Vol.2. Biology of thermal potentiation of radiotherapy. VSP BV Utrecht, The Netherlands (1992, 1989); Hyperthermia and Oncology – Vol.3. Interstitial Hyperthermia: Physics, biology and clinical aspects, VSP BV Utrecht, The Netherlands (1992) Hyperthermia and Oncology – Vol.4. Chemopotentiation by hyperthermia VSP BV Utrecht, The Netherlands (1994)
1179. Overgaard J (1978) Effect of Local Hyperthermia Alone, and in Combination with Radiation, on Solid Tumors. In: Streffer C, vanBeuningen D, Dietzel F et al (eds) Cancer Therapy by Hyperthermia and Radiation, Urban & Schwarzenberg, Baltimore, Munich
1180. Szász A (2007) Hyperthermia, a Modality in the Wings. J Cancer Res Ther 3(1):56–66
1181. Fiorentini G, Szasz A (2006) Hyperthermia today: electric energy, a new opportunity in cancer treatment. J Cancer Res Ther 2(2):41–46
1182. Brunner G (2006) Zellulare Elektrohyperthermia-Effecte in vitro in einem Progressionsmodell des plattenepithelkarzinoms. Hyperthermie Symposium, Cologne, 22–23 September 2006
1183. Kodama K, Doi O, Tatsuta M et al (1989) Development of postoperative intrathoracic chemo-thermotherapy of lung cancer with objective of improving local cure. Cancer 64(7):1422–1428
1184. Sneed PK, Gutin PH, Stauffer PR et al (1992) Thermoradiotherapy of recurrent malignant brain tumors. Int J Radiat Oncol Biol Phys 23(4):853–861
1185. Szasz A (2006) Physical background and technical realization of hyperthermia. In: Baronzio GF, Hager ED (eds) Locoregional Radiofrequency-Perfusional- and Wholebody-Hyperthermia in Cancer Treatment: New clinical aspects, Ch. 3., Springer, New York, NY, pp 27–59
1186. Oleson JR, Calderwood SK, Coughlin CT, Dewhirst MW, Gerweck LE, Gibbs FA, Jr. Kapp DS (1988) Biological and Clinical Aspects of Hyperthermia in Cancer Therapy. Am J Clin Oncol 11:368–380
1187. Ganong WF (1997) Review of medical physiology. Appleton & Lange, Stamford, CT
1188. Vander AJ, Sherman JH, Luciano DS (1994) Human physiology: the mechanisms of body function. McGraw-Hill, New York, NY
1189. Joy MLG (2004) MR current density and conductivity imaging: the state of the art, engineering in medicine and biology. IEMBS 26th Annu Int Conf IEEE 2:5315–5319
1190. Suk, Hoon Oh (2003) Conductivity and current density image reconstruction using harmonic Bz algorithm in magnetic resonance electrical impedance tomography. Phys Med Biol 48:3101–3116

1191. Bondestam H (2007) New requirements for magnetic field measurements. Conformity, pp 30–35
1192. RF Radiation and Electromagnetic Field Safety ARRL handbook for radio amateurs. American Radio Relay League Inc, USA, 1997
1193. Institute of Electrical and Electronics Engineers: Safety Levels with respect of Human Exposure to Radiofrequency Electromagnetic fields, 3 kHz to 300 GHz, IEEE C95.1-2005
1194. ICNIRP (1998) Guidelines for limiting exposure to time-varying electric, magnetic and electromagnetic fields (up to 300 GHz). Health Phys 74:494–522
1195. Renner H (2006) Radio-(Chemo)-Thermo-Therapie. Kasuistische Erfahrungen bei lokal fortgeschrittenen Kopf-Hals Tumoren. Hyperthermia Symposium, Cologne, September 22–23
1196. Herzog A (2008) Oncothermia applications, Presentation, Baden Baden October 31.
1197. Sahinbas H, Grönemeyer D (2002) Local and regional deep-hyperthermia in combination with radiation- and chemotherapy for advanced tumors. 20th European Society for hyperthermic oncology, Bergen, Norway, 23–25 May
1198. Sahinbas H, Baier JE, Groenemeyer DHW, Boecher E, Szasz A. (2006) Retrospective clinical study for advanced brain-gliomas by adjuvant oncothermia (electro-hyperthermia) treatment. www.gimt-online.de/uploads/media/Therapieergebnisse_Giloma_Studie_01.pdf
1199. Szasz A, Dani A, Varkonyi A (2004) Az elektro-hipertermia eredményei nagyszámú beteg retrospektív kiértékelésének tükrében Magyarországon. Magyar Klinikai Onkológiai Társaság III. Kongresszusa, Budapest, Hungary, 17–20 November 2004
1200. Wismeth C, Dudel C, Pascher C et al (2010) Transcranial electro-hyperthermia combined with alkylating chemotherapy in patients with relapsed high-grade gliomas – Phase I clinical results. J Neurooncol 98(3):395–405
1201. Sahinbas H (2004) EHT bei Kindern mit Hirntumoren und nicht-invasive Messverfahren am beispiel von Hirntumoren. Symposium Hyperthermie, Cologne, 15–16 October 2004
1202. Lang I, Piko B, Juhos E, Hitre E, Szucs M, Zsalek J, Csejtey A (2005) Case report on heavily pre-treated advanced breast tumor. Symposium Cologne, 2005
1203. Kongsgaard UE, Werner MU (2009) Evidence-Based Medicine Works Best When There is Evidence: Challenges in Palliative Medicine When Randomized Controlled Trials are not Possible. J pain Palliative Care Pharmacother 23:48–50
1204. Bruera E (2006) Process and content of Decision making by advanced cancer patients. J Clin Oncol 24:1029–1030
1205. Tassinari D, Montanari L, Maltoni M et al (2008) The palliative prognostic score and survival in patients with advanced solid tumors receiving chemotherapy. Support Care Cancer 16:359–370
1206. Pauling L (1989) Biostatistical analysis of mortality data for cohorts of cancer patients. Proc Natl Acad Sci USA 86:3466–3468
1207. Rosenbaum PR, Rubin DB (1983) The Central Role of the Propensity Score in Observational Studies for Causal Effects. Biometrika 70:41–55
1208. Rosenbaum PR, Rubin DB (1984) Reducing bias in observational studies using subclassification on the propensity score. J Am Stat Asso 1984 79:516–524
1209. Williams PL (1996) Sequential monitoring of clinical trials with multiple survival, endpoints. Stat Med 15:2341–2357
1210. Whitehead J, Donaldson AN, Stephens R, Machin D (1993) A simulated sequential analysis based on data from two MRC trials. Br J Cancer 68:1171–1178
1211. Altman DG, Machin D, Bryant TN, Gardner MJ (2000) Statistics with Confidence. British Medical Journals Book, British Medical Journal, London
1212. Swinscow TDW, Campbell MJ (2002) Statistics at square one. British Medical Journals Book, Br Med J, Bristol
1213. Juckett DA, Rosenberg B (1993) Comparison of the Gompertz and Weibull functions as descriptors for human mortality distributions and their intersections. Mech Aging Dev 69:1–31

1214. Ricklefs RE, Scheuerlein A (2002) Biological implications of the Weibull and Gompertz models of aging. J Gerontol A Biol Sci Med Sci 57:B69–76
1215. Tyurin YuN, Ykovlev AYu, Shi J et al (1995) Testing a model of aging in animal experiments. Biometrics 51(1):363–372
1216. Vanfleteren JR, De Vreese A, Braeckman BP (1998) Two-parameter logistic and Weibull equations provide better fits to survival data from isogenic populations of Caenorhabditis elegans in axenic culture than does the Gompertz model. J Gerontol A Biol Sci Med Sci 53:B393–403
1217. WeibullSopo, Case LD, Morgan (2003) Design on Phase II cancer trials evaluating survival probabilities. BMC Med Res Methodol 3:6–18
1218. Work of two Hungarian institutions are involved: HTT Med Day-clinic, Budapest, Hungary (investigators: A. Dani and A. Varkonyi). Referred to in the text as HTT. And Peterfy Hospital, Budapest, Hungary, (investigator: T. Magyar). Referred to in the text as PFY.
1219. Wismeth C et al (2006) Loco-regional hyperthermia in patients with progressive astrocytoma WHO III or glioblastoma WHO IV (RNOP-10) – a prospective single arm phase I/II study; EANO
1220. Wismeth C et al (2009) Transcranial electro-hyperthermia combined with alkylating chemotherapy in patients with relapsing high-grade gliomas – Phase I clinical results. Expanding the Frontiers of Thermal Biology, Medicine and Physics. Annual Meeting of Society of Thermal Medicine, Tucson, USA, 3–7 April 2009
1221. Greig NH, Ries LG, Yancik R et al (1990) Increasing annual incidence of primary malignant brain tumor sin the elderly. J Natl Canc Inst 82:1621–1624
1222. Karatsu J, Ushio Y (1977) Epidemiological study of primary intracranial tumors in elderly people. J Neurol Neurosurg Psychiat 63:116–118
1223. Scott CB, Scarantino C, Urtasun R et al (1998) Validation and predictive power of Radiation Therapy Oncology Group (RTOG) recursive partitioning analysis classes for malignant glioma patients: A report using RTOG 90-06. Int J Radiat Oncol Biol Phys 40:51–55
1224. Stupp R, Dietrich P-Y, Kraljevic SO et al (2002) Promising survival for patients with newly diagnosed clioblastoma multiforme treated with concomitant radiation plus temiozolomide followed by adjuvant temozolomide. J Clin Oncol 20:1375–1382
1225. Stupp R, Mason WP, van den Bent MJ et al (2005) Radiotherapy plus concomitant and adjuvant temozolomide for glioblastoma. N Engl J Med (352):987–996
1226. Brandes AA, Vastola F, Basso U et al (2003) A prospective study on glioblastoma in the elderly. Cancer 97(3):657–662
1227. Scott JN, Rewcastle NB, Brasher PMA, Fulton D, Hagen NA, MacKinnon JA, Sutherland G, Cairncross JG, Forsyth P (1998) Long-term glioblastoma multiforme survivors: a population-based study. Can J Neurol Sci 25:197–201
1228. Chen JW, Lin J, Madamanchi N, Trier TT, Campbell G (2000) Apoptosis occurs in a new model of thermal brain injury. J Biomed Sci 7:459–465
1229. Fuse T, Yoon K-W, Kato T, Yamada K (1998) Heat-induced apoptosis in human glioblastoma cell-line A172. Neurosurgery 42:843–849
1230. Andocs G, Szasz A: Preliminary results of Oncothermia induced apoptosis in nude-mouse inoculated human-glioblastoma cells, Results of hyperthermia, Seminar, St. Istvan University, Aug. 24., 2005. (in Hungarian)
1231. Vogel P, Dux E, Wiessner C (1997) Evidence of apoptosis in primary neuronal cultures after heat shock. Brain Res 764(1–2):205–213
1232. Sahinbas H et al (2006) Retrospective clinical study of adjuvant electro-hyperthermia treatment for advanced brain-gliomas. Deutsche Zeitschrift fuer Onkologie 39:154–160
1233. Hager ED et al (2008) Prospective phase II trial for recurrent high-grade malignant gliomas with capacitive coupled low radiofrequency (LRF) deep hyperthermia. ASCO, J Clin Oncol, Annual Meeting Proceedings (Post-Meeting Edition) 26:2047

1234. Dani A, Varkonyi A: Electro-hyperthermia treatment of malignant brain tumors, Results of hyperthermia, Seminar, St. Istvan University, Aug. 26–27., 2003. (in Hungarian)
1235. Szasz A (2009) Brain glioma results by oncothermia, a review. Expanding the Frontiers of Thermal Biology, Medicine and Physics Annual Meeting of Society of Thermal Medicine, Tucson, USA, 3–7 April 2009
1236. Douwes F, Douwes O, Migeod F, Grote C, Bogovic J (2006) Hyperthermia in combination with ACNU chemotherapy in the treatment of recurrent glioblastoma. http://www.klinik-st-georg.de/pdf/hyperthermia_in_combination_with_ACNU_chemotherapy_in_the_treatment_of_recurrent_glioblastoma.pdf
1237. Hager ED et al (2003) The treatment of patients with high-grade malignant gliomas with RF-hyperthermia. Proc ASCO 22:118, #47;Proc Am Soc Clin Oncol 22: 2003
1238. Fiorentini G, Giovanis P, Rossi S, Dentico P, Paola R, Turrisi G, Bernardeschi P (2006) A phase II clinical study on relapsed malignant gliomas treated with electro-hyperthermia. In Vivo, 20:721–724
1239. Szasz A., Sahinbas H., Dani A.: Electro- hyperthermia for anaplastic astrocytoma and glioblastoma multiforme ICACT 2004, Paris, 9–12. February, 2004
1240. Sahinbas H., Grönemeyer D.H.W., Böcher E., Lange S.: Hyperthermia treatment of advanced relapsed gliomas and astrocytoma, The 9th International Congress on hyperthermic oncology, St. Louis, Missouri, ICHO, April 24–27, 2004
1241. Hager E.D.: Response and survival of patients with gliomas grade III/IV treated with RF capacitive-coupled hyperthermia, ICHO Congress, St. Louis, USA 2004
1242. Hager E.D.: Clinical Response and Overall Survival of Patients with Recurrent Gliomas Grade III/IV Treated with RF Deep Hyperthermia – An Update, ICHS Conference, Shenzhen, China, 2004
1243. Sahinbas H, Szasz A: Electrohyperthermia in brain tumors, Retrospective clinical study, Annual Meeting of Hungarian Oncology Society, Budapest, November 3–5, 2005
1244. Medical Research Council Brain Tumor Working Party (2001) Randomized Trial of Procarbazine, Lomustine, and Vincristine in the Adjuvant treatment of High-grade Astrocytoma: A Medical Research Council Trial, J Clin Oncol 19:509–518
1245. Sahinbas H (2004) EHT bei Kindern mit Hirntumoren und nicht-invasive Messverfahren am beispiel von Hirntumoren. Symposium Hyperthermie, Cologne, 15–16 October 2004
1246. Fisher PG, Buffler PA (2005) Malignant gliomas in 2005. Where to GO from here?, Editorials, JAMA 293:615–617
1247. Bodoky Gy (2003) Experiences with the treatment of advanced pancreatic cancer in Hungary, in Hungarian. Hungarian Oncology 47:10–13
1248. Dani A, Varkonyi A, Magyar T, Szász A (2008) Clinical study for advanced pancreas cancer treated by oncothermia. Forum Hyperthermie 1:13–20
1249. Douwes F (2004) Thermo-Chemotherapie des fortgeschrittenen Pankreaskarzinoms. Ergebnisse einer klinischen Anwendungsstudie. http://www.klinik-st-georg.de/publikationen/pdf/thermo-chemotherapie_des_fortgeschrittenen_pankreaskarzinoms.html
1250. Douwes F, Migeod F, Grote C (2006) Behandlung des fortgeschrittenen Pankreaskarzinoms mit regionaler Hyperthermie und einer Zytostase mit Mitomycin- C und 5-Fluorouracil/Folinsäure. http://www.klinik-st-georg.de/pdf/pankreastherapien.pdf
1251. VeraMed Clinic, Meshede, Germany; Dr. M. Kalden, unpublished data, private information
1252. Renner H, Albrecht I (2007) Analyse der Überlebenszeiten von Patienten mit Pankreastumoren mit erfolgter kapazitativer Hyperthermiebehandlung. Internal Report of Praxis in Klinikum Nord, Nurnberg, Germany, (Prepared by: Mr. Mirko F; May 2007)
1253. Roazzi P, Capocaccia R, Santaquilani M, Carrani E (2003) EUROCARE Working Group; Electronic availability of EUROCARE-3 data: a tool for further analysis. Ann Oncol14(Suppl 5):v150–v155.
1254. Hager ED et al (1999) Deep hyperthermia with radiofrequencies in patients with liver metastases from colorectal cancer. Anticancer Res 19(4C):3403–3408
1255. Ferrari VD, De Ponti S, Valcamonico F et al (2007) Deep electro-hyperthermia (EHY) with or without thermo-active agents in patients with advanced hepatic cell carcinoma: phase II study. J Clin Oncol 25:18S, 15168

1256. Panagiotou P, Sosada M, Schering S, Kirchner H (2005) Irinotecan plus Capecitabine with regional electrohyperthermia of the liver as second line therapy in patients with metastatic colorectal cancer. ESHO, Jun.8–11, Graz, Austria
1257. Fiorentini G, deGiorgi U, Turrisi G et al (2006) Deep electro-hyperthermia with radiofrequencies combined with thermoactive drugs in patients with liver metastases from colorectal cancer (CRC): a Phase II clinical study. ICACT 17th, Paris, France, Jan 30–Feb 2 2006
1258. Vigvary Z, Mako E, Dank M (2002) Combined radiological and interventional treatment of non-operable rectal tumors and their liver metastases. Regional Radiology Conference, Maribor, Sept. 19–20, Slovenia
1259. Szasz A (2009) Clinical studies evidences of modulated rf-conductive heating (mEHT) method. Paper presented at the 25th Annual Meeting of the European Society for Hyperthermic Oncology, ESHO, Verona, Italy, 4–6 June
1260. Aydin H et al (2003) Strahlen-Hyperthermie bei Lebermetastasen und bei therapieresistenten Knochenmetastasen. Hyperthermia Symposium, Cologne, Germany, 25–26. October
1261. Feyerabend T, Wioedeman GJ, Jaeger B et al (2001) Local hyperthermia, radiation, and chemotherapy in recurrent breast cancer is feasible and effective except for inflammatory disease. Int J Radiat Oncol Biol Phys 49:1317–1325
1262. Dani A, Varkonyi A, Magyar T, Szasz A (2010) A retrospective study of 1180 cancer patients treated by oncothermia. Forum Hyperthermia accepted (pp. 1–11)
1263. Renner H. (2003) Simultane RadioThermoTherapie bzw. RadioChemoThermoTherapie, Hyperthermia Symposium, Cologne, Germany, October
1264. Szasz A et al (2005) Retrospective analysis of 1180 oncological patients treated by electro-hyperthermia in Hungary. Jahreskongress der Deutschen Gesellschaft für Radioonkologie, DEGRO 11, Karlsruhe, 26–29 May 2005
1265. Bogovic J et al (2001) Posttreatment Histology and Microcirculation Status of Osteogenic Sarcoma after a Neoadjuvant Chemo- and Radiotherapy in Combination with Local Electromagnetic Hyperthermia. Onkologie 24:55–68
1266. NCBI, MeSH. http://www.ncbi.nlm.nih.gov/sites/entrez?db=mesh. Cited 02 Sept 2010
1267. Feynman RP, Leighton RB, Sands M (1963) The Feynman lectures on physics. Section 44-3. Addison-Wesley, Massachusetts
1268. Feynman RP, Leighton RB, Sands M (1963) The Feynman Lectures on Physics. Section 46-5. Addison-Wesley , Massachusetts
1269. Shannon CE (1948) A Mathematical Theory of Communication. Bell Syst Tech J 27: 379–423; 623–656
1270. Jaynes ET (1965) Gibbs vs Boltzmann Entropies. Am J Phys 33(5): 391–398
1271. Kuhn H (1992) The origin of life – A task in Molecular Engineering. Molecular Engg 1:377–399
1272. Rottenberg H (1979) Non-equilibrium thermodynamics of energy-conversion in bioenergetics. Biochim Biophys Acta 549(3–4):225–253
1273. Taylor CR, Weibel ER (1981) Design of the mammalian respiratory system. I. Problem and strategy. Respir Physiol 44(1):1–10
1274. Labra FA, Marquet PA, Bozinovic F (2007) Scaling metabolic rate fluctuations. PNAS 104(26):10900–10903
1275. Calder III WA (1984) Size, Function and life history. Dover, Mineola, NY
1276. West GB, Born JH (2000) Scaling in biology. Oxford University Press, Oxford
1277. Savage VM, Gillooly JF, Woodruff WH et al (2004) The predominance of quarter-power scaling in biology. Funct Ecol 18(2):257–282
1278. Levine HJ (1997) Rest heart rate and life expectancy. J Am Coll Cardiol 30:1104–1106
1279. Glazier DS (2005) Beyond the '$^3/_4$ power-law': variation in the intra- and interspecific scaling of metabolic rate in animals. Biol Rev 80:611–662
1280. Glazier DS (2008) Effects of metabolic level on the body size scaling of metabolic rate in birds and mammals. Proc Biol Sci 275(1641):1405–1410

1281. Gazit Y (1996) Fractal Vasculature and Vascular Network Growth Modeling in Normal and Tumor Tissue. PhD Thesis, MIT, Cambridge, MA, USA
1282. Sabo E, Boltenko A, Sova Y et al (2001) Microscopic analysis and significance of vascular architectural complexity in renal cell carcinoma. Clin Cancer Res 7(3):533–537
1283. Mancardi D, Varetto G, Bucci E et al (2008) Fractal parameters and vascular networks: facts & artifacts. Theor Biol Med Model 5:12
1284. Baish JW, Jain RK (2000) Fractals and Cancer. Cancer Research 60(14):3683–3688
1285. Bauer W, Mackenzie CD (1995) Cancer detection via determination of fractal cell dimension. Eprint arXiv:patt-sol/9506003. Cited 02 Sept 2010
1286. Lopez-Urrutia A (2003) Allometry: How reliable is the biological time clock? Nature 424:269–270
1287. Clarke A, Fraser KPP (2004) Why does metabolism scale with temperature? Funct Ecol 18(2):243–251
1288. Clarke A (2004) Is there a universal temperature dependence of metabolism? Funct Ecol 18(2):252–256
1289. Gillooly JF, Allen AP (2007) Changes in body temperature influence the scaling of VO2max and aerobic scope in mammals. Biol Lett 3(1):99–102
1290. Craig RW, Seymour RS (2003) Mammalian basal metabolic rate is proportional to body mass$^{2}/_{3}$. PNAS 100(7):4046–4049
1291. West GB, Brown JH, Enquist BJ (2001) A general model for ontogenetic growth. Nature 413(6856):628–631
1292. West GB, Brown JH, Enquist BJ (2004) Growth models based on first principles or phenomenology? Funct Ecol 18(2):188–196
1293. Banavar JR, Damuth J, Maritan A et al (2002) Modelling universality and scaling. Nature 420:626
1294. West GB, Enquist BJ, Brown JH (2002) Modelling universality and scaling –Reply. Nature 420:626–627
1295. Guiot C, Degiorgis PG, Delsanto PP et al (2003) Does tumor growth follow a “universal law”? J Theor Biol 225(2):147–151
1296. Bertalanffy von L (1957) Quantitative laws of metabolism and growth. Q Rev Biol 32(3):217–231
1297. Deisboeck TS, Guiot C, Delsanto PP et al (2006) Does cancer growth depend on surface extension? Med Hypotheses 67(6):1338–1341
1298. Deisboeck TS, Wang Z (2008) A new concept for cancer therapy: out-competing the aggressor. Cancer Cell Int 8:19
1299. Jonscher AK (1983) Dielectric Relaxation in Solids. Chelsea Dielectrics Press, London
1300. Foster C, Schwan HP (1996) Dielectric properties of Tissues. In: Polk C, Postow E (eds) Handbook of Biological Effects of Electromagnetic fields, Chapter 1, CRC Press, Boca Raton, FL, pp 25–102
1301. Pauly H, Packer L (1960) The Relationship of Internal Conductance and Membrane Capacity to Mitochondrial Volume. J Biophys Biochem Cytol 7(4):603–612
1302. Grodzinsky AJ (2003) Fields, Forces and Flows in Biological Systems. http://ocw.mit.edu/OcwWeb/Biological-Engineering/20-430JFall-2004/LectureNotes/index.htm
1303. Cole KS, Cole RH (1941) Dispersion and absorption in dielectrics. I. Alternating current characteristics. J Chem Phys 9:341–351
1304. Cole KS, Cole RH (1942) Dispersion and absorption in dielectrics. II. Direct current characteristics. J Chem Phys 10:98–105
1305. Osypka M, Gersing E (1995) Tissue impedance spectra and the appropriate frequencies for EIT. Physiol Measure 16:A49–A55
1306. Tremoliers J (1978) Effects biologiques des champs electromametiques non ionisants. Electron Appl 7:71–77
1307. vonEuler H (2002) Electrochemical treatment of tumor. Doctoral Thesis, Swedish University of Agricultural Sciences, Uppsala

1308. Blank M, Soo L (1992) Temperature dependence of electric field effects on Na,K-ATPase. Bioelectrochemis Bioenerget 28(1–2):291–299
1309. Varma R, Selman JS (1991) Techniques for characterisation of electrodes and electrochemical processes. John Wiley & Sons, New York
1310. Jacobsen T, West K (1995) Diffusion Impedance for Planar, Cylindrical, and Spherical Geometry. Electrochimica Acta 40(2):255–262
1311. Robinson FNH (1962) Noise in Electrical Circuits. Oxford University Press, London
1312. Kittel C (2004) Introduction to solid state physics. Wiley, New York, NY
1313. Nigam NC (1983) Introduction to Random Vibrations. The MIT Press, Cambridge
1314. Cramer F (1995) Chaos and Order (The Complex Structure of Living Systems). VCH, Weinheim, New York
1315. Peng CK, Buldyrev SV, Hausdorff JM et al (1994) Fractals in Biology and Medicine: from DNA to the Heartbeat. In: Bunde A, Havlin S (eds) Fractals in Science, Springer, Berlin, pp 49–87
1316. Szasz D (1994) Boltzmann's Ergodic Hypothesis, a Conjecture for Centuries? Lecture given at the international symposium in honour of Boltzmann's 150th birthday. The Erwin Schrodinger International Institute for Mathematical Physics, Vienna, 24–26 February 1994
1317. Sneddon I (1955) Handbuch der Physik Bd. II. Springer, Berlin
1318. Gillespie DT (1992) Markov processes. Academic, San Diego, CA
1319. White DC, Woodson HH (1959) Electromechanical energy conversion. Wiley, New York, NY
1320. Lanouette W (1992) Genius in the Shadows. Macmillan, London, pp 350–361
1321. McNamara B, Wiesenfeld K (1989) Theory of stochastic resonance. Phys Rev A 39(9):4854–4869
1322. Smoluchowski M (1998) Barrier fluctuations and stochastic resonance in membrane transport. Chaos 8:549
1323. Smith GF (1971) Constitutive equations for anisotropic and isotropic materials. Int J Engg Sci 9:899–916
1324. Anderson DA, Tannehill JC, Pletcher RH (1984) Computational fluid mechanics and heat transfer. McGraw-Hill, Washington, DC
1325. Lupis CHP (1983) Chemical thermodynamics of materials. Elsevier, Amsterdam
1326. Dewey WC (1994) Arrhenius relationships from the molecule and cell to the clinic. Int J Hyperthermia 10(4):457–483
1327. Perez CA, Sapareto SA (1984) Thermal dose expression in clinical hyperthermia and correlation with tumor response/control. Cancer Res 44:4818s–4825s
1328. Separeto, S.A., Dewey, W.C (1984) Thermal dose determination in cancer therapy. Int J Oncol. Biol. Phys 10:787–800
1329. Thrall DE, Rosner GL, Azuma C (2000) Using units of CEM43°C T90 local hyperthermia thermal dose can be delivered as prescribed. Int J Hyperthermia 16:415–428
1330. Bhowmick P, Coad JE, Bhowmick S et al (2004) In vitro assessment of the efficacy of thermal therapy in human benign prostatic hyperplasia. Int J Hyperthermia 20(4):421–439
1331. Leophold KA, Dewhirst MW, Samulsky TV et al (1993) Cumulative minutes with T90 greater than temindex is predictive of response of superficial malignancies to hyperthermia and radiation. Int J Radiat Oncol Biol Phys 25:841–847
1332. Thrall DE, LaRue SM, Yu D et al (2005) Thermal dose is related to duration of local control in canine sarcomas treated with thermoradiotherapy. Clin Canc Res 11:5206–5214
1333. Dewhirst MW, Vujaskovic Z, Jones E et al (2005) Re-setting the biologic rationale for thermal therapy. Int J Hyperthermia 21:779–790
1334. Delannoy J, LeBihan D, Hoult DI et al (1990) Hyperthermia system combined with a magnetic resonance imaging unit. Med Phys 17:855–860
1335. van Rhoon GC, Wust P (2005) Introduction: non-invasive thermometry for thermotherapy. Int J Hyperthermia 21(6):489–495

1336. Foster KR, Lozano-Nieto A, Riu PJ et al (1998) Heating of Tissues by Microwaves: a model analysis. Bioelectromagnetics 19(7):420–428
1337. Eckert E (1949) Warme und Stoffaustausch. Springer, Berlin
1338. Hobbie RK (1988) Intermediate Physics for Medicine and Biology. Wiley, New York, NY
1339. Hodgkin AL, Huxley AF (1952) A quantitative description of membrane current and its application to conduction and excitation of nerve. J Physiol 117, 500–544
1340. Spanner DC (1954) The active transport of water under temperature gradients. Symp Soc Exptl Biol 8:76–94
1341. Sackmann E (1995) Physical basis of self-organisation and function of membranes: physics of vesicles. In: Lipowsky SE (eds) Handbook of biological physics, Vol I. Elsevier, Amsterdam
1342. Ricker DW (2003). Echo Signal Processing. Springer, Heidelberg
1343. Hamada N, Kodama S, Suzuki K et al (2003) Gap junctional intercellular communication and cellular response to heat stress. Carcinogenesis 24(11):1723–1728
1344. Vigh L, Maresca B, Harwood JL (1998) Does the membrane's physical state control the expression f heat shock and other genes? Trends Biochem Sci 23(10):369–374
1345. Measurements were made in Laboratories of Professor HH Kampinga. Department of Cell Biology, Medical Center Groningen, University of Groningen, Groningen, unpublished
1346. Bini M, Ignesti A, Millanta L et al (1985) An unbalanced electric applicator for RF hyperthermia. IEEE Trans Biomed Eng 32(8):638–641
1347. Feynmann RP, Leighton RB, Sands M (1964) The Feynman lectures on physics volume 2: mainly electromagnetism and matter. Addison-Wesley Publishing, Reading, MA
1348. Spector WS (1956) Handbook of biological data. Saunders. Philadelphia, PA
1349. West GB (2004) Brown JH. Life's Universal Scaling Laws. Phys Today 57:36–43
1350. Bellissant E, Benichou J, Chastang C (1996) The group sequential triangular test for phase II cancer clinical trials. Am J Clin Oncol 19:422–430
1351. Evans CH Jr, Ildstad ST (eds) (2001) Committee on Strategies for Small-Number-Participant, Clinical Research Trials, Small Clinical Trials: Issues and Challenges. National Academy Press, Washington, DC, http://www.nap.edu/catalog/10078.html
1352. Aczél J (1960) Vorlesungen über Funktionalgleichungen und ihre Anwendungen. DVW, Berlin
1353. Wilson DL (1994) The analysis of survival (mortality), data: fitting Gompertz, Weibull and logistic functions, Mech. Aging Dev. 74:15–33
1354. Piantanelli L (1986) A mathematical model of survival kinetics. I. Theoretical basis. Arc Gerontol Geriatr 5:107–118
1355. Economos AC (1982) Rate of aging, rate of dying and the mechanism of mortality, Arc. Gerontol. Geriatr 1:3–27
1356. Weston CL, Douglas C, Craft AW, Lewis IJ, Machin D (2004) Establishing long-term survival and cure in young patients with Ewing's sarcoma. British J Cancer 91: 225–232
1357. Kolmogorov NN (1937) On the statistical theory of the crystallization of metals, Bull. Acad. Sci. UssR, Math. Ser. 1, 355
1358. Joshnson WA, Mehl PA (1939) Reaction kinetics in processes of nucleation and growth. Trans Am Inst Min Engrs 135:416
1359. Avrami MA (1939) Kinetics of phase change I–III. J Chem Phys 7:1103
1360. Cope FW (1977) Detection of phase transitions and cooperative interactions by Avrami analysis of sigmoid biological time curves for muscle, nerve, growth, firefly, and infrared phosphorescence, of green leaves, melanin and cytochrome C. Physiol Chem Phys 9: 443–459
1361. Cope FW (1980) Avrami analysis of electrical switching in hydrated melanin suggest dependence on a phase transition. Physiol Chem Phys 12:537–538
1362. Cope FW (1976) The kinetics of biological phase transitions manifested by sigmoid time curves: a review of approaches. Physiol Chem Phys 8:519–527

1363. Ricklefs RE, Scheuerlein A (2002) Biological implications of the Weibull and Gompertz models of aging. J Gerontol A Biol Sci Med Sci 57:B69–76
1364. Glantz SA (1997) Primer of biostatistics. McGraw Hill Health Professions Division, New York, NY
1365. Sposto R (2002) Cure model analysis in cancer: an application to data from the Children's Cancer Group. Stat Med 21:293–312
1366. Sposto R (2007) Criteria for Optimizing Prognostic Risk Groups in Pediatric Cancer: Analysis of Data From the Children's Oncology Group. J Clin Oncol 25:2070–2077
1367. Frankel P, Longmate J (2009) Parametric models for accelerated and long-term survival: a comment on proportional hazards. Stat Med 21:3279–3289
1368. Dani A (2004) Electro-hyperthermia for advanced pancreas tumors. Deutscher Kongress für Radioonkologie, Strahlenbiologie und Medizinische Physik, Erfurt, 10–13 June
1369. Dani A, Varkonyi A, Magyar T, Szasz A (2010) Clinical study for advanced non-small-cell lung-cancer treated by oncothermia. Forum Hyperthermie; accepted for publication
1370. Douwes FR (2006) Thermochemotherapy of the advanced pancreas carcinoma. Biologische Medizin 35:126–130
1371. Szasz A (1991) An electrically driven instability: the living-state (Does the room-temperature superconductivity exist?) Physiol Chem Phys Med NMR 23:43–50

Index